Human Factor and Reliability Analysis to Prevent Losses in Industrial Processes
An Operational Culture Perspective

Human Factor and Reliability Analysis to Prevent Losses in Industrial Processes
An Operational Culture Perspective

Salvador Ávila Filho
Federal University of Bahia, Salvador, Brazil

Ivone Conceição de Souza Cerqueira
Federal University of Bahia, Salvador, Brazil

Carine Nogueira Santino
Federal University of Bahia, Salvador, Brazil

ELSEVIER

Elsevier
Radarweg 29, PO Box 211, 1000 AE Amsterdam, Netherlands
The Boulevard, Langford Lane, Kidlington, Oxford OX5 1GB, United Kingdom
50 Hampshire Street, 5th Floor, Cambridge, MA 02139, United States

Copyright © 2022 Elsevier Inc. All rights reserved.

No part of this publication may be reproduced or transmitted in any form or by any means, electronic or mechanical, including photocopying, recording, or any information storage and retrieval system, without permission in writing from the publisher. Details on how to seek permission, further information about the Publisher's permissions policies and our arrangements with organizations such as the Copyright Clearance Center and the Copyright Licensing Agency, can be found at our website: www.elsevier.com/permissions.

This book and the individual contributions contained in it are protected under copyright by the Publisher (other than as may be noted herein).

Notices

Knowledge and best practice in this field are constantly changing. As new research and experience broaden our understanding, changes in research methods, professional practices, or medical treatment may become necessary.

Practitioners and researchers must always rely on their own experience and knowledge in evaluating and using any information, methods, compounds, or experiments described herein. In using such information or methods they should be mindful of their own safety and the safety of others, including parties for whom they have a professional responsibility.

To the fullest extent of the law, neither the Publisher nor the authors, contributors, or editors, assume any liability for any injury and/or damage to persons or property as a matter of products liability, negligence or otherwise, or from any use or operation of any methods, products, instructions, or ideas contained in the material herein.

ISBN: 978-0-12-819650-2

For Information on all Elsevier publications
visit our website at https://www.elsevier.com/books-and-journals

Publisher: Susan Dennis
Acquisitions Editor: Anita A. Koch
Editorial Project Manager: Lena Sparks
Production Project Manager: Bharatwaj Varatharajan
Cover Designer: Mark Rogers

Typeset by MPS Limited, Chennai, India

Contents

About the authors ix
Preface xi
Acknowledgment xvii

1. Introduction 1

1.1 **A brief discussion** 1
    1.1.1 The organization inserted in the social environment 1
    1.1.2 Conceptual and mathematical models in human and operational reliability 1
    1.1.3 Risk management in complex processes and environments 2
    1.1.4 Competency analysis and task planning 2
    1.1.5 Analysis and diagnosis of human factors 3
1.2 **Discussion timeline and schools** 5
    1.2.1 A historical vision about schools related to origin of human reliability 5
1.3 **Worker role in job and society: human error** 8
    1.3.1 Human activity 9
    1.3.2 Role at work and society 9
    1.3.3 Departments, functions, and human reliability 11
1.4 **Risk management on material losses and operations** 19
    1.4.1 Material loss risk 19
References 23

2. Human reliability and cognitive processing 25

2.1 **Human reliability** 25
    2.1.1 Why study human reliability? 25
    2.1.2 Classic concepts of human reliability 28
    2.1.3 Modeling—first, second, and third generation 31
    2.1.4 Standardized plant analysis. risk-human reliability analysis method and a case in a chemical facility 36
    2.1.5 Human reliability involved in the cultural link 38
    2.1.6 Operator discourse—fuzzy 41
    2.1.7 Third-generation application in the calculation of organizational efficiency, Oil & Gas 48
2.2 **Human reliability and cognitive processing** 52
    2.2.1 Cognitive processing 52
    2.2.2 Introduction to cognitive processing, cases, failure, and skill knowledgement, rules 52
    2.2.3 Cognitive functions and decision processes 60
    2.2.4 Learning and skill 64
    2.2.5 Motivation and decision 65
    2.2.6 Decision-making process 66
    2.2.7 Cognitive model discussion 71
    2.2.8 Human behavior dynamics in the company 74
References 82

3. Factors affecting the performance of tasks 85

3.1 **Human and social typology** 85
    3.1.1 Human typology 86
    3.1.2 Social typology and archetypes 88
    3.1.3 Classification of human error 90
    3.1.4 Concepts and the investigation of latent failure 93
    3.1.5 Socioeconomic environment: human reliability analysis 95
    3.1.6 Investigation of socioeconomic-affective cycle and cycle of thinking and decision making—utility 101
    3.1.7 Executive function analysis 102
    3.1.8 Environments in human and operational reliability: human error 104
    3.1.9 Diseases, bad habits and cognitive academy 105
3.2 **Task assessment** 106
    3.2.1 General 106

         3.2.2   PADOP—environments and cognitive aspects in task preparation-execution   113
         3.2.3   Tools for planning the standard task   122
         3.2.4   Decision analysis under stress: emergency simulation   136
         3.2.5   Application of PADOP for the sulfuric acid plant case   138
   3.3   Discussion about API 770   143
         3.3.1   Concepts and assessment from API 770   143
         3.3.2   Analysis of the API 770 survey   144
   References   148

4. **Process loss assessment**   **151**
   4.1   Context   151
   4.2   Competencies to assess process losses   155
         4.2.1   Premises and competencies   155
   4.3   Losses in the process industries   157
         4.3.1   Process losses in the oil industry   159
         4.3.2   Process losses in gas industry for energy   163
         4.3.3   Process losses in the biofuels industry   164
         4.3.4   Process losses in the petrochemical industry   166
         4.3.5   Process losses in the chemical industry   167
         4.3.6   Process losses in the polymer industry   167
         4.3.7   Process losses in the metallurgy industry   168
         4.3.8   Risk of loss due to technology   168
   4.4   Diagnosis of process losses   169
         4.4.1   Introduction   169
         4.4.2   Knowledge of the production process   172
         4.4.3   Collecting data—inputs   172
         4.4.4   Measuring results—outputs   176
         4.4.5   Introduction to tools and methods   185
   4.5   Cases: diagnostics with quantitative and qualitative analysis   194
         4.5.1   Chemical and polymer case: diagnosis based on operator's discourse   195
         4.5.2   Case of metallurgy based on manager discourse and technical issues   207
         4.5.3   Discussion   211
   References   211

5. **Learned lessons: human factor assessment in task**   **213**
   5.1   Routine, environments, human types, and class of errors   213
         5.1.1   Routine management case GR: director's behavior (511—GR1)   214
         5.1.2   Technical and operational culture: solution for waste in the reaction (512—CTO2)   215
         5.1.3   Emergency case: situation in reaction stoichiometry (513—ER3)   215
         5.1.4   Practical skills case: perception and monitoring (514—HP4, HP5)   216
         5.1.5   Routine management case: meeting to change time in shift group   216
         5.1.6   Operational—process control: investigation of the process and wastewater (516—COP7)   218
         5.1.7   Problem analysis: diagnosis and process mapping (517—AP8)   219
         5.1.8   Problem analysis: negotiation for preventive action (518—AP9)   220
         5.1.9   Operational safety case: about safety culture (519—SO10, SO11, SO12)   221
         5.1.10 Inappropriate design and operation: technological solutions (5110—PJ13)   223
         5.1.11 Organizational change: change in practice without consulting past ritual (5111-PR14)   223
         5.1.12 Routine management, technical-operational culture: bias in execution (5112—VIO15/19)   224
         5.1.13 Accident cases in contractors: inadequate standard for services (5113—AC20 a 21)   227
   5.2   Routine learning: guidelines for human reliability   228
         5.2.1   Learning points   228
         5.2.2   Route of human, group and organizational error   234
   5.3   Lessons learned and validation of the guidelines   235
         5.3.1   Cognitive and behavioral academy: routine and program friends of emergency pool   237
         5.3.2   Application of tools for archetype analysis and executive function in the industry   248

		5.3.3	Communication in routine—environmental accident with HCl (Souza et al., 2018b)	251
		5.3.4	Investigation of technical failure and human error in the sulfuric acid plant	255
		5.3.5	Industry alarms and shutdown (Ammonia, HDT, Cyclohexane): H_2 and CO compressors	261
		5.3.6	Task complexity, low efficiency, and accident investigation	269
		5.3.7	Just culture in metallurgy and oil industries	273
	5.4	**Human reliability, sociotechnical reliability, culture of safety demands**	**277**
		5.4.1	Chemical industry and electricity distribution	278
		5.4.2	Petrochemical industry	286
		5.4.3	Onshore offshore oil and gas industry	287
		5.4.4	Fertilizer industry and refining units	294
		5.4.5	Metallurgical industry and chicken manufacturing	296
		5.4.6	Public security agencies: security, mobility and health, firefighter—PuA	300
		5.4.7	Context conclusion: research versus society's demand	304
	References			304
6.	**Human reliability: SPAR-H cases**			**307**
	6.1	**Introduction**		**307**
		6.1.1	Reviewing the chapters	307
		6.1.2	Human errors in the context of critical activities	309
	6.2	**Concepts and SPAR-H calibration**		**310**
		6.2.1	Operational context	311
		6.2.2	Calculation of human error probability	314
		6.2.3	SPAR-H calibration	314
		6.2.4	Discussion of performance shaping factors and calibration	315
	6.3	**Case studies**		**324**
		6.3.1	Chicken industry	325
		6.3.2	Uranium industry	328
		6.3.3	Chemical industry	331
		6.3.4	Refining industry	333
		6.3.5	Fertilizer industry	337
		6.3.6	Coconut industry	342
		6.3.7	Packing list services in the manufacture of sports products	344
	6.4	**Comparative analysis**		**347**
		6.4.1	Recommendations	349
	6.5	**Integrated reliability: the beginning**		**349**
		6.5.1	Control and metrics to achieve integrated or sociotechnical reliability	351
		6.5.2	Simulation of the application of the integrated reliability method	353
		6.5.3	Results	354
	References			356
7.	**Human reliability: chemicals and oil and gas cases**			**359**
	7.1	**Methodology description**		**359**
		7.1.1	Guiding the algorithms to apply technological tools and social environment	359
		7.1.2	Concept, tool, and procedure for technical, social, environment, and human typologies	361
	7.2	**Chemical industry case application**		**391**
		7.2.1	Abnormalities inventory	392
		7.2.2	Abnormality logic in complex processes	393
		7.2.3	Aliphatic amines	395
		7.2.4	Aromatic amines	396
		7.2.5	Polycarbonates	401
	7.3	**Oil and gas case application**		**404**
		7.3.1	Context identification: company, experience, rituals and organizational culture	405
		7.3.2	Survey of technical, human and social data	408
		7.3.3	Abnormal event mapping, signs, and failure mode (MEA and FMEA)	412
		7.3.4	Process analysis, logistics, operations, maintenance and safety (T3—AEP, T4—EVA)	414
		7.3.5	Analysis of the task and results	423
		7.3.6	Analysis of human and social data in the work environment	435
		7.3.7	Competence analysis and results	441
		7.3.8	Standard for behavior analysis	445
		7.3.9	Abnormal event cluster analysis (T14)	449
	7.4	**Qualitative results: chemical industry cases**		**451**
		7.4.1	Management aspects for decision	451
	7.5	**Quantitative results: oil and gas case**		**457**
		7.5.1	Failure energy analysis	457
	7.6	**Future work: task cross-assessment based on particle swarm model**	**468**
		7.6.1	The path of workers	469

		7.6.2 The bridge to the future	470
	References		471

8. Conclusion and products — 473

- 8.1 Conclusion — 473
- 8.2 Future book: human factor routine and emergency analysis — 475
- 8.3 Products in general — 476
- 8.4 Product 1 (Chapter 4) — process loss mapping — 478
  - 8.4.1 Introduction and methods — 478
  - 8.4.2 Methods — 478
  - 8.4.3 Discussion — 479
  - 8.4.4 Calculation based on metallurgy case (Section 4.5.2) — 480
- 8.5 Product 2 — task assessment — PADOP — 480
  - 8.5.1 Introduction — 480
  - 8.5.2 PADOP standard and review — 481
  - 8.5.3 Task failure assessment — 488
  - 8.5.4 Task emergency assessment — 490
- 8.6 Product 3 — cognitive quality — 491
  - 8.6.1 Introduction — 491
  - 8.6.2 Cognitive quality elements, functions and subfunctions — 492
  - 8.6.3 Static cognitive quality — COGNQe — 494
  - 8.6.4 Discussion about dynamic cognitive quality — COGNQd — 494
- 8.7 Product 4 — human reliability SPARH — 495
  - 8.7.1 Operational context — 496
  - 8.7.2 Performance factors assessment — 496
  - 8.7.3 Calculation and calibration — 497
- 8.8 Product 5 — social-technical reliability — 498
  - 8.8.1 Culture and manager profile — 498
  - 8.8.2 Complexity of parametric equations — 498
  - 8.8.3 Individual reliabilities — 499
  - 8.8.4 Social-technical reliability calculation — 500
- 8.9 Product 6 — operational-technical culture and prediction — 500
  - 8.9.1 Introduction — 500
  - 8.9.2 Methodology and products — 502
  - 8.9.3 Management programs for human and sociotechnical reliability — 503
- References — 506

Annex — 509
List of abbreviations — 541
Glossary — 543
Index — 545

About the authors

Salvador Ávila Filho—Chemical Engineer (UFBA) and Petrochemical Process Engineer (Petrobras)—has extensive knowledge in the chemical industry. He is specialized in statistical techniques (CQE/ASQ) to investigate abnormalities in the industry and work as an organizational consultant (UCSal) for transformation of culture. He has developed auditing techniques, effluent control, and fugitive emissions to reduce environmental impacts in the industry. He has also developed a technique for analyzing the perception of the operators regarding routines that have an impact on safety, energy, and environment (UFBA). As a manager of industrial facilities, he created a model for clean management, including human behavior and technologies. He has worked as a professor in graduate courses in administration, chemical engineering, and industrial engineering at different universities, especially in the subjects of risks, reliability, and human factors. Meanwhile, human aspects are investigated in the courses and practices of psychoanalysis allowing to enter by the cognitive fault in the behavior, in the society and the work. He has partnerships in the private initiative in reliability centered maintenance and human reliability. doctoral research in the human and organizational cultures area and provides concepts, techniques, and methods to keep industrial processes under control. He has presented articles were presented in the areas of human factors, risk, process safety, energy and water efficiency and currently in culture and behavior change. He has also work in the energy sector and developed a solution for productive arrangements. He also contributed to reduce the load of nitrogen in the effluent of the fertilizer industry with statistics, process studies, equipment technology, investigation of procedures, and educational campaign (Friends of the Lagoon). As a professor and researcher of UFBA in the department of mechanical engineering, he conducts research and services in the areas of risk management, organizational culture and human reliability, loss of process, energy, and water savings. Currently, he has cooperation with RLAM (Refinery) in energy efficiency, UOBA Active North in the area of reliability and risk, Secretariat of Public Security in the area of risk in mega-event and behavior of the police force of the State of Bahia. In addition, he has worked on tools and projects of public safety for preparedness for emergency to avoid disasters.

Ivone Conceição de Souza Cerqueira—Graduated in Nutrition from the Federal University of Bahia. She has been a Federal Public Servant for the Ministry of Education—Federal University of Bahia, since 1982 and current coordinator of the nucleus of follow-up of undergraduate courses assessments—pro rectory of undergraduate—UFBA. She is specialized in hospital management by the Institute of Collective Health/UFBA, 1996, in clinical nutrition by the University of Navarro and in people management by the School of Administration, UFBA. She holds a master's degree in medicine and health, Faculty of Medicine, UFBA. A member of the Communication and Health Research Group—UFBA Institute of Collective Health. She has experience in international hospital accreditation in the areas of education and professional qualification and facilities security. She has published articles in security, risk in mega-event, human factor, leadership, and human reliability and was a doctorate student in industrial engineering—Polytechnic School—Federal University of Bahia-2017.1–2020.1.

Carine Nogueira Santino—Graduated in Production Engineering from Salvador University and master's in industrial engineering from UFBA: study in loss mapping in the metallurgical industry. Doctorate student in Industrial Engineering/UFBA: study in process optimization. She has professional experience in quality checking in steel industry, involving audits and Six Sigma project; she has 4 years of experience in human resources in the area of people management. Member of the Research Group GRODIN—Polytechnic School-UFBA—research group at dynamic risk, and topics that address reliability safety and reduction of environmental impacts. She has published articles in the areas of process safety, human reliability, and loss mapping.

Preface

This book aims to discuss human reliability and the most suitable working environment in the industry to avoid human error. We will address the following issues:

1. General considerations on human reliability analysis
2. Understanding operator discourse from routine data collection
3. Cognitive processing and the possibilities for human error
4. The role of human in the execution of the task and aspects related to safety
5. Discussion of events, incidents, and accidents in the chemical industry based on operational culture
6. Human reliability method [Human reliability method (SPAR-H)] and human factor analysis in industry
7. Case discussion of the LPG industry for operational culture by applying the tool, operator discourse analysis, for human factor adjustment

1 Paradigms

The theoretical and practical knowledge on human reliability and human factors are approached progressively from Chapter 1 to Chapter 8, as shown in Fig. 1. After reading the book and noticing the techniques applied in each topic, it is concluded that due to cultural changes that change human behavior, it is essential to review tasks and investigate human factors. It is important, if necessary, to return and revise the concepts in the discussion of previous chapters, as Chapter 1 and to advance upwards in the spiral shown in Fig. 1, toward operational excellence, organizational resilience with the maximum level of reliability, in the sociotechnical production systems.

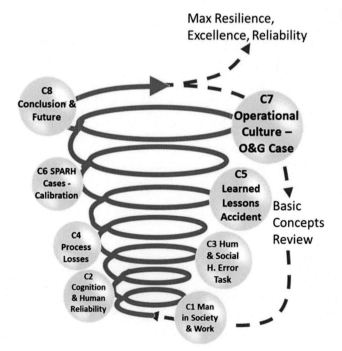

FIGURE 1 The upward spiral of knowledge and applications.

2 Book structure

This book breaks new ground when discussing the relationship between social, organizational, and operational cultures for inferring about unexpected future behavior. It attempts to include, in conceptual models, an intuitive, affective learning with cognitive processing. This subjectivity should be recognized through tools that assist in the investigation of task failures. These aspects are presented in Fig. 2.

The relationship between process losses, accidents, failures, and deviances indicates a possible complexity of the failure including the resulting human performance factors and technologies, processes, and equipment. Several cases are discussed qualitatively but, also, quantitatively for the calculation of human reliability.

The book presents methods and techniques to mine data, measure organizational efficiency, and reduce the number of data that represents the operating culture. It also attempts to relate which are the priority human and organizational factors to, through adjustments, return the organizational efficiency to the region of normality.

In the Preface, the subjects discussed in the book are presented. Each practical scenario confirms the conceptual models and makes the reader thinks of ways to avoid the accident. Here we present also acknowledgment, glossary, list of abbreviations, and summary of complete book.

In the introduction, Chapter 1, human reliability is discussed regarding multidisciplinary aspects such as the following: (1) organization within the social environment; (2) conceptual and mathematical models in human and operational reliability; (3) risk management in complex processes and environments; (4) competency analysis and task planning; and (5) diagnosis of human factors. Multidisciplinary sciences, academic schools, and private and public initiatives participate in the discussion on human reliability, from 1990 to 2016, and their bases inserted in the complexity of human factors in the industrial environment.

The importance of human's role in organizational efficiency includes a discussion of his roles in society, his departmental functions, and the possibility of causing human error. This recognition about multiple connection of human, from society to economic outcome, indicates the need to investigate risks in a dynamic format.

Chapter 2 deals with the themes of human reliability and cognitive processing, discussing the challenge of mathematical models in the three generations. In the case of the SPARH technique, basic concepts about human factors that indicate the level of reliability are presented. In the case of the third generation, we indicate the use of fuzzy mechanics,

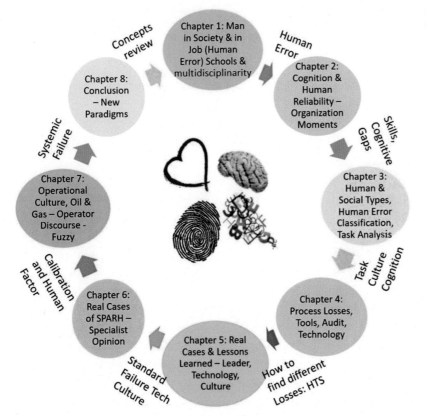

FIGURE 2 Concepts, subjects, paradigms, and results throughout the book.

citing a case presented in this book in Chapter 7 that uses the operator discourse analysis technique, which signals the resulting operational culture. Analysis of the operator discourse in a productive environment involves human performing the task in productive systems. This topic also discusses ways to collect data for the production analysis in complex systems and the connectivity analysis between factors in production systems.

It is also in Chapter 2 that the theme cognition and human error are presented through the lens of different authors that suggest schemas, mental maps, and cognitive models. Cognitive processing is important for researching about the causes of human error. Many human profiles have their psychological functions altered due to a stressful environment that causes nonidentification between individual and organizational values. Defects in cognitive processing cause human errors in the execution of tasks. Cognition is involved with psychological functions such as memory and attention and is influenced by the work environment and family. The perception of signals in cognitive processing is discussed in activities involving task control. Based on authors of cognitive psychology, some rules are discussed about the memory and attention functioning, which influences learning. The cognitive models suggested by Hollnagel and Avila are presented and adapted with guiding examples. In the conclusion of the chapter, the psychological functions are presented in a dynamic way by enabling the analysis of the task. A logic diagram is proposed for the application of the techniques in this chapter.

Chapter 3 deals with the identification of the human and social typologies that may face stressful or routine environments, which have an impact on team performance. The characterization of human types and the forms of individual or group decision making are addressed. By understanding the type of team reaction in routine, it interconnects with human error through different event classes that are derived from the various causes. Techniques for task planning, execution, and control are discussed to achieve greater effectiveness in factory operation and better safety in industrial facilities. The task is discussed relating to possible cognitive gaps, different complexities, behavioral variations, and the trigger of human errors.

Task analysis begins with the verification of human-centered environments. Regional culture and behavior of the groups and of the leaders are assessed. Mass culture also produces abnormalities such as memory gaps and attention difficulties. The loss of affective bonds, considered as a consequence of cultural biases, can also initiate human errors and be part of the accident construction. Human performance factors have impact on emergency, operational, tactical, and strategic situations.

In Chapter 3, analysis of human profiles and cultural aspects occurs routinely. Human typology helps classify human errors into various classes, and the convergence of these failures and errors receives the elaboration of the derived human error.

Rasmussen describes the disciplines to build a safety culture and accident investigation. Its models help in the adequacy of environments to avoid stress of people and, consequently, reduce equipment risks to failure.

Unfortunately, contrary to popular belief, redundant safety devices may bring new chances for the accident to occur through the phenomenon of "cognitive laziness."

API 770 is a material worked on in the book, used as a reference standard for human reliability that includes questionnaires from each area of knowledge, allowing the diagnosis of situations that most affect human error. A logic diagram is proposed for the application of the techniques in this chapter.

The issue of process losses is discussed in Chapter 4, and it is concluded that the main causes of low reliability-quality and increased input consumption are human errors. The main difficulty is the low visibility of the root cause of problems. This happens because operational risks are considered as dynamic.

Industrial competitiveness in a globalized world does not allow for the existence of failures that cause losses. Process losses are directly related to the control of operations and require appropriate methods and techniques for root cause identification. Each technology has its peculiarities that must be identified before investigation because of the need to treat data and information, often not available in the technology description. It is intended to apply techniques to identify, measure, and treat process losses by discussing the definition of losses, peculiarity of each industrial segment in continuous processes, competencies for analysis of process losses, data collection methods and techniques, and the identification of losses and respective diagnostic tools.

In Chapter 5, lessons learned or routine issues in the chemical process industry are discussed, by indicating aspects of human reliability such as cultural environment, leadership profile, decision model, possible human errors and hits, mind map construction, organizational culture, and others. Routines may be related to safety and operational culture, such as auditory perception and plant start-up, or quality of communication during shift change. Some situations discussed may be in the area of technology, such as emergency due to uncontrolled chemical reaction, failures in diagnosis and process mapping or, yet, change in the O-ring diameter of alternative pumps type Alfa Laval. However, it is known that many routine issues involve the managerial profile and attitudes in the safety area such as director's centralized behavior at plant start-up, considering that the latter used to be operator, supervisor, and engineer. It is observed that, apparently, there is repetition of pattern for abnormalities in the various dimensions to occur.

Chapter 6 brings together research work on human reliability of the continuous process and manufacturing industry. The steps, difficulties, and results are discussed regarding the refining, food, uranium, and fertilizer industry cases. The reason to use SPARH in calculating the likelihood of human error is the simplicity of applying workstation issues to the team of executors and managers. The application of SPARH does not require database and need expert opinion. This discussion generates a calculation that is compared to economic losses in operations and may recommend method calibration or if checked, may recommend change in human factors. Discussing SPARH is important to demonstrate through management, the best indicators, the best regions, and functions for investment in risk management. Another important work is the introduction of indicators that include human, process, operations, and equipment reliability, that is, sociotechnical reliability. This chapter presents a simplified script for SPARH application and its relationship with process loss and cognitive gap assessments.

Chapter 7 summarizes the methods and techniques already discussed for calculating human reliability based on human errors, human factors, and process loss auditing to diagnose operational culture. Chapter 7 discusses the application, in the LPG industry, of concepts, techniques and procedures for maximum organizational efficiency in production. Management programs and tools to application in the prediction of failure situations are proposed; thus, the steps for the implementation of the methodology developed in industrial plant are presented. An activity roadmap for investigating technical–operational culture at an oil and gas industry workplace is presented.

Consequently, this discussion brings the definition of new criteria for human machine interface design, methods to measure human reliability considering the theory of fuzzy mechanics, and guidelines with primary and secondary indicators to the adjustment of human factors.

A nondeterministic method that begins with identifying signals to measure emerging hazard energy can identify and evaluate parallel abnormality chains to better install the safeguards. Turning large numbers of data into smaller quantities with principal component analyses has enabled the graphical view of optimal point approximation of organizational efficiency using the scatterplot.

Chapter 8, the conclusion of this work, reviews the paradigms that are being broken by new conceptual models and algorithms for calculation. Risk management will be punctuated with the practical exercises and the results of applying these techniques. It is also intended to include prior discussion on the measurement of human factors, networks, concept of hazard energy enabling elements, and calculations for integrated or sociotechnical reliability. Chapter 8 includes a brief discussion about a future book and indicates the relationship between the chapters and respective products,

Finally, Appendix 1 presents several materials related to data collection with respective application forms to be applied in a real case from the reader based on the techniques and concepts discussed. Also, in this topic, there is a database organization and questions for experts for the examination of operational culture.

In summary, we present, in Fig. 2, issues approached throughout the book:

- The discussion about the role of human in society and how his imbalance can become human error (Chapter 1) helps to understand the models of cognitive processing inserted in cultural environments (Chapter 2).
- Management decision depends on human reliability indicators calculated after the analysis of human factors and recovered in Chapter 6.
- Concepts related to the decision and execution of planned actions are included in Chapter 2, indicating the birth of skills and cognitive gaps. On the other hand, they facilitate the discussion of human types (characteristics of the worker in carrying out industrial tasks) and social types (influence of culture) to devise procedures for classifying human error as indicated in Chapter 3.
- From the evaluation of human error and the task, it is possible to discuss the origin of process losses and their consequences, and these aspects are discussed in Chapter 4.
- The questions of the operational routine facilitate the investigation of the patterns that lead to systemic failure, Chapter 5.
- The applications of SPARH in Chapter 6 meet the demand for indicators on human reliability to direct interventions and apply resources.
- And the application in oil and gas, in Chapter 7, discusses the technique of the third generation in human reliability related to technical–operational culture and the analysis tools on the operator's discourse.
- The conclusion indicates the importance of this new methodology and unfolds the perspective on new concepts to measure human factors and find the regions with the highest intensity in danger energy, to optimize the use of scarce resources.

3 Products related to the chapters

The products related to the chapters are discussed following Fig. 3, step by step, including what is needed in guiding the researcher/student/manager/engineer in the investigation journey about human factors to avoid process losses. This explanation shows basic concepts about the products offered in this book with guides: (product 1) process loss mapping; (product 2) task assessment; (product 3) cognitive quality analysis; (product 4) applying the human reliability SPARH method; (product 5) sociotechnical reliability application; and (product 6) operational culture assessment.

Knowing more about human errors allows opening a "range" of opportunities and applications of techniques for solutions in the industry. This book covers an introduction to the study of human error, and how it presents itself in the roles of society, work, and family [1.3]. Human error can be considered as a generic term, in which it encompasses all occasions when a planned sequence of mental or physical activities fails to achieve its goal. The study of human error in Chapter 1 introduces the addressing of human trust [2.1] in industrial settings. The human reliability analysis (HRA), discussed in Chapter 2, can be defined as a measure of probability for the malfunction of a human system. In this way, in the work environment, the failure derived from the action of human can generate several losses, among which the most impactful is the accident. Therefore it is essential to identify the reliability of the operational team, knowing the task scenario and an organizational structure that comprises it.

For the researcher/manager/student to assess human reliability, it is necessary to identify the type of risk in the industrial environment, which in this book is discussed as dynamic or stable. Dynamic risk can be characterized by the instability of variables in a process, such as the task routine, equipment, staff, and management. Stable risk, on the other hand, is a more predictable risk, with controllable variables, such as in super-automated processes. After defining the type of risk, this is the moment when you, the researcher/manager, will identify the type of risk in your process and follow in the tutorial the suggestions for starting an investigation.

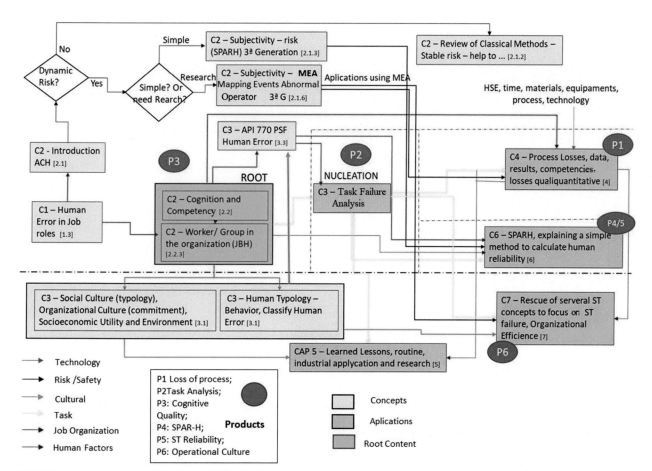

FIGURE 3 Tutorial about chapters, investigation process, and products.

The book's approach is around processes with dynamic risks. Therefore, for the development of an investigation with dynamic risk processes, two investigation methods are indicated: SPARH [2.1.3; 4], which is a method for assessing human reliability and human performance factors, and has easy application of data collection, such as questionnaires and interviews; and the abnormal event mapping [2.1.6; 4; 7], which is based on the records of signs of noncompliance or signs that indicate intermediate states for noncompliance. It is worth mentioning that the stable risk approach was carried out in Chapter 2 with a review of classic methods, such as the technique for human error-rate prediction (THERP) method.

To understand how errors and losses occur in a dynamic risk environment, it is necessary to have knowledge in the cognitive dimension [2.2], that is, in the cognitive system of human; and also understand how it behaves in groups in an organization [2.2.3]. An individual's cognitive system involves the individual's affective, social, and technical relationships. The behavior of human in the work environment is an important aspect that is inserted in matters of occupational safety and health. On the other hand, productivity requires movements of the worker that integrate perception, memory, and attention. The ergonomic design for perceiving process signals, adjusting the control devices, and movements define basic criteria to achieve success in performing routine tasks at the workstation. With these important questions, it is necessary for the researcher to assess the cognitive quality of the team involved in the task, thus recognizing the difficulties and opportunities to be explored in this team.

Acknowledgment

First and foremost, we would like to thank all the family members for their valuable support throughout the project.

We also acknowledge the support of José Rafael Nascimento Lopes and Luiz Fernando Pelerine Pessoa at all the levels of the project, especially at the beginning and during meetings and for providing important contacts for the preparation of the initial chapter, which we will certainly count on in future projects.

In addition, we acknowledge the financial support of CNPq, National Council for the Development of Research and Technology, via a research grant for technological development to the book project leader, Dr. Salvador Ávila Filho, registered as DT2 in the period 2017−20 and 2020−23. We also recognize the financial support for research, through CNPq's Universal Notice in the period 2015−18 for the construction of the project.

In addition to the partnerships the FAPESB and CAPES research agencies have played an important role in funding scholarships such as master and doctorate scholarships for industrial engineering program for Carine Santino as the author. Scientific initiation for the graduate student in Production Engineering, PIBIT, Lucas Menezes, worked as translator collaborating on the project.

We also thank the industrial engineering program of the Federal University of Bahia, which brought together the group of authors, Carine Santino and Ivone Cerqueira, under the leadership of Dr. Ávila Filho during the graduate program.

We would like to thank the undergraduate and graduate students Jade Ávila, Júlia Ávila, Rita Ávila, and Lucas Menezes for their help in editing the book, elaboration of figures, and supervision. We especially highlight the important role of the undergraduate Lucas in translating the materials for this project.

We are grateful to all the partner institutions, National and International Universities (Texas AM University, Poznan University), Brazilian and International Industries, Companies in general and Government Agencies in particular Public Security, for enabling the validation of the concepts, tools, and procedures presented in this book.

We thank Vivien Susan MacIver for the English language review.

A special message to the authors: *"The persistence and acceptance of the challenge of this project, despite the restrictions in Economics and in the Management of Educational Institutions, is proof that Brazilian Culture produces high-quality characters worldwide!"*

Postmortem message: *"The union as a couple gave comfort to the author Ivone to be in this project. The lack of Gil reinforced the importance for its completion, and wherever he is, we want a peaceful rest and the certainty that he was also part of the construction of this book."*

Chapter 1

Introduction

1.1 A brief discussion

The discussion on human factors that can, directly or indirectly, cause errors in the execution of a task, is multidisciplinary and current, intended to provoke the curiosity of readers to search for more information on the subject. It is important to deal with the complexity of working with human factors in dynamic social environments.

The topics in this book include discussion on the following: (1) the organization inserted in the social environment, (2) conceptual and mathematical models in human and operational reliability, (3) risk management in complex processes and environments, (4) analysis of competencies and task planning, and (5) diagnosis on human factors.

1.1.1 The organization inserted in the social environment

Companies (here considered with the same meaning as "industries" and "organizations") are profit-seeking institutions, which transform materials into ready-to-use products. These Companies also execute activities in assembly, product manufacturing, maintenance services, and chemical transformation processes. These products and services are for the use of society, complying with the specifications for use and following the legislation determined by the social environment, altering demands according to the local requirements and in accordance with global market laws, based on economic and environmental restrictions.

Social demands are presented to the Company by the legislation that is, generally, altered in a reactive manner; through compulsory standards that require standards of behavior by the Company and its employees; and from demands of the community that emerge in a latent way and reach the workforce.

Changes in the economic environment influence the social environment by altering human behavior during the accomplishment of the task. Individuals tend to commit human errors such as slips, communication mistakes and inability to work under high levels of stress.

These human characteristics confront the need for competencies to act in complex processes and with legal constraints, such as the availability of water and energy that define the challenges for future organization.

Regarding the possibilities of managerial actions to avoid failure, Silva (2015) discusses how routines work in the Company, which includes aspects of formal and informal leaderships. Managerial decisions should follow standardized procedures as to the choice of alternatives with less social and environmental impact.

However, it is not always possible to achieve this goal, especially when the actions are of urgent demand. Formal leaderships or managers should be sensitive to fast and continuous changes, that society and the market require from this organization. These changes demand transformation of the work team behavior and intuitive emergency actuation skill, which was normally not valued in the past and continues not to be valued to date.

Some cultures inherit inadequate traces of their history (paternalism, avoiding conflicts, centralization), that are transferred to the company's work in the form of vices that must be identified and neutralized.

Rasmussen (1997) indicates the need to understand the various subjects of human knowledge to plan the task and meet the laws and social rules. One of the reasons for omission in the communicating of failures in the steps of the task or procedure is the necessary social adjustment, but not authorized by the company, because of the "false" risks that consider the rigidity in writing the procedure.

1.1.2 Conceptual and mathematical models in human and operational reliability

The difficulty in predicting failure in the dynamic social environment requires scientists to prepare conceptual and/or mathematical models that attempt to simulate the behavior of man in accomplishing the task (Hollnagel, 1993; Swain and Guttmann, 1983), based on expert's opinion and on databases use. Some models are deterministic and need adaptations to withstand oscillations of behavior (Pallerosi, 2008). For deterministic models, generalization difficulties occur.

Other models use fuzzy logic to translate the industry operator's discourse into values that alter the probability of failure (Ávila, 2010; Mosleh and Chang, 2004), or even to avoid failures from heuristics in routine activities such as truck loading of HCl (Hydrochloric acid) or the welding of equipment in maintenance (Moré, 2010).

Ogle et al. (2008) discusses a model that deals with the influence of the automation level in the industry, on the possibility of occurrence of human error, and analyzes the severity of tasks when considering manual, semiautomatic, or automatic activities.

On the other hand, local, regional, and global bad habits are discussed and addressed after understanding the conceptual models indicated by Marais et al. (2006) that suggests the weakening of archetypes to avoid social failure and human error.

1.1.3 Risk management in complex processes and environments

Risks exist because dangers can turn into an abnormal event with different impacts, from deviance to accident with sick leave, or even a disaster. The dangers are hidden in complex processes (Perrow, 1984) that hinder the definition of barriers or safeguards, such as managerial actions and technical devices to avoid accidents.

The difficulty in visualizing the processes and their dangers can be a consequence of the type of control and the respective level of process automation. This low visibility indicates that the complexity of the productive system can induce latent failure in the industry, which is difficult to find, such as "a needle in a haystack." In addition, the social environment with its increased complexity, caused by communication noises, can be an inducer of failures in the task.

Llory (1999) discusses about latent aspects of failure in the organizational movements, due to lack of alignment by leaderships, incorrect managerial decisions and the various human types in shift group leading to the error. Llory (1999) questions the current model of accident investigation that does not consider the reality in fact and does not identify the root cause of accidents.

Industries must discuss the social movements, the formation of leaderships and the risk perception. In fact, companies only care about objective evidence. They only generate in their analysis of failures, incidents and accidents, observations on the events that happened. Companies do not detect emerging social or individual movements to predict future failures.

Dodsworth et al. (2007) and Bevilacqua and Ciarapica (2018) discussed the relation between safety culture and human factors in the task and investigated possible results in terms of accident reduction. Schönbeck (2007) presented a method to revise the safety integrity level after analyzing: organizational and human aspects, energy of failure in the classic model of risk management and analysis of events that compose the accident. Lees (2005) indicates the importance of the signs in the analysis of the failure anatomy and Leveson (2004) proposes a technique to construct this anatomy.

Ergonomics studies the workstation and both physical and cognitive environment, where the n intervention of man on the machine occurs, looking for, through analyses, improvements to avoid discomfort (Iida and Buarque, 2016). This science is concerned with metrics, but it also inserts aspects of social relations at work.

The human reliability improvement programs measure the rate of operational failure, the probability of human error and the mean time between failures (MTBF), besides studying social relations at the workstation and the impact on the success of the task.

Daniellou (2007) discussed the philosophical approach to Ergonomics by introducing the discussion on which is the most appropriate pattern for the work. Carvalho et al. (2006) present that the observations of the control room are important aspects, such as small emergencies. They also discussed the level of adherence between the formalized procedure and the work performed in nuclear power plant operations.

1.1.4 Competency analysis and task planning

Complex processes require higher perception of events and on the respective anatomy of the failure; therefore, they demand new task planning competencies. Muchinsky (2004) discusses the planning of personnel selection and development activities to achieve a maximum level of effectiveness in the execution of tasks. The projection of people development in task performance is not being achieved due to various factors, for example, workers do not decide with efficiency in stress environment, do not have capacity to innovate in problems that happen in complex processes, do not have capacity to work in group, or do not cooperate in the team routine.

Considering research with oil industry staff groups, such as engineers or psychologists, it is observed that there are various organizational factors that make people development difficult. Among these factors, we can discuss inadequate measurement of knowledge in personnel selection; inadequate training programs to achieve the goal of maximum effectiveness in the task; team's lack of commitment in learning; and changes in technology imposing high amount of information to be learned at the same time.

Moreover, social relations are not considered during the development of the routine task, regarding the leadership aspects of the team and the quality of communication in the group. Beyond these human factors, technical information and the company's policies are also part of a technical-operational culture that needs to be understood. Valle (2003) states that there is a technical culture for each type of technology and company. This culture can reduce or increase the team's performance in relation to the expectation on the competencies for the task.

The economic and social oscillations demand new skills that are dynamic and that help in establishing new standards to maintain industrial processes under control. The work team has blocks of fixed concepts used to accomplish the task. Added to these concepts, there is new information regarding appropriate social relations at work.

The lessons learned must be transferred to the operational mass by indicating regions with higher probability of socio-technical failures, for example, due to corrosion, fatigue and human errors. Therefore, the ways that failures may occur are widespread, indicating only what are the critical equipment and processes and what are the possible superficial relation between e human errors and the technical failures.

Nowadays companies should value a broader training for technicians, beyond the fixed concepts acquired in classical courses, also the best way to operate an industrial plant. It is important to know how to relate the causal nexus of operational and of equipment failure with social and organizational factors.

Task planning is successful when procedures are elaborated and effectively performed according to its written design or sequence. Similarly, aspects of risk on task regarding the failure of equipment, products and people should obtain consensus in the implementation of the new procedures in shift teams (Embrey, 2000). In this way, the definition of procedures, requirements, goals and their auxiliary documents (checklists, bookcases, drawings, and others) transform the task in successful activity.

According to Lees (1996), the competencies demanded by the task planners are different in relation to the competencies for their validation during execution. In planning, a broad knowledge base is required, while in the execution/review of procedures experience and skills in the routine performance is required. For the sake of cost reduction, the company may erroneously assign the planning, execution and revision of procedures to new operators/engineers, which bring possibilities of failure in the task due to non-acceptance of prior operators, misuse or even use of inappropriate task in the operational scenario.

1.1.5 Analysis and diagnosis of human factors

Some pathologies resulting from human relations conflicts in the social environment are repeated at the workplace. It is important to understand the main psychopathologies (symptoms of neuroses and psychoses) to learn about the types of human errors that are committed in this environment.

On the other hand, information, and knowledge about the functions inherent to cognitive processing (Sternberg, 2008) are sought, such as perception, attention, memory, and the mental map to improve the analysis of social fault in the accomplishment of the task.

The definition and identification of human types in the team assists in the improved allocation of people in their respective tasks. Fadiman and Frager (1986) and Jung (2002) present human types with the emergence of their respective human errors in social activity, and that, eventually, may influence the performance of individuals and groups in the work routine.

Disturbances, originated from specific personality traits, may be associated with organic pathologies and a link connecting the type of affection, behavior, and possible diseases in the body (Haynal et al., 2001). This knowledge resulting from psychosomatic medicine can help the area of occupational health, operation and company safety, regarding the best mode of routine management.

It is important to talk about the characteristics of these affection networks, the consequences in the body and the behavior for diagnostic work in human reliability. On the other hand, information based on interviews or polls may have low quality because it does not represent the human type present in the workplace.

The study of human error demands new techniques that are more assertive. It is necessary to provoke stress, and investigate its effect (Souza et al., 2002) on behavior and on organic disturbances, and measure the impact by stress level, directly and indirectly, in the decisions to carry out emergency tasks (Ávila, 2010).

Data collection to study behavior and to diagnose human types and their social relationships (social and human latent situations) is based on business techniques that involve subjects such as:

- behavior under stress in emergencies (Ávila, 2010),
- formal and informal leadership based on the FIROB (Fundamental Interpersonal Relations Orientation-Behavior) technique,
- group aggregation or disaggregation,
- commitment analysis,
- quality in formal and informal communications, and
- self-analysis of the psychological quality of the worker.

Pasquini et al. (1997) analyze different types of human machine interface, in control panels, considering the quality of the team's cognitive processing and the quality of oral and written communication. Sternberg (2008) presents intrinsic factors in the analysis of cognitive processing when verifying the possibilities of human errors, the impact of its consequences and the suggestion of barriers for its prevention.

Reason (2003) and Dekker (2002) discussed the study of human error and the treatment to prevent its recurrence, in the company. These authors affirm that companies treated human error by attributing guilt to the injured parties, and currently, they analyze the error in search of the environment promoter of this event. The definition of the modes of occurrence of human error is fundamental for the preparation of algorithms that exercise the prediction of failure (Hollnagel, 1993; Mosleh and Chang, 2004; Pallerosi, 2008).

The diagnosis of cognitive gaps and human factors aims to identify human and environmental characteristics that induce failure and that can be avoided, mitigated, or have their effects mitigated through specific programs. The emotional balance in the shift team of an industry can potentiate situations of comfort, create situations unfavorable to cognitive traps, and increase the quality of cognitive and intuitive processing (Ávila, 2010).

The discussion about stress is made by Lorenzo (2001) when he presents the external and psychological stressors; Souza et al. (2002) when identifying psychopathologies and behaviors arising from stress; and Pallerosi (2008) in an attempt to quantify the level of stress at work.

The analysis of failures in the chemical industry demanded the preparation of a standard for managerial action. The API770 Guide (Lorenzo, 2001) indicates situations of this industry that can provoke human error and treats the influence of stressful environments or stressful situation that causes the beginning of the process of failure.

Ávila (2011) discusses situations of changing patterns in the chemical and petroleum industry and indicates ways of analyzing human factors for environmental impacts, accidents, operational failures, and equipment in general. In some cases, the preparation of educational and motivational programs is indicated as a managerial action to change old bad habits installed in the operational culture and/or in the behavior of people and groups (Fig. 1.1).

After this discussion, it is perceived that all topics cited—items from (a) to (e)—have significant importance for good conduction and execution of activities and in the risk, control previously known by the company. In your opinion, which of the topics would be the "top" of the iceberg to unleash human error? Moreover, why is that?

FIGURE 1.1 BOY 1 time to think!

1.2 Discussion timeline and schools

The human reliability theme emerges based on various movements of schools of administration, security engineering, in the areas of production administration and psychology. This movement induces research on human factors in the analysis of the task.

1.2.1 A historical vision about schools related to origin of human reliability

The description of the schools and activities in human reliability is based on an interpretation without scientific basis, therefore, an intuitive analysis of the findings and practices performed. Fig. 1.2 presents institutions, segments of the economy and schools involved with techniques in the areas of administration, engineering, ergonomics, psychology, and safety guided to human reliability.

In terms of countries dealing with these issues, the United Kingdom should be remembered with its mass production, the United States with Fordism and Japan with Toyotism and Total quality control. In the Social area, the psychology of American-English work and the psychodynamics of work in France. In ergonomics, the French cognitive-social discussion and the American objective-anthropometrical ergonomics for Interface and control room projects. Finally, the subject of human reliability led by the nuclear energy area at the United Nations, represented by the Committee on the Safety of Nuclear Installations (CSNI) and the analysis of the task in the chemical-oil industry in England (Manchester).

Some cognitive models explain human error, as by Hollnagel in Nordic countries, such as Sweden and Norway.

Some Economy Sectors included human factors discussion and human reliability, task analysis: American Aviation introduced the Reliability Centered Maintenance (RCM) and discussed human reliability; American Chemical and Petroleum Industry analyzed stressful environments (Lorenzo, 2001); Rail Transportation Department in the United States and South Korea; Commercial Maritime Transportation in the United Kingdom, discussed task and just culture (Seahorse Project). Finally, the analysis of medical errors has been intense in the United States, in hospitals and universities within health sciences.

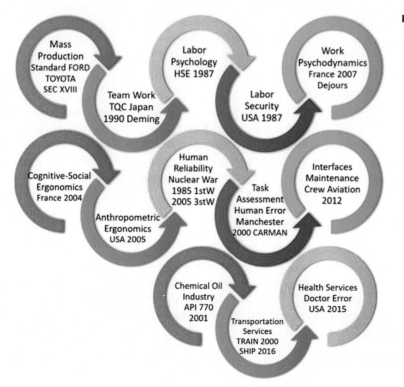

FIGURE 1.2 Schools related to human reliability.

1.2.1.1 Mass production in the United Kingdom and Fordism in the United States

The techniques were born in the Eighteenth Century, during the Industrial Revolution, aimed toward the specialization of certain functions to increase production. In this period, there was the development of steam machines, but the pace of work was intense and there was no concern about the impacts of the activity on health and safety. In continuity with the phenomenon of mass production, the American industry started Fordism by overloading the work team and achieving consecutive production records. In these two cases, in England and the United States, there was a greater effort to organize the task in production cells with their respective jobs. It was a period of learning and applying patterns in the routine to achieve quality in the final product.

1.2.1.2 Toyota, total quality control and total productivity maintenance in Japan

Although the initial period of mass production had been successful, there were many incidents and productivity losses. Due to these product quality and productivity aspects of the team, the revolution of quality arises through Deming (1990). The Japanese industries have applied these principles and statistical techniques for the economic recovery of the country following the Second World War.

The Japanese people had cooperative culture and found that most of the deficiencies in technology were discovered and mitigated through developing teamwork. Thus systems such as TQC (total quality control) and TPM (total productivity maintenance) were created that focus their actions on teamwork. An example of the economic resumption in Japan is the Toyota Philosophy, where the mapped production cells possess operators with specific characteristics. There was a need to develop techniques to increase the efficiency of teamwork through leaders with specific profiles. Despite the positive results in this period, there are events that escape the control of the TQC or TPM, which is the analysis of the environments on the worker's behavior during the accomplishment of the task.

1.2.1.3 Psychology and safety of work, discussion originated in the United States and Europe

Human errors in production influenced the quality of the products and the productivity of the team. In this scenario, American manufacturing entrepreneurs hired research to diagnose the reasons for these process losses. The question was the loss of motivation for the work, and the level of teamwork involvement in the production task (Herzberg, 1987; Maslow et al., 1987).

On the other hand, the increase of accidents concerned European and American Companies with the need for new guidelines including risk management. At this stage, the mobilization was around the supervision of labor activities. The work is organized according to rules that must be followed. In the case of non-conformity in safety, there would be the attribution of guilt and penalty. Labor standards were altered to investigate accidents by assigning the fault mainly to the injured party and not analyzing the work environment with the respective production factors.

1.2.1.4 Psychodynamics at work in France and England

The Industrial Revolution initiated in England generated physical efforts to achieve the goals for production, generating dissatisfaction of workers and political-labor movements in the unions (Marxism). In France, humanitarian movements were initiated after the Industrial Revolution that reflected in the work environment. Studies and applications of work psychodynamics were done to mitigate the impacts on health, safety, and quality of life. The relations between health, work and life are discussed by Dejours (2007) based on a dynamic vision, the French psychodynamics. It is possible to perceive the possibilities of the work as a psychic structuring factor, in which the possible referrals of learning toward pleasure and health occurred.

1.2.1.5 Cognitive ergonomics (France)

French ergonomics was born based on the psychological, cognitive and social discussion, while American ergonomics was built to define physiological patterns that avoid discomfort in the operations performed at the workstation, named as Anthropometrical and physiological analysis. Despite the effort to reduce discomfort, it was not possible to explain the causes of fatigue and human error without discussing the factors related to the planning and execution of the task (Daniellou, 2007).

Cognitive ergonomics was born from the discussion about the relations between capital and work and the measure of suffering or satisfaction in relation to the work performed. These philosophical discussions did not describe the efficiency and reliability of the work activities and did not mediate the failure rate or service of the man machine interface.

Social relations in the workplace are approached in French cognitive ergonomics, in which the role of man was analyzed considering the possibility of exceptions in relation to the standard; hence there are no major concerns about the measurement, then delivering specific patterns for each type of installation for each region.

1.2.1.6 American anthropometrical ergonomics

Anthropometry is the study of distances, colors, times, figures to compose the design of computer screens in the industrial control rooms to facilitate the human machine interface. American ergonomics defines metric standards to enable the design of equipment at the workplace (Pheasant and Steenbekkers, 2005).

These projects are prepared based on the average criteria of the local culture (height, distance, color, timbre) without considering the worker's physical type exceptions. The exceptions considered in this discussion can avoid major accidents in the industry. Thus the design criteria cannot be very rigid, thus admitting certain variability for the adopted parameters.

1.2.1.7 Human reliability in nuclear power plants (CNSI/ONU)

The CSNI began to worry about Human Factors and scale the projects of future nuclear plants. This occurred after accidents where human and managerial error were part of the main causes.

The type of operation and nuclear fission reactions can bring chain consequences and affect the local and global community in an impactful way. The industries in this area induced academic researchers to work in modeling events and avoid their repetition. These industries demanded the formulation of conceptual and mathematical models in research in the academic community to increase knowledge about the human factor in the accomplishment of the critical task and that can cause the accident. A database of fault in critical task was constructed.

Human reliability is withdrawn from the philosophical discussion and goes to the practice considering the possibility that exceptions cause the accident. The schools that developed models, between the end of the 20th century and the beginning of the 21st century, for the analysis of human reliability were concerned with collecting the specialists' opinion. One of the main flaws commented in the reports of the nuclear industry is the quality of the data that does not reflect all the task situations due to the variability of human behavior.

1.2.1.8 Analysis of the task in England, Manchester, and Sheffield

The analysis of critical tasks is important to avoid possible process losses including, mainly, accidents. It is known that a better understanding of the tasks reduces the loss of image and capital and increases time for accidents to happen; decreases material losses and bad image in environmental accidents; it also decrease loss of energy and a better image with the lower fuel consumption and decrease in CO_2 emissions to the atmosphere.

Schools in Manchester and Sheffield have worked in the field of industry task analysis, but without success in correlating it with their safety culture. The authors of the task planning and analysis area are from Embrey (2000) and Lees (1996) Schools. Embrey (2000) presented that consensus with the staff is very important for the elaboration of safeguards in critical tasks involving technologies, people and equipment.

1.2.1.9 Cognitive models about human error in Nordic countries

Suggested models for task planning and the best decision against crisis scenarios facilitate standardization, training and selection of appropriate people for leadership in these activities. The models for task analysis must communicate with the cognitive models for decision-making or for the automatic execution during emergency. Human types are not discussed in these models. Human reliability invests in the discussion about the analysis of the task by considering the cognitive process of each team in the company.

1.2.1.10 Human error analysis in US aviation

An airplane crash has a high social impact due to the amounts of human losses. Thus the work of Human reliability began in the following areas: in the projects of the aeronautical industry; in the control of airspace; in the aircraft operation in terms of pilot-crew communication; and in aircraft maintenance to increase availability. The high impact of poorly performed tasks by the supervisory bodies controlling operations on airplanes, led to detailed studies of Man Machine Interface in the control of aircrafts.

1.2.1.11 Chemical industry and oil industry in the United States

Oil and chemical industry accidents affected Society by bringing negative consequences for public health. Society demands solutions to reduce the risk of accidents. The result of this pressure is the publication of the API770 (Lorenzo, 2001). The environmental impacts discussed in the legislation are not detailed because of failures from human factors. Currently, Human reliability advances in relation to security since it considers that human error is the result of the environment of the task, the organizational environment and the socio-economic position of the Company and the employee.

1.2.1.12 Rail transportation in the United States and maritime transportation in the United Kingdom

Rail and maritime transport services make concern the society due to the economic and social losses caused by accidents due to human and organizational factors. In the United Kingdom, there are Human reliability Programs for maritime transport. In the state of Virginia, in the United States, there is a guideline for human reliability with the analysis of critical tasks in the railway network.

1.2.1.13 Medical error in the US health services

Universities (North Carolina) and hospitals have jointly initiated research in human reliability. The services in which the decision model is studied in the health area are anesthesia, surgery, diagnosis of serious diseases and nursing (Fig. 1.3).

Schools composed an evolutionary movement in the concepts of human reliability and the failures investigation derived from the influence of human factors. The access to the true root cause of complex events is not satisfactory due to almost no investigation performed in subjective dimensions. Given this, what is your suggestion to investigate the latent failures?

1.3 Worker role in job and society: human error

To achieve higher human reliability, it is necessary to understand the importance of Man for organizational efficiency. It should be asked,

What is it to be efficient to the organization? What is the role of the workers? How?

Nowadays, only profit and being positioned in the economic market is not enough, to maintain sustainability. Therefore, we seek to understand the demands of society, nature and economics that define the rules of action. Accordingly, the company's performance is evaluated not only on economic issues, such as paying taxes or wages, but also on natural and social environments.

Organizations are inserted in a context with various stakeholders. If the team is not dynamically involved, the rules followed "are changed" and companies lose the control of socio-technical processes over time. Socio-economic and

FIGURE 1.3 BOY 2 time to reflect!

environmental sustainability depends on achieving satisfactory results (profit), considering their strengths and the interests from the scenario in which they are inserted.

Every company, even if it is an industry with a high level of automation, consider Man as its fundamental part, whether involved in the technology, the design, the assembly, the operation, the movement of materials and even in the marketing area. So, it is important that the company's work team is focused on achieving success in the execution of their tasks.

1.3.1 Human activity

The worker to be focused on his/her tasks in the workstation must perform activities that bring satisfaction and emotional balance and feel included in society. Therefore, the analysis of the roles assumed by the worker it is essential for the work team, Roles vary according to dominant culture and sometimes, change with seasonal effects. As human reliability involves business organizations, it addresses two strands or segments: Man's role in society and Man's role in the organization. This analysis includes the workstation project and the execution of the task scheduled in the routine, including behavior in emergencies.

As shown in Fig. 1.4, the role of Man at work influences his role in society and in the organization, according to his level of performance and satisfaction. The operator can perform different functions in their job, such as:

- Perform practical tasks or in areas of automatic control.
- Plan tasks to be done with the respective tools.
- Lead teams to develop tasks and organizing working groups.
- Search systems, changing patterns, testing situations, and tracking changes.
- Represent the organization interests in external means.
- Define and implement strategies to enable sustainability in the company.
- Deal with high-risk situations that require specific profiles to work in the field.

1.3.2 Role at work and society

Chiavenato (1993) describes Man's role in society (Lewin's theory) and Man's role in the organization by indicating the importance of discussing the appropriate managerial profile to delegate activities in production. According to Lewin (1965), in field theory, the worker/man is influenced by his roles in society (family, school, leisure, profession, groups and politics). To maintain efficiency in accomplishing the task, the worker's emotional balance needs to be reached.

A field of forces that moves Man's attention to each of the areas (society and organization) is established. In certain situations, there is no way to choose where to direct more attention, so there may be an incorrect division of attention. In the analysis of human and social typology, the learning processes with emotional, intuitive and cognitive bonds with the possibilities of occurring human error are discussed.

Family. The role of man in the family is linked to affective and learning bonds in the education of parents and children, or in their miseducation, possibly caused by current mass culture (example: TV and internet). Affective bonds with the family may indicate that dominant (global or regional) culture influences education. Emotional stability may result from family stable bonds that lead to conformant behaviors in relation to the expected task execution. The inverse

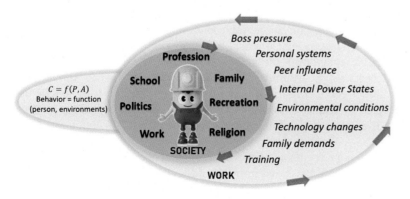

FIGURE 1.4 Relations of man in the society and in the job. *Adapted from Chiavenato, I., 1993. Human Resources. São Paulo, SP.*

can also happen, when workers who are unable to maintain links can bring lack attention and memory problems to their work.

School. Learning levels that depend on the educational process, promote facilities or difficulties in creating new concepts, or in analyzing causality. The abstraction, logic thought, phenomenology investigation and the analysis of the movement are the result of educational processes. These processes should not always be static and substantiated on the past. In the formation of competences or dynamic skills, efforts are mobilized to apply knowledge and skills through the creation of new concepts.

Leisure. When work is assumed compulsively, the accumulation of tensions from vigilance actions and use of attention continuously can cause fatigue in the worker. Physical or mental fatigues also emerge from irritated behavior. Irritation affects people in their professional conviviality and provokes emotional imbalance. Compulsive work is dangerous in terms of the possibility of causing human errors. Thus seeking emotional balance is important to share work time and leisure moments. Everyone has a preference on how to spend time considered suitable for leisure.

Profession. Working brings the benefit of social and economic inclusion and helps to maintain emotional stability in social relations. The work demands from Man, an attention level appropriate to the continuous exercise of decision and leadership in social environments. This attention demand is reflected by his behavior in family life and, on the other hand work activities influence behavior in family routines. Not fulfilling the performance criteria in the company can create a sense of exclusion in the workplace and, consequently, social exclusion.

Groups. Citizenship is important for the stability of Man at work. Social, political, and religious groups generate opportunities for the worker to act as a citizen and strengthen his/her role from the humanitarian point of view. This strengthening helps to maintain emotional balance, broaden the horizons of knowledge, and motivate Man to act more actively in society.

Role in the organization. More than half of the individual's time is spent in an organization for production. Knowledge of power relations and their demands move strategies and define this individual's behavior at work. Therefore, the Organization demands attention to strategies in the direction of achieving efficiency and this will result in a higher motivation of the worker.

States of internal energy. The experience in childhood and the generation of paradoxes of each learning phase provoke individual history marks that define current traits of behavior that cause psychological stress, unstable internal energy (Fadiman and Frager, 1986). Internal energy states are able to favor or harm the position of the worker in the organizational environment. Moreover, they may be the reasons for unexpected coping of the authority (executive) or the creation of insecurity situations in emergency scenarios (memory or causal nexus blackout).

This discussion is also valid for routine situations, where constant anguish or anxieties may occur, or even depression, cyclical humor that damage the sociability and functionality of the worker. Emotional stability depends on states of internal energy and hinders predictability, so it is appropriate to maintain stable social environments so that individuals have basic humor without oscillations.

Environmental conditions. Observation and attention on physical and cognitive signals emitted in the workplace indicate risks and opportunities that change organizational competitiveness. When risks are offered, psychological pressure can be caused on the worker in decision-making and, thus, generating negative biases. In complicated decision-making processes, it is important to adopt techniques for neutral decisions. The emotional balance resulting from stable environments through the practice of leisure and the magic of family relationships becomes an increasingly demanded quality in organizations to avoid the stressful environment.

Training programs. The company demands continuous improvement of its skills, and, for this, continuous learning programs are developed. The development of skills and specific knowledge for new situations is the activity required by the organization to be accomplished by the workforce. Knowing the level of competence "installed" and "applied" after certain training is not a simple activity and requires the application of dynamics rather than static tests.

Pressure from the manager. It is important to adopt a management model that is not centralized and enables leaders of smaller groups to make decisions. If the model is centralized, the inhibition of initiatives for the development of competencies will occur, considering that the Manager solves everything. The good employee is the one who meets the pressure form the superior leader and captives his/her confidence in the desired results, freeing him from routine decisions. The questioner employee is not yet accepted by managers and supervisors influenced by regional culture with centralizing profile. The conflict of generations is a behavior resulting from the non-acceptance of centralizing leaders at the workstation.

Influence of colleagues. Paternalism is a trait of some regional cultures and can generate workers' dependence on the leaders. The globalized and media culture propels selfishness and provokes difficulties in cooperation for accomplishing the task. A negative influence of colleagues can motivate decisions with loss of perception of a current scenario, causing a climate of internal dissatisfaction. These are uninteresting situations for teamwork. On the other hand, there is a positive influence when seeking to improve organizational efficiency while maintaining quality of life. At this moment, we need to forget paternalism behavior and work in a cooperative manner.

Personal systems. Understanding the organizational values and the group values is important to establish a psychological contract. Keeping this contract maintains the commitment in a continuous manner, which positively affects Organization results after performing the tasks. Thus, personal systems consisting of criteria and heuristic rules weigh in worker's decision-making and influence on the development of the shift team, of their competencies (Ávila et al., 2008) and on the level of commitment with management and the organization.

Changes in technology. Changes in technology influence the role of Man in the company. Automation, if not well done, has a double sense of either stabilizing or disorganizing the relationship between individual and organizational values. Man may feel threatened or prestigious with automation, depending on how it is deployed. Changes in the process that lead to the increase in production scale pressure the work of Man in the search for results and depends on the learning in relation to the new technology. Therefore, all technological changes, including new types of products, bring stressful movements that need to be mitigated in order not to create an anomaly in the quality of task execution [Center for Chemical Process Safety (CCPS), 2008, 2013].

Family demands. As previously mentioned, the family has contributed to man-worker in the company, in a double sense: (1) promoting emotional stability, which develops an appropriate profile for the task or even emergency situations; and (2) generating instabilities in the basal mood resulting from financial and affective demands not found in the family environment and that affect the work activities.

1.3.3 Departments, functions, and human reliability

The particularities in a company operation that has risky activities, and its organizational and functional structure division should be considered for the construction of a human reliability program. The discussion about the functions and objectives for the departments helps the implementation of this program and the elaboration of procedures, such as the classification and the analysis of human errors. Some departments analyzed in the company regarding the subject of Human and Operational Reliability are Business; Projects and Technology; Production Management; People Management; Production; Health, Security and Environment (HSE); Logistics; and Commercial, as seen in Fig. 1.5.

1.3.3.1 Strategic area—business environment

The Strategic Function of Project and Business Administration is responsible for the risk analysis of the project in its multidisciplinarity. Business management deals with the environments, stakeholders, legislation, aspects related to the community and aspects of local culture. This strategic function also discusses the market of products and the labor market related to the available offer of skills and specialties.

Rasmussen (1997) cited that the legislation represents population's desire and delimits forms of action in the economic environment. Knowing the limits imposed by the law and the flexibilities for the environmental, social and occupational aspects allows to assemble strategies for the sustainability of the company, including to define patterns of behavior that increase human reliability.

The traditions and customs of the region must be respected to avoid the occurrence of failures and accidents. Strategic decisions cannot force a situation that intimidates the region's culture. Unfortunately, this situation can be transformed into latent difficulties not foreseen in the project. The dynamism of the mode of failure that includes human errors transforms future behavior into uncertainty. Therefore, strategic management should understand the environments that provoke the variability of human behavior.

Although the phenomenon of deviance normalization presents situations in which, the beginning of the failure with low impact on the operation is not a priority in the company's decisions, suddenly, the deviance can turn into an accident.

The lack of respect for population stereotypes indicates biases, such as risk aversion, opportunism and normalization of deviances that need to be identified and treated. This treatment is done in a respectful manner and requires the publication of organizational actions for the internal and external community.

To exemplify, the Human Element program of a multinational chemical industry indicates actions to balance the demands from society and those from the company in a given geographic region. The actions chosen and followed, measured as reducing deviances, were visits of relatives to the company and active participation of employees in neighboring community as leaders of activity for the community or active participation in college courses.

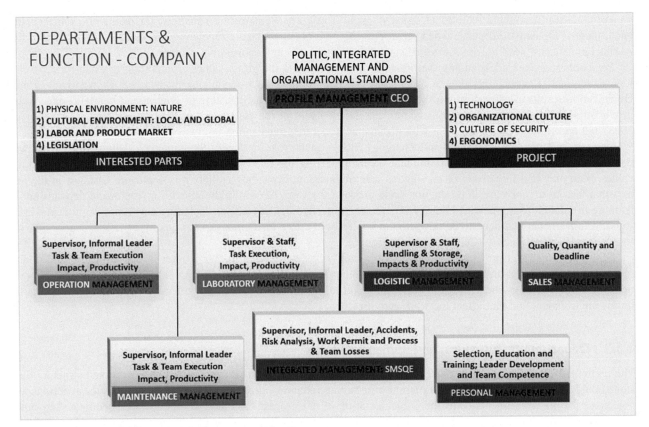

FIGURE 1.5 Human reliability analysis in departments and functions.

The environmental and occupational policies built from restrictive standards makes the operation of complex industrial processes (e.g., chemicals) arduous. The investigation procedures on the accident demand different outlooks of the same problem and are not always available in the organization. It is important to define appropriate internal procedures and standards, analyzing the level of adjustment possible in revisions, to build systems of Management in Safety, Health and Environment for the routine of operations.

Certain emerging events may signal the need to improve internal-external communication and the representation of the organization with the neighboring community. The environmental diagnosis involves the identification and treatment of risk situations that affect the community, making it difficult for the company to be present. This movement can achieve uncontrollable risk or latent resistances.

Knowledge management must be installed in the company, as a strategic activity for sustainability. It is essential to know ways to attract dynamic skills and develop essential competencies to implement the human reliability programs.

The development of competencies may not be a priority in organizations that have a centralizing managerial profile. Accordingly, some modifications are necessary in the organizational culture to allow changes toward a new reality. Only in this way can organizational resilience be achieved.

1.3.3.2 Technology and projects

The technologies elaborated in Research Institutions may not consider the dynamics of human factors in the operation, following incorrect definitions that cause human errors in the assembly or operation of industrial plants. In detailed designs, failures may occur due to inadequate tools, reduced attention, or inadequate budgeting, cutting money from human error barriers investments. The failures in the detailing step can function as a domino effect, inducing the low quality of materials and services.

The considerable approximation of the project area with the operational history of the industrial plant mitigates serious errors of unsolved specifications in the routine. Thus the project team should be composed, also, by experienced technicians of the operation, who contribute toward avoiding serious errors that consider incorrect specifications.

In new plants and equipment, it is common to find situations that are not acceptable as new pumps operating in cavitation, control meshes by by-pass, compressor trip without explanations, and others. The criticality in handling products resulting from the increase of operational load or new environmental laws, as well as the lack of willingness or motivation of the panel operators, cause failures in the operational routine.

Serious technological errors induce repetition cycles in the failure anatomy, during the attempt to recover the state of normality. Poorly structured technology has difficulties in controlling intermittent process states and those resulting from production capacity changes and raw material supplier's changes. These issues should be analyzed in the project as operational risks. Some recommendations are present in the strengthening of the operational environment (human factors) or even in new devices for the control of processes through automation. This technological team must have operational and human expertise, to identify the risks of failure and accidents, mainly during changing of parameters, equipment, and workers.

Although ergonomics deals with issues related to Human Reliability, such as task analysis and workstation design, this science does not treat the variability of intrinsic behavior, caused by conflicts in social culture.

Ergonomics is related to the projects and operation of man machine interfaces, equipment, process and the task. These interfaces are integrated with automated systems, such as supervisory or PLC and are inserted in the discussion about comfort during the execution of the task.

Some design rules and systems are important to avoid human error as equipment specifications; programmable logic controllers; supervisory; and specifications for computer screens.

The cognitive limit affects the ability to treat alarms and are part of a cast of standards adopted in the design of control systems in the industry of continuous and intermittent processes. On the other hand, the analysis of regional influences that transfer cognitive gaps and cultural bias in behavior are fundamental to insert specific safeguards for the local working group.

1.3.3.3 Organizational culture

Organizational policies should be analyzed solidly to avoid conflicts with the practices. Written policies, standards and procedures should not disseminate guidelines that are impossible to practice, causing organizational discrediting.

When there are distortions between organizational policies and practices, the company's employees normally do not externalize the problem; only indirectly, demonstrating discredit that leads to non-acceptance of certain routine procedures. The latent positioning of the team, classified by Reason (2003) as a resident pathogen, can be transformed into failures and accidents.

The team's resistance in relation to the organizational culture can happen because there are different cultures in the same company, as the desired culture and the installed practical culture. The best for the company is that the desired culture is installed. The culture diagnosis intends, by interviewing the organization team, to detect the difference between policies and practices and the level of impact on human error. It can be affirmed that the organizational culture and safety culture has a guiding role to the company's safety standards and worker behavior in general.

1.3.3.4 Production management profile

In Production Management, it is important to define the criteria based on human reliability to allocate the worker at the workstation, also, considering restrictions on social relations and knowledge of technology. The manager with an adequate profile should know how to act in routine or emergency situations in search of the best organizational efficiency (Handy, 1978; Motta, 2004).

Individuals have developed and offered competencies for the necessary functions.

The worker's commitment from a strong organizational culture and the appropriate managerial profile will work to transform the competencies available in the workstation to the desired one. People management and production management together bring individual values closer to the organizational values, promoting training and adopting dynamic criteria in personnel selection.

Decision-making in Production Management involves knowledge, global vision, emotional balance, risk identification and analysis of the objective function in the activity. Technical and behavioral aspects are part of the knowledge used in the management work.

A global vision enables an understanding of the environments that comprises decision-making. The emotional balance makes it possible to make decisions in a neutral way, without biases or pressures from internal-external requirements and family.

The analysis of the objective function considers the benefits or financial burdens of the alternatives in decision-making. The objective criteria of costs and benefits, measured and calculated, give the condition to choose the best alternative after the maximization or minimization analysis of this function. This care intends to avoid cultural biases at the time of decision.

The criteria for cost, profit, quality of life and impacts on the business or the activity are verified. Knowing these criteria allows for developing competencies to choose the best alternative in the decision-making process in production management.

Cost criterion involve: circulating materials, depreciation of equipment, utilities, and expenses with own and outsourced personnel, the business result of the company, payments of fines and others.

Staff allocated to the workstation is also considered due to the difference in team's behavior and productivity. The delegation of the task can generate savings in the result of the company.

Multinational companies tend to select managers who have a systemic and generalist view. These managers should also have intuitive decision-making capabilities and not just methods that rationalize the decision (Motta, 2004). This indicates that there are two ways of thinking and acting in management, and what is sought is an equilibrium of both: cognitive and intuitive-affective thinking.

Knowing how to delegate and trust is a managerial art. Knowing how to select in an intuitively cognitive way also depends on practice. The manager should be sure regarding the obliquity in decision-making and adopt support techniques such as statistics or decision theory (Simões et al., 2002) to avoid rework. Management documents must be clearly written, and if there are questions, a draft must be presented for a conference of understanding. The Manager is responsible for making it clear to the team which pattern is desired and the reasons for preventing each one from adopting a different pattern. Rules of conduct (standard behavior) must be written with limits and examples of acceptable and unacceptable situations.

1.3.3.5 People management

Production management is concerned with leading the team to achieve results in operation, maintenance, and laboratory. On the other hand, People Management has the role of coordinating or guiding important activities such as selection, training, education, leadership development and group relations.

The selection process is the People Management role that is directly related to the Production Management. The psychological tests do not guarantee the results of the professional's performance in the work routine (Muchinsky, 2004), demanding the Production Manager participation in the interviews and group dynamics. People Management usually offers training, but the knowledge multipliers are usually from the production team. The identification and development of informal leaderships are performed in the production environment, during routine and emergency situations. The Educational Programs aim to internalize new competencies and motivate the change of standards.

1.3.3.6 Selection by function

Subjects related to human reliability include concepts of production management and aspects on success in the task. The competence desired for the task are from map features, technology, process and product and in the operational conditions delimited by the legislation for economic activity (Rasmussen, 1997). These mapped characteristics are compared with the analysis of human errors and the investigation of abnormal events to confirm the skills to be selected or developed. This map of cognitive gaps guides the necessities of knowledge and skills avoiding failure. This is the necessary guidance to prepare the training and development program, in which knowledge becomes practical application in routine.

Man, or the worker act, mainly, as a group in organizations and, according to psychodynamics; the worker can be positioned as leader, supporter, planner, creator, executor, critic or another role in the workstation. Thus the profile of the individual to be selected should be analyzed, projecting his/her behavior in the work reality.

Hollnagel (1993) states the need to separate decisions by type of activity and by impact level, if human error could occur. The selection by function also depends on the type of activity. What is the kind of professional we are looking for? Who has qualities for emergency behaviors or routine behavior, or able to construct mind maps in tasks or complex activities to investigate problems?

1.3.3.7 Education and training

Analyzing the basic knowledge inventory and its application rate at work is an activity for People Management, but it requires the knowledge and participation of Production Management. These basic knowledges are fundamental "rocks" for the formation of competences. Thus depending on the quality of primary education of the individual, there may be restrictions to develop certain activities at work.

The investigation of processes depends on workers with a broad knowledge base. The good base allows abstraction, logic and causality, which facilitate the construction of a mind map. This fact helps enrich the troubleshooting process. Eventually, the professional has good performance in abstraction, but does not advance in the extrapolation of real-life situations. The knowledge base is necessary to acquire competence in transforming plans into action.

Having a set of skills for the work does not mean being available to apply. It is important to take into account the need for attitude to enhance this skill in routine. This statement is based on the relationship of the power of information in the work environment. The managerial profile and the type of organization can induce selfish behavior at work, without democratizing key points to solve problems in the operational routine. This is not good!

Training programs are part of the People Management activities and depend on mapping the needs (demands of technology and organization) to format the skills offered in training. Personnel allocation and competence development are recommended action toward team performance adjustments. A good practice of the Production Manager is to prepare training multipliers who develop basic material on process failures and lessons learned. used in routine training, increasing skills in the administrative routines and during shifts.

1.3.3.8 Leadership

The identification of informal leaders to perform task forces is healthy for the exercise of delegation of activities with moderate risk. The informal leaders break the centralizing structures in production management and help the processes of improvement that require technical [Center for Chemical Process Safety (CCPS), 2008] and Organizational [Center for Chemical Process Safety (CCPS), 2013] changes. Often the professionals who resist the desired pattern of the organization are potential and strong leaders and, if there is negotiation about individual behaviors and values, the achievement of trust occurs and becomes an ally in the managerial processes. Many informal leaders find themselves "hidden" or inhibited by not finding a chance to externalize their positions.

With the identification of these leaderships, it is possible to start the development through tests, simulations, and decisions in normal operation situations. The People Manager organizes the training of leaders and applies certain external trainings; the Production Manager puts the leader in position of action, accompanies his performance and corrects errors during the training. These informal leaders can be transformed into task force leaders in production or, even, substitutes for formal leaders in shifts. His critical vision assists in increasing human reliability.

1.3.3.9 Operation profile

The function of performing material processing tasks or raw materials for the purpose of adding value to the final product is the end-activity assigned to the operation. Regardless of the technology, scale and level of automation, the operator must perform fluid transfer and processing activities, chemical reaction control, and state change of some materials. The mental map built to accomplish these tasks has low complexity and requires simple and fast thinking, as well as understanding the steps, requirements, monitoring points of intermediate states and consequences after the realization of task.

The variability of processes, the quality of the materials and the reliability of the equipment and management systems cause the need to revise procedures or standards. This review motivates the constitution of operation team prepared for the realization of changes or improvements of the task.

Worker productivity depends on environmental considerations and aspects of the task, such as the availability of tools at the workstation and the understanding of the steps to be performed. The possibility of performing the task in less time leads to the belief that there are more dynamic ways to operate than the previous ones. Knowing the operational practices and analyzing which are the best behavior patterns in performing the task (Embrey, 2000; Lees, 1996) depends on the overall vision of the process, including the impacts to the environments of influence on human performance.

Since mass Production is designed to run 24 hours per day throughout the year, the operation is distributed in shifts, intending to avoid fatigue or stress due to overwork. The shift scheme is designed to avoid exaggerated effort at work. Excess of double shifts ($>25\%$) in which the operator performs two consecutive shifts to attend to the Company, brings financial benefit to the employee, but, if it is done repetitively, means overwork and possibility of fatigue. This situation can cause negative effects on the performance of the shift and, if it is not noticeable, it becomes latent. The impact can be indicated by increasing the number of deviances that do not indicate an immediate negative consequence.

Today, we discuss the phenomenon of normalization of deviances [Center for Chemical Process Safety (CCPS), 2018] that is part of the research on the operational culture.

When the task is planned and practiced (Lees, 1996) many factors must be considered: the text, the steps and the sequence, the requirements, the consequences, the final process states, the responsibilities, and the task failure analysis. The procedure before being executed in practice must go through the imaginary state for the composition of the mental map and the list of requirements and steps to be fulfilled.

To improve human reliability in shifts, it is necessary to analyze the effectiveness of the task, the signs of failure and quality problems in the process/product. Common performance factors (Hollnagel, 1993) are elements that influence human behavior and that can lead to error. These factors are related to environments, structures and processes that have non-desirable characteristics for process technology, tasks performed by man, technical control, and managerial control.

The responsible for Routine Management in turn should perceive how these factors fluctuate from one state to another state, and how it can cause the loss or increment of human, operational and equipment reliability. This analysis is usually performed on visible failure, such as the inadequate way to operate, the rupture of seals and the cavitation of bombs. The operation Manager should observe possible vices on his team and that are considered as latent failures (use of by-pass, customize preferences, accumulate off-spec fluid to not allow close batch in shift time).

The analysis of past events considered as an industrial emergency or the risk analysis of fires or leaks requires knowledge and skills that may not be present in the shift team, especially with regard to the ability to withstand stress. The Operation Manager must map the team to develop appropriate profiles that can handle emergencies. The coexistence of the leaders on the factory floor enables the identification and development of the team in the emergency treatment.

By initiating risk management in the industry, the operation offers confidence to investors and the community regarding the treatment of emergencies, since the risks of larger events are diminished based on the actions that treat these events. One of the consequences in the business dimension is the improvement of the company's image and the reduction of the cost of renewing insurance.

1.3.3.10 Maintenance profile

The availability and the efficiency of the machines depend on a good work of equipment reliability. The maintenance and the reliability area are committed to controlling the performance of this critical equipment but are also prepared to meet the demands of the industrial plant's day to day. If the staff and the production team have not solved the various causes of abnormalities in the industrial area, the maintenance will perform excessive corrective activities making it difficult to meet the preventive activities. In other words, there will be a lot of reworks to ensure the return of productive systems (machinery, instruments, and equipment). These systems suffer damage resulting from latent failures causing an increase for services, and not performing root cause diagnoses postponing the solution of the problems. This phenomenon is nicknamed as a cycle of non-quality in production.

The maintenance function gives support to the work of the operation when the process is in stable state, performing services to keep lubricated rotating equipment, aligned systems, available engines, and valves in its best control. These services are manual and depend on human work with experience and knowledge. If the backlog is high, maybe the quality of the service decreases!

Maintenance activities may be specific, dependent on specialists, or may be cooperative, demanding study of spare parts flow, times, and deliveries by professionals of different function. Therefore, both operation and maintenance depend on professionals who are committed to the reestablishment of the systems in minimal time with their guaranteed functions.

The commitment to the activities carried out by the maintenance area depends on the identification of the team with the organizational values, as well as the role assigned and performed by the operation. The operation team, when it believes that there is a lack of competence in maintenance team to solve process problems, can create a distrust and state of irritation in relation to the maintenance professionals. This situation disrupts the compromise between the parties and changes the application rate of knowledge and skills on routine tasks and special maintenance tasks. In this way, an increase in failure rates can occur, with the return of less reliable production systems.

Maintenance routine tasks include planning, writing, and movement of equipment and spare part, near target equipment. The clarity of writing is essential. Poorly controlled operation or project failures can generate situations of process instability, human errors in the operation, material changes, and inadequate adjustments in critical equipment making it difficult to achieve quality in maintenance services. Thus the control indexes of the maintenance and reliability of the service will be outside the expected standard, among them: use rate of spare parts, MTBF, average maintenance time and efficiency/effectiveness of the equipment after start-up.

In the maintenance area, similarly to the operation area, it is important to discuss the appropriate job profiles for emergencies (equipment, and plant shutdown), routine situations (such as lubrication), and specialized work (maintenance studies that require dimensional and logical analysis to improve performance of complex equipment in operation).

1.3.3.11 Laboratory profile

Maintenance and the laboratory are part of the production, together with the operation. Communication is expected to flow between the sectors clearly (without noise) and without biases (power in the organization). When the support areas rely on the demands requested by the operation, they feel part of the organization and meet the demand in a more committed way.

In the laboratory there are complex procedures to generate decision information that is used by the operation to maintain or adjust the quality in the process. The knowledge of these procedures and their execution should avoid the behavior of excessive self-reliance and the use of outdated standards.

The laboratory performs several methods of detection in repetitive activities and assists the operation with the interpretations of the results. Thus, cooperative behavior is an important feature in the production team that includes operation, maintenance, and laboratory.

1.3.3.12 Health, security and environment profile

The HSE sector (safety, health and environment) is in a prominent position, included in the Organization's investments discussion, and addressing local and global demands of society. In Integrated management systems, the HSE sector has standards to be achieved that depend on the commitment of the production team. Thus the HSE professional must have technical skills to develop diagnose tasks, as well as knowing how to motivate workers to follow the standards established by the legislation. This requires a specific profile for sensitization, motivation and training regarding the forms of action.

Health, Security and Environment relies on the operation team to maintain stable processes and prevent increased release of hazardous contaminants to the environment. This uncomfortable situation may not allow the achievement of planned goals, and HSE sector, in situations of environmental accident, becomes responsible for the news about fines, and negative image with the external community. The impacts of an accident in the industry brings to the production team, a feeling of discouragement and incompetence. This fact may lead to an attempt to migrate responsibility for the causes of problems, involving, operation, maintenance, management, and other departments.

The HSE sector is responsible for the communication of the company with the demands of the stakeholders and for answering the constraints required by the legislation. However, the posture of assigning guilt and inspect workers makes it difficult to predict a risk situation, since the operation team will not be motivated to signal deviances to the HSE sector, indicating the points where the risks are increasing in the technical and social concerns.

1.3.3.13 Logistics

The logistics activities, until recently, were function of the operation. The increase in the production scale and the large movement of materials resulted in the transfer of this storage and transport function, to this new sector, logistics. This sector, in the continuous process industry, is represented by the production control and programming area.

The final storage tanks bring together production batches sent by the operation, which are transferred to the next actor in the productive chain, which is the customer. Thus, all the effort made by the production to keep the machines operating satisfactorily, to sustain the processes in the projected boundaries, and to meet the demands of the commercial area, can be lost in improper handling of materials by the effect of errors in the logistics task. Another important activity of the logistics area is to monitor transfers in large volume, through direct supply lines to customer, by pumps or compressors.

There are risk operations in the use of transportation and loading stations of hazardous products. Each transport option is involved with corresponding volumes: transport by ship means transferring high volumes of products (e.g., LPG); transport by train has volume limitations and requires traffic care due to accident possibilities; and transport by trucks means transit of toxic goods in medium volume, with possibilities of tipping.

Logistics operations are susceptible to human error and there may be loss of product batch, due to lack of attention in manual alignment, as well as lack of commitment in loading times, which generates delays and fines in the final delivery. The biggest risk is the event of accidents with high impact, such as the release of flammable gases to the atmosphere or toxic liquids in rivers or seas. The fatigue is a human common problem involved in all transportation equipment.

1.3.3.14 Client

Certain guidelines or decisions in the commercial area provoke stressors that generate difficult schedules to perform, and that may require complex technical adjustments. Thus, the commercial area can be a great motivator to initiate human error in production.

The customer is the reason for the existence of the company and is one of the main interested party in its operation. Some essential product characteristics required by the customer as quality, quantity, and price and delivery time need control. All these items are under the Worker's field of action in production work or logistics.

The quality can be impaired, where good batches can be transformed into off-specification batches in the mixing procedure. The laboratory may err in false compliance due to incorrect analysis procedure for the issuance of product reports. The quantity not attended due to unavailability of equipment critical in the operation, or due to the lack of parts (error in supply or inspection) or by difficulties in "maintainability."

On the other hand, the operation can practice serious errors generating system downtime and lead to the event of accident or incident, which would prevent the formation of the desired batch by the customer. As the price is established by the market, unplanned downtime, and errors in handling of finished products is capable of causing major damage. Production and marketing costs may not be remunerated, and value aggregation will be insufficient for the company.

1.3.3.15 Summarizing

Human error affects all areas of industrial installation. The root cause of a problem can be initiated by the dissatisfaction of performing tasks that will cause impacts in the future. At this time, the affective bond with work is fragmented, weakening the success rate in the task. Human error can be caused by technical issues (fatigue resulting from the rework of services to control unstable processes) or by the pressure to meet the requirements of the legislation. The community acts in keeping with the history of the industry, perceiving a positive or negative image of the business. This image is regarded as a reflex of environmental and labor liabilities, which can hinder the operation of the company indirectly through the representations of society.

1.3.3.16 Example of the possibility of the same error caused by different functions

Reason (2003) presents an example of human errors in different functions. In a certain factory that operates in the market for 20 years, there is an apparent integration between the various areas, such as strategic management, production management, operation, project, research and development of new technologies. The product developed has physicochemical properties, which at low temperatures can freeze and consequently cause severe accident by obstruction and explosion. This example reveals that the areas can make mistakes at different levels due to the poor realization of the function. Lorenzo (2001) also discussed this example in API770.

In the case approached, human error is a consequence of incorrect actions in the various functions in consecutive or alternate way (Fig. 1.6):

1. In technology, where the chemist failed by not warning to the production team, that the product freezes at temperatures close to the operation.
2. In the detailing project, the engineer failed to specify the vapor trace to keep the operating temperature, reaching its freezing temperature and the consequences.
3. In the assembly project, where the contracted company failed to install the vapor trace—the steam purge is not specified, leading to lack of temperature stabilization.
4. In the maintenance, the ironmonger failed to change the vapor trace from triple to double, decreasing the heating capacity.
5. In the operation, the operators failed because they did not know the need to align the steam to the pipeline through which the product passes, moreover, these activities were not written in the procedure and were not cited in the routine training.
6. In the production management, to save energy due to the unavailability of water and difficulties in the generation of steam, decided to reduce the steam for some systems considered "risk free".

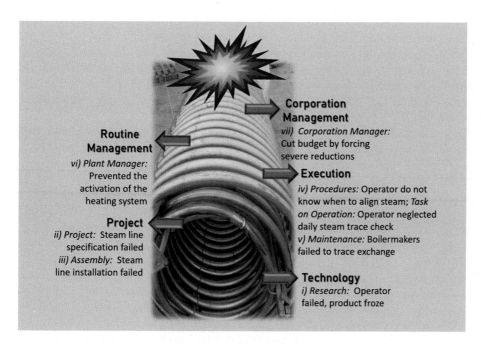

FIGURE 1.6 The failure initiation by different functions. *Adapted from Hollnagel, E., 1993. Human Reliability Analysis Context and Control. Computers and People Series. Academic Press Inc. San Diego, CA. Lorenzo, D.K., 2001. API770 — A Manager's Guide to Reducing Human Errors, Improving Human in the Process Industries. API Publishing Services, Washington.*

7. In corporate management, there was a decision to cut the budget, precluding investments in the utilities area or in the maintenance of production costs in general, affecting the heating of the line that froze the product and caused the explosion in this Industrial facility.

It is important to know how to differentiate the "criticality" of systems and the relative cost of low quality in production. Since there are human errors in different functions, it is important to evaluate the organizational environment or economic position of the company to avoid more serious problems, such as a major event that becomes an accident, or, even a disaster.

1.4 Risk management on material losses and operations

1.4.1 Material loss risk

The importance of analyzing process losses related to low human reliability is explained in an introductory way in this topic.

Industry should avoid process losses (tangible and intangible factors) due to current conditions of competition in the globalized market. The causal nexus of operational failure is not always visible, which induces inadequate decision-making and does not allow the process loss to be avoided.

Activities to reduce process losses have superior benefits in relation to certain types of investment in the production process. This fact is evidenced from the measurement of the results after the implementation of a process loss-avoiding program that uses the information collected in the routine of the shift for investigations.

Process losses can be of various types: material, asset, equipment, quality, financial among others. Human reliability has an intangible element that influences the process loss: low motivation. This element generates human errors that can cause the losses. Individuals with low motivation may incorrectly perform routine tasks and, consequently, cause low productivity. This fact causes excessive costs of materials and inputs, or the non-framing of product quality indices.

The occurrence of losses can be present in every department in industrial process. The variability in the execution of the procedures, in the quality of the raw material, or in the motivation of the workers affect productivity. Changes in management policies and practices can affect the team, especially in industries that depend heavily on human intervention.

The control of process parameters such as pressure, temperature, flow, and the necessary knowledge of the operator change the index of losses. Thus, a program to reduce loss in the process by applying specific techniques to investigate and map critical points can increase the visibility of abnormal occurrences in the routine, which are usually, not signalized by fault indicators, such as alarms or trend-charts.

In the continuous process industries, there is a high movement of materials, involving activities such as sampling, storage and transportation to the final destination. These activities mainly use human labor, in addition to some automated systems (process controllers). For the accomplishment of the activity, it is necessary to perform operational procedures that guide the operator in the execution of the activity. The procedures are fundamental in any type of applied technology, as they guarantee greater reliability in processes with moderate level of automation.

In Risk Management, the integration of the project activity (operational specialist inserted), the operation with sufficient knowledge to maintain the routine of the productive system (ability to adjust the controls and the reactions) is verified and maintenance centered on preventive function, with the necessary support to avoid risky situations.

As the equipment is assembled and starts the pre-operation, the achieved state of the industrial plant does not always reach the state expected by the project. The measurement of control items is important for the delivery of the technology, and mainly for the measurement of production capacity, quality, and environmental impacts.

Analyzing plant conditions during start-up is essential to avoid premature adjustment of social systems in the organizational environment, in addition, it avoids the growth of latent failures, which are initiator events of future material losses.

Knowledge about latent failure depends on: (1) the sensitivity in reading the operational routine; (2) the recognition of delimiters of human performance that depends on the chosen technology, the design and assembly; and (3) the managerial style that makes up the organizational culture under construction. The delimitating factors of human performance are environmental aspects that can induce the operation or production team to commit human error.

Human error is the result of worker's intervention in the execution of the task by inserting subjectivities that are not readily identifiable. Research on human error mechanisms demonstrates whether they are based on knowledge, ability, rule, memory difficulties and attention fixation. The identification of the human (individual) and social (team) typology makes it possible to visualize whether or where the induction of failure occurred from stressful environments leading to unsafe actions.

The analysis of the processes, the history of accidents and incidents in the industry, the investigation of failures encompass the entire production system depending on the causality analysis of human factors in the organizational systems. The task performed by the operator suffers the action of environmental stressors that demarcate their performance (API770). When they reach a level of hazard energization in the failure process, they can cause abnormalities that lead to accidents or material losses.

The company's safety database and respective safety reports discuss accident scenarios for each kind of task. This information may recommend defense mechanisms to be installed to avoid the transformation of risks into accidents, but without assurances as to their effectiveness.

The scenarios built from the failure mode must be based on socio-technical aspects that undergo change and not only in the static analysis of past accidents. At this time, the reading of the operational routine together with the knowledge about human error allows to increase the certainty regarding the need for changes in the risk management guidelines. It is important that routine management of industrial operations be close to the dynamic risk analysis. The risk is not only symbolized as a probability calculation, but the dynamic environments that alter behavior should also be considered.

Designing industrial or service systems that are intended to achieve goals require accumulated knowledge of the history of technology and of the region (cultural and physical aspects). This story can be universal or specific when referring to the accumulated experience of corporations in processing product, of operating equipment in specific unit operations and of performing specialized services.

The expertise required for the process design depends on the technical history that may be, but is not always, documented. A good project begins with the documental definition of production patterns in current systems, which depends on the good circulation of information regarding best practices, and the effective communication implemented in the routine of production. The knowledge of the practice, the characteristics of the products and the processes, besides the difficulties of executing tasks by the team facilitates the elaboration of standards. This expertise assists in the preparation of new projects of industrial plants or specific systems for plants in operation.

The project requires equipment calculations, testing and use of simulators. However, knowledge of the practice improves the quality of the result because it assists in defining the criteria of the project. The competency allocation analysis, for example, is as important as choosing devices to avoid explosion in the storage section of process fluids. Therefore, the social and technical aspects should be considered in the project phase because they are at the same level of importance.

The investigation of operational abnormalities that leads to process losses makes possible the analysis of unsafe actions in the routine of production. Certain details of equipment can generate situations of physical discomfort or low reliability in the representation of data for the production. We need to perceive and correct these situations.

Aspects of the leadership or vices of the operation team can influence the control of systems using the practice of block electronic signal. On the other hand, some individuals are unable to process the same amount of information when confronting supervisory alarms patterns in stressful situations.

Several processes require local operational control based on tests and inspections that if not performed can generate risks or process losses. This fact indicates that the role of the field operator is as important as the supervisory operator.

Additionally, some features of the team may make it possible to increase operational abnormalities and energize a failure process. Unsafe actions inserted in the productive process during the accomplishment of tasks generate incremented scenarios of risk, where danger approaches its objective, that is the accident or the plant shutdown.

The analysis of these scenarios allows, with the knowledge of the operator discourse and investigation of the Occurrence Register, take preventive or corrective actions to adjust the environments and systems. Thus, barriers against accidents or losses in general are constructed.

The operational failure signals indicate if the risk of accident or system downtime increases. These warnings or signals are, normally, not quantified, due to the difficulty of interpreting complex system information.

Some aspects that exist in the workplace may become conducive to avoiding or facilitating human or operational errors. API770 (Lorenzo, 2001) and Hollnagel (1993) signal the existence of common performance factors in the workplace that can cause human error or operational abnormality.

Among these factors, we can mention the following:

- deficient procedures, resulting from inadequate writing or failure in the logic of the mental map;
- inadequate instrumentation that leads to failure in controls due to incorrect signals or actuations failures in the control loops;
- insufficient knowledge, resulting from unrealistic demand for training or inadequate analysis of knowledge inventory;
- conflicting priorities due to difficulties in leadership and organization of inflexible groups in their positions;
- inadequate identification of resources for the execution of the task as equipment, people and, procedures;
- conflict between organizational policies and the practices, due to aspects related to credibility; and
- inappropriate feedbacks, which derive from the use of incorrect language in people management processes.

According to Lorenzo (2001), other factors that can lead to human error and consequent reduction in organizational efficiency are the following:

- poor availability or unavailability of equipment, low quality of facilities of the man/machine interface, and bad maintenance;
- difficult communication resulting from lack of integration between people, that unbelieves in the organization's policies;
- sensitive control and excessive vigilance with the excessive use of automation, measurement and control in non-adapted production systems;
- excessive mental and physical burden for the task;
- inadequate organization and order in which the form indicated by the management is not the most appropriate to facilitate the accomplishment of the task in the field;
- failures in PLC control and lack of competence completely developed for the accomplishment of the task;
- unestablished or inadequate physical constraints for risky tasks; and
- excess of functions may result in overlapping of roles with inefficacy in the action around the task.

The high-risk industry can reduce process losses by deploying a human reliability program and critical task analysis. The most impacted areas in reducing losses are production, costs, personnel, image and sales. A good projection of process loss reduction based on human reliability programs is presented in Fig. 1.7 for the following items: overtime, plant shutdown, reprocess, off-spec production, customer loss, spare parts, loss of Materials in reaction, effluent and through gaseous emissions, increased load, and wastewater treatment (Fig. 1.8).

Risk management aligned with process loss control provides an operational area more integrated with operators and leadership. Process losses affect various sectors of the organization. Through a survey in the Chemical, LPG and Energy industries (Fig. 1.7), loss reductions were observed using human reliability programs. Perform a critical evaluation on the percentages proposed in Fig. 1.7, relating the human factors (API770) that influence each loss.

FIGURE 1.7 Recovering process losses.

Chemical Process Industry

A_{QP}: Extra maintenance and operation time
30% about 5%

B_{QP}: Loss of production
30% of 10 hours/m

C_{QP}: Cost of reprocess
30% about 1% reprocess

D_{QP}: Loss Price Product offspec
15% about 2% offspec

E_{QP}: Customer loss (term or quality)
Recover 30%

F_{QP}: Spare parts
Avoid Spending 8%

G_{QP}: Raw material and input
Reduce 20% reaction loss
Reduce 20% losses in the effluent
Reduce 5% of emissions

H_{QP}: Load
Load increase 0.4%

I_{QP}: Effluent treatment costs
10% cost reduction

LPG Processing - Gases

A_{GLP}: Volume increment processed
15% increase

B_{GLP}: Probability of disaster/accident
10% reduction

C_{GLP}: Reduction of deviations
20% reduction

D_{GLP}: Electricity consumption reduction
10% reduction

Electric Power

A_{EP}: Accident cost reduction
12% reduction

B_{EP}: Reduction in power interruption
1 hour every 14 hours (7%)

C_{EP}: Reduction in activity costs
10% about 1% extra hour

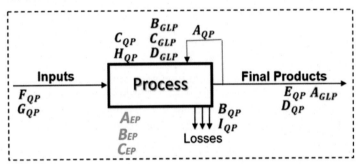

FIGURE 1.8 BOY 3 time to evaluate.

References

Ávila, S.F., 2010. Etiology of Operational Abnormalities in Industry: Modeling for Learning (Doctoral thesis). Federal University of Rio de Janeiro. Rio de Janeiro, RJ.

Ávila, S.F., 2011. Dependent layer of operation decision analyzes (LODA) to calculate human factor, a simulated case with PLG event. In: Proceedings of Seventh GCPS Global Conference on Process Safety, Chicago, IL, March 13–17, 2011.

Ávila, S.F., Pessoa, F.P., Andrade, J.C., Figueiroa, C., 2008. Worker character/personality classification and human error possibilities in procedures execution. In: Proceedings of Third International Conference on Safety and Environment in Process Industry. CISAP3, Rome, Lazio, Italy, May 11–14, 2008, pp. 279–286.

Bevilacqua, M., Ciarapica, F.E., 2018. Human factor risk management in the process industry: a case study. Reliability Engineering & System Safety 169, 149–159.

Carvalho, P.V.R., Santos, I.L., Vidal, M.C.R., 2006. Safety implications of cultural and cognitive issues in nuclear power plant operation. Applied Ergonomics 37, 211–223.

Center for Chemical Process Safety (CCPS), 2008. Guidelines for Managing of Change for Process Safety - MOC. Wiley, New York, NY.

Center for Chemical Process Safety (CCPS), 2013. Guidelines for Managing Process Safety Risks during Organizational Change. American Institute of Chemical Engineers, New York, NY.

Center for Chemical Process Safety (CCPS), 2018. Recognizing and Responding to Normalization of Deviance. Wiley, New York, NY.

Chiavenato, I., 1993. Human Resources. São Paulo, SP.

Daniellou, F., 2007. The ergonomics in the conduction of project designs of systems and work devices. In: Falzon, P. (Ed.), Ergonomic. Blucher. São Paulo, SP.

Dejours, C., 2007. The Human Factor. FGV, Rio de Janeiro, RJ.

Dekker, S.W.A., 2002. Reconstructing human contributions to accidents: the new view on error and performance. Journal of Safety Research 33 (3), 371–385.

Deming, W.E., 1990. Quality: The Management Revolution. Marques Saraiva, Rio de Janeiro, RJ.

Dodsworth, M., Connelly, K.E., Ellett, C.J., Sharratt, P., 2007. Organizational climate metrics as Safety, Health and Environment performance indicators and an aid to relative risk ranking within industry. Process Safety and Environmental Protection 85 (1), 59–69.

Embrey, D., 2000. Preventing Human Error: Developing a Consensus Led Safety Culture Based on Best Practice. Human Reliability Associates Ltd, London, GB.

Fadiman, J., Frager, R., 1986. Personality Theory. HARBRA, São Paulo, SP.

Handy, C.B., 1978. How to Understand Organizations. Zahar, Rio de Janeiro, RJ.

Haynal, A., Pasini, W., Archinard, M., 2001. Psychosomatic Medicine: Psychosocial Approaches. Medsi, Rio de Janeiro, RJ.

Herzberg, F.I., 1987. One more Time: How do You Motivate Employees. Harvard Business Review, Watertown, MA.

Hollnagel, E., 1993. Human Reliability Analysis Context and Control. Computers and People Series. Academic Press Inc, San Diego, CA.

Iida, I., Buarque, L., 2016. Ergonomics: Design and Production, 3. (ed.) Blucher, São Paulo, SP.

Jung, C.G., 2002. Psychic Energy, 8. (ed.) Vozes, Petrópolis, RJ.

Lees, F.P., 1996. Loss Prevention in the Process Industries: Hazard Identification, Assessment and Control. Butterworth-Heinemann, Great Britain, GB.

Lees, F.P., 2005. The hazard warnings structure of major hazards. Trans Ichem E 60, 211–221.

Leveson, N.G., 2004. A new accident model for engineering safer systems. Safety Science 42, 237–270. USA.

Lewin, K., 1965. Field Theory in Social Science. Pioneira, São Paulo, SP.

Llory, M., 1999. Industrial Accidents, the Cost of Silence. Multiação, Rio de Janeiro, RJ.

Lorenzo, D.K., 2001. API770 – A Manager's Guide to Reducing Human Errors, Improving Human in the Process Industries. API Publishing Services, Washington.

Marais, K., Saleh, J.H., Leveson, N.G., 2006. Archetypes for organizational safety. Safety Science 44, 565–582. USA.

Maslow, A.H., Frager, R., Fadiman, J., et al., 1987. Motivation and Personality, Third Edition Harper and Row, New York, NY.

Moré, J.D., 2010. Human reliability analysis in an oil refinery. Use of fuzzy methodology. Cuadernos del CIMBAGE 12, 71–84.

Mosleh, A., Chang, Y.H., 2004. Model-based human reliability analysis: prospects and requirements. Reliability Engineering and Sistem Safety, USA 83, 241–243.

Motta, P.R., 2004. Contemporary Management: The Science and Art of Being a Leader, 15th (ed.) Record, Rio de Janeiro, RJ.

Muchinsky, P.M., 2004. Organizational Psychology, 7th (ed.) Pioneira Thomson Learning, São Paulo, SP.

Ogle, R., Morrison, D.T., Carpenter, A., 2008. The relationship between automation complexity and operator error. Journal of Hazardous Materials 159 (1), 135–141.

Pallerosi, C.A., 2008. Human reliability: new methodology for qualitative and quantitative analysis. In: Proceedings of sixth International Symposium on Reliability, Florianopolis, SC, Brazil, May 22–24, 2008.

Pasquini, A., Pistolesi, G., Rizzo, A., Risuleo, S., Veneziano, V., 1997. Reliability analysis of systems based on software and human resources. IEEE Transactions on Reliability 50 (4), 337–345.

Perrow, C., 1984. Normal Accidents: Living with High-Risk Technologies. Basic Books, New York, NY.

Pheasant, S.T., Steenbekkers, L.P.A., 2005. Anthropometry and the design of workspaces. Evaluation of Human Work. Taylor & Francis, pp. 715–728.

Rasmussen, J., 1997. Risk management in a dynamic society: a modeling problem. Elsevier Safety Science 27 (2/3), 183–213.

Reason, J., 2003. Human Error. Cambridge University Press, Cambridge, UK.
Schönbeck, M., 2007. Human and Organizational Factors in the Operational Phase of Safety Instrumented Systems: A New Approach (Master's degree thesis). Eindhoven University of Technology (TU/e), Trondheim, Sør-Trøndelag.
Silva, M.C., 2015. Application of best practices in projects management though identification of organizational culture de Charles Handy. In: Eleventh National Congress of Excellence and Management, Rio de Janeiro, RJ, August 13–14, 2015.
Simões, C.F., Gomes, L.F.A.M., Almeida, A.T., 2002. Management Decision Making. Atlas, São Paulo, SP.
Souza, A.D.D., Campos, C.S., Silva, E.C., Souza, J.O.D., 2002. Stress and Work. Monograph submitted at University Society Estácio de Sá, Campo Grande, MS.
Sternberg Jr, R., 2008. Cognitive Psychology. Artmed, Porto Alegre, RS.
Valle, R., 2003. (Org) The Knowledge in Action. Relume Dumará, Rio de Janeiro, RJ.
Swain, A.D., Guttmann, H.E., 1983. Handbook of Human Reliability Analysis with Emphasis on Nuclear Power Plant Applications. (NUREG/CR-1278, SAND800 200, RX, AN), Sandia National Laboratories, Albuquerque, NM, August.

Chapter 2

Human reliability and cognitive processing

2.1 Human reliability

2.1.1 Why study human reliability?

The objective of the human reliability (HR) study is broad and complex because it reaches all areas of human activity in critical activities. This study involves different modes of operation, both routine and emergency, and meets the demands of functional hierarchies, from operating mode to managerial and strategic.

It is intended to evaluate elements of the project and the operational control of plants, when discussing technologies and operations management. Design criteria should be related to worker development, critical equipment details, automation, operational controls, and task planning-execution-review.

One of the challenges facing the chemical and manufacturing industry is the area of human behavior, linked to the possibility of human errors at work. Control of operational discipline (CCPS, 2011) in times of technological or organizational change (CCPS, 2008, 2013; Gerbec, 2017) has been inefficient because it has failed to advance on improving operational culture and technical culture (Ávila, 2010). This phenomenon is challenging scientists to understand how to avoid normalization of deviances (CCPS, 2018).

The complexity of the HR study, as indicated in Fig. 2.1, requires investigating situations that have occurred, and the reasons for technical failures in combination with human errors. The description of the operational context should be added to the investigation of technological and organizational reasons.

The resulting organizational–social and natural environment at the workstation can trigger communication noises that impact the outcome of technology system operations, initiating the failure cycle. Human error may have strong personal and group characteristics or be linked to failures in operational control mode.

Formal and informal workstation rules that depend on the type of production, shift and management profile indicate that the confinement and isolation of the workforce can increase stress and the likelihood of human error.

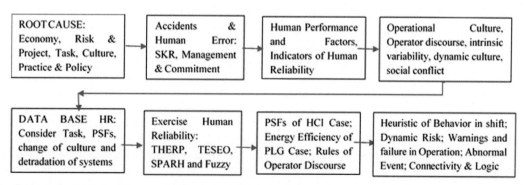

FIGURE 2.1 Gold map for the study of human reliability and cognitive processing.

Llory (1999) studied the great industrial accidents in history, in terms of human and organizational factors. The author analyzed the chain of events of the nuclear industry and raised the following points for the Chernobyl case:

1. The organizational culture and management of large industrial enterprises (Electricité de France, EDF) stems from the physical (technology) and administrative (management) sciences and does not include human factors in the operational routine;
2. Lack of articulation between technical culture (Engineering) and the operational culture (task);
3. The human factor is no longer seen as negative but as a preventive factor, positive;
4. The operator seems lost in a huge, sophisticated and very complex technical scenario.

For the case of Three Mile Island, Llory (1999) raised the following aspects:

1. The accident occurred following a series of policy violations and successive disconnection of most safety systems;
2. Engineers know utility systems and do not know how the core works;
3. During the test, load reduction with disconnected emergency cooling system—professional training inefficiency occurred;
4. Lack of knowledge of external requirements for the test; and
5. Fatigue and nervous tension installed in the operation group.

Souza et al. (2018) investigated the causes of accidents, failures and incidents in more than 20 events in industrial activities related to leadership aspects, organizational environments and human errors (Ávila, 2011b; Carvalho, 2016; Ávila et al., 2018b; Ávila et al., 2018c). Perceptions and actions were performed by the organization, operator or engineer, all in the production line and technology adjustments. The following aspects were examined: Management, Skill, Perception, Knowledge, Rule, and Commitment indicated in Table 2.1. The root cause of the problems may involve human factors or situations influenced by the economic, technological or natural environment in the industry.

It can be noted in Fig. 2.2 that the factors listed as major causes for accidents are: conflicts between policies and practices; the influence of cultural traits; task planning events; inadequate design criteria, static risk management, and economic biases. Depending on the amount of hazard energy on failure, the main event can be classified as accident or incident; the events with average hazard energy are already the various technical, human and social failures; and the events with low hazard energy are warnings of failures or deviations that are not properly investigated, due to the omission caused by the possibility of punishment.

Performance factors alter operator behavior leading to human errors such as slips and forgetfulness. The consequences have impacts on critical tasks, breaking safety standards, quality and production. The Human Resources research seeks to identify cognitive gaps (Ávila and Costa, 2015c) and devices that induce human error. Some of these issues relate to psychological factors, facility design, and cognitive ability to recognize and treat alarms.

High risks of events that release hazardous products and expose workers to hazardous processes demand automatic safety mechanisms and make human errors more visible. The increasing complexities in production systems and organizational formats demand greater insight into the environments that initiate and develop human error. Maybe the operational decisions are made in shorter times and with antagonistic objectives in the managerial aspect, why? Technologically, the quality control and process safety mechanism configurations are more intricate and can account for the great quantity of hazard energy circulating in the process.

A paradox arises, when increasing the level of automation to reduce human error can inhibit initiatives to recover the normal state of industrial plants. This inhibition develops a picture of loss of skills and decreased HR.

Ineffective process safety mechanisms may indicate that there is no human error proof automation. Safety barriers based on Safety Instrumented Systems are important investments, but they do not guarantee that the accident will not happen due to negligent operator vigilance behaviors regarding the process. Safety barriers may be less redundant if parallel HR increases.

Process losses caused by human error are described in Fig. 2.3 through the explanation of Johnny Big Head (JBH) operator activities that were presented in the book introduction, and refer to the following production components: materials, time (plant shutdown), rework, reprocessing, image (accident and environmental impact), assets and energy losses, which affect the objective function of the business. Impacts are on the cost, profit, revenue and useful participation of operator JBH on Society and Economy.

Thus the individual can assist the company to achieve better performance through alert, safe and resilient behavior (Ávila et al., 2018b) in activities avoiding accidents, environmental impacts, energy losses, and asset failure. If this man is not in tune with the organizational orchestration, the result will be human error.

TABLE 2.1 Management analysis, skill, perception, knowledge, and rule (Ávila, 2011b).

	Event
Management	Centralization of the director in the start-up of the plant causes inhibition of the shift activities.
	Manager pressure for unethical environmental solution (drum of chemical), viable operating solution.
	Manager change that does not respect ritual group integration, instability, absenteeism and accident.
	To protect group errors and failures in the event of solvent loss to the reaction, group omission was planned to avoid blame—underreporting occurred.
	Electric damage on forklifts during waste transference. With rain, toluene leakage from tanking from dike to courtyard: dike fault, rain, fire and death.
	Communication interrupted at shift passage—Education: Pass the complete information to the colleague (pass the round ball to colleague)
Knowledge	Improperly specified flow controller acquisition bringing stoichiometric uncontrol events and shutdown. Field laboratory assembly and operational adjustment test.
	Vacuum oscillations in distillation and the presence of benzene in the effluent in the chemical industry. Report with hypotheses that led the group to think that there was a hole in the exchanger shell.
	Difficulties in improving control of the stripper column due to delayed analytical results. Installation of polymer plates to be a visual indicator of solvent in the effluent.
	Deviation logging allows routine mapping, statistics, and causal nexus suggestions through signals, warnings, and faults. Construct abnormalities chain for analysis.
	Alpha-Laval pump O-ring changed from initial start-up to adapt operation. This situation remained for a long time generating environmental impact.
	Reactor release without proper inspection increases the risk of new holes. The rush was avoided with increased redundancy in manual inspection and risk recording to other inspection.
Skill	Individual perception could prevent loss of vacuum—tuberculous ear, operator nickname.
	Staff instruction for new standards, negotiation, adjustment. Listening to experienced people and tuning good setup on the panel.
	Uncontrolled drainage of the refinery wastewater and oil passage to the sea, lack of attention.
	Errors in TDI tank release leads to fatal crash. Skills and knowledge.
Rule	Inappropriate use of PPE standards for boots. We performed an interim audit in job restroom, damage to feet.
	Do not report = do not be guilty, no info to EHS—glove misuse.
	Effluent outflow out-of-range PH and framed apparently. Field research, not accepted.
	Director violates by spraying solvent into effluent—health impact.
with	Initiative to identify hazards in the area assuming paternity to clean and repair.

Diagnostic questions in Annex 2, relation to Chapter 3, Worker Performing Task, and Chapter 4, Process Loss Assessment.

What are the causes of prosecution in this incident?

- Contamination by effluent or air pollution
- Occupational accident
- Process accident
- Failure of critical equipment for quality and production
- Low productivity

    What are the roots or root causes that most influence this incident?

- Low sales with budget cuts? Economic bias
- Task Planning

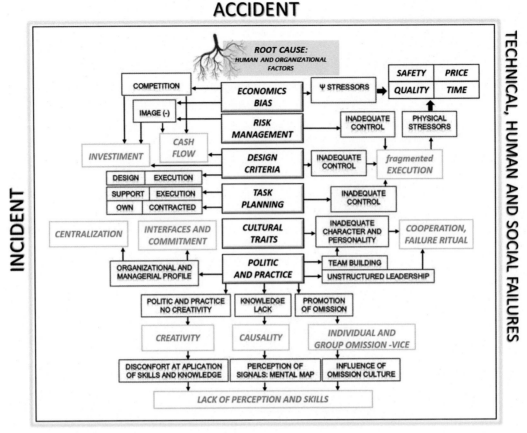

FIGURE 2.2 Root cause and elements for building accidents. *From Ávila, S.F., 2011b. Assessment of human elements to avoid accidents and failures in task perform, cognitive and intuitive schemes. In: 7th Global Congress on Process Safety, Chicago, IL.*

- Cultural traits that affect the perception of risk, opportunism
- Design criteria
- Risk management
- Project criteria: technology and workstation
    Which cause the following cognitive gaps:
- Memory operations failed
- Incorrect perception of risk, reality, and event connection
- Priority and attention calibration
- Difficulty creating hypothesis
- Defects in communication by social conflict
- Leadership failure in elaboration of competence
- Lack of social inclusion and organizational integration
- Lack of sense of justice and clarity

2.1.2 Classic concepts of human reliability

2.1.2.1 The worker is guilty!!!

Hollnagel (1993) defines HR as the probability measure for human system malfunction, as well as the average time before this system fails again (mean time between failures, MTBF). Both definitions theorize that man is a machine whereby observing events, knowing patterns and collecting data, it is possible to calculate the reliability. From these assumptions, the author assumes that people's behavior does not change over time and that HR is a measurable and predictable issue.

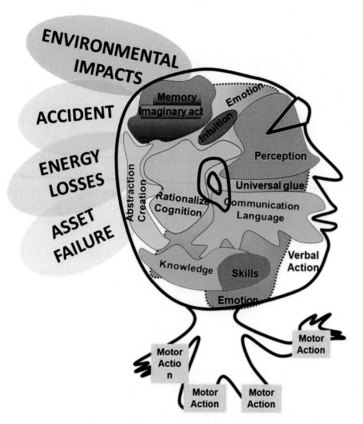

FIGURE 2.3 Losses of process by Johnny Big Head. *From Ávila, S.F., 2011b. Assessment of human elements to avoid accidents and failures in task perform, cognitive and intuitive schemes. In: 7th Global Congress on Process Safety, Chicago, IL.*

A case published by Stanton et al. (2001) uses the HR analysis (HRA) method to choose the best commercial alternative supervisory system integrated with the management system in order to reduce human error and production cost in an Oil & Gas processing unit. Management models are proposed for training, communication, and staffing organization, at the workplace. The HR indicator used to compare the results of alternatives is the mean time between failures (MTBF). We would like to ask if these parameters from models are dynamic or fixed on time?

HR is a complex indicator based on a series of observations and measurement of variable of technological and cognitive processes that aim to obtain similar results with the repetition of the measurement. It is intended to reduce the undesirable consequences of human systems by amplifying the analysis of stressful environments and their relationship to the stability of industrial processes. There is a natural difficulty in obtaining results that follow a single standard, so it is necessary to study the human performance shaping factors and environments that affect the equilibrium (Hollnagel, 1993; Lorenzo, 2001).

How to identify the impact of the operational culture of a given industrial unit on performance shaping factors, and how to know the impact of management profile on the occurrence of human error?

According to Hollnagel (1993), psychological factors that cause human error are discouragement, complacency and insecurity about the future. Physical factors, tangible, that cause human error include: risk of chemical or physical agents; the analysis of the task during its execution; and inadequate physical conditions in the anatomy of the task, generating discomfort, fatigue, irritation and premature completion of the task. Cultural factors, intangible, social customs that affect the mind map.

The team's discouragement may be due to people's discredit of the organization because of conflicts between policies and practices. In complacency, uncritical acceptance is the result of myths in the social culture allowed by the company. Worker's insecurity about the future may be related to the economic situation of the company or the social position of the individual in the company. This creates a sense of exclusion and impacts the quality of decisions and actions.

HR increase indicates positive consequences for industrial process safety. Cognitive processes have a direct and "organic" relationship to process control and safety devices for critical chemical industry activities. We define organic relations when we do not find direct cause-consequence, but, indirect events link, of influence, oftentimes, of nonunderstandable situations.

Human factors can induce a comfortable environment and safe behavior for the repetition of cognitive controls in the task. This comforting environment exists only under a communication-based culture. Thus unconscious and conscious factors are present in the discussion about HR.

Reason (1990) states that:

> *Human error is a generic term used to cover all occasions when a planned sequence of mental or physical activity fails to achieve its intended purpose, and when these failures cannot be attributed to the intervention of any other agent.*

In this context, human error or wrong acts occurs when the individual, even in the face of a systematic procedure, takes actions that lead to a state of alert or an incident or accident. According to Reason (1990), it is observed that the attitude taken by the individual is associated with their performance in relation to the system that operates. On the other hand, it is observed that there are external factors that influence the accomplishment of the task in the workplace and may lead to a human error, such as stress environment, excessive alarms, some family situations and requirements, excessive number of hours worked and others.

Reason (1997) discusses the two phases of accident analysis, of human error investigation in Occidental Culture. The first investigates human error by analyzing the responsibility of the individual who initiated, developed or was impacted by the failure cycle until the occurrence of the top event. When this way of studying the failure mode is adopted, blame for the accident is worked on, following the line of civil law and the Blame Culture where:

2.1.2.2 The injured is guilty!

The second phase seeks physical-cognitive—organizational learning about the environment where the accident occurred. This environment has changed the stability of decisions by triggering human error and the failure cycle with respective events.

Human error may contain cognitive gaps (Ávila and Costa, 2015c) caused by noise in the communication between manager, planner, executor and researcher. Or, due to complex operating situations and interpretation difficulties, the incorrect decisions may occur, resulting in human error and failure.

Dekker (2006) points out that human error is conceived as a derogatory action. In this context, it is explicit that the cause of the failure mode is the nature of the human being. Thus the Blame Culture is reinforced and presents the omissions as a consequence.

In this perspective, it is necessary to format a scenario of the whole event to evaluate the occurrence and the reasons that led to the act. Thus it is possible to identify the actions and decisions taken at a given time and avoid the search for the culprit.

When the investigation mode is adopted to indicate the responsibilities for the accident, a database is needed to discuss the failure rate per critical task step or process control system. Thus the best safeguards are established, ensuring that there are no omissions that interfere with the development of barriers that, certainly, will prevent new events.

The following questions come up:

Is HR related to psychological quality?
How do the functions of perception, attention, and memory work?
Is HR calculation related to the measure of whether people and groups are reliable? or if the organizational aspects are compatible with practices?
How to keep reliability if groups change their behavior dynamically?
What if undetectable omissions or commissions occur?

These questions related to the individual's behavior in task performance are answered in Section 2.2 on cognitive processing, gaps, and human errors. As well as the discussion about the relationship between culture and behavior in a social work environment.

HRA in high-risk activities following the classic THERP (Technique for Human Error-Rate Prediction) method requires extensive data collection on deviations in operation while performing the normal operating procedure.

The discussion about subjectivity resulting from behavior was brought to the spotlight during the training of Nuclear Power plant staff. The new classifications of human error and the proposition of HR methods that observe the characteristics of Operational Culture motivated the training team to a strong conclusion.

Annual data collection, using the same HR method (THERP), with operational events registered from the same facility and, the human aspects identified in this period—do not guarantee that different German and American specialists, for example, have the same recommendation for human systems, in this industry.

It can be assumed that managing people, culture, and technologies to increase HR in organizations is a very important activity and has high subjectivity! Mainly because complex production systems have poor visibility into the causes and intricate relationships that cause the incident and the accident.

The individual, during the execution of the task, can influence his peers, with vices and qualities acquired in the organization and the culture itself. In the company, there are specificities, power relations, between departments or functions, that can change HR indicators. It is important to recognize these specificities to implement appropriate programs and procedures to avoid human errors.

Answer with your own words. Why don't Americans and Germans make the same recommendation for the same production system using the same data with the same HR modeling? The problem is more complex than it sounds. Again, why?

2.1.3 Modeling—first, second, and third generation

System reliability is estimated from the Industry's use of models and techniques to calculate the likelihood of a human error occurring. The nuclear industry motivated the academy (NUREG-6883, 2005) to create techniques and began testing the use of reliability calculations for the human case. We can list the fault tree analysis (FTA); Failure Mode and Effects Analysis (FMEA); and cause tree adapted for human injury accidents such as the Accident Sequence Assessment (ASEP).

Particularly, we applied FTA to Oil & Gas Industry including human error, named as FTAH. We applied FMEA to analyze the work quality of operator activities and of maintenance, such as the discussion about turbine speed control, we named it FMEAH. We worked with risk assessment and defining of barriers using Bowtie H for entertainment megaevent, like carnival in Salvador—Brazil, for example.

Other methods were created based on the occurrence of past events for failure prediction; human error assessment and reduction technique; human cognitive reliability; operator action tree, or decision and action mind map (operator action event trees, OAET), and human error assessment matrix. HR can also be measured by:

1. Probability of failure of an event (human error) = (number of human errors/number of opportunities to make mistakes);
2. The probability of a failure resultant from a tree or chain of events leading to a top event;
3. Failure rate measured over time = (number of human errors / total task duration);
4. The average time taken before the human system fails again, or MTBF; and
5. Process losses = number of losses ($) due to human error in the construction of the top event.

The quantification in HR occurs in this model, with the calculation of the human error probability (HEP), probability that a human error occurs in the accomplishment of a task. For example: within a week, a given task was performed 16 times successfully, and 4 times unsuccessfully.

$$HEP = 4/20 = 0,2$$

In an attempt to develop models that represent reality and assist in proposing safeguards for future production, the nuclear industry and the CSNI stimulated modeling and testing that resulted in three generations of HR estimation methods. In Fig. 2.4 the three generations are demonstrated. In a simplified form, the first generation equates human performance with machine performance, using data not always available for quantitative assessment.

In the second generation, the search for a model for human failure rate continued, with specific environments that alter the result. The need for databases remains to ensure the HRA, considering that the risks are operational and dynamic (Ávila, 2015b). The third generation is a method of analysis that considers cultural influences on the way the task is performed. The result is performance clouds with varying risks (A, B and C), considering the principles of Fuzzy Logic.

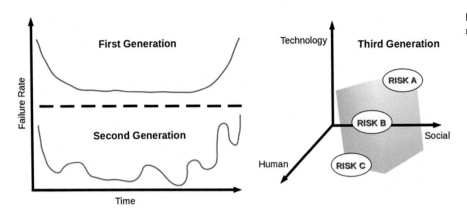

FIGURE 2.4 Generations of human reliability.

TABLE 2.2 Probability data—THERP.

Type of activity	Discussion	Probability
Error of omission of 12th item from list	Although there are very disciplined operators and teams, I believe that human errors for the chemical industry are of around 5%–10%	0.0004–0.005
Align remote and manual valve		0.0004–15
Wrong control selection		0.001–0.01
Identify correct controls and close valve		0.002–3
Monitor processes or surveillance, inspection		0.015–0.019
Operator, procedure, and auto check error	Maintenance and operating errors of 5%–10%—here with smarter increase	0.003–0.03
Clock instrument Reading		0.08
Operator error supervision error	Supervisory errors 10%	0.1
Errors in high stress or dangerous activities	High stress 20%–30% but can reach 70% as indicated in LODA	0.2–0.3
Improper disposal check	Check for inadequate disposal, 90%?	0.1–0.9

Source: Adapted from Lorenzo, D.K., 2001. API770—A Manager's Guide to Reducing Human Errors, Improving Human in the Process Industries. API Publishing Services, Washington, DC.

The TESEO (Technique Empirica Stima Errori Operatori) Tool and the THERP Tool predict the probability of the unsafe act. Failure measures are estimated in proportion to the lack of knowledge about system operation. The result is the level of reliability or risk associated with the system.

2.1.3.1 First generation—Technique for Human Error-Rate Prediction

In the first generation, probability function models follow reliability-like characteristics for machines in HR considering specific environments. The purpose of first-generation methods is to calculate and reduce the likelihood of human error and its consequences. All steps of a task are considered to be performed through predefined procedures and completion of each sub step is essential to its success. Human error may be due to omission or commission. Operator actions can be estimated at the same level of success or failure as a pump or valve and can occur as a result of an unsafe act. THERP is a first-generation method, and depends on an established database for the probability of human error for each step of a task, as indicated in Table 2.2.

HRA is performed in the following steps:

1. Collect information, visit the plant to know the systems, procedures and tasks;
2. Break down tasks into steps (hierarchical analysis of tasks): actions taken (equipment and systems), performance information and identification of probable errors at each step (omission), considering the factors that affect performance;
3. Develop the event tree or fault tree from the listed errors;
4. Determine the nominal probabilities of occurrence of omission errors (a database with description);
5. Estimate the effects of performance shapping factors (PSF) on nominal probabilities and modified;
6. Determine the effects of recovery factors, including these factors increases the likelihood of success; and finally,
7. Total probability calculation.

According to Bayma (2019) each node represents an action and the ramifications are human errors or successful execution of the task, all based on the event tree. Given the use of the tree in binary event format, it is observed that this method does not meet the conception of the relationship between event and performance factors. Thus its application does not meet the expectation of a more careful evaluation, therefore it is not a sustainable HRA for possible changes in the system and in the human performance factors (HPFs).

Silva et al. (2017) propose an estimation of the effects of unsafe acts (human error) on events where failures occurred in order to evaluate HR Analysis methods. According to the author, it is necessary to present what should be changed in the system, to perform a new calculation of the probability of failure. It is noteworthy that the systems are dynamic and can suffer a variability that interferes with the individual's performance. This makes a deterministic calculation of the probability of failure and the result of the HRA unfeasible.

The requirement for a quantity of documented material to perform the analysis makes this method questionable, allowing gaps for close-to-reality analysis. It is important to highlight that the probability of failure is directly linked to human behavior. Thus the variability of behavioral factors, such as stress, directly influence worker performance, although the system and technical knowledge may seem sufficient for good performance and promote a lower probability of failure. This performance needs to be ensured by safeguards based on corporate culture and safety.

The assessment of cognitive processes depends on events in the timeline. Terra et al. (2019) cite the diagnosis of an alarm system during an accident, a timeline base investigation with necessity of interpretation about the sequence of alarms. In high complexity process and task, THERP is not much recommended. In fact, the binary tree of events and performance factors makes it difficult to get an answer. Additionally, as the behavior of the individual is not predictable, and the demand of many branches representing the actions in the fault tree can cause human error. This methodology is not appropriate to represent complexity and indicate a diagnostic.

The existence of the database is not sufficient for analysis due to the possible inconsistency of the results. In the HRA method, a series of variables that motivate or not the individual and teamwork is verified.

In this way, the data resulting from dynamic behaviors are worked and modeled to avoid failure, either from the system perspective or the attitude of the individual.

2.1.3.2 Proposed exercise

In this exercise, we consider propane alignment with the possibilities of human error. Simply put, it was divided into four phases:

Phase 1—Gather information for event tree

- Define a possible system failure
- Event started—Which are the equipment and human action failures that can lead to system analysis failure?
- Different types of letters for human failure branches in relation to equipment failure

Phase 2—Analyze task steps and develop event tree

- Draw the event tree with failure branches and success branches, only failure branches

Phase 3—Qualitative analysis of the branches

- Which first branch only with human failure? Does training resolve?
- Which first branch only with equipment failure? Does preventive maintenance resolve?

TABLE 2.3 THERP exercise.

Failure description and estimated probability	
(1) Operator fails to close propane valve first—0.07	(3) Propane output valve fails to open lock—0.004
(2) Propane inlet valve fails to open lock—0.001	(4) Operator fails to detect locked valve—0.019
	(5) Operator chooses to close coolant valve for leak prop—0.27

Phase 4—Quantitative analysis (HEP, estimating PSF effects, assessing dependencies, and determining failure and success probabilities)

- The probability of a branch is the multiplication of all probabilities attached to this branch—failures and successes
- The probability of total failure is the sum of the probabilities of each branch.
- Finally, the probability of failure analysis is multiplied—first event (Table 2.3)

2.1.3.3 Second generation—TESEO e CREAM

The second generation covers HR techniques that include corrections to task performance. This model includes the variation of behavior in the face of changes in social and organizational rules. Second-generation methods incorporate aspects of human cognition, ergonomics, psychology, and aim to identify actions that require important cognitive activities, determining the conditions and actions that may constitute a source of risk; incorporate knowledge related to the interaction between user and system; identify and model omission errors.

TESEO is a technique for estimating the probability of operator error by combining factors during the application of the following parameters (Kotek and Babinec, 2009) that are related to human errors:

$K1$—nature of activity, routine or nonroutine that requires attention: $0.001-0.1$;
$K2$—temporary stress factor for routine and emergency activities: .5-10; .1-10;
$K3$—Operator qualities: selection, training, and expertise: 0.5-1;
$K4$—Activity anxiety factor, which depends on the situation and emergency level: $1-3$;
$K5$—Ergonomic activity factor, microclimate quality, and plant interface: $0.7-10$.

Silva et al. (2017) proposes to estimate the effects of unsafe acts on the events that occurred, their failures, and indicates that necessary changes must be made in the production system and new probability of failure calculations.

In the TESEO method, it is observed that the analysis is based on the individual's technical performance, the environment in which their tasks are performed and the psychological conditions to which these workers are subjected during the performance of their duties.

2.1.3.4 Exercise

A chemical plant that produces isocyanate constantly has short stops caused by human error. A reactor start-up procedure has brought poor reliability in the reaction industry. After the analysis of the 5 factors,

$K1 \times K2 \times K3 \times K4 \times K5 = 0.1 \times 5 \times 2 \times 2 \times .8 = 0.16$ or 16%, considered high percentage.

Conduct training to reduce emergency anxiety by bringing 2 to 1 on K4 and improve training to increase knowledge and skill by bringing 2 to 1, so the probability of human error goes to 4%, a very reasonable value.

Answer with your own words the following questions:

- How to state that applying techniques with these characteristics can lead to the presentation of a HRA?
- Assuming that the individual is the agent performing the task, is it possible to measure behaviors that may be linked to culture, stress, motivation, profile, commitment, commitment?
- How to state about the difference between the history and the realization of the actions. What is the expected performance of the task and what can happen if environmental differences occur?
- Is there an absolute value for system reliability? And human?
- How to reach this value since there are a multitude of factors that need to be perceived, tracked and measured?
- What is discrete modification in the odds? How to do the analysis by the foggy context of answers?
- What is the explanation, what, how and when do unsafe acts happen?

The Cognitive Reliability & Error Analysis Method (CREAM) is part of the second generation. It is a method that analyzes cognitive processing error and aims to evaluate human performance in system security (Hollnagel, 1998). It presents a structure that allows the interaction of the human cognitive process and the action performed. There is the basic approach and the extended one. The extended version produces an analysis based on the results of the basic approach to provide more realistic information (Akyuz and Celik, 2015; Liao et al., 2016; Wu et al., 2017; Wu and Chen, 2017). This analysis is based on Contextual Control Model of Hollnagel and Woods (2005), which is also used for emergency analysis.

The application of CREAM is divided into the following steps: (1) Construction of the sequence of basic events (performed in the Hierarchical Task Analysis); (2) Evaluation of Common Performance Conditions; (3) Adjustments taking into account dependencies; (4) Indication of probable control mode; (5) Adequacy of the organization; (6) Human–machine interface (HMI) and operational support; (7) Availability of procedures and plans; (8) Adequacy of training; (9) Quality of Team Collaboration.

According to Hollnagel (1998), the distribution of the control mechanism meets the performance evaluation of a group or an individual, keeping the same characteristics of the method application.

Like the TESEO method, the procedures analyzed by CREAM remain delimited, which can lead to false conclusions, especially if this model is used for emergencies. In these scenarios, there is an intense demand for technical and psychological factors that interact with contingency actions.

2.1.3.5 Third generation—Using fuzzy logic to change performance factors

In the third generation, technical and human characteristics are standardized and localized for the accident, as well as the management actions needed to adjust the PSFs. Third-generation methods incorporate aspects of human cognition, ergonomics, psychology, process technology, and environments, transforming discourse and subjectivity into a quantitative language that reveals values for calculations.

Hollnagel (1998), Ávila (2010), and Boring and Gertman (2005) adopt Fuzzy mechanics for HR modeling. These researchers indicate that the solution sets for the problems are dynamic. There is not only a single answer or an absolute number, but a range of numbers for causes that promote unsafe acts. The routine when it is close to the region occupied by low organizational efficiencies in the safety area indicates the proximity of the accident and requires corrections in the human error inducing factors (Fig. 2.5). Thus HR is guaranteed at high levels.

Hollnagel (2017) presents the Functional Resonance Analysis Method to model the functions performed in the routine, as well as to obtain safe practices and to avoid process failures. The author explains that specific events are evaluated to demonstrate how activities can be connected, and how changes in daily life can interfere with an individual's performance positively or negatively. It is a method for use in complex and dynamic socio-technical systems (Ávila et al., 2019a, 2019b).

According to Zadeh (1988), the Fuzzy philosophy brings the discussion about the Fuzzy Logic. Even in advanced computer-monitored technology, situations that express ambiguity do not follow the rules of Boolean logic, which requires precise definitions as yes or no.

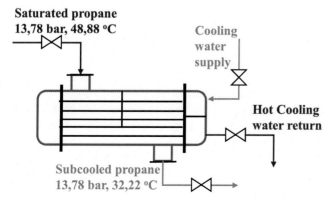

FIGURE 2.5 Propane condenser alignment failure.

Expressions that define high or low may also be average; full or empty, can be Half. Therefore operator discourse (OD) may reflect information rich in a variety of possibilities, and thus the indication of the appropriate way to formulate HRA becomes more complex.

Example 1—The rules of belonging are elaborated from the behavioral regions; thus the Fuzzy Logic indicates the possibility of inference about the risk.

IF the adequacy of the organization is inefficient AND the working conditions are compatible AND the availability of procedures and plans is acceptable AND the suitability of the HMI and operational support is tolerable AND the number of concurrent objectives is greater than actual capacity AND the time available is adequate AND the time of day is daytime AND the appropriateness of training AND experience is highly adequate AND the quality of team collaboration is efficient, THEN the operator would act in random control mode in a timely manner.

Example 2—Prediction of failure based on organizational efficiency or method for estimating and decreasing the probability of failure by measuring organizational efficiency and adjusting PSFs.

Socio-technical behavior is the result of intricate relationships promoted by a set of specific heuristics in the workplace. Processes, people, and procedures produce discrete and continuous variables that explain the phenomena of production. The use of variance, similarities, and dissimilarities allows the number of variables that indicate the performance result on each day to be reduced. The chosen variables (data mining) are integrated through the socio-technical causal link, which is not always easily explained. Organizational efficiency mapping is a technique that analyzes the history of the routine over a long period of time indicating the rules, variables, and conditions for approximation or distance from comfort regions in the industry.

2.1.4 Standardized plant analysis. risk-human reliability analysis method and a case in a chemical facility

The studies of human factors permeate the analysis of the individual in several aspects. The relationship of employees with the company and its organizational culture will reflect on how the employee validates and understands the mission, vision, and values of the company, affecting its performance and satisfaction. Experienced leaders aligned with the organizational culture are able to make assertive decisions in a short time. A mobilized and committed team is able to self-manage and achieve the set goals.

The organizational climate can lead to the strengthening or weakening of employee performance and their link with the environment and co-workers. Stress caused by an unstable environment promotes emotional changes, causes fluctuations in employee judgment, attention, productivity, decision-making, and creates environments for human error to occur. Ávila (2010, 2012a, b, 2013) carried out the human-social risk analysis in an Oil & Gas production plant, aiming to verify the risk level of aspects related to human factors regarding (1) inadequate mental model, (2) low commitment to results, (3) difficulty in dealing with teams, (4) low perception of failure and others. The results indicate high values of risk of problems with leadership, commitment, and mental map. Additionally, actions were proposed to mitigate the problems identified.

HR methods work to ensure proper operator performance in performing the task. The objective is the nonoccurrence of failures for a certain period. The psychological aspects of employees are considered factors that may impact on the execution of the task and, consequently, on the prevention of failure. An individual's personality in shock at potential workplace emits leads to issues related to assimilation of content, steps, and patterns for proper task execution, and social interaction problems with other team members. Along with poor learning, individual emotional fluctuations that affect attention and perception for potential risk situations, memory failures, slips, and problems with accepting the authority of current leadership (Ávila, 2010).

SPAR-H, which is a HRA method, is based on the assessment of PSFs. For analysis purposes, this method divides into steps and task groups that may impact on process safety such as: ergonomic aspects referring to the quality of the HMI; aspects of the task, such as time available, training, work processes, complexity, quality of procedures and aspects inherent to the worker such as stress and aptitude for service and experience.

The use of SPAR-H begins with the analysis of critical tasks that can cause accidents from 8 factors that induce human error in diagnostic and execution activities. Accidents are the result of incorrect decisions that cause human error, failure and abnormal events. Human error, on the other hand, results from cognitive failures such as low perception of risk, failures in memory operations, and inadequate workstation interpretations.

Diagnostic tasks depend on the level of knowledge and time-intensity of experience to understand existing conditions, plan and prioritize activities, and determine appropriate courses of action (NUREG-6883, 2005). In summary, when using SPAR-H, the analysis team presents decisions related to the function of a particular activity to perform the diagnosis or action. The steps for analysis include:

1. Separate the probabilities of misdiagnosis or failure of action;
2. Assess the context associated with human failure events using PSF;
3. Use probability values of human error and predefined PSF following NUREG-6883 Guide;
4. Dependency analysis between tasks. If the task is the first in the sequence, dependency analysis does not apply, considering that there were no previous activities. In this case, spreadsheets developed specially to ensure a consistent analysis are used (Ferreira, 2018).

The method recognizes the reliability of the system, but highlights the assessment of human factors influencing cognitive procedure that can be flagged if there is stability, and anticipates procedures that can prevent a human error, an accident. Chapter 6, Human Reliability: Spar-H Cases, presents 7 cases of SPARH for HR application in the manufacturing and continuous process industry.

The HR methodology using the SPAR-H proposes equations for calculating the probability of human error, associated with the analysis of performance factors, defined by Eqs. (2.1) and (2.2), however, if at least three evaluated aspects have a value greater than 1, it is necessary to correct the value of the probability of error through Eqs. (2.3) and (2.4). Once the errors for the evaluated aspects of the activity (action and diagnosis) are calculated, Eq. (2.5) is used for evaluation of activity error. Equations are used to determine the probable human error (HEPDiag) and action-related probabilities (HEPAtion), and the total human error probability is defined as the sum of the two factors. During the case study performed, the performance factors used were: time available, stress, task complexity, experience, training, procedures, ergonomics, work fitness, and work relationships.

$$HEP_{Action} = 1 \times 10^{-2} \times \prod PSF \tag{2.1}$$

$$HEP_{Diagnoses} = 1 \times 10^{-3} \times \prod PSF \tag{2.2}$$

$$PSF_{Composite} = \sum PSFr \tag{2.3}$$

$$HEP = \frac{NHEP \cdot PSF_{composite}}{NHEP \cdot (PSF_{composite} - 1) + 1} \tag{2.4}$$

$$HEP = HEP_{Diag.} + HEP_{Action} \tag{2.5}$$

$$R_{hu} = 1 - HEP \tag{2.6}$$

The application of the case of the Chemical Industry in Camaçari is presented here for discussing the basic concepts of SPARH. Interviews were conducted with field operators as well as area managers. The results were compiled and analyzed, assigning levels to each of the eight criteria that the method works on. These levels are related to numerical values, which will compose the calculation of the probability of human error, which leads to the HR value. The HCl loading task was analyzed as a diagnosis and/or action (Ávila et al., 2018a).

The work process affects individual or group performance, just as the task is planned, communicated and executed influences HR. In this case study, the work process is considered good due to good communication, well-understood policies and good group relationship. The suitability for the activity, related to physical and mental fitness indicated nominal level, the operator is able to perform the tasks, but does not surprise with preventive actions.

Ergonomics and HMI indicate that there are low quality and quantity of instrumentation and operation information to perform critical tasks such as the HCl loading study case. This situation of poor visualization and difficulties in action by the control panel requires extra communication between the various areas. The procedures have appropriate guidelines for carrying out the activity. It was indicated in interviews that it would be possible to perform the task even if there was no previous experience. The procedure for the diagnosis was considered incomplete, as there was a lack of specific information about the process that would be needed in the decision-making.

Training time and operator experience for the task were considered high for 90% of operators. The stress level on the task was considered high since the task is performed with PPE throughout the unloading causing discomfort. Another factor noted is the variation of the tanker dome—in some situations, it is very small—which makes the task of connecting the hoses ergonomically uncomfortable. Normal stress is considered positive to keep the operator alert and motivated. High stress levels impact the operator's ability to complete an activity as they disrupt attention, physically and mentally affect the operator (excessive heat, noise, poor ventilation, uncomfortable positions).

The complexity of accomplishing the task depends on the context. The risk of human error increases with ambiguity in decision-making and when there is greater physical or mental strain. The evaluation of this criterion for the case study indicates that the complexity of action and task diagnosis is moderate due to the existence of parallel loading and unloading activities. In addition, HCl is a dangerous, volatile product that releases toxic vapors and causes damage to the mucosa, eyes and skin.

The time the operator has to diagnose or respond to an adverse situation is considered nominal for rapid action that includes triggering pushbuttons or radio communication to the brigade team. The time was considered extra because in abnormal situations safety devices will act and bring the plant to a safe condition. Then a diagnosis can be made after returning to a safe condition for leak containment or pressure control.

The HR calculated from the study performed on an HCl pumping system using Eq. (2.6) was 71%. This value includes filtering historical data and calculating equipment reliability (Ávila et al., 2018a).

2.1.5 Human reliability involved in the cultural link

2.1.5.1 Human design and reliability

Knowledge and concepts related to risk activities are transformed into criteria and factors in production system designs and used in the design, equipment assembly and industrial operation stages in Management of Dynamic Risk paper at GCPS (Ávila et al., 2016a); and Reliability Tool paper at AHFE and TAMU (Ávila et al., 2019a, 2019b). Due to a lack of HR knowledge, designers add capacity to equipment to cover possible human errors and lost time due to operational failures. This is called engineering ignorance coefficient.

Another situation occurs when project specifications differ from those used in the operation, and if the expected discipline of routine activities differs from reality during plant operation. In such cases, there will be greater process and service effort to compensate for lost time and quality failures. This "nonquality cycle" (Ávila, 1995) increases plant capacity and adds new human errors and failures to the production system.

2.1.5.2 Structures, processes, and environments

Ávila (2010, 2012a, b) considers that production systems are formed by structures, processes, and environments. The social structures of personality-character-work ability, the physical structures of facilities and equipment are input resources for the functioning of productive systems. Social processes are divided into learning and management, while industrial processes include the flow of fluid and its transformation. Social processes also include decision-making and standardization applied to routine and crisis management.

Well-designed structures and processes indicate stable patterns, reduce the occurrence of deviations, human errors or production failures. Often the changing natural-organizational and social environments threaten the balance of industrial activities. Organizational changes, social phenomena and climate change impact on the workplace.

Human errors are caused by a lack of competence due to training failures and misdirected learning. Human errors occur during the intervention of human structures on equipment and processes. Bad habits are caused by cultural biases that are in the social environment and generate negative archetypes for quality or safety. Incorrect decisions are the result of mismanaged processes in organizational change.

The harmonization of structures, processes and environments makes the production system more reliable and resilient. Culture and technology design promotes balance, indicates high production and safety performance and is demonstrated by mapping the reliability of the socio-technical system (Ávila and Bittencourt, 2017a; Ávila and Dias, 2017b; Ávila et al., 2019a, 2019b).

2.1.5.3 Operational failure exergy

Ávila (2012a, b) analyzes the exergy of operational failure in socio-technical production systems and seeks to discuss the phases of the project, from the project to the operation. Checks the possibilities of failure/hit, or total loss as success of failure, as well as the crisis with the possibility of accidents or mass disasters, both social and technical origin.

The Production System, which includes technological and human resources, is open with continuous migration of hazardous and physical energies into and out of the system or equipment. This flow of energy with a physical and social dimension follows patterns of technology or the functioning of society. Thus deterministic phenomenology loses importance in the analysis of human error integrated with technical failure.

The industrial process can be positioned between normality and abnormality, with or without the presence of deviations and failures. The normal state scale shows how much hazardous energy can be released and what its impact is.

Successful process control indicates a normal state with few deviations. The absence of operational deviations is almost impossible considering the complexity of the chemical industry, for example. Above a certain level of deviations, the existence of a cultural event of deviation normalization and the release of hazardous energy above the normal state is considered.

Human error caused by mind map noise (Ávila, 2010), caused by the difference between concept and practice, occurs in the various working pairs: manager and planner, planner and executor, maintainer and operator, among others. This human error is transferred to the technological production systems through the actions of the operational team.

The severity of human error indicates the intensity in the release of hazardous energy. The transfer of energy from the socially hazardous body (noise in the mind map) to the technically hazardous body (abnormal phenomena in the production process) is discussed in the human error materialization analysis tool in Chapter 3, Worker Performing Task.

This energy of social danger migrates into the equipment and process, becoming technical danger and promotes a state of deviation, failure or incident. Minimum reliability is related to system shutdown with maximum hazard energy release, and maximum reliability is when the amount of shift deviations does not indicate the presence of emergent faults.

Understanding acceptable noise in social processes, such as standardization and operational deviations in industrial processes, will indicate the acceptable level of energy in the social and technical work of the production system, that is, the failure exergy scale (Ávila, 2012b).

2.1.5.4 Human reliability database

Behavioral changes that occur in the production team may be a consequence of changes in the social, physical and technological environment. These transformations affect productivity as well as the economic and social results of the company. Threats induce behavior change by requiring unique team characteristics of resilience, flexibility, and detection of technical failures or emerging human errors.

A social, generation, gender and multicultural conflict may have a short timeline considering history today. This means that behavioral changes took 20 years before media culture and the internet. Today it occurs in 5 years. Thus the human error database is dynamic and exchanges old data for new data over a 5-year cycle. Difficulties in measuring subjective variables require a great deal of data mining and outlier analysis indicating emerging failures.

2.1.5.5 Fuzzy mechanics in human reliability

HR calculation includes factors that affect human performance and human error rate in performing the critical task. Thus it is possible to obtain a behavioral prediction of the individual and to plan the respective safeguards.

Classical HR (first and second generation), although indicating methods for its measurement, may not be effective. Again, the analysis depends on comparison with fluctuating patterns based on environmental changes. That is, fuzzy measurements of factors with fuzzy patterns are compared. Therefore it is not a deterministic science.

HR also refers to the study of psychological quality. It is important to know how cognitive processes work and the influence of unconscious factors on worker behavior. These factors result from a stressful environment and are responsible for accidents from unexpected behavior (Lorenzo, 2001).

Ávila (2010) and Hollnagel (1993) point out that the modes of operation are field-influenced by cultural forces and can trigger human error. The resulting behaviors may be as expected or may be affected by cultural biases and archetypal actions of bad habits.

Ávila discusses the concern with building a common mental map or schema for staff, management and execution functions. The author highlights the organization, the flow of information and the means of communication to avoid the low visibility of the operational routine and, consequently, the omissions or commissions.

Carvalho (2016) points out how the operational culture resulting from this field of cultural forces indicates the culture level of guilt (Fragoso et al., 2017) in operational practices. The author states that it is important to measure the direct or indirect variables of human behavior (stress level vs doubling shift work) that have managerial and technological impacts on critical industrial activities.

Experience from the study of accidents indicates that failure is a function of context, and variability is the greatest indicator of the onset of the accident. This is the most sought sign (Hollnagel and Woods, 2005), rather than the probability of human error. Therefore care should be taken when using probability tables for human tasks in the industry as it does not always work for every industry, every region or every time.

Memory fragments, fact making, memory storage for work, psychosomatic learning, the relationship between affection and commitment merge with cultural influences to result in the rules of routine, that is, heuristics of behavior.

Macro-cultural rules can transfer characteristics for or against safety in the task. The adjustment of macro-cultural rules (social, regional, and global culture) induces archetypes that impact production, quality, and safety issues. Latent and intangible movements considered human errors (Reason, 1990) are studied in the first phase of human error classification (Reason, 1990; Dekker, 2006), and blame the performer of the action.

Micro-cultural rules can be a consequence of personal characteristics, leadership influence and events relating to the company and the employee's personal values. The company designs a certain organizational and safety culture that can succeed and inhibit personal or group movements toward the accident. If the company is a subsidiary, the implementation of the matrix culture may not respect local population rules. When mixing, cultural rules including safety and organizational culture, operator behavior can arise that does not reflect the projected organizational culture. This is the operating culture.

Routine or heuristic rules result from the fusion between micro-macro cultures and confirm the technical (types of technology and equipment) and operational (operations and people) culture with their accepted and biased formal or informal standards. Hence the importance of developing new techniques for HRA based on OD and observation of the critical task routine. For simplicity, the resulting culture on the factory floor we will call operational culture.

Ávila (2010) points out the importance of performing bottom-up analysis and avoiding only top-down observation. It is necessary to account for data on tasks performed during the operational routine and to value intuitive actions of production management, including the subjectivity common to the technical-operational culture in the organization.

Performing tasks that are useful to society and the economy is a continuous act of activities, actions by function, flow of information and communications, between staff and shift workers. Communication includes: discrete, automatic patterns, feedback, auxiliary memory, media, redundancies, interfaces, procedures, signals and controls (Drigo, 2016).

Human communication depends on the facilities granted by the organization, favored by management and dependent on the team's spirit of cooperation. Thus the resulting HR depends on the communicative process. Most human errors are caused by noise in communication and induce different mental maps for decision-making and action. Certainly, clear and fair communication confirms an effective standard.

Noises in communication affect the physical process, and the transfer of this noise, referred to as a mental map defect, materializes in equipment and processes. Operating mode, routine situation or operational emergency may cause high release of hazard energy. Often the alignment of various technical and human problems causes an accident or even a disaster.

In the HR study, we need to be sensitive to study social data that may affect the technical environment. Some social phenomena enter the industrial environment and create noise in labor communication, conflicts between different generations, genders, and cultures within the shifting environment. This can lead to omissions of the affected individuals. Therefore dynamic risks should be considered in HRA through variables representative of the socio-technical system (Ávila et al., 2019a, 2019b).

Reliability increases with the control and prediction of human behavior at work and the identification of new heuristics from new environmental and cultural threats. Increased reliability raises the level of system resiliency, and prevents loss of operational control as well as re-entry into abnormal states and their repetitive chains. The safeguards for threats that impose new heuristics on routine decisions are in the area of team communication and harmonization. Informal or heuristic rules indicate danger regions that need to be tested to confirm task reliability.

Ávila (2004) developed the technique that uses OD to mine data to process control, and manager discourse to explain heuristics. The author presents concepts of the OD analysis technique and develops a case of the Oil & Gas industry in Chapter 7, Human Reliability: Chemicals and Oil & Gas Cases.

The socio-technical system (Ávila et al., 2019a, 2019b) is highly likely to initiate high-impact, low-visibility faults (Perrow, 1984) represented by the digital systemic failure model (Ávila and Dias, 2017b). Appropriate safeguards for this failure depend on tools that handle complexity. Ávila (2010) suggests and applies the following techniques in the chemical and petroleum industry cases that will be presented in Chapter 3, Worker Performing Task, (Oil & Gas, Polycarbonates, Toluene Diisocyanate (TDI), Sulfuric Acid, and others): logical failure analysis, chronology analysis and release of danger energy, connectivity diagram, study of parallelism and dominance in critical task and others.

In summary, Ávila suggests that the study of operational culture is a third-generation HRA method and will measure the vector and cylindrical distance between the normal state with the least danger energy being released and the current state (Ávila, 2010). This distance or height presented by the cylindrical failure model will indicate the proportion that represents the socio-technical or integrated reliability. The graph (Figs. 2.11 and 2.12) indicates regions from most comfortable to near discomfort. This is the mathematical way of calculating reliability by the distance from the center of mass of these regions, as well as by the distance between the current hazard energy of cylinder events and the total amount of hazard energy at system shutdown.

2.1.6 Operator discourse—fuzzy

2.1.6.1 Operator discourse

OD analysis is a technique for interpreting latent information to maintaining control of the production system. The OD can be observed in texts and behaviors, recorded or not: operation and maintenance occurrence books; technical and supervisory reports; specific procedures and checklist records; process and production data in the operation routine. The operator oral speech allows interpreting information through: coexistence in the workstation; interviews and dynamics for emergency decisions; questionnaires for the survey of the organizational climate and the level of perception of risk. According to Ávila (2010) the operation shift book analysis helps in making more assertive decision-making.

The production manager speech transfers the summary of the operation's work to the organization. This transcript is shrouded in political issues reproduced by the operation manager's apparent profile. The manager's oral speech in stressful situations reflects his management profile, not necessarily reality scenarios.

To manage risks in the industry, it is important to increase visibility into failure processes by seeking signals from knowledge built in practice and following the interaction of social and technical systems. Risk management is qualified when technology, tasks, individuals with their characteristics and skills are known.

The functionalities and commitments of the groups will be validated when the OD can be turned into signals that indicate the most likely forms of operational failure to occur. This will enable prevention work in organizational environments.

OD analysis is a Bottom-Up tool and discusses the culture (regional, organizational and safety) at its source. This discourse characterizes the type of group relationship, the economic positioning, the level of stress resulting from the company's image in the occurrence of accidents, and indicates the average human type of the shift group.

The OD that represents the socio-technical system and may also indicate the following information: type of human error practiced in the routine environment; operator behavior related to task types; differences between writing and task interpretation to enable its analysis, execution and revision; group patterns resulting from the dominant culture; level of complexity of process and safety interconnections; possibility of failure due to nonrespect of population stereotype; lack of coexistence between various cultures at the workstation; influencing environments or performance factors that promote fluctuation in operator and manager behavior; measurement of demanded and offered skills; and biases in decision-making.

Paul Sharratt of UMIST in 2006 states regarding the OD technique:

> *Key to human error research is the OD technique that understands the complexity of operations and can integrate the task performed with the appropriate safety culture. The complexity of operations involving different processes interconnected via utilities or ancillary systems, the complexity of processes interconnected through vices or positive characteristics of individual or group operators that repeat "modus operandis" to different process points are printed on the information fragments. which are in the context of the OD analysis tool.*

The application of OD analysis is important for the elaboration of the abnormal event map (AEM), and the relationship of causal connections to the root cause in the birth phase of the operation failure (Ávila et al., 2006, 2008).

The quantitative study that turns discourses into values or probability calculations uses fuzzy mechanics that translates this information into a set of rules to be analyzed. If the operators express (linguistically) their experiences during the work shift and if there is something that needs attention to avoid the error, then conditioning rules IF ... THEN. Thus it will be possible to formulate an algorithm as a result of the expressions. The result of this transformation of the subjective into numbers is a rule-based inference system, in which the fuzzy set theory presents as a mathematical tool suitable for pointing out the answers that can serve as parameters for the best decisions.

These answers are not given deterministically, as stated earlier, there will be a range of responses that indicate the variability of errors and hits. The closer the numbers to a comfortable range, the greater the reliability and the smaller the range of accidents or human error incidents.

2.1.6.2 Dynamic risk management and operator discourse

Dynamic risk management identifies the chain of events during the operational routine to avoid the accident or incident. Prediction and decision-making around the root cause of problems depend on the identification, treatment, and interpretation of the operator's speech in conjunction with the manager's speech based on statistical validation and phenomenological calculations.

The risk of loss is manageable when social and technical processes are visible in the production system. Preestablished technology through projects provides information to enable systems to be operated. At the operation stage, the insertion of social systems into technical systems requires adaptations to achieve the normal operating state and the integration of different teams with formal and informal leaders.

Risk management that uses perceived signals in practice builds appropriate knowledge for organizational resilience by integrating social systems with technicians. Risk management is qualified when technology, tasks, individuals with their characteristics and competencies, groups with their functionalities and commitments are known. Risk management is validated when OD can be transformed into signals that indicate the most likely form of operational failure, enabling prevention work in organizational environments, equipment and the workplace.

At this time, different levels of information that feed risk management intersect and become routine management. At the strategic level, the need to estimate insurance against the risk of material and human loss is matched by the company's market value. At the tactical level, the validity of the organizational guidelines and their credibility with the production team make the organizational climate reliable and bring the closest operators to the desired standard for the organization and society. At the operational level, knowledge of the task and its risks leads to safer action and clearer visualization of possible signs of operational failure, or, in other words, the departure from the best production standard. At this level, information from the operator, staff and management discourses circulates, leading routine management to appropriate operating standards and understanding of the intrinsic variability of the socio-technical system (Ávila et al., 2019a, 2019b).

Special attention should be given to situations of risk close to major losses such as major accidents or even disasters. In emergency situations, teamwork gives way to individual leadership skills for contingency or technical specialization. Therefore operators must have contact (simulated training) with risk situations to make the right decisions in real situations in the face of high stress.

Emergency situations are outside the rules of routine management and also depend on warnings visibility for quick action. Often the biggest problem is not the availability of signals but their excessive quantity in random format. In this context, the lack of knowledge about the causal link in the timeline and over-automation leads to cognitive gaps (Ávila and Costa, 2015c). This fact creates the illusion that it is possible to design industrial systems without human error.

In risk management, the technical systems are analyzed taking into account the history of the facilities in the "said" range, but it is also important and essential to learn how to read the Operator's Speech, that is, the "unsaid." "Thus it is possible to bring into the manageable area actions not yet thought out, which can avoid the realization of risk and the transformation of danger into major accidents."

Thus knowing how to read the operator's speech is the difference to achieve competitiveness and becomes the primary competence for risk management.

Dynamic risk research relies on the recognition of information embedded in the DO that represents the socio-technical system (Ávila et al., 2019a, 2019b) involved, this private label with the history of the failure is named as the operational failure fingerprint (Ávila and Bittencourt, 2017a). These include: the type of error commonly practiced in that environment in the routine, operator behavior according to different types, differences in writing and task interpretation to enable its analysis and execution, group patterns and dominant culture complexity of the process interconnections and the safety involved (including redundancies), the possibility of failure due to nonrespect of the population the fluctuation of operator and manager behavior and, finally, the measurement of demanded and offered competences and decision-making biases.

2.1.6.3 Assumptions for investigation of operator discourse
2.1.6.3.1 The phenomenon of omission and commission (reason)

Operator inaction or lack of action performed is a type of human error called omission. This may be intentional or unintentional and may happen partially. Failure to complete the task may also be an omission. Incorrect performance of a task or action is a commission error. Operators perform correct actions according to their understanding and current knowledge of the system and its behavior. However, the system may be in a state where a correct intention to operate is not appropriate. The discovery of the reason for the omission is in the study of OD.

The type of behavior depends on aspects related to human typology and human reaction to environmental stressors (Chapter 3: Worker Performing Task). The result is hits, mistakes, vices, unexpected situations, scheduled situations, group relationships with their respective roles and leadership positions. All these aspects occur during the operational routine and influence the company's results.

2.1.6.3.2 Operational mode differences for task analysis

Task-related situations depend on the operational mode involved, whether in routine or emergency situations, whether they are execution, strategic, tactical or planning tasks. Thus the type of language adopted and the use of knowledge and skills depend on each case.

2.1.6.3.3 Patterns and culture

Standards adopted by the operator (in fact) may not be standards desired by the organization (from the desired culture). Thus it is necessary to understand the difference between organizational and individual values to identify patterns in the application.

2.1.6.3.4 Process interconnects

The number of connections as well as energy and mass recycle in process interconnections can make it difficult to perform the task. The network logic in the failure of process and product processing needs to be identified through the logic, connectivity and materialization diagrams discussed and applied in Chapter 3, Worker Performing Task. Therefore it is possible to analyze the factors related to the industrial process (equipment, controls, and materials) and operational failure.

2.1.6.3.5 Multicultural

The effectiveness of the task depends on the multicultural level involved, so people from certain regions behave in a typical way as a result of local history. The aim is to break down barriers and develop a common place where operators, despite having different origins, will work in group cooperation with staff for the correct execution of the task. Operators from Bahia will act in tune with the strategies led by the management of Rio Grande do Sul, and the guidance of Rio's staff, within an environment of social and labor justice.

2.1.6.3.6 Fluctuation of behavior

Failure to guarantee the stability of human behavior can trigger socio-technical events that may initiate a latent process involving human error in conjunction with technical failure. The characteristics installed on personnel and equipment structures can trigger the failure cycle when stressful environments and behavioral oscillations are present. Human error due to this unexpected fluctuation of behavior makes the accident or incident possible.

2.1.6.3.7 Skills analysis

The difficulty in identifying the skills demanded and offered [cognitive gaps (Ávila and Costa, 2015c)] is due to the complexity of working with socio-technical systems (Ávila et al., 2019a, 2019b). Therefore OD analysis is part of HR studies. It is necessary to analyze the competencies offered using the OD and to analyze the complexity of the technology from the investigated process and equipment abnormalities.

2.1.6.4 Abnormality mapping and operator discourse

The transcription of the operational abnormalities recorded in the occurrence book is fragmented in event trees (Map) and after their coding, the events are reorganized in a causal network format. The mapping of abnormalities builds the history and enables the analysis of operational factors as well as the identification of the root cause of problems in the socio-technical system (Ávila et al., 2019a, 2019b). Thus it is possible to translate the complexity of discourse information to be used in production control.

2.1.6.4.1 Complexity

Complex production systems (Perrow, 1984; Ávila, 2012a) have parts that are not in a logical sequence and may have physically close modes of connection. Auxiliary (inserting, sealing, lubrication, off-gas) and heat exchange (steam and water) systems add technical complexity to the chemical process industry. Another important aspect is the existence of multiple feedback loops with a ball shape in the process systems. Complexity can be inserted when information is fed back into different points of the environment under study in a similar manner to individual, managerial and organizational systems.

Complexity is also a result of nonvisible interactions between control parameters through intervention in the organizational and economic environments that alter the behavior of man in the execution of the task. Complexity can also stem from changes in the natural environment (temperature or rainfall) over the resulting process equipment.

Perrow points out that several subsystems interconnected through pipelines and thermal systems or through organizational links of information. Ávila (2012a) presents complexity reducing its visibility regarding the industrial and organizational process states. This may result in loss of control and operational abnormalities. When structures are interconnected through processes and environments, it is necessary to be aware of the information circulating in the routine to monitor apparent normality in industry equipment or service activities.

2.1.6.4.2 Failure logic in complex systems

In complex systems with influences of multidimensional factors in industry or services, it is necessary to review methods that analyze process failures, including human and organizational factors in conjunction with known technical factors. Thus failure analysis includes multiple dimensions (technical, organizational, human and managerial) in factors that have different functions in failure.

Some observations can be made: the causes can be multiple and of different levels (root cause, secondary cause) and dimensions; the consequences can also be multiple and of different levels (primary, secondary, and tertiary) and dimensions; the disruption of the production system can be classified into different dimensions (technical, organizational, managerial and human); migration of fault energy that is in one dimension to another occurs, for example, from communication noise to excess pressure in the equipment; branching of the fault energy also occurs; the emission of fault signals does not develop work, that is, there is no apparent damage to the production system; failure energy retention occurs at certain times, promoting delays in the failure passage between equipment and people; corrective and preventive actions can drain the energy of the failure until it disappears.

2.1.6.4.3 Influence of technology on human error

Depending on the type of technology there may be different human errors. Lack of practical knowledge can cause inaccuracy (technology) and change the quality of task planning and execution. Procedure writing may be inaccurate and may cause equipment or process failure. Incorrect language or incorrect sequence of steps indicates that there are difficulties in knowledge/skills, or difficulties in drawing-up the mind map to plan the procedure. Sometimes improper communication of work patterns can lead to problems in performing the task. This causes technical failures or human errors caused by discomfort or fear of acting.

Control system monitoring or sensing of errors can cause process fluctuations, create unusual situations that the operator may not know how to handle and generate system shutdowns. Automation and control technology, therefore, is important for developing proper workplace expertise.

Sometimes deficiencies installed in the interfaces between man and machine (alarm management) do not allow actions to be taken to avoid human failure, which links the lack of technical memory with the deconstruction of the mind map. Or, there may be a lack of visibility of process variables in the HMI that does not allow correct planning of the task.

The use of additional chemical and safety knowledge is required when there is a product or process with high toxicity and flammability. Human error exists when the required expertise for the task does not exist or is being poorly used.

2.1.6.4.4 Operational factors for failure

The causal logic in terms of normal operation and the model regarding the types of operational factors involved in failure is developed for the execution of activities and material transformations across industry and services. Each factor has a specific function with rules and behaviors that can be achieved depending on the operational and human failure scheme. The explanation of operational factors begins in the normal process and comes to an abnormality or abnormal event with the disturbance or noise caused.

2.1.6.4.5 Normal state of process

In Fig. 2.6 solid formation in the reactor takes the process out of normality to initiate the failure cycle. In 48 hours, the pump cavitates. In 150 hours, the reactor temperature control valve changes opening, and finally, after some time (e.g., 240 hours), the reactor overflows. After this event, there will be a need to stop the industrial plant.

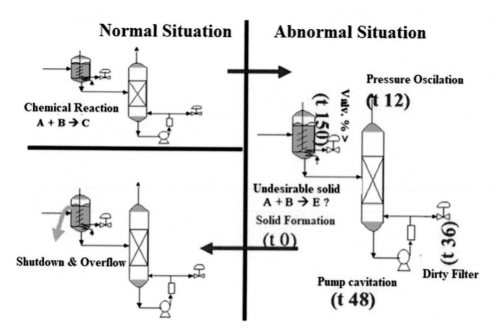

FIGURE 2.6 Normality and abnormality in process states. *From Ávila, S.F., 2010. Etiology of Operational Abnormalities in the Industry: Modeling for Learning (Thesis; Doctor's degree in Chemical and Biochemical Process Technology). Federal University of Rio de Janeiro, School of Chemistry, Rio de Janeiro, 296 p.*

Process inputs are specified by technological standards, material transformation and information that are part of management processes and systems. Operational factors involve: tasks; management guidelines; technical data of operation; legislative information; market requirements; changes in nature; equipment and process specifications; modes of quality control in processes, inputs and products; installed skills; and organizational policies.

The production process is framed when operational factors are specified and controlled. Between the optimum condition (100% normal) and the plant shutdown condition (0% normal), there is a cycle that can continuously signal failure and affect production costs. Thus each time there are signals and events happening in the failure life cycle this ends in an undesirable event.

2.1.6.4.6 Process disorders

When there are environmental/structural/procedural changes, the disturbance may remove the process from normality. Some of these changes may be: behavioral fluctuations, managerial/organizational changes; changes in environmental and/or occupational legislation; changes in business requirements; and changes in the states of nature.

To return the process, product, equipment, and people back to the normal state it is important to: (1) understand how the characteristics of the elements in production are favorable to the onset of failure; (2) identify or have visibility of what is the root cause of the disturbance in the production process; (3) check whether there are potential operating factors to increase or decrease the strength of the failure in its life cycle; (4) identify monitoring points or signals that indicate the level of hazard energy released in the operational failure; (5) identify hazard energy feedback points; (6) identify drainage points of this force/failure energy that are corrective and preventive actions; (7) know the physical and human structures to avoid entering the risk zone where the failure could "explode," resulting in system breakdown; (8) verify how the fault circulates in its multiple dimensions; (9) identify the interconnection between different abnormality chains through common operating factors (using the connectivity technique) and other important actions.

2.1.6.5 Identification of types of operational factors in failure

2.1.6.5.1 Cause, signal and consequence

Cause: these are factors involved with the onset of failure (Fig. 2.7). In the cylindrical model, it means breaking inertia by initiating latent failure or visible failure. Causal factors can be classified as the root of the problem. These are primarily responsible for feeding the energy of the fault and causing the inertia to break. Operating factors classified as secondary causes are not reasons for initiating the failure. A series of events comprising operating factors indicating a possible source of failure and its final consequence.

Signal: These are indicators of the presence of the fault, emitted by nearby equipment or the equipment themselves that will suffer the consequence of the failure. The greater the number of signals, the more visible the fault becomes and the faster one can act to correct it.

Consequence: These are operational factors resulting from the work of the failure and which may be classified as primary, secondary, tertiary and even quaternary. These factors only happen when the fault accumulates enough power to damage the facility.

2.1.6.5.2 Potentiating factor, false factor, and anticipated factor

Potentiating factors are operational factors that contribute geometrically to the growth or decrease of the failure. It is important for action and the adequacy of environments.

False factor is a disconnected factor that arises during the interpretation of the causal link by logic and chronology. It indicates the existence of another chain of abnormalities or incorrect construction of the mind map.

Anticipated factor occurs when the operational factor is likely to happen in advance rather than the current position, there may be a low-visibility fault feedback ball giving the disconnected appearance in this case.

2.1.6.5.3 Failure energy feedback

This operating factor is represented by material or power recycling lines in equipment and information recycling streams in cases involving human, social and organizational dimensions.

2.1.6.5.4 Corrective action and preventive action

With system shutdown, corrective actions are taken to drain the failure bottlenecks and restart production. This energy drainage may not necessarily be performed at the root cause but at the consequence or secondary cause. If production is restarted, in a short time e the same problems resurface.

After verifying the root cause and secondary causes, with the plant and the process still operating, certain actions are taken to prevent the failure from increasing energy and avoiding consequences for the production system (Fig. 2.7).

2.1.6.6 Failure analysis in production systems

The failure investigation follows the cylindrical model indicated in Fig. 2.8 and is performed in three levels or phases: (1) low-visibility inertia and deviations (latent failure); (2) failure development with its signals, failures and incidents in the operating routine; (3) image risk and gross profit (critical or emergency situations). The multi-causality of failure makes it difficult to identify the root cause, with multiple causes crossing the path and sometimes confusion about position and function, whether it is the cause or consequence.

There is a chronology (Fig. 2.9) for failure to manifest in industrial sectors, either serial or parallel, and depending on the retention time per system (failure time to exceed sector or equipment boundaries). During fault analysis, it is important to know how to differentiate signals (they do not have the power to cause/feed the fault), causes (factors that initiate or contribute to initiate the fault), consequences (fault work). Knowledge of how to classify causes and consequences as primary or derived. Knowledge of how to identify which causes potentiate the energies of failure in production systems. Knowledge of how to define corrective actions on the consequence of the failure and preventive actions on the cause of the failure.

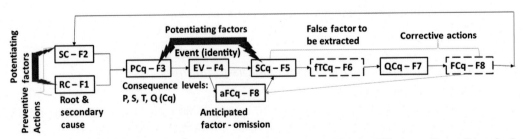

FIGURE 2.7 Studies of operational factors. *From Ávila, S.F., 2004. Methodology to Minimize Effluents at Source From the Investigation of Operational Abnormalities: Case of the Chemical Industry (Dissertation; Professional Master's degree in Management and Environmental Technologies in the Productive Process). TECLIM, Federal University of Bahia, 115 p.*

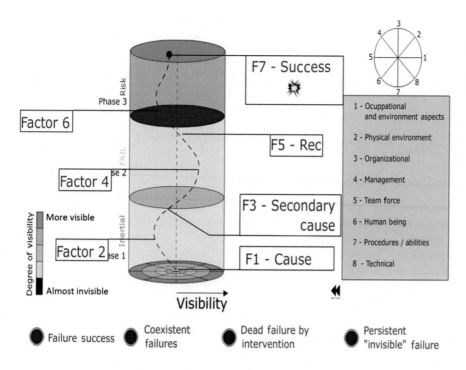

FIGURE 2.8 Cylindrical failure model. From Ávila, S.F., Dias, C., 2017b. Reliability research to design barriers of sociotechnical failure. In: Proceedings of Annual European Safety and Reliability Conference (ESREL), CRCPRESS, Portorož, p. 27.

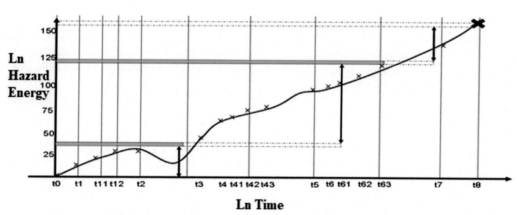

FIGURE 2.9 Failure timeline. From Ávila, S.F., 2010. Etiology of Operational Abnormalities in the Industry: Modeling for Learning (Thesis; Doctor's degree in Chemical and Biochemical Process Technology). Federal University of Rio de Janeiro, School of Chemistry, Rio de Janeiro, 296 p.

2.1.6.6.1 Identify the moment of failure materialization

Manager's speech for action (on task) and transformation into processed material. Knowledge of how to identify nodes with failing branches for events of various dimensions. Knowledge of how to define corrective actions on the consequence of the failure and preventive actions on the cause of the failure.

When operator speech in conjunction with equipment performance and task complexity is analyzed it is possible to view the fault in complex mode with multiple dimensions as in Fig. 2.10.

Often, as indicated in Fig. 2.10, the fault has several dimensions. Blue failure is due to human fatigue, green failure is due to improper rotary lubrication, red failure is due to improper pump operation and consequent cavitation. It was not necessary to identify all the root causes in the configuration shown in the figure. When complex (multidimensional) architecture is understood, acting on operational factor 4 becomes viable, as an intermediate event linking the three dimensions of the same fault.

Failure investigation, analysis and decision-making are related to HR and the abnormal chains identified from the OD. Knowledge, behavior, safety culture are factors that influence human performance and depend on the availability

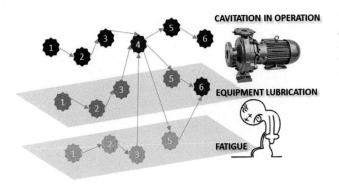

FIGURE 2.10 Interconnection in systemic failure. *Adapted from Ávila, S.F., 2010. Etiology of Operational Abnormalities in the Industry: Modeling for Learning (Thesis; Doctor's degree in Chemical and Biochemical Process Technology). Federal University of Rio de Janeiro, School of Chemistry, Rio de Janeiro, 296 p.*

of the cognitive machine. The application of fault detection and analysis tools is complementary to the integrated model of dynamic risk management (that includes human error). This chapter discussed the diversity of HR content, with emphasis on OD in adjusting the safety culture. This tool allows you to increase the visibility of potential latent failures and maintain process control in the production system.

In this context, we will begin the discussion of a case of analysis of HR (third generation) for Oil & Gas, whose application will be detailed in Chapter 7, Human Reliability: Chemicals and Oil & Gas Cases. Then we will study Cognitive Processing that presents ways to correct HPFs intrinsic to individual-group behavior. It is important to know the psychological functions in the workplace and in relationships with society.

2.1.7 Third-generation application in the calculation of organizational efficiency, Oil & Gas

The HR technique uses OD to integrate the task with the culture by predicting future behavior. It is intended to anticipate the future failure situation by studying filtered and mined variables in socio-technical systems. The selection of valid data representing operational and technical culture is defined through task analysis, culture analysis, high-stress decisions, human and social typology, process behavior, product and effluent quality, and maintenance and safety studies.

2.1.7.1 The method

A three-month survey of variables representative of the operational-technical culture will help to construct the organizational efficiency map. Daily, the number of data is reduced through the principal component analysis (PCA) to plot a point which along with other points represent different classes of variance. Each point represents one day, with its organizational efficiency with shared weights between production, safety, and cost. If the chosen variables really represent the technical-operational culture, the points released will indicate regions of organizational efficiency in the form of curves or clouds. Curves should be calibrated to confirm where the future outcome is located and to review human and organizational factors through the heuristics transcribed in fuzzy modeling.

The resulting per production day indicates the level of organizational efficiency. Effective corrective and preventive actions can be taken by identifying characteristics of human–organizational–technical factors because they can occur in the physical–functional region of hazard energy concentration. When the data are used in a simulation of the future situation, the position on the efficiency curve and the offset calculation between the current point and the center of mass of the maximum comfort region are obtained, which means a low amount of danger energy.

Direct variables are not always measurable, sometimes indirect variables that may represent socio-technical behavior are necessary. For example, stress is a variable that is difficult to measure for cultural, educational and personal reasons. In addition, high stress causes a reduction in routine task efficiency (Muchinsky, 2004) and generates human errors that can lead to accidents (PEH from 10% to 70%) due to incorrect decisions (Ávila, 2011a). Thus indirect variables are sought to represent the stress level.

The type of problem analyzed indicates the objective function or the objective functions with the resulting weighted average for organizational efficiency. The main objective of companies can be to increase profit as well as reduce events with an impact on safety and the environment. The AEMs to be analyzed depend on which objective function is being tracked. Although accident reduction is sought, the number of shift deviations is used in the composition of the objective function.

The elaboration of the AEM has a technical−human and cultural basis. This map and the technological information gathered (process, event, task and technology statistics) allows us to choose the discrete and continuous variables that will form the clusters to apply in the Principal Component Analysis (PCA) technique.

Human behavior is explained from the analysis of socio-human typology and varies with different personality traits. Human-type research is important when specific functions control large amounts of hazardous energy, for example, flight controller and nuclear power plant operator. The aspects analyzed are:

- Application of competence;
- Occurrence of slip in the task;
- How the individual or group processes information;
- The form of the causal nexus for problem-solving;
- The way the individual or group decides on alternatives to action.

These human behaviors are influenced by natural (temperature and climate) and social (organizational culture and level of cooperation) environment variables to compose decision heuristics. Changing environments, people, and equipment-operation setup may change the reduced values in the scatterplot as a result of PCA.

Traditional methods for reliability analysis may indicate wrong regions of the cause of human error where hazard energy is located. The complexity of factors and their connections to failure indicate a universe of possibilities for the construction of unsafe acts. Therefore, for high-impact and low-frequency systems, it is necessary to expand the research dimensions. The OD plays an important role in the reliability analysis as it indicates opinions in the different social and technical dimensions.

The risk of poor worker performance exists and is related to what is said or not said, described in the OD. Omission or slowness is a projection of fear related to informing facts that may conflict with expected behavior (Drigo, 2016). This situation makes data collection complex and interferes with the quality of task performance as well as the likelihood of human error occurring.

In this uncertain scenario, Ávila (2010) and Luquetti and Rodrigues (2011) discuss the application of a method that includes the requirements and contributions of the 3 generations to meet the variability of the nondeterministic factors used to evaluate HR.

2.1.7.2 Generic algorithm

HR techniques adopt a conceptual algorithm and statistical tools that include PCA, as indicated in Fig. 2.5. The model aims to estimate the prediction of behaviors resulting from the variables on the organizational environment. This calculation results in regions with their organizational efficiencies, where the goal of maximum efficiency is sought.

After identifying the types of technical, human and social behavior in the workplace, it is possible to suggest individual and group actions, bringing the human response closer to the best execution of the task.

Parameters that integrate the groups of variables are chosen for PCA regarding the process and operator behavior. The resulting calculation or estimate indicates a tendency to control/control and depends on the direction of the artificial variables in relation to the goal or pattern of optimal organizational efficiency.

Regardless of the results of approaching or departing from the target organizational efficiency, the records are transferred to the organization's memory. The memory of the past influences the current control of production systems.

The process efficiency (organizational efficiency) influenced by the failure acts as concentric contours in relation to the target cloud (maximum efficiency), moving away when the hazard energy increases with the appearance of the failure, dimensionless variable. The alteration of the main component is made through: (1) environmental change, (2) different social human typologies (Chapter 3: Worker Performing Task) and (3) application of competencies. Thus the resulting PCA is adjusted and can direct the value to the goal or the opposite direction.

2.1.7.3 Situation simulation

Simulating a situation: from the choice of measuring results as reducing the frequency of accidents that can affect the community, the chains of abnormalities involved with this possibility are chosen, and discrete and continuous variables are extracted. These are indicators of the uncontrolled system toward the accident. Technical variables are grouped by similarity in variability and PCA artificial variables are calculated. Two groups of chosen variables indicate approaching or departing from accidents in which, when grouping these variables, a scatterplot is drawn up: group 1—specific pressure and temperature and time to perform the critical task; group 2 contaminant in the raw material and number of times of filter cleaning in the shift.

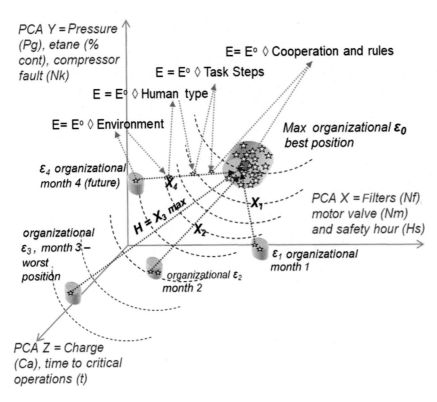

FIGURE 2.11 Conceptual model about the failure. *Adapted from Ávila, S.F., 2010. Etiology of Operational Abnormalities in the Industry: Modeling for Learning (Thesis; Doctor's degree in Chemical and Biochemical Process Technology). Federal University of Rio de Janeiro, School of Chemistry, Rio de Janeiro, 296 p.*

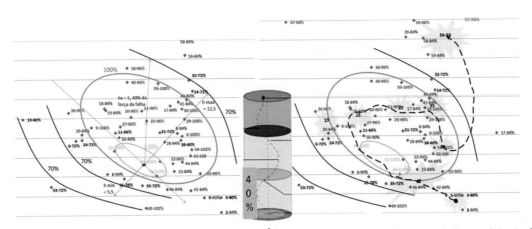

FIGURE 2.12 Fault model application for Oil & Gas. *Adapted from Ávila, S.F., 2010. Etiology of Operational Abnormalities in the Industry: Modeling for Learning (Thesis; Doctor's degree in Chemical and Biochemical Process Technology). Federal University of Rio de Janeiro, School of Chemistry, Rio de Janeiro, 296 p.*

It is noted that some aspects of human partners are included in the chosen variables of the groups and are presented in the graph of artificial variables of the PCA. Knowing the organizational efficiency results for this variable, we classify the grouping of data derived from the PCA artificial data, according to the point of accident (maximum failure) or without accident (maximum efficiency in the organization).

This history-based mapping indicates hot and cold cloud regions for the crash. This process uses historical data mined from the process and plotted on the graph through actual measurements and projects the future result at a new point on the graph. This real situation refers to new points resulting from the artificial variables of the PCA that measure the deviation from the optimal point. Then changes are made in the preprogrammed setups that lead to corrections through the heuristics, as to the team, tools, environments, and tasks. How organizational efficiency mapping works depends on understanding the complexity that begins with OD.

Fig. 2.11 presents the conceptual model that indicates the organization's characteristic curves for operational failure, as well as projecting the future failure condition based on measurements of PSF changes.

The verification of the risk level aims to change situations or PSFs, for reduction. The use of simulations avoids real tests in industrial plants and helps to identify abnormality maps and choice of discrete and continuous variables to prepare preventive programs and increase HR.

In Fig. 2.12, the PCA results are plotted for data collected from an Oil & Gas plant. Each point represents a production day with its organizational efficiencies. Organizational efficiencies are classified into three main ranges. In the red lines, low-income ranges (70% to 84%), the green circle middle income range (100%) and the yellow circle in the center, you can reach 120% of organizational efficiency.

Several points are found to be outside the ranges defined by the curves or circles and it is believed that there is a need to improve data quality or data processing. There are ways to try to improve results, such as: increasing the quality of the data collected; choose other related data to test the analysis that was performed in two dimensions; perform the analysis with three groups of variables; and analyze the results considering other compositions of organizational efficiency.

As indicated in Fig. 2.12, the maximum efficiency is located in the center (yellow circle) where the point of maximum efficiency (center of mass) is marked and a line is drawn to the red curve representing low organizational efficiency. This distance that links the point of maximum efficiency (maximum processed volume, no abnormalities and low technical idle) to the minimum indicates an uncomfortable situation or the near incident or serious accident.

When measuring the process by mapping the organizational efficiencies posted for exercise purposes, it is noted that point 25 with 96% organizational efficiency has traces of failure and could be closer to the 120% organizational efficiency range. When measuring the distance from this point, passing a perpendicular to the center of mass of the maximum h point and leading the line (purple) to point 25, it is noted that the fault is 40% (as indicated on the cylinder of the failure) of the near miss, leading the organization to invest more intensively in the study of the problems.

For the purpose of testing the procedure, and knowing that this is a method for analysis of socio-technical systems (Ávila et al., 2019a, 2019b), it is considered successful in the demonstration. The mapping of socio-technical behavior, after the adoption of simulation data for production planning, allows locating future performance in regions of organizational efficiency. Thus it is possible to change the production schedule.

Grouping movements around organizational efficiency is a technique of social displacement of particles or particle swarm (Fig. 2.12), where: (1) The direction of the resulting collective displacement of particles depends on the sequence of points and the density of the data; (2) There are points that do not fit the displacement indicating that there are different collectivity rules or processes that require debugging data and techniques.

Diagnostic questions Annex 3, relation Chapter 6, Human Reliability: Spar-H Cases (SPARH) Chapter 7, Human Reliability: Chemicals and Oil & Gas Cases (OD & operational culture).

What are the critical tasks and procedures?

- Number of steps and necessity of memory
- Risk of Steps to leakage, time loss or other impacts
- Different people doing a Task from different departments
- Migration of attention, disperse to focused
- Parameters and Variables in different functions at Production
- Others

What are the performance shaping factors more critical?

- Level of stress
- Planning and execution of procedures
- Organization of job
- Ergonomics
- Attention and memory quality
- Process, task, and social relation complexity
- Others

What are the principal heuristics?

- Procrastination
- Risk aversion and failure
- Negative position
- Hero

- Suggestible
- Flexible in excess
- Obsessive in excess
- Paternalism and no critics
- Others

2.2 Human reliability and cognitive processing

2.2.1 Cognitive processing

Human cognition is related to the psychological functions essential for work and for maintaining efficiency in the accomplishment of the task. This efficiency in the task also depends on the workers' state of emotional equilibrium, where a good affective relationship and commitment join to enhance cognitive functions. These functions directly related to the task are: memory, attention, logical operators, perception, decision, translation of the task into action, critical sense, and a sense of utility in relation to the task's ultimate goal.

Cognitive processing is a science that involves objective aspects of Man at work and the subjective aspects of Man in society, as discussed in Chapter 1, Introduction. In this introduction, we discuss the concepts presented by Reason (1990, 1997) and Rasmussen (1997) and from other authors.

The behavior of Man in the workplace is an important aspect of occupational safety and health. On the other hand, productivity requires worker movements that integrate perception, memory, and attention. The ergonomic design for process signal perception, the adjustment in control devices, and movements of workers and machines set basic criteria for achieving success in workstation routine tasks.

Reason (1990) presented three main views for the discussion and study of human error. The behavioral psychological view, the cognitive-informational view, and the natural-organic view. This discussion formats the theoretical basis for definitions of the sharpness of understanding the mind map or working scheme.

Reason (1990) and the Psychosomatic Specialists (Franke, 2015; Aisenstein, 2018; Peterson, 2018) establish two types of knowledge in human learning, the representational and the presentational. Representational language is used where one seeks to unite the word (signifier) with its meaning (Lacan and Miller, 1988). For people of different cultures who perform the same task, communication difficulties may occur due to the existence of different signifiers for the same meaning.

In the process of socialization of individuals, words come to life, incorporate affection and can generate different meanings. In this sense, the level of harmonious socialization is essential for people working in teams, trying to reduce the number of signifiers for the same meaning.

The "Presentational knowledge" of the body that speaks, in its present form indicates that body diseases can be activated from specific affections or even specific words. Thus increasing HR generates emotional balance and body health; just as communication without noises transfers affection to the word and body assertively. In just culture at organizations, one signifier is suitable for only one meaning, or vice versa.

Human labor errors can be integrated with occupational disease because they are "Presentational knowledge." Meanwhile, human errors integrated into communication can indicate an unrealistic situation, transferring different signifiers to the same meaning, and creating the phenomenon of underreporting or misreporting in the safety culture. This effect of signifier and meaning is taken into account in interviews and behaviors, in the record of occurrences and in the relationship between the leaders and the followers.

In understanding the cognitive apparatus, it is necessary to use new tools to understand the correct meaning of words in the occurrence book, interviews and observation of routine work.

Cognitive functions are closely related to affection, imposing the need to discuss perception, attention, and memory while discussing team commitment and cooperation. The quality of cognitive and affective functions enables the link between the task, the behavior and the respective actions that can lead to human error. Each role, process, element and event is important in building an operating scenario demonstrating critical human–organizational factors and the influence of individual and team activities.

2.2.2 Introduction to cognitive processing, cases, failure, and skill knowledgement, rules

Human elements, functions and factors will be discussed and exemplified in routine industrial situations. The intention of this discussion is to anchor the biases (gaps) of cognitive processing with the operational routine. Perception, attention and memory are closely related, but with moments of signal processing, information and memories (auditory, visual and sensitization). Interpretations, concepts and discussions are reviewed through real-world cases of industry situations.

2.2.2.1 Case 1: Divided attention in the oil industry

Divided attention causes errors by the disconnection of perception of nonpriority events. Several activities performed at the same time require auxiliary memory, auxiliary processing and signaling to activate attention at critical moments. The control that filters information is flexible and can be somewhat restrictive. Interference at different stages can impair perception and change reality.

In the multichannel processor theory, any complex task will depend on the number of specialized and independent processors. The quality of perception is diminished due to the lack of auxiliary tools to memorize events and fix attention, thus making it possible to perform several tasks of low quality at the same time (Fig. 2.13).

In case 1, the field operator was transferring water from the organic effluent to the sea and then transferring the oil from wastewater back to the tanks. He was absent to solve family problems and, due to the good (not expected) performance of the "frog" pump, the transfer was completed to the sea, including the oil that was supernatant in the effluent. Simple event with high environmental severity.

2.2.2.2 Case 2: Illusions of memory in the workplace

Memory can create conscious or unconscious pitfalls (Ávila et al., 2016b). Memory illusions are creations of the imagination not lived in the past. These creations can be based on unconscious will, when there is no desire to remember risky events of the past or, perhaps, even a conscious will. The illusions of memory can be the invention of false reality, such as the lack of memory breaking the logical segment of past events or the strangely altered memory in relation to the truth registered in groups or through irrefutable documents. Thus it can be seen that memory can be constructed with low reliability, hence the discussion of memory combined in Figs. 2.20 and 2.21.

Media and Internet-based learning (Fig. 2.14) can form hyperactive personalities, which is a current feature of society. Education without the presence of parents, as a reference, does not encourage the establishment of bonds, that is,

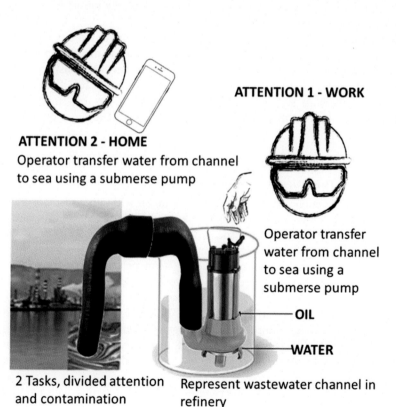

FIGURE 2.13 Operator, 2 tasks, divided attention cause contamination, wastewater to sea from refinery.

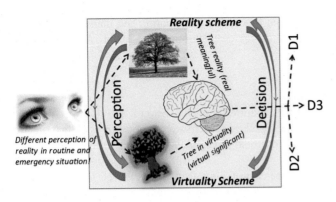

FIGURE 2.14 Perception of the virtual in substitution to the real and decision. *From Ávila, S.F., Fonseca, E.S., Bittencourt, E., 2016b. Analyses of cultural accidents: a discussion of the geopolitical migration. In: Proceedings of Annual European Safety and Reliability Conference (ESREL), CRCPRESS, Glasgow.*

facts and lived experiences do not add value during learning. The experience with images and sounds of the media, as well as the experience of close and distant friends, changes the perception of routine. This is the effect of the social phenomenon of virtuality related to distant references.

Routine is perceived by a dichotomy between the true image of reality or the virtual image of reality. Thus to understand if the tree, before decision-making, will be a single definition image in relation to reality, or in relation to virtuality, or a dynamic that varies according to stress and level of importance.

The processing of this image in the face of a situation of decision confusion between signifiers (Sr ↔ Sv) indicating different meanings (sr≠sv). Thus in the workplace, while performing a critical task, this phenomenon can alter the decision regarding the accident. Some results for the event of these phenomena are: reality is perceived in a modified way; the feeling of virtuality offers a high risk of decision comfort and the final decision being based on virtual games.

Accidents and serious failures in the industry create a tense climate under pressure from senior management to identify causes and their actions that ensure nonrecurrence and urgent performance improvement. The Company with a reactive culture centralizes decisions and pressures everyone on the team as a result of not identifying the root cause. This company puts pressure on the "culprit" for the accident and its work team. In this scenario, the virtual images and fantasy messages advance, occupying the real space in the routine decision.

In the workplace, combining organizational pressure, the behavior of internet and media culture in the virtual-reality and the lack of commitment due to differences between individual-organizational values, has made the worker unconsciously claim to have performed actions that in fact did not happen. Or, the operator and his team consciously perform the group omission defending the instinct for work survival. Both are serious cases and occur in real life. As an example, a real case where the group agreed to tell the staff an "invented story" that was not the real story of the environmental accident, where methylene chloride was released to wastewater in great quantity. The sense of reality-responsibility must be maintained in the production team.

2.2.2.3 Case 3: Primary memory capacity and memory with association

The primary memory, here considered as immediate perception memory, without association with other memory fragments, has a limit on the number of memorizable items. Tests in groups (Muchinsky, 2004) show that the maximum number of nonassociated memorizable items is 7. It is understood that association occurs when the similarity of the object or the remembrance of affective bonds occurs, in relation to what is reserved in memory.

Other tests carried out in courses for the petroleum industry show that if words, items or objects are associated, 24–28 items can be memorized. In a simplistic way, it is known that procedures above 28 associated steps need auxiliary memory, such as a checklist.

The limit of memorization gets complicated when the cognitive effort for this task becomes higher due to the need for complex calculations and comparisons depending on external information, the areas of support of logistics, laboratory or maintenance. In this case, in addition to the checklist for procedures above 28 associated items, it is important to provide algorithms for mathematical processing in an automatic way, as an aid to control of the process on chemical plants, for example.

But, even with a checklist, it is not appropriate to perform complex procedures of the plan starting with over 60 steps without a cross-conference.

2.2.2.4 Case 4: Mind map, perception and mnemogram

The theoretical constructs of memory (mnemograms) and of perception ("percept") are interconnected and condensed into a single system known as mind map or mind scheme (Sternberg, 2008). Sensations, thoughts and data are processed in subroutines in the mind map to support decision and action. This mind map quality impacts on the performer's behavior at work, and also codifies and stores the information for the action. The process of coding (perception) and memorization (Mnemogram) may be weakened according to the instabilities: in the mind processes (through learning inadequate concepts) and in the decision-maker's emotional balance (due to high stress).

The mind map or mind scheme can be an activity roadmap, action plan, or prototype to be tested. The type of personality influences the development of knowledge and skills from the construction of long and short-term memories, respectively. The development of skill and knowledge includes interpersonal relationships that facilitate group inclusion and organizational integration. Thus the realization of complex activities depends on factors that are not only cognitive in the routine of the operation.

2.2.2.5 Case 5: Slips, organizational and individual values

Slips (Reason, 1990) represent the verbalization of words inappropriate to the pre-established rule in the social or work environment. These words are true in the human being's intrinsic thinking. Embrey (2005) considers the slip to be an error made in performing the task, that is, it is a failure in the context of cognitive processing. Likewise, the pattern written in the form of procedure indicates how to act correctly and for a certain reason, something happens and triggers a human error.

Slip prevention includes establishing a sense of reality in performing critical tasks, making it clear what the task is for planners and workers. Attention should be calibrated to tasks according to: risks (using exercises in a sense of reality), complexities (information flow), modes of operation (emergency; routine execution; managerial and strategic actions). Slip prevention is also addressed when team cooperation reaches the level of establishing effective communication and when seeking to understand the meanings of the procedure.

2.2.2.6 Case 6: Defragmenting the operator's discourse to investigate accident

The perceived reality from observations and reading of documents or occurrences does not exactly represent reality. The reality perceived by the senses, and especially the sense of vision, is added to the memory and associated feelings. They are part of the perception and (in spite of the attempt) they do not represent exactly the true reality of each part constituted (object, sensation, and movement). The affective perception is different from the rational perception which differentiates the fact constructed from the fact itself. To find out what the real fact is, it is important to fragment reality and remove the emotional bonds that turn into biases involved in the scenario.

Society's desire at work and in the family adopts a behavior expectation that may be different from the actual scenario. Society does not want to know exactly the true fact, but the fact expected in the standard behavior of social life. In this way, the reality turns into gradations of approximation to the expected pattern and the construction of memory has a conjugated memory configuration.

The operator's speech wrote in the shift book does not exactly represent actual events because of the loss of reliability and the gradations of the truth admitted by the author in this record. The discourse, therefore, is fragmented and often meaningless. Tools should be developed to defragment the discourse or record of occurrences and unite these fragments of information with other perceptions of different dimensions, in an attempt to validate a real interpretation based on reality.

The individual discourse does not represent exactly the fact as occurred, and its modification can be explained according to the rules of socialization of work, formal or informal. These rules may disagree with the practices that occurred, then causing the record to be different from the fact. Teamwork can change written and spoken messages when they are involved in biased environments in a political direction and not coincident with practice. The situation makes it difficult to investigate losses in the process, such as occupational or process accidents.

The analysis of the OD was discussed by Ávila (2010) and investigated by Drigo et al. (2015), Drigo (2016) within a broader view on organizational communication (Fig. 2.15).

2.2.2.7 Case 7: Human error, automatism and excessive self-confidence

Skilled people have activity schedules with memorized and automatically repeated steps to perform tasks. It is important not to rely on automatism because of unconscious forgetfulness that can occur in a psychological stress environment or in a low workload environment (technical idleness). Thus skills do not replace rules. They are complementary forms that allow the action/task to be done. The skill allows to create the rules that are validated with team consensus, says Embrey (2000).

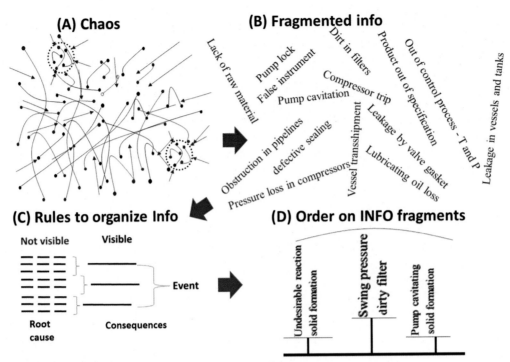

FIGURE 2.15 Operator discourse, facts, gaps and process variables. *From Ávila, S.F., 2004. Methodology to Minimize Effluents at Source From the Investigation of Operational Abnormalities: Case of the Chemical Industry (Dissertation; Professional Master's degree in Management and Environmental Technologies in the Productive Process). TECLIM, Federal University of Bahia, 115 p.*

Excessive self-confidence behavior can cause human error due to nonconsultation of the procedure, the rule, the checklist, and lastly, of the task. The most experienced people often believe that time, experience and skill resulting from the repetition of activities make it possible to skip steps, dispensing consultation of the written procedures. It is essential to have simplicity and care to complete the task through the checklist query.

This discussion is important because there may be situations in which the correct is not to follow any procedures, in case these are outdated due to changes between the actual state of the equipment in relation to the state considered in the planning. The procedure goes into discredit and different classes accomplish the task as they understand it.

2.2.2.8 Case 8: Gender conflict and ethnicity prejudice in shift operator accident

A fatal accident occurred in a refinery unit, located far from the oil industry Decision Center, within a public economic group. This accident drew attention due to the human and social aspects involved.

Before this occurrence, two other fatal accidents occurred with contracted workers indicating issues related to the safety culture and organizational culture. This unit has operations with new and old equipment, with or without automation. After this occurrence, an employee with a management position suffered an accident during the investigation of the previous event (operator death). These events occurred in a 3-year period and are related to each other.

This description is fictitious, and it is prepared to offer an example of the consequences of the lack of social inclusion and lack of organizational integration in the refinery unit, but was written on the basis of actual facts with complementary elements for discussion.

A national competition for hiring process operators attracted people from all over Brazil and also from regions with differentiated population characteristics. In the case, a black woman process operator, migrating to regions with cultural resistances to these characteristics. The feeling of not being inserted in the group during a period of over 2 years, probably brought-about by differentiated social values between company and employee. Personal depression characteristics led to actions that initiated the accident.

A drain valve from the boot of a decanter, requiring additional torque to be closed due to the defect caused by the routine. This torque was not executed and the drainage of inorganic product continued with hydrocarbons in large quantities. The mentioned operator continued her routine activities. The unit was old and conduits passed through paths on the factory floor, underneath metal protections. One of these conduits was in an electrical short circuit, causing fire and burning to the operator and causing her death. A clear conclusion is that population stereotypes cause accidents with fatality (Fig. 2.16).

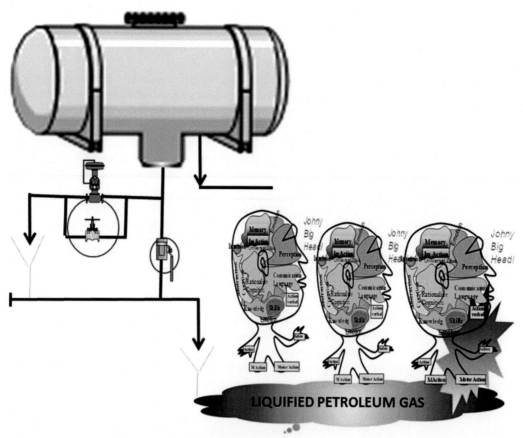

FIGURE 2.16 Drainage of inorganic product, accident, and population stereotype.

2.2.2.9 Case 9: Conflict of generations and supervision of shifts

A HR expert was invited to give a motivational lecture in a quarterly meeting with the manager, the engineering team and the shift supervisors of a polymer production unit in southeast Brazil. The main topic discussed was the causes of human errors resulting in production stoppage.

At this time, a lecture on human factors, talking about the subjectivity of behavior at work, the influence of stress level, emotional balance and population stereotypes were presented as causes of human error and were mainly related to local and global cultural issues.

One of the supervisors who was from the petrochemical unit of Camaçari industrial site, after hearing about Sennett (2005) who wrote the book Corrosion of Character, remembered how the shift worked 10 years ago and compared it with the current situation, complaining about the profile of the new operator. The supervisor of the company stated that nowadays he needs to argue for the operator to go to the field to open or close a valve, which is an action required for the procedure. He remembered that 10 years ago things were more practical and straightforward and that today, the boys are very full of rights and justifications.

2.2.2.9.1 The supervisor asked the expert how to proceed in this situation of generational conflict?

Initially, the expert affirmed that the treatment of this social phenomenon is a responsibility of the organization that should instruct leaders for necessary cultural adaptations. Then he passed the unfortunate news that this conflict is not considered a bad habit or human error, but new profiles that appear in the workstation. That is, in fact, the supervisor's argumentative s' power should improve during the routine so that in emergency situations there is a trust in leaders related to quick decisions. The supervisor was surprised. What do you think about it?

2.2.2.10 Failure analysis with cognitive nature cause

In the human body, in an organic system of information flow, there are restrictions on the transmission of messages through the neuronal network. This organic defect added to cognitive failures in perception, decision-making, and/or motor action, can reduce performance or even block information processing.

Information processing in a rational way may be surrounded by the possibility of failure. False perception can induce decision failure. The decision can be impulsive, which generates uncertain motor actions. Perception can cause altered memory, or even require attention, which is currently failing and without focus. The change in the level of attention may take time, and may not always be available, causing the worker to commit failures.

Automatic processing does not involve rationalization. It leads to impulsive action, often necessary for emergency situations, automatically repeating what is done in simulated exercises. The phenomenon of normalization of the deviance (CCPS, 2016) in the operational context may be involved with the automatism of the actions. This automatism happens due to cognitive failures caused by negative archetypes in the safety culture.

Incomplete cognitive processing occurs due to: attention and perception failures; incorrect memory bindings; and the complexity of logical operations. Automatism in actions occurs when workers assume that the recent work memory based on the skill represents the current situation.

By neglecting attention on the here-and-now, the perception of the current environment is not valued, relying on the repetition of previous perceptions and practices. Thus excessive self-confidence can lead to automatic information processing, and the result of this phenomenon is the rapid acceptance of truth without critical analysis, based on memory and pre-formatted actions, leading to human error.

Often cognitive failure is not in the routine, but in unusual situations that require new concepts for control actions. Thus the analysis of human errors involves two situations: normalization of deviances in routine and lack of knowledge and skill for dealing with unusual phenomena that require new attitudes.

By processing information that involves new events and new environments, there is the possibility of inference of the conditioning of the action, undoing old concepts and creating a new line of thought. The cognitive vision intends to open the relationship between the old and the new concepts. The discussion of resistance regarding the creation of new concepts is essential to improve learning.

This dialectical discussion allows the fluctuation between antagonistic concepts, without the immediate formation of opinion about a specific reality of the industrial plant and society itself. The cognitive structures that construct knowledge and skills depend on the functions established and have differentiated weight for the steps of analysis, planning, and execution, all depending on assertive decisions.

Knowledge base allows the search for solutions through reasoning. The skill base allows the coexistence with practical situations and the adoption of automatic behaviors and not always safe, authorizing the decision by alternative action. Skill, knowledge and experience work together to better tailor decisions.

Planning of tasks depends on specific socio-technical competencies to avoid human error, which is a consequence of cognitive gaps (Ávila and Costa, 2015c). The procedure is the result of planning the task in the written format, so it is necessary to allow some flexible space in the objectives of the procedure, in the expectation that the fluctuation of the operation (Rasmussen, 1997) occurs in the routine, which facilitates reaching established guidelines, although many experienced oil industry operators affirm the contrary.

According to Rasmussen, the final adjustment in the action is performed by the operator in the field with changes in requirements and new process states, that are similar to the original one. By personalizing the actions, the centralization, that can be dangerous, occurs due to the possibility of environmental oscillations, Man's mood, and also the mood of the team and of the manager in the workplace.

Human errors caused by unconscious demands are reflections of an internal world that does not have rational access. Internal perception transforms into action without cognitive reflection. The other possibility happens when the individual perceives external movements and then triggers logical operators of comparison between the memory and the fact that happened.

Improper behaviors can be mapped by the identification of behaviors and psychological aspects of Man and the group. It is observed that the unconscious aspect arises without warning, and it is not possible to trigger preventive measures. When this worker possesses unique functions of hazard control, the causes and possibilities of unconscious human error should be analyzed.

The development of competences for these singular functions should be dealt with in a special way because of the possibility of spurious human errors to happen. Thus training and development programs for operators operating the control of nuclear fission, airspace and catalytic cracking in the refining industry must include psychological analysis and behavior analysis in situations of high stress, for performing selection and development of management processes.

In the theory of expected utility (Cusinato, 2003), the utility function or reward analysis, in the labor environment, allows in a nondeterministic way to anticipate the view of the available alternatives in decision-making. Based on the individual's expected attitude in a given environment, the reality is extrapolated, and incorrect or correct attitudes that will be later performed are anticipated.

The decision is made through the use of clearly established scenarios and the maximization of the expected utility. When the operator makes routine decisions automatically, the analysis of subjective utility becomes unfeasible. In this perspective, the responsibility for the result is transferred to the trainings that simulate situations through scenarios. Training is often based on accidents or failures that, unfortunately, in the future, when demanded, the procedures will not occur as planned, they are going to take place in different areas, which demands differentiated competencies and concepts.

2.2.2.11 Case 10: Culture biases and population stereotypes

Cultural aspects, traditions and archetypes interfere in the behavior of people and groups by altering the final result, then indicating imperfect rationality. These traditions and rituals are valued by the team and are the result of the local dominant culture. Cultural biases should be considered in the analysis of HR and in the construction of mind maps.

The culture indicates scenarios with behavior trends of operators who are actors or executors of a plot. These actors of the failure scenario should be analyzed in terms of: knowledge; environment and workplace; information flow; access to procedures; and level of commitment in treating critical tasks.

The organization's criticism of the regional behavior pattern (regional culture) of the worker with the respective group beliefs leads to prejudices, team disintegration and difficulties to meet the production and safety goals.

Identifying how the failure or the operational problems (as a result of archetypes) occurs allows changes in actors' mind maps in the scenario. The study of mind maps and failure modes guides the project of preventive and corrective actions on task failure and defects in organizational environments, such as the incorrect selection and/or allocation of personnel to the function. Mind maps are graphic schemes that facilitate the elaboration of preventive programs.

2.2.2.12 Case 11: Emergency, Three mile island (nuclear fission versus utilities), and Fukushima (tidal wave)

The nuclear industry has a high risk of contamination as long as an accident happens, especially when it affects the nuclear fission reactor. In the case of the accident with the Three Mile Island plant, the studied accident scenarios for risk analysis and respective safeguards involved only more severe cases but did not include accidents with contamination due to failures in the energy utilities, such as water and steam. In Fukushima, something similar happened. The scenarios studied included external tidal events of up to 17 m high and what happened, in reality, was double that height.

What can be done when the events occurred do not coincide with: the technological knowledge, the level of stress expected, the training to correct emergencies and the scenarios built beforehand?

_____
_____
_____
_____

It is known that it is virtually impossible to predict all sorts of events. Thus knowledge, skill and intuitive learning for high-stress situations should be joined to structure an appropriate leadership for quick decisions.

What is lacking is the training based on operative groups (Pichon-Rivière, 1988) and on progressive stress (Ávila, 2011a) to allow thinking outside the box and elaborating solutions besides the obvious ones offered by the organization and the history of accidents. Another important consideration to be analyzed is how to treat the unusual situations. Vygotsky (1987) presents the need to elaborate pseudo concepts to define new future concepts, still without a public domain.

The construction of an unprecedented accident, with the release of fireball in a region close to community (800 m), was elaborated by Ávila et al. (2015a) and tested through a tool that measure the dynamics of stress to conclude about the behavior of leaders, in similar situations to the reality, and based on local discourse, equipment, events and people.

2.2.2.13 Discussion on skills, knowledge, and rules

Mental schemes direct the skills to perform actions through patterns or instructions for the task. The highest level of skill and experience and a stable environment, without cultural bias in technical and organizational changes, diminish the variability in the operator's behavior and action.

It is important to evaluate the influence of local-regional heuristics in the accomplishment of the tasks. The logical—chronological analysis and connectivity of factors and events interconnected in the task that cause failures indicate the exact point of corrective-preventive-mitigatory action (Chapter 3: Worker Performing Task). The effectiveness in the application of the skill to accomplish the task depends on tools related to formal and informal rules, and in tune with new concepts for unusual situations.

The established behavioral rules and operational standards try to maintain stability in industrial processes. This task planning interconnects the detection of the fault, its causes, its consequences, the safeguard actions and the revision of the operational standards. It is important to remember the importance of including the operator's emotional stability as an essential requirement for the critical task and that impacts the main psychological functions, such as perception, attention, and memory.

The team's appropriate skills and emotional balance in a stable production system with updated standards and procedures keep the facilities controlled in the operating routine. Unfortunately, new situations can emerge without historical recognition, demanding new actions and procedures, that is, new knowledge.

Previous knowledge is used to generate new knowledge in new situations based on hypotheses generated in the routine. It is important to analyze the old concepts and visualize the possibility of creating new concepts through intermediate elaboration phases of Pseudo-concept. New situations are achievable, in which actions are planned in real time through the use of conscious analytical processes to extrapolate/interpolate toward the new possibilities of corrective-preventive-mitigating action.

After understanding the importance of the Rasmussen (1983, 1997) and Reason (1997), skill rule and knowledge (SRK), the level of approximation between good practices and written procedures is discussed (Embrey, 2000). The procedure is the best way to maintain the desired "*meta*-state" in the industrial process. Its activation, analysis, planning or execution depends on the level of need for maintaining the pattern. The observation of the action accomplished indicates requirements-steps-goals-impacts of the task that are compared with the best practice and with written procedures. The observation of the action must confirm the written procedure and the best practice.

The interpretation of the needs for each step and the measurement points to verify and evaluate the efficiency and effectiveness of the procedure occurs through selected control items. The selection of goals for each step and the realization of the procedure is essential in the operational decision. Writing the procedure is performed after field test, joint interpretations and putting this procedure in practice (Rasmussen, 1997).

The evolution of cognitive processing directs questions about cognitive functions in the decision-making process. Models are discussed to evaluate the functioning of the decision and to understand where the main reasons for the incorrect decision are.

2.2.3 Cognitive functions and decision processes

The study of cognitive processing indicates psychological and psychic functions for the analysis of human error. There are functions that influence cognitive processing efficiency and the environment where actions are performed. The emotional balance that affects memory and attention is highlighted. Another important logical operator is the reward analysis and/or fulfillment of expectations in balancing the company's and employee's values, which directly affect the complying with the rules and applying knowledge and skills.

Analysis of historical and environmental factors may indicate the reason for the unexpected or emergent behavior, then triggering the accident. The effectiveness of performing the task is not only restricted to cognitive factors, but also includes intuitive-instinctive factors such as environmental-historical knowledge and the emotional-organic issues.

Current studies in the areas of human factors and HR discuss cognition and tasks without understanding the subjective aspects of man related to affection, unconscious personality movements and organic-motor adequacy in the

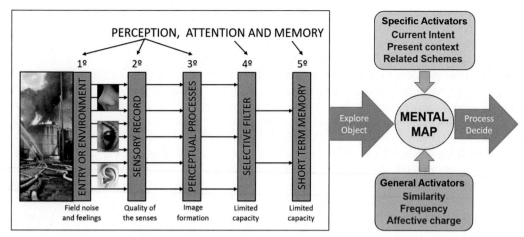

FIGURE 2.17 Cognitive processing and psychological function for decision. *Adapted from Sternberg Jr, R., 2008. Cognitive Psychology. Artmed, Porto Alegre.*

execution of the task. A case of the effectiveness of the task may be related to the level of acceptance of authority at work, based on the operator's family history, implying losses from process when managerial guidelines are applied.

The operator in his work routine manages to distinguish and meets the company's expectations regarding the accomplishment of the task by living with psychological discomfort. At a certain moment, even in an unconscious way, the defense begins in relation to the coexistence of the manager (symbolic for father) and of the worker (symbolic for son), through the action of denying compliance with the set of procedures and rules. Thus unconscious aspects of the past induce human error.

The study of cognitive processing indicates alternatives to the decision in the seek for the best socioeconomic-affective utility (Chapter 3: Worker Performing Task). Concepts about cognitive functions are discussed considering a decision environment that suffers dynamic cultural threats. These concepts are discussed including the role and operations in cognition, memory, learning and decision-making. These cognitive models (Hollnagel, 1993; Reason, 1990) assist the identification of gaps causing human errors that can be potentiated by aspects raised in the dynamics of psychological functions in an organizational environment (Ávila, 2011b).

Cognitive processing involves logical-rational operations and respects (or is influenced by) cultural and organizational aspects. Cognition relates the operations of comparison of scenarios perceived with old memories, retrieval of concepts and similar experiences, extrapolation to new concepts, decision-making, and analysis of priorities regarding the future reward.

Some of the indirect functions of cognitive processing for the execution of the task are: intuition, emotion and motor impulsiveness. Attention, memory, and perception are functions directly involved in cognitive processing.

Attention is the taking of possession by the mind in a vivid and crisp way from one of several objects or lines of thought to initiate the cognitive process or the act of thinking. Attention implies moving away from situations to deal with others, that is, a cognitive filter (Sternberg, 2008).

Attention works as a way to concentrate limited mental resources for certain information and trigger highlighted cognitive processes at a given time. In an operational routine situation, a perceived strong noise can direct the operator's attention to identify the occurrence of a possible explosion.

The development and activation of the procedure require attention focused on cognitive processes. The same can be affirmed regarding the surveillance task and its follow up, although the attention cold be scattered or divided. The intermediate and final goals are compared with the measurements to confirm the priority level and initiate corrective actions. The sense of usefulness is an integral part of cognition through the analysis of achieving the necessary goals and compliance with adjustments. Fig. 2.17 is an adapted model for risk awareness training in the industry, which exemplifies the functioning of cognitive functions. Deutsch and Deutsch (1963) and Norman (1980) indicate the existence of perception levels, attention filters and the moment of memory activation for decision-making. In the discussion about attention, emphasis is given to the most important classes related to the role of the operator performing the operational procedure, but the activation of the mental scheme and decision-making is also discussed.

The phases indicated in the model of Fig. 2.17 are Perception (P1, P2, P3), Attention (A), Memory (M), Mental scheme activator (M2), and Decision (D).

Initially, the equipment and processes emit noises, odors or signs (P1) that can be captured by the human senses (P2) as shown in Fig. 2.17. In this case, the noise, the fire scene, and the burning smell are part of the signals emitted that indicate a serious event.

Then, capturing the signals occurs through the organs of the senses (P2) that should or must be available. These captured signals may mean some events related to the safety goal being reached. The ear, the eyes and the smell capture these signs (P2). Failure to capture these signals (emission or organic capture) can change the decision. The failure in the individual decision is riskier because there is no redundancy in the capture of the signals or in the construction of the image. The individual failure in group decision considers that redundancy occurs, which decreases the probability of failure, therefore, the failures of capturing of the visual and audible signals for group decision are damped.

The recognition of the sensory signal is part of the perception (P3), as shown in Fig. 2.17. Life experience assists the recognition of signals through the comparison of the current situation with past moments. Thus the framework of experienced moments is verified as useful information for the current cognitive process. This comparison is instantaneous, but it is not considered cognitive yet.

After the signal is emitted by the equipment and captured by the organs of the senses (individual or group), it goes to the perceptual processes. There the work of reasoning begins, reaffirming the sign as a component that is part of the decision or constitutes variable for the follow up of motor action. Part of the signals is discarded depending on the focus (attention to certain aspects) and another part follows as a component in the cognitive process. This third phase of perception is indicated in Fig. 2.17 through P3.

The filtration of constructed image-signals is part of the second phase, which works with the attention (A) function, in which the level of risk and urgency for action are considered important criteria. In the third phase, the long-term memory for use, as working memory (M in Fig. 2.17) is taken to compare with the perceptual image and initiate decision-making with the choice of the alternative and the control of the action. Memory operations, mental map activation, and interpretation for decision and action are discussed in cognitive models at 2.2.6.

Failure in retrieving knowledge (long-term memory) for use in the current work may be caused by oversight of past facts (no affection or compromise involved in memory construction) or due to exceeding, taking into account the limited capacity of working memory. The cause of this failure is due to the presence of voids or substitute memories (personality), or still resulting from a state of stress and emotional imbalance, which causes these gaps. The built scenario is probably different from the current reality. To avoid memory failures [cognitive gaps (Ávila and Costa, 2015c)], auxiliary, sequential, or parallel memories are used to accomplish the task. These auxiliary memories and activators help the construction of mind maps to allow robustness in decision-making.

Comparisons trigger logical operators and actions with results that modify the state of the task until they reach the goal. The mind map activation to decide what to do about the scenario and what the best alternative is in phase 4.

Hollnagel (1996) indicates that the activation of maps or mental schemes depends on general and/or specific aspects. Among the general aspects or activators, the following ones are discussed: similarity in relation to past facts facilitates the retrieval of memory by association; frequency of events makes the memory more vivid; presence of elements that impact the environments, the organization and on the task. Finally, the high affective load activates the map for decision and future action.

Among the specific aspects or activators, there are: the seek for the group's goal, in which the current intention can activate the construction of a mental scheme; the present context with signs of abnormalities that show the need to elaborate mental schemes for decision-making, thus leaving from the state of surveillance to the state of search; the related schemes lead people to believe that there are things going on that need to be investigated with new mental schemes. Something different happens in the production system.

When activating the schemes or mind maps, the outputs are: images of the action before it happens (prediction) and its possible consequences; actions to be performed for validating the existence of this map; words that allow applying tests for validation and communication of interpretations to the related community; perceptions with specific information on signs of abnormality; and feelings and affections related to the mental scheme that indicates the need for deciding what is the priority action.

The exploration of the object is necessary for the mental scheme to work in the decision-making process and is part of the fifth phase of Fig. 2.17. Neisser (1976) proposes that from the exploratory investigation, the sample of data should be performed, representing the environment. With the available information, it is possible to choose the object for the cognitive process analysis. Data processed on the chosen object generates information for decision-making.

This research proposes the elaboration of a mental scheme that facilitates decision-making and subsequent action. The information that resulted from the action modifies or constructs the mind map. It is also influenced by memory.

The mind map formed after recognizing the object of analysis closes the cycle and directs the exploration of environments. This leads to specific searches for acting and not only surveillance. The mental scheme directs the cognitive work, thus reducing the sensory effort in the perception, which indicates the most probable alternatives in the decision.

In the second phase of the model adapted by the authors (Fig. 2.17), care should be taken with certain phenomena that can generate human error. Divided or dispersed attention in the case of the operator performing parallel tasks, causes the redirection of limited attention from a high-risk task to a task with lower risk.

In this scenario, the drivers' behavior who speaks while driving is exemplified. Limited attention can cause failure. A case is presented for the operational context in an industry:

In a given shift group, an operator was absent in the team and the field operator performed sampling and started, in parallel, the loading of a 35% HCl truck. At the end of the maneuvers, he returned to the loading station and then found the HCl truck overflowing. This environmental accident indicated that certain tasks are priority, which demands concentrated attention.

How to avoid this human error based on cognitive gap that might had cultural influence?

_____
_____
_____
_____
_____

The surveillance state aims to detect whether an emitted signal is perceived and what the meaning in a larger area is, where various equipment and similar systems should be monitored. Surveillance allows action quickly if a sign of abnormality appears. Surveillance does not seek a specific location or action but rather the general situation.

2.2.3.1 Routine of a field operator in chemical industry

In a bomb park, the operator carries out his/her general check routine. In the domestic residence, vigilance is made to the smell of gas when water is heated due to past occurrences of leaks with current possibilities.

The surveillance activity is constant in the control room of industrial facilities because of the automatic control panels that need supervision. To follow operations there is a roadmap following the priorities in terms of safety and quality. When deleting important sections or controls, the operator loses the chance of correction.

In monitoring the control panel, the operator may perceive abnormalities through alarms and signaling in the operation of bombs. The same surveillance is important in the field, where the operator uses a roadmap to be followed.

In a few moments, the operator performs specific operations with focused attention or search activity. The active searches are performed by determined stimuli, the surveillance may initially occur regarding the smoke that appears and the search for the cause of fire in a given location. In the operation, the cause for the bomb to have disarmed is searched. Thus the operator heads to the specific location and collects information about the equipment and the related data.

Selective attention is a process of choosing some stimuli and ignoring others. If the focus of attention is concentrated at a given point, it improves the ability to manipulate stimuli for other cognitive processes. Thus when the company develops preventive work involving statistics and mapping of abnormalities, based on the shift deviance signals, it is possible to direct efforts toward a given location to investigate the reduction of losses, for example, losses of methylene chloride solvent in polymerization reaction systems to produce polycarbonate

In the third phase of the model adapted by the authors (Fig. 2.17), M, care should be taken with certain phenomena that can generate human error from memory, a function of cognitive processing. The current perception is compared with past experiences and the result of this analysis between pattern and reality, and after the application of logical operators, the result is reached on the action of control or execution. This result depends on the environments that involve a decision. Memory does not always represent reality itself, but it indicates a transformed reality, showing the existence of a conjugated memory different from reality.

In an accident investigation, operators confirm the realization of activities that would block failures, then avoiding the accident. But the accident occurred, why?

The memory trap occurred, thus indicating that actions were not performed, but were registered as performed by the operation. So, the individual memory does not correspond to the actual memory of the facts.

How to avoid this type of cognitive event?

_____

Memory is also important to apply old concepts through actions or factors and create new concepts to explain a new event. The new facts and observations are compared with the previous ones. The signs of the current reality are compared with the situations experienced in the past.

Reading texts indicates past experiences or the imaginary of the local culture, which represents the past, in the educational process. This difference between the new and the old can indicate an emerging phenomenon that was dormant or simply a new phenomenon that needs to be explained through a new concept.

In the construction of human and social typology in Chapter 3, Worker Performing Task, one of the main aspects analyzed is memory. In the analysis of regional culture and organizational culture, there are aspects registered in the community or in the organization. These aspects need to be rescued to investigate social behaviors with tendencies to depression or low motivation. In mass culture, there is the characteristic behavior of porous memory and the lack of affective bond between people themselves or between people and organizations (Sennett, 2005).

2.2.4 Learning and skill

Learning depends on the memory retrieval of past experiences (related to old concepts) that return to be used at work, and after several applications, they can become current skills (short-term memory), or if not applied, returns as the basis of knowledge (such as long-term memory). These memory operations are constant when the cognitive processing in performing the task demands the complete scheme SRK, from Rasmussen (1983): skill (working memory or short-term memory), knowledge (long-term memory) and rules (established standards).

As shown in Fig. 2.18, the skill is formed from continuous and consecutive accomplishments of the task. The skill effectiveness is reduced as the frequency that the task is performed decreases. The insertion of new knowledge blocks focused on human factors and lessons learned from failure processes reinforces decision effectiveness. In the current culture, multidisciplinary, interdisciplinarity and transdisciplinary are broadly discussed, but, in fact, it is not put into practice, probably due to limitations in the development of skills.

The intensity in the practice, the commitment, and knowledge about the causes of human errors motivate the group of operators, experienced and beginner ones, to achieve and maintain a standard of excellence through the increase in the learning speed.

The inverse phenomenon may happen for experienced operators, in the situation that the excess of repetition makes the work tedious, which decreases the speed of learning. The learning limits and challenges indicate skills quality. A good strategy is to adjust responsibilities to restart the learning process in areas with new knowledge and organizational challenges (Muchinsky, 2004).

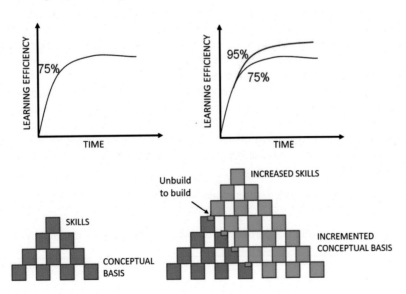

FIGURE 2.18 Learning capacity: concepts and processes.

In HR, number of tests or times that the task was accomplished to increase the effectiveness at levels above 90% (Fig. 2.18) is questioned. As commented, only the will is not sufficient for learning. There must be a compromise, measured through a satisfactory psychological contract, which serves both parts and allows the effective application of skills in the accomplishment of the task. The very automated systems, without the occurrence of small failures, can limit the learning process and the development of solutions.

The construction of skills implies the understanding, from the practice, of two important processes that are part of this learning: adaptation and habituation (Sternberg, 2008). The responses that involve physiological adaptation occur in the sensory organs and create better conditions to receive the signs from the environment, whereas the answers that involve cognitive habituation occur in the brain and are part of the learning process.

Adaptation is not accessible to conscious control. An example is the perception of odor or perception of light intensity change. This learning process is linked to the intensity of stimuli (the intensity of light influences the senses for its adaptation). The adaptation is not related to the number of repetitions, duration and period of exposure to light (Sternberg, 2008).

Habituation is accessible to conscious control and is not linked to the intensity of the stimulus, but rather, linked to the number of tests, the duration and the recent character of previous exposures. The night operator can operate on the daytime shift after a period of habituation (is it possible to speed up this habituation?). Another example is presented: it quickly gets used to the sound of an "electric trio" with increased exposure to sound more frequently for longer periods and on more recent occasions.

2.2.5 Motivation and decision

The logical operations of memory start and maintain the accomplishment of the task until the goals are reached. In nonstandardized situations, the best alternative is sought through the decision-making process, triggering the respective cognitive functions: attention, perception, memory retrieval and utility analysis of the task. Nonmotivation at work makes it difficult to perform these functions, especially in unusual cases involving intense memory retrieval and a large amount of energy for the creation of new concepts.

The cognitive process is initiated after confirming the usefulness of the task for society, the operator, and the usefulness of work, intrinsically, for the worker-citizen. Confirmation of the usefulness is made by comparing the possible state of being achieved and the established goals. Confirmation of organizational efficiency through numerical results is an important element for the motivation at work and indicates the level of approximation or getting apart to meet the business objectives.

High motivation ensures that the operator/executor efficiently performs the task in the organizational environment. The action initiated depends on an analysis of the comparison of the expectation and reward at work, of the discomfort caused by decision-making, and also the return to internal equilibrium after the decision. The action initiated also depends on the variation between these states.

Cognitive processing is "operated" in routine decision-making with the environments and possible alternatives, but always considering the presence of biases resulting from the pressures of external reality (management) and internal reality (reward relations and work satisfaction) (Chiavenato, 2004; Sternberg, 2008).

A cyclical movement occurs with the operator during his/her coexistence in the work routine, which involves balance and imbalance phases in the decision process. At the time of the decision, the operator's internal equilibrium is momentary, because there are stimuli generated by the need to perform actions and that will meet the industry demand on the production line.

The operational control to keep the industry running is carried out by the work team and depends on the level of motivation. Some routine or emergency work are performed after reaching the team's necessary state of motivation. In the continuous process industry, operations are discussed, such as maintain the reactor temperature at the setpoint; loading with the final product; quality control and quantity produced of high-pressure vapor; vibration reduction of large rotating equipment; or avoid the beginning of a fire in the industrial area.

From the operational need, a tension field is created, which starts the decision-making process. In it, analyses of alternatives and consequences are made, and in a group or individual way, the task is accomplished through parallel or serial actions. An important factor for an effective decision is related to the compatibility between the worker's competence and commitment besides the function to be performed at the workplace. The benefits achieved with the realization of the action authorize the reward and the operator's satisfaction, then the return to the state of internal equilibrium maintains the motivation. Frustration can cause a loss of motivation.

Frustration occurs after the impact of an incorrect action. In this case, the return to internal equilibrium is slower, and the discomfort of a nonaccomplished task is perceived. Other derivative behaviors can be activated to compensate for the frustration of the incorrect action on the task. It is important to analyze these possibilities to improve the operating standards, in which the written procedures equate the good practices for the routine (Embrey, 2000).

The motivation for action alters or maintains the activity patterns at the workplace, in the execution of the task, in achieving individual goals, such as: social inclusion; satisfaction of family desires; and meeting financial needs. Motivation may result from the perceived relationship between the productivity performed and the accomplishment of individual goals at work. This result indicates the effort required to meet the company's expectations or individual expectations.

2.2.6 Decision-making process

The classic models of the decision-making process follow fixed steps. The construction of the decision depends on criteria that do not consider cultural biases—classical decision. There are cultural heuristics that cause decision priorities to be changed. These models involve changes of rules in the internal environment (individual) or external environment (economy and nature) in the decision matrix; this is another proposed model—decision by priority orbitals.

In the classical decision, the decision-making processes performed during operational routine activate psychological functions in consecutive steps, such as: perception of signs; comparison between perceived facts and working memory; construction and activation of a mind map for the task; control and decisions, requirements, steps and goals; Type of communication in the steps; and carrying out action.

The decision-making process includes actors with different roles: the organization works as a promoter of policies that enable economic results in production. The physical-cognitive and organizational environment demands studies on organizational and safety culture. The decision-making model in the factory operation is a result of the managerial style (manager) adopted in the organization. The worker and his/her interpersonal relationships participate in this process through the application of knowledge (expert) and skills (executor) in specific areas. The efficiency of decision-making depends on aspects of cooperation, leadership (leader) and its relationship with people. The type of task involved, the process technology (continuous or intermittent) and the type of product are factors that impact the decision model.

The Failure Risk analysis, according to Table 2.4, performed by specialists in human factors in industry and provoked by functions, factors and actors in the decision, indicates that the greatest impacts are, in actors-factors: the leader (26) and the task (22); and in functions: communication (23) and attention (22).

Fig. 2.19 illustrates the classic decision-making process involving the following elements applied in each step: (1) the environment or workplace; (2) causal factors related to the events happened (backcasting) and the projection of future (forecasting); (3) cognitive processes (comparison with utility criteria, conditioning factors, extrapolation of situations, and memory retrieval); (4) the moment of decision-making (time and clarity); (5) the analysis and discussion of alternatives and their consequences; (6) choice of the best alternative and (7) action after decision.

A decision-making process is discussed in the case study of a sulfuric acid reconcentration plant operation. The decision begins on the alternatives for cleaning the tank of the finished product. This decision involves aspects of safety, equipment availability, and people for the task. In the reconcentration operation, after evaporation, the salts

TABLE 2.4 Impact of failure in function and relation with actor & factor (red high, medium, green).

Actors & Factors Functions	Organization	Environment	Manager	Leader	Executor	Specialist	Task	Product	Process	Risk Assessment
Perception	0	3	3	3	1	2	3	3	3	21
Attention	3	2	2	3	2	3	3	2	2	22
Work Memory	0	0	0	3	3	1	2	2	2	13
Long Term Mem.	3	2	2	2	1	3	2	1	1	17
Cognition Causal N.	2	2	3	3	1	3	3	1	1	21
Mental Map	0	2	3	3	1	3	2	0	3	17
Decision Process	3	2	3	3	1	3	1	2	2	20
Communication	3	3	3	3	2	2	3	1	1	23
Action	3	1	2	3	3	1	3	1	1	18
Risk Assessment	17	17	21	26	15	21	22	15	18	

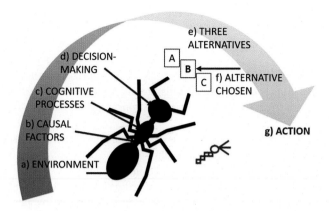

FIGURE 2.19 Decision-making process and types involved.

resulting from corrosion accumulate in concentrated acid tanks to the point of obstructing lines and making it impossible to operate transfer pumps. Tank cleaning is scheduled annually according to specific procedures and variations depending on environmental and organizational issues.

Description of the steps of this decision-making process for the sulfuric acid plant with the use of the classic model:

1. The environment

    The following items are taken into account when defining the context for this activity: time, occupational risks, availability of people, the possibility of contamination of the tank and what is the current organizational moment.

2. Causal factors

    Corrosion is the main reason for the procedure to be carried out. The moment of stopping the plant operations for cleaning depends on the accumulated number of salts in the equipment, the consequences of this accumulation and the production planning. Since there is no direct evidence to indicate the problem, some indirect evidence is sought. Postponing and deciding mean major future damage involving the loss of assets. There are alternatives to the moment of tank release and the way to perform this procedure.

3. Cognitive processes

    The perception of the current state of corrosion of the tanks indicates the right moment. The most appropriate pattern depends on comparing the state found with the history. (1) Comparison with criteria based on usefulness, speed of corrosion (mpy) and estimation of precipitated salt; pump yield; inspection of corrosion in lines; history of equipment release in acidic areas; and the ways of executing the task. (2) Conditioning factors: time to perform the work; possibility of accident and the existence of support for risk analysis; ability to accomplish the task. (3) Extrapolation of situations: the scenario is elaborated from the preparation of the procedures and from the definition of the moment for executing the activity. (4) Memory retrieval from reports and information on other cleaning and corrosion processes.

4. Analysis of the alternatives and consequences

    If the cleaning is not done in time, there will occur total failure of the pump and the possibility of a crack in the tank. Cleaning is accompanied by inspection and service requiring urgency. If the service is in the daytime, the release of gases may disturb the administrative staff. If the service is carried out without technical or safety monitoring, the opportunity to perform other repairs and inspect the tank status can be lost. The people chosen to accompany and perform this service must have specific skills to handle acids safely, in addition to skills in releasing equipment for services, that is, monitoring of PT.

5. Moment of decision

    Once the important information is gathered, the following should be defined: time, duration and clarity of the procedure. The decision will be shared between operation, safety, and maintenance.

6. Action after the decision of the chosen alternative

(1) Service: Tank cleaning, inspection and small boiler services in the tank and auxiliaries; (2) Staff: two operators, two boilers, a welder, a safety support and an engineer for monitoring, independent of the shift team; (3) Cleaning staff: four people; (4) Materials and tools: neutralizing, handling of solids and acidic liquids; (5) Time and day: On Friday night, opening and withdrawal of solid material with neutralization; (6) On Saturday morning: inspection and quick weld services; (7) Emergency and safety: use PPEs and have the provision of first-aid services.

Priority Orbital Decision is a proposed model to explain the dynamic behavior of the individual or group in decision-making, and the risks of committing individual or group failures, which affect the organization's efficiency. This model uses the classical decision steps in an environment that changes the criteria dynamically, altering the objective function and consequently the chosen alternatives.

The decisions revolve around the individual representing the decision matrix (Fig. 2.20). This matrix consists of three areas: (1) information and Action Processing Center (Perception and action); (2) historical traits inherited by the individual and by the environment of collective unconsciousness; (3) conjugated memory consisting of original or actual memory modified by defense mechanisms, indicated in Figs. 2.20 and 2.21.

The information processing center is responsible for triggering the causal nexus for decision-making by action. The identification of the decisions to be taken and the priority analysis is also processed in this center. The motivation to analyze the situation and the motivation to act in verbal or motor form is a process performed in this center.

The historical traits are inheritances recorded and used by the individual (archetypes). The processed data takes into account intuitive aspects (historical and environmental trends), archetypes based on the collective unconsciousness discussed by Jung (2002) and instinctive aspects such as physical survival, movements in the seek for pleasure and preservation of the species.

The conjugated memory consists of records from the present that are altered to meet some uses in individual and social relations (psychic survival and social inclusion). Based on the explanation above about what the individual matrix is, around which the decisions by priority orbital are made, two types of action are defined: impulsive action and rational action.

In the impulsive action, the individual's memory is the basis for the action. There is no cognitive processing. The cycle of thinking does not occur and the action after perception is impulsive, thus not considering any analysis of the alternatives and impacts.

In rational action, development depends on the cognitive apparatus. Action is based on perceived signs in the inventory of affective or rational memory, in bonding relationships with the groups that compose society. The decision to act, therefore, depends on conscious and unconscious aspects, triggering the action according to the energization of the decision-making process.

The individual matrix is composed of diverse materials of the conscious and unconscious type. These materials are recorded in the past or present time and can infer about future time. In the individual matrix, there are memories of the affective experiences in crossing with memories about rational experiences. Part of what is recorded in memory that has sufficient bonds to form facts and, part of it constitutes the recorded traits of the routine that are used for experience-based decision.

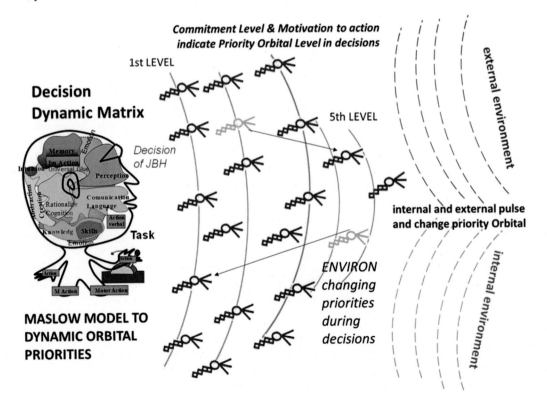

FIGURE 2.20 Decision model by priority orbitals.

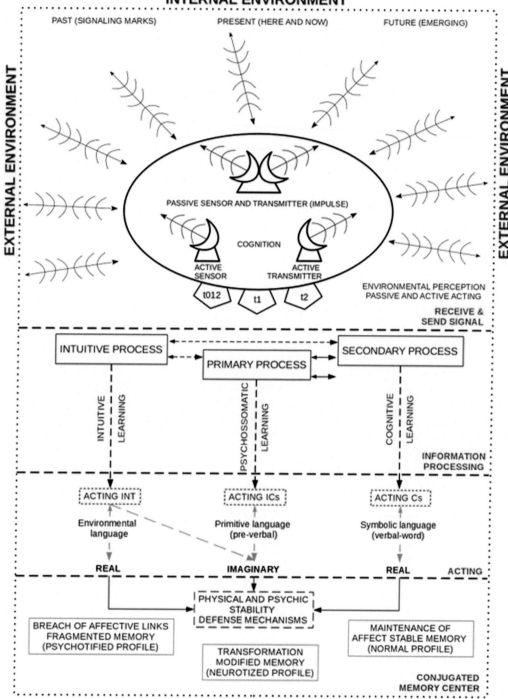

FIGURE 2.21 Information and action processing center. *Adapted from Ávila, S.F., 2010. Etiology of Operational Abnormalities in the Industry: Modeling for Learning (Thesis; Doctor's degree in Chemical and Biochemical Process Technology). Federal University of Rio de Janeiro, School of Chemistry, Rio de Janeiro, 296 p.*

Rational memory can have a valuation for society. Thus it is applicable to the work environment. When this rational memory constitutes the basis for labor activity it is called knowledge and when this memory constitutes the basis for action on the environment it is called skill.

Professional competence depends, therefore, on the inventory of skills and knowledge, but also depends on the inventory of social experiences in the work environment. Thus memory about a social relationship is part of the matrix, in the conjugated memory, and makes up the competence for the work.

Being a leader or being led, acting as a process manager formally or informally, possessing specific functions in the complex social relationship, possessing responsibilities or authority on administrative or technical processes: an are competencies that are recorded in memory increasing the professional's value in the market.

Fig. 2.20 presents the model of decision by orbitals of priority, developed by Ávila (2010). The priority is not static but dynamic, thus it suffers the action of the internal and external environment. The socioeconomic reality is understood as an external environment, which due to globalization, can be changed in a short period of time. Environmental changes affect priority orbitals by altering the causal and organic links that trigger the decision-making process. In the case of internal environments, the energy waves of this environment cause energized orbitals (with priority change) or from orbitals, that is, alteration of priority and also of the decision-making process.

A repetitive movement of energization or de-energization influences the change of priority of the alternatives and subjects in the orbitals. Priorities are dynamic and change over time, including criteria, alternatives, weights and information quality. It is necessary to study the internal (behavior) or external (natural and organizational environment) to verify whether the team has trends of changes in their values and whether they are discordant to the company's culture, which unfortunately is static.

In other words, Maslow's Hierarchy of Needs (1987) changes the priorities in the workstation, changing the position of the alternatives; after 6 years, a biodiesel unit that had its entire model of decision based on the raw material starts to divide their concerns about water availability in their objective function for decision-making, an external and natural environment.

The topographic model of the information and action processing center is shown in Fig. 2.21. This center is divided into four functional units: (1) capture and emission of signals; (2) information processing; (3) action or performance; and (4) center of conjugated memory.

In the capture and emission of signs (1) perception and action can happen cognitively or also impulsively. Signals involve external environments (industrial scenarios) and indoor environments (demands for new emerging social values). When the environment is internal there is no time measurement because the processes are primary. The signs may come from individual or technical records from the past. They can also be a perception from the internal or external present, or even, they may be signs of emerging trends of technical or social behavior.

The processing of information (2) occurs in three dimensions: intuitive-instinctive process, secondary-cognitive process and primary process based on the individual and collective unconsciousness. In this process, heuristics of behavior and tendencies are used for changes in the social typology that is in the collective unconsciousness, but there will be no discussion about the logic of the primary process.

The motivation, decision, and processing of information will be treated in the analysis of the task, analysis of human factors and possibilities of human error always in the cognitive basis as a secondary process. But issues that may involve rare risks and behavioral changes will use secondary data with group responses changing decision criteria based on priority orbitals.

The intuitive and instinctive learning is in the behavior when emergencies are faced and is based on techniques that transform the affective dimension in the operator's decision (Ávila, 2011a). Psychosomatic learning is not an object of study in this book, but it influences decisions in the model of priority orbitals. The focus for the routine analysis is in cognitive learning. Learnings come together in the routine, whereby it is not possible to separate it from the reality in the subjective analysis. In the quantitative analysis, including the subjective, it is attempted to verify the trends through behaviors that tangent the real scenario leading to other hypotheses for future emergent behavior.

In the same way that the capture of signals and their transmission is discussed for the internal and external environment, since the processing of information is multidimensional, the performances work accordingly.

High performance is considered a result of cognitive processing, as in the example of industrial tasks. But also, the actuation is real and resulting from instinctive survival actions and historical archetypes (intuitive actions). Action can also be from the imaginary dimension, possibly confusing the memory of the real when there are defense mechanisms.

Conjugated memory is the result of the recording of the actions performed according to the level of affection that can circulate the fragmented events until they become psychic fact or fragmentary memory without retrieval.

Through organizational language, the memory of learned lessons depends on the commitment to the task. Successful and unsuccessful experiences are means of learning. Through psychosomatic language, affection will surround, according to psychic energy, the event in the body causing a neuronal fixation. By the intuitive language, the repetition of emergency simulates and the perceptions of the environment will indicate the way to perform actions by impulse.

The worker's physical or psychic stability promotes a state of emotional-affective equilibrium that demands fewer defense mechanisms in the cognitive, motor and psychosomatic performance. This balance maintains the memory and cognitive processing stable by responding with assertiveness and the agility demanded.

The breakdown of labor, social and family affective bonds, which can be caused by lack of feedback, lack of commitment, lack of genuine leadership, lack of accredited organizational values, generate imbalances that promote fragmented memories, which makes decision-making difficult. These neurotic profiles can cause variations in memory, in various moments of planning, diagnosis and execution of the task, provoking human errors.

Borderline profiles, currently very common, are today on the threshold between normal and psychotic behavior. They are selected for work and can have a "fanciful" memory effect which puts themselves over the memory of real psychic facts. This is a serious situation in which workers who control or supervise high-risk activities (airplane pilot) can freak out and replace reality with fantasy (affective bonds interrupted), then causing major accidents. My wife doesn't give me attention, then I can die with all the passengers.

2.2.7 Cognitive model discussion

Cognitive, motivational and decision-making models (Hollnagel and Woods, 2005) do not answer all questions about the cause of human errors in performing work activities. At this step of the work, it is intended to confirm the heuristics in addition to indicate where and how cognitive gaps (Ávila and Costa, 2015c) occur to suggest preventive actions for human error. The cognitive models from Hollnagel (1993), Reason (1990), Ávila (2011b) are adjusted to inhibit the recurrence of human error.

2.2.7.1 Rules, gaps, and questions

At this moment, questions are presented about how the cognitive gaps (Ávila and Costa, 2015c) that cause human error occur. With the knowledge of this initial reading, answer the following questions.

When work activities are individual-based, how to ensure that the perception builds reliable image in relation to reality?

___

How to migrate between focused attention and dispersed attention in environments that require this dynamism?

___

How do memory operations work between retrieval and deposit and what human errors are involved in memory failure?

___

Do behavior models always indicate a truth or are there variations in the rules?
Is there any weight for mind map activators?

___

Does the motivation to act depend on personality or decision environment aspects?

___

What is the relationship between cognitive gap and human error?

How do you know that the standardized procedure has lost validity and it is necessary to test a new alternative in the decision on the action?

Cognitive models attempt to explain how the operator performs the task and the possibilities of causing human error. The first model will discuss memory based on Reason's discussion. The second model will indicate the information flow concentrated in the control room; and the third model will present the importance of integrating cognitive processing, gaps, and errors, with the fundamental aspects in the construction of the task.

1. MEMORY. Cognition from reason and Sternberg
2. SUPERVISOR & SUPERVISORY SYSTEM. Decision of observer
3. COGNITION AND TASK. Cognition model.

2.2.7.2 Model I—Cognition and memory operations

Memory is a limited function and works with codes, symbols, and words not to spend excess energy on cognitive processing. These words or signifiers associate and retrieve other memories stored, bringing meaning to the facts. Memorization is a nonrational process and can be complex. Memory retrieval is done from fished words, and with their contents, and the relationship with other words and contents, and there may be many loops of concepts, promoting complexity in certain tasks (Sternberg, 2008).

Short-term memory has a higher priority level than long-term memory. It takes up more space and can be retrieved rapidly because it has relation codes and by the fact that it is linked to more logical operations. Nonuse of memory on the work environment or family routine authorizes a reclassification from short to long-term memory. Certainly, this also depends on the affective bonds and the level of priority for the family and for work. To make the reverse change, that is, leaving from long to become short-term, repetition and organizing information are important, that is, the mooring of words with codes of relationships and logical operators in the decision-making process (Sternberg, 2008).

Signals from the external environment or sensory inputs of the cognitive machine are introduced through the information and actions in the processing center (Fig. 2.22). These inputs are consciously introduced to working memory and in an unconscious manner, forming the knowledge base. Working memory or short-term memory is conscious, selective, and has limited resources. Therefore not every signal goes through sensory inputs. Working memory requires more effort, working intermittently or in blocks, and has computational power (Reason, 1990).

On the other hand, the knowledge base is long-term memory that harmonizes similarities and is formed unconsciously, through the sensory input of signals from the external environment. It is a seemingly unlimited and fast memory with parallel processing, thus requiring less effort. There are two basic heuristics: the knowledge base harmonizes signs as they happen; and this harmonization occurs in favor of the most frequent items (Reason, 1990).

Energy consumption for memorization or memory retrieval process is controlled according to the need. Situations involving obsessive behavior or major concerns in terms of imminent danger cause psychological fatigue. Memory retrieval can be an exhaustive operation, trying with increasing efforts, to recover something that is still lost, a gap to be filled-in. Sometimes our mind deceives us and fills the gap with untrue but comfortable memory for the inclusion or for the relationship of internal satisfaction. This is discussed in the conjugated memory model (Sternberg, 2008).

The memory retrieval process can be self-finalized in case of overspent power, like an interlocking of the memory retrieval process. Memory can be redeemed with some gaps. But there may also be the filling of these gaps with appropriate content for the moment. Thus memory distortions are real possibilities and, depending on the individual (human typology), the work team (group typology) and the relation with the organization (expectation and reward), the operations that cause distortion in memory may be multiple and may be very common in everyday life. Human and social typology are addressed in Chapter 3, Worker Performing Task.

Some memory failures occur in the worker's routine: transience where it disappears quickly; distraction where you forget what was accomplished; psychological block where it does not remember something that threatens psychic life; misattribution in relation to what was said or heard; suggestiveness, when people accept as truth what is suggested;

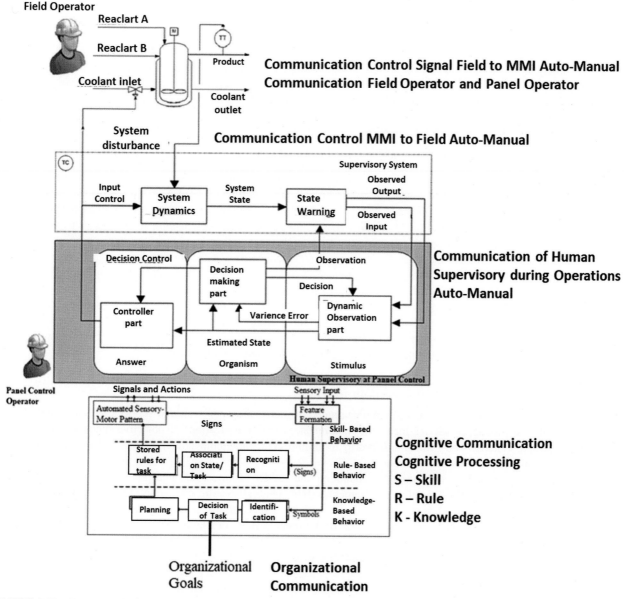

FIGURE 2.22 Observer/controller, system, and SKR. *Adapted from Reason, J., 1990. Human Error. Cambridge University Press, Cambridge; Hollnagel, E., 1993. The phenotype of erroneous actions. International Journal of Man-Machine Studies, 39 (1), 1—32.*

cultural biases in decisions, which indicates substitution of reality by accepted memory; and persistence where the population remembers more of the failure than successes (Sternberg, 2008).

To avoid a threat situation due to memory failure, practices are required to ensure control of a productive system. The practice of communication during operational routine (operational communication), so that signals are sent to the control panels. Communication from the HMI, which is a signaling and uncertainty procedure, should stimulate changes in human supervision. The process of technical—human—organizational communication is another phase that aims at control and, finally, organizational communication, which contemplates the whole process of Communication.

2.2.7.3 Model II—Decision model for the observer and the controller

The production system is dynamic. Operational controls maintain processes at specification limits which are technically and commercially accepted standards. Pipes, raw material and utilities participate in these processes. In the first phase of control, operational communication occurs, in which signals and words are transferred from the field and are recognized in the panel for keeping standards. In a second phase, control is sought on the HMI in distributed systems with complex processes or the control of critical equipment. The signal resulting from process state measurement and the equipment is observed and manipulated on the control panel. This is the Man—Machine communication phase, also called control supervising phase (automatic or manual) (Fig. 2.22).

Dynamic observation shows the difference in relation to the standards. This difference will indicate decisions to be made. The analysis of alternatives and the way of decision-making is a responsibility of an organism that is part of the human supervisor and that is in contact with the supervisory through HMI. When comparing the input with the output, there are differences to be adjusted through human decisions implemented by the supervisory. The instances involved in this cognitive model are the supervisory system and the human supervisory system. In the human supervisory system, there are the stimulus, the organism, and the response.

The input of the productive system and its output, after the control actions, are compared to generate new controls. This is necessary because the measured state is different from the expected standard state. This difference is the stimulus for changes in control, which are performed in the communication of human supervision, the third phase.

Following, thus, the process of technical (based on knowledge) — human (based on skills) — organizational (based on rules) communication, is the fourth phase also called cognitive communication stage, where Reason's SRK Model (1990) is discussed, including commitment in this scheme, the SRKc. The organization then transfer expected goals, procedures and expected behavior patterns in the routine, through organizational communication, the fifth phase.

In the supervisory system, and the HMI, the practice of communication, whether in cognitive, organizational or operational processing, influences the performance of the production system through standards set by the organization, the attainment of objectives and the recognition of the competencies that they produce. play. involves attitude, commitment, skill. In this context, Model III presented below highlights routine procedures in the execution of production operations, the validity of compliance with written procedures (a communication format) and its revision, in order to evaluate the state of the process, without changes or indicating abnormalities. This interpretation allows the retrieval of memory and its knowledge content, considering that memory influences the perception, planning and execution of the task.

2.2.7.4 Model III—Simple model for cognition

The study on the physicochemical—mechanical phenomena of industrial production and services indicates that the best way to elaborate the product or service meets the maximization (or minimization) of objective functions. The optimal points for objective functions are: maximum profit, minimum cost, maximum quantity produced, maximum quality, minimum accidents, and environmental impacts.

The accomplishment of these operations in production and services follows repeated procedures during the production schedule in the daily routine, meeting requirements, steps, general/specific goals, and acceptable level of impact. Therefore written procedures are elaborated to be accomplished. A schedule for reviewing the procedures (with the control of goals and objectives) is made after the analysis of the process state, in which some ranges are accepted as normal and other values may show abnormalities.

As indicated in Fig. 2.23, planning, control, and diagnosis on the task and the review of procedures are interconnected with the cognitive ability to perceive, memorize, interpret, decide and act in each function of the process.

In the Hollnagel model (1993) described in Fig. 2.23, technical data, measured variables, production information, and events are inputs of the cognitive model through the function of perception. These observations make it possible to interpret events and to get to know what is needed to maintain control. When planning the accomplishment of tasks, goals are established and actions are chosen. The execution of the actions is compared with the data entered. This is how the cycle ends.

In this model, interpretation influences planning, enables memory retrieval and the generation of a knowledge repository. Planning, on the other hand, influences interpretation. Memory changes perception, modifies task execution, and influences interpretation and planning.

Thus we can affirm that the lack of operator preparation can lead to human errors. It is understood that low memory quality, lack of knowledge, lack of experience, failure in the knowledge repository, bad habits and errors in decisions indicates a lack of operator preparation and/or cognitive gaps (Ávila and Costa, 2015c). These gaps have a differentiated impact by function, type of technology and role performed on the task: planning, diagnosis, execution, or project.

2.2.8 Human behavior dynamics in the company

Cognitive processing, widely discussed in this chapter, will be approached here by considering psychological functions at different times in the organization. Thus the discussion is divided into: classification of psychological functions;

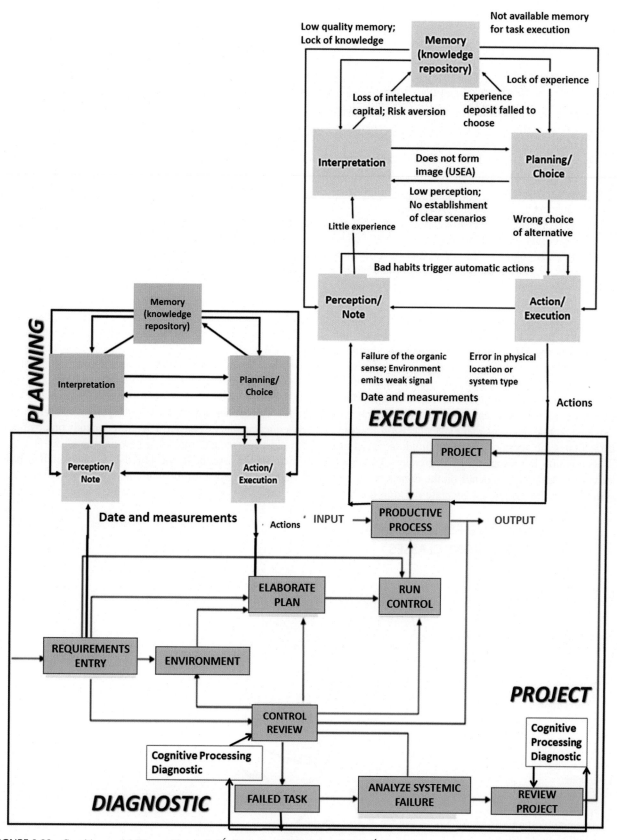

FIGURE 2.23 Cognitive model III: cognition, task (Ávila et al., 2015d). *Adapted from Ávila, S.F., Sousa, C.R., Carvalho, A.C., 2015d. Assessment of complexity in the task to define safeguards against dynamic risks. Procedia Manufacturing 3, 1772–1779; Hollnagel, E., 1993. The phenotype of erroneous actions. International Journal of Man-Machine Studies 39, 1, 1–32.*

functions and behaviors in the organization; the movements of these functions in organizational moments; and the JBH life cycle in the organization. Functions analyzed include cognitive processes, but extrapolate cognition by advancing through intuition, language, communication, and analyze the fundamental role of emotional balance in organizational processes.

2.2.8.1 Classification of psychological/psychic functions in the informational process

The operator JBH in Fig. 2.24 shows the role of psychological functions and their classification. JBH represents human complexity in performing serial and parallel cognitive processing through multiple functions involving cognition, emotion balance, intuition, language, competency construction, and motor activities.

Psychological functions are divided into classes to facilitate the analysis of human errors. Thus subjectivity is present in all initiatives, including the operation. Innovation is a current need for the sustainability of the company. Thinking is a movement harmonized with individual objectives. Learning is continuous and involves informational aspects, concept formation and extrapolation of situations. Socialization is the current reason for the disintegration of teams due to a lack of leadership and lack of social inclusion in the companies.

In subjectivity, three important functions are analyzed: emotion, memory and imaginary action. Emotion works like glue that keeps cognitive, intuitive, and memory functions active in the socialization process, in a balanced and harmonic way. As a result of this balanced subjectivity, skills and knowledge have consistency, communication is satisfactory and motor action is concatenated. Working memory and knowledge base are formed in subjectivity during learning at school with the help of family education. They are important functions in the professional aspect and allow, through long-term memory encodings, the formation of skills. Another important function in subjectivity is the imaginary action that allows predicting what future tasks will look like. This facilitates the construction of mind maps for decision and also performing risk analysis or task analysis during planning.

Innovation is considered a competitive differential for sustainable companies, therefore, an important class of functions that need to be identified. In the selection of personnel, one should identify the creative human types and predict the analysis of unusual situations. According to Motta (2004), intuition is currently valued for management and leadership roles in the work environment. The operator, through the correct use of intuition, can perceive environmental movements that qualifies and makes decision-making and consequence analysis easier. Creation depends on abstraction. It facilitates the operator's work in building causal nexus from old concepts and including new concepts.

Perception is present at the beginning of psychological movements, except in the impulsivity, in which it has no influence. Perception depends on the organs of the senses that are not always entirely available. Disorders in the hearing system or in the vision can generate noise in the informational processes. Cognition is resulted from the application of

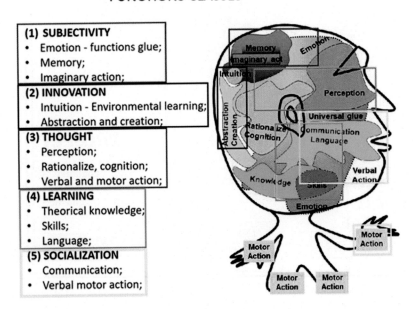

FIGURE 2.24 Psychological classes and functions.

logical operators to get the idea that will feed the decision-making process. Some of the logical and/or cognitive operators are: comparison, induction, union, intersection, interpolation, extrapolation, inference and preference.

The nonlinear sequence of cognition (complex mode) is completed in mind map formation, in the consequence analysis and the decision for action. Then, as parts of the thinking process, there are verbal and motor actions that make it possible to act objectively in the work environment or in society. Action should be monitored, as it can be carried out aggressively with the intensity scaring the interested community. Action is the continuity of thought and must be under its control. Impulsive action escapes from control of the cognitive process, thus requiring special attention in specific moments of the organization.

Important functions that depend on cognition and memory are present in learning. The construction of knowledge, the construction of specific skills, and the formation of language, a factor directly related to the dominant culture. Knowledge is the result of concepts formed through cognitive operations between memories and new signals in the search of solutions for tasks demanded by society.

The concept forms the basis for the meaning of living (usefulness in society and at work). The restrictions on disaggregated knowledge are diverse. They cannot construct a base to develop skills, thus discarding the need for application in the workplace. The skills depend on the disruption of inertia (attitude) and the realization of motor actions in accomplishing tasks.

Learning new skills is broad at the beginning and the same tasks are repeated. Thus auxiliary knowledge is added to enable better performance in the application of skills. Language is a complex operation that completes learning, with a reflection on the society of the communicated fact. The difficulties involve cultural differences and language conflicts, such as the existence of several words for the same meaning, or the case of various meanings for the same word.

In the socialization class, communication, motor and verbal actions are involved after deciding about the task. Communication noises should be identified and neutralized by leaders in the work routine. Action is also a form of body communication that can be represented by words.

2.2.8.2 Movements: functions and behaviors in the organization

Movements in organization and society activate perception and cognition through their logical operators, memory, and depend on emotional stability (Fig. 2.25). Each movement triggers specific psychological functions to achieve certain goals that can be conscious or unconscious. When working in the labor environment, most actions and purposes are conscious. Thus brief considerations about each movement are presented.

At the movement of free-thinking (FT), moments of contemplation are necessary without requirements regarding the usefulness of tasks for the company or for society. FT involves functions such as perception, memory, emotion, and imagination (or mental action).

Survival (S) is a constant need. It is driven by individual demands for psychic survival (social insertion) or physical survival (maintenance of life), thus avoiding psychological damage (group experience), physical damage (accident) and loss of organic stability (feeding). In the organization, the struggle for survival involves fiefdoms, powers, and influences that make the routine full of defenses and exercises, so we appoint as survival and power in labor relations. The psychological functions involved are: perception, memory, emotion, cognition, imagination (mental action), communication/language, and motor action.

Experience and concept is an input for the professionalization process, without which, although operators have skills, they may have extrapolation difficulties in understanding problems in industrial processes. The experience is attached to the conceptual basis, thus bringing competence to work. The functions involved in this case are: perception, cognition, emotion/ memory/ imagination, verbal/ motor action, and skills/knowledge.

The thinking, deciding, and acting movement. The moment for the decision is based on the formation of concepts and the memorization of tasks. This allows the construction of dynamic skills (Prahalad and Hamel, 1990). The moment for decision-making requires intuitive environmental knowledge. For acting, it is necessary to think and decide. In this case, the related functions are: perception, cognition, memory, imagination, knowledge, verbal/motor action and skills.

The movement for comparison (almost impulse) and act (CA). In this case, although cognition is not the strongest process, organizational movement involves quick and impulsive cognitive actions, based on experience. Thus the functions involved are: perception, memory, emotion (balance), cognition, imagination, cognition, motor action, skills, communication, and language.

Impulse action or movement. In this case, habituation through the repetition of emergency tasks allows the application of practices when the situation is real (fire and serious leaks or even explosions). The functions involved are: intuition, emotion (balance), skill, verbal/motor action.

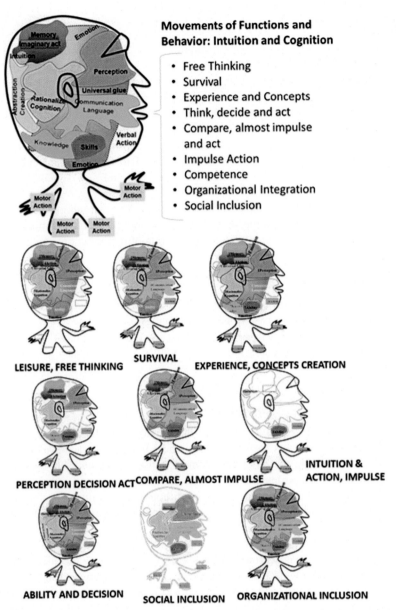

FIGURE 2.25 Johnny Big Head's movements of functions and behavior.

Competence training movement: the cycle of create, think, decide and act. In this cycle the operator is involved with the formation of competence, starting from abstraction for rationalization followed by the action. It is more complex and has several psychological functions, which are: perception, abstraction, emotion (balance), memory, imagination, cognition, skills, verbal/motor action.

Organizational inclusion is a movement in which organizational values must be integrated into the work team. The intentions are to reflect on/think about the group of workers and what are the organizational practices accepted by the company's directors and managers. Organizational inclusion involves incoming ritual presenting the company's vision, mission and values to employees. The functions involved are: internal and external perception, memory, emotion, imagination, cognition, verbal/motor action, communication/language, skill, and knowledge.

Social inclusion values man through work and allows social and financial activation. Therefore tasks running in a production environment means value in market and emotional balance. Financial balance is a health-friendly environment, although many workers have depressive traits despite not being unemployed. Main activated functions are: internal and external perception, cognition, memory, emotion, communication/language, imagination, verbal action, motor action, and skills.

2.2.8.3 Organizational processes

Mental movements or processes involve different actions, and this may be a decision in the cognitive region (conscious and objective). It can involve impulse (unconscious) or intuitive (semiconscious) sensations. These intuitive sensations may still be complemented with cognitive processing (objective) or impulses in action (unconscious). For each type of structure (psychological profile), there is a dominant behavior with certain movements of psychological functions in relation to others. Another important variable is the pressure of the external environment and the emergence of impulses from the internal environment.

At this moment, it is intended to focus on (1) the movements in the cognitive processes, external pressures and critical profiles for this situation; (2) the movements involved in unconscious processes that can generate impulses, promoting imaginary, verbal or motor action, giving life to fantasy; and (3) the movements involved in the intuitive processes that come from environmental information and are processed unconsciously or consciously, and then reaching final conclusions.

2.2.8.3.1 The moment of reality and cognitive processes

Cognitive processing is based on the individual's memory, which depends on his/her emotional balance. Psychological functions, highly known to be involved in this processing are: memory, rationalization, abstraction, and knowledge constructing. Emotion affects cognition when it aggregates or disaggregates memory, which causes problems in the steps of rationalization for decision-making. Perception is related to cognition once it involves the external environment, but when there is emotional instability, it is possible to affect cognition with false signals, and then the internal environment invades the perception for decision-making. Skills are constructed based on objective aspects, therefore, an important result of various cognitive processes of thinking, to choose action or even, FT, or experimenting to save in skill form. During cognitive processing, there may be thoughts about the actual action, and this may be involved with imaginary action, retrieving memory and constructing a mind map.

2.2.8.3.2 The moment of elaboration of causal nexus

The dominance relation depends on weights attributed to each situation and the type of model most commonly used by psychological profile. When the preference is to use objective movements, the cognitive process should tie causal nexus and put the battle on the force ground to infer about the outcome in decision-making.

2.2.8.3.3 The moment of fantasies and unconscious processes

Multiple recycles of the perceive, think, analyze, create, infer has as a consequence the imaginary action (fantasy), which if out of control advances through motor or verbal action (conscious and objective action). Fantasy represents a story created from various imaginary actions that, at first, do not go beyond the motor and conscious field.

Past and more recent memory, and environments of influence (emotion, feeling, physical survival, psychic and social inclusion) promote the creation of stories, novels that can be downloaded in the oniric/verbal form (dreams) of production, or through motor production (acting), or even through production from the body (psychosomatic diseases). Thus when imaginary actions are in the cognitive field, they are considered as included in thought. And when they are in the oniric field, they are included in fantasy.

2.2.8.3.4 Thought and fantasy can provoke verbal and motor actions

If there is no habituation (simulated), impulses and almost impulses can relate to imaginary actions in the unconscious field. Many human failures or errors occur without logical reason. These human failures can occur by workers who have skill and knowledge, in a favorable organizational/economic environment and the Manager has a profile according to the worker's ego. In this case, human errors are difficult to treat, as they move through the unconscious environment.

2.2.8.3.5 The moment of the organic nexus

It is difficult to affirm where or how the life of failure began, as multiple recycles of information, material and energy occur in processes of action of the individual in society or at work. This confusion of causalities and potentiating events leads to failures in interpretation with several false and anticipated factors.

In the organic nexus, the important thing is to analyze connectivity in the failure life cycle and try to track unconscious movements based on the interpretation of fantasies. Thus they are important factors in the definition of organic nexus and in predicting individual failure: psychological profile with specific psychological movements and functions, environments influencing failure including movement for social inclusion and psychic/physical survival, influence of regional values and organizational values when the individual acts in a given work environment. Based on individuality and organizational/regional/ social values, a field of forces is developed, in which bonds and commitments are built or destroyed, generating the probable scenarios of human action.

2.2.8.3.6 The moment of causal and organic nexus

Due to the presence of factors that occurred, studied and related, some historical processes are objective, and their study is possible through causal nexus, simple cause/consequence relation. Thus it is more appropriate to study the causality of intuitive processes than processes in the order of fantasy (with unconscious languages). On the other hand, some historical factors are unknown and move latently. There is involvement with sensations, emotions through these operational factors. Some of them can be considered false (indicate nonexistent stories). This possibility promotes the existence of multiple information and energy recycles, leading to organic nexus. Information analysis is required in complex systems.

2.2.8.3.7 The moment of information processing, memory and action

Signal perception and transmission are the first fundamental psychic functions for intuitive, cognitive and unconscious processing. This perception is based on historical memories (external environment), memories of the emotional experience (internal environment), emerging movements without early perception and passing through conscious perception apparatus.

2.2.8.4 Johnny Big Head's life cycle in the organization

JBH goes through different phases, all indicated by Fig. 2.26. In his life cycle in society and in the organization, divided as the figure shows, there are: construction of knowledge base, learning for and at work, maturity at work and retirement. During the day, we separated these into work and rest. Each division has psychological functions activated differently and strongly.

In the formation of the knowledge base, it is important to create concepts, survival and social inclusion. In learning for work, the main moments are: organizational integration, social inclusion, creation of concepts, leisure, decision based on knowledge and action of almost impulse. In maturity for work we can describe: social inclusion, organizational integration, almost impulse actions, impulse actions—automatism, knowledge-based decision, and skill-based decision. In retirement: leisure and social inclusion.

Diagnostic questions in Annex 3, related to Chapter 3, Worker Performing Task (Human and Social Typology).
What are the external influences on decision-making, thinking and actions?

- Archetypes result of regional culture biases, global culture biases, leader profile in decision
- Nature phenomena and climate changing stress level
- Process state, equipment and task causing difficulties to achieve the target
- Economy characteristic, work charge, profit expectation
- Socialization process, linguistic and expected patterns, social conflicts
- Information flow in process control (external flow)
- Others

What are the internal influences on decision-making, thinking and actions?

- Emotional Balance: overconfidence, commitment, attention calibration
- Intuitive perception
- Internal cases: divided attention, memory illusions taking decision, memory capacity, quality of mind map, fragmented discourse and accident, impulse action, memory operations,
- Quality of memory operations and cognition
- New concepts, motivation, actors and risk to decision
- Dynamism of Memory including affective learning and intuition, priority orbitals
- Conjugated memory, reception (perception), transmission of data (actuation)
- Effect of virtualism, multiculturalism, generation and gender social phenomena on decision
- Information flow in process control (internal flow)
- Others

What are the organizational influence on decision-making, thinking and actions?

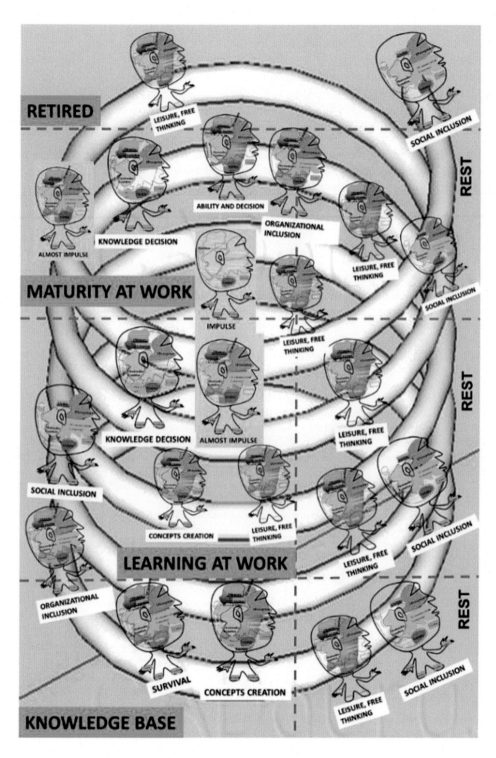

FIGURE 2.26 Organizational movements and behaviors. *From Ávila, S.F., 2011b. Assessment of human elements to avoid accidents and failures in task perform, cognitive and intuitive schemes. In: 7th Global Congress on Process Safety, Chicago, IL.*

- Emotional balance: responsibility, function, attention calibration
- Objective function, target
- SRK target and human error
- Communication quality in task planning & field execution
- Communication quality in interface, screen
- Quality level of psychic, intuitive, affective functions to workers, gaps, human error
- Complexity level in task

- Requisites and impacts on task (hollnagel)
- Classes to results: subjectivity, innovation, thinking, learning, socialization
- Integration and inclusion process
- Psycho, physical surviving and protection and justice sense of leader
- Blame culture and just; sub reporting and misreporting
- Level of cognition: quality, time, impact, frequency, visibility
- Old concept accuracy, new concept near reality, experience and competence level
- Leisure time in organization
- Organizational moments: reality and cognitive process; causal nexus elaboration; fantasies processes; cognition and fantasies cause actions; organic nexus; causal and organic nexus; processing information;
- JBH life cycle: knowledge base; learning at work to routine and emergency; maturity at work; rest and retired time.
- Information flow in process control (internal flow)
- Others

References

Aisenstein, M., 2018. The unconscious in work with psychosomatic patients. On Freud's The Unconscious. Routledge, pp. 203–215.

Akyuz, E., Celik, M., 2015. Application of CREAM human reliability model to cargo loading process of Oil & Gas tankers. Journal of Loss Prevention in the Process Industries 34, 39–48.

Ávila, S.F., 1995. Changing organization. Quality Control Magazine. Banas, São Paulo.

Ávila, S.F., 2004. Methodology to Minimize Effluents at Source From the Investigation of Operational Abnormalities: Case of the Chemical Industry (Dissertation; Professional Master's degree in Management and Environmental Technologies in the Productive Process). TECLIM, Federal University of Bahia, 115 p.

Ávila, S.F., 2010. Etiology of Operational Abnormalities in the Industry: Modeling for Learning (Thesis; Doctor's degree in Chemical and Biochemical Process Technology). Federal University of Rio de Janeiro, School of Chemistry, Rio de Janeiro, 296 p.

Ávila, S.F., 2011a. Dependent layer of operation decision analyzes (LODA) to calculate human factor, a simulated case with PLG event. In: 7th GCPS Global Conference on Process Safety, Chicago, IL.

Ávila, S.F., 2011b. Assessment of human elements to avoid accidents and failures in task perform, cognitive and intuitive schemes. In: 7th Global Congress on Process Safety, Chicago, IL.

Ávila, S.F., 2012a. Failure analysis in complex processes. In: Proceedings of 19th Brazilian Chemical Engineering Congress—COBEQ, Búzios, Rio de Janeiro.

Ávila, S.F., 2012b. The exergy of operational failure, a philosophy discussion with practice base. In: Proceedings of 19th Brazilian Chemical Engineering Congress—COBEQ, Búzios, Rio de Janeiro.

Ávila, S.F., 2013. Review of risk analysis and accident on the routine operations in the oil industry. In: 5th CCPS Latin America Conference on Process Safety, Cartagena das Indias, Colombia.

Ávila, S.F., 2015b. Prevention of operational and dynamic risks in the process industry, an exploratory discussion. Blucher Chemical Engineering Proceedings 1 (2), 7786–7793.

Ávila, S.F., Ahumada, C.B., Cayres, E., Lima, A.P., Malpica, C., Peres, Drigo, E., 2019a. Management tool for reliability analysis in social-technical systems—a case study. In: 10th International Conference on Applied Human Factor and Ergonomics and the Affiliated Conferences, Orlando, FL, July 21–25, 2018.

Ávila, S.F., Ávila J.S., Drigo, E., Cerqueira, I., Nascimento, J.N., 2018c. Intuitive schemes, procedures and validations for the decision and action of the emergency leader, real and simulated case analysis. In: 8th Latin American Conference on Process Safety, Buenos Aires, Argentina, September 11–13, 2018.

Ávila, S.F., Ávila, J.S., Drigo, E., Cerqueira, I., Nascimento, A.J., 2019b. Analysis of reliability mapping in refining industry: identification of critical regions and interventions in complex production systems. In: 22nd Annual International Symposium, College Station, TX, October 22–24, 2019.

Ávila, S.F., Bittencourt, E., 2017a. Operation risk assessment in the oil and gas industry—dynamic tools to treat human and organizational factors. In: first ed. Olson, K.F. (Ed.), Petroleum Refining and Oil Well Drilling, vol. 1. Nova Science Publishers, New York, pp. 1–125.

Ávila, S.F., Carvalho, A.C.F., Portela, G., Costa, C., 2015a. Learning environment to take operational decision in emergency situation. In: 6th International Conference on Applied Human Factors and Ergonomics and the Affiliated Conferences, Caesars Palace, Las Vegas, NV, July 26–30, 2015.

Ávila, S.F., Cerqueira, I., Drigo, E., 2018b. Cognitive, intuitive and educational intervention strategies for behavior change in high-risk activities—SARS. In: International Conference on Applied Human Factors and Ergonomics, Orlando, FL, July 21–25, 2018. Springer, Cham, pp. 367–377.

Ávila, S.F., Costa, C., 2015c. Analysis of cognitive deficit in routine task, as a strategy to reduce accidents and industrial increase production. In: Safety and Reliability of Complex Engineered Systems, London, pp. 2837–2844.

Ávila, S.F., Dias, C., 2017b. Reliability research to design barriers of sociotechnical failure. In: Proceedings of Annual European Safety and Reliability Conference (ESREL), CRCPRESS, Portorož, p. 27.

Ávila, S.F., Ferreira, J.F.M.G., Sousa, C.R.O., Kalid R.A., 2016a. Dynamics operational risk management in organizational design, the challenge for sustainability. In: 12th Global Congress on Process Safety, Houston, TX.

Ávila, S.F., Fonseca, E.S., Bittencourt, E., 2016b. Analyses of cultural accidents: a discussion of the geopolitical migration. In: Proceedings of Annual European Safety and Reliability Conference (ESREL), CRCPRESS, Glasgow.

Ávila, S.F., Pessoa, F.P., Andrade J.C., 2006. Dynamic analysis of human reliability. In: Proceedings of Brazilian Congress of Chemical Engineering, Santos, Campinas, 16, 2006.

Ávila, S.F., Pessoa, F.P., Andrade J.C., Figueiroa, C., 2008. Human reliability risk management system (SGRCH). In: Proceedings of Brazilian Congress of Chemical Engineering, ABEQ, Recife.

Ávila, S.F., Ramalho, G., Ramos, A., Costa, I.P., Souza, M.L., 2018a. Analysis of the reliability model for interdependent failure mode, a discussion of human error, organizational and environmental factors. In: 10th National Congress of Mechanical Engineering (CONEM), Rio de Janeiro.

Ávila, S.F., Sousa, C.R., Carvalho, A.C., 2015d. Assessment of complexity in the task to define safeguards against dynamic risks. Procedia Manufacturing 3, 1772−1779.

Bayma, A.A.C., 2019. Analysis of Human Reliability in the Emergency Evacuation of an Aircraft (Doctoral thesis). University of São Paulo, Brazil.

Boring, R.L., Gertman, D.I., 2005. Advancing Usability Evaluation Through Human Reliability Analysis. Idaho National Laboratory (INL).

Carvalho, A.C.F., 2016. Diagnosis of Gaps in Team Competencies for Coping With Situations Outside the Routine (Master dissertation). Post-Graduate Program in Industrial Engineering—UFBA.

Center for Chemical Process Safety (CCPS), 2008. Guidelines for Managing of Change for Process Safety. New York.

Center for Chemical Process Safety (CCPS), 2011. Conduct of Operations and Operational Discipline for Improving Process Safety in Industry. Center for Chemical Process Safety, American Institute of Chemical Engineers, New York.

Center for Chemical Process Safety (CCPS), 2013. Guidelines for Managing Process Safety Risks during Organizational Change. Center for Chemical Process Safety, American Institute of Chemical Engineers, New York.

Center for Chemical Process Safety (CCPS), 2016. American institute of chemical engineers. Center for Chemical Process Safety. Guidelines for Safe Automation of Chemical Processes. John Wiley & Sons, Hoboken, NJ.

Center for Chemical Process Safety (CCPS), 2018. Guidelines for managing of change for process safety. Recognizing and Responding to Normalization of Deviance. Wiley, New York.

Chiavenato, I., 2004. Organizational Behavior: Success Dynamic for Organizations. Pioneira, São Paulo.

Cusinato, R.T., 2003. Decision Theory Under Uncertainty and the Expected Utility Hypothesis: Analytical Concepts and Paradoxes (Master's degree in economy). Federal University of Rio Grande do Sul, Porto Alegre, 115 p.

Dekker, S., 2006. The Field Guide to Understanding Human Error. Ashgate, Burlington.

Deutsch, J.A., Deutsch, D., 1963. Psychol Rev, 70. Stanford University, Stanford, CA, pp. 80−90.

Drigo, E., Ávila, S.F., Sousa, C.R.O., 2015. Operator discourse analysis as a tool for risk management. In: Podofilini, et al., (Eds.), Safety and Reliability of Complex Engineered Systems. Taylor & Francis Group, London.

Drigo, E.S., 2016. Analysis of the Operator's Speech and His Communication Tool Between Shifts as a Tool for the Management System (Master's degree dissertation). Federal University of Bahia, Salvador, Bahia.

Embrey, D., 2000. Preventing Human Error: Developing a Consensus Led Safety Culture Based on Best Practice. Human Reliability Associates Ltd, London, p. 14.

Embrey, D., 2005. Understanding human behavior and error. Human Reliability Associates 1, 1−10.

Ferreira, J.F.M.G., 2018. Analysis of the Reliability of the CO Compression System in a Petrochemical Plant Considering the Technical-Operational and Human Factors (Dissertation from post-graduation program of industrial engineering). Federal University of Bahia, Salvador.

Fragoso, C., Ávila S.F., Cerqueira, I., Sousa, R., Pimentel R., Massolino C., 2017. Blame culture in workplace accidents investigation: current model discussion and shift requirements for a collaborative model. In: International Conference on Applied Human Factors and Ergonomics, Los Angeles, CA, July 17−21, 2017. Springer, Cham, pp. 318−326.

Franke, F., 2015. Is work intensification extra stress? Journal of Personnel Psychology .

Gerbec, M., 2017. Safety change management—a new method for integrated management of organizational and technical changes. Safety Science 100, 225−234.

Hollnagel, E., 1993. The phenotype of erroneous actions. International Journal of Man-Machine Studies 39 (1), 1−32.

Hollnagel, E., 1996. Reliability analysis and operator modelling. Reliability Engineering & System Safety 52 (3), 327−337.

Hollnagel, E., 1998. Cognitive Reliability and Error Analysis Method (CREAM). Elsevier Science Ltd, Oxford.

Hollnagel, E., 2017. FRAM: The Functional Resonance Analysis Method: Modelling Complex Socio-Technical Systems. CRC Press.

Hollnagel, E., Woods, D.D., 2005. Joint Cognitive Systems: Foundations of Cognitive Systems Engineering. Taylor & Francis, Boca Raton, FL.

Jung, C.G., 2002. Psychic Energy, eighth ed. Vozes, Petrópolis, Rio de Janeiro.

Kotek, L., Babinec, F., 2009. Použití metody human hazop při redukci chyb operátorů. Časopis, AUTOMA.

Lacan, J., Miller, J.-A., 1988. El seminario de Jacques Lacan: Libro 7, La ética del psicoanálisis, Paidós, 1959−1960.

Liao, P.C., Luo, X., Wang, T., Su, Y., 2016. The mechanism of how design failures cause unsafe behavior: the cognitive reliability and error analysis method (CREAM). Procedia Engineering 145 (6), 715−722.

Llory, M., 1999. Industrial Accidents, the Cost of Silence. Multiação, Rio de Janeiro.

Lorenzo, D.K., 2001. API770—A Manager's Guide to Reducing Human Errors, Improving Human in the Process Industries. API Publishing Services, Washington, DC.

Luquetti, I., Rodrigues, P.V., 2011. Ergonomics in the licensing of nuclear installations. Ergonomic Action Magazine 2 (1).
Maslow, A.H., 1987. Motivation and Personality, Third ed. Harper and Row, p. 335.
Motta, P.R., 2004. Contemporary Management: Science and the Art of Being a Leader, 15th ed. Record, Rio de Janeiro.
Muchinsky, P.M., 2004. Organizational Psychology, seventh ed. Pioneira, Thomson Learning, São Paulo.
Neisser, U., 1976. Cognition and Reality Principles and Implications of Cognitive Psychology. WH Freeman and Company, New York.
Norman, D.A., 1980. Errors in Human Performance. Center for Human Information Processing, San Diego, CA.
NUREG 6883, 2005. United States Nuclear Regulatory Commission Regulation—NUREG—The SPAR-H Human Reliability Analysis Method. United States Department of Energy, Washington, DC.
Perrow, C., 1984. Normal Accidents: Living with High-Risk Technologies. Basic Books, New York.
Peterson, C., 2018. Stress at Work: A Sociological Perspective. Routledge, Abingdon.
Pichon-Rivière, E., 1988. The Group Process. Martins Fontes, São Paulo.
Prahalad, C.K., Hamel, G., 1990. The Core Competence of the Corporation. Harvard Business Review, Cambridge, pp. 81–82.
Rasmussen, J. 1983. Skill, rules and knowledge, signals, signs, and symbols, and other distinctions in human performance models. In: IEEE Transactions on Swems, Man and C—Hernetics SMC-13.
Rasmussen, J., 1997. Risk management in a dynamic society: a modelling problem. Safety Science 27 (2–3), 183–213.
Reason, J., 1990. Human Error. Cambridge University Press, Cambridge.
Reason, J., 1997. Managing the Risks of Organizational Accidents. Ashgate, Aldershot.
Sennett, R., 2005. Corrosion of Character. Record, Rio de Janeiro.
Silva, B., Grazielly, J. et al., 2017. Human reliability: a current approach to human error. In: Proceedings of IX SIMPROD, Sergipe, Brazil, November 28–December 1, 2017.
Souza, M.L., Ávila, S.F., et al., 2018. Failure analysis cases in operational control and recommendations for task criteria in the man-process interface design. In: International Conference on Applied Human Factors and Ergonomics. Springer, Cham, pp. 317–329.
Stanton, N.A., Ashleigh, M.J., Roberts, A.D., Xu, F., 2001. Testing Hollnagel's contextual model: assessing team behaviour in a human supervisory control task. International Journal of Cognitive Ergonomics. Mahwah 5 (2), 111–123.
Sternberg Jr, R., 2008. Cognitive Psychology. Artmed, Porto Alegre.
Terra, S.X., Paula Neto, J., Rosa, A.F.P., 2019. Human reliability: a comparison of methods. <https://even3.blob.core.windows.net/anais/120615.pdf> (accessed 07.09.19).
Wu, B., Chen, X., 2017. Continuance intention to use MOOCs: integrating the technology acceptance model (TAM) and task technology fit (TTF) model. Computers in Human Behavior 67, 221–232.
Wu, B., Yan, X., Wang, Y., Soares, C.G., 2017. An evidential reasoning-based CREAM to human reliability analysis in maritime accident process. Risk Analysis 37 (10), 1936–1957.
Zadeh, L.A., 1988. Fuzzy logic. Computer 21 (4), 83–93.

Chapter 3

Factors affecting the performance of tasks

The role of people in the society, as discussed in Chapter 1, Introduction, indicates the possibility of the occurrence of human errors due to biases in labor and emotional imbalance in the family environment. In Chapter 2, Human reliability and cognitive processing, the indicators regarding the level of human reliability in carrying out critical tasks are discussed. In calculating these indicators, we present human factors (HF) in the task and the influence of work organization. Thus the cognitive processing for decision making depends on basic functions such as attention, memory, and perception. Some aspects considered are directly related to the task, whereas other factors such as social influence and the individual's own personality and character are indirectly related to performing critical activities. These classifications and investigations are described as human and social typology when carrying out procedures, one of the subjects covered in this chapter (Fig. 3.1).

Elements that impact people performance in carrying out tasks are addressed, which must comply with safety and production standards and goals, based on a careful assessment of the types of human profiles that form the groups responsible for the procedures applied in the work routine. Situations that promote failure and accidents caused by human errors are carefully avoided. In this context, the human error class, the human-social typology, the task analysis, and this error and failures in the task evolution are evaluated. Work environments are discussed from the API770 document on HF management, to establish a culture of strong security. This discussion is nondeterministic and requires tools for the research and diagnosis of social groups for the management of the critical task.

This chapter deals with human and social typology, classes of human error and how these occur in the routine, socioeconomic environment and research on human error considering the cycles of decision and feedback in the task, task analysis and respective tools that deal with complexity, time, risks, and functions. Finally, the elaboration of a routine operational environment with the appropriate requirements for the construction of a safety culture is also presented. This discussion is nondeterministic and requires a robust technique for diagnosing social groups, including the one already discussed and an analysis of the operator's discourse.

3.1 Human and social typology

The shift crews include several human types, defining the final typology of the team (Fig. 3.2) (Ávila et al., 2013b). Some incompatibilities must be avoided in the composition of the class so that the performance of tasks remains in line with expectations. Therefore the following points are recommended:

1. Avoid excessive obsessive types in the same shift crew, to avoid the risk of noise in the construction of the mental map that decides critical situations in the group.
2. Avoid establishing standards without considering the social environment and available profiles, as the result of the standard depends on the human-social type installed in the shift.
3. Avoid the presence of porous or depressed profiles that hinder decision making.
4. Balance the proactive and the reactive. Excessive proactiveness causes greater argumentative discussion making it difficult to command in emergency situations. Excessive reactiveness does not develop solutions for unusual situations.
5. Establish a balance of formal and informal leadership. Formal leaders can have experience for emergency or practical situations, and informal leaders develop better solutions, in addition to better polarizing tasks in the search for a successful outcome.

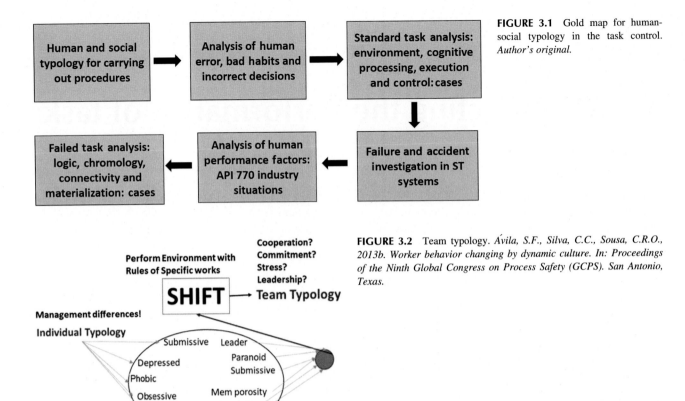

FIGURE 3.1 Gold map for human-social typology in the task control. *Author's original.*

FIGURE 3.2 Team typology. Ávila, S.F., Silva, C.C., Sousa, C.R.O., 2013b. Worker behavior changing by dynamic culture. In: *Proceedings of the Ninth Global Congress on Process Safety (GCPS)*. San Antonio, Texas.

3.1.1 Human typology

In shift and administrative work regimes in the production area, supervisors and managers are aware of certain characteristics that should be avoided. Experience indicates that the way of being and acting of some individuals can be more of a hindrance than helping in group work.

Thus the HF analyst for the critical task must discuss the concepts of psychology—psychoanalysis to be applied in the work environment. There are different forms of individual and group actions that are prone to human errors in the routine.

3.1.1.1 Types of organizational behavior

This proposition is just a reference, considering that certain people have a mixture of human types in their routine performance. The classification is based on both social and organizational behaviors. Thus when starting each discussion, we consider the advantages and disadvantages of each profile. Individual typologies (Ávila et al., 2008) are divided into organizational behavior to play roles in teams and social behavior that hinders or facilitates work occupation in performing the group function.

In this discussion, the comparison of energy of motivation and goal established for each human type are analyzed in a figurative manner. The internal energy must be sufficient to surpass the organizational goal and the internal perception of that goal, bringing the result of action and breaking inertia. And so, following each organizational or social profile and relations of motivational energy. Initially, the organizational & paper and then social behaviors are discussed (Ávila et al., 2008).

Behavior for organizational roles is divided into reactive, proactive, and leadership human typologies. Some of the characteristics of these types of behavior are outlined in the following.

1. Reactive behavior presents itself, performs, and participates when provoked, with no initiative of its own for innovations. It follows what the rule defines and has low motivation energy. The personal goal for efficiency is low and action is taken when motivated by the external environment. With each new situation presented by the external

reality, a new motivational energy goal is established and action is taken. The individual needs to gather factors in the analysis of the alternatives for action, sufficient to move the libido toward the action and the object of action. The reactive individual works in response to the external environment, which may require greater efforts at every moment. If this behavior exceeds the limit of common sense, it can be considered a threat to the environment, if motivated for aggressiveness, it needs to be evaluated clinically (Aplin et al., 2014).

2. Proactive behavior can sense environmental changes and, with this perception and compared to personal values, history, and the recorded, the individual takes the initiative before events occur. The worker is predisposed to act at any time, regardless of the demand for the internal or external environment; it can be said that this individual is self-motivated. The greater the difference between the current motivation and the goal, the greater the probability for action within the shortest period (Huang et al., 2014).

3. The leader has the ability for moving people around organizational objectives, which occurs when there is an individual—organizational rapprochement. This profile is prepared to act in organizational changes. The target energy to activate the action is below the motivating and ascending energy. The leader is always willing to act and seeks to bring the team together regardless of the period or environmental influences. This profile is also self-motivated, that is, it consciously or unconsciously develops sufficient bonds to maintain the will to surpass the goal and remain active. The leader may decrease activity momentarily, but after reviewing the new positioning of objectives, he returns to the normal state of activity and moves according to environmental risks. And if he is proactive, he will always be in a state of activity even if he is apparently in inertia. Think before you act, or let environmental aspects emerge for more coherent action (Day et al., 2014).

Behavior for social roles at work discussed in the typology includes the submissive, obsessive, cyclic, depressed, phobic, porous, and paranoid types:

4. The submissive type is dominated by internal factors that reflect in situations of external reality. The submissive individual has fixation in phases of control in the development of personality. This profile has a low energy of motivation and works on the dictates of authority. The submissive worker is one in which the unconscious energy of motivation is always below the goal. He is motivated to act. Depending on the character traits, it is sought to transform the submissive profiles into reactive ones through educational campaigns and then, anticipating future scenarios, to proactive profiles (Catarino et al., 2014).

5. The obsessive type always acts rationally and does not attach value to emotional issues. It feeds on work in an unhealthy way without valuing the social environments outside the organization. Obsessive individuals can create rituals to replace past memories. Workaholic individuals can have obsessive traits. At certain times, he wonders about the need to act in the work environment and the fatigue in which he finds himself. After a while, he finds that there are no such strong motivations for him to continue acting. It can be compared to the feeling of extreme passion (Astakhova, 2015).

6. The cyclic individual has an oscillating basic mood and volatile behavior, being able to feel moments of high and low motivation. The target motivational energy is also variable, reflecting on the unconscious portion of this energy. This profile is suitable for activities in the commercial area. There is a coexistence with depressive moments and moments of full activity. At each moment, the motivation may have values higher than the target, pointing to the possibility of action or, it may have values lower than the target that indicates the condition of passivity. Failure to set standards shows no resolution of the paradoxes that arise for the individual. Doubt is constant and the establishment of bonds and commitments is difficult.

7. The depressed individual has an average low mood. This personality tends toward depression and has no disposition for any professional activity to be developed. The target energy to initiate the action is variable and is always above the motivation energy; thus the state of passivity is constant. Depressed mood accompanies moments of loss of the individual's libido object. This object can be positioned inside or outside itself. Thus the grief of a loved one, the loss of a job or the loss of a limb can cause a depressive state. The inaction that comes from depression occurs because the energy of motivation fails to exceed the goal for action to take place. The unconscious energy of motivation is decreasing and the greater the distance from the goal, the more difficult it is to restore action. The leader must seek to rescue the trust of the followers who are in a depressive situation, to avoid functional loss (Bhui et al., 2012; Fleck et al., 2014a,b).

8. In phobic individuals, the impact of past events involving humiliation and aspects of the external environment makes the individual scared and impairs certain functions in his performance. Fear is constant and hinders rational action causing inappropriate behavior to work, such as the fear of heights. The phobic individual is inactive in the initial phase and then develops increased activity when he avoids the moments of phobia. The environment that has aspects reminiscent of his phobia causes states of inactivity, varying with states of activity. The phobic does

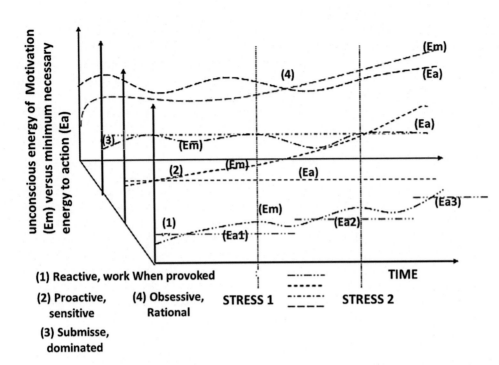

FIGURE 3.3 Individual motivation energy and goal by human type. *Author's original.*

not normally adopt antisocial behaviors. In the workplace, phobia is manageable by activity leaders, removing moments of anguish where the motivation energy does not exceed the goal, causing inactivity (Guimarães et al., 2015).

9. The porous individual has a fanciful behavior because of nonfunctioning memories and sensations with gaps. The lack of attention and memory are constant in this profile. The aspects of creativity and innovation in the organization may require some porous profiles. The porosity of perception and memory causes states of inertia or depersonalization (strangeness) that can cause fantasy. The established energy goals of unconscious motivation are failures, leaving the individual to standing still due to lack of knowledge of what to do, due to the loss of reference of the bonds (psychological contract). In this profile, there is no increase in the unconscious energy of motivation, as it is not possible to establish a strong bond in relation to social and organizational activities.

10. In the case of the paranoid individual, with fantasies that cross the cognitive mind, these can motivate action. The problem occurs when fantasies come to life and, through persecution, invade the individual's routine in society and at work. Persecution is a fanciful state that harms work by disrupting the bond. Although there can be growing fear, the employee's social function is not lost. If this happens, it is no longer a trait and becomes effectively a condition of illness. After a certain moment, the activity is greater than the passivity and of the persecutory state and the search for results at work are established.

In Fig. 3.3, the relationships between individual energy -motivation unconscious and energy perceived individually with the organizational goal are presented. This behavior follows different levels of stress.

After discussions about personalities and behaviors that influence the work environment, especially in group activities, we emphasize how much social typology, through environmental factors, can influence or change behavior. Initially, a discussion on the social, economic, and organizational environments, followed by an analysis of the bad habits of Brazilian culture and global culture. The types of organizational culture as well as how to measure organizational efficiency are discussed. How does the social-economic cycle affect the thinking and deciding cycle, in addition to the various levels to investigate human error in the production environment? To deal with environmental factors and the influence on workers' behavior, the types of Human Error, due to human and social typology are contextualized, as discussed below.

3.1.2 Social typology and archetypes

The concepts of each profile are based on the analysis of archetypes, which influence the maintenance of safety, as well as the development of human error. Archetypes can dynamically model behaviors incompatible with isolated and group decisions, which affect the organization's security (Ribeiro, 2012). The Analysis from the historical-regional point of view indicates that a group of people is influenced by habits in the corporate work environment, by the action of characters who act repeatedly, which are here named as archetypes in the decision of the industrial routine.

According to Jung (2002), the collective unconsciousness is a reservoir of primary images that represents the development of the human psyche. This collective unconsciousness influences the functioning of the cognitive machine. The virtual image based on the memory environment becomes reality when it identifies targets belonging to the scene. Thus these inherited or learned images are part of archetypes of the collective unconsciousness.

The archetype comprises the set of a common language that helps in the development of cyclical actions. The scenario composed of archetypes can summarize the type of decision and future action considered inappropriate for an environment that requires strict standards for safety, quality and production.

The knowledge of desirable profiles in workgroups can avoid human error through automaticity, affecting the stability of the organization that operates in a shift regime with individuals that have intrinsic behaviors under the influence of a social environment, which features a behavior discussed herein as social typology.

The identification of archetypes and their influence can avoid human error while performing the task. Consultation with interface users (panel operators), industrial risk analysts, process safety specialists and field operators on typical shift behavior, allows the identification of patterns that are repeated and the corresponding phrases that direct to name the archetypes working, for example, in the control room.

These cyclical behaviors can result from historical influence, economic and social movements, regional customs or specific strong leadership. These behaviors can block wrong actions (positive archetypes) as well as motivate wrong actions (negative archetypes). It is important to understand which human type and situation is in line with which different archetypes, to make it possible, from this knowledge, to design safeguards against this movement or to transform the movement into a main rule for maintaining operational discipline.

Organizational behavior may result from individual or combined constructs. The fusion of hero and devil constructs results in a great leadership. These archetypes and derivatives can contribute to addictions and bad habits that lead to failure and cognitive gaps in the operational routine.

Some mechanisms of action from the archetypes require attention in organizations: the reinforcement of bad habits that turn into addictions; the barrier in counterpoint to these vices; the ritual and "mass" to cancel the "seduction" in this reinforcement; the connection process; the transfer between characters; the migration from the real world to fantasy and vice versa.

These mechanisms can cause conflicts and cultural reactions: comparison of the operational routine with valued population stereotypes; the difference between organizational and individual values, which affect commitment because there is not a reward; disagreement between policies and practices; conflict between safety and production objectives; isolation of individuals or groups; and a group feeling of hopelessness in the future.

These conflicts promote discomfort, fear, and stress and affect cognitive decision, reducing the efficiency of routine tasks. In a simplified form, archetypes represent characters that have preformatted behavior that can affect industrial safety.

The archetype analysis methodology (Ávila and Menezes, 2015) indicates how to identify factors that influence the social performance of the work following the steps outlined in Fig. 3.4:

1. consultation to security experts on aspects of human reliability using as basis the API770 questionnaire;
2. identification of archetypes and mechanisms that can strengthen safety, generate unsafe situations, and mitigate those situations;
3. analyze the influence of archetypes on decision making and the occurrence of human error;

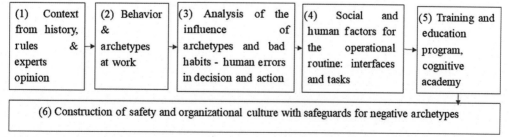

FIGURE 3.4 Methodology for archetype analysis. Ávila, S.F., Menezes, M.L.A., 2015. Influence of local archetypes on the operability & usability of instruments in control rooms. In: Proceedings of the European Safety and Reliability Conference (ESREL), Zurich.

4. consider social and human aspects (types and situations) that cause changes in the control of MMI, processes and machines, which are the cultural paradigms to be investigated;
5. use the human intellectual capital for training and education; and
6. establish a safety culture based on behavior.

To exemplify the archetype methodology, a real case of a chemical industry is discussed in three steps. In the first step, context, the observation for six months of the work processes in the control room of a chemical process plant (industrial gas), with little employee turnover, allowing the analysis of influence of the archetypes. The consultancy was carried out with operators and the respective managers, and then some safety activities were delegated. The operation team controlled the variables and indices in these areas.

In a second step, good and bad behaviors observed from operators or groups in control room are shown as follows:

1. do not accept the guidelines with resistance in the implementation of organizational goals;
2. believe in negative results for the tests or changes in the control room;
3. are very smart in solving problems with causal knowledge;
4. isolated and with partial participation indicating difficulties to socialize or to integrate in the group;
5. create ironic situations for reality, which makes it difficult to achieve the objectives;
6. understand or seem to understand the reality and industrial risk scenarios but in practice behave in a dispersed manner;
7. understand in detail and like to teach the group what it means;
8. have physical layout for field performances; and
9. have a positive position regarding challenges and changes.

It is important to confirm that these often-antagonistic behaviors happen in the routine of operators in the shift with impact of the actions in the control room.

In the third step, archetypes and mechanisms that influence the decision are the following:

1. rebellious—resistant to organizational goals, take informal positions when they were leaders, hinder the work and decisions. It is necessary to convince this group by demonstrating safely the benefits of actions;
2. pessimistic—the behavior can influence the group, which lowers the morale of the team, increasing the dispersion and decreasing motivation for the work;
3. smart—help in preparing the best alternatives for troubleshooting and allow the study of signs in the routine for taking preventive action;
4. excluded—can promote isolation and depression, affect the communication process and cause loss of attention, affecting the health and promoting fatigue;
5. cynic—this behavior may indicate that the actions are being carried out, but not in authentic way, promoting omission and making it difficult to control the task;
6. teacher dispersed—shows knowledge, but in action or communication commits failures, require some redundancies to avoid human error;
7. strict teacher—can transfer knowledge and skills to the group and perform when there is understanding, interesting for the task of planning and its dissemination;
8. athletic—physical layout for performances in the field and proper body care helps to balance the team in performing the task;
9. optimistic—sees things in a positive way can motivate everyone toward the goals, but can also lead to overconfidence.

Chapter 6, Human reliability: spar-H cases, learned lessons, case studies related to tools in this book, including archetypes analysis, classification derived from human error, analysis of the executive function (EF) (Ávila and Costa, 2015) and others, describes what are the causes for some important human errors. In these cases, we will discuss two situations, archetypes that reinforce safety and production, in addition to situations that induce failures in the control room and respective losses in the process.

3.1.3 Classification of human error

Research on human reliability carried out by Ávila and Dias (2017) considers the intrinsic aspects of human types, influences of socioeconomic movements, and social types of behavior with the possibility of bad habits, by including the impact of organizational changes due to geopolitical-natural factors and the impact of technological and security changes. The organizational and safety culture must be prepared with appropriate barriers to control the risks of possible accidents.

Ávila (2010) worked with these different aspects in a research that resulted in a doctoral thesis and the recommendations for the diagnosis of cognitive gaps for the case of the oil and gas industries located in Northeast Brazil, close to the sea with logistics based on pipe, ship, and road transportation modes. Owing to the complexity of the chemical industry and the dynamics of the risks involved, we sought to define a procedure for the classification of human error through dominant factors that induce failure mechanisms, thus creating the procedure for the classification of the derived human error.

Ávila et al. (2008) address the types of human errors based on the personalities of workers with different types of behavior, stating that by knowing the types of human behavior and how technology works during the routine of the operation and the influence of environments on the work of the group, it becomes possible to identify the root cause of latent failures.

Even without identifying the root cause, because of the complexity of multiple material and energy relationships and the flow of information in the chemical industry (Perrow, 1984), corrections are identified in the competence program to avoid the appearance of secondary causes that draw operational routines in the direction of accidents.

Several authors have discussed human errors to improve managerial decision, including rationalizing the financial and human resources for preventive and corrective actions. The identification of the human performance factor mentioned in) and mentioned by Hollnagel (1993) indicates the best actions based on the history of the error.

The analysis of the operational routine is carried out by mapping abnormalities, over a long period of time, in the shift, with information from the organizational-human and technological dimensions. This analysis makes it possible to build the connection between apparently disconnected events, bringing the need to study human error in a multidimensional way, where different classes are possible to form and build the failure caused by man and by the technological-organizational environment. The methodologies available for the analysis of the factors that promote human error, fail to advance in this complexity, settling in specific dimensions bringing incomplete solutions (Hollnagel, 1993).

Lees (1996) presents factors that influence human performance in the chemical industry, indicating regions of possible action in case of error. Reason deepened the discussion on human errors on the cognitive side, where he states that errors are based on skill, knowledge, and rule. The skill is part of a behavioral discussion and includes aspects such as omission or commission at work; errors in not performing according to the rule, exist on a contextual level, according to language, word association and memory; at the conceptual level, knowledge, where cognitive mechanisms and the possibilities of errors are related.

Hollnagel (1993) presents a classification for under-specification errors in the task related to human behavior: word identification, recalling a list of verbs, category generalization, language slips, action slips, and failures in prospective memory and others. Sternberg (2008) explains, based on psychology, that human error has the following types: capture errors, omission, repetition, and correct actions on the wrong object, incorrect data from the sensory, associative activation and loss of activation to complete the task.

Based on the various models of classification of human error, derived human errors appear, depending on the various aspects of knowledge, skill, technology, personality, decision making, tools, and rules. Fig. 3.5 shows the classification of human errors to systematize the classification of errors derived from humans based on research on the origin of the failure in the operation (Ávila, 2010).

The perception of failure signals or warnings indicates when and where the release of technical or social danger occurs and depends on access to the flow of technical and social information in the operational routine. In this first moment, the signs (1) of abnormality are recognized, which represent intermediate states of failure in the processes, equipment and behaviors, which, in the absence of proper intervention, take the structures to the nonstandard state.

Signs of abnormalities are indications of the existence of a latent (not visible) fault that can become an active or visible fault. It is known that sensor control instruments can cause the failure in the factory operation, by degradation of equipment materials, and by human errors in the process control. When deviations in the task are detected in relation to the intended goal, it is recommended to calculate the objective function for the conformity analysis (2). In reality, it is still not considered that the problem exists when signs of failure occur, but the tendency of these signs should be noted to extrapolate sociotechnical behaviors and identify the regions, functions, structures, and processes, which are likely to be part of incidents involving critical tasks.

The urgency to solve problems depends on the impact of future events as follows: (1) noncritical events and potentially low-probability, archive research; (2) noncritical event with high potential, involve an investigator (3) to find root cause; and (3) when current and future losses occur, critical events, a (4) multidisciplinary task force is necessary.

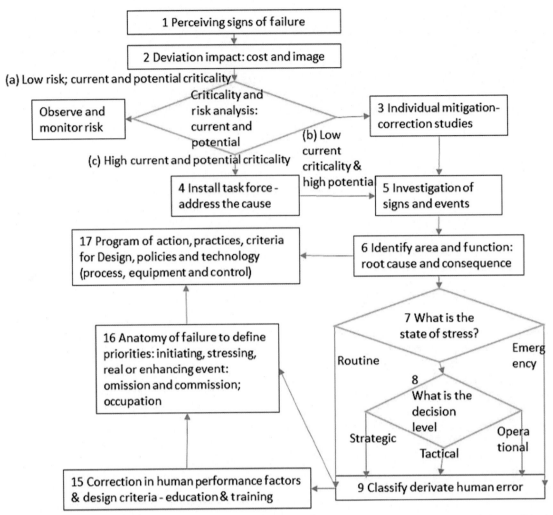

FIGURE 3.5 Preliminary action in classification of human error. *Adapted from Ávila, S.F., 2010. Etiology of operational abnormalities in industry: a model for learning. (Doctoral thesis), Federal University in Rio de Janeiro. School of Chemical, Rio de Janeiro.*

The investigation of the signs (5) requires the mapping of the abnormalities and their treatment to direct or differentiate what is hypothetically the cause or consequence. In this same investigation of the signs, the areas (6) where the symptoms and consequences occur are located, or yet the areas that cause these events, that is, where the root cause is located.

In the operation, the stress level alters perception, memory functioning, and cognitive processing, which makes it important to check the state of stress (7). The events may be taking place in a routine situation (that is easy to investigate) or in an emergency (risk of fire, explosion, and contamination; organizational changes, change of shareholder; and different times in the company calendar).

Failure process may eventually occur in the organizational dimension (8), that is, a strategic decision, or even in the managerial dimension, where there is a tactical failure in the manager's guidelines, or behavioral failure when leading the team. In addition, it may only be an issue related to the operational aspect. Part of the problems of oscillation of the operator's behavior, are related to organizational issues, a failure may possibly have organizational (pressure or climate) and execution dimensions, human error in the execution of the task. The manager's attitude also has a considerable influence in the context of tactical decisions associated with aspects of task execution.

This analysis of the classification of human error that begins with the identification of how the failure process occurred requires the cognitive-affective counterpart discussed in the EF tool in Section 3.1.7 herein.

After the initial aspects to be considered about human behavioral error or human error in technical environments (operational failure), the effort to classify (9) the error by class begins; thus they are divided into cognition, design, behavior, task, and environments (Fig. 3.6).

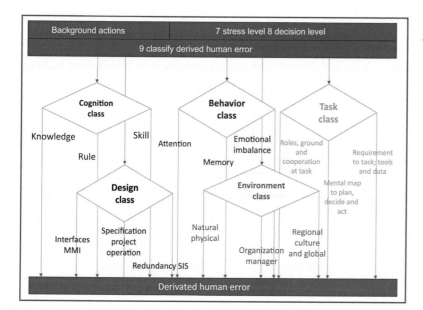

FIGURE 3.6 Classification of human error. *Adapted from Ávila, S.F., 2010. Etiology of operational abnormalities in industry: a model for learning. (Doctoral thesis), Federal University in Rio de Janeiro. School of Chemical, Rio de Janeiro.*

In the cognition (10) class of human error, it is intended to verify aspects related to knowledge, skills, and rules. Another item is worth adding, that mixes with human errors in the behavior class, depending on memory and attention: it is cognitive processing.

In the design topic (11), aspects to be analyzed in human errors are taken into account, such as operation premises for the project specification, details of the human-machine interface (HMI) and aspects regarding safety systems, including redundancies.

In the behavior class (12), depending on the human type, human errors can be classified based on memory, attention, and behavior in general, due to emotional imbalance. Behavior is intertwined with aspects of cognition and suffers the action of environmental aspects.

In the task topic (13), aspects such as group formation are inserted to perform the task with the respective roles and types of communication. Some aspects are the consequence of globalized culture, such as the level of cooperation or even the isolated work traits, harming group work. Also, in this human-error class, there is the definition of requirements for the task, what tools are used and the possibilities of involvement with the technological-operational failure and also, checking the consistency of the data that allows the passage from one step of the task to the next.

The environment topic (14), related to human-error investigation, can be considered as the main initiator of the failure. The organizational environment is the main cause of failures (Llory, 1999), and the managerial profile strongly influence the frequency of failure. The physical environment changes the intensity of the failure, through rain, humidity and other elements of nature. The environment inherited by regional or global culture can also be a trigger for failure.

When the failure is identified and it can be classified into several competing types to study its anatomy, there is a need to classify failure as a derivative (15) from previous origins.

Thus the separate classes are joined in the form of a derived or compound error, facilitating easy monitoring. In the classification of failure, knowledge of policies, people and technologies allows testing and modifications in environments (16) to correct performance factors, thus initiating the preparation of a program (18) to avoid losses through process stability, changes in tasks and factors that alter human reliability. The anatomy of the failure (17) is now validated after tests and confirmation of causality.

Human error often begins with gaps in cognitive processing during normal operation controls, where cold cognitive components occur, and also affective or warm components. This discussion will be held in Section 3.1.7.

3.1.4 Concepts and the investigation of latent failure

Visibility of the failure in the sociotechnical system depends on technology (complexity), operational context, and culture. Complexity depends on the type of technology and according to Perrow (1984), it depends on energies and masses that are "hidden" in the process. The operational context depends on the number of steps, presence of auxiliary memory or auxiliary processing of information, or even on the norms referring to the standardization of procedures. The way in which human error is treated can be related to the culture of guilt, which indicates the tendency to underreport.

Visibility of the event is directly related to the impact and the type of failure. Engineers have been trained to discuss different processes and equipment failures, but the analysis of complex operational failures and other human errors are extremely difficult. A complexity is the result of establishing connections with different dimensions participating in the same chain of abnormalities. Therefore in this haystack, it is difficult to find a needle.

Latent failure, as indicated in Fig. 3.7, is part of a hypothesis raised in regions where there is high uncertainty. The question is what are the real operational and HF inserted in the chain of abnormal events that lead to accidents. These hypotheses are tested by crossing the events recorded in the shift book, the process variables, the manager's speech and the experts' opinion.

HF that were discussed deal with interactions in the work environment between people, facilities and equipment, and management systems. Just observing that included among these characteristics are:

1. equipment and facilities (equipment design, control systems, control center design, remote operations, workstation design, HMI, emergency safe locations, and equipment identification);
2. people (training, communications, design and use of documentation, environmental factors, workload and stress level, shift work, and material handling);
3. management systems (safety culture, safe behavior, planning and executing projects, procedures, maintenance, work practices, work permits, hazard and risk analysis, safety systems, competence management, emergency response, and investigation incidents).

Part of the management systems is closely related to what other authors classify as organizational factors (Ávila et al., 2018b,c), these being leadership, culture, vision, values, mission, talents and innovation indicating company value, changing group behavior and individual behavior.

During courses on human reliability for the oil & gas industries (from 2009 to 2013), other aspects influencing the loss of performance and that cause human error were questioned. The students, operators and engineers answered: administrative factors to work organization; social and geopolitical culture; traces of regional culture with presence of strong archetypes; emergency communication to the internal and external community; human behavior; inadequate health and social behaviors (alcoholism); relations with the hired employee; relationship with government authorities; and stakeholder analysis.

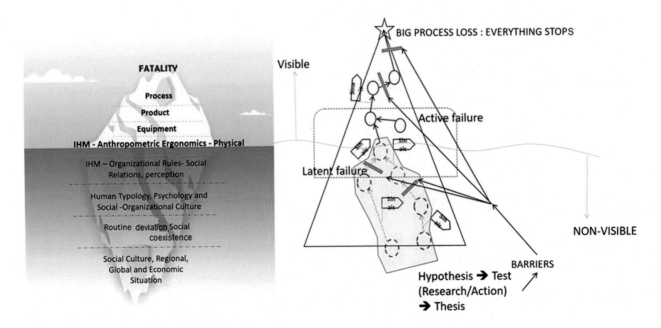

FIGURE 3.7 Method for human error validation and operational research. *Author's original.*

When human and social typology are discussed, it is possible to go further in relation to HF and into the discussion on the intrinsic issues of team building with behavioral characteristics that can hinder the maintenance of safety due to low operational discipline.

When discussing social typology, we advance the list of archetypes and group movements that promote phenomena such as altered communication and possible group omission. These phenomena involve violations, human errors and bad habits, and which affect the organizational culture. The group can make internal agreements regarding the official facts to be communicated, changing the essence of the events.

The classification of the derived human error identifies, based on the events, the probable classes that promote this error. The explanation of how the error occurs is provided by the discussion of the EF in performing the task (Ávila & Costa, 2015; Barkley, 2010; Malloy-Diniz et al., 2008), demonstrated below. The trends of global and regional culture are the topic of item 3.1.5. All these issues presented are essential before analyzing the critical task (Section 3.2) and before going on to Chapter 4, Process loss assessment, which highlights how to avoid process losses through abnormalities map, process mapping, and the operator's and manager's speech.

3.1.5 Socioeconomic environment: human reliability analysis

In a globalized environment, news that affects the stability of the company travels extreme distances instantly, and restructuring to overcome the resulting challenges occurs continuously. On the environmental side, natural systems have lost the ability to adapt to the presence of intensive economic activities, thus any accident involving industry causes major impacts to nature and can generate changes in local legislation and restrictions to the presence of this industry in the location, region or country. The stock market responds radically to these movements, resulting in the loss of value of the company in the market, and this may even reach its collapse, as happened with Union Carbide (Broughton, 2005; Llory, 1999) or can bring disastrous economic consequences, such as the largest oil spill in history in the Gulf of Mexico, involving British Petroleum.

Companies in the globalized world aim to "play" to achieve their structural stability by changing their missions, such as the oil industries that diversified their activities by expanding to renewable energy. All of these organizational movements bring about changes in policies including increased ecological awareness.

An environmental accident on the American coast by British Petroleum caused a decline of the stock market to the point of reducing to half the value of the company and generated in the world, including Brazil, a change in environmental rules for offshore oil operating projects, which affected the presalt installation schedules due to changes in rules by IBAMA.

Another move that has been less valued by some organizations is in relation to the society's behavioral changes, which affects the way groups are formed and the establishment of bonds for their maintenance. The Industrial Companies receive human resources from the society with great changes when compared with 20 years ago, leading to additional efforts so that there is not a loss in the process of skill development due to increased turnover. Workers are more vulnerable when they conclude that individual and organizational values do not coincide.

Fig. 3.8 illustrates the capacity that the environment exerts on the individual and the existing relationships resulting from globalization, where society, the economy, the organization, and the workplace interfere in the human performance of the activities.

The natural environment is part of the discussion with climate change at high speed, affecting the ambient temperature and humidity. The impacts of effluents and residues from the activities of society, industry, services and the attitudes of Man in society cause contamination of drinking water and river water, reducing the availability of this precious cooling and process fluid, generating the urgent need for alternatives and cleaner technologies for the maintenance of the industry.

A discussion about the vices and biases of Brazilian culture is necessary to structure the safety culture and organizational aspects of a company. Group formation rituals (Handy, 1978; Lapassade, 1989) are important, enabling the strengthening of organizational structures and allowing the formation of affective bonds between company and employees or between leaders and followers. These relationships inside an inappropriate environment with Brazilian culture vices (Motta and Caldas, 1997) can cause sick states in the Organization as shown in Fig. 3.9.

Formalism adopted by leaders inhibits the initiatives of the team and informal leaders. Paternalism around the team or people, leads to undue protections for maintaining power (Handy, 1978). Loyalty without criteria, that is, "blind" loyalty causes excessive dependencies of the team in relation to the leader and is contrary to self-management (Lapassade, 1989; Motta, 2004; Renesch, 1993). Excess flexibility, avoiding conflict situations at any cost (leave it at that), leads the followers to "weak" positions and an expectant attitude.

Barros and Prates (1996) claim that national traits work as part of each individual's unconsciousness and, therefore can be reflected in Brazilian behavior. Some of these traits would be the privilege of relationships between friends and

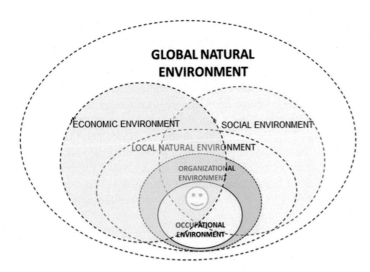

FIGURE 3.8 Environments that influence people in their tasks. *Ávila, S.F., Silva, C.C., Sousa, C.R.O., 2013b. Worker behavior changing by dynamic culture. In: Proceedings of the Ninth Global Congress on Process Safety (GCPS). San Antonio, Texas.*

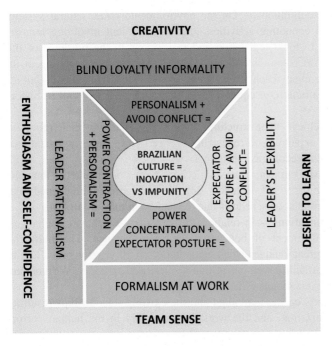

FIGURE 3.9 Brazilian cultural type and organization. *Adapted from Motta F.C.P., Caldas, M.P., 1997. Organizational Culture and Brazilian Culture. Atlas, São Paulo, 325 p.*

the appreciation of the Brazilian "way to do things" and the thought of little effort to achieve success. Other typical behaviors by Brazilians (Fig. 3.9) are flexibility, creativity, authoritarianism, and a paternalistic way of leadership, ambiguity, egocentrism, self-confidence and enthusiasm.

Stefano and Maia (2015) state that, when analyzing the aspects of "productivity" and "innovation," the national results are inferior to those of the United States and Japan because of low educational development, in addition to other factors. They highlight the low level of education (on average 7.5 years) and the low percentage of university education of the population (around 11%) and in companies (around 13% in large companies).

According to Ávila et al. (2015b), the typical situations of Brazilian culture coexist with global needs and characteristics imposed by economic activities, through the influence of media, to change behavior in the search of the compulsive buyer that feels calm when purchasing the goods that symbolize social insertion. From this confusion between global and local, addictions can affect the work environment. Some of these described Brazilian and global characteristics happen in the routine and affect operational discipline, for example, the behavior of normalization of deviances is derived from these defects from the culture.

The discussion of the cause and consequence of these speeches is part of the strategic planning for the construction of the organizational culture. We draw attention to these phrases that turn out to be heuristics and improper behavior in Table 3.1.

TABLE 3.1 Phrases and heuristics that induce archetypes and behaviors.

1.	Order those who can and obey those who has any sense	It reflects the classic position that for not complying with the rules, punishment occurs. Therefore there is no room for discussion. Currently, there is the opposite behavior that in some companies is said to be the right of refusal in the workplace. Thus excessive rules and inflated ego are behaviors that cause low productivity in the work environment.
2.	Do not change anything on a winning team	In the current economic dynamics, the challenges are continuous, and creativity is a necessity in the profile of the body of work. In many regions of Brazil, conservatism is striking while maintaining the known hinders the treatment of new situations. Thus it seeks to incorporate innovation into the organizational environment by proactively mobilizing the organization to prevent problems arising from "pulsating" social, natural, and economic environments.
3.	Gerson's Law: I like to take advantage of everything	Cooperation is a group characteristic that is fundamental when looking for a team result and clashes with the increasingly fierce competition between companies and which is translated as competition between teams. By adopting self-centeredness and selfishness as the reference behavior of society, the interaction between people with common goals is lost. This situation finally causes nonfidelity, high turnover, and difficulties in competence formation.
4	Murphy's Law: If something wrong can happen, it will certainly happen.	Thinking about the right destination and negativism in relation to work activities leads to the pursuit of reduced goals and, in the presence of small restrictions, the rapid abandonment of reaching it. The activity must be developed without creating cultural restrictions and with teams that have enough affection to move (break the inertia) and enough reason to analyze the stopping point, so as not to spend excess resources on actions without expected results.
5.	The important thing is to compete.	The unique image of competition does not bring effective results for the company. Today, with global movements constantly changing the rules of the market, the sustainability of the business is more important than ever, and even large corporations break down. This statement is validated when analyzing the situations of large chemical companies such as Union Carbide (Bhopal) and Oil Cia such as BP (Gulf of Mexico) in the respective environmental accidents.
6.	Here I earn little, but I am happy! Social activities…	This position makes the body of work feel very comfortable, and their motivation for innovation decreases. Thus job stability is a positive factor, but it causes stagnation in the work team. It is necessary to have low turnover to favor the affective bond, but to move in the form of a task force, breaking inertia, leading to greater competitiveness…
7.	I have stability, so I only do what I want	Public service stability. The effect of job stability leads to addictions linked to inertia or even to political union issues making it difficult to install a policy of consequences.
8.	Let us keep this in the informal way…	Informality brings comfort in relation to punishment and allows illicit financial and political gains to be increased. This informality often generates acceptance of unethical behavior, which can lead to violation in the workplace.

In discussing the research, Ávila et al. (2015b) state the need for adjustment in the worker's profile in a search for better emotional balance, which improves attention, memory, and mental map for decision, with the consequent quality of communications in the routine and good spatial location.

This diagnosis (Ávila et al., 2015b) points out that the reallocation of personnel and training in specific skills and abilities can contribute as instruments of performance correction. For group action, it was observed that cultural changes in society affect factors related to the level of selfishness and self-centeredness, which make it difficult to reach ideal standards for the formation of cohesive groups, and these challenges are for managers who need to deal with heterogeneous teams in terms of age, educational background, experience, regional culture, etc.

3.1.5.1 Social culture, globalization, and human types of the society for work

Social culture was discussed during the period in which psychoanalysis (Mezan, 1986) was introduced. According to Freud, thinker of culture, culture is based on a patriarchal family where education is guided through discipline. For every mistake made according to the expected standard, punishment is given according to its severity. The coexistence within the secondary process of guilt for the mistake made leads the individual to operate the memory and behavior to cushion the pain of the "executioner" who demands the price for the guilt. Diseases are the results of culture.

In this period of the beginning of psychoanalysis, as stated by Bonfim (2014), the Order of the Phallus reigned and many of the diseases that arose from a mental or bodily order were the result of living with the Father who represented the persecutor of order in the family. The Father, God, the State, are all responsible for punishing those who do not follow the pattern of behavior expected by society and dictated by the paternal will. However, the new Society is experiencing the fall of the "phallus" order bringing new profiles of mental illness.

Some classic diseases arose because of culture based on the resolution of the Oedipus complex and the castration complex. Some psychosocial phenomena have consequences such as violence by state disintegration, advance of mass culture, family without parental reference, appearance of different diseases related to self-estrangement, and depersonalization (Lacan and Miller, 1988).

The family was left without the role of the father and mother who prefer not to assume the responsibilities that are encumbered on them, seeking in an inadequate way to occupy the place that the media imposes on them, freeing their children to suffer direct media action, a new phase in social culture begins, as shown in Fig. 3.10.

It is intended to insert, through advertising and soap operas, the desire to acquire objects compulsively and, therefore bring to this process the dissatisfaction for not reaching the enjoyment promised by the media, which generates an internal conflict of the body between the ideal of the ego produced by the media and the missing ideal ego resulting from family reference.

In this way, the strangeness and consequent illnesses, which were an exception in the past, emerge with high frequency. The classical psychoanalysis cannot act with free association because, these diseases are not in the field of the symbolic, apparently, because it appears in the nonverbal phase, in which the child lives with the informational vectors of the media, with the absence of the father and, with weak links in relation to the symbiosis with the mother.

The new diseases, in fact, have existed since the time of Mezan (1986) to a lesser extent and were also studied by Lacan and Miller (1988) and Klein (1987). These diseases can be classified in the borderline region, but are directly linked in the performance by the body without involving the symbolic. Some personality characteristics are the result of mass culture, excess of image value, and excess of criticism regarding not meeting the standards, lack of affective bond and others, as shown in Fig. 3.10.

1. mental conflicts involving the being, its internal and external self-image (relationship with community standards);
2. demand for socioeconomic inclusion that depends on meeting the wishes of the community in relation to possessing goods and influence through social relationships;
3. instability of humor facilitating the dispersion of attention on objects or people, causing a lack of emotional bond (appealing media with erotic, instant, pulsating calls);
4. the family is disaggregated and is no longer a reference for their children, and "media education" assumes and forms compulsive personalities;
5. a picture of emotional instability and the breaking of emotional bonds is installed in society, in other families of workers;
6. diseases intensify with "schizoid-depressive" characteristics, that is, close to psychotic conditions (borderline);
7. escaping through drugs, by incorporating 'images and objects that speaks into the body (piercings and tattoos) as not being able to match the ideal presented by the media;
8. when recognizing the strangeness between the ego ideal and the current ego, compulsive and deconstruction mechanisms of the ego appear, resulting in bulimia, anorexia, and obesity;
9. the recognition that there are no parental references in the training itself, generates anguish and causes depression with the shattering of the personality; and

FIGURE 3.10 Man seen as mass culture product. *Author's original.*

10. this limit situation will cause a short circuit transferring the discharge of psychic energy to the body causing the appearance of bodily illnesses or fluctuations between mental and bodily illnesses;

3.1.5.2 Influence on the workplace

The global culture impacts environments that involve Industry: in the economic environment, where new technologies include new risks from equipment and processes, and where old technologies need to be better operated to avoid material and energy losses; in the occupational environment, where new requirements from legislation generate fines that obligate changes in the industry behavior; in the social environment, where conflict occurs between different generations, cultures, customs and genders (Sennett, 2005); in the environmental trend that calls for a reduction in product consumption; and in the mass culture chain that causes shopping compulsion.

Within the work environment, situations of uncontrolled tasks appear, resulting from the stress caused by new paradigms in the environmental, economic and social areas. Souza (2002) claim that tensions in the workplace reduce people's efficiency and consequently their productivity, thus generating haste, interpersonal conflicts, demotivation, aggression, isolation; a destructive human environment with possibilities of delay, strike, absenteeism, violations, high turnover, weak link between people and relationships characterized by rivalry, distrust, disrespect and disqualification.

This unstable situation causes changes in psychological functions, generating situations that are prone to human error in the work environment, with possibilities of accidents and abnormal events that lead to increased process losses.

Thus the conflict between local and global cultures, challenges to sustainability and personalities with fluctuating behaviors indicate that the selection and development of competences are, increasingly, strategic activities for the industry.

3.1.5.3 Organizational environment: human reliability and efficiency

The organizational environment (Handy, 1978) is analyzed in relation to the type of culture and the implications regarding the possibility of team failure in carrying out the task. For each type of culture, there is a behavior adopted in relation to the team and the routine.

In the culture of power, the "bosses" centralize decisions, reducing the need for skills in the team and increasing the possibility of resulting errors, causing a low morale and low efficiency of the organization in a globalized world.

The procedures are valued in the culture of the papers, the departments are strong, allowing for a more agile production although, with slow changes due to the centralization in decisions that is still present. The roles in the job are more important than individual life, generating undervaluation and causing conflicts in reward analysis. In crises, managers fall and in situations of stability, they are valued.

In the culture of the task, the mobilization of the teams is around the project. The power is found in the network between the various projects, therefore decentralized. The culture is adaptable, ideal for branches in foreign countries although the morale of the groups declines in political strife. Controlling organizational efficiency is difficult and depends on project leaders.

In people's culture, individual values are important and the base of power is specialists, suitable for high-tech businesses such as consulting. Individual goals are above collective goals and the influence on corporate decisions is shared.

Organizational culture in industry is the result of the fusion of the types presented in different situations and specific regions. The culture of power facilitates quick decision making, but regardless of the organization's wishes for its mission, the body of work is now more critical, with top-level operators questioning action with feasible arguments. Emergencies, on the other hand, must be dealt with established and strong leaders, with centralization and authorities to avoid delay in taking action. Another important aspect about organizational culture is due to the size of the companies that makes it difficult to import managerial models of process security and that are applicable in the local technical culture.

As shown in Fig. 3.11, organizational efficiency (Handy, 1978) depends on the development of work by the team and on factors that make up the psychological contract between the company and the employee. This psychological contract promotes the worker's commitment to perform the task where, some premises can be adopted by the worker and guide his tasks, increasing the probability of making him more efficient. The organizational climate and the harmony between individual and organizational values (Ávila, 1995) influence the quality of the bond

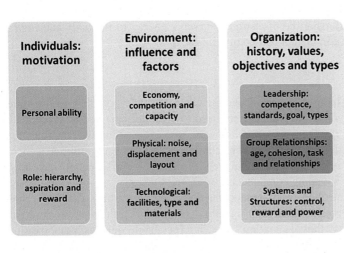

FIGURE 3.11 Components of organizational dynamic efficiency. *Author's original.*

between the Company and the employee, thus generating a situation of union or fragmentation. The credit given by employees to the organizational policies, mission and vision directly influences the way in which individuals and teams act. The management system, which must be in line with the company's organizational guidelines, influences the positioning of employees and teams in relation to the policy decided by the organization. Thus depending on the business culture and depending on the type of business and the respective technology, there may be a propensity to human error.

Drucker (1967) was a pioneer in suggesting a management model that meets the needs for organizational efficiency. He has defended the possibility of formatting effective managers despite personal differences and changes in the socioeconomic environment. Leadership over teams is also defended, making them productive through the effective use of time and the right decisions that leads to the organization's sustainability in economic issues and linked to the natural environment. Drucker does not go further on group relationships and intuitive-affective aspects of production management, which are current demands of new businesses in this globalized world.

When discussing the role of man and social relations, we conclude about factors that influence organizational efficiency. In the upper part of Fig. 3.11 there are aspects related to leadership, group relations, the roles of man in the organization and the ability to meet organizational demands.

Leadership includes specific skills and knowledge that can be inherited from experience. The leadership style depends on personality characteristics, including the reaction under stress conditions. The leader can adopt organizational standards following the Company's guidelines and values, or the leader can row against organizational standards, without entering into the judgment of what is correct. The manager's biggest challenge is to place all leaders (formal and informal) believing in organizational values, resulting in the transformation of policies into practice. The type of task to be performed by the team influences the leadership style. Leadership relationships can be genuine, superficial, or Machiavellian, or the three alternately.

Some aspects influence group or social relationships: the size of the group, member's age, and the level of cohesion. There is a possibility that the group will not be cohesive and adopt positions of apparent consensus despite isolated actions, and each worker will act individually and differently from the consensus. The transparency of group goals defines organizational efficiency. In addition, the truthfulness of the relationship intergroup and intragroup can bring positive results for the organization.

Some characteristics of Man's role at work, offer resistance to the achievement of organizational efficiency. The personal situation of values in imbalance; the hierarchy not accepted by the group or by the worker; the vices to publish results, not actually achieved; the work rewards not accepted on the balance sheet (expectation × reward); and inadequate time for the execution of tasks or to reach the reward. The motivation of the worker toward work depends on dynamic capacities and competence (application of knowledge including efficiency on interpersonal relationships) to meet the proposed role.

At this time, some factors that depend on the organization and are offered to the workforce are analyzed. The administrative systems that act on the control of human resources can inhibit the worker's practical skills in the subjects: control of materials, organization chart, matrix of responsibilities, authorities, and the relationship between positions and salaries in the company. Financial stability allows comfort to increase Human Reliability because there is no stressing effect on negative cash flow. The physical environment presents anthropometric or thermal stress, ergonomic aspects which influence efficiency achieved in the workplace.

3.1.6 Investigation of socioeconomic-affective cycle and cycle of thinking and decision making—utility

The organizational and social routines alter the individuals' and group' configuration of cognitive processes. To explain these social movements, organizational, economic and individual routines are presented on job completion cycles with feedback confirmations about the task: the cycle of socioeconomic-affective utility (SEAU), the thinking and the deciding- cycle.

3.1.6.1 Cycle of socioeconomic-affective utility

The work in the organization exists to meet the need for developing products and services that meet the demands of society. These demands vary according to social class, country, region and time of the year. In the SEA utility cycle (socioeconomic and affective), the worker performs his/her function properly, which activates social and economic satisfaction and balancing the role of this individual, within the family, thus serving the affective part. This utility occurs only due to the existence of knowledge, skills and facilities of group activity, thus achieving the intended goals of this work within a larger group.

If the goals achieved with the completion of the work does not meet the expectations of society and the organization demanding this service, the imbalance of SEA cycle occurs with the shutdown of this worker. In this case, there may be impacts on the individual organic system, on the cycle of thinking and on the emotional balance at his/her home, therefore; the social function is lost shortly.

Satisfaction must be in carrying out the task and meeting the individual's functions in society and in the family, thus facilitating the insertion, and utility sense of SEA, with socioeconomic affective satisfaction. In the case that the worker operates and satisfies the demands of work and society, the activation of thinking can be considered of higher quality, compared with the imbalance of the SEA cycle, inadequate execution of the task and the disruption of the function.

3.1.6.2 Thinking cycle

When the cycle of thinking is stabilized, it maintains coherent linguistics, enabling verbal and written communication between workers and the respective leaders. The lack of affective experience of the worker can generate the lack of installed concepts, which generates difficulties in the investigation of problems and in communication between people.

The new concepts are more difficult to be inferred when emotional imbalance and loss of confidence occur with loss of commitment to the company. Thus thinking is activated inappropriately. The images present in the memory may be altered or modified or missing, which makes it difficult to establish the mental map. Dynamic images and their signals are not perceived properly. If the mental map formed has low quality, it may happen from the attempt to correct it through the consecutive activation of the thinking cycle, then imprisoning the worker in an infinite inertial stage in time.

If the psychological functions are balanced, even in the cases in which the economic function is not being satisfied, there is the discernment and formation of good quality mental map, thus activating the decision cycle and taking the worker to the choice of alternative correct action. If the cycle of thinking has low quality, the abstraction is inadequate, the mental map is made up of unreal arguments, and the quick decision leads to an unsuccessful SEA utility returning almost instantly to the thinking cycle.

3.1.6.3 Decision–action cycle

Thus when deciding, an instantaneous utility analysis is performed regarding the efficiency of this action, and returns to the thinking cycle, with balance or imbalance, thus affecting the quality of the mental map.

In the decision, it is important to activate the memory and rescue whole or partial facts, movements, whole or partial objects so that, together with the external information, a consequence analysis is carried out on each possible alternative of action to verify whether the impact is as expected when compared with the target. In this context, decisions and actions are taken along the rational line, motor and verbal action. After the decision is complete, some consequences occur in the process state that should be analyzed and compared, which brings the success or failure in relative levels, and feeding the thinking cycle, then favoring the stability or otherwise of the mind map. Fig. 3.12 illustrates the affective cycle and the decision–action cycle.

The feedback is cyclical and it manifests in each cycle as information and response to the steps that aim to prepare the resource-structured mind map, either technical and/or human, and develop the goals set and assertive decisions.

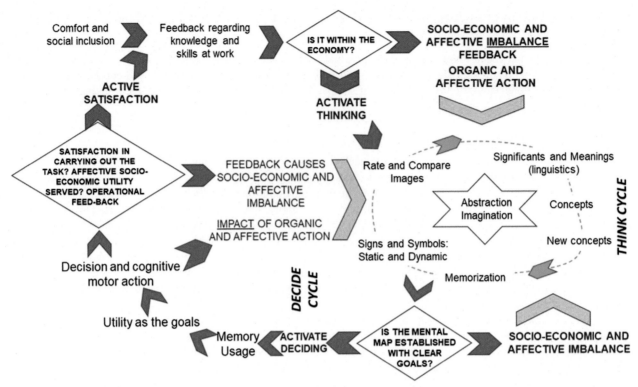

FIGURE 3.12 Thinking-decision-action cycle and socioeconomic-affective utility: feedback. Ávila, S.F., Silva, C.C., Sousa, C.R.O., 2013b. Worker behavior changing by dynamic culture. In: 9th Global Congress on Process Safety (GCPS). San Antonio, Texas.

Satisfaction, responsibility, skill, knowledge and understanding about the importance of the thinking and acting moment, appear as devices for carrying out operational procedures.

3.1.7 Executive function analysis

The EF according to Barkley (2010) is a set of neurocognitive processes that enable the organization of the individual's behavior to achieve goals, in harmony with the environment and professional activity. For the mental map to aggregate the maximum amount of information for the performance of operational activities, the EF must contemplate indicators of leadership and operator behaviors, which can interfere in the production and maintenance of safety, reducing failures and human error. The interpretation of the role of EF and its performance, allow to analyze cognitive impairment. Motor, verbal, cognitive and emotional activities, as well as work memory, planning and action are called EF.

A study by de Oliveira points out cold components of EF that represent movements of learning, perception, and knowledge, in direct stimulation to perform the task. Among the cold components are cognitive processes. And the hot components activate the action, relate of affection behaviors and motivation. When it comes to the work environment, affection can be attributed to the behavior that unites the commitment to responsibility, and the motivation is given by the effort in carrying out the task.

EFs play a fundamental role in routine activities that require organization of tasks to achieve the desired objective. There must be harmony to contain impulses and regulate, balance motivation in the face of a relevant purpose, and yet, to choose a decision centered on what is happening now and what can happen as a consequence of the decision made. These skills lead to multiple behaviors that are essential for the correct execution of the task.

The EF evaluation intends to investigate the performance in the production task, in a highly complex environment. In this way, the analysis of procedures is anticipated, including probable human flaws and errors. To achieve this goal, it is necessary to understand the cognitive planning that defines and structures actions to accomplish the task.

Due attention must be applied to the EF of the critical task by the operators. Unfortunately, the operational team does not properly train on how to carry out written instructions in routine failure situations. Ávila et al. (2015a) and Ávila and Costa (2015) comment that the exercises performed in simulators do not stimulate the behavior for decision making in a real abnormal situation, or even during a probable accident. Such fact can lead to inadequate decisions in emergency demands and keep the operators with low experience. The EF represents the performance control functions during the task. The excess of information and the complexity, including high automation level hinder the activity of the operation in the control of production and of security systems. Our target is that individual and group cognitive processing allows the performance of the task reach the expected standard after adjust the EF (Boyeur and Sznewar, 2007).

The analysis of the physical and cognitive ergonomics of operations allow that procedures performed in the process industry achieve the targets. The Task perform efficiency depends on the ergonomic requirement from that Company. Some characteristics of workplace ergonomic requirements, do not, necessarily, need fixed standard, allowing variability that is natural when working with different culture:

1. availability of suitable tools;
2. enough space to perform task (plan and control);
3. correct review of procedures available for group work;
4. well projected safeguards to prevent worker's injuries or equipment damage;
5. clear responsibility delegation of the operator;
6. good communication tools;
7. daily presentation using charts about targets with the operation routine;
8. a team with good perception for signs that can indicate equipment-process abnormalities; and
9. priorities organization for process control alarms.

In this scenario, Ávila and Costa (2015) propose a new model for the analysis of EF, from the "calibration" of the appropriate attention so that the possible flaws are pointed out before occurring and through the investigation of the operational environment, or workstation. Operations that are repeated daily, must be carefully observed. These procedures can create bad habits, such as overconfidence, inattention, neglect regarding actions and consequences. The flaw needs to be realized for the action to have positive consequences. Procedures are adjusted after continuous task failure assessment. Therefore the elaboration of a schedule of tasks with their specificities needs to be connected with a description that indicates how to proceed in a situation other than the routine.

Another aspect, which must be taken into account, is the need for communication between operators in the different areas. The occurrence of individual cognitive failure tends to decrease when there is a good group interaction. The synergy of operational activities and between operators in the work environment is essential for resilience and return to normality after crises.

The cognitive limits in operational and process control must be recognized and treated to keep the operation in normal mode; some strategies are planned and applied in the routine. Developing a visual map that identifies the stages of the process and their risks can avoid forgetting actions as well as the weaknesses of the controls (cognitive flaws). The triggered alarm must be identified and properly treated, when it is quickly recognized which are the causal factors. Although there are procedures that automatically manage the priority of alarms to be treated, accidents and failures continue to occur in the industrial sector. Thus it appears that HF determine the EF compliance flow.

The importance of HF in the execution of operations, challenges management as to the coexistence of exact sciences (engineering) and human sciences (psychology) in the control of the factory floor. In this context, the analysis of different perceptions and interpretations is performed to indicate the best solution to avoid any accident or the loss of production. Thus programs for adjusting skills are implemented according to human typology and technological characteristics that may be related to social culture in the work group.

The role of the operator and the group is analyzed through its EF where it is sought to diagnose the routine where operational failures may be occurring. The threat of the organizational environment alters the quality of cognitive processing.

According to Ávila and Costa (2015), human error in the operator team is derived from a multicausal phenomenon of interaction between the technical, organizational and psychosocial dimensions. The EF analysis is the representation of a study of the operational context that involves the influence of the organization, the workers, the company culture and those who operate in the different production scenarios. It also involves the appropriation of cognitive processing during the performance of the task, an understanding of the economic and psychosocial factors that impact the task, and the identification of flaws that may occur as a result of cognitive gaps of operators and leaders in the production area.

Fig. 3.13 illustrates the model for analyzing the EF, based on concepts in the literature and based on the Ávila and Costa (2015) research on human and sociotechnical reliability. Human error associated with cognition, behavior, and the organizational-social environment is discussed analytically, including the task steps and the reflection on cognitive processing.

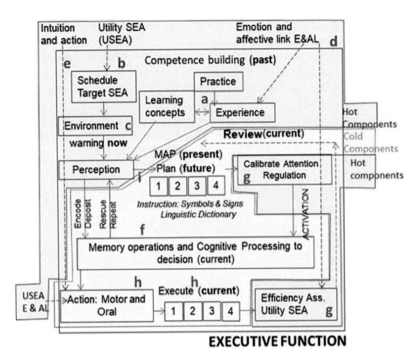

FIGURE 3.13 Executive function analysis. Ávila, S.F., Costa, C., 2015. Analysis of cognitive deficit in routine task, as a strategy to reduce accidents and industrial increase production. Safety and Reliability of Complex Engineered Systems, London, p. 2837–2844.

Thus cognitive failure or gap can be caused by the reasons shown in Fig. 3.13:

1. inadequate competence;
2. nonrecognition of the agenda and task goals;
3. low quality perception (*c*, *i*) of environmental signals;
4. prejudices that hinder the decision;
5. difficulty with memory storage and memory retrieval, activation of cognitive system is delayed or inadequate, and low rescue memory, or low integration of long-term memories as well as external memory and of the expert group;
6. attention or prioritization or inadequate concentration due to the change of affection weights in alternative, and attention or prioritization or inappropriate concentration allow the cross imagination with real perception, not creating real alternative in decisions;
7. lack of motivation to action, locating errors in steps, and the task of the effectiveness of the task, and lack of feedback without the behavior correction and planning and task scheduling; and
8. inadequate techniques to plan the task and Errors in the building of the map and mental task of reviewing.

Chapter 6, Human reliability: spar-H cases, will discuss the application of an exercise performed in a chemical industry, which will demonstrate the analysis of EF.

3.1.8 Environments in human and operational reliability: human error

The organizational, social and economic environments affect human error, which can cause process losses. In the organizational environment, some aspects can be analyzed to avoid the low reliability of the sociotechnical systems: the organizational climate generating different types of culture; the managerial profile, with the respective policies and practices and levels of commitment differentiated by task or activity; and the efficiency of the company's communication tools and processes.

Some social factors can cause human errors, such as:

1. ritual or myth of local society classified with low importance creating population stereotypes;
2. regional aspects of communication and linguistics that hinder the task;
3. types and respective relationships between formal and informal leaders;
4. aspects related to the power and movement of the leaders, leading to the formation of informal organizational charts overlapping the formal ones;

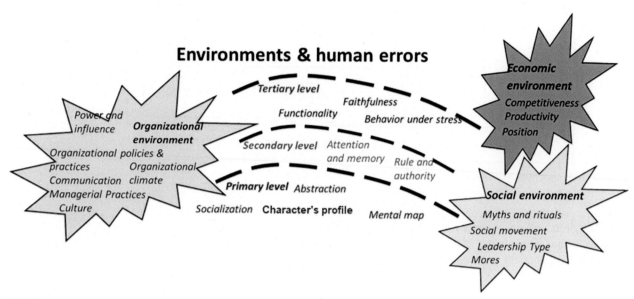

FIGURE 3.14 Levels for investigating human errors and environments. Ávila, S.F., Pessoa, F.P, Andrade J.C., Figueiroa, C., 2008. Worker character/personality classification and human error possibilities in procedures execution. In: Proceedings of CISAP3–Third International Conference on Safety and Environment in Process Industry, Rome, pp. 279–286.

5. local customs for food, leisure, physical activities, parties that can increase efficiency performance; and
6. political and social movements in the region.

Human error is analyzed at three levels (Fig. 3.14) that suffer from the environmental influence of the organization, the economy and society. At the first level, or primary level, the formation of the personality structure and socialization processes take place. The individual profile in coexistence with the society results in a behavior with peculiar characteristics, indicated as character traits resulting from behavior. Faced with this scenario, learning processes occur that enable or hinder the construction skills.

The second level includes research on the reasons for human error, such as lack of memory and difficulties in focusing attention, acting with authorities, accepting rules, building the inventory of knowledge and skills, decision-making processes for execution of the task.

The third level presents the organizational measurements related to: the level of loyalty (psychological contract) and functionality, the resulting productivity in the task that is influenced by social relationships and the behavior of the employee under psychological stress.

3.1.9 Diseases, bad habits and cognitive academy

3.1.9.1 Case study—sulfuric acid and oil industry

Fig. 3.15 shows the activities of research on the effect of culture on the dynamic behavior of the worker. After structuring the necessary analysis (dominant culture types and dynamic), data are included for further analysis of the company (technology, organization, routine, and interviews) to perform the first hypothesis. This analysis indicates the possibility of human error and verifies social priorities (priority vaccines to cognitive diseases). Finally, it is possible, using the technique action research to operate on cultural adjustments, people training, and projects of the organization to avoid the presence of oscillation behavior of the team.

In technology data, we discuss how the process works, products, materials, and limitations. Subsequently, we discuss the routine from observation of specialists and discussion with managers, supervisors and operators. After this, it is necessary to understand what kind of factors affect the team behavior.

The rules, procedures, and patterns of organization are discussed in guided interviews. These aspects allow the elaboration of other hypotheses about behavior change. What are the origins in culture, human types, and features of culture and why changes occur with time or with organizational-technological-natural modifications. After this, we consider the necessity of doing tests with organizational and operational-technical culture.

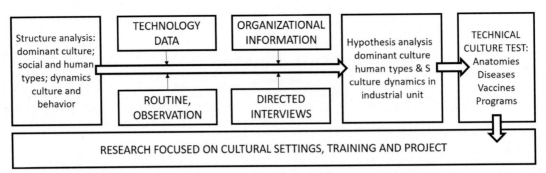

FIGURE 3.15 Activities for research on cultures, behaviors and barriers. Ávila, S.F., Silva, C.C., Sousa, C.R.O., 2013b. Worker behavior changing by dynamic culture. In: Proceedings of the Ninth Global Congress on Process Safety (GCPS). San Antonio, Texas.

These tests indicate programs to be implemented in the organizational culture to change the behavior of operator. In Chapter 6, Human reliability: spar-H cases, on learned lessons, we will discuss a case of a sulfuric acid plant and the cognitive diseases to program vaccines after understanding the social & human anatomy of failure, as presented in Fig. 3.16A,B.

3.2 Task assessment

The authors (Ávila, 2010; Embrey, 2000; Hollnagel, 1993; Lees, 1996) from the task analysis area discuss methodologies and techniques to standardize, control, and review the task that may be involved in the failure processes. Task investigation requires the interpretation of signs, warnings, and symbols from the shift report. The behavior of people in the task is also analyzed, in routine or emergency, and the risks that can lead to major accidents, using the tool named as dynamic of stress, and after, applying the Sociotechnical Failure Tree Analysis ST-FTA (Ávila, 2011b).

After checking the psychological movements of the worker in the industry, indicating the importance of understanding how cognitive processing works, analyzing the physical-natural and cultural environments that can alter human behavior, the discussion begins in the task analysis field.

The task consists of a sequenced or parallel order of instructions, which, when carried out, allows the production process or service to reach the desired state. Thus the following items are considered: (1) logical organization of the task steps; (2) authorizations to perform the steps; (3) goals to be achieved; and (4) tools and information necessary for carrying out the procedure.

3.2.1 General

The discussion on task analysis will start with Hollnagel (1993) model named as COCOOM for decision making during the task execution, and then with the goal and requirements analysis technique for the task (GMTA). The views of Embrey, Ávila and Lees are discussed as to the premises and characteristics for the planning, training and execution of the task. From these discussions, numerous tools help the manager and staff to format and test standard task. Between these tools will work with standard task construction, assess of the failure in the task during the routine, and analysis of the behavior of the teams in an emergency.

Based on Hollnagel's interpretation of the task analysis, the distinction is made regarding the emergency and routine aspects of the task. Decision making depends on the level of urgency to reach the expected states of the process and depends on the type of communication adopted. The results of the decision depend on the choice of the best alternative of action with less probability of accident and lower cost in production.

Production activity reaches a minimum cost value, when planned, and reaches a maximum cost value, when it is unusual. Hollnagel classifies the decision on the task in four levels applied in cases of Brazilian industry in this book: strategic, managerial, operational routine and emergency situations (Vil and Rodrigues, 2016).

At the strategic level of decision making, policies guide technical and behavioral standards. These policies are motivated by the vision of the future and the mission perceived by employees, shareholders and society in relation to the company. Deadlines are longer for strategic planning although requiring a quick response to urgent socioenvironmental and economic challenges.

At the managerial level of decision, the aim is to bring policies closer to the resulting routine practices. The administration's spokesperson needs instruments to make the team's management feasible, so planning is essential to avoid

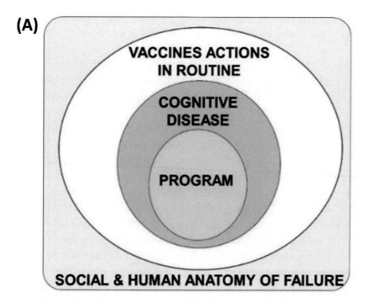

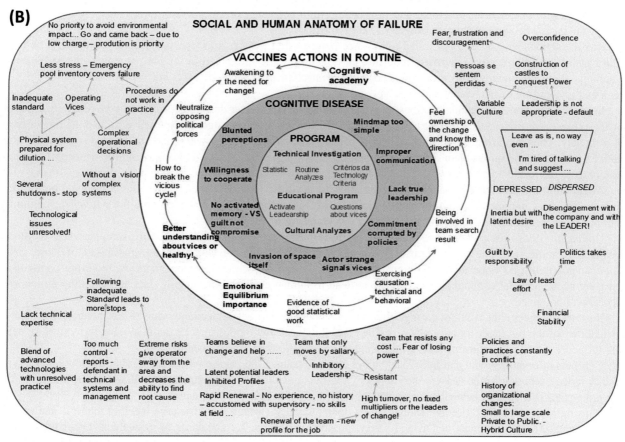

FIGURE 3.16 (A) Program to adjust operator behavior after dynamic culture impacts, general. Source: (B) Program to adjust operator behavior after dynamic culture impacts, detail. *(A) and (B) Adapted from Ávila, S.F., Silva, C.C., Sousa, C.R.O., 2013b. Worker behavior changing by dynamic culture. In: 9th Global Congress on Process Safety (GCPS). San Antonio, Texas.*

unattainable policies for achievable practices. The time for planning and for carrying out management guidelines is shorter when compared to the time for strategic decisions, although both can be confused at certain periods of high pressure from society.

The management guidelines, with countless standards and intended behaviors of the team, are broken down into operational routines. Accordingly, the operational procedures are designed to bring the desired final state to the processes and add value to the business.

The outputs, products and image, tangible or intangible results, are possible, after written documents become practical activities. The Manager and staff need to list the operation standards as values of process variables, level of system reliability, patterns of behavior of the individual and the group. A daily operation instruction that the Manager writes for the team indicates special and new procedures to specific activities and routine procedures.

The planning of tasks avoids environmental, occupational, and production risks bringing the balance between the costs of the activity and the impacts on occupational health and the quality of processes and products.

In the emergency decision mode, there is a mixture of intuitive, instinctive and cognitive behaviors, making it necessary to evaluate the objectives, to have an idea of the time horizon, to understand what the expected results are, and how to evaluate the priorities to guarantee a maximum level of organizational resilience.

The analysis by decision mode discussed in the COCOOM tool by Hollnagel (1993) and adapted by Ávila et al. (2015c) verifies the managerial and routine aspects to make strategic decisions possible. It also analyzes the weight of planning on emergency issues that require a subjective measure of impact, in the short timeline. Each level of decision described, strategic, managerial, routine, and emergency, is important in carrying out the task in industry and services.

When comparing routine procedures with emergency ones, regarding aspects of the decision, it is noted that the routine is usually planned, and using cognitive task analysis while the emergency action depends on an intuitive moment and with little planning. Training to act in emergencies requires "hot simulation," which are close to reality, and transfer skills to leaders, in quick actions, here and now, taking into account cultural characteristics. In these cases of emergency, the basis of communication is oral (Ávila et al., 2015a).

Routine actions are planned and tested in advance before being put into practice, enabling adjustments to be made. The basis of communication is written, requiring care regarding the interpretation of the procedure due to misunderstandings about documents. The production costs are controlled in scheduled tasks, but on the other hand, costs are high and uncontrolled when emergencies occur, such as accidents.

In the management of routine situations, the risk of an accident is lower due to the safeguards installed after analysis of the project risk. Emergencies can trigger accidents due to lack of knowledge or experience to lead actions to return to normalcy. Thus due to the unprecedented nature of the situation, it is not known which are the best solutions or alternatives to solve a certain problem. Maybe the operator does not have a correct mind map and the decision becomes a lottery as to the results.

Attention to the task must be continuous, even after the necessary pauses that can cause requirements to be forgotten to avoid failure. Continuous attention requires an additional memory effort, such as recording the state of the equipment and processes before the interruption to avoid errors when resuming the action.

Hollnagel (1993) discussed aspects that can lead to failure in carrying out the task, for example, when people are involved in more than one task, requiring a quota of additional attention and with the possibility of lapse errors. Another possibility is, when tasks are interconnected through points of multiple dependencies, requiring conferences, measurements, records, communication, which can lead to human errors in exchanging data and adopting an incorrect decision alternative.

Rasmussen states that it is not strange that the procedure undergoes adjustments due to the noncontinuity of the availability of the resources initially planned. The control devices and modes may be degraded due to weather or external influences, requiring adjustments to avoid accidents, changing the wrote procedure.

Conflicts of organizational and individual values can also change the priority of actions during the routine. The variability in the execution of the task can be caused by the fluctuation of human behavior that alters the motivation to perform the task. This state of behavior can reduce creativity in the decision to take action.

Thus the writing of the procedure should not be done "literally", which would require meeting multiple and specific requirements. The worker must understand safety, production and quality standards to meet them. Other nonessential standards bring difficulties to the work of the operator, who will have to change the stages, several times, during their performance. It ends up taking the focus of the aspects that can generate the accident by giving attention to document controls. This bureaucracy hinders production and still has the possibility of the operator not communicating the change in the practiced procedure, in the case of deviation normalization.

On the other hand, for other critical system situations, flexibility is avoided in complying with procedures. For this specific technology, the way of acting is unique, and any deviation can cause accident with fatalities and loss of property. In systems with extremely high risk, such as nuclear fission control, air space control and catalytic cracking control, procedures are not flexible, according to the opinion of experienced professionals.

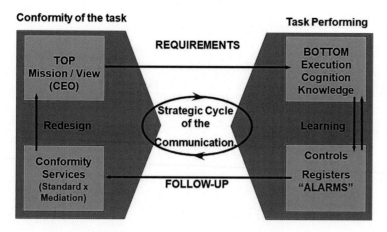

FIGURE 3.17 Organizational complexity in task execution. *Author's original.*

3.2.1.1 Task control complexity

Monitoring and controlling the task involves information flow with organizational, managerial and human activities (Fig. 3.17). This flow depends on the proper functioning of the technology in terms of process control. The company's Mission expresses desires for profitability and sustainability that depend on customers, the community, and government and worker satisfaction. Thus task requirements are defined based on legislation and on the nonexplicit demands of workers and the community close to the task. The aim is to produce a product that meets the demand without impacting the neighboring environment and keeping a good quality of life of employees.

Meeting requirements is the first action to achieve task compliance. Task Planning indicates a series of steps in the procedures that demand, on the shop floor, knowledge and skill through cognitive and motor processes, the execution of each step demands conferences that are part of the strategic communication cycle in the complexity of the task control.

When carrying out and checking the result of the procedure, a learning process begins in the here and now, where managers must be sensitive to promote the increase of competence of their team and their leaders.

The action controls, perceptions, checking requirements and identifying the state of normality are learned and performed. In practice, the suggested alarms and expected patterns are compared with the routine alarms, with signals emitted by the interface, and with the signals emitted by the equipment itself. These alarms and signals are compared with the state of the intended and achieved process and with the quality of the intermediate or final product. These conferences increase the perception of the production team, where the repetition of this cognitive processing increases the skills regarding the identification of process states.

The follow-up of these signs of abnormality shows the need to check the desired pattern with the state of processes reached, the state of services and the respective transformation of materials into products. Often a mediation is made between the ideal intermediate pattern and the one that can be reached at the moment, to finally achieve success at the end of the task. And so, these gaps are dealt with the necessary revisions of the technological project and organizational culture.

3.2.1.2 Methods in task analysis

Several authors have developed methods for analyzing the task with different objectives: (1) the CARMAN created by David Embrey, discusses how to prepare the manual of good practices and its importance for safety; (2) Frank Lees discusses the task in the routine of operating industrial facilities, mainly chemical processes; (3) Hollnagel discusses the importance of the goals, requirements, impacts for carrying out the procedure; and (4) Ávila et al. (2015c) includes a discussion about the environment in which the task is performed (physical-cognitive-organizational), presenting the most appropriate type of cognitive processing and how to measure its efficiency and effectiveness.

Table 3.2 indicates the opinion of an oil operation team (Embrey, 2000) about reasons to not comply with the planned and written procedures. We show the principal impact on production and some recommendations.

The improvement of the safety culture by transform practices into procedures, respecting safety and technology aspects to adapt to cognitive limits and specific and regional HF. Embrey makes some suggestions for the success of this methodology: the operation's staff should participate in the task's risk assessment; operational experience is valued

TABLE 3.2 Reasons for operator not to follow procedure, CARMAN—impact Interpretation.

Reasons for oil industry	Impact on production	Recommendations
The procedure needs to be revalidated for current situations where deficiencies in production management occur	Uncontrolled production; human error and deviation	Review procedures from task nonconformity
There is no possibility to perform the procedure with the tools, data, environments, and times available (lack of knowledge in the preparation);	Uncontrolled production; occupational and process accidents	Select the task's elaboration and review team with appropriate criteria
There are restrictions for performing some of the steps in the procedure such as difficulties in: building the causal link for mental map; and management deficiency in building/maintaining skills	Errors in interpretation of procedures; uncontrolled production; deviation normalization.	Select the task's elaboration and review team with appropriate criteria
The Individual practices are used in place of Best Practices, then, operators do not perform the best task. The inadequate way of reviewing, looking for the best place and not fitting for the best procedure. Existence of excessive self-confidence	Rising production cost	Observe and follow best practices to update procedures. Task analysis—good practices
The number of connections between different procedures with multiple requirements, which can also occur, in the steps of the same procedure, lead to a complex situation. The complex causal nexus require new auxiliary tools, nonexistent, yet	Inability to recognize the relationship between different procedures; uncontrolled production; negative image	Perform analysis of complexity of process, task and social relationships at work in conjunction with the risk analysis (frequency and impact)
There is a lack of information to carry out the procedure and a lack of resources for emergency situations. Certain safety rules are circumvented due to production being a priority in relation to safety. The organizational policies different from practices change the steps of the procedure. The management has deficiencies in the translation of policies or in the establishment of basic operating guidelines	Deviations, omission and discredit; increases the likelihood of accidents and production losses; equipment or product quality failure	Implement human reliability program
The procedure is not easily recognized, leading to "assuming" what should be done without due consultation to document. There are deficiencies in the management system, in locating and updating critical procedures. The ISO system is too bureaucratic transforming in a system disease, difficulty to establish standards	Knowledge-skill deficient for not accessing the task's proper form. Production out of control, and accident	Critical and complex task review Implement Human Reliability Program
The warning that should trigger use of the procedure, is not clear, leading to no use, or misuse of procedure. Improper communication in procedure step	Doubtful task communication—deviation, failure or accident	Improve communication tools in routine and emergency—auxiliary memory
The limited knowledge of the operation, indicate that the procedure will not be done on time, if it is done to the letter. The steps and times are changed on their own	Production invoicing changes the way the task is performed—accident	Human reliability program and adjustment of human performance factors
The time of experience and the criticism of how to develop and revise standards lead the operator to assume that he knows the best way to perform the task. Happen excessive self-confidence	Human error, deviations, failure and a greater likelihood of an accident	Human reliability program and adjustment of human performance factors

(Continued)

TABLE 3.2 (Continued)

Reasons for oil industry	Impact on production	Recommendations
The operator over-relies on his skills, leading to a lack of commitment in relation to the recommended way of carrying out the task. Normally this situation results in do not follow the recommendations from risk analysis in the production routine	The "super-operator" is convinced that the best way to operate is his, aversion to risk and lack of cooperation—diversion, failure and possible accident	Human reliability program and adjustment of human performance factors

to develop best practices; facilitators are used to convince that best practices should be practiced; support best practices with tools indicated by the workforce; develop training systems to provide risk management skills.

In a simplified way, Embrey (2000) states that the Best Practices Manual consists of: listing of tasks, identification of documents or procedures used and respective characteristics of the operator; analysis and inclusion of aspects of technology; analysis of the dangers and consequences if the task is not carried out properly; inclusion of specific aspects of the legislation and external references regarding the dangers and consequences in the task; and inclusion of aspects related to reference procedures.

3.2.1.3 Requirements for task analysis

Task planning is based on project data and operating system specifications. The procedure designer has previous experience and knowledge about the production system. In the case of the industries in operation, we seek to observe the task, to improve their writing, to change some of the requirements. These adjustments allow the adjustment of cognitive processing time and, avoid the standardization of the scenario based only on nonvalidated facts, on "*I think that*," after all, there are no arguments against facts. The survey of the previous data allows the binding of criteria for the decision during the performance of the task. Historical events help to change procedures, but they often bring human errors, replacing past memory for present requirements.

In planning the task, Hollnagel suggests and Ávila discusses the following checks: the workload analysis from the level of physical and cognitive effort; the analysis of the chronology in the task indicating the probability of failure by action alternatives in the timeline to change the tools to reduce the time of decision and execution of the task; identification of the logical sequence of steps trying to reduce repetitions and increase redundancies for high risk actions; identifying the critical points of the task in relation to safety, cost, fatigue and other aspects.

In this direction, important aspects are considered in the task planning: the available time and the realizable time of the activities in the task; sequentially or parallelism in carrying out the steps of the task-procedure or between simultaneous tasks; the understanding of the situation of normality and abnormality in the process, in the people and in the equipment, expected standards and observation of the apparent current standards; the availability of data and support for decision and action; the emotional stability of the task participants; discernment as to the type of action, emergency (control or contingency actions) or routine.

3.2.1.4 Knowledge base and skill in planning and adjusting the task

Lees (1996) points out the importance of planning operational procedures for writing, through planners who have a good knowledge base. During the period of performance of the task, an evaluation of the results is made by comparing the intended goal with the achieved goal, thus some requirements are changed and the analysis of the consequence after the task is revised. It can be concluded that, after the first performance of the task in pilot format, a final configuration of the operational procedure is reached. In preparing the procedure, it is important to have a good knowledge base. The procedure is reviewed to achieve the goal of best practices in relation to safety, cost, quality of life, and environmental impact. Changes are made to pre-existing procedures based on the skill of experienced people.

3.2.1.5 Means and goals in task planning

The GMTA is a method of analyzing the task created by Hollnagel. The main feature is to define the overall goal and partial goals for each task or for a series of steps within the task. For each step of the task, the goals of the process/equipment/product and people with the respective requirements, impacts, side effects and conditions for their accomplishment are tied. The goal is achieved by performing the steps and may need to meet previous requirements. Thus the map of relationships between the stages of the task is constructed. After testing with the GMTA, the following algorithm were developed, already validated in the chemical and petroleum industry:

TASK PLANNING

1. Establish the objective of the procedure as to its main function
2. Analyze the requirements of legislation for product, process, work, and natural environment
    a. Establish general goals and conditioning indicators for compliance with the procedure
    b. Establish the goals and conditioning indicators of the fulfillment of the steps in the procedure
    c. Perform a hierarchical Analysis of the task and divide the steps by hierarchy
    d. Check the level of experience, knowledge and skill in the task for training or selection
    e. Communication style, decision stress, cognitive load for alarm normalization

TASK EXECUTION

1. GENERAL REQUIREMENTS, OBJECTIVES AND GOALS
    a. Step 1
        i. Requirements
        ii. Activities
        iii. Local and Global Impact
        iv. End of Step 1 indicator, rework where necessary
        v. End of Step 1
    b. Step 2
        i. Requirements
        ii. Activities
        iii. Local and Global Impact
        iv. End of Step 2 indicator, rework where necessary
        v. End of Step 2
    c. Step 3
        i. Requirements
        ii. Activities
        iii. Local and Global Impact
        iv. End of Step 3 indicator, rework where necessary
        v. End of Step 3
        vi. Step 3 Conference
    d. Step 2 Conference
    e. Step 1 Conference

END OF PROCEDURE—FINAL CONFERENCE
TASK CONTROL

1. GENERAL AND SPECIFIC GOAL CONFERENCE
    a. Control of the Procedure
    b. Compliance with Requirements
    c. Registration of Control Indicators
2. COMPLIANCE WITH CHECK-LIST

TASK REVIEW

1. CRITICAL TASK ANALYSIS WITH SUGGESTED REVISIONS
    a. Present Training, Project, Requirements
        i. Requirements Review
        ii. Procedure Review
        iii. Task Review Design
        iv. Technological Design Review
    b. New Training, Project, Requirements

END OF TASK ANALYSIS

3.2.2 PADOP—environments and cognitive aspects in task preparation-execution

The Frontier of Technology to achieve competitiveness is in control over human behaviors and machines in high-risk tasks, in this situation it is intended to transform individual skills into productivity for the team in the operation routine. This requires knowledge about the training mode of the operator, about the reward analysis, and about the understanding about commitment level from the comparison between individual and organizational values. In addition, the choice of the best rituals to establish good social relationships in the task execution environment and in the job. Therefore the task analysis is not limited to questions of cognitive processing, expanding this view to the knowledge of work patterns from the various actors in the organization (Ávila, 2011a; Ávila et al., 2015a,b,c).

This Methodology called PADOP presented in Fig. 3.18, has the following steps: (1) IAT—task environment identification; (2) PCET—cognitive processing and task execution; (3) ATEE—task analysis and its effectiveness.

The first stage, task environment (Zardasti et al., 2017) identifies the ideal state based on process technology and human aspects inserted in the local culture and in the organizational culture. The verification of the knowledge and skills inventory of the operator and the group makes it possible to schedule the training. Organizational values are inserted and reaffirmed in the work team through educational, integration and social inclusion programs. The forms of group action are the result of managerial profiles and the role of informal leaders in acting in task forces to solve problems. These group actions can be improved through training and socioeducational programs.

3.2.2.1 IAT—task environment identification

To analyze the task environment (Fig. 3.19), it is important to know how the work group carries out the activities, the level of cooperation, and how individual initiatives work in terms of creativity and socialization of knowledge. Identify the way in which the worker and the group decide, to achieve resilience in adaptations that improve the results in carrying out the task. The ideal state of the task depends on technological and social aspects. The domain of technology indicates which are the indicators to decide on the state of the process and what level of knowledge is required about the product, process, task, controls and equipment. In social terms, there must be good organizational integration and social inclusion at work.

Organizational incorporation is achieved when the group's role is recognized at work, where the level of commitment is reached for the high-risk activity. This level of commitment depends on the acceptance of organizational policies, the managerial profile, insertion in group relations and the acceptance of leaders. The group beliefs established

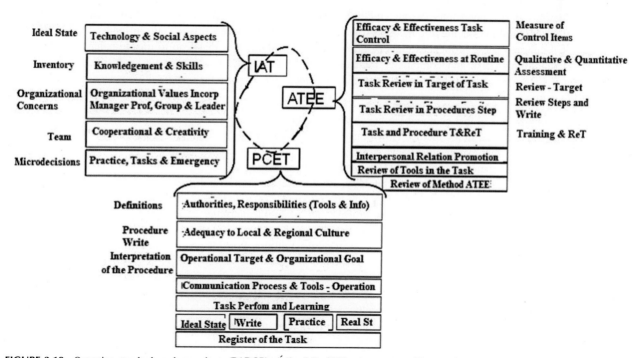

FIGURE 3.18 Operation standards and procedures (PADOP). *Ávila, S.F., 2011a. Assessment of human elements to avoid accidents and failures in task perform, cognitive and intuitive schemes. In: Proceedings of the Seventh Global Congress on Process Safety, Chicago.*

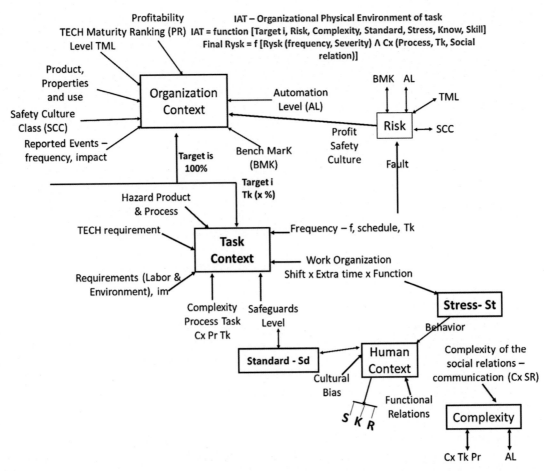

FIGURE 3.19 Context in the task environment IAT. *Author's original.*

toward productivity and a good interpersonal relationship in the group (without conflicts), results in a good level of communication that facilitates achieving the effectiveness of the task. Routine decisions are influenced by regional culture and global culture. These influences must be addressed to avoid changes that lead to unsafe behavior in the workplace. The calculation of risk and complexity indicators can be developed for preventive analysis of task failure as shown in Fig. 3.19. Some questions are asked for the *task environment analysis* in the context of company, task and person:

Company context

1. What is the company's global goal and the benchmark position?
2. Are there events reported in the media? How many a year? What impact?
3. What are the factors that classify the safety culture? What is the class of safety culture?
4. What are the characteristics of the product? Physical-chemical properties
5. Factors that classify the maturity of the technology. How mature is the technology?
6. What is the level of profitability and where is it in the ranking?
7. What is the company's automation level? For process and product technology...
8. Level of SKR demanded by the company, basis for training and selection of the company
9. Position of the company as a class of safety and profitability culture. For this process and product technology.

Task context

1. What is the goal of the task and how much does it represent in relation to the company's goal?
2. What is the danger of the product? Flammability, toxicity, flash point, chlorine, GHG emission...
3. What is the danger of the process? Critical temperature or pressure conditions? The material hazard is the result of cyclic movements that can cause fatigue?
4. What is the technology restriction? Materials? Corrosion? Capacity? Temperature? Mechanical shock?
5. Dangerous transport? Accident with impact? What is the impact and frequency?

6. What is the quality of the product, raw material and processes? Required by the market and hit by the task? How much is due to the task?
7. High cost due to time... Quality?
8. Who plans, elaborates, executes and evaluates knows the restrictions of the technology? The team have good Knowledge Base and Skills;
9. What are the environmental and occupational restrictions on events in this branch of technology? In this region? With this amount? And for the ministry of labor? What is the impact and frequency?
10. How often is the task performed?
11. What is the work organization? Are there double shifts? Frequency? Generating what level of stress?

Context of the person

1. What are the barriers to human error?
2. What are the standards? Are they flexible? Have they been agreed?
3. What are the cultural biases? And the group? Clear archetypes? Which are?
4. What level of knowledge is offered? What skill level is offered? What is the commitment to the rules? And with sincerity?
5. What are the functions and functional relationships? What are social relationships? How does functional and social communication work within the company? On task? And in personal terms?
6. How is routine behavior? In the Emergency? And in unusual situations?
7. Are there double shifts that cause stress or even an extreme demand for the result?

3.2.2.2 PCET—cognitive processing

The PCET is a stage that works with the elements to carry out the task in the routine. They avoid emotional bias in the choice of people to carry out activities and to fulfill them. When detecting that the group has difficulties in communication through the study of the task environment (IAT), this problem can be corrected through actions in the perception stage, in the data collection, or even in the construction of the mental map for the Task. PCET includes *cognitive processing and task execution*.

When identifying the Environments and Standards to establish the Normal State of the process, the PCET begins with the writing of the procedure, taking into account aspects of the local culture. At this stage, the procedure is interpreted in relation to operational and organizational goals. These procedures, depending on the complexity, need to be incorporated through Communication for the operation group and only then is the task or procedure executed, with an analysis based on the observation of the procedure being performed.

The PCET defines the Authorities and Responsibilities. In this stage, it is intended to analyze the knowledge inventory, the technical / personal skills, the level of commitment (rate of application of these inventories) with the execution of the task to reach the ideal state.

We divided the PCET into: (1) planning the task, (2) elaborating the procedures, (3) designing the workstation, (4) chronology and criticality analysis.

1. *Planning the task*

To Plan the Task with Hierarchy Assessment (Embrey, 2000; Lees, 1996) and with *identification of work environment* (Ávila et al., 2015c), it is necessary to explain the physical and social phenomenon that occurs and the respective actions and responsibilities in the field and on the control panel. Technological restrictions and environmental impacts indicate the main failure mode and the probabilities of events occurring ends each sub step.

Previous knowledge allows to structure the task indicating how to build the manual of good practices, the procedures by hierarchy, the steps and routines that are repeated and the possible emergency. The types of activities (search-action, surveillance, planning and decision) indicate the best auxiliary tools to control the task.

The records from the activity, monitoring of indicators and of control items help to indicate periodic measurements of process variables, production parameters, team participation by function, equipment availability, occurrences observation, and follow-up of corrective and preventive actions indicating efficiency level.

In structuring memory and auxiliary processing, the detailed procedure, checklist, booklet, drawing, auxiliary processing, specific memory, digital tools are taken into account. In addition, it is intended to analyze the risk in the task including technical and social aspects. It also seeks to structure training by complexity and level of competence demanded versus installed. When setting up the task with its chronology, the task hierarchy must be tested (Table 3.3).

TABLE 3.3 Task planning questionnaire.

PCET	Cognitive processing and execution	Task planning	Explaining the task
The phenomenon: causality projected in the process to identify how many steps are necessary, which equipment, which materials circulate and which are associated risks. Identify the process and task actions and field or panel responsibilities, which are the risks and goals in general. Identify and reference technology restrictions, possible impacts and related standards. Describe the cause mode with the deviations and the consequences indicating what measures are taken in the project (project barriers). If the barriers fail, present the consequences and impacts.			
PCET	Cognitive processing and execution	Task planning	Structuring the task
Locate the Task in the Operation or Best Practices Manual and who would be the facilitator. Check constraints, goals, process status for the larger task group. Detail the task within the larger group with the general requirements and approvals to start it—discuss possible failures. What are the procedures that will be performed in this planned task? What is the memory? What is the risk? Communication? Initial barriers? Automation level? How many steps are foreseen per procedure in general and within it if there are repetitive routines? What are the emergency situations and procedures to interrupt operations avoiding the release of danger… Are there any barriers in the emergency initially designed? Which are these? For what plant capacity?			
PCET	Cognitive processing and execution	Task planning	Activity type, registration, and control
Through the explanation of the phenomena and the classification for the manual of good practices it is possible to suggest the type of activities to be developed by the operator in the task. Suggest the competency requirement according to the type of activity for this task, process and relationships within the work environment. Define which types are most common and which activities will be in the respective classes: (1) Search and action; (2) Surveillance—checking; (3) Diagnosis; (4) Planning and decision.			
PCET	Cognitive processing and execution	Task planning	Risk assessment
Design failure scenarios and analyze how the hazard can lose contention in the task. Check frequency and severity. Analyze the influence of undue communication. Know how to identify initiating factor or root cause. Know how to identify factors of human performance. Analyze cost of failure. Analyze save failure. Define ways to measure the accomplishment of the task.			
PCET	Cognitive processing and execution	Task planning	Auxiliary memory
Perform the risk analysis (frequency, severity and complexity—based on the number of steps) to choose the type of auxiliary memory. List specific technical items, details on equipment, devices and processes that can cause major impacts. Check if the established routines are for the control of the continuous process or for the purpose of diagnosis. Choose the appropriate auxiliary memory for each workstation level (control room, field, and equipment). Check the level of general knowledge and compare with the demand for knowledge, add auxiliary processing or even digital tools. Detailed procedure, checklist, booklet, drawing, auxiliary processing, specific and auxiliary memory, and, digital tools.			
PCET	Cognitive processing and execution	Task planning	Structure training
Based on the knowledge inventory and the demand for general and specific knowledge and skills; Based on the complexity of the task, processes and social relationships; Based on failure mode and failure mode documentation for training; Based on the risk of an event occurring at any stage of the task; Structure the training: concepts, skills, failure process, communication, requirements for the task, connectivity between tasks and actions, emergency situations, normal situations, management guidelines and strategies that influence the work of the operation.			
PCET	Cognitive processing and execution	Task planning	Task hierarchy
Structuring the manual of good practices with facilitator. Have a general and specific idea and set up the hierarchy for future application and corrections. Predict task failure mode to understand how to write the procedure. Indicate in which stage the auxiliary memory will be used and in which parts of the procedures training is required. Indicate the responsibilities and competencies for carrying out this task. Suggest care for very complex tasks, number of steps and missing information.			

2. *Development of the procedure*

The procedure writing is important so that the requirements and actions are fully understood. The elements of writing, including interpretation of linguistics, type and quality of measurement and the expected results of the task indicating that the ideal state has been reached, which includes the proper use of financial, material, and human resources.

The criteria used to assess writing are: requirements for task control; beliefs, and biases; regionalization of words; mental representation of the action; action value by writing; artifice in the construction of phrases that change the procedure; omission in writing; substitution in writing; repetition in writing and; management guidelines on writing procedures. The Interpretation of the Procedure, its rereading by executors and the verification of errors in the reading or interpretation are activities carried out. There is a need to agree on what to do during the operational shift (Table 3.4).

3. *Workstation design*

The importance of discussing physical ergonomics to avoid fatigue and discomfort in the work environment is known by the industry and the area of technical and administrative services. The areas of logistics, transport, storage, and production also demand solutions for the design of the workstation in the industrial area. Logistics in the

TABLE 3.4 Questionnaire for the elaboration of the procedure.

PCET	Cognitive processing and execution	Elaborate procedure	General actions
\multicolumn{4}{l}{Write based on the description of the phenomenon and on the structure of the goals, requirements, and general status for the beginning of the procedure. Indicate the final state (and the final goals) to be reached in relation to quality, safety, time, environmental impact, energy, commitment, reliability, savings (costs, quantity). For each stage established indicate the time spent and the possibility of deviation varying the time. For each block of stages indicate which are the variables, time, quality etc., to follow. The approvals for each stage or block depend on reaching the submissions and reaching intermediate states of the process, thus requiring approvals. Each block of steps also fits the specific observation of the environment and the confirmation of the accomplishment with the operators. The identification of risks in each stage or block of stages allows identifying when to perform redundancy of people. Analyze the local and global impact after performing steps in the procedure. Write each step of the process based on the phenomenon described.}			
PCET	Cognitive processing and execution	Elaboration procedure	Procedure writing
Final goal and submissions, final state and intermediate states of the process established, results = performance criteria. Quality, safety, time, environmental impact, energy, commitment, reliability and economy (Ql, Sg, Te, Am, Em, Cp, Cf, Ec). Establish necessary knowledge to perform the task. Check if the state of the process after intervention can have consequences that make it difficult to achieve the company's vision, including laws. Write the task step, type of activity, final risk (Ri and Cx). Establish the appropriate change in the operating condition, indicate how to perceive this change also in the product and equipment. Establish intermediate control items. Write task steps in sequence (procedure) until completion.			
PCET	Cognitive processing and execution	Elaboration procedure	Validation procedure
Establish the necessary skill to validate the procedure and to perform in the field; establish points of detection of the need for action, that is, the alert to indicate what is happening. Through comparisons and processing of the data resulting from the performance of the written procedure, step by step, it is verified and the tendency is to reach the goal state passing through the intermediate states. If there is a deviation from the expectation of the procedure when in the elaboration phase, review the intermediate steps and include conditions or new steps necessary to reach the target state; review the task and requirements, benchmarks, and intermediate goals and process states; review the written procedure with the respective auxiliary memories, auxiliary processing and training needs. Repeat the revised procedure until it is verified that the target state will be reached in compliance with the established requirements.			
PCET	Cognitive processing and execution	Elaboration procedure	Barriers and safeguards
Failure, Initial barriers, Final risk, Review of barriers. Identification of failure modes. The knowledge of the causal link of the functioning of the process allows to analyze whether the initial barriers are sufficient to avoid the risk.			

operational routine includes the movement of materials, machines and people, both in the control panel and in the field. Locations for functional meetings and operations control require visibility, access and availability of procedures and documents for the various functions of the job.

Functional relations are accompanied by responsibilities and authorities in the control of resources and factors of production. Each function serves the production areas for correction and prevention regarding routine and critical operations. Strategic activities work in conjunction with managerial actions and production control.

The availability of physical tools and methods on the workstation facilitates operational control within an expected time in the company's mission. In this context, it is sought to avoid deviation and its acceptance as a normal situation (the normalization of deviation). We seek to understand the actions to maintain operational control, avoiding the abnormal state.

The treatment of deviations in uncontrolled processes intends to return the current conditions back to the normal state. Understanding failure warnings makes it possible to build scenarios that indicate future deviations (prediction). From these scenarios, preventive actions are designed to prevent the real states from being abnormal. Hence the importance of the standardization process of critical procedures and the respective tools for work and social communication. The status of the process depends continuously on safe, alert, and resilient behaviors (Ávila et al., 2018b) that are achieved through the Cognitive Academy, discussed in Section 3.1.9, where the planning of rituals (daily safety dialog) and educational campaigns are recommended.

The workstation must function organically balanced in terms of the flow of information and the availability of data to decide and act. The quality of data and information defines the results of controlling critical tasks.

The operation activity map contains the physical and cognitive limits of man and the physical limits of equipment and processes. Each task is performed in an orchestrated way in relation to the previous and the subsequent ones, therefore they follow hierarchy and repetition. Responsibilities and competencies are also part of the task requirements and the data available for the decision.

Fig. 3.20 describes the complexity of sequential and parallel tasks that demand the discussion of competencies and skills in the operation. An agenda of activities must be carried out to carry out the production. This agenda has a specific order or sequence of realization.

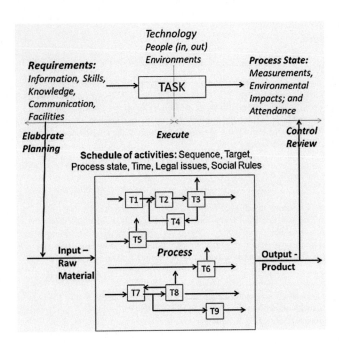

FIGURE 3.20 Schedule of activities. *Author's original.*

Each task has intermediate goals or, for a task block, final (global) goals. It is important to define the intermediate process states and the direction for the failure or normality. The time analysis allows to verify production restrictions; therefore the tasks must be timed. Legal labor and environmental aspects are considered essential to define limits in carrying out the task. Regional or organizational social rules must be known to avoid the low effectiveness of the task, indicating ways of acting more appropriate for each case.

The workstation (Fig. 3.21) becomes a comfortable environment when adopting appropriate criteria (Ávila et al., 2019) for its design and adjustment in the accomplishment of the routine tasks. Organizational and cognitive design deserves attention of entrepreneurs to avoid human error and failure. This design is based on discussions about cognitive processing, standards with risk analysis and guidelines for good practice. The writing and the execution of the procedures demand revisions to the control of the tasks and their conformity in the multidisciplinary aspects. Failure on the task analysis due to human and organizational factors requires investigation of the chronology, logic and materialization of the failure indicating the need of revision the culture design and the process.

The workstation project includes the planning of adequate and available skills to perform the tasks. This set of characteristics adds to the comfort of controlling systems and processes using technological interfaces. The safe, alert and resilient behavior of the worker is the result of intervention on human performance factors (Table 3.5).

4. *Chronology and criticality analysis*

Chronology and criticality analysis are addressed in the discussion of the tools in Section 3.2.3 and are techniques already mentioned by Embrey (2000) and Lees (1996). Here we introduce some characteristics and steps of these tools a priori. Chronological or temporal analysis will also be discussed in emergency situations regarding the rate of release of hazardous energy.

When planning the task, a sequence of actions is established to be performed in a certain time (expected standard) and which are part of procedures connected in series or in parallel (Fig. 3.20). Depending on the level of reliability of the equipment and processes, there may be redundant systems presenting the operator with the option of choice. This means that, redundancy increases reliability, but not necessarily in the same proportion between "brother" systems.

The choice of the system to be operated, therefore depends on the analysis of the probability of failure and the type and risk of the activity. It becomes possible to check the time bottlenecks and the risks of not fulfilling the task, treating the problems in a preventive way. Based on probability and time, the best routes for the use of equipment and systems are chosen.

The criticality analysis of the activity in the task requires knowledge of the type, physical displacement and difficulties, thus defining the physical and cognitive efforts. This analysis also allows preventive actions to avoid human errors or technical failures with a review of the planning based on the previous diagnosis.

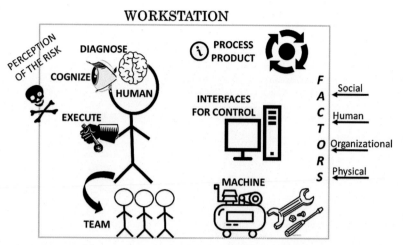

FIGURE 3.21 Workstation criteria, cognitive failure, and poor perception of risk. Ávila, S.F., Mrugalska, B., Wyrwicka, M.K., Souza, M., Ávila, J., Cayres, E., 2019. Cognitive and organizational criteria for workstation design. In: International Conference on Applied Human Factors and Ergonomics. Springer, Cham, pp. 161–173.

Another important aspect is related to the activities carried out in parallel by the same person or by different people. This issue will also be addressed in Section 3.2.2, using the risk dominance and parallelism tool. When analyzing that excessive effort can cause human error, the strategy of reinforced communication or other automation options is established.

Criticality analysis also works with indicators such as task fulfillment and team behavior analysis; relationship between level of automation in relation to critical manual activities; proportion of main and auxiliary systems in the project; level and attention and human types; memory level and human types; ability to withstand stress in emergency situations (Ávila et al., 2015a); task complexity level to predict future failures. Criticality analysis is influenced by process and organizational technology. Thus the complexity of the process (Perrow, 1984), the task and social relations increase the complexity (Ávila, 2012, 2013b) and, respectively, the Criticality of the Task.

3.2.2.3 PCET—task execution

The goals of the critical task include different dimensions of compliance and require adequate communication processes for the functioning of the routine and the correction of abnormalities; the execution of tasks by skilled operators; maintaining compliance through task adjustments; the registration and analysis of indicators, facts, and variables to anticipate future failure.

Each communication process needs to be studied and standardized, admitting some flexibility and some informality. Communication processes are dynamic and divided between formal (standards and procedures) and informal. Feedback after operations is considered an informal communication process and helps to detect trends in human error. Communication in the execution of the task is studied to maintain its control. It is done through the following characteristics: working hours; type of communication (written, spoken, and gesticulated); quality of communication, in strategic terms and as to the speed and behavior of the communicator; actors involved in communication; and communication tools-devices for each routine or emergency case.

When comparing the real process state and the intended-ideal state, a big difference is noted, demanding the accomplishment of the task to transform the process state. In the execution of the task, tools for control action (manual or automatic) are used. These tools need maintenance to prevent failure. The concern is to carry out the activities correctly and correct the flaws in the control system.

Often, with the degradation of technological and human systems, it is necessary to carry out corrective or preventive maintenance on process control and safety control devices. But it is also extremely important to review the procedures.

Task planning authorizes the writing of the procedure that must be controlled through goals such as quantity produced, quality, and safety. These technical-social parameters are registered through records that are part of the operational standard.

Task analysis is performed during its execution or, historically, when collecting data and resulting information. The analysis of the task is performed from the registration of the process parameters, the conditions of the task, the status

TABLE 3.5 Questionnaire for the workstation project.

PCET	Cognitive processing and execution	Workstation project	Behavior SKR C
External perception allows the formation of an image for decision and action processes. Skills initiate behavior with actions learned in practice. From these experiences, actions are carried out. In the absence of references from the experience, the rules and standards transferred from the system to the individual are sought. Then new filters and standards are formulated and accepted to initiate action. If perception does not indicate the presence of recognized signs or patterns, then the knowledge base is sought, from the causal nexus, the rules are reformulated, new standards are accepted and actions are taken. And the commitment? Operates the psychological contract for utility in the task...			
PCET	Cognitive processing and execution	Workstation project	Skill SKR C
Perception is an activity resulting from the sense of usefulness of the task and the role established at work. Perception would be off without utility and clear role. The sensory input (perceiving the external signals of the process, people, and equipment). Some filters are established and some patterns are recognized (signs). Are there archetypes that define behavior? Are there resistances or acceptance of population stereotypes? What is the quality of image formation with sensations and elements and recovery of stereotype? Preconditioned actions are performed automatically based on experience and recognition of signals. But, the perception of signs may not go through recognition, in this case, the actions are on impulse.			
PCET	Cognitive processing and execution	Workstation project	Rules SKR C
The signs were not recognized for immediate or almost immediate action. Commitments to organizational values indicate the need to follow an established standard. Rules-based behavior begins... the signals indicated in the perception (formed image) are recognized in the rules or standards. A series of actions (steps) are established to reach a goal state. A number of procedures are related, including what was recognized for the action—there is an association, requirements, an order... Signs that were previously accepted based on experience are now accepted from the experience of task planners, or that is, if I accept the signs inscribed in the procedures...			
PCET	Cognitive processing and execution	Workstation project	Knowledge SKR C
If there are no similar previous experiences. If there are no procedures related to what was perceived, that is, the image of the equipment and processes and the demand for actions, there is no reference in the standards and procedures. We are going for knowledge-based behavior... The unusual requires a consultation of the broader knowledge base or hypotheses that enable tests and generations of new concepts. A causality of factors needs to be systematized based on safety, production and quality goals. Planning what to do from the discussion of the phenomenon. Establish a procedure and accept it as a sign for action.			
PCET	Cognitive processing and execution	Workstation project	Commitment SKR C
Be careful with impulse actions, including emergency situations. Commitment results from the recognition of individual values within the organizational scheme through feedback. Comfort for cognitive processing and rescue / operations of perception and memory. The team needs comfort for discussing and confirming new concepts. Draw up a clear mind map for the decision. Provide feedback to maintain the expectations of the organization and define standards of behavior. Feel inserted in society through the role in the company. Accept the dynamics of continuous improvement of competence for continuous improvement of productivity. Beware of aspects related to labor legislation. Establish these criteria in behavior based on skill, rule and knowledge.			
PCET	Cognitive processing and execution	Workstation project	Interface man procedure
Establish the roles of operators on the shift and related activities. Availability of documents at each workstation: good practices, procedures, documents and records. Provision of information via system, telephone, radio and decision reports (micro and macro). Facilities and devices available (physical). Description of the established functions and types of common activities in the procedure. Establish glossary (dictionary) to facilitate errors due to linguistics. Check the forms of communication at the entrance: information, guidelines, parameters, states and goals. Check the forms of communication at the exit: variables, occurrences, states and goals achieved.			
PCET	Cognitive processing and execution	Workstation project	Interface —ManM and ManPr
To Provide manuals for machines, processes and safeguards. Alarm and audit management system. Communications and documents involved with the machine and the process—drawings, booklets, and notices. Indication of the conduct in routine and in emergencies. Establishment of the main modes of failure in the process and equipment. Indication of the main possibilities of failure in the process and equipment safeguards. Be careful about behavioral biases such as using a bypass valve or not handling an alarm. Schedule of routine activities with established frequency, severity and complexity. Guide on signs of abnormality or deviations to increase the perception of the beginning of the failure by the shift.			

indicators of the technical process, and the status indicators of human-social processes. The collected records make up the efficacy-effectiveness report of the actions resulting from the Operational Routine. We summarize the task execution as:

1. perform the steps and measure the results;
2. continue the steps with confirmation of the goal status of the process, service and product;
3. compare standards, expectations, with performance, and correct if there are deviations;
4. record occurrences for feedback on technical and social behavior and to investigate failure in the task;
5. be available to learn while performing and to record routine learning;
6. check the need to review the procedure while it is being carried out and give feedback;

7. check the stability of the barriers and anticipate the possibility of failure through preventive actions;
8. always consult the knowledge base and practice respecting the experience with the equipment, processes and services; and
9. establish ways to quickly consult the formal and informal rules that make up the knowledge.

3.2.2.4 ATEE—assessment of the task, efficacy, and effectiveness

The third step, ATEE, involves the measurement of control items that indicate the efficacy and effectiveness of the procedure after being performed. With the group formed opinion, it will be possible to perform the correction in the various fields of task execution: review of the process goal, correction in the writing of the procedure, training and retraining, activities to promote interpersonal relationships, review of the tools to be used in the task, and review of the methodology for analyzing the effectiveness and effectiveness of the task.

The results achieved with the implementation of PADOP, cover the entire production area, and thus increase the motivational level, decrease human errors due to the adaptation of human-organizational values, decrease errors due to the adjustment of leadership when dealing with difficulties group relationship, and facilitates the identification of the best standard for the operation team.

The control and verification items indicate the state of the technical and human-social process for reviewing the control actions and modes. Changing these parameters and variables in production is expected to correct the process state and prevent the real state from being different from the ideal. At certain times, certain control items, especially human processes, are difficult to measure, so we seek to make a correlation with results in behavior that indicate a level of motivation, therefore we seek the correlation between the operator's speech and the action in the process. Hence the importance of investigating the operational routine based on the operator's speech, as shown in Fig. 3.22.

The task control items can be considered extremely important for: cost (time, people, materials used); quality (conditions of the product, process, and effluent); human (motivation and fatigue); environmental-occupational aspects (impacts, health, fines, effluent, and losses); frequency and quantity of product processed with the task; and the image preservation.

From the previous analyses (PCET and ATEE), the conclusion is reached on the level of efficacy and effectiveness of the task, therefore starting its correction. It may be that, due to different factors, it is also necessary to review the

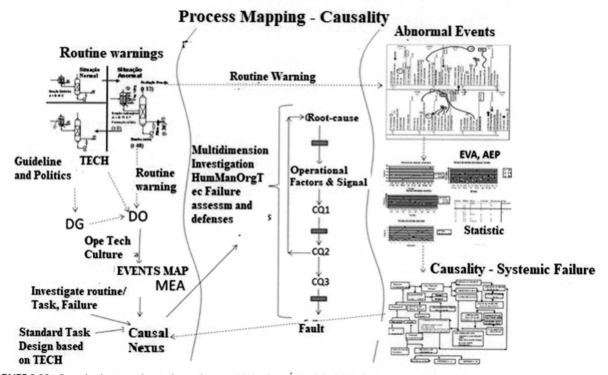

FIGURE 3.22 Investigation to review tasks, projects, and behavior. Ávila, S.F., 2012. Failure analysis in complex processes. In: Proceedings of the Nineteenth Brazilian Congress of Chemical Engineering. Búzios, Rio de Janeiro.

ideal state to be reached in the industrial process. Among these factors are the following: changes in load (increment); change of raw material supplier or catalyst; more restrictive legislation increasing controls; restrictions in process systems (dirt in exchanger and others); and restrictions in terms of personnel (previous operator leaving). Another item that can be changed is in the content of the task, with its activities, times, tools and others. This may be due to early recognition of an ongoing failure seeking mitigation.

Production management requires the training of multiplier agents for routine control. Training modules need to be prepared for the dissemination of Lessons Learned by experienced operators. Normally, human errors caused by low quality in communications are inserted in a low integration social environment, where leaders and teams' live psychodramas in their daily routine. One way to recognize this lack of integration and preparation of the team and leadership is through the application of the Operative Groups technique developed by Pichon-Rivière (1988) and adapted by Ávila (2011b) in the LODA tool.

By developing a context of psychodrama with high and progressive stress, one can recognize the social actors who operate the group, increasing friction, worsening communication and causing the accident. Just as the opposite can happen with the work of social actors who operate the group, reducing friction through better communication, which avoids the accident.

LODA (Ávila, 2011b) was applied to industrial units that process Liquefied Petroleum Gas where, in a dynamic scenario, failures occur with progressive release of danger.

By approaching the process analyst with the task scenario and the team of executors through his speech, Pandora's Box opens up regarding the deviations that were in the informality. Information flows and it becomes possible to correct the task to unify the operator's performance during the shift.

3.2.2.5 PADOP—implementation process

Implementation of the PADOP method (Ávila, 2011a; Ávila et al., 2015c) is discussed in Fig. 3.21 and detailed in the case of the Oil and Gas Industry, in Chapter 7, Human reliability: chemicals and oil and gas cases, of this book. Input information for analyzing the task, such as technology design data and environmental threats, indicates the compliance level of the constructed Scenario. The investigation stages on the Task include the Identification of the environment, the elaboration of the scenario, the planning of the task (writing and procedure), the execution and control of the critical activities described in the procedure. Requirements and data can be revised based on practice or external threats, creating new conditions for the environment, planning, execution and control.

While the environment is analyzed by the contexts of company, task, and person to compose the manual of good practices, planning follows the hierarchy in the task and the analysis recognizes the importance of the means, goals and steps to avoid the accident. Good practices (Embrey, 2000) require the presence of facilitators to avoid noncompliance with the procedures. The physical-cognitive project combines with social-human content and rules for the execution of the task. Revisions can be made directly in the execution and in the planning, stage as shown in Fig. 3.23, which represents the PADOP method.

The intimacy between the operator and the industrial plant becomes a priority so that the learning results in procedures compatible with the best practices. At this point, Ávila contributes to the discussion of human behavior in the task of communication and feedback in the safe, alert and resilient system (Ávila et al., 2018b,c), only developed in "Just Culture," with a focus on the development of the task in communication environment.

When measuring and comparing practices with behaviors and production variables, it is possible to state about the conformity of the task or to demand minor revisions (procedure) or major revisions such as the technological and organizational redesign (equipment and information flow).

3.2.3 Tools for planning the standard task

The definition of the standards to be adopted for carrying out the task requires specific studies and analyses that were discussed in PADOP (Ávila et al., 2015c) and in the introduction of this Section 3.2.2.

In the architecture of the task, causal relationships between steps, procedures and tasks are indicated, thus presenting auxiliary tools to increase reliability. The task project seeks to facilitate the development of procedures through issues related to its overall form and specific details.

The dominance studies of the risk, the parallel analysis of the steps, the implications of physical and cognitive effort indicates the need for establishing new criteria for the preparation of the task, including the suggestion of an important

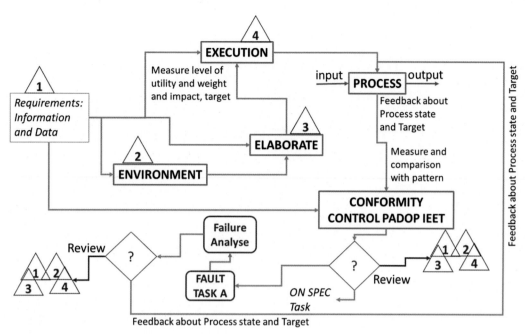

FIGURE 3.23 Process implementation of PADOP. *Author's original.*

tool for memory and cognitive processing. The decrease in physical and cognitive effort may require changes in the ergonomic and control aspects of the process.

The complexity analysis of the task (Ávila, 2012, 2013b) allows checking the level of automation, suggesting auxiliary systems, reducing physical-cognitive efforts and reviewing the type of action to avoid human errors and failures in critical procedures. The recognition by types of action and by the place indicates the need for safeguards because of the low or high automation level. The analysis of the timeline shows the most time-consuming steps and the actions involved, seeking to change the technologies to adapt the task in a more comfortable zone.

1. *Task architecture*

In task architecture (Fig. 3.24) a format is suggested to build the manual of good practices or best practices. What is the hierarchy, how many tasks are there to control the operations and what are the procedures and related records? In each procedure, it is important to check the steps and the cognitive ability with the respective controls and safeguards.

At that point, task controls and compliance analysis are defined. The standards are accompanied by written procedures with the respective checklists, readings of specific data by equipment, booklets for quick reference and drawings of processes and equipment. These are mnemonic instruments to prevent human error.

Hierarchy is preserved by classifying the procedures in the Task into levels, primary (main tasks), secondary (related procedures) and tertiary (steps in each procedure). All levels with respective panel, field, equipment and specific system records.

2. *Task project*

In the task design, general aspects are considered for all procedures, specific aspects per procedure, and aspects related to automation and safety in the task. Regarding the general and specific analysis, the following are discussed: title, objectives, numerical goals, requirements, risks, authorizations for sequencing the procedures and steps in the task. In the case of functional analysis, security barriers and the level of redundancy are identified, among other aspects of automation and security.

In the case of critical equipment, signals that indicate the main failure modes will be analyzed to avoid process abnormalities. Routine care becomes a criterion in the project and enables future operation with guaranteed conformity in terms of safety, quality, quantity, times, and control of human resources. The types and quantities of alignment for fluid flow and the function of the control loops show the level of comfort offered by the process design to achieve the optimum level of reliability.

3. *Risk dominance, step parallelism, physical, and cognitive efforts*

In the dominance analysis (Fig. 3.25), the steps of the procedure being executed are allocated between greater and lesser impact on security and cost. The stages are classified in series, parallel and transversal when related to other

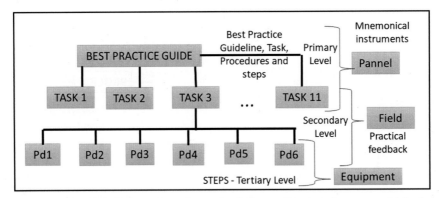

FIGURE 3.24 Task architecture. Ávila, S.F., 2010. Etiology of operational abnormalities in industry: a model for learning. (Doctoral thesis), Federal University in Rio de Janeiro. School of Chemical, Rio de Janeiro.

external tasks. This classification allows to confirm how the information flow in the group will happen and what type of safeguard will be necessary in cases of high risk. Physical or cognitive effort can alter the level of attention and difficulties in recovering memory. In this way, it is possible to study "bottlenecks" in the procedure, which can cause loss of time, work accidents, process accidents with or without environmental impacts.

In Figs. 3.23, 3 graphs are presented, representing 5 steps of the procedure, where steps 2 and 3 are in parallel, and step 4 has a high level of risk 3. In step 4 there is a high physical effort and low cognitive effort adding up to value 4, an uncomfortable situation (risk 3) that can cause the accident. The total duration of this procedure is of up to 7 hours. The greatest physical-cognitive effort has a value of 6 and is found at the medium risk level, 2 in step 3.

4. *Task complexity* (Ávila, 2012, 2013b)

The complexity of the task and its impacts is part of all the concepts and tools discussed in the book. Directly, this discussion becomes more evident through the tool that analyzes the dominance of risk concomitantly with the parallelism of the stages, and continues to be detailed in the discussion in Table 3.6. Other information helps us to analyze the complexity: number of automatic actions in relation to manual activities; number of auxiliary systems in relation to the main ones; number of procedures followed to the letter in relation to those that are not followed, which admit the creation of steps depending on the current scenario; level of calibration of attention, whether focused or dispersed to perform, respectively, a single step or if it is divided into several steps, procedures and tasks. This knowledge facilitates the definition of auxiliary tools to keep tasks under control and helps to define indicators on the level of complexity of the task.

5. *Place and task type*

The type of activity in the task depends on characteristics such as: level of attention demanded; use of main and auxiliary memories; need for a knowledge base (mind map in problem solving), and need for specific skills (in emergency situations). Task planning depends on the type of activity on the task (surveillance; search/action; and planning/decision) and the location of the task. The location will indicate difficulties in transporting operators, machines, and materials.

Activities related to the monitoring of process variables (temperature, level or pressure), operational parameters (cycle time or quantity produced), signals perceived in the field (smoke or noise) are considered surveillance (V), with no established focus of action. The activities directed to a specific location and with specific action are identified as search for equipment with respective action (BA), as is the example of drainage actions in certain equipment using specific valves.

Activities that involve planning for subsequent decision making (PD) require greater cognitive effort with higher use of memory and construction of mental maps. These can be considered complex activities.

The definition of the task's location helps to establish the physical effort of displacement, indicating the need for remote operations (automations). When the locations between steps of the procedures are close, there are no large displacements or the need for constant use of communication devices.

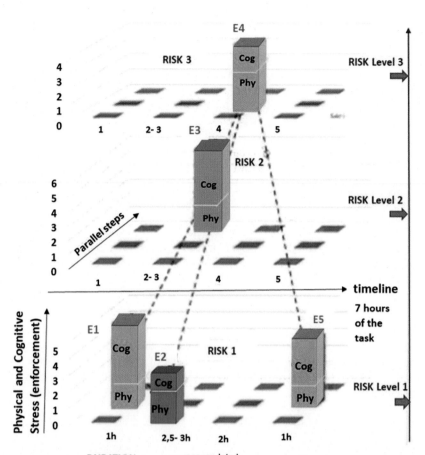

FIGURE 3.25 Risk dominance and parallelism of task steps. Adapted from Ávila, S.F., 2010. Etiology of operational abnormalities in industry: a model for learning. (Doctoral thesis), Federal University in Rio de Janeiro. School of Chemical, Rio de Janeiro.

Step	Time	Risk	time	Physical	Cognition	Total	Serial-parallel
1	1	1	1	1	4	5	Serial
2	3	1	4	2	1	3	Serial-parallel
3	2,5	2	3,5	4	2	6	Serial-parallel
4	2	3	6	3	1	4	Serial
5	1	1	7	1	3	4	Serial

TABLE 3.6 Task assessment: complexity to startup an important pump.

Step	Physical effort	Cognitive effort	Task complexity			
			Auto/ mannual (%)	Attention, memory (%)	Principal/ auxiliary (%)	Official/ alternative (%)
1. Turn on pumps to transference	1	2	30	50	30	70
2. Start pumping	2	1	40	Cavitation possibility 60	50	50
3. Follow transference by pressure and level	2	2	50	Test of flow and level—70	50	70
4. Normal shutdown of transference	1	2	40	50	50	80
5. Emergency shutdown of pumping and transfer	3	3	10	80	40	30

3.2.3.1 Practical case—fertilizer industry

In the fertilizer industry, the alternative high-pressure pump (180 kg/cm^2) injects liquid ammonia into the urea synthesis reactor. If there is a back pressure, to preserve the ammonia pump, the raw material feed block occurs, but the pump starts the flow recycle. These actions were performed manually, bringing the operator's exposure to equipment and events that, in failure mode, can dismantle and launch projectile (cylinder) to the environment and leak chemical. When analyzing the risks of accidents and process losses, the recycle valve started to be activated remotely, bringing comfort to the operation and reducing the process failures.

6. *Timeline analysis*

The chronology analysis mentioned by Embrey (2000) and Frank Lees (1996) helps to find time bottlenecks and improve the availability of systems, initially presented in Section 3.2.1. In Fig. 3.26, the steps of the task are discussed by type and time of completion. The reliability-availability analysis of the main and alternative systems for critical tasks depends on the control over the time spent, the type of action and the place where the action is being carried out.

In this tool, it is also indicated whether the activity involves consultation of data and information about the normal process, both on the panel and in the field. This consulted data and information feeds the decision and action process. The various action probabilities can be launched to calculate the probability for the occurrence of the desired end state.

3.2.3.2 Case study—oil and gas processing

This discussion is exemplified in the procedure of drying LPG columns in a Petroleum plant. The total time to perform this procedure is 124 minutes or approximately 2 hours. The actions are field or panel, the square indicates start of operations, the black circle indicates actions taken and the white circle indicates checks of data or process-production variables. The timeline is in minutes and when there are two alternative systems or equipment, the aim is to discuss the probability of success in the reliability assessment.

3.2.3.3 Dynamic investigation of task in failure

From the operator's speech (Chapter 1: Introduction, and Chapter 2: Human reliability and cognitive processing) recorded in the operational routine, the analysis of the task failure begins, indicating points where preventive/corrective measures can be taken with a greater probability of success.

Among the techniques, the study of the phenomenon in the operation indicates the volume of control where the events related to the failure occur with the path of the "body of the failure." The failure logic indicates the functions, sequentially, size of the failure and level of risk that cause financial impact and impact on the Company's image.

The chronology of the fault indicates the release of hazard energy, or size of the event, through a dimensionless parameter in conjunction with the time and phases involved in the operation mode.

The materialization of the failure refers to the analysis of the migration of the failure energy (or danger energy released in the event) that goes from the dimension of the manager's speech/operator's speech to the dimension of physical alteration of equipment, processes and products. Or, in the opposite direction, it migrates from the material or physical dimension, such as the presence of solids circulating in the process to the dimension of behavior as a result of the

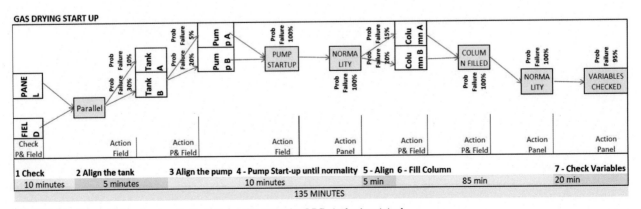

FIGURE 3.26 Temporal-functional analysis, drying, and regeneration LPG. *Author's original.*

operators' fatigue. The connectivity analysis indicates the points of connection between common factors in different failures.

1. *Cylindrical models for operational failure*

The explanation of the origin of the failure and its development is formulated from cylindrical models. The failure may result from the relationship between structures, processes and environments, as shown in Fig. 3.27. The plot of the failure history is described in the cylindrical model, with a multidimensional nature and presented in Fig. 3.28. We describe the interconnected operational factors, through the causal link and that have different levels of danger energy during the occurrence of these events. The cylindrical model of failure is presented in an ascending level of accumulated energy, from bottom to top, until the complete degradation of the system, considered successful in the failure process.

As shown in Fig. 3.27, there are inputs, outputs and recycling currents, both in technical structures (production processes and equipment) and in human structures (individual and group). In the technical structure, it is the circulation of materials and energy. The functioning of the human and social structure in management processes involves inputs and outputs for the formation of competence and for the accomplishment of the task. In the case of building competence, previous learning is the input and the aptitude for work is the output. In the case of carrying out the task, it refers to the ability to plan the task using imagery, abstraction and the construction of the mental map where, the input of information, with the previous knowledge of the patterns brings as a physical result, the transformation of the product to a higher added value, without generating waste, translated into output information.

The processes are technical, when they include input-output of materials through the structures. The processes can be human-social, which are the learning and management processes that facilitate the correct application of the skills to perform tasks on the equipment. The reward processes for work are also managerial and are related to the approximation of individual values in relation to organizational values.

The natural, occupational, organizational, social and economic environments pulsate energies for the occurrence of the accident. It is up to the company and the groups to perceive and install physical and communication safeguards. Information that flows dynamically threatens the stability of behavior, causing failures in the design and operation of industrial facilities.

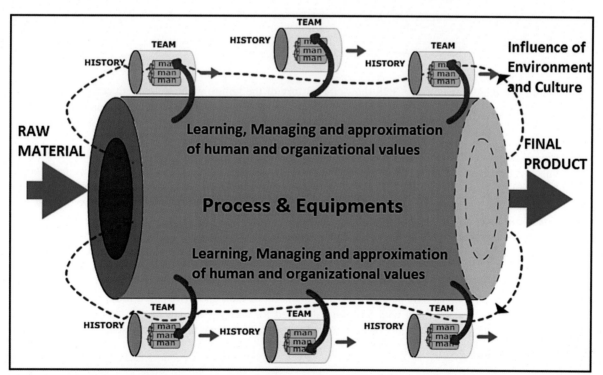

FIGURE 3.27 Dynamics in the production system. *Ávila, S.F., 2010. Etiology of operational abnormalities in industry: a model for learning. (Doctoral thesis), Federal University in Rio de Janeiro. School of Chemical, Rio de Janeiro; Ávila S.F., Dias, C., 2017. Reliability research to design barriers of sociotechnical failure. In: Proceedings of Annual European Safety and Reliability Conference (ESREL), 27, 2017, CRCPRESS, Portoroz.*

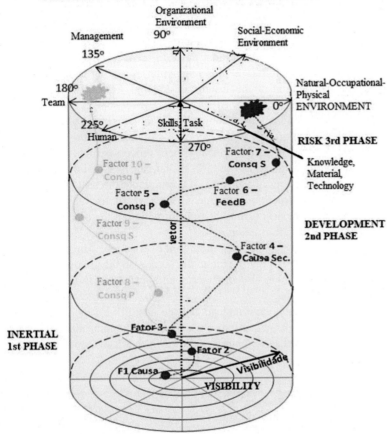

FIGURE 3.28 Failure of cylindrical model. Ávila, S. F., 2010. Etiology of operational abnormalities in industry: a model for learning. (Doctoral thesis), Federal University in Rio de Janeiro. School of Chemical, Rio de Janeiro.

The cylindrical model shown in Fig. 3.28 shows the history of the failure with the inherent complexity of sociotechnical systems. The cylindrical coordinates represent the quality of the events involved in the anatomy of the failure (operational factors), where, the same factor can have multiple dimensions involving the technical-human and organizational dimensions.

The various factors involved in the failure anatomy can be classified in different zones of danger energy release, in the failure-accident process, which starts in the inertial zone, passes through the failure-accident development, and can reach the level of maximum risk, the success of the failure. Each of these zones has specific characteristics.

In the inertial zone, visibility is low (radial axis) making it difficult to take actions. In the development zone, the complexity resulting from multiple extensions of hazard energy release events (failure process) can occur. In this case, it has complex characteristics with fault energy feedback points. Still in the development zone, it is possible to detect signs of failure in the technical-social dimensions, making it possible to monitor them in sociotechnical systems.

In the risk zone, the consequences have greater impact and maximum visibility (cylinder shell). It seeks to eliminate or mitigate failure through intervention on the cause or consequence. Technical and social devices are used to drain energy from the failure. The axial axis indicates the accumulation of fault energy released in the chain of abnormalities indicating the power level of the fault, with position calculated for each operational factor, and height as the result of the scalar sum of the release of this hazard energy, indicating the failure strength and its consequences (failure work).

2. *Phenomenology of task failure*

After analyzing the abnormalities of the shift routine (DIPEA—Dissecting the Process to Investigate Abnormal Events), using the abnormal events map (MEA), registered through the abnormal events report (REA) and discussed in the descriptive statistics of process variables (AEP) and analysis of abnormal events (EVA), the phenomenology of the

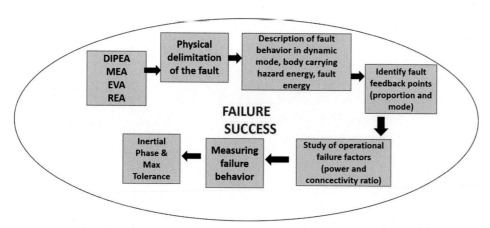

FIGURE 3.29 Failure phenomenology. Ávila, S.F., 2010. Etiology of operational abnormalities in industry: a model for learning. (Doctoral thesis), Federal University in Rio de Janeiro. School of Chemical, Rio de Janeiro.

failure is recognized (Fig. 3.29), as well as its physical delimitation, indicating the areas in which investigations on path of the "body of the failure" in equipment and production staff. This topic was started in Chapter 2, Human reliability and cognitive processing, and continued in Chapter 4, Process loss assessment, also will be discussed in the last Chapter 7, Human reliability: chemicals and oil and gas cases.

The failure energy feedback points are identified, thus presenting a complex and "organic metabolism" anatomy. It is necessary to study the operational factors through new failure analysis tools in a sociotechnical system to make the intervention more assertive. Multidimensional analysis includes the identification of logic and connectivity in systemic failure.

The location of the danger energy release regions indicates points for measurement for the strength of the failure that will cause consequences. The distancing of the source in the failure process from its maximum risk indicates the path of total energy release. The location of the hazardous energy release level may indicate the likelihood of the system losing utility. The height of the cylinder indicated in Fig. 3.28 will describe the objective function ranging from the null process loss conditions to the maximum process loss, which may be the risk of loss of revenue above the maximum value, risk of increasing the cost of production above maximum point, and the high risk of accidents occurring.

3.2.3.4 Case study—oil and gas processing

The "failure bodies" initiated through noise in the mind map in the decision are materialized (Figs. 3.29 and 3.32) and circulate through equipment in the processing of LPG as indicated in Table 3.7. These bodies, in the process, have the following forms:

a. presence of humidity and air in the storage of pressurized LPG and LPG for burning;
b. soot solids (when burning), the flow of LPG (process) in heating the bed through the oven;
c. presence of ice and hydrates in the cooling and separation of water from the heated LPG; and
d. uncontrolled pressure in the tanks, which, at specific times, can limit drying rates.

The study of this phenomenology is restricted to drying and regeneration operations, with the volume of control and circulation of the fault body restricted to related equipment.

3. *Fault logic diagram*

The occurrence of each operational factor in the cylindrical model leads to the accumulation of fault energy (with probability of danger energy being released) with a positive sign when it is classified as the cause, in this case, power supply. The signal of the fault energy is negative when the operational factor is classified as corrective action, draining the fault energy. The type of energy is related to the quality of the corresponding operational factor, which can be structural (technical and human failure), environmental (failure in the organizational and physical environments) and procedural (management and learning).

The Diagram for the Logical Analysis of the failure (Fig. 3.28) presents operational factors classified in the dimension technology, man, management and organization covering the structures, processes and environments of the dynamic model of the productive system (Fig. 3.25). The energy of the failure (or release of danger) in the production system can be translated into work in the form of signaling or warning (low energy), or consequence (medium energy), or even accident (high energy) from operational failure.

TABLE 3.7 Failure energy-carrying body in the task and failure phenomenology.

Body type and body in process	
Solid	Solid at furnace; solid on level valve
Water, humidity and hydrate	Excess moisture; water in the boot drain; hydrate in refrigeration
LPG flow and pressure process	
Flow rate—LPG pressure on flare	

Source: Ávila, S.F., 2010. Etiology of operational abnormalities in industry: a model for learning. (Doctoral thesis), Federal University in Rio de Janeiro. School of Chemical, Rio de Janeiro.

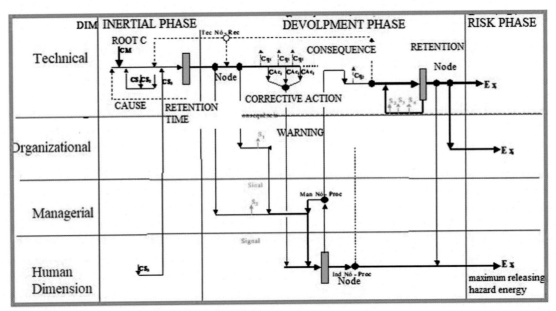

FIGURE 3.30 Failure logic diagram. Ávila, S.F., 2010. Etiology of operational abnormalities in industry: a model for learning. (Doctoral thesis), Federal University in Rio de Janeiro. School of Chemical, Rio de Janeiro.

The ability of physical structures to tolerate failure without causing the maximum loss value indicated in the objective profit function, is translated as system tolerance or capacitance, and indicated by the height of the risk zone in the cylindrical model in Fig. 3.28. Each system has specific fault tolerance, and depends on time-varying aspects, such as: the type of abnormality, the environmental influence, the capacity of the sociotechnical system to withstand stress-fatigue and the existence or not of safety barriers and respective efficiencies. The tolerance of the system is maintained and increased when the project is changed with technical changes: larger dimensioning of instruments, installation of other instruments for safety, alteration of internals, or even alteration of the equipment capacity.

As shown in Fig. 3.30, the failure often does not go forward in the sequence of the process because it is stuck at a certain point in the production system, accumulating energy. This means that the transfer of hazardous energy failure occurring, depends on the saturation level of the system. In the case of complex systems, the fault has a network-like architecture, which can be investigated through this logical diagram. Operational factors can divide the energy flow of the fault, opening a new branch forward or backward (feedback), feeding or re-feeding the hazard energy in the fault process.

The human network node occurs when the procedure or behavior of an individual causes the fault to be fed at different points or gives rise to different fault paths (common procedure). The organizational node, based on norms, occurs when undue disclosure is made about ways to behave in the company. The managerial network node occurs when the manager transfers his addiction and causes stress at a different time in the production team or even at the same time during the event of a certain failure. The technical node occurs when it involves the process, unit operations, equipment and even the operational procedure, through lines that interconnect different systems, transferring specific problems to several different points.

The logic diagram of systemic failure represents the complex mechanism that involves the possibilities of failure in the production process with the occurrence of several chains of abnormalities based on signals or operational factors. In the logical analysis of the failure in the task, as previously mentioned, we seek to define the causal nexus by indicating probable causes, the intermediary operational factors, the consequences and the interconnection between chains of different abnormalities through common factors.

This analysis seeks to define the regions where barriers are installed to prevent failure factors from continuing to work, generating major incidents or accidents. In the current case, the analysis refers to the chains related to the drying and regeneration task in oil and gas plants. The abnormalities analyzed are part of an investigation already carried out, where the logic diagram is used to mitigate or eliminate failure. It is noted that some of the failures have their origin in organizational issues.

3.2.3.5 Case study—oil and gas processing

In Fig. 3.31, the factors are coded, facilitating the visualization of the relations and their interpretation. Man's fatigue resulting from excess manual services, with high risks of working with LPG, ends up returning and feeding abnormalities. The consequences can be summed up by the loss of productivity in the case of consecutive stoppages, due to oven interruptions and in events of greater impact, the explosion of the oven with an internal atmosphere rich in LPG and the lighting of the oven. If, through preventive actions, the failure processes in: 1Fn4, 7F3, 78nBp, 9/10mB, 23C1 are isolated, the six aforementioned events are simply deactivated.

In Fig. 3.29, the causal factors are divided by size and location, in relation to the phases of the operational failure. Note that in the inertial phase (with low visibility) the following factors are found: 1Fn4 (does not allow pilot lighting), 7F3 (very high process temperature) and 23C1 (improper maintenance). It is noted that in the development phase, factor 78nBp (inadequate flow control) is located and in the risk phase, the operational factor 9/10mB (passage of water for drying or sphere). As for the dimension of these operational factors, we have that: 1Fn4, 7F3, 78nBp and 9/10mB are technical factors, while 23C1 is a factor related to the task.

The root causes are described as follows: inadequate maintenance contract management; lack of commitment to the task; improper operation and maintenance; nonfunctioning of safety instruments for the operation of a LPG vessel; Inadequate CV of control valves since the project (requiring studies on the reasons); load fluctuation in refrigeration; fatigue of operators and maintenance technicians.

4. *Chronology of failure*

3.2.3.6 Case study—oil and gas processing

The occurrence times for each operational factor of the task failure are estimated from data and events in the plant's history, from the discourse of experienced operators and from the analysis of external accidents.

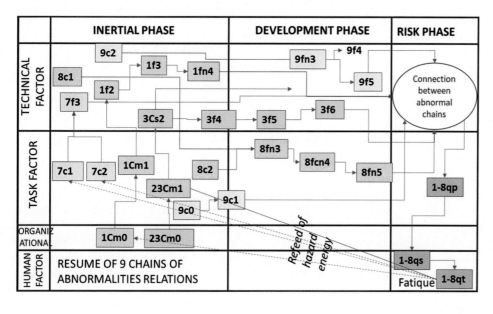

FIGURE 3.31 Logical diagram of the failure involving LPG regeneration. Ávila, S.F., 2010. Etiology of operational abnormalities in industry: a model for learning. (Doctoral thesis), Federal University in Rio de Janeiro. School of Chemical, Rio de Janeiro.

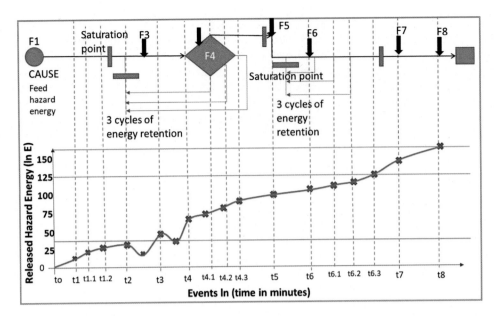

FIGURE 3.32 Chronology and functionality of operational failure factors. Ávila, S.F., 2010. Etiology of operational abnormalities in industry: a model for learning. (Doctoral thesis), Federal University in Rio de Janeiro. School of Chemical, Rio de Janeiro.

The measurement of the hazard energy (strength) released is relative and is related to the approach of the maximum energy point of the failure in the task, for example, a fire-ball that affects the neighboring community. Thus the maximum strength of the failure is indicated when some continuous and discrete variables approach the state of total lack of control, with the fireball licking the houses of the community.

The failure scenario analyzed by the chronology technique (Fig. 3.32) aims to verify times and operational factors that increase or decrease the strength of the failure, and also the format of the failure with time. An exercise is performed, simulating a situation prepared based on events measured with the oil and gas company staff, simulating the joint occurrence of the factors and the estimated times.

The events that took place in the drying and regeneration sections were chosen to illustrate a simulated explosion situation, from the signs of abnormalities, such as the absence of a flame, to corrective measures to reduce the strength of the failure, reaching the explosion of the oven and of equipment through a "domino effect." These events were based on the operator's speech (occurrence book) and on hypotheses of abnormalities.

Due to the specific behavior of the failure in its final stage, where in a very short time, the strength of the failure (dimensionless) increases geometrically, leaving the classification of signals for process uncontrolled and then error in the task, and hence medium and serious events that involve a Bleve (Boiling Liquid Expanding Vapor Explosion). Thus to facilitate the construction of a graph, values of energy released from danger (or failure energy) were used, as a result of the discussion with experienced operators, with the respective logarithm (ln) of the hazard energy released in the failure, against the logarithm time (ln) as indicated in table and figure discussed in Chapter 7, Human reliability: chemicals and oil and gas cases, in oil and gas case.

At the beginning of the diagram, the danger release energy is low over a long period where the failure is represented by the operational deviation factors. At the end of the diagram, the hazard release energy is very high in a short hazard release period, in the accident-disaster phase. It is difficult to carry out preventive or corrective actions in a quick period of time where the most appropriate action is mitigation with the removal of people close to the fireball in a Brazilian village.

5. *Materialization of the Failure in the task, migration of social and technical energy in the task*

In Fig. 3.34 the diagram for the analysis of the materialization of the failure is presented and in Fig. 3.33, the scheme showing the passage of the failure energy from the manager to the operator and to the physical area in the process. As shown in Fig. 3.33, the failure task reveals the social body of that failure as the occasional mental map resulting from the manager's speech, the acceptance by the operator that is influenced by certain organizational environments and regional culture.

Company pressure can generate addictions or inadequate ways of operating (the social body of the failure—the imaginary of the operator) resulting in the low efficiency of the process (physical body with danger). These noises acquired through the social body of the fault (improper communication) are materialized in the panel and field operations, generating process losses.

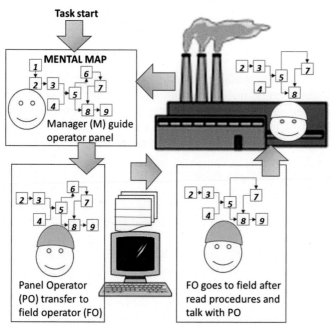

FIGURE 3.33 Failure of social body and physical body. Ávila, S.F. 2013b. Review of risk analysis and accident on the routine operations in the oil industry. In: Proceedings of the Fifth CCPS Latin America Conference on Process Safety. Cartagena de Las Índias.

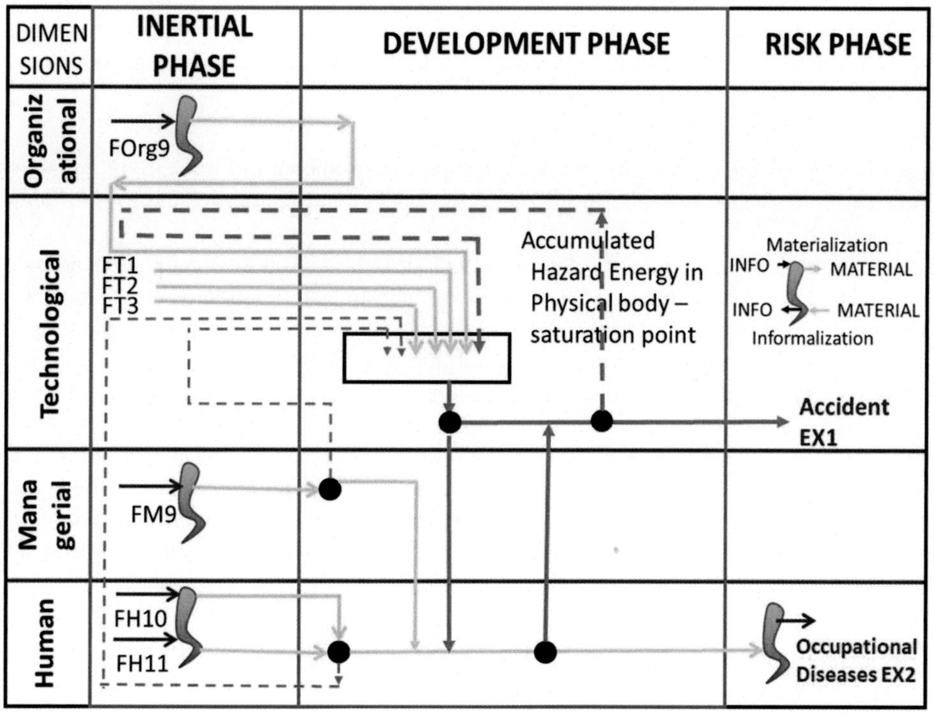

FIGURE 3.34 Materialization and informalization of operational failure. Ávila, S.F., 2010. Etiology of operational abnormalities in industry: a model for learning. (Doctoral thesis), Federal University in Rio de Janeiro. School of Chemical, Rio de Janeiro.

The materialization diagram in Fig. 3.34 indicates the multiple dimensions of the operational factors in the failure process and where the migration of the social body (green) to the physical body located in the equipment and process lines occurs. The danger energy when installed in the social body is caused by the manager's speech, organizational culture or intrinsic variability of human behavior.

In the risk phase, it may be that the danger energy of the physical body that causes uncontrolled product quality, for example, is dematerialized to the image of the company that is considered a social body with danger. Thus the failure's success, regardless of the phase, can be in physical or social body. Preventive, corrective, and mitigating actions must be planned in the correct dimension of the cause causing this failure energy.

3.2.3.7 Operational failure connectivity

According to Perrow (1984), the complexity of the processes is in the multiple process pipelines that connect recycled material or in the multiple interconnections that return or send energy, material or information. When it is not visible, this migration makes it difficult to identify the factors that promote the state of the process. Unlike the methods already discussed, where the task is studied and then the possibilities of failure are verified, this method seeks to identify the signs that indicate the presence of operational, technological and human failure through the operator's discourse and indicated by critical process variables.

To enable the investigation of complex processes, the following techniques and models were presented: (1) analysis of the production system based on the influence of environments (organizational, socioeconomic) on structures (equipment, man, and team) and on managerial and educational processes; (2) analysis of the hazard energy present in the operational factors that make up the fault, its visibility and quality; (3) analysis of operational failure by the logic of its operation within the network topology.

Still in the investigation of complex processes, other complementary analyzes are introduced. The failure analysis by chronology (4) where, for each operational factor the current strength of the failure and the time for this factor to manifest are identified. On the other hand, Materialization Analysis (5) where failure energy migrates from the information present in tasks or standards to changes within industry process equipment, where the appearance of the fault body is shown (fluctuation of pressure, contaminant, reaction imbalance) or in the reverse process, with the informalization of the physical aspects of the operational failure, bringing the result to the informational energy (for example: stress of the operator).

It is worth mentioning that, in a failure process, the manager's discourse transmits the desired standards to be performed, to be transformed into work in the structures of technology (machines and processes). This work is materialized in the process equipment. From the speech of the manager and the leaders, the operator performs the task on the equipment and processes and returns to record his speech on task control in the shift occurrence book, the main database of this methodology.

3.2.3.8 Principles and concepts of connectivity analysis

The connectivity analysis (6) that investigates the complexity of the failure process is shown in Fig. 3.35, in a practical example of failures in the cooling and brine water systems. In this case, the operational factors related to the failures are identified, classified, listed and studied.

The following aspects of the failure are analyzed: type of relationship classified as causal or organic, depending on the level of complexity (number of automatic meshes per manual, number of main lines in relation to utility lines,% of use of the procedure in practice) and the level of involvement of the social group in the technical failure indicating the classification of the failure as human error; type of operational factor with the respective causes (root and secondary) and consequences; points of materialization and informalization of the failure; fault recirculation points, energy, material and informational feedback loops; identification of the signs; existence of the fault; identification of the failure force measurement points; identification of corrective actions (which act on the consequences and not on the causes) and preventive actions (which should act on the root cause); identification of the main operational factors in the process steps and in auxiliary systems with the identification in block diagram. The maximum state reached in the failure process leading to the accident or any other major event where there is a high economic impact or quality of life is also identified in the connectivity diagram.

3.2.3.9 Case study. Cooling system

Failures in the cooling water system can alter the normality of processes causing impacts in the areas of quality (off spec product), safety (obstructions with leakage of flammable product) or deviations in production control (decrease in the quantity produced) (Fig. 3.35).

The type of relationship in the cooling system is classified as causal and organic nexus where human errors that affect the physical system occur. The complexity level is average, resulting from the number of automatic meshes per manual, number of main lines in relation to utility lines, % of use of the procedure in practice. This complexity indicates the presence of all factors: root cause, signal, primary, secondary, and tertiary consequences in addition to a top event. The causes of failure events in the cooling water system are the presence of solid and low flow in the system. The failure is materialized at high CW temperature and high temperatures in the process fluids in the purification and in the caustic scrubber.

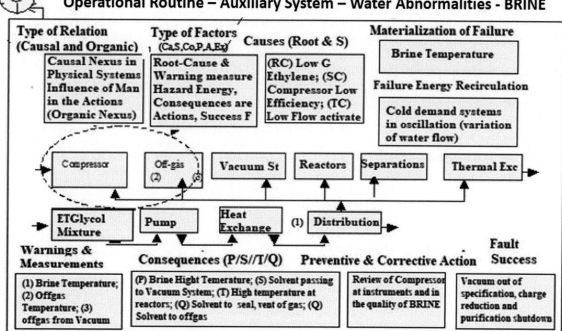

FIGURE 3.35 Analysis of connectivity between operational abnormalities. Ávila, S.F., 2010. Etiology of operational abnormalities in industry: a model for learning. (Doctoral thesis), Federal University in Rio de Janeiro. School of Chemical, Rio de Janeiro.

The failure recirculation points involve the circulation of the process solvent requiring cooler to reach the standards. The signs that indicate the degradation or improvement of the situation are in the temperature. Corrective actions are in improving the quality of cooling water and the success of the failure occurs when the situation causes plant shutdown. These factors are intertwined with the chilled water system. The main operational factors in the process steps are flaws in the quality of CW that affect separation, caustic washer and compression operation.

3.2.4 Decision analysis under stress: emergency simulation

The behavior of the operation team is different between routine moments and emergency situations, where extreme stress occurs, and its consequences for the task, and for the individual, for example, an emergency stop after obstruction with organic (dinitro toluene) in the feed supply pipeline in a sulfuric acid reconcentration plant, in the production process of toluene diisocyanate. The impact can be occupational accident or low availability of plant.

Thus it is proposed to analyze the behavior of operators in high-risk conditions, in search of answers on how to adapt their knowledge and skill in teamwork under this stressful situation.

The methodology is proposed to control the effects of the high state of stress on the body and the cognitive decision. It is intended to analyze whether the knowledge installed in the team is sufficient to resolve the emergency. This method involves tools from the risk area, HF and the role of production in an emergency situation all published in different safety process conferences. Tools:

1. the analysis of the accident by the team based on the routine allows investigating behaviors and management mechanisms, stress dynamics;
2. risk analysis using the failure tree with the calculation of the frequency of human errors, technical factors, accident and disaster after chain reactions;
3. study of defense mechanisms to be installed in equipment, processes to reduce the risk of accidents; and
4. study of safeguards to define the appropriate leaders in emergency and disaster mitigation situations.

The application of this tool requires the construction of an accident and disaster situation from events sequenced in progressive levels of stress, as shown in Fig. 3.36 during a certain period of time (2–4 months). Operational decisions are made for each group of events classified by the level of consequence and stress. These decisions are made by the operation at different levels:

1. ineffective group research on deviations from the routine (here referred to as signs of failure);
2. uncontrolled process that require diagnosis and actions, which in this case, were also not efficient;
3. degradation of equipment caused by the incomplete or deficient task, which unfortunately was ineffective;
4. accident caused by consequences of inappropriate decisions to deal with leaks in equipment that cause explosion and fire. Corrective action not effective; and
5. and finally, with the chain reaction, and errors in the contingency actions, it ends up causing disaster and many fatalities in addition to high property losses.

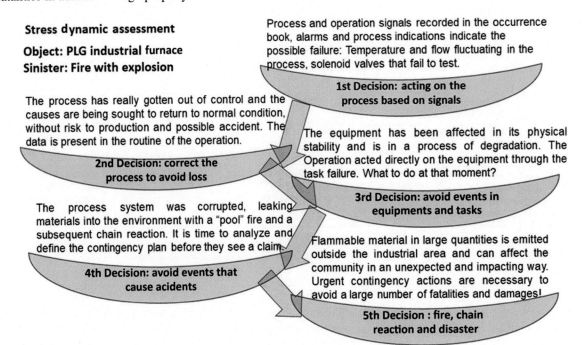

FIGURE 3.36 Risk analysis based on 5 levels of stress and dynamics. *Ávila, S.F., 2010. Etiology of operational abnormalities in industry: a model for learning. (Doctoral thesis), Federal University in Rio de Janeiro. School of Chemical, Rio de Janeiro.*

The first tool aims to measure the behavior, reaction of the team, and also the quality of the proposed actions, to avoid the continuity of the accident process. The second technique aims to calculate the frequency of the accident based on the frequency of the events estimated in the first, to analyze the consequences of the accident for the company and the community. The third and fourth tools, proposes the installation of barriers to prevent accidents at the managerial level and with technological safety devices for industrial installations. It also considers recalculating the frequency of accidents.

The technique for the elaboration of procedures and standards (PADOP) presented by Ávila (2011a), performs the analysis of the critical task added to the environment, the task's functionality, instruments used in production and the barriers intended for safety to perform the task.

This tool has already been applied to the behavior analysis of the emergency team in the following cases: (1) Leakage of LPG with a fireball in Brazil (Ávila, 2011b); (2) Damage caused by gasoline leak in pipes in the United States with group dynamics at the CIDEM Congress in Feira de Santana; and (3) Group Dynamics in the Polytechnic School Building Fire Case at UFBA.

The Scientific and Academic discussion on this subject includes: LODA Tool (Ávila, 2011b) that involves LPG leakage; LESHA tool (Ávila, 2013a) that involves Fatigue in the Mining Industry; and Accident Analysis to investigate cognitive and behavioral gaps in leadership and emergency Staff (Ávila et al., 2018a).

3.2.4.1 Organization of techniques for task review

The construction of standard procedures and the review of the task requires a previous analysis regarding the level of risk and complexity (as indicated in Fig. 3.37) through indicators proposed by Ávila (Ávila, 2011b, 2013a) and by Ávila and Barroso (2012a,b), and by Ávila et al. (2013a), based on information from the Technological-Organizational Project and operational history of the plant. This identification will direct the appropriate tools for the critical task Project and for the review of this project in case of preventive identification of environmental contexts or even after the occurrence of safety, quality, health and production incidents-accidents that occurred in the operational routine. The tools used for failure analysis and the need to review the task are divided into types: Preliminary analysis (risk and complexity as in Fig. 3.37); Context Analysis (definition of general and specific requirements—consider regional aspects); Pattern and Procedure Analysis (cognitive processing with methods to maintain quality in the routine); Task Failure Analysis (in case of failure in a critical task that has a multidisciplinary nature); and Pattern Control involving safety, reliability, production, environmental impact, communication as showed in Fig. 3.38. The numeration described in the list of tools will be applied in chapter 3.2.7 for application in an acid plant. All these analyses are in a dynamic and operational risk context (Ávila et al., 2016a).

Preliminary analysis phase:
Preliminary Analysis of Risk and Complexity to define Criticality: APR, HAZOP, LODA.
Context analysis phase:

1. context analysis in the task environment, requirements—PADOP;
2. human and social typology: cultural bias, archetype analysis, and EF.

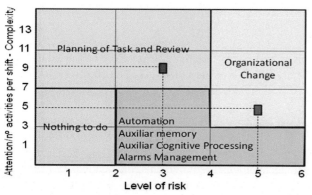
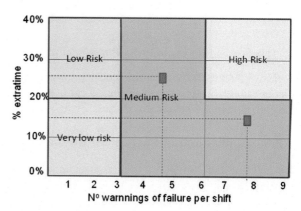

FIGURE 3.37 Risk map operation and task. *Ávila, S.F. 2013b. Review of risk analysis and accident on the routine operations in the oil industry. In: Proceedings of the Fifth CCPS Latin America Conference on Process Safety. Cartagena de Las Índias.*

Control	Measurement	Acceptance Level	Non Acceptance Level
Efficiency	% Target	Above 80%	Below 70%
Safety	P (accident)	Below 10^{-5}	Above 10^{-3}
Reliability	Failure rate	1 per 2 weeks	4 per 2 weeks
Time	Minutes	T activity < 20% T total ; DP/M below 20%	T activity > 50% T total ; DP/M above 50%
Quality	Adequate Purity	90%	70%
Requirement	% attended	Above 80%	Below 60%
Communication	% understanding	> 80%	< 70%
Environmental im.	Quantity* Risk	Below 10	Above 70
Effectivity	No times	Success in 95%	Success in 90%

Control Measure			Review Tipe 1–5 → except	Consensus	Barrier
ACCEPTABLE	INTERMEDIARY	NON ACCEPTABLE	90/100% AC = w1	All	Project
			80/90% AC = w2	Experts	Review
			70/80% AC = w3	Experts	Memmory
			30/40% INT = w4	Supervisor	Personal
			40/50% INT = w5	Manager	Organizational
			NAC Safety = w6	Supervisor	Trainning
			NAC Leak = w7	All	Organizational
			< 60% AC = w8	Experts	Process
			40/50% INT = w5	Manager	Organizational

Control: Efficiency, Safety, Reliability, Time, Quality, Requirement, Communication, Environmental IMP, Effectiveness

FIGURE 3.38 Indicators of behavior in task acceptance levels. Ávila, S.F., Sousa, C. R., Carvalho, A. C., 2015c. Assessment of complexity in the task to define safeguards against dynamic risks. Proc. Manuf. 3, 1772–1779.

Standard analysis and procedure phase:

1. task architecture, hierarchical analysis and CARMAN;
2. risk analysis in the sociotechnical system of the task: PADOP;
3. review of steps, goals and threats: GMTA;
4. chronological analysis of the CARMAN task; and
5. analysis of complexity, dominance and PADOP parallelism.

Failure analysis phase:

1. failure analysis: logic, materialization and connectivity: PADOP safeguards;
2. failure analysis: stress dynamics, failure chronology and LODA failure tree.

Standard control phase:

1. Standard and review in PADOP task control.

Fig. 3.39 presents criteria for choosing the appropriate tools to review the task that this involved or caused the deviation or failure. Scripts are established according to the complexity and risk of the failed task. The 10 review tools are in the context, failure and control phases. The tools are referenced from numbers 1–10.

3.2.5 Application of PADOP for the sulfuric acid plant case

3.2.5.1 Introduction

PADOP (Ávila, 2010, 2012; Ávila et al., 2015a,b,c) checks the technological requirements, the complexity of the task, discusses the risk and barriers in industrial environments. However, the main constraint is not the validated standards for quantitative analysis of risk and complexity. Ávila indicates the need to consider the environment dynamic, where nonvisible new threats appear and impact cognitive processing, altering the efficacy and effectiveness of this task. The

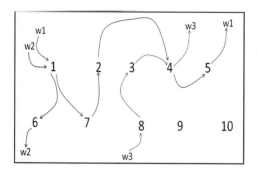

FIGURE 3.39 Route for revision of critical task. *Ávila, S.F., Sousa, C. R., Carvalho, A. C., 2015c. Assessment of complexity in the task to define safeguards against dynamic risks. Proc. Manuf. 3, 1772–1779.*

Risk Assessment on the Environment of the task, analyzes the environmental requirements for the task, from the tooling, standard, knowledge and commitment of the operator.

The fundamental requirements are discussed for the elaboration of control techniques applied in the execution of the task: level of physical conditioning, type of information for authorizations in the partial stages and the task as a whole. The ideal state to be reached by the task includes aspects of process, production, quality, health, and time.

The standards should be revised from environmental indicators that alter the achievement, risk and complexity, requiring new levels of attention and increasing physical and cognitive efforts. The responsibility for the control, monitoring diagnosis and planning the task depends on the level of competence and emotional stability. A requirement for the working efficiency is a good level of communication and cooperation within the group and between groups. The evaluation of cognitive aspects such as perception of risk, quality of cognitive processing, priority in the choice of alternatives and avoiding cultural biases at the moment of decision is essential for positive results in the production.

The goal is to discuss and define the tools to control the process and the devices to control security that can control-avoid the event of the accident, production stoppages, and also the loss of quality of the product. The discussion of the task's functions indicates what are the probabilities of failure of the safety, process and production safeguards.

The safeguard project depends on the level of complexity and the organizational-human risk. The informal and formal rules established in routine depend on restrictions from the process, equipment and product. Safeguards are designed after understanding and complying with technical and social restrictions, such as stress. The objective of the critical analysis of the task is to indicate the risk of inefficiency based on the complexity of execution and control of the procedure. This complexity is analyzed by Ávila (2010) in the analysis of complex processes of tasks and their social relations (Ávila, 2012). To establish the appropriate barriers, we seek to understand the influence of physical and cognitive efforts on human error. This analysis aims to establish a minimum level of energy attention to accept project decisions. Achieve this minimum quota may be impossible with the parallelism high, high risk and high physical and cognitive effort. Finally, after this analysis, safeguards are established in the areas of automation, HMI, redundancies and management.

In Fig. 3.37 a risk map is presented, which incorporates performance indicators and compliance analysis as to acceptable levels for different types of tasks. The result is presented in the form of matrices that indicate (1) the level of technical risk from the level of attention and complexity estimated by the number of steps of the task and (2) the number of deviations in the 8-hour shift from the percentage of extra work time in production.

As shown in Fig. 3.38, the indicators for task compliance can include: efficiency, safety, reliability, time, quality, customer requirements, communication, environmental impact and effectiveness. For each of these efficiency performance indicators, there is a standard adopted for "acceptance" and "nonacceptance." In the case of percentage of requirements met, with 80% acceptance and 60% nonacceptance.

As shown in Fig. 3.39, for each set of performance indicators a type of review is chosen, decision-making consensus and what barriers to avoid repetition. If the initialization procedure of the sulfuric acid stripper, is below 90%, for example, it will follow the route w1 of application of the tools mentioned in 3.2.6 through the following script: $1 => 7 => 2 => 4 => 5$. If it were route w2, the revision includes tools 1 and 6. If it were route w3, the revision includes only tools 8, 3 and 4.

3.2.5.2 Operational context of the case of sulfuric acid facility

The sulfuric acid plant case (Fig. 3.40) was based on a real application considering the need to review the procedures due to equipment failures and constant production shutdowns. The complexity of routine operations was discussed. The purification of a liquid stream of sulfuric acid containing about 83% acid, 14% water, 2% dinitro toluene and 1%

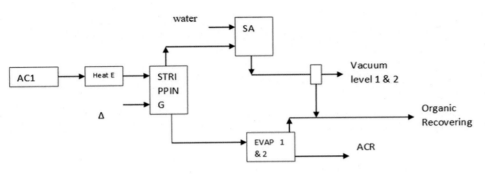

FIGURE 3.40 Sulfuric acid plant. Ávila, S.F., 2004. Methodology to minimize effluents at source from the investigation of operational abnormalities: case of the chemical industry. Dissertation (Professional Master in Management and Environmental Technologies in the Productive Process) — Teclim, Federal University of Bahia. 115p.

nitrous gases is done by evaporation, in successive steps, leaving the acid purified (96%). The sulfuric acid purification process involves high temperature, vacuum, field work and control room adjustments (Ávila, 2004).

The hazards associated with materials and products are:

1. risk of thermal or mechanical shock with damage to glass equipment;
2. blockage of organic and iron sulphate glass tubes in pumps and tanks;
3. risk of bodily injury in direct human contact with acid and risk of pollution with nitrous gases;
4. obstructions and holes in the equipment and leaks in the pumps;
5. damage to vaporizer column fillers; and
6. damage to product tanks with an environmental impact on effluents.

The project did not provide devices to treat problems in the effluent due to equipment failures. The managerial profile is centralized and the maintainers do not have an industrial culture, but a banking background. The lack of feedback causes omissions in critical procedures. There are occupational problems, since the effort is high and the risk of acid burn as well.

The position of not investigating the effect of variation of the raw material in the breaks of the glass equipment was adopted. There is no practice of collecting historical data for operational research events, and no similar records in other installed plants. No analysis of the root cause of equipment breaks. The use of inadequate Personal Protective Equipment (PPE) and the consistency of breaks promote an environment with a sense of fear due to the burning of chemicals. The lack of trust between the executive group, employees and managers makes it a difficult environment to research the causes. A learning cycle has not been established.

The excess of redundant systems without addressing the cause of downtime increases operational complexity, which makes it difficult to solve the problem. Although the level of automation was average for the simplicity of the process, the existence of contaminants generated a new situation that hinders the establishment of a standard based on best practices. The procedures were not followed due to variations in the quality of the raw material. Booklets with drawings that simplify the action are not used. Failure to follow the same processes during periods of normality.

Main restrictions for performing the task with sulfuric acid: constant stops, lack of presence of the operator for field actions; and high contamination of processing materials. Organizational restrictions are conflicts between policies and practices in high and middle management and no social rituals were performed to aggregate the team.

The impact of the hot acid accident is serious and can cause fatality. The time available to carry out control actions is reduced. Intuitive actions did not bring results and were not applicable because there were constant fluctuations in the quality of raw material and the process. The procedures only occurred in normal hours. These factors generated a state of fatigue of the operators, hindering analysis and decision making as to the best actions to be taken.

3.2.5.3 Task and barriers analysis

The task analyzed in this case is the startup of the stripping column in the sulfuric acid plant. For correct execution, a requirements analysis for this operation is necessary, with the gutting column starting. In addition, there are requirements that must be met, such as: training on care in operating with sulfuric acid and the use of appropriate PPE, training in product handling and the possibility of thermal shock. It is worth mentioning that a checklist should be established to improve visual inspection. This checklist should be leaner compared to the procedure.

The Planning and Action of the procedure depend on the analysis of chronology, function and type of activity taking into account variations due to contamination in the raw material. At the startup of the column, alignments and sampling are discussed due to the presence of organics with periodic inspections of the equipment, process and product.

In the case of task architecture for the preparation of the good practices manual, specifically for this purification column, the operation involves the removal of organic gases and direct contact with steam and evaporation of water.

The procedures should address each operation: (1) handling requirements and care; (2) startup; (3) normal operation; (3) shutdown; and (4) emergency stop. In addition, checklists are needed to make decisions about the suitability of the product at each stage of the process.

In the task design, the steps are established in the application order: (1) water alignments for exchangers and systems; (2) condensate drain; (3) flow and temperature control considering "manual" and automatic interlocks; (4) operation of the acid pump and recirculation of the tank; (5) inspection of equipment and instruments; (6) adjust the manual flow control valve and monitor leaks or fluid dynamics; (7) adjust the steam valve manual and follow events; (8) adjust the temperature from the preheater to the stripper; (9) increased acid flow; (10) gradual adjustment of the steam flow to the stripper; (11) temperature control at the cooling outlet; (12) adjusting the temperature profile; (13) sample collection; and (14) alignment for purification.

In the design of the equipment, limitations on the perception and speed of human action should be considered in the state of normality and in relation to the state of abnormality where the control of anomalies is sought. Based on this context, it is recommended to reassess the HMI, the procedures that involve human perception (visual booklets) and the review of priorities in alarm management. The investment in personal protection achieved an improvement in the behavior of the operation team who were afraid of personal damage in the control of the process and equipment. Adequate training for the level of automation of routine and emergency controls; review of organizational rituals; feedback and communication to increase commitment to the result.

The Task in the design phase is critically reviewed considering general and specific requirements, partial goals, detailed and simplified steps, barriers for the control and safety of processes and care to avoid abnormalities. Redundancy protection systems are reevaluated to reduce contamination of the raw material. The use of appropriate PPE and routine plant inspection can decrease occupational events at the plant. The investigation and recording of sulfuric acid processing, relating past events of leaks, obstructions and breakages of equipment, can assist in the future decision. Specific actions are taken: delegation of responsibility; acquisition of new technological tools for monitoring; managerial actions to increase the team's level of commitment and cooperation; new rituals to integrate the team and reduce omissions.

In the task environment, it is important to: implement motivational campaigns to communicate events; establish criteria for team priorities; attraction of informal leaders to improve operational decision. In cognitive processing, try to check the task agenda to suggest ways to prepare the mind map for the performer. Study of interpretation of the procedure from a sample in the routine—check understanding.

Problem Analysis in a clear and consistent manner and the risk analysis in the procedure allowed to reduce the fear of accidents. The effectiveness of the task comes from a review of the quality criteria, timing and impacts of the events. There must be an analysis of compliance with indicators that guide the best revision to be made after the failure.

The investigation of the signs of failure in the routine enable the study of the failing task through techniques developed by Ávila (2010) such as chronology analysis, functional analysis, logic, connectivity and materialization of the failed task. These techniques were applied experimentally to discover the root-cause of major glass tube breaking events in this purification.

3.2.5.4 Critical analysis

An increase in attention directed to product quality, safety and handling requirements will bring comfort to the team in carrying out the tasks. The systematic study of procedures, technical standards and organizational norms complemented by the respective training will facilitate the understanding of the requirements. The review of the stages of the starting procedure of the sulfuric acid unit, gutting column, in consensus with the formal and informal leaders of the shift crews, considering the situations of normal and emergency operation, brought greater comfort to the team regarding the routine standards with significant improvement in the organizational climate.

The analysis and discussion of the impacts of the work of manual and automatic actions on equipment and products indicates the importance of maintaining control safeguards (process, quality, safety) in normal operation. The constant consultation of feedback from the operation imposes radical changes in the communication structure of the leaders. The analysis of the complexity in the task environment indicates the strong influence of cultural traits, the tendency to disaggregate values, the failures in leadership regarding the use of incorrect signifiers for failure meanings. This linguistic phenomenon that is part of inadequate archetypes for operational discipline does not guarantee the effectiveness of the task and promotes failure.

The influence of factors that affect human performance for the acid case is summarized in Fig. 3.41. These include: (1) lack of security; (2) pressure to perform the task; (3) stress; (4) lack of competence; (5) failure to understand the root cause of the problems; (6) lack of commitment; (7) lack or diversion of attention; (8) difficulty of meeting the agenda; and (9) low perception. Organizational and workstation factors also influence this mechanism. These factors depend on the company's management profile or history: (1) number of accidents; (2) inadequate PPE; (3) frequent equipment failures; (4) budget cuts; (5) lack of investment; (6) difficulty in receiving the supplier, due to import delays;

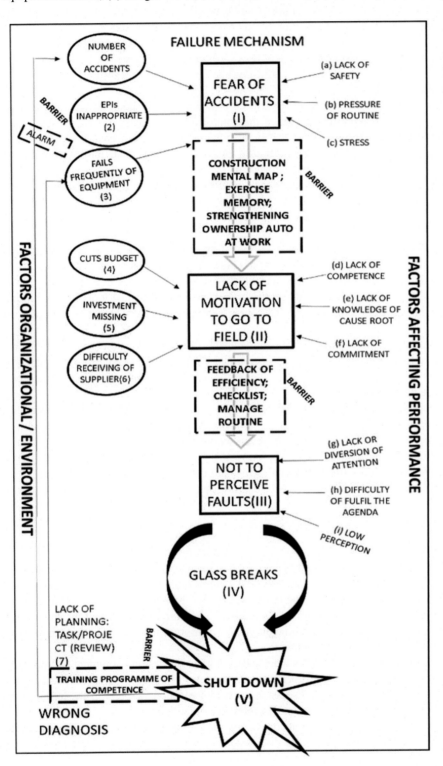

FIGURE 3.41 Systemic failure mechanism in sulfuric acid plant. Ávila S.F., Santino, C. N., Santos, A.L.A., Fonseca, M.N.E., 2016b. Analysis of cognitive gaps: training program in the sulfuric acid plant. In: Proceedings of the European Safety and Reliability Conference (ESREL), 25–29 September 2016, Glasgow.

(7) lack of planning for the task and the project (review). These cognitive and organizational factors cause the plant to stop, generating operational costs and, with the familiar cognitive gap, deliberate on diagnostic errors, activating the failure cycle again (Ávila et al., 2016b).

3.3 Discussion about API 770

The previous discussion on human and social typologies, study of the environment, analysis and task control in industrial facilities with a high level of risk and complexity, concluding with the real case, already published and investigated by the authors, brings up the discussion on what are the guidelines, used by the chemical industry, to prevent accidents caused by human errors.

The document API 770 presents the discussion on the factors of human performance and indicates a script of studies to identify which are the main dangers that can be released in the organization, technology, group and individual set.

In this chapter, we intend to discuss the concepts of API 770 related to the previous concepts and retrieving cases already discussed in the chemical industry and related services. In the end, after the critical discussion, the themes were related, thus building an "anatomy" about causes and consequences involving the failure process.

3.3.1 Concepts and assessment from API 770

Accidents in chemical and oil installations contaminated the environment, killed hundreds of people and caused damage to the enterprises. Thus studies and guides dealing with this subject must take into account human errors and human reliability analysis (HRA). Human errors are present in the design, construction, operation, maintenance and administration of companies and are generally the main causes of product quality problems, production losses, energy losses, and accidents.

Managers believe that just by training workers, the operation of any operating system will be adequate and that human errors are the result of carelessness or ignorance. Consequently, to reduce human errors it is necessary to blame and discipline.

On the other hand, managers who incorporate the topic of human reliability in their management know that most mistakes are made by skilled, careful, productive, and well-intentioned employees. Therefore they try to identify the essential causes of error in the work situation and implement the appropriate corrective actions.

Human factor engineering, and ergonomics is an area of science and technology that includes what is known and theorized about human biological and behavioral characteristics that can be validly applied to the specification, design, evaluation, operation and maintenance of products and systems (Attwood et al., 2007). It complements other engineering disciplines that aim to improve equipment performance without regarding how the equipment is operated and maintained.

3.3.1.1 Are managers capable or not?

Managers are capable or able to use techniques to improve human performance, through the identification and elimination of situations with probable errors that may not be so obvious because they are not visible.

The strategy is to reduce the frequency of human errors by applying HF engineering principles: equipment, tasks, and the work environment. For this approach to be successful, it is essential that the workers themselves be involved in the design process. From the management point of view, it is necessary to consider the knowledge of workers, which is a valuable resource that already exists within each company. The strategy will be most effective when workers and HF experts are involved in the project. This strategy can be difficult if Regional and Organizational Cultural Biases are strong.

One of the tools that managers can use to reduce failures is the Human Reliability Analysis (ACH). Similar to other assessment tools (for example, fault tree analysis), ACH can provide both qualitative and quantitative information. Qualitative results identify the critical actions that a worker must take to perform a task correctly, identify erroneous (unwanted) actions that can degrade the system, identify likely error situations, and identify any factors that could mitigate human errors. Quantitative results are numerical estimates of the likelihood that a task will be performed incorrectly or that unwanted actions will be performed. What is the best way? Quantitative or qualitative?

According to API770, the factors that influence the environment in which tasks are performed are divided into two groups: (1) situations that alter the individual's cognitive processing and decision and (2) alter the performance of the task on equipment and processes. In this document the main situations that can cause human error are presented

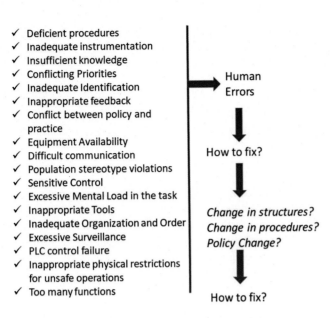

FIGURE 3.42 Situations that cause human error. *Adapted by Ávila, S. F., 2010. Etiology of operational abnormalities in industry: a model for learning. (Doctoral thesis), Federal University in Rio de Janeiro. School of Chemical, Rio de Janeiro.*

(Fig. 3.42) and in the manuscript, we will criticize and confirm this description. API770 did not present ways to correct these errors and mentioned that changes in structures, procedures and policies are needed.

Situational characteristics are workstation architecture; physical environment risks; hours of work versus interval; organization of shift and administrative work; availability and adequacy of equipment, instruments and supplies; organizational structure with authorities, responsibilities and communication channels; actions developed by people, by type of function and by interested parties; preparation of standards; company policies; and human resources organization.

The characteristics related to task planning and performance of procedures on processes and equipment are the following: written and unwritten procedures; written and unwritten communication; working methods; needs for greater perception and attention; physical needs for speed and strength; need for previous actions; interpretation for execution and decision; complexity regarding the amount and type of information retrieval; memory operations in the short and long term; auxiliary cognitive processing through auxiliary calculations; types of HMI; characteristics for control from sample (information, situation, or text); feedback quality; critical task analysis capacity; frequency, repetition and learning when performing the procedure.

In Fig. 3.42, the main situations that cause human error are suggested. For each situation, we consider the size, frequency, impact, complexity, and investment required considering the risk analysis for two plants already presented, the sulfuric acid purification and the oil & gas processing plants.

Risk analysis considers the classification of frequency and impact for each plant and then squared when complexity occurs (low and high) in Table 3.8. The frequency and impact values range from 1 to 3, and when the plant has complexity in the process, task and social relations (Ávila, 2012), the result is squared. Then letters are launched that represent investment level at three levels, low when revisions are needed (B), medium when there is a change in Technology (M), and high when it involves changing the organizational and security culture (A). The yellow color refers to characteristics of the sulfuric acid plant and the orange color refers to characteristics of the oil and gas plants. The human error situations are related to performance factors and which have high investment are for the sulfuric acid Industry (1, 3, 6, 7, 9, 10, and 12) and for the oil and gas industries (6 and 10). The risk matrix is constructed by product type and has values between 1 and 81, represented by Fig. 3.43, where the highest risks are in the Acid Plant due to impacts, these being Factors 6, 8, 9, 10, and 14.

3.3.2 Analysis of the API 770 survey

The standard is divided into sections and appendices. Section 3.1 discusses the importance of improving human performance, and Section 3.2 defines human error and the most common causes. Section 3.3 identifies specific factors in the workplace that increase the likelihood of human error and discusses ways to improve human performance. Section 4 describes several ACH techniques available for a numerical estimate of the likelihood of human error and how these

TABLE 3.8 Characteristics of risk, complexity, and investment in barriers.

Human error	Frequency	Impact	Risk	Complexity	Power		Investment
1. Deficient procedure	3	3	9	B	B	9	4	A	B
2. Inadequate instrumentation	2	3	6	B	B	6	4	B	M
3. Insufficient knowledge	2	3	6	A	A	36	9	A	B
4. Conflicting priorities	3	3	9	B	B	9	4	M	M
5. Inadequate identification	2	1	2	B	B	2	4	B	B
6. Inappropriate feedback	3	3	9	A	A	81	36	A	A
7. Political & practice conflict	3	2	6	A	A	36	16	A	M
8. Equipment availability	3	3	9	A	B	81	2	M	M
9. Difficult communication	3	3	9	A	A	81	4	A	M
10. Violation of populational stereotype	3	3	9	A	B	81	6	A	A
11. Very sensitive control	1	1	1	B	B	1	4	B	B
12. Excessive mental workload	3	2	6	A	B	36	4	A	M
13. Inappropriate tools	2	2	4	A	B	16	2	B	B
14. Undue order and organization	3	3	9	A	B	81	4	B	B
15. Excessive surveillance	2	3	6	B	B	6	2	M	B
16. PLC control failure	3	2	6	B	B	6	4	B	B
17. Inappropriate physical restriction	2	2	4	B	B	4	4	B	M
18. Over-function	2	2	4	B	A	4	16	M	M

	Sulfuric acid	PLGas	Sulfuric acid	PLGas	Sulfuric acid	PLGas
HIGH RISK (60-81)					6, 8, 9, 10, 11	
MEDIUM RISK (22-59)			3, 7	6		
LOW RISK (1-21)	11, 17, 18, 5	1, 2, 4, 5, 9, 11, 12, 14, 16, 17; 8, 13, 15	2, 4	3, 7, 18; 10	1	

FIGURE 3.43 Risk matrix for sulfuric acid and oil and gas plants. *Author's original.*

techniques can be used. Finally, Appendices 1 and 2 contain self-assessment questionnaires, and Appendix 3 provides an example of HRA. The main topics covered in the API770 self-assessment questionnaires are discussed.

3.3.2.1 Policy analysis and guidelines

Analyzing management's commitment to workers' health and safety requires knowledge about the tools, policies, practices, and the team's perception of risks. Organizational policies generally clearly communicate the strategic guidelines; however, in the management of the production routine, the need to meet billing goals can generate ambiguities in the management's discourse. Thus the team's commitment and confidence in management can be diminished. This is a latent movement and difficult to be perceived, which makes it difficult to define programs for adaptation.

The group of executors and leaders generally believe that the company has constancy of purpose (Deming, 1990) in terms of prioritizing safety over production. If this trust does not occur, much information does not flow in a true way and a false environment appears. This false projection of reality serves to control the risks of loss of power in organizations.

Integrated safety and health management uses procedures and records to achieve the objectives of regulatory requirements. However, these patterns can work anomalously by directing attention to nonexistent dangers without prioritizing the real dangers that are not perceived in the routine. Near misses may not be discussed if the perception of danger is not being trained.

A major difficulty in industries is to recognize the importance of studying the cognitive process as much as studying the industrial process. Both are part of the technology. Thus the job post is analyzed for knowledge by the level of training required and the adaptation process. The analysis of decision making under stress and the stability of the psychological contract over time are not considered. If attention, memory, perception, mind map, stress studies existed, it would be possible to remedy deficiencies based on HF engineering.

In new projects, normally, there are no criteria that address the factors that alter human performance in carrying out the task. The parameters are functional and consider that the risk of man-made mistakes are minimal, even in the face of a universe of digital information invading a simple mind map. Companies routinely are not concerned with having among their specialists a team that deals with HF in cognitive processes, which makes the analysis of causes diffuse, before and after operational failures. In some cases, there is concern, but there is no periodic review of standards related to HF.

The need to prepare specialists in HF, including presenting how cognitive engineering is integrated into the design of productive processes as well as the elaboration of procedures, is latent. These professionals would also contribute to the demand for technicians who can identify error situations from the signals emanated from the process and the team

It is noted that the culture of guilt is still dominant in management practices. Errors are discussed reactively and not naturally. Workers' failures are treated as evidence of incompetence, and some requirements of the legislation indirectly promote operators' distrust when investigating serious failures. When the investigation does not address the various HF involved, there will be a superficial analysis without reaching the root-cause that is in a latent area, that is, of low visibility.

Data on human errors in the operation task are not collected widely. The observations are centered only on the location of the incident and do not cover the volume of control of the investigation in the face of a social bond that involves the event. Data are investigated, usually, after the accident and leads the manager to make decisions that are neither efficient nor effective.

3.3.2.2 Analysis of assignments in tasks

The start-up of an industrial plant is carried out following procedures indicated by the project area and adapted by the operation's staff. After a certain period of operation in the industry, some tasks are identified as critical, as they can cause incidents/accidents with negative environmental and occupational impacts. It is of utmost importance that a detailed analysis of these tasks be carried out, both in physical and behavioral aspects.

Actions or devices to reduce the likelihood of failure or to mitigate the consequences of human errors can be suggested. Operator attention is an extremely important aspect, so double shift and overtime hours must be well managed. The rotation in carrying out the task serves to level the workloads and to increase the experience of the workers.

Another way to classify the activity as critical is based on intensity, repetition, and cases where the lack of history does not allow a more careful analysis. Qualitative analysis does not offer certainty regarding the definition of the level of risk. It is important to measure the workload for each environment. It is also relevant to map the behavior of the operation's personnel regarding the effect of stress on task effectiveness.

3.3.2.3 Analysis of the human machine interface

HF engineering and cognitive engineering analyze the interface instruments between man and machine, enabling solutions for inadequate postures and movements (discussed in criteria of anthropometric ergonomics). The analysis regarding the usability of the instruments in the console (layout, type of indication, sequence, grouping, and cancellation of the action) and the flow of people in emergency situations are important for the control of industrial processes.

Generally, the control center project does not balance the number of signals offered by the project to control the processes, the number of signals that can be analyzed by the team, and the number of signals needed to investigate the problems of the process.

The process control project needs to be analyzed: accessibility, ease of use, respecting regional characteristics, manual adjustments in routine and emergency. The necessary changes to adapt the instruments for use must occur at the design and preoperation stage, without which there will be problems in the control of the process.

The automatic safety devices installed since the project must be reconciled with the decision model and the possibility of cognitively monitoring the actions. Without reconciliation, failures in the automatic system can lead to serious problems. Usually, the interconnections of the control systems make the flow of information so complex that it is impossible for the operator to understand. The usability of process and safety control instruments requires procedures and tools for: repair, deactivation, calibration, alarm settings, and controls.

The physical risk of accidents with products and equipment is reduced with the proper installation of safety barriers and with the appropriate training for their use. In the chemical and petroleum industry, it is often overlooked that operators are "afraid to go out into the field," both in routine tasks and in emergency situations. A preventive action to mitigate the risks of improper handling is to maintain the appropriate signage for equipment and accessories.

The design of the communication tools in the workplace is important to ensure the effectiveness of the team task at different times: in the passage of service between shifts, in the instructions between the staff and the group of the relay shift, in the instructions between different departments, during emergencies (evasion, community, authorities). The knowledge base of the team in the shift may not be sufficient for the decision making of critical situations. The communication process is important to guide consultation with staff, manager, and specialists. In this way, unnecessary stops resulting from emergencies not properly treated are avoided.

3.3.2.4 Analysis of written procedures

A set of updated procedures must be available to deal with risky events that impact safety, quality and the environment. Updating the procedures is part of the operation's work, since the processes and the functioning of the equipment are dynamic. A concern in the analysis of the procedures is regarding the perception of signs for unknown events and the lack of perception of signs that are classified as of low importance because they are known and repetitive events.

When there is an incompatibility between the complexity of the procedure and the workers' knowledge, it is necessary to have changes in writing due to lack of understanding. This incompatibility can be in several areas: language codes, established concepts, available experience, and others.

The technical culture established in several chemical process and manufacturing industries, is not to observe the written procedures in their entirety. Therefore the review must be carried out and can be demanded by the team or from a pre-established agenda. The lapse of a procedure usually occurs when changes to processes or equipment are made. Some procedures do not need to be detailed due to the level of competence of the team and for being routine activities.

A commonly used procedure is permission to work on equipment. The risk analysis of all work permits is considered tedious work, generates inattention and the possibility of accidents. A different posture makes it possible to judge the need for risk analysis with two possibilities: (1) the analysis would not be done because it is a routine action and with established competencies and commitments, or (2) the risk analysis would be done because it considers the service to be critical with possibilities of causing accidents. An uninteresting situation is when the contractor is afraid to perform the service after the work permit is filled out.

In the project of the procedure, it is necessary to analyze whether there are parallel actions and which manual and automatic controls are necessary. In each critical procedure, additional tools can be used for verification and control, or for auxiliary memory. Procedural steps or phases are planned through sequence analysis, authorizations, requirements, goals and impacts.

3.3.2.5 Analysis of the worker

The schedules for rotation of the shift are adjusted to minimize the interruption of circadian rhythms. This planning for maintaining the circadian cycle is lost when there is unscheduled work, or double shifts for participation in social activities. Thus the fatigue of workers can occur without due analysis.

It is important to carefully relate the demands for the job to establish the metrics of the tests and selection criteria of the workers. Among these criteria are skills, aptitudes, experiences and others. To increase the performance of these selected workers, it is important to establish an adequate training program with aspects related to concepts, failure processes, phenomenological principles and aspects of social relations at work.

The monitoring of the consequences of stress on health should be as important as the functional monitoring of workers' health in the company, after all it is part of the context of the individual's mental health status and the consequences. Thus periodic examinations would indicate the need to adjust the level of stress, preventing emotional imbalance and low psychological quality of the worker, which impacts production and the safety of the company and of the workers.

References

Aplin, L.M., Farine, R.D., Mann, R.P., Sheldon, B.C., 2014. Individual-level personality influences social foraging and collective behavior in wild birds. Proceedings of the Royal Society B: Biological Sciences 281, 1789. Available from: https://doi.org/10.1098/rspb.2014.1016.

Astakhova, M.N., 2015. The curvilinear relationship between work passion and organizational citizenship behavior. Journal of Business Ethics 130 (2), 361–374.

Attwood, D., Baybutt, P., Devlin, C., Fluharty, W., Hughes, G., Isaacson, D., et. al., 2007. Human factors methods for improving performance in the process industries. CCPS, Center for Chemical Process Safety.

Ávila, S.F., 1995. Organization under transformation. Quality Control, São Paulo, 33, pp. 44–46.

Ávila, S.F., 2004. Methodology to minimize effluents at source from the investigation of operational abnormalities: case of the chemical industry. Dissertation (Professional Master in Management and Environmental Technologies in the Productive Process) – Teclim, Federal University of Bahia. 115p.

Ávila, S.F., Pessoa, F.P, Andrade J.C., Figueiroa, C., 2008. Worker character/personality classification and human error possibilities in procedures execution. In: Proceedings of the CISAP3–Third International Conference on Safety and Environment in Process Industry, Rome, pp. 279–286.

Ávila, S.F., 2010. Etiology of operational abnormalities in industry: a model for learning. (Doctoral thesis), Federal University in Rio de Janeiro. School of Chemical, Rio de Janeiro.

Ávila, S.F., 2011a. Assessment of human elements to avoid accidents and failures in task perform, cognitive and intuitive schemes. In: Proceedings of the Seventh Global Congress on Process Safety, Chicago.

Ávila, S.F., 2011b. Dependent layer of operation decision analyzes (LODA) to calculate human factor, a simulated case with LPG event. In: Proceedings of the Seventh GCPS Global Conference on Process Safety, Chicago.

Ávila, S.F., 2012. Failure analysis in complex processes. In: Proceedings of the Nineteenth Brazilian Congress of Chemical Engineering. Búzios, Rio de Janeiro.

Ávila, S.F., Barroso, M.P., 2012a. Human and social risk preliminary analysis (HS-PRA), a case at LPG Site. Global Congress on Process Safety, Houston.

Ávila, S.F., Barroso, M.P., 2012b. Social HAZOP at oil refine industry. Global Congress on Process Safety, Houston.

Ávila, S.F., 2013a. LESHA—multi-layer progressive stress & impact assessment on health & behavior. Global Congress on Process Safety, San Antonio.

Ávila, S.F. 2013b. Review of risk analysis and accident on the routine operations in the oil industry. In: Proceedings of the Fifth CCPS Latin America Conference on Process Safety. Cartagena de Las Índias.

Ávila, S.F., Pessoa, F.L.P., Andrade, J.C.S., 2013a. Social HAZOP at an oil refinery. Process Safety Progress Journal 32 (1), 17–21. Available from: https://doi.org/10.1002/prs.11552. March 2013.

Ávila, S.F., Silva, C.C., Sousa, C.R.O., 2013b. Worker behavior changing by dynamic culture. In: Proceedings of the Ninth Global Congress on Process Safety (GCPS). San Antonio, Texas.

Ávila, S.F., Costa, C., 2015. Analysis of cognitive deficit in routine task, as a strategy to reduce accidents and industrial increase production. Safety and Reliability of Complex Engineered Systems, London 2837–2844.

Ávila, S.F., Menezes, M.L.A., 2015. Influence of local archetypes on the operability & usability of instruments in control rooms. In: Proceedings of the European Safety and Reliability Conference (ESREL), Zurich.

Ávila, S.F., Carvalho, A.C.F., Portela, G., Costa, C., 2015a. Learning environment to take operational decision in emergency situation. In: Proceedings of the Sixth International Conference on Applied Human Factors and Ergonomics and the Affiliated Conferences, 26–30 July, 2015, Caesars Palace Hotel, Las Vegas, Nevada, USA.

Ávila, S.F., Carvalho A.C.F., Kalid, R., Sousa C.R.O., 2015b. Influence of Brazilian culture. In: Proceedings of ABRISCO Conference, 23–25 November 2015, Rio de Janeiro.

Ávila, S.F., Sousa, C.R., Carvalho, A.C., 2015c. Assessment of complexity in the task to define safeguards against dynamic risks. Procedia Manufacturing 3, 1772–1779.

Ávila, S.F., Ferreira, J.F.M.G., Sousa, C.R.O., Kalid, R.A., 2016a. Dynamics operational risk management in organizational design, the challenge for sustainability. In: Proceedings of the Twelfth Global Congress on Process Safety, Houston.

Ávila S.F., Santino, C.N., Santos, A.L.A., Fonseca, M.N.E., 2016b. Analysis of cognitive gaps: training program in the sulfuric acid plant. In: Proceedings of the European Safety and Reliability Conference (ESREL), 25–29 September 2016, Glasgow.

Ávila S.F., Dias, C., 2017. Reliability research to design barriers of sociotechnical failure. In: Proceedings of Annual European Safety and Reliability Conference (ESREL), 27, 2017, CRCPRESS, Portoroz.

Ávila, S.F., Ávila J.S., Drigo, E., Cerqueira, I., Nascimento, J.N., 2018a. Intuitive schemes, procedures and validations for the decision and action of the emergency leader, real and simulated case analysis. In: Proceedings of the Eighth Latin American Conference on Process Safety, 11–13 September 2018, Buenos Aires, Argentina.

Ávila, S.F., Cerqueira, I., Drigo E., 2018b. Cognitive, intuitive and educational intervention strategies for behavior change in high-risk activities-SARS. In: Proceedings of the International Conference on Applied Human Factors and Ergonomics. 21–25 July 2018. Springer, Orlando.

Ávila, S.F., Souza, M.L., Ramalho, G., Passos, I.C, Oliveira, A.R, 2018c. Analysis of the Reliability Model for interdependent failure mode, a discussion of human error, organizational and environmental factors. In: Proceedings of the Tenth National Congress of Mechanical Engineering (CONEM). Rio de Janeiro.

Ávila, S.F., Mrugalska, B., Wyrwicka, M.K., Souza, M., Ávila, J., Cayres, E., 2019. Cognitive and organizational criteria for workstation design. International Conference on Applied Human Factors and Ergonomics. Springer, Cham, pp. 161–173.

Barkley, R.A., 2010. Differential diagnosis of adults with ADHD: the role of executive function and self-regulation. Journal of Clinical Psychiatry 71, e17.

Barros B.T., Prates M.A.S., 1996. The Brazilian style of managing. Atlas, São Paulo.

Bhui, K.S., Dinos, S., Stansfeld, A.S., White, P.D., 2012. A synthesis of the evidence for managing stress at work: a review of the reviews reporting on anxiety, depression and absenteeism. Journal of Environmental and Public Health 2012. Available from: https://doi.org/10.1155/2012/515874.

Bonfim, F.G., 2014. Perspectives on Lacan's writing: "the meaning of the phallus.". Revista de Psicanálise. Analytica 3 (5), São João del Rei, Minas Gerais. Brasil.

Boyeur, G.C., Sznewar, L.I., 2007. Enation and work process: an actuationist approach to operative action. Gestão & Produção, São Carlos 14 (1), 97–108.

Broughton, E., 2005. The Bhopal disaster and its consequences: a review. Environmental Health 4 (1), 6.

Catarino, F.B., Gilbert, P., Mcewan, K., Baiao, R., 2014. Compassion motivations: distinguishing submissive compassion from genuine compassion and its association with shame, submissive behavior, depression, anxiety and stress. Journal of Social and Clinical Psychology 33 (5), 399–412.

Day, D., Fleenor, J.W., Atwater, L.E., Sturm, R.E., Mckee, R.A., 2014. Advances in leader and leadership development: a review of 25 years of research and theory. The Leadership Quarterly 25 (1), 63–82.

Deming, W.E., 1990. Quality: The Management Revolution. Marques Saraiva, Rio de Janeiro, RJ.

Drucker, P.F., 1967. Out of the Crisis. Guanabara, Rio de Janeiro, 166p.

Embrey, D., 2000. Preventing Human Error: Developing a Consensus Led Safety Culture Based on Best Practice. Human Reliability Associates Ltd, London, p. 14.

Fleck, M.P.A., Lafer, B., Sougey, E.B., Porto, J.A., Brasil, M.A., 2014a. Guidelines of the Brazilian Medical Association for the treatment of depression (complete version). Brazilian Journal of Psychiatry 25 (2), 114–122.

Fleck, M.P.A., Lafer, B., Sougey, E.B., Porto, J.A., Brasil, M.A., 2014b. Guidelines of the Brazilian Medical Association for the treatment of depression (complete version). Brazilian Journal of Psychiatry 27 (11), 24.

Guimarães, A.M.V., Neto, A.C.S., Vilar, A.T.S., Almeida B.G.C., Fermoseli, A.F.O., Albuquerque, C.M.F., 2015. Anxiety disorders: a prevalence study on specific phobias and the importance of psychological help. Undergraduate Notebook-Biological and Health Sciences-UNIT-ALAGOAS, 3, 1, pp. 115–128.

Handy, C.B., 1978. Understanding organizations. Zahar, Rio de Janeiro, 498p.
Hollnagel, E., 1993. Human reliability analysis context and control. Computers and People Series. Academic Press Inc, San Diego, CA.
Huang, J.L., Ryan, A.M., Zabel, K.L., Palmer, A., 2014. Personality and adaptive performance at work: a *meta*-analytic investigation. Journal of Applied Psychology 99 (1), 162.
Jung, C.G., 2002. Psychic Energy, 8th (ed.) Vozes, Petrópolis, Rio de Janeiro.
Klein, M., 1987. Selected Melanie Klein. Simon and Schuster, Juliet-Mitchell.
Lacan, J., Miller, J.-A., 1988. El seminário de Jacques Lacan: Libro 7, La ética del psicoanálisis, Paidós, 1959−1960.
Lapassade, G., 1989. Groups, Organizations and Institutions. 3rd (ed.) Francisco Alves, Rio de Janeiro, 316p.
Lees, F.P., 1996. Loss Prevention in the Process Industries: Hazard Identification, Assessment and Control, 2. Butterworth-Heinemann, Great Britain, GB.
Malloy-Diniz, L.F., Sedo, M., Fuentes, D., Leite, W., 2008. Neuropsychology of executive functions. In: Fuentes, D., Malloy-Diniz, L. F., Camargo, C. H. P., Cosenza, R. M. (Eds.), Neuropsicologia: Teoria e Prática. Artmed, Porto Alegre.
Mezan, R., 1986. Freud Culture Thinker, 4th (ed.) Editora Brasiliense, 680p.
Motta F.C.P., Caldas, M.P., 1997. Organizational Culture and Brazilian Culture. Atlas, São Paulo, 325 p.
Motta, P.R., 2004. Contemporary Management: Science and the Art of Being a Leader, 15th (ed.) Record, Rio de Janeiro, 256p.
Llory, M., 1999. Industrial Accidents, the Cost of Silence. Multiação, Rio de Janeiro, p. 333.
Perrow, C., 1984. Normal Accidents: Living with High-risk Technologies. Basic Books, New York, NY, p. 453.
Pichon-Rivière, E., 1988. The Group Process. Martins Fontes, São Paulo.
Renesch, J., 1993. New Traditions in Business. Cultrix, São Paulo, p. 246.
Ribeiro, A.C.O., 2012. Quantification of the Impact of Human and Organizational Factors on Human Failure Probabilities used in Probabilistic Security Analysis. (Doctor's thesis). Federal University in Rio de Janeiro, RJ, Brazil.
Sennett, R., 2005. Corrosion of Character. Record, Rio de Janeiro, 204p.
Souza, AD, 2002. Stress and work. Course Conclusion Paper (Specialization in Occupational Medicine). University Society of Estácio de Sá. Campo Grande, Mato Grosso do Sul, Brazil, 68p.
Stefano, F., Maia Jr., H., 2015. Brasil takes beating from the United States in productivity. <http://www.exame.abril.com.br>.
Sternberg Jr, R., 2008. Cognitive Psychology. Artmed, Porto Alegre.
Vil, S.L., Rodrigues, S.M., 2016. Leadership of Industry Leadership.15(1), 147−154.
Zardasti, L., Yahaya, N., Valipour, A., Rashid, A.S.A., Noor, N.M., 2017. Review on the identification of reputation loss indicators in an onshore pipeline explosion event. Journal of Loss Prevention in the Process Industries 48, 71−86.

Chapter 4

Process loss assessment

4.1 Context

The industry encompasses many processes and operations that require cost control and loss reduction. Competitiveness and sustainability are stimuli for these issues, resulting in process loss control. The application of these methods and techniques is possible, after the root causes of problems in the human and technological dimensions is investigated. Each loss in the industrial environment involves increased production costs, environmental impact, and social conflict in the operation and support team (Sartal et al., 2017).

Process loss management is the responsibility of the operational team. Turning experiences and skills into specific knowledge helps shape production systems and can result in improved productivity. Often, specific technical skills are not available, and existing ones depend on the motivation to be transformed into productivity gains and for reducing the production costs.

Reliability analysis improves plant productivity by identifying the root cause and its ramifications. Thus it is possible to change technologies, products, and processes to meet the demands of new environmental legislation that include the challenges of saving water, using renewable energy sources, and reducing overall losses.

Productivity improvement programs help frame critical variables in the 99.95% range, between the maximum and minimum allowable limits. Cleaner production programs and social responsibility programs have driven the industrial sector to improve its processes. The chemical and petroleum industry are being required to implement the programs, such as responsible care, to meet society's demands.

Framing the process and the product variables depends on investigating the causes of human errors and applying techniques that enable greater effectiveness in performing critical tasks. Process losses should be discussed from technical, economic, and human factors to propose objective functions followed by each tool or method. Several losses in industrial environment need investigation:

- Leaks from during processing products or in final product;
- Damage to assets due to uncontrolled proceedings or due to improper operations;
- Plant shutdown leading to loss in profit and increased production cost;
- Uncontrolled process and operations leading to poor product quality;
- Difficulties in meeting production on time;
- Loss of company image because of accidents with environmental or occupational impact, thus affecting the value of the company in the stock market;
- Billing losses due to product price reduction resulting from poor quality and delays, and;
- Absenteeism due to occupational diseases.

These factors and technical-economic impacts were the result of human action inserted in an organizational environment. The causes of loss may involve aspects such as poor risk perception, incorrect cognitive decision, or omission resulting from high stress situations.

Productivity of the industry depends on the reduction of losses present in the operational routine involving cultural, social, and human issues with low visibility. Current management models make it difficult to visualize the chains of abnormalities embedded in complex models. This chapter describes how to identify fragmented data over time or in production areas to avoid process loss.

Statistical monitoring of events and critical process, production, and maintenance variables may indicate possible abnormalities. Its mapping enables the indication of the appropriate location and barrier to be installed to avoid routine losses. Process monitoring is part of the loss investigation methodology and describes data from the operation routine to reduce production variability and the occurrence of abnormal events. According to Rasmussen (1999), social variables involve the analysis of the task and its procedures regarding requirements, steps, memory, training, and flexibility level accepted in the standard to avoid failure.

Process losses can be prevented through advanced knowledge of the causal link that includes intermediate factors until operational failure occurs. It is necessary to identify the obvious and, also, the latent causes related to the operation of equipment, organizations, and people that participate in the construction of the process loss.

What are process losses?

Process losses can be of several types: material, equipment, time, image, sales, and motivation. Some of these losses are described and evidenced as losses that lower human reliability, that is, that are not easily measurable, because they are located in the latent failure zone. Reasons for human error are interconnected with subjective issues which are not easily recognizable; thus not every company knows how investigate.

Process losses can be classified as materials, assets, time, quality, image, occupational events, and human errors. Thus these are discussed in detail:

1. Losses of materials may result from product leaks in equipment: in the production area; from truck tipped over; and from inadequate operating conditions causing loss of pressurized gases or spill of liquid;
2. Loss of assets, as in the case of accidents or incidents in the industry, cause the stopping of large compressors or even rupture of vessels or tanks. Loss of assets is considered when equipment or facilities are damaged in incidents and accidents;
3. Loss of time, when there are shutdowns of industrial units due to leakage, accidents, or failure of equipment. This loss of time can also be caused by delays in the release of equipment or its return from maintenance, and may also result from failures in the performance of the task;
4. Losses in the quality of the product and difficulties in meeting the deadline contribute to reducing billing and to the increase of fines, reduction in the final price or, in the worst case, loss of customer;
5. Loss of the company's image due to accidents that end up being reported by the media. It is also considered a loss of image when the employee is dissatisfied with the company, generating labor actions, or "negative campaigns" of the employee or partner in relation to the company where he works; (Fig. 4.1)
6. Losses related to the worker's absence due to occupational diseases and accidents.

Fig. 4.2 indicates the process losses classified in omage, production, and personnel fronts. After the implementation of this program, losses can be reduced in chemical, oil & gas, and electricity power industries.

Human reliability, as discussed in Chapter 2, has an important and intangible element of loss. Low motivation generates human errors that can cause process losses and promote the incorrect way of proceeding in the routine of the operation. This fact leads to low productivity and along with the lack of a diagnosis that visualizes the failure in the routine, causing excessive energy (physic and psychic) expenses and increasing production cost.

Answer the questions... Did you forget any process loss? Do you disagree that these are Process Losses?

_____
_____
_____
_____

Why investigate the routine to avoid process losses?

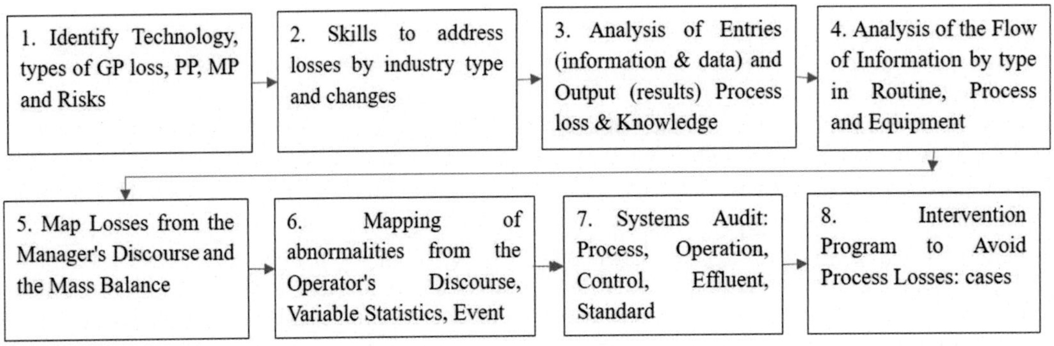

FIGURE 4.1 Gold map to process loss investigation. *Author's original.*

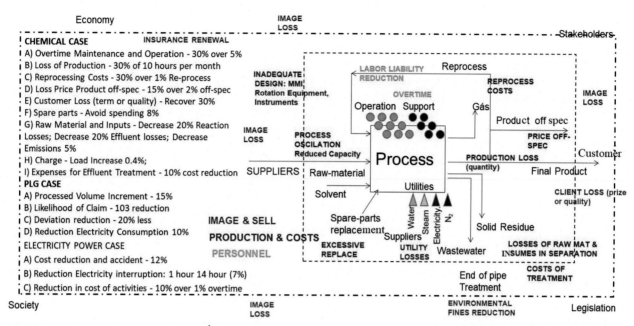

FIGURE 4.2 Losses by class in the process. Ávila, S.F., Santos, A.C.N., Pessoa, F.L.P., Andrade, J., 2011. Risk management and prevention program to avoid loss of process (LP3) based on human reliability, practical application and measurement. In: 7th Global Congress on Process Safety, Chicago.

The industry should avoid process losses due to the current conditions of competition in the globalized market. Process losses can comprise tangible and intangible factors that subtract the company's results. Tangible factors allow the quantification of losses, and intangibles are more difficult to measure and treat, such as human motivation. The causal nexus of operational failure is not always visible and can direct work in areas of low impact on loss reduction. Investigation of process losses becomes clearer when operational routine information is verified (process variables and shift book) to achieve better economic-financial and image results.

Answer the questions... In your opinion, what is the importance of routine for studying the causes of process losses?

_____
_____
_____
_____
_____
_____
_____

Do process losses reach any type of industry? Are there differences by industry segment? Do industries in all basic, intermediate, and manufacturing sectors need to assess their process losses and seek corrective and preventive measures?

Every industrial process has the possibility of losses due to changes in working standards or the quality of raw materials. Another possibility that can lead to process loss is the change in management policies and practices caused by organizational changes. Any industrial segment is prone to loss if it does not adjust its processes to environmental-social fluctuations and changes in the characteristics of the materials being processed, as well as the materials of the equipment with their degradation. The unstable process certainly causes losses of various origins in industrial sites.

The possibility of process losses changes depending on the type of fluid, the value of pressures and temperatures required, and the relationship between each other, the variation and equipment limits. The level of complexity of certain technologies may not make it possible to control actions due to improper interpretation of the procedure. Specific knowledge and skill requirements limit the availability of people to perform activities, so the lack of competency can lead to human error and loss.

Process losses also depend on the organizational climate. Interpersonal relationship difficulties can lead to serious losses, such as a fatal accident. This event can occur due to multiple omissions, poor communication quality and even the presence of dangerous product, service or process.

Differences between organizational policies and operational practices indicate the frequency of deviations in the routine. Company standards can differ from your team's willingness to design and deploy solutions.

Communication and feedback within the organization are critical to aligning the operation with management guidelines and organizational policies. The organization must be in line with the demands of local legislation and internalize as requirements in operating procedures. Finally, industries from all sectors need to assess their process losses and seek corrective and preventive measures.

Answer the question… In your opinion, which types of industries have the most losses? What losses?

_____
_____
_____
_____
_____
_____
_____

What is the relationship between process losses and the execution of operating procedures? Technical solutions (process control devices or equipment in the technology) "guarantee" the non-existence of process losses.

Operating procedures guide industrial operation and enable the training of operators, regardless of the type of technology. Industrial plants have a high level of automation, so they are less likely to have a human error. On the other hand, they have high impact due to dealing with a greater amount of hazard. Industrial plants with low level of automation, low quantity of product and higher probability of human error, normally have lower severity in events. Human errors can trigger process failures and losses, in a large-scale and high complexity industry, that can cause incidents, accidents and even disaster.

Some labor-intensive processes require simple procedures such as logistics, transportation, storage, and sampling. All liquid material transformed in the industry are transported to vessels through rotating equipment and piping. Solid material is loaded by conveyor belts or trucks to the silos. Transport, storage and sampling are labor-intensive operations despite the use of automated control devices. The probability of human error should be analyzed in the manufacturing, assembly, production, operation, maintenance and project design processes (which requires specific skills in applying methodologies and modeling). It is noteworthy that the influence of human activity on processes will depend on the type of technology in each industry and its state of degradation.

The automated industry decreases the possibility of human intervention. Maintenance activities and projects are labor intensive. According to Mosier and Skitka (2018), the automated operations in plants controlled by few people with high competency and great responsibilities has low quantity of human errors, on the other hand, another reality is that human errors in automated plants with large production scales have major impacts when process losses occur.

The industrial plant designer adopts various criteria and assumptions to avoid process loss events. The experience of having worked in the operation, operating in the same region with local characteristics makes it easy to understand the application of intrinsic criteria in the project. The participation of the operation team in Project design is essential for adjusting equipment, process control and safety technologies, increasing the likelihood of success in the operation phase.

Part of the "unspoken" knowledge of routine when perceived and transformed into environments with good plant operational standards enables adjustments to the acquired technology. If the adjusted operating standards are not compatible with local reality, the performance achieved by the industrial plant may not reach the state expected by the project. The team mastering of some important operational aspects permits the delivery of technology. Then, the measurement of control items, the productive capacity confirmation, and good result from quality audit, the routine events check, and the validation of no occurrence of environmental and occupational impacts are confirmatory aspects of technology delivery.

Known plant conditions in the startup process prevent the emergence of latent failures due to lack of adjustment in the socio-technical systems in organizational environments. The growth of these latent failures can initiate unknown future process losses.

Mastery of technology from the project and operational knowledge allows the best operability of the equipment. It will be possible to propose programs focused on reliability, predictive maintenance, predictive and specialized studies of maintenance engineering. In addition, we will implement programs that will control human factors to reduce the probability of errors in critical activities in order to avoid the process losses.

Answer the question... According to your experience or knowledge, indicate a critical procedure and its process losses...

_____
_____
_____
_____
_____
_____
_____

How to detect, evaluate, or diagnose process losses? Are there any specific techniques?

The way to detect process loss will indicate the effectiveness and efficiency of the treatment to reduce it. Specific investigation and process mapping techniques need to be applied to increase the visibility of abnormal routine occurrences that are not flagged by fault indicators such as alarms or trend charts.

Failure mapping in the industrial process is a diagnosis that depends on the type of technology, the team involved in production and the environment in which the industry operates. By drawing failure process mapping through chains of abnormalities at industry, it will be possible to initiate an audit & action program to prevent process loss, including human aspects that affect task effectiveness.

Answer the question... What are the appropriate diagnostics for preventing and correcting process losses?

_____
_____
_____
_____
_____
_____
_____

4.2 Competencies to assess process losses

The industry is always searching for programs to reduce process losses, new sources of energy and water, as well as reuse of water and materials. The low availability of nonrenewable resources due to natural or geopolitical reasons, cyclical climate movements, and the possibility of the greenhouse effect drive the industry in search of solutions.

The best programs to prevent process losses are guided toward achieving the requirements from international standards through techniques and practices in the industry. Authors (Ahidar et al., 2019; Neto et al., 2019; Poveda-Orjuela et al., 2019; Tavares, 2019) discussed these requirements and applications of ISO Standards in Quality (9001), Environment (14001), Safety & Health (45001), Risk (31000) and Energy Efficiency (50001) with these objectives.

The knowledge necessary to develop and implement this process loss program in the routine are:

- Understanding the tools to identify, prevent, and avoid process losses;
- Statistical monitoring of quality in processes, utilities, effluents, and products;
- Management of corrosive processes;
- Mathematical tools to predict future failure situations;
- Simulation of processes;
- Energy and mass integration with data reconciliation;
- Reliability centered maintenance;
- Human reliability analysis;
- Production planning and control analysis;
- Analysis of the critical task (operational procedures to reduce the risks of accidents);
- Projects for reuse and reprocessing of water currents.

The expected result with the incorporation of this knowledge is to avoid process losses and achieving increased productivity, increase in gross profit, and improvement of the company's image for stakeholders.

4.2.1 Premises and competencies

The competency to perform activities in the programs that intend to avoid process losses depends on technicians that have had coexistence with the "factory floor", besides having the capacity to systematize data to prepare diagnoses that allow assertive actions (Cerezo-Narvaez et al., 2017). The mapping of failures can be difficult when there is no knowledge about the tools to investigate the failure in the industry. Mainly, when it involves multivariate statistics or even lack of registration of the occurrences of the operation. The lack of experience in the interpersonal relationship makes it difficult for the operation team to implement the actions on the factory floor.

A program to prevent process losses should be prepared based on information gathered in the areas of process monitoring/control, reliability of critical equipment and human reliability. The method depends on the assertiveness with which the industrial processes are mapped regarding the failure. The mapping of failures is a dynamic process diagnosis and depends on the type of technology, the team involved in production, management, and the industrial environment.

After the diagnosis, the program to avoid process losses can work in a standardized way. Diagnosis and implementation require the participation of consultants to establish and adjust the appropriate tools and standards for the work.

In a scenario of industrial competitiveness, improvement must be continuous to follow technological changes and the resulting social conflicts. It is necessary to discover new competency gaps in order to, constantly, empower teams, from the operational to the managerial level. The development of skills is essential in the industry, to the extent that it contributes to the training of people and the change in attitude toward work practices, or even to the perception of the new reality considering dynamic risks. This fact adds value to the organization and the teams.

As indicated in Fig. 4.3, the competencies and experiences required for the mapping of the technical-social process are as follows:

- Ability to perceive signs & warnings of technical and human processes in the industry routine;
- Multidisciplinary knowledge in the technical (engineering) and human (management and psychology) areas;
- Knowledge of restrictions regarding competitiveness in the industry;
- Experiences in the management area;
- Knowledge and practice in human reliability (human factors) and processes;

Executors of the process loss program need the following skills to perform the required actions:

- Knowledge of tools for data collection and measurement of results;
- Good interpersonal relationship;
- Knowledge and experience in contract administration;
- Awareness of the technical team;

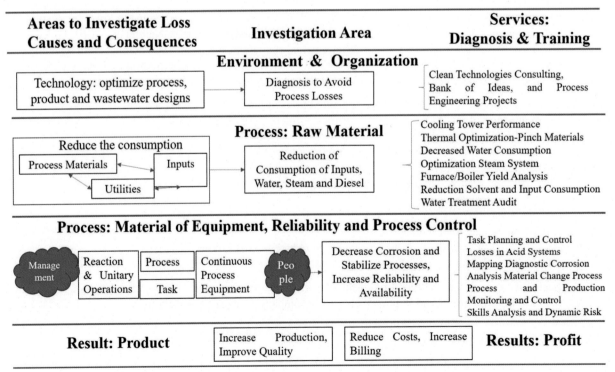

FIGURE 4.3 Process loss skills. *Author's original.*

- Knowledge and experience in writing procedures;
- Knowledge in the area of statistics;

4.3 Losses in the process industries

The discussion of process losses by industry type enables us to anticipate the data to be collected, the diagnoses to be made, and the results to be measured. These industries are of continuous process and most involve chemical processes. General considerations are made depending of industry types and what are the potential for process losses. The priority will be to describe the characteristics in the continuous industrial plants, in the segments of the oil, gas, chemical, petrochemical, biofuels, polymers, and metallurgy industry. It is noteworthy that the cases in this Chapter were based on private chemical and metallurgical industries and public oil industries.

The raw materials are different in physical state, corrosion level, toxicity and production scales for each industrial segment. Industrial process characteristics may influence the type and amount of process losses (Jarvis and Goddard, 2017). For example,

- The level of specialization of the worker;
- The quantity of automation involved;
- The impact on the environment and society;
- The type of equipment, its operation and construction materials;
- The organizational culture and the managerial profile that acts controlling the production;
- Operational procedures and team behavior in emergencies.

We will begin with an initial discussion about these technologies classified as of continuous process and based on physical and chemical transformation.

The oil and gas industries include processes with fluids in liquid and gaseous phase, oil-water separators, and high flows in refining production areas. The scale of oil production depends on the size of the exploration field. In the gas industry, the analysis of losses of industrial gases such as nitrogen, oxygen, carbon monoxide, hydrogen and hydrocarbonates are considered of utmost importance. These losses directly influence the economic and environmental aspect of the industrial gas business.

Naphtha derived from oil is the main raw material for the petrochemical industry. It is cracked and separated into several other streams to supply the petrochemical, chemical and polymer industry. In the petrochemical industry, BTX derivatives are sent to the chemical, polymer and paint sectors, among others. The petrochemical process involves flammable gases in which any leakage can cause fire and explosion. Therefore, these plants usually are classified as highly hazardous.

The active principles manufactured in the chemical industry are used in cosmetics, food, cleaning, and other industries. Polymers such as polyethylene and polypropylene are prepared from petrochemical processes to be used in the transformation industry to produce plastic bags and bottles.

These bottles are used for liquid contents prepared for the cleaning industry, for example. Although there are processes that handle solid and gaseous products, this industrial chain works mainly with the liquid phase. Thus transport, storage and processing consider the need for level measurement, pressure and temperature monitoring. The main fluid conveying equipment is the centrifugal pump, although centrifugal and reciprocating compressors are used in many industrial plants.

Some of the polymerization reagents have high toxicity, requiring escape plans and care to avoid problems with human health and, as already seen, impacting on company image losses. From the moment petrochemicals turn into polymers, hazardousness decreases and care to avoid process losses focuses on the handling of solids, batch control, logistics, nonconforming product control and reprocessing. Products handled in the chemical industry may have high toxicity compared to the polymer industry; having a high flash point, which indicates high flammability and probability of fire explosion.

The chemical and physical transformation industries produce fewer toxic products when compared to the basic chemical industry. Material losses in cleaning, cosmetics, paints and pharmaceuticals can be caused by uncontrolled reaction stoichiometry and by inadequate automation.

The processes with medium pressure ($5-10$ kgf/cm^2) have possibilities of leaks and the products vaporize quickly. The teams have variable skills and need to adapt quickly to new challenges. The economic situation makes the chemical industry smaller and more vulnerable to market fluctuations. This scenario indicates discomfort for production managers in cash flow management, due to constant corrective actions for business maintenance.

The oil-petrochemical-chemical-polymers supply chain is inserted in a socioeconomic-environmental condition. Product and energy inventories in the oil chain were reduced to give way to the distribution of electricity from wind and solar energy. The Petrochemical Industry seeks solutions to replace the mineral matrix (oil and coal) by animal and vegetable matrix (renewable) as indicated in Fig. 4.4. These solutions involve new logistics to face geopolitical, social and economic conflicts.

The metallurgical industry has high work temperature and needs detailed procedures and with signaling of critical activities. Metallurgy offers several benefits to society, producing consumer goods and civil construction products.

In the development of process loss diagnostics, it is necessary to discuss the risks, technologies, regional culture and types of expected losses. Presented below are some questions to contextualize the discussion on losses in the technological chain of oil, bioenergy and metallurgy. The risks of losses are transformed into losses during the description of the technologies.

Questions by subject:

Can the risks of failure in industrial activity and critical tasks, the impacts of hazardous product handling and process controls to prevent loss of standard in quality and safety create a context of high stress? Consequently, is there a possible task environment with scenarios prone to human error? Give an example of this situation.

_____
_____
_____
_____
_____
_____
_____
_____

There is a situation in the operational routine, where managers do not know the reality of their team? Where the team doesn't get involved because they are composed of different groups? Different people mean different tribes, or even people without tribes, where, workers not adopted by the classic tribes of the routine can cause human error. What do you think? Can, this situation prepare a context of low confidence in the group work, or even, no equivalence

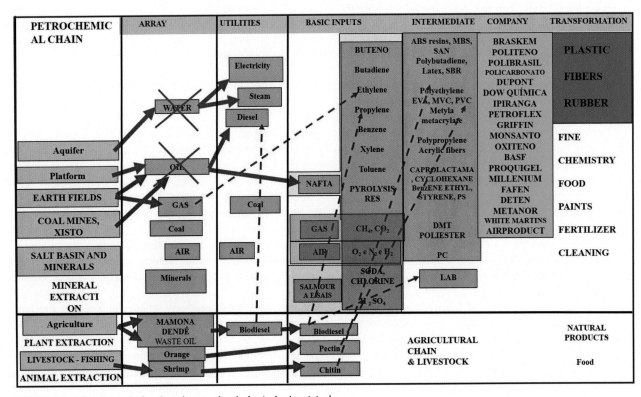

FIGURE 4.4 Mineral matrix for plants in petrochemicals. *Author's original.*

between planners and executors, making it difficult to clarify the shift changeover? Present how this situation can cause an accident in Petrochemical Technologies or in others.

Do the environmental challenges lead the operation to work with new technologies and new conditions with reduced availability of energy, water and with restrictions on the quantity and quality of products? Do these restrictions, promoted by the non-availability of production factors, lack of new knowledge, necessary for changes in the middle of the production process, affect the operational routine. How? Develop a scenario of this type...

We can analyze a situation where strategic leaders (with good and bad habits) are very different from operational leaders. This situation brings a distance between beliefs and makes it difficult to compromise between the organization and the individuals within critical environments. In the case of oil refining, with high flows and temperatures, and the possibility of emissions of greenhouse gas: to what extent can you live in a situation of this type within an industrial environment? Is this a matter for the production or management of the company or only for the supervisors? Do you have examples of this situation, and in the metallurgical industry, high temperatures at workstation, how would it be? Discuss...

4.3.1 Process losses in the oil industry

Modern civilization began industrialization at scale due to the availability of energy from oil, coal, and gas. After 200 years of this beginning, population growth and the unbridled consumption of these fossil fuels significantly reduced mineral energy stocks. The maintenance of current levels of oil and gas consumption in contrast to the reduction of available inventories of this raw material and energy (geopolitical and environmental issues) raises the importance of energy supply activity and demands of maximum reliability and availability of assets.

Process loss in the mineral energy industry is inconceivable due to the decrease in inventories and the increase in production costs. Loss in the oil industry means less availability of naphtha for the petrochemical and chemical chain, or even diesel for the transportation sector (Jarvis and Goddard, 2017). The incorporation of renewable energy matrices and the concern to save energy help the management of scarcity crises of water, energy and nonrenewable materials. Countries have sought to incorporate renewable energy sources into their energy matrix, as well as to implement activities to avoid process losses.

Oil production and extraction divided into offshore (sea), onshore (land), and subsea (under the sea) as shown in Fig. 4.5. Oil industry processing is divided into downstream (field extraction), midstream (refining and storage), and upstream (distribution). Oil extraction is a "dirty" process with separations of water and oil, presence of sulfureous molecules, presence of toxic elements, difficulties in water treatment, and transportation through oil and gas pipelines, requiring specific unit operations to avoid contamination.

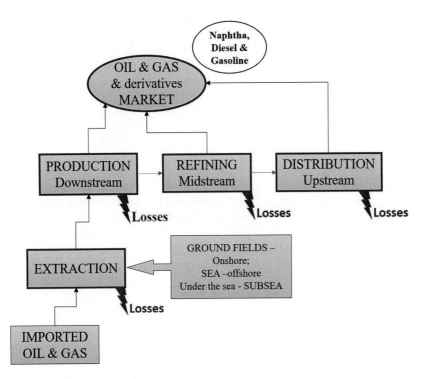

FIGURE 4.5 Oil industry classifications. *Author's original.*

The frequency of leakage in offshore oil extraction and production has decreased. On the other hand, the severity has increased. The same scenario occurs in the transportation of oil using vessels in import-export operations. The environmental implications are less severe and more frequent in onshore production. The physical area where leaks may occur is large due to the long extension of fluid transport from the field to refining. Pipelines cross various agricultural properties, with risk of environmental contamination and damage to population.

Oil production (downstream)

Oil fields can be located close to urban or rural areas in onshore production. The production fields need a certain uniformity in relation to operational safety. The company's image before the community demands technological and political compliance, therefore, it seeks to: meet the demand for energy products and the petrochemical chain; be aligned with new technologies to avoid environmental impacts and energy losses; and work routinely to reduce production costs. In order to avoid the occurrence of probable events with high impact, it is important to have the environmental awareness of the teams and the knowledge about the technological tooling. Nowadays, we also consider the ethical positions toward population demands.

Onshore fields operated by small private companies are more likely to have environmental impacts due to human error although less severe than offshore fields.

As the physical distance from operational managers is large, the level of employee commitment to work or to the company may be low. Therefore, the outcome of the work also depends on the motivation to meet the delegated function.

The level of automation in mature fields is low (onshore), and several operations are carried out manually following old procedures and operational instructions. The number of human errors that occur onshore and downstream is probably higher than the number of errors that occur in midstream refining.

Crude oil is a viscous product, with high saline content, and mixed with water. Handling this product requires careful alignments, separation operations and corrosive processes. The higher efficiency of oil separation in the field provides better quality raw materials for refining, as well as decreases the corrosive attack on product transfer pipelines and decreases the likelihood of leaks.

Case study

The onshore maintenance team travels 8 hours in a small airplane or in a large car to embark to an oil production field, in the interior of the state. The maintenance team travels 6 hours by ship or 4 hours by helicopter to embark on offshore oil production platforms or FPSO ships. The rules of embarkation, the requirements for embarkation of these teams, the delays and preferences result in the separation of the production groups employed by the company from the maintenance groups that are mostly subcontracted.

Does this "caste" separation situation have consequences for the work environment that requires large displacement? Is it possible that a worker who has to dismantle a large valve will suffer an accident as a result of failure to consult documentation, perhaps due to the stress of non-rest that is a requirement or rule for starting activities in onshore production? What is your perception of this situation taking into account the risks of process, product, materials and equipment? What are the types of losses and their causes for this situation?

_____
_____
_____
_____
_____
_____
_____
_____

Refining (midstream)

The risk in oil processing at refineries is high due to the possibility of fugitive emissions, combustion, explosion and contamination. High temperature and pressure are physical properties of many unit operations.

The refining units indicated in Fig. 4.6 have high flow rates where most of the processes are technologically known. The level of automation is high and the possibilities for human error are smaller. Unless unexpected events and consecutive failures of the control instruments occur.

Residual or output currents at the bottom of secondary distillation columns have high viscosity. LPG (liquefied petroleum gas) and gasoline currents are highly volatile, and can cause fugitive emissions due to maintenance fault in gasket installation and/or wrong gasket specifications in rotating equipment. The petrochemical industry, which will be addressed later, presents similar problems to those that occur in oil refineries, such as: explosion, fire and toxicity, but with less product circulation (lower flow) than a refinery.

The main reasons that cause the oscillation in oil quality are the following: (1) the change in the type of oil used in production, which depends on the origin; (2) changes in the efficiency of separation operations in the field; and (3) contaminations caused by the transfer. The equipment and operations are common and the process operator has a similar background to act between the oil and petrochemical areas. In these industries, there are many gaseous currents and many viscous liquids. Thus specific types of rotating equipment are gear pumps.

In the case of centrifugal or alternative compressors, the presence of liquid is prohibited in gaseous current, and may impair compression efficiency, or even cause problems that affect sealing systems and promote excessive vibration in rotating equipment. On the other hand, in energy cogeneration turbines from heated gases, the lack of control of the turbine speed can cause this large equipment to stop. In the case of gear pumps, small fluctuations in viscosity can generate reasonable changes in fluid transfer capacity. The existence of solid residues in these pumps can provoke wear in

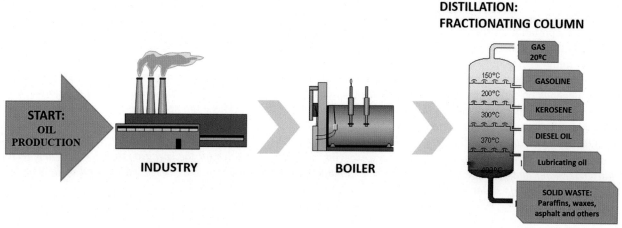

FIGURE 4.6 Oil production and refining. *Author's original.*

the components that reduce their capacity. Problems with the gear pump can cause leaks and dirt in the areas near this equipment.

High flow rates lead to the possibility of turbulent flow, if the change of rotating equipment is not accompanied by changes in piping. This type of flow causes mechanical stress on the pipes and flanges, which can cause leaks or fugitive emissions. Liquids and gases in the refining industry are highly flammable; consequently, the risk of explosion for fixed volume equipment is high.

Based on experience and perception, what is the relationship between rotating and static equipment with the knowledge of the team for the internals of such equipment. What is the relationship between complexity in process control and safety with human errors and losses in the refining unit operation? What is the relationship between a large refinery, its losses and the impact on the neighboring community where most direct employees do NOT live in the region? Pass on your perception...

_____
_____
_____
_____
_____
_____
_____
_____

Storage and distribution (upstream)

Logistics in the oil industry demands a large quantity of tanks, pipes, and rotating equipment. Because of the variation in the quality of crude oil due to different fields of origin (onshore), it is necessary to carry out cleaning in the pipelines that serve the various products. This cleaning procedure has high risks of environmental impact and material losses.

When the finished product batch is formed in the tank, the oil goes to the refining units. The storage operation must follow specific criteria to avoid the decantation of solid materials and corrosion due to the acidity of the product. Depending on the distances involved, there may be a need for a booster in order for the product to reach its final destination. The visualization of possible leaks is often a difficult identification and depends on the external community, which is not always available. Therefore, the pipelines need monitoring for pressure drop and loss of metallic material due to corrosion with leakage as a consequence. These pump stations have operators who work in isolation from the rest of the organization. This fact requires care regarding the aspects of communication and appreciation of the work of the operator, in order to reduce the risks of human error. The worker during the embarkation period, in most cases, lives at the pumping station itself, which increases the possibility of making mistakes by slipping or inattention.

Overpressure is one possible operational event that can cause overflows or major leaks. Therefore, the entire upstream must be monitored with level and pressure instruments. The quantity of finished products from petroleum refining is very large, and may be products with reduced viscosity such as gasoline, gaseous products (LPG) and even high viscosity such as lubricating oils, diesel and paraffin. These products make up lots with controlled quality and transported in various ways to tanks, with the objective of redistributing them to consumers, such as industry or the community.

The transportation of finished products in the oil industry is an issue that also deserves attention from environmental agencies, due to the number of trucks that travel long distances on roads with variable quality.

Case study

A very long stretch of pipeline (5000 km) that requires continuous monitoring during the transfer of oil and its derivatives depends on the community to realize the onset of losses to avoid future disasters. This situation requires scouts who accompany the operational teams and who have the spirit of safety. These are sent directly from top management to the plant floor; understand differences in communications, cooperation, and compromises of micro cultures.

What type of management model is appropriate to maintain the initial psychological contract, without the degradations inherent to physical distance, will the very experienced technician solve all the specific issues? Will a new technician understand and perceive the demands of the communities where this pipeline passes? What kind of human error mostly happens in this situation? Describe your perception...

_____
_____
_____

4.3.2 Process losses in gas industry for energy

Natural gas extraction follows operations and procedures similar to downstream oil extraction, such as drilling, separation, storage, and transfer through pipelines. The increased presence of sulfureous gases requires a reduction in the acidity of the gas prior to transfer. Corrosion in pipelines is a current problem and drives research at universities and oil companies.

Natural gas can be used as raw material for the petrochemical industry in the production of polyethylene. Certainly, the quality of the gas required to burn in domestic or vehicular use is different from the quality for use in industry that demands greater control of operations.

The transport and storage of gases are more expensive than liquid, due to the inherent risks of pressure, temperature, leaks, and explosion. The gas coming from the field has contaminants, which can be removed in logistics and industry activities. There is a need to monitor the thickness of the pipes in the gas pipelines, as well as the pressure in the transfer of this material. Concerns regarding human error aspects are the same in relation to the oil industry.

Other gases, such as nitrogen, oxygen, hydrogen, and carbon monoxide, are used as inputs in the chemical and petrochemical industries. The purification and suitability processes for transporting or storing these gases involve cryogenic temperatures in units known as "cold boxes". Any variation in the quality of the gases, or in the efficiency of cryogenics, can cause losses, which are sometimes not detectable. Hydrogen processing is very dangerous and is surrounded by safety devices. In this type of product, it is always necessary to have a specific follow-up in order to detect possible faults that could cause leaks.

Gas production for energy

The production of gas in oil and gas fields has operations with possible losses that can be very large. These losses depend on pressure fluctuations, very low pressure, presence of water in the gas, presence of CO_2 and other gaseous components, such as sulfur, which are extracted together with natural gas. Crude gas can have contaminants that influence the occurrence of process losses.

The existence of liquids in the gas flow brings harmful effects to the pipes (erosion and corrosion), which can cause leaks. Natural gas can contain sulfur, a toxic substance that generates acids in contact with water. Therefore, the transport natural gas over long distances is not indicated, due to the corrosive aspects of these substances.

The losses that occur in the extraction of gases in the field have similar causes to the extraction of oil. Among the reasons, one can mention the physical distance of the team in relation to the manager (the lack of commitment, the lack of competency to develop the extraction services); possibilities of corrosion and erosion problems; and dirt in the pipes.

Transport and storage

Aspects relating to the transport and storage of gases are more complex than those relating to liquids, such as oil and its derivatives. In the case of gases, compressors are used for transfer and storage. This equipment increases the pressure of the gas to keep it in liquid form and stored in spheres. Fugitive emissions can cause losses that affect the areas of integrated management, safety, environment and occupational health.

The transportation of natural gas for energy or for inputs in the polymer industry is done using trucks with cylindrical loads and curved ends at high pressure. This transportation is carried out with the contribution of safety systems to prevent exhausted pressure and to detect leaks. Another widely used means of transportation is the gas pipeline, which can be sent continuously or in transfers to intermediate storage between spheres.

The risks of gaseous emissions generate a hazardous environment, that is, a place not suitable for work or housing. Concerns increase in cases of sparks, avoiding smoking and any action that may cause sparks. The situation in the field becomes more critical due to the possibility of impacting on neighboring communities.

Employees working in gas storage and pumping units should be prepared to deal with emergencies with the possibility of explosion, and know the importance of keeping the safety systems in good working order, such as relief valves that are calibrated to open at a certain pressure.

Refining of gases for energy and chemical inputs

Rectification of light and heavy contents, such as the removal of liquid contaminants, may be necessary for natural gas. Refining should be carried out close to the extraction to avoid transport of gases with contaminants, and eventually, cause corrosion or erosion.

Gas separation involves cryogenics or membranes. The most commonly used process is cryogenics, using liquid nitrogen. The monitoring used to prevent cold leakage and loss of efficiency is carried out through changes in the levels of columns that have a reduced diameter. When natural gas is available for use in the petrochemical industry, it is separated by membranes. This process provides quality to promote the necessary chemical transformations for the generation of the polyethylene or polypropylene polymer. In the case of industrial gases, extraction is made from atmospheric air (nitrogen and oxygen), or through chemical reactions, as secondary products. These products are separated and reused in the industry.

There are two different situations in the refining of industrial gases: (1) high purity gas derived from atmospheric air, which needs to correct moisture aspects and traces of contaminants; (2) gases resulting from reaction, which need to be purified due to the possibility of the presence of process contaminants. In both cases, the level of competency of the operation must be high, as must the level of automation. The prevention of process losses requires attention to warn management, the opening of control valves, the level of cryogenic columns and the pressure differential in the case of reverse osmosis membranes.

In refining, highly flammable gases are isolated, and if compressor seals or pipe flange gaskets fail, leaks, fires and the possibility of explosion can occur. Therefore, industrial plants must be monitored with indicators of the presence of flammable gases (explosimeter).

4.3.3 Process losses in the biofuels industry

The substitution of materials and energies from the mineral matrix to the agriculture-livestock matrix has been encouraged as a solution for limitations in the availability of nonrenewable energy.

Government programs subsidize the production of biofuels, such as biodiesel and ethanol that drive the regional and rural economy. The biodiesel industry is new, while the ethanol industry is older.

This work presents processes for the production of oil, biodiesel, sugar and alcohol. The similarity between biodiesel and diesel produced in the oil chain is known. In the vegetable oil chain, the products are also obtained through chemical reactions for the production of biodiesel and glycerin. It is known that some oilseeds are part of the food industry (soybean and sunflower oil), and the same happens with the economic chain that produces alcohol (sugar market).

The losses in the biofuel industry are diverse, including the agricultural stage, where there is a loss of productivity such as the appearance of pests until the industrial stage with the loss of processes. The current work will focus on the analysis of agribusiness up to the production of biofuels.

The handling of solids for the extraction of oil (biodiesel) and liquid (alcohol) indicates precautions to be taken at the stage of vegetable collection, with the objective of avoiding acidity and other contamination. This care should be taken mainly with regard to biodiesel, which, since it comes from various sources, may have its productivity modified due to changes in its physical and chemical properties.

Biodiesel

The price of vegetable oils fluctuates according to the time of year and according to the demand for the food area, considered more profitable for the supplier than the energy area. This means that the biodiesel production campaign must be well scheduled and executed to avoid losses, where reliability must be high and so must safety standards. The choice of oil blends with specific reaction and separation conditions will indicate the best performance for the process. At the same time, alcohol and oil recycling should not fluctuate too much in load or composition.

For each different oleaginous base, there is a specific operating condition with concentrations, balance conditions, final biodiesel quality, waste streams, quantity and quality of by-products and their respective recycles. Depending on the size of the plant and the technology that can use heterogeneous catalysis, we named this industry as Biorefinery.

As in the oil extraction industry, the laboratory teams that perform chemical analysis and those that operate biodiesel plants come from the rural region or city of the interior. The possibility of low technical competency requires additional preparation of personal stuffs to avoid process losses.

Other factors that can contribute to operational failures and that promote process losses are: quality of operations; forms of process control; process logistics; qualities of raw material, product and effluent; tools available for the task; oral/written communication; and many others. Legislation in the area of biodiesel requires appropriate performance to meet the final product quality standards required by energy regulatory agencies. Some mill owners use the biodiesel produced as fuel for their own truck fleet, not meeting the same product quality control required by the Agency for external sales.

The destination of biomass (oilseed cake) for fertilizer or feed and the production of by-products (glycerin) are important steps, but not valued in biodiesel plant projects. This means that the inadequate processing of biomass and glycerin implies a loss of competitiveness, which generates waste that cannot be used and the lack of revenue from these co-products.

The methanol used in the chemical reaction is highly toxic, and if there are losses due to fugitive emissions, they directly affect the worker, as they amplify the negative effects on occupational health. Ethanol, another input used in the transesterification reaction, is separated by azeotrope distillation with possible quality problems and reduced efficiency in the reaction, through the alcohol recycling chain.

The production manager of the biodiesel plant may be isolated from the advances in new technologies. This situation can generate loss of time and discredit of the team in relation to the management. Another problem that affects the industry may be related to the effects of climate change that decrease the availability of water for the operation of its cooling system (this is an actual case).

Sugar and alcohol

Sugarcane comes from the field transported in large trucks and material losses are visible due to the characteristics and balance of the vehicle. Thus the distance between the plantation and the mill should be as short as possible. The proportion of material to be milled will indicate whether there is adequate loading for the mill as well as the proportion of water to be fed. The processes are simple, but they require attention as to the load of solid material. The juice resulting from washing and milling the cane goes into decantation, evaporation, cooking and crystallization operations. These are operations of sugar preparation and deserve care in the quality of the processes, since they involve heat in the form of steam and temperature control.

Sugar production is automated and brings positive results when the load is constant and the chains of inputs have adequate proportions. By following this pattern, the industry will certainly achieve the planned productivity, reducing process losses. Other concerns are the behavior of workers when they adopt risk aversion.

The raw material to produce alcohol is the same as the sugar production, and the amount of molasses for alcohol may decrease or increase depending on the load used in sugar production, that is, the market demand for food. Thus the load tables, procedures and controls depend on the sugar and alcohol market and the respective regulatory rules.

The complete shutdown of a production system affects the loss of reliability of rotating equipment. Upon the return to production, adjustments will be made with loss of time and re-service of maintenance and operation.

The quality control of the process is fundamental for the molasses to be at the appropriate level to start the fermentation and formation of the wine. The content of heavy components in this kind of wine is separated in the form of vinasse and goes to fertilization in the fields. The losses should be minimal in vinasse and sludge to avoid the use of excess water in the extraction of the juice. Then, the alcohol goes to the distillation and production, mainly of anhydrous, dehydrated and neutral alcohol.

In this process, there are static and rotating equipment. Although there are no major corrosion problems, the initial handling of solids requires great mechanical effort, especially in the mill. Another important detail is the adequate removal of moisture from sugarcane bagasse. If it is not adequate, the burning in the boiler will have low efficiency generating less energy and steam.

The alcohol and sugar mill, from sugarcane bagasse, can be automated with centralized control, which reduces the risk of human error through manual procedures. The fermentation operation in the production of alcohol from the juice depends on the control of the biochemical reaction. In these reactions, adequate acidity and temperature conditions must be maintained so that no waste of time or materials occurs. Molecular sieve operations require special attention for some procedures with potential for loss. An alcohol production scheme is indicated in Fig. 4.7.

The site of the mill must meet certain management requirements and may have biases resulting from the regional culture. These aspects need to be addressed and incorporated into company policies in order to avoid losses due to low labor productivity.

Transport and storage

Alcohol, the plant's final product, is highly volatile and can ignite. The transportation of this product requires appropriate procedures, safety systems and equipment to avoid losses due to fugitive emissions. It is important to specify the design for parts, gaskets and mechanical seals.

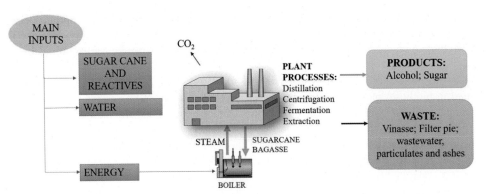

FIGURE 4.7 Alcohol production scheme. *Author's original.*

Operations that transfer volatile and flammable fluid are considered high risk. In this way, it seeks to create an appropriate technical culture for the non-occurrence of accidents. Every accident can cause loss of material, time and motivation of the operation team. The displacement of fluids must be done carefully, with the specified materials, and the operators trained.

The rules of the legislation for the transportation of alcohol must be established and complied with to avoid problems such as tipping over, contamination, and even the possibility of explosion. In certain cases, the transport of alcohol in quantity is done through pipelines (long distances). Similarly, to oil production, the issues of distance and isolation make it difficult to meet the management guidelines for the maintenance and possible contamination of these systems.

4.3.4 Process losses in the petrochemical industry

The petrochemical industry works with gaseous products and processes that use heterogeneous catalysis technology at high pressures. This type of process requires a high level of automation that is reflected in closed process control loops, safety systems, and preventive systems related to process failure. On the other hand, the flows in this industry are high, and the equipment has more noble materials and accessories to avoid corrosion and leaks.

The main raw material of the petrochemical industry is naphtha, and its cracking occurs through catalytic reactions. This process can have oscillations originated in the quality of the naphtha and make the operational control difficult. These transient states require greater attention from the operational team to define the appropriate human intervention for the return of process stability.

As the flow rates are very high, any process oscillation can generate off-spec product that is sent for firing or reprocessing. Both considerably reduce industry gains.

The resulting stress level of the routine in this system is high, for both the operator and the manager. This stress derives from the need to avoid oscillations of the process variables, since any minute of material loss can be a great loss for the productive system.

The production of petrochemicals requires controls that are adjusted in the process to avoid energy and water losses, inputs that are scarce in production schedule and control. The solution to reduce water and energy losses involves the reuse of available energy through heat exchangers and intermediate treatment of water for internal reuse. The reuse of energy, water, and materials is only valid when it does not influence the quality of the products and the stability of the processes. The reuse must have an indication of levels that are suitable for process safety, without which it becomes a low-quality solution. Therefore, the stability of the process variables is a preponderant condition to start a loss reduction program. This stability in organizations is achieved through quality and engineering tools.

Large technological systems (oil refining and production of petrochemical intermediates) require from the group of operators the necessary technical knowledge and skills for situations in a state of normality and for emergency in industrial facilities.

When individual knowledge is strongly rooted in personal protective characteristics, to maintain power, there is a decrease in the consultation of group knowledge. In other words, excessive self-confidence influences the lack of communication and discussion on technological solutions. Under these conditions, the risk of the operator assuming that his technical position is the most correct one increases. There will be no concern of argumentation or consensus with the team. There are now several ways to operate with the possibility of high omission of routine records. This can compromise the investigation of process issues to avoid the accident.

It is worth mentioning that the oscillations of composition or load in large equipment can generate excessive vibration in rotating equipment, or even in static equipment. In rotating equipment, such as pumps and compressors, excessive vibration affects the lubrication and sealing system, in addition to promoting displacement in the operating curve, which increases the consumption of electrical energy to perform the same service. In static equipment, there may be, for example, falling plates in distillation columns, fatigue of materials, displacement of internals in vessels (demister and chicanes) and, in worse situation, packaging or displacement of catalysts causing high loss of load and or preferential paths.

4.3.5 Process losses in the chemical industry

The control of the chemical reaction is fundamental in this type of process, the variations of which in stoichiometry imply great differences in the reaction environment with the appearance of inadequate secondary products. The uncontrolled reaction mass can generate, for example, solid material or alter physical-chemical properties useful in the application in formulations of the fine chemical industry.

Visual monitoring of the reaction is possible in some processes and products of the chemical processing industry. Changes in color and appearance may indicate the onset of lack of control in stoichiometry. Thus with due operational ability, the generation of solid and liquid residual contaminants can be avoided.

In the chemical industry, although there is closed-loop control, the level of automation is lower than in the petrochemical and petroleum industries. The operational control depends on Man in a more intensive way and depending on the production management mode, many human errors can occur. The quantity of circulating product is lower than in petrochemicals and as the process control is less robust, the reduction of the risk of events that cause losses depends on the control of the operations.

The presence of solids in the reaction that circulate through the pipes and pumps brings negative impacts to the process equipment, which affects the tightness of the sealing systems and can cause obstructions in low-speed regions.

Human contact with finished products or reaction intermediates can generate complex metabolisms in the body with serious long-term consequences. Occupational diseases affect the image of the company and cause loss of health and motivation in the team, with reduced productivity. Contact with the finished product can happen inside the factory in the handling of the products in the tank or in the production process, or outside the factory in the stage of use of the product by the customer, and can generate by warning, health problems by contamination of chemical products.

The standard adopted by management and transferred to the production team can generate credits or discredits of the organization before the society in a timely or definitive way. For example, by prioritizing actions that meet the expectations of production and cost reduction in absentia to meet safety and environmental requirements may make the sale of industrial assets unfeasible due to environmental and labor liabilities.

4.3.6 Process losses in the polymer industry

Various types of polymerization reactions have varying criticality in terms of solvents, pressures, or even high toxicity products. The polymerization that uses solvent and reaction terminator have critical points that must be analyzed in order not to provoke losses of process, as:

1. Purity of the raw material;
2. Contaminants in the catalyst;
3. Pressure oscillation in gaseous input;
4. Pressure in the oscillating sealing liquid;
5. High plant load;
6. Bottlenecks in specific process and rotating equipment points, which causes low efficiency and mechanical and process problems;
7. Uncontrolled polymerization flows, which generates solids and/or corrosive products;
8. Standards not properly established for operational control, generating different states after the execution of the task;
9. Off-spec product in reprocess; and
10. Possibility of rework in operation or maintenance, which generates low efficiency of the Systems.

Polymers are raw materials for the manufacture of plastics. The polymers, after being produced, are sent to the physical area of the industry to be added and extruded (extrusion), enabling them to achieve appropriate physical-chemical characteristics for the market. If mechanical problems occur in the extrusion equipment, there can be losses in the production stop of the pasta (polymer cooled in the form of strips). The performance of the product in the application is important in the polymer industry, because the plastic transformation is a physical operation that reproduces the intrinsic characteristics expected by the production. The polymer cannot be contaminated to the extent that it cannot be reused. The reuse can be internal (in the plant itself) or external (off spec-sales).

Some precautions should be taken to avoid customer complaints due to polymer end-use problems. These are considered commercial or technical assistance losses in the polymer and plastic industry and may reduce negative image billing of the commercial team.

4.3.7 Process losses in the metallurgy industry

The metallurgy industry's main objective is to extract and manipulate fused metals and generate metal alloys. The raw material of a metallurgy is always some kind of ore, a rock that mixes the desired metal with oxygen and other impurities. One of the main processes of elimination of impurities called "reduction" consists of removing the oxygen atoms that make up the ore. To perform the reduction, it is necessary to melt the raw material at high temperatures in a furnace, emphasizing that each metal is extracted at a certain temperature. In the furnace, the desired mixture of materials is processed at temperatures of up to 2000°C.

The losses in this process can occur in several operations, such as in the processing of the materials, as to the loss of energy for not reaching the ideal temperature in the mixture; losses related to solid waste, such as slag and fine slag; time losses, due to rework; loss of the final product, among others. A large amount of energy is used to produce the slag, and it also involves a large number of materials in a generally low yield processing. The metallurgical sector is essential for the development of society, since it feeds the production chain of metal transformation.

Activity in a material melting industry can take place in batches or in continuous production. Normally, the processes in this branch of industry use human labor intensively, which makes the process vulnerable to losses. A case will be explained in the following item 4.5, addressing a methodology to map and identify process losses.

4.3.8 Risk of loss due to technology

A series of questions are asked to prioritize the areas of investigation and intervention for process losses. These questions are related to the type of technology, type of losses, risk of loss and components of losses. By answering the questions, it is possible to draw up the Fig. 4.8 by technology and compare technologies and products.

Technology. What kind of technology? Energy? Chemicals and petrochemicals? Polymers? Bioenergy? Metallurgy? Mineral extraction for oil or for minerals? Agribusiness? Distribution? Gas, liquid, or solid processing? Process in industrial site, field, or sea? Physical or chemical transformation?

Type of losses. What type of loss? Materials? Production? Image? Or time? If the loss is material: processing materials or equipment materials? Type of equipment, static or rotating? Is the loss on the equipment components? Which one? If the losses are from production, what are the present types? Stop—start, reserves, sales, reprocessing? If the losses are of Image? Accident, labor claim? Pressure from the company? In addition, if the loss is of time?

Risk of loss criteria. Losses can happen in the following items: BL—Big losses, SL—Small losses, AL—Average losses. The risk classification considers the loss as being, high risk, BL = 4 and red color, medium risk, AL = 2 and blue color, low risk, SL = 1 and green color. How do you classify the risk of loss by type of loss and technology?

Quantitative risk of loss by industrial segment. In Fig. 4.8 the risks of loss by industrial segment, type of loss, and part of the process are analyzed. It is noted that the field activities of oil (2.2), refining (1.8), and cellulose (1.8) are the most impactful compared to the oil distribution, which is 0.7. In terms of losses, it can be stated that the most impactful types of losses are of failures in equipment components (2), re-service (1.7) and accidents (1.7). In second level of impact are raw material (1.6), social impacts (1.6) and loss of motivation (1.6).

Quantitative Risk of Process Loss by Type. Here we will make a sum of the individual risks and a normalization for four individual risk components. When analyzing the risk by type of loss the waste of time is still considered as the main avoidable loss with a 100-risk score, next is image with 91, followed by material losses 88. In terms of individual losses, maximum scores are related to rotating equipment (27), spare parts (stock and quality—26), operating and maintenance rework (26) and overall waste of time (25), Fig. 4.8.

INDUSTRIAL SEGMENT	Materials				Production				Image			TIME Motivation	Class
	MP-Input	E. Static	Rotary	Piece	Stop	Reservice	Sell	Reprocess	Accident	Employee	Society		Loss
ENERGY	11→1,6	6→0,9	10→1,4	14→2	11→1,6	12→1,7	6→0,9	6→0,9	12→1,7	9→1,3	11→1,6	11→1,6	119→17
Oil - Field	AL	SL	AL	BL	BL	AL	SL		BL	AL	BL		26→2,2
Oil - Distribution			AL	AL	SL	SL	SL		SL				8→0,7
Oil - Refining	SL	SL	SL	AL	SL	AL		BL	AL	AL	SL	BL	21→1,8
Gas	SL	SL	SL	SL	SL	SL		SL	AL	AL	BL	AL	17→1,4
Biodiesel - Field	BL		SL	AL	AL	BL	AL			SL	SL		17→1,4
Biodiesel - Extraction	AL	AL	SL	AL	AL	SL	SL		SL	SL		SL	14→1,2
Production/distrib.	SL	SL	AL	SL		SL	SL	SL	AL	SL	SL	BL	16→1,3
PETROCHEMICAL	2→0,5	2→0,5	10→2,5	6→1,5	6→1,5	6→1,5	2→0,5	5→1,3	2→0,5	3→0,8	5→1,3	7→1,8	56→14
Basic	SL		AL	AL	SL	SL		AL	SL		AL	SL	13→1,1
Intermediate		SL	BL	SL	SL	SL		AL		SL	AL	AL	16→1,3
Chemical		SL	AL	AL	AL	SL		AL	SL	SL		AL	13→1,1
Polymers	SL		AL	SL	SL	AL	SL	SL		AL		AL	13→1,1
TRANSFORMATION	3→1,5	1→0,5	2→1	1→0,5	3→1,5	3→1,5	3→1,5	2→1	3→1,5	3→1,5	1→0,5	3→1,5	28→14
Chemical Process	SL	SL	SL		SL	SL	AL	SL	SL	SL	SL	SL	12→1
Physical Process	AL	SL	SL	AL	AL	SL	SL	AL	AL		AL		16→1,3
CELLULOSE/PAPER	8→2,7	2→0,7	5→1,7	5→1,7	3→1	5→1,7	4→1,3	2→0,7	7→2,3	5→1,7	7→2,3	4→1,3	57→19
Field	BL		SL	SL	AL	AL	SL		BL	AL	BL	SL	22→1,8
Cellulose	AL	SL	AL	AL	SL	SL	SL	SL	SL	SL	AL	AL	19→1,6
Paper	AL	SL	AL	AL		AL	AL		SL	AL	SL	SL	18→1,5

FIGURE 4.8 Quantification of losses and risks by type industrial segment. *Author's original.*

4.4 Diagnosis of process losses

4.4.1 Introduction

The diagnosis of process losses depends on the level of knowledge, availability of technical and social data (human factors), and the response in variables, or in different process states. Thus it should be considered that engineers and technicians determine a good diagnosis of loss when they:

- Understand the phenomenon in the functioning of the equipment, in a practical and theoretical way, regarding the model and of their relations;
- Identify where to collect the technical, human and social data and the quality of them to estimate the level of uncertainty;
- Understand from the relationships that variables will be modified and process states to follow the outputs, variables, states, parameters, the various dimensions also, technical, human, social and organizational.

The collection of data and measurement of results depends on the knowledge of the production process and the respective flow of information that enables the control of production (cost, people) and of the process (equipment, products, and Safety). The flowchart of Fig. 4.9 indicates the flow of information from the data collection to the measurement of results that are present in the management reports and in the statistics of critical variables.

Knowing the processes means understanding the operating patterns of equipment and operations. Understanding of how data flows for, automatic and manual, system control. This information analysis allows us to indicate whether the best decision to resume the normality of the process is in the correction of variables, in the review of procedures, in the monitoring of consumption and losses, or in the analysis of the discourse related to routine. This decision will result from the utility between the cost and benefit of the method or steps chosen.

The information for the analysis of process losses is of technical dimension (which indicates the state of normality of the technology in operation) and social (which present the organizational environment including management, policies and culture). This information is available in documents/records/parameters and specific variables in the various areas or departments of the organization. Some of these documents and data are considered input data for the analysis of opportunities to avoid process losses, other documents and data are the result of process state, measurements, before and after the intervention of the group that deals with process losses. The comparison between the results indicates the

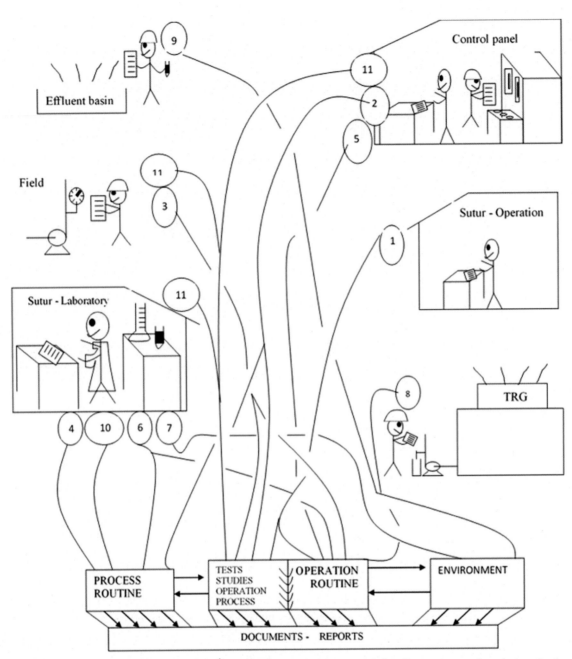

FIGURE 4.9 Data and information flow in production. Ávila, S.F., 2004. Methodology to minimize effluents at source from the investigation of operational abnormalities: case of the chemical industry. 2004. 115 p. Dissertation (Professional Master Management and Environmental Technologies in the Technological Production Process)—Federal University of Bahia.

gains in the program. The areas consulted to form the knowledge base in the implementation of the process loss program are: production, engineering, integrated management (safety—health—environment), commercial, people, and institutional relations.

Production is the area responsible for transforming materials into finished products. The main possibilities of occurrence of process loss events are in production. Engineering provides the technological support so that it is possible to perform the intended transformation of materials into products. Statistical studies and quality control are part of this area. Integrated Management (Health Safety Environment—HSE), on the other hand, brings together concerns about compliance with voluntary requirements (International Systems—ISO) and compliance with legislation on occupational and environmental aspects.

In the commercial department, there are results from the production work regarding the quality and quantity of the product, considering the technological base of engineering, and the rules and standards established by the safety and operational culture. The result of the production work is the delivery of finished products and services that are considered satisfactory for use by customers and the market. Thus any errors or failures in these operational areas may indicate non-compliance of the product in the commercial area. The importance of the activity of purchasing materials and inputs for the operation of machines, equipment, processes, and operations is worth mentioning.

The operation of machines and equipment has different dependencies in relation to the work of the operator. The level of automation and the technical-social complexity of the process influence decision-making in the routine and indicate the normal state for the operation of the processes. The role of Man is important in all areas; therefore, it is necessary to establish the requirements for the allocation of the worker in the workplace, the appropriate training and the type of internal communication in the company. These are extremely important issues and are dealt with in the Management of People and of the operations.

The quality of the practice is a result of the business vision for the routine activities and the established operational-technical culture. Practice is analyzed from the project, the assembly, the operation until the delivery of the finished product to the client. The area of institutional relations involves the creation of an organizational culture that keeps the company's image positive "outside" and "inside". The resulting culture and practice depend on the level of employee satisfaction.

The terms used in this chapter of process losses is comparable with those used in quality control techniques (Juran, 1988, 1992). The knowledge of the productive process is a basic element to evaluate the statistical parameters and represents the state of the process, so the recommendation is to involve the internal team of execution of the task in the work of improving the quality of the process-product, the behavior of the worker and the performance of the equipment.

The data used for the qualitative mapping of processes and for the analysis of operational abnormalities are also statistically processed. This data is considered as control items that, together with the actions of planning and changing patterns, are inputs for continuous improvement. The items that indicate the results of the interventions on the industrial plant are the verification items used in the statistical control program.

The classification of data and information by area, used in process control, have the following formulation:

- In 1, management guidelines, the control of the legal requirements, and the writing of the shift report that summarizes the state of the process, process variables and production parameters, in its shift group and time.
- In 1, the information is circulated for the control of people, whose objective is to maintain the availability of cognitive and motor skills for work. This is a direct function of the supervisor.
- In 2, related to information from the control panel, together with information from actions and field tests organized to improve processes;
- In 3, there is the activity of the field operator, who monitors the indications of the process state in the equipment, verifies the appearance of the product, verifies the state of the equipment and monitors the services in the area, as is this case of monitoring the maintenance work. The variables and parameters in the field control are complementary to the panel control. But we can also consider redundancy, when in the panel occurs the automatic controls, in the field occurs the comparison with reality;
- In 4 and 10, there are the analysis of products in processing and final, carried out to indicate, or assist in indicating the state of the process during the operational routine. Although the process variables are framed, it is important to monitor the qualities through chemical and physical analysis;
- In 5, the control panel is the place where the operators meet to keep control of the processes and where the field operators and maintenance support teams, safety, environmental technicians pass through. Much data and information, technical and managerial, circulates in this environment. Depending on the state of the process, the level of stress may be moderate or high.
- The panel operator (5) is the central communication between the execution of actions and the flow of information for production management, quality, integrated management, where the operational routine is controlled;
- The laboratory, 6, handles information to keep the routine of the process and operation under control;
- The laboratory, 7, also works to keep the environment framed;
- In 9, data is collected from the effluent, care is taken to avoid contamination and meet the needs of the environmental agency;
- In 11, the control panel also centralizes information for tests in normal operation, tests of production capacity and tests of new process systems;

- In 12, there is field monitoring for the operational routine in water quality.

4.4.2 Knowledge of the production process

The knowledge of the production process involves the technology of reaction, unit operations, production control, logistics, process control, standards, and operational procedures. A list of recommended knowledge and documents is presented in Table 4.1.

Auxiliary systems also help stabilize reactions, operations and fluid transfer. To understand the origin of process losses and their role in the chain of abnormalities (cause or consequence), it is important to analyze the fundamental principles that describe this event, as well as the influence of auxiliary systems that are not adequately monitored.

The study of critical systems indicates the possible process deficiencies that occur in the complex mode and that involve equipment, auxiliary systems, process control and operational procedures. These studies help to identify the interconnection with other processes in the internal or external supply chain. The analysis of complex systems demands the redesign of flowcharts and diagrams, from the reinterpretation of the production process. This analysis is carried out during the processing of data from documents and routine operation reports.

The knowledge of practices and standards (operating procedures) allows clarifying the origin of process losses in the production routine. Therefore, it is important to analyze the task and monitor its accomplishment during the shift to identify the respective failure modes. This monitoring functions as a participatory audit where the auditor, in a transparent manner, presents to the operators his interpretation of what was seen in the shift.

Management style, type of organizational culture, policies, and the integrated management system should also be investigated and interpreted to study the origin of the process losses (it may be related to behavioral or attitude aspects at work). Process losses can be a consequence of the operator's dissatisfaction with being an employee of the Company and resulting in a poor performance of the operation in the execution of the task.

4.4.3 Collecting data—inputs

Data collection for an action program that intends to avoid process losses must be performed by the team who knows the processes, has skills to perform the tasks in the routine, knows the human relations in the shift, understands about the organizational demands and reworked the flowchart of the main process with the inclusion of auxiliary systems. Data collection involves several techniques, such as interview, reading of documents, reports, records, field audits, and questionnaires.

TABLE 4.1 Summary of knowledge and documents.

Department	Topic or subject	Document or record
(1) Production	Operational procedure	Related records
	Assumptions and criteria	Reports—Management practices
(2) Engineering	Machines and systems	Engineering flowchart
	Technology book	Mass balance
	Process equipment	Instrumentation and control
	Sampling methods	Methods of analysis
	Process control	Report monitoring process
(3) HSE	Integrated management system	Records: ex risk analysis
(4) Commercial	Policy	Customer records, non-conformity reports
(5) Personnel	Policy	Health-related statistics
(6) Institutional Relationship	Culture	Socioeconomic environment
	Organizational standards	Integrated management system

The types of data handled for verification regarding process losses are continuous number (quality and quantity), discrete number (spare parts replace), descriptive (routine relational facts), tangible (concrete), intangible (something abstract is measured, expert opinion), primary (measured directly at source), and secondary (inferred from direct measurement).

The areas where data will be collected in the industry to avoid process losses are the same areas where the knowledge is analyzed for the mapping of process losses, these being: production, engineering, integrated management, commercial, people and institutional relations.

The production area includes operation, maintenance and management of production. The documents and records are researched, and the workplace is also observed to verify the behaviors. From the listening and reading of the discourse of the operator (OD) and the discourse of the manager (MD) an analysis is made. Engineering includes the area of monitoring processes and operations in addition to the industry conceptual project. Information about the process technology and the quality of the product/process is obtained in these areas. Integrated management brings together the areas of safety, environment and health. This area analyzes the impacts and represents the company in terms of legal or voluntary issues, as well as discussing ethical issues related to the use of data to avoid process losses.

The monitoring of non-conformities and characteristics of relationships with suppliers and customers indicates information to be used, which are obtained from the commercial area. Personal data is also confidential and may not be used politically or with preferences in the decision. The personnel area is in possession of the employee's psychological profile, which may indicate the greater or lesser facility to fulfill certain functions. On the other hand, the training program prepared to address the deficiencies of the teams may indicate a tendency for error, or correct operational procedures. Another important factor in the human resource (HR) area that indicates actions to avoid process losses is the type of communication and relationship that occurs in the work environment.

The institutional relations are the result of the present and past policies for Management in the company. It is intended to act in the company as it is written in the policy and reproduced for management. The external image is built from the relationship with the community by means of communication instruments, the treatment of events that may affect society, the relationship with customers and stakeholders, in short, the way to manage the business.

Documents for data collection by area

Each area has documents, reports, data (alarm list) and records used to perform diagnostics to avoid process losses. Routine procedures, the report from the production manager, the reports from the process engineers of each area and the reports from the board of directors present the technical data flow for the control of production, and how communication occurs between people, producing the technical culture.

The quality of communication tools and the quality of social labor relations in the shift groups are visible from some documents and reports, such as:

- Operating instructions available for the shift work;
- Report of occurrences written in the shift;
- Reports of incidents in production
- Occurrences in the field and the style of participation of the team in the meetings

The data used in the mapping of process losses are collected, mainly (50%–60%) from the shift occurrence book. This record describes technical, human and group aspects of production operation. The occurrences or events resulting from process control and operation reported in this shift report may or may not control the process.

In order to identify the organizational climate and the type of culture that encompasses the organization, it is necessary to investigate the management systems in the areas of quality, environment and safety. The organizational standards define the desired type of management. The management performed may differ from the desired management. Despite the policy transferred to the management guidelines, both in the standard and in the routine, there may still be practices that are at odds with the policy resulting from the balance of values between individuals and the organization. This data defines the psychological contract and the employees' links with the manager and the task. The non-conformity reports and the audit of the integrated system make it possible to know the main "bottlenecks" in this industry.

The measurement data of critical process, product and effluent variables are interpreted together with the facts that occurred and can indicate the causes and interactions between operational failures and deviations that occur in routine production. It is important to analyze the variability of the process, the main loss indicator parameters, the best operational conditions to maintain the quality of the raw material, inputs, finished product, co-products and liquid effluent. The evaluation of the quality of the process depends on techniques such as the audit of the opening of control valves and the use of bypasses in the field and in the control panel. The interlocking system indicates the critical points for the occurrence of safety failures.

Routine and emergency operating procedures, assumptions and criteria used to control the plant for quantity, quality and production cost are basic information that feeds the investigation of process losses. Some information about worker and the group help the investigation about the failure, such as:

- What are the expected psychological profiles of the employee;
- What are the appropriate profiles for the job;
- Evaluation of the performance of managers and employees;
- The history of the employee inside and outside the company in terms of interpersonal relationships;
- The desired training and training undertaken by the employee in the job and what courses were taken;
- Specific performance in operational incidents.
  - From the records of employee's behaviors, level of knowledge and speed of learning can be projected.
  - From experiences, similar failure situations are projected in the future to verify the respective behavior.

The selection and training policy indicates the criteria used for the absorption and development of the company's employees. The evaluation of the inventory of knowledge and skills is represented by exams taken in courses, school records, training and others. The employee's file indicates the occupational history that indicates the behavior in situations of process losses. Included in this data are:

- Involvement in accidents and incidents;
- Information on occupational health based on examinations;
- Information on occupational health based on anamnesis;
- General information on occupational health;
- Information on contamination caused or perceived by the employee;
- Behavioral problems regarding the use of PPE (Personal Protection Equipment) or in meeting the company's standards in the area of safety, health and environment.

The technology documents (process manuals, operation and process control manuals) indicate basic information of critical systems, according to the vision of the technology supplier and which are confirmed by the Production Reports. The observation and monitoring of the:

- Technical environments, in the planning and execution steps of task assessment;
- Socio-technical environments in the execution of the task;
- Human Performance related to skills in production team indicate information from/about:
  - How the discourse and action works;
  - Who has written the procedures;
  - What does the technology supplier indicate as recommendable?

The commercial records indicate the quality of the lots of finished products and the service by the sales team. These reports may be included in the quality system, but the impact on sales is found in business reports that contain statistical studies of purchases and sales. In purchasing, especially spare parts, it gives us an idea of what part of the production system needs technical or operating changes.

The data are collected from records and documents in different sectors of the manufacturing unit. For each sector the importance of the documents is described, to extract the measurement items that will be used in the analysis.

4.4.3.1 Production area

This involves operation and maintenance of production processes. This area is primarily responsible for carrying out activities that promote physical or chemical transformation adding value to the product. The main document used for the study of process loss is the shift book or the occurrence book of the operation. Operational factors are identified and the etiology of the technical failure with the respective chains of abnormality is made.

The description of how the tasks should be performed from the operational procedures is located in the operation. The way of working is guided by assumptions and criteria to be adopted in the operation and maintenance work. Operational failure can have several consequences, including operational incidents that affect the quality of the product and the process, not necessarily causing risks to man or machine.

The probability or possibility of the productive system failing depends on the level of competency of the operation and maintenance team, which is also the result of the inventory of available and applied knowledge and skills, which also depends on the level of commitment. Some non-conformity reports are generated from the work of the operation and can have an impact on the cost of production, in the same way that risk analysis can be done to avoid process losses, and not only prevent accident possibilities.

Product quality, process, and occupational safety management systems directly influence the assumptions and criteria for production management to occur. Recording the behavior of the team per shift and of the manager helps management to assign responsibilities, and indicates trends regarding the construction of the group failure. The acceptance of organizational norms by the manager indicates the ease with which he transmits and represents the organization to the team.

Finally, the knowledge of the organizational norms and the observation of the managerial practices are indicators that allow the knowledge about the construction of the group failure. Information on quantity produced, quality, personnel and production can be extracted from the production manager's report. This information indicates the criticality of the work in the production area facilitating the calculation of the failure in mathematical model.

4.4.3.2 Project and process engineering area

The monitoring of processes, products, and effluents indicate technical solutions to adjust the production, by means of methods or industrial projects. In process control, the variables are measured directly in the equipment to be indirectly corrected through other manipulated variables. The existence of critical process variables based on product quality, environmental impacts, the possibility of accidents and production costs is considered. These variables are monitored statistically and some are part of closed control loops or cycles. Some valve openings can be important indicators in the trend of loss of control for continuous processes.

The monitoring of the quality of the finished product is carried out by means of batch control with the respective sampling and analysis procedures. The follow-up of the effluent quality is of great importance to investigate the contamination at the source of the processes. These records are part of the input data in the qualitative analysis of the operational failure. However, this "world" of data needs to be "filtered" to consider only those that are statistically significant in the relationship between facts and data.

Variable identification cannot be based solely on records and data from digital systems; you should look at the productive system to check the reasons for process variations. These observations facilitate the analysis of risks in the production system. Data on process, operation, safety, analytical methods, equipment and many others can be found in the instruction books and flowcharts provided by the supplier of the technology. This material, considered the technology "book", is important to understand how the equipment was originally structured to work. Another important activity in the design phase that must be carried out prior to the operation of the industrial plant is the risk analysis of accidents and operational incidents, with an impact on safety and security. This analysis can help with information for the qualitative-quantitative assessment of operational failure.

The quality of the finished product continues to be essential for sales, which indicates the importance of the quality management system and ISO-9001 certification (quality standard in the ISO international series). In this system, the knowledge of the nonconformity reports and the analysis of the risks of the product going out of specification (process variables, failures in operations and equipment) help in the study of the operational failure.

4.4.3.3 Integrated management, occupational, and environmental area

The occupational area involves the areas of safety, ergonomics environment, hygiene, and occupational health. The occupational history of the team at the industrial plant indicates the result of the organizational climate, level of knowledge, commitment and acceptance of the interface between the operator and the equipment or product. Thus it is important to recover this history for analysis of the occupational environment. The field audits on the use of PPE are important to evaluate influencing factors on the behavior. Correlation analysis between items such as type of leadership, dismissal risks, team fights and operational incidents or accidents can help in the construction of influencing environments for the failure.

The way in which the organization views, writes, and executes safety-related priorities reflects the credibility of management systems. It is important to review safety audit reports to assess aspects of belief, writing and practice.

Field observation of the criticality of tasks is essential to assess influencing factors on behavior. Often great difficulties are solved with simple ideas and still favors the performance of the operator's work.

Records from ergonomic studies facilitate the verification of similar cases and measure the influence of these cases on the occurrence of fatigue and operational failure. The historical record of worker contamination indicates the chemical and physical risks to which the operator is subject. Field observation of the appearance of operators is important to assess these influencing factors on behavior.

Confirmation through the audit is essential to identify the level of importance in terms of occupational hygiene in the organization. The risk framework can indicate the relationship with influencers on the possibility of failure. The environmental issue is a strategic part of maintaining the company's business, just as product quality is fundamental to marketing, competitive pricing and customer satisfaction. Knowing the environmental and quality management system with the respective audit results helps to identify other factors that influence the issue of process loss.

4.4.3.4 Commercial

Non-conformities in product quality and delivery time are the result of flawed processes that can be related to the entire production module. Commercial records may indicate changes in the quality of the intrinsic variables of the product, because of different ways of operating or degradation of process equipment. Thus purchasing control is important to evaluate which spare parts are being purchased in excess, and which equipment is being changed quickly, either by corrosive processes or wear due to erosion or fatigue situations (dynamic systems).

4.4.3.5 Personnel area

Selection, training, body health, and psychological activities are the direct and shared responsibility of the personnel area. The area of selection and follow-up of people in the company has the obligation of knowing the psychological profile, verifying the character traits of each individual in the organization. There are some profiles that should be avoided for specific functions, such as people with an aggressive profile in production, or non-communicative people in management/leadership.

Performance evaluation, according to Muchinsky (2004), may not confirm the performance indicated by the psychological profile. Thus in order to find more correlations that are efficient, both the profile of those who initiate the work activities and of the performance of the task after the training, need to be measured.

The history of the interpersonal relationship allows defining the roles of each one in the team, including the leaders. The training transcript and the analysis of the individual's behavior in each subject can give indications of the existence of characteristics for the beginning of process losses or failure. The quantitative result of the training performed is analyzed and influences the failure rate based on knowledge and ability.

The people who make up the production team take exams at entrance, during the work period and at dismissal from the company. Therefore, there are variables that indicate the possible presence of individual failure. Some workers' diseases are related to somatization resulting from stress and workload in the performance of routine or emergency tasks.

4.4.3.6 Institutional relations area

The area of institutional relations involves the socioeconomic environment. Social culture, customs, religious issues, myths and rituals need to be understood for the analysis between individual, social and organizational values. It is also important to verify by means of tests, which are the socioeconomic emergencies that occur in the current time. Indirectly, the stakeholders also influence the construction of the loss of process.

Table 4.2 presents a summary of the data used for the analysis of process losses.
In which department is the data located.
Which document or record should be consulted?
What type of data, whether continuous (C) or discrete (Di)?
What is the description of some fact (DF)?
Whether it refers to something tangible (T) or something intangible (I).
Besides the classification of whether the data is primary, that is, measured directly from the source, or if it is secondary, which indirectly represents the source.
The weight to determine process losses classified as high (A), medium (B) or low (C).

4.4.4 Measuring results—outputs

The results of this process loss program may involve activities from all of the areas of the company. The identification of critical factors to reduce process losses guide the correct measurement of results. It is important to minimize effluent

TABLE 4.2 Data entry summary.

Department	Document or record	Type			Variables, indices, yields	W
(1) Production	Shift book	NA	DI	P	Events and variables out of control	A
	Operational incident (RNC)	NA	DI	P	History of incident—probable causes & actions	B
	Management practices	NA	DI	P	Registration of acceptance of practices audit	B
	Management Report	NA	DI	S	Writing and practice. Results and goal	B
(2) Engineering	Critical variable—Statistics	C	T	P	Temperature, pressure, ratio, vacuum, flow	A
	Complex processes	Di	T	P	Number of meshes closed per operator	B
	Control valve	C	T	P	% of modulating valves; 0%; 100%	B
	Effluent quality—Statistics	C	T	P	Historical analysis with related events	A
	Product quality, raw material	C	T	S	Purity and contaminants—statistics	A
	Steam and cooling system	C	T	S	Losses; % in relation to standard; performance	B
	Correlation analysis	C	I		Relationship between critical variables	A
(3) HSE	Occupational H.: Safety, ergonomics	Di	DI	P	Accidents or incidents: criticism	A
	Audit: Field observation	NA	DI	P	% in relation to the standard	B
	Correlation analysis	C	I	S	Relationship accident/incident—climate/event	C
	Identification of scenarios and risks	C	DI	P	% in relation to the standard	C
	Environmental audits	C	DI	P	% in relation to the standard	B
(4) Commercial	Non-conformities	Di	I	S	% in relation to the standard	A
(5) Personnel	Psychological profile	C	I	S	Skills and features: memory and attention	C
	Learning performance	C	I	S	% of standard	B
	Interpersonal relationship, behavior	NA	DI	S	Conflicts; role; leadership	C
	Inventory knowledge and skill	C	I	S	Measurement in quantity and usage rate	B
	Training results	C	I	S	% of standard	C
	Examinations	C	DI	S	Group trends: punctual	C
	Health history	C	DI	S	Group trends: history	B
(6) Institutional relations	Social relations	NA	DI	S	Image measurement: % of standard	C
	Organizational climate	NA	DI	S	Climate report: Type	C
	Type of culture	NA	DI	S	Culture report: Type and practices	C
	Environmental conditions	NA	DI	P	Stakeholder risk analysis and competition	B

generation, improve the quality of the finished product as measured by the number of batches that follow the specifications, increase the capacity and availability of the industrial plant, increase production, and reduce consumption rates.

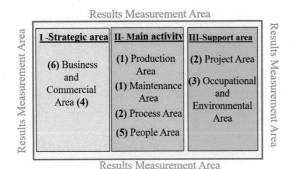

FIGURE 4.10 Results measurement. *Author's original.*

In production, operation, maintenance, and management results are collected. The measurement of the results is performed in the same area in which the collections happen. Some variables may also result from the process quality area, product quality, and process control. The results may be in the occupational and environmental areas, such as accidents or leaks. The results of the commercial area are related to product specifications and deadlines. Other results also are found in the areas of human resources and business (institutional relations).

The areas are classified as indicated in Fig. 4.10: (1) Strategy—commercial and institutional relations area; (2) End activity—human resources and production area; (3) Support—areas of integrated management with occupational and environmental impacts (HSE) beyond engineering and project.

(I) Strategic area: institutional, business, and commercial relations

The improvement of the company's public image occurs because of transparent actions and depends on the level of knowledge on industrial operational failure. Improving a company's public image in terms of order delivery, finished product quality assurance, and internationally compatible prices will result in higher customer satisfaction levels. Thus controlled processes, without surprises, indicate stable variables and indexes that are traced to avoid process losses. This monitoring of results is performed in the strategic area.

4.4.4.1 Satisfaction and loss of customers

Several factors influence customer satisfaction and are controlled through the quality plan and in the HSE. Product quality should be as agreed. The agreed delivery deadline should meet customer demand without any surprises. Price stability and courteous treatment/reception should be maintained during the relationship and avoid switching representatives during negotiations. Therefore, this index represents the overall customer satisfaction level measured by: number of customers per product per month; the number of customers who stopped buying per month; and percentage of customer satisfaction through a market survey.

4.4.4.2 Revenue

Revenue is the result of sales, service, but also primarily represents the maintenance of quality, time, and price in the timeline. Quality and quantity, on the other hand, depend on production and engineering. This result is measured against total monthly sales by product and by customer.

4.4.4.3 Sales amount

Higher sales volume does not mean that profit has increased, or that high revenues were achieved. It depends on the production costs and the final price of the product. To have a competitive price, it is important to reach: a good sales team, good public image of the products and, especially, reduced production cost, which indicates a decrease in process losses. This measurement is made by cubic meters per month, or the number of pieces per month, or even in tons per month.

4.4.4.4 Fines

A disorganized company with lawsuits has a negative public image for customers and partners. Fines result from different areas of impact:

- Environmental area, where some events such as toxic gas leakage occur;
- Services area, in which the social and economic impacts from such leakage on the neighboring community are assessed;

- The handling of social demands is not satisfactory;
- Commercial area, where the contract sales rules not complied with, causing conflicts with the customer;
- Occupational area, when involving inappropriate actions from the company in relation to regulatory norms; and in
- The labor area, where nonconformities with the labor legislation (CLT-Consolidation of Labor Laws) occurs. The measurement made in the number of fines per type per year, fine value per type per year.

4.4.4.5 Image

The satisfied customer increases their purchase at a reasonable price, thus increasing the supplier's revenue. If the supplier has a controlled production cost and few contractual fines, then positive results will be obtained. This means that companies with few process losses have a positive image. The image is defined by approximation to the standard (percentage).

4.4.4.6 Insurance

Risks are minimized in a company where losses are controlled; therefore, the renewal of insurance related to these risks has a reduced value in comparison to the market average. The measurement is performed in the same risk rating by dividing renewal value by the total of annual sales.

(II) Core activities: process and design engineering

The stabilization of process variables results from the identification and mapping of interconnected factors and events through causal relationships, which can be described through the statistical process and production parameters, including recognition and revision of standards/procedures, action plans and field modifications.

Improvements are measured by framing the statistical parameters to 6σ, that is, 99.99% of the operational parameter measurements within the statistical limits. Effluent quality is also improved from the framing of variables and the speed at which solutions are given when the effluent is contaminated. Unit operations, reactions, machines, product quality and quantity, and consumption of raw materials are indexes of major interest for the Process Engineer.

The comparison between the current state of the process and the projected requirements to be adjusted and discussed on the system's reliability level.

4.4.4.7 Process variables

Variables can be investigated using descriptive and multivariate statistic methods to stabilize industrial processes by adjusting old standard to new conditions. Process and production variables indicate whether the target state has been reached, that is, whether the process state is within the default values. The parameters to be modified are chosen through process improvement. The parameters that indicate the improvements are analyzed through the variability factor, inverse Pearson factor (1896); and Cpk (process capability) per month (Juran, 1988). In the case of discrete variables, the machine cycle and the production schedule are analyzed.

4.4.4.8 Yield

Boiler burn efficiency (Carbon in CO_2) or engine yields (electricity at work) for strategic equipment in terms of process losses are continuously monitored. This equipment is large, with the possibility of operating with non-standard conditions, which causes high consumptions of diesel or electricity. Yield is calculated in percentage by comparing the startup standard and actual consumption.

4.4.4.9 Unit operations efficiency and chemical reaction

The structure of the industrial unit considers certain conversion and selectivity for the chemical reaction. Processing capacity generally depends on the composition of process streams relative to the standard range. The aim is to maintain the quality of incoming currents, the physical condition of the reaction system and the knowledge of procedures to keep the specified reactor operating. The reaction efficiency is measured by converting the raw material into the product on a mass basis. The separation operations follow the same type of control, where to keep the separation within the standard; the yield is given by the separation efficiency and depends on the internal hydrodynamics where the accessories have specific utility that can change the efficiency. The efficiencies of unitary separation operations are measured by the amount of specific material separated from the total amount of incoming material.

4.4.4.10 Product quality

The final product is the result of the process and operation controls of the industrial plant, including the quality condition of the raw material and the inputs. The way it operates, the environmental condition and the work stress can change the quality of the finished product measured in terms of purity (percentage), contaminants (percentage or ppm) and physicochemical characteristics (approval or percentage in terms of standard).

4.4.4.11 Consumption index

This item is considered in the core activity because of procedures, controls, and quality of inputs. It is measured by dividing the quantity in tons of material consumed by tons produced in the analyzed period. Consumption rate depends on reliable measurement of levels in process vessel, however in some production systems this reliability cannot be guaranteed.

4.4.4.12 Effluent quantity and quality

Effluents are measured by analytical methods and are monitored by the engineering area. Effluents should be controlled, and if possible minimized, as they affect the institutional relationship with the community. This control fits the measured values to the standards required by the legislation. It is followed by the HSE. Effluent generation is measured by parameters such as flow (m3/h or m3/month), COD—chemical oxygen demand (ppm or ton/month), pH and contaminant (ppm; ton/month). The monthly cost of treatment can also be tracked ($/month).

Industrial plant load (capacity) and continuity (availability) result from knowledge of technology, well-designed procedures, the reliability of systems including equipment and workers, and the availability of these systems. The load is expressed by the percentage in relation to the standard. Continuity is measured as a percentage from the hours worked divided by the total hours available.

4.4.4.13 Industrial plant load (capacity) and continuity (availability)

The load and continuity result from knowledge of technology, well-designed procedures, the reliability of systems including equipment and workers, and the availability of these systems. The load is expressed by the percentage in relation to the standard. Continuity is measured as a percentage from the hours worked divided by the total hours available.

4.4.4.13.1 (II) End activities—production: maintenance

Equipment availability is increased when downtime for preventive or corrective maintenance is decreased. The implementation of causal identification and mapping techniques, process statistics, normal operation testing, adequacy of standards and procedures, localized action programs, training and educational campaigns can lead to a reduction in equipment downtime and increased availability for the operation. With the implementation of these techniques, the consumption of spare parts is also reduced. These items are possible outcomes in the production area, specifically due to improvements in operation accompanied by maintenance. Statistical monitoring of the spare parts replaced in the maintenance intervention is performed by the procurement, purchasing or maintenance area. After establishing the causal nexus of the process loss, reviewing the procedures and the facilities for failure detection will make it possible to reduce overtime and maintenance costs.

4.4.4.14 Maintainability and availability of critical equipment

Equipment with high potential of being related to process losses should be monitored in aspects of maintainability and availability. Some equipment stop and may be involved in accidents or leaks/breakages, which increases the likelihood of remaining stopped or underused. Equipment that can cause load reduction or discontinuity above 5% is considered critical and should be monitored in terms of downtime and availability for operation in standard conditions. It is no use just working; it is important that it works well. These results are measured as follows: maintainability—time between large equipment downtime/period; availability—hours of equipment operating/total hours available for operation.

4.4.4.15 Spare parts costs

The "cold" statistical treatment in purchases, based only on history, without adding technical information as the main function, does not result in diagnosis or recommendations to avoid losses due to lack of understanding of what is happening in the process. To understand the reasons for the excessive spending on parts, it is necessary to know the area, the sector, the material and the equipment where the parts are consumed. In addition, it is necessary to have full knowledge of the type of service that this material is serving, the type of fluid, the type of mechanical stresses, equipment history and the way it works. The results measured from the consumption of critical parts/period with the history of facts in industrial plants allow the diagnosis regarding the cause of problems or loss of performance.

4.4.4.16 Overtime—extra hour

Operational problems require a high workload and attention from the service area to avoid: system downtime (accident or incident), load reduction, high consumption, or loss of end-product quality. In addition to the negative economic factor in overtime use, there is also the human fatigue of being overworked during downtime. These results are measured as follows: overtime divided by total available time.

4.4.4.17 Maintenance re-work

Another factor that creates fatigue in executing services is performing the same service several times by repeating the same procedure. Thus it is known that although the equipment is available for use, it is being misused, so that there are many stops for the same reason. Rework generates conflicts between maintenance and operation personnel, leading to the transfer of responsibility for the error from one team to another and vice versa. These results are measured as follows: hours spent on services already performed divided by total hours of services.

4.4.4.17.1 (II) End activities: production, operation and management

This is the main area where actions to reduce losses will be carried out; therefore, the area where the main data is collected and where the normality of the process is observed. Normality depends on process states that can be classified as normal transients for normality, transients for abnormality and abnormal process states. The results in the area of operation and management may indicate, through historical statistical parameters, the fit process normality state. These results come from the management and operation area, and are measured through the following items: plant load (measured capacity), continuity (availability), production losses due to downtime or waste generation, reprocessing, consumption rates, overtime and the re-work of the operation. Some of these items involve the entire plant and have been mentioned before. For these cases, specific comments are presented to the production area and, particularly, management and operation.

4.4.4.18 Load (capacity) and continuity (availability)

Capacity can be considered as a control or verification item. When the load is predefined from production scheduling, which is based on the sales schedule, the load is considered as a fixed parameter or goal to be met. A number of conditions can cause low stability in process variables, which leads to load reduction in specific campaigns, quality problems in materials (also raw materials) and services. It may also happen that due to operational difficulties, it is necessary to reduce the plant load, thus becoming an item that results from the production process. The continuity of the plant is always a consequence, that is, a check item regarding whether or not to reduce process losses.

Continuity is measured in percentage because of the number of hours worked in relation to the number of total hours available under normal conditions, excluding scheduled downtime. The load is measured in percentage of standard and can be considered as Conceptual Design or Revised Design. The load depends on equipment, chemical reaction, separation operations, and fluid and heat transfer operations. The resulting process load depends on a number of operating conditions and their combination to avoid process fluctuations. Continuity measurement depends on the operating condition of the equipment and operations. Equipment hydrodynamics influence load and continuity through individual yields and efficiencies, which results in overall process efficiency.

4.4.4.19 Production

The amount produced depends on the factors reported, the efficiency of the chemical reaction, and the efficiency of the separation or purification operations. Production also depends on factors not directly related to the material being processed, such as the condition of the equipment and local occupational and environmental laws. Some equipment are considered critical, such as: compressor, charge/load pump and cooling system. Some operations may incur in fines that are the result of contamination by leakage and gaseous or liquid emissions.

Certain operations may cause lost time accidents with the subsequent need to reorganize production activities. Statistical trends are used to evaluate yield in intermediate and final plants. At first, the critical variables are related to the load, the continuity of certain equipment and plants, besides some important process variables.

The quantity produced depends on the capacity, availability, and final quality of the product. There is a goal to be achieved and some of these parameters may correlate with others. This item is measured in specified tons produced per month.

4.4.4.20 Quantity and reprocessing costs

When the process is out of control or the operational activities are not properly performed, the final product that cannot be sold is considered material to be reprocessed. The quantity and cost of reprocessing are parameters for verifying quality in production or in process and operational control. Reprocessing also affects the quality of the product and the efficiency in unit operations, when the return of unreacted raw material or other products ends up impairing selectivity in the chemical reaction, as well as reheating the material. Increasing processing quantities will result in uncontrolled separation and transfer operations due to a change in speed and in flow rate.

4.4.4.21 Consumption index

Already mentioned in item (II) end activity is process engineering. However, it is noteworthy that although it is difficult to measure instantly, due to fluctuations in process vessel levels, it is necessary to define the consumption index based on accumulated values for three days.

4.4.4.22 Extra hour/overtime

The reasons for working overtime are related to the recovery of process normality. Extra hour/overtime is spent on actions to recover finished product quality, plant continuity or load, availability of certain equipment, unsafe conditions or real leakage risks, process investigation and completion of hazardous tasks. The way to measure overtime is the same as that mentioned in the maintenance part.

4.4.4.23 Repeated process and efficiency of operating procedures

The reworking of the operation in the process and the efficiency of the operating procedure are related to each other, and when there is low operating procedure efficiency, the re-execution of the procedure may be necessary. Failure to reach the desired goal state in the procedure may require repeating the procedure. The way to measure the re-execution is the same as for maintenance.

4.4.4.24 Product quality

Item discussed in (II) End-Activities: Process Engineering, which is also directly related to production.

4.4.4.25 Effluent quantity and quality

Item presented in (II) End-Activities: Process Engineering and detailed treatment in (III) support areas. It is noteworthy that this item is also the result of a good job in production.

4.4.4.26 Turnover

Followed by personal or human resources management, but intrinsically linked to production planning.

4.4.4.26.1 (II) End-activities: people or human resources

By implementing an action program to prevent process loss, problems affecting staff productivity become known and overtime work is reduced. As a result of this implementation, there is a decrease in absence from work, higher participation of operators in training and leisure activities promoted by the company, which shows increased motivation. Another good result of the program to avoid process losses is the reduction in the number of incidents and operational accidents when the comfort at work was increased. People, when involved in improvements, get motivated and improve the quality of work. Attention is drawn to common goals. Thus there is a decrease in turnover and a reasonable improvement in the internal work environment.

4.4.4.27 Absences at work

The effect of absenteeism at work is severe when there is no immediate replacement of the absentee because the teams are "lean" (the number of resources—people—is calculated for maximum productivity). The justified absence at work, although supported by official rules (laws), may indicate lack of motivation in the company and at work. Thus the lack of health problems, family problems and education problems could be indicators of lack of motivation at work. The measurement is made considering the number of absences per employee or by sector, by period.

4.4.4.28 Turnover

The result from the policies and the organizational climate in the production area is seen through turnover indexes. The intensity of satisfaction interferes in the willingness of the worker to seek other jobs. In industrial plants, physical and psychological stress may be manageable; there may be low turnover. Decision-making regarding the process requires higher efforts from the employee and affects the entire team on the motivational aspect, thus causing a negative multiplier effect. The number of dismissals/total number of employees/year/by productive capacity measures turnover.

4.4.4.29 Image on the workplace

The employees take home the company's internal image or internal climate, which positively or negatively reproduces the company's performance in the social environment. The transfer of this image affects the company's image to society. The satisfaction level is measured through the answers to surveys, in which the valuation of this intangible aspect is the result of checklist (% of deviation from standard).

4.4.4.30 Motivation

The measurement of the motivational level is indirect and results from the organizational climate, from the operator's success in performing the task under his responsibility and from the team's success in performing group tasks in which the worker is part. Thus the motivated worker participates in the voluntary activities carried out in the company. Measurement of motivation level is intangible and depends on scores to be given, so that is possible to define the percentage in relation to standard motivation for the company, function and type of business.

4.4.4.31 Accidents, incidents and morale changes

Accidents and incidents within the company may be unintentional. Such a problem usually occurs at random moments and, therefore, it is difficult to analyze its root cause. It is easier to blame the injured person than to analyze and investigate the setting and environment of the event. The search for the culprit is a process that causes internal conflicts in the organizational structure. Accident or incident investigations should be transparent to avoid causing a "distrust and hunting" climate. If the investigation is superficial, the responsibility may be focused on only one person, then possibly leading to a change in team morale and lowering staff productivity. The measurement of this item occurs through an increase in the annual accident or incident rate, a description of coincident events, and a correlation with the average team productivity.

4.4.4.31.1 (III) Support areas: integrated management and project engineering

Monitoring health, safety and environment requires a structured team that in many companies make up the area of the HSE. This team provides support for the action in the production area on factors that impact on occupational and environmental aspects.

Techniques for planning industrial plants range from the basic design stage to equipment assembly. This planning indicates control actions that are based on experience, knowledge and available technology.

The manner of operating the industrial plants and the type of technology defines the possibility of leaks and the need for monitoring and control in the operation practice. To get to know about the real risks, it is necessary to operate the industrial plant and to verify the equipment working with the respective chemical products load.

Some basic items are checked to verify the results of the process loss program, among them: number of accidents and incidents in the safety and environment areas; effluent characteristics; characteristics (quantity and quality) of the generated industrial waste and needs for treatment; fines in the occupational and environmental area.

Due to incorrect assumptions in the operation of industrial units, projects are prepared, but the acquired equipment is not suitable for use in the real operating conditions, that is, there are costs in the project engineering area that can be reduced when the engineering area is near the production area. This subject will be studied because of the process loss program.

4.4.4.32 Accidents and incidents

Some comments have already been made on human issues related to incidents/accidents and the change in the team's morale. Techniques that analyze accident risks that consider environmental variations in production are discussed. When discussing environmental variations, it also involves physical (climate), organizational (politics and culture), occupational (new legal requirements) and socioeconomic aspects (societal and market demands). In risk analysis, therefore, there are some initial considerations about the environment where the failure or loss is present. It is necessary to revise the risk monitoring and points for control. The ways to measure accidents and incidents were also commented in the discussion related to People.

4.4.4.33 Effluent and waste treatment

In the results related to process engineering and the industrial plant, considerations were made about the generation of organic effluents. The adequacy of the effluent to the legislation standards requires physical, chemical and biological treatment. This adjustment is repeated for waste to comply with the legislation, and involves treatment, transportation, disposal and registration. The measured characteristics of solid waste indicate whether the action program to reduce process losses is meeting its goals, such as reducing quantity and toxicity. The measurement of solid waste should be monitored monthly. Trends should be analyzed according to actions to decrease process losses. In the case of gaseous effluents, control and measurement are more difficult and are estimated by difference in global mass balance.

4.4.4.34 Effluent quality and quantity

Organic effluent is monitored in quantity and quality. The factory operation is responsible for keeping the systems watertight without liquid leaks from the various process points, for example: pump sealing, washing of equipment and places, aqueous streams for process separator, heating systems and nozzles, purging contamination or samples. Therefore, it should be noted that most of the organic wastewater generation can be avoided when adopting operating standards with cleaner procedures.

4.4.4.35 Fines

The reduction in fines shows the reduction in process losses. The reduction of environmental fines and reduction of fines in the safety-related labor area expressly show this trend. The reduction of fines or contractual claims between the employee and the company also presents this trend. The reduction of environmental and occupational fines is an indicator of a better process status. The measurement of this result has also been described before and involves the number of fines per month, during each 3-month period, and the number of fines per period as well.

4.4.4.36 Costs from inadequate projects

The difference between the real operating condition and the designed condition depends on the equipment, in which the reaction, separation, heat and mass transfer occur. The equipment used to perform the service is designed considering the standard process state assumption. Thus if what is considered in the project is not the same as the operating condition, there will be problems in load, continuity, quality, and cost. The production team's non-participation in the design stage can lead to discrepancies. The measurement, therefore, considers the percentage (%) of the difference of the actual project cost in comparison to the expected.

Final comments: Many variables, indexes and yields are summarized in Table 4.3, which indicate the units, areas and sectors where measurements should be made. There may be a relationship between them, which facilitates measurement and interpretation.

TABLE 4.3 Variables and Indexes for measuring results to avoid process losses.

Department	Sector	Variables, indexes, yields	Unity
	General		
	Effluent	Flow rate; COD; Ph; Poisoning; Value	(m3/h; m3/month); (ppm, ton of solid); (ppm, ton/month); ($/month)
	Product	Purity; Poisoning; Feature	(%); (% or ppm); (no per year, %)
	Production	Charge; Continuity; Amount	(%); (%); (ton/day, ton/month)
	Consumption	Index	(Ton raw material/ton of product)
(1) Production	Maintenance	Availability; Maintenance; Parts; extra hour; Re-service	(%); hours between failures; (%); (%)
	Production	Charge; Continuity; Amount	(%); (%); (ton/day, ton/month)
		Re-Process; Re-Work; Quantity	$/ton; (ton/month, %)
	Consumption	Index	(Ton Raw Mat /ton product)
	Operation	Extra Hour; Re-service; Efficiency Procedures—%; %, %	
(2) Engineering	Process	Variability	Variability factor Pearson; Cpk
	Process	Yield	% Over standard
	Process	Reaction and separation efficiency	% Converted or separate
	Product	Quality; Purity; Contaminants;	%; (% or ppm); (no per year or %)
	Consumption	Index	(Ton Raw Mat/ton product)
	Effluent	Quality and Quantity	(m3 /h, month); (ppm, ton/month); (ppm, ton/month); ($/month)
	Project	Inadequate project cost	% Over standard
(3) HSE	Events	Accidents; Incidents; Disease	# Of events, criticality, area
	Effluent	Treatment, quality and quantity	Ton; $ /month; (m3/h, month); (ppm, t/month); (ppm, t/month); ($/m)
	Fines	Safety; Environment	Fine/(month, year); Fine/type/(month or year)
(4) Commercial	Client	Satisfaction; Loss; Quality; Price; Term; Treatment	Client/prod; %satisfaction; Dissatisfaction /month;
	Sales	Revenues; Sales amount	Billing/month/(product, total, customer); (pieces, ton, m3/month)
(5) Staff	Motivation	Missing work	Number of absences/periods
		Staff turnover from standard	%
		Internal image, employee and family	Class, %
		Motivation as standard deviation	Class, %
(6) Institutional Relations	Fines	Labor; Safety; Environment; Commercial; Compare with pattern	Fine/(month, year); Fine/type/(month or year); %
	Safety	Annual renewal	S/Volume of Sales/Class

4.4.5 Introduction to tools and methods

The processing step includes techniques that process qualitative and quantitative data, which allow the construction of the causal nexus for the occurrence of operational failure that causes the process loss. The mapping of abnormalities and the correlation between descriptive and numerical data define what the roles of the operational factors are.

Identifying the causal nexus requires the definition of process states from normality to abnormality. Certain operational factors may indicate the existence of failures, contrary to what is indicated by the records and the discussion of official documents, which can show the adoption of labor standards in production.

The analysis of the collected and processed data will indicate an action program to be validated in the field, in this case, after the analysis of the Operator's Discourse (OD) and/or the Manager's Discourse (MD). The abnormal events map composed by OD and MD indicates the critical process variables to be tracked as presented in Chapter 2. After interpretations, relationships and correlations are made, some changes should be suggested and then validated.

Audits confirm the need for monitoring and making changes to variables, procedures and standards. Man's role and the influence of the organization on problems involving process loss are evaluated to define the causal nexus with technical problems.

For each specialty, a specific diagnosis is used and then a set of actions for the multiple areas is indicated. The order for the actions depends on the priority in terms of process losses.

These diagnoses will be validated through the follow-up of indexes and results, confirmation by the production team, revise of procedures, standards, and testing to verify the benefits of change.

The techniques for constructing the system failure diagnosis are procedural. They depend on: Technology and operating standards, Managerial characteristics, Organizational characteristics, Order of use, Time of use, Period, Parameters, Type of altered correlation.

The methodology is procedural because the data and means of data collection also depend on the types of technological/technical systems, the type of social system and worker profiles, as well as organizational and socioeconomic changes.

The choice of results to be monitored depends on the technical and social types involved in process losses. The main techniques that make up the diagnosis for process losses are:

- Analysis of the causal nexus in the construction of events;
- Process and abnormal event mapping
- Mass balance with interpretation of the process variables
- Intervention program based on the operator's discourse and manager's discourse;
- Analysis of organizational and human influence on process losses, and;
- Production audits to confirm the relationship between action and resulting variable.

4.4.5.1 Abnormal event qualitative-quantitative map based on operator's discourse

The area of operations is accustomed to classify states as normal or abnormal, but there are actually intermediate states prone to abnormality or to normality.

The operational culture present in production does not value the signs of failure perceived directly or indirectly by the operator through the sense organs, aligned with the operator's difficulties in building a mind map, which is a result of current perception and poor quality of the memory from personal/team. These cultural events indicate a high possibility of latent deviances and emergent failure processes that can lead to process loss. Thus there is a number of low visibility information not valued on the process and operation state and could assist in solving routine problems. The information recorded in the shift occurrence book indicates events classified as normal when in fact they are intermediate, and then if analyzed can detect the emerging fault. On the process loss program, that information and warnings collected and posted to the relational database, with the abnormality level classification, about the fact under discussion.

The quantity produced and the purity of the finished product are indexes that result from the process state and operating condition of industrial plants. Therefore, these process and production data records are added to the abnormal and intermediate states record.

The level of abnormality or the abnormal state of the process, operation and management are identified by comparing the current state to the best process standards. These standard states are identified through the history of the industrial plant. If this history is not known, the process analyst who is implementing the loss prevention program should (based on his experience and external references) define which abnormal and intermediate states are in addition to the default normal state. The state of abnormality for the process area is based on the plan of quality, which includes the limit values for the plant to operate normally. The state of abnormality for management is based on the team's work schedule, their respective roles and the monitoring of team and individual productivity.

It is important to demystify some items from the operational routine that if not considered among the abnormal states and which may indicate the point where process loss begins, so it is necessary to analyze the frequency of these deviances to investigate process loss.

Based on routine knowledge in chemical, petrochemical, polymer and transformation plants, it is suggested that the technical-operational culture should not ignore the following signs of failure:

- Dirty filter, dirty disperser, and pump tripping;
- Valve opening condition, variable mean change;
- Line noise, vibration, small leaks;
- Samples indicating abnormality;
- Level changes in intermediate vessels;
- Altered steam quality and cooling water return temperature on exchangers;
- Trends in change of mean and standard deviation, even if they are framed;
- Trends in production and quality, even if the items are framed;
- Staff irritation, cut-off communication, excessive aggressiveness or passivity;
- Obstructions or holes in reserve lines;
- Flange leakage and sealing of pumps, which have spare;
- Bypass operation;
- Valves operating on 0% or 100%;
- Manual operation of control valves;
- Effluent visual aspect, presence of water hoses;
- No registration of the team presence;
- Persistent health problems;
- Use of refurbished equipment and parts;
- Oscillations in recycling.

The preparation of an abnormal event map (MEA) (Fig. 4.11) in a process is based on records of nonconformity or signs indicating intermediate states of noncompliance. These signs detailed and confirmed from interviews with shift team leaders and from the interpretation of the production process are components of MEA. The MEA tool already detailed in Chapter 2, which presented the MEA methodology alongside the OD. However, it is worth highlighting some essential points for the application in the case study at the end of this Chapter.

4.4.5.1.1 Period for investigation

A period of 2–5 months was chosen to perform the mapping, in which production was involved with a number of abnormal events of various types, which caused low production, reduced productivity and environmental and occupational impacts. During this period, it is important to check process variables that have fluctuated or varied excessively, indicating loss of control in certain loops for process control. In addition, the systems and equipment that suffered oscillations was checked. It generated solid and liquid waste.

4.4.5.1.2 Process abnormalities

Operational events or facts that may be abnormal, intermediate, or normal are taken from the shift book. Critical production and process variables are extracted from process and production reports. These facts are processed and the resulting information indicates the probable causality between events in an attempt to identify the root cause of abnormal cycles.

Some factors that contribute to the process loss can be misinterpreted, which causes a classification of false or anticipated factors in relation to the problem. The facts may not be in explicit or visible form. They were constructed from inferred hypotheses from the data and occurrences recorded in the shift book.

It should be observed that there is much omission on the records of occurrences due to shift "protections" from administrative staff when it involves blame for failure or accident, as well as aspects in production, which may indicate incorrect decision-making with process losses. In addition to the omission, there is the delayed effect of the process problem transferring to the next shift, as in the following case: the high level in the process vessels is adopted to avoid transferring the nonconforming product in the shift, making it possible for the next one. After analyzing this situation, it

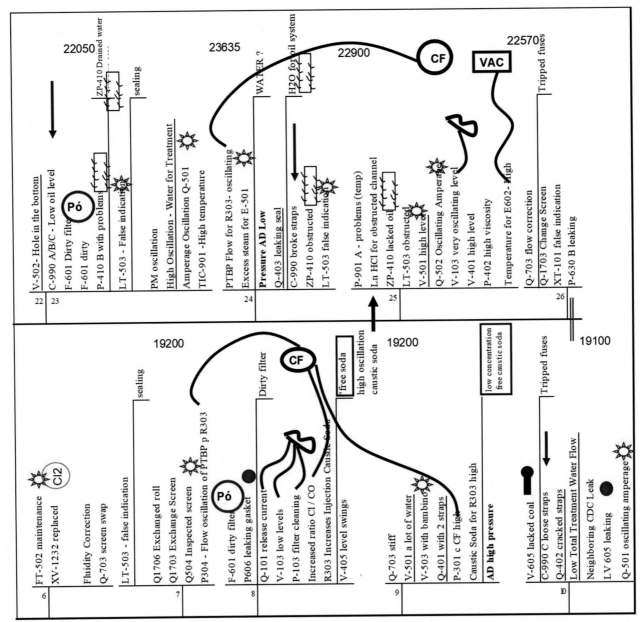

FIGURE 4.11 Abnormal event mapping. Ávila, S.F., 2004. Methodology to minimize effluents at source from the investigation of operational abnormalities: case of the chemical industry. 2004. 115 p. Dissertation (Professional Master Management and Environmental Technologies in the Technological Production Process)—Federal University of Bahia.

is clear that the operation tends to transfer the problem with low quality of information, thus generating high rework and reprocessing for the following shift that starts at 15:00 hours, considering 8-hour shifts.

The construction of the MEA is based on data relationships from specific criteria obtained during the study of the process and interviews with operation representatives. After the map is prepared, it is important to use the competency to infer events not visible in the production, process, management and people areas. This inference needs to be validated through the use of abnormal events statistics and critical variables.

The inference about the causal nexus of the abnormalities is performed through the use not only of the knowledge base, but also: the resulting mind map of alternative flowchart practices that include utilities and recycling lines, then resulting in the abnormality chain composed of operational factors.

4.4.5.1.3 Operational factors

Each chain of abnormalities has causal factors with different levels of failure energy or physical energy that increase process losses. When the operational factor is a cause and has high influence for the process loss to occur, it is named "root cause." There are other causes, which have lower influence in the generation of losses, which are called "secondary causes."

The consequent factors are those caused by the causal operating factors. The consequences generate losses at different levels. Some consequences only happen after others occur, leading to a sequence of events after the failure occurred.

When process loss has not yet occurred but the signaling of a problem occurs, the latent failure is said to occur. This operational factor is considered as a sign of failure.

Failure feedback with an increased likelihood of process loss can occur through recycling, winding, or sealing lines, which link the end of the process to its beginning. The engineering staff do not properly monitor the recycling lines. This lack of follow-up can often increase the power failure, reaching their limits, which increases process loss and causes great damage to the company.

Corrective action is taken after the loss of process occurs and without the real cause detection. Corrective action allows operations to resume, zeroing power of failure, while not ensuring that the problem does not recur due to root cause resolution. The preventive action depends on the real knowledge of the causal nexus of the failure, being an operational factor that depends on the clear and objective visibility of the failure process.

Corrective and preventive actions are active operating factors because they depend on the decision and action of the management and of executing bodies. Causal factors and their consequences are passive operational factors because they are not thought voluntarily to initiate failure. The signs of failure are passive. Decisions and related actions that are made based on knowledge and skill try to keep operations and processes in control. These decisions and actions are active operating factors and depend on human action and the construction of a mind map.

4.4.5.1.4 Description of failure & history resulting in process loss

The chain of abnormalities aims to describe the latent facts, which tries to demonstrate which activation points of the fault and which structural (technical and human), managerial and organizational characteristics promote its growth. An important factor to note is that there is a possibility that several abnormal chains occur in parallel with coincident factors, which leads to a complex structure. The complex structure can make it difficult to identify the root cause because there are multiple causes, factors resulting from certain abnormalities, being cause for others, or the material recycling, fatigue recycling lines, which generates new abnormalities. On the other hand, a root cause, if identified, can, when corrected through actions, disrupt the intricate network of various abnormalities.

4.4.5.2 Mapping losses

Mapping losses resembles the waste stream mapping discussed by Kurdve et al. (2015). The mapping lists the losses by type, which can be material (M) and energy (E). This Process Loss Mapping analysis uses the MD to gather information through interviews and data (Kipper et al., 2011; Santino, 2018).

Santino (2018) suggest applying the tool to identify Losses through Mapping of process, with these characteristics. The Frequency (F) with a range (from 1 to 3) that is based on the number of losses occurrences or failures in a given scenario. The Social, Economic and Environmental Impacts (I) is related to the causes of failure s in the process, and is resultant from the frequency type (1 to 3) for each of the three types of impacts. The Urgency (U) for action is graded as 1 to 3. For each factor, there are assigned criteria of quantitative and qualitative levels, according to Table 4.4.

TABLE 4.4 Loss assessment.

Value	Frequency	Value	Impact	Value	Urgency
3	Daily or >3 times week	3	High impact	3	High urgency
2	1 to 3 times a week	2	Medium impact	2	Moderate urgency
1	Up to 4 times a month	1	Low impact	1	Low urgency

This mapping helps to identify critical losses. The critical loss is the priority 1 loss in the treatment through the program against losses. Loss prioritization (Pp) is the result of the product/multiplication of the factors mentioned above. The result of prioritization suggests that the higher the value, the higher the priority for the search, according to the equation: $Pp = F \times I \times U$

In addition to the priority classification, to understand how to reduce losses, one must understand which variables influence the problem through root cause investigation.

The urgency can be classified as follows:

- High urgency—Loss has no action problems, endangers operator's life, or adds unnecessary high costs to the company;
- Moderate Urgency—The loss was predicted with some precision. Some actions were already planned or are in progress;
- Low Urgency—Losses do not cause moral or economic damage and there is already an action plan in progress.

4.4.5.3 Mass balance

The mass balance in the industrial environment may not be equivalent to what happens in the plant reality. Many volumetric measurements or samples do not accurately reflect the compositions and quantities of what goes through the pipelines and process equipment. Mass balance is a mathematical tool that guides decisions but does not show true reality.

Mass balance techniques are used in the methodology to avoid process losses. The purpose of using this tool is to issue/deliver/address actions to change or confirm components, characteristics and properties in the reaction, in unit separation operations, cooling and industrial effluents. Trends are confirmed from flow indicators, field tests and chemical analysis results from laboratory methods.

The mass balance is both an investigative technique to get to know what is happening in the process, and a learning technique that puts the analyst in contact with flows and compositions, thus indicating what is apparently happening in the equipment, but always considering other possibilities due to uncertainty level in the chemical industry.

Investigation of process losses may require an overall plant-wide mass balance that calculates inputs and outputs. Mass balances can be used for finding parameters of both system and cooling tower in order to estimate evaporated quantities or the balance by contaminant for the organic effluent to verify where contamination occurs. A widely used technique is the Data Reconciliation, in which, based on the quality of information, the original balance is adjusted because it originally did not appear to be correct, that is, the output flow didn't match the input.

The water balance helps in the investigation of excess water consumption and effluent generation. The water balance is related to the currents that involve the presence of more or less water in the industrial plant, both in the purest quality and wastewater. Water sources can come from process, rainfall, rotary machines cooling, liquid-phase cooling tower (clarified and chilled water), and steam systems. Other sources of purer water are demineralized water and potable water. At the plant outlet, there are aqueous streams with different qualities, classified: organic, inorganic and rainfall effluent. Material losses can contaminate the effluent, the atmosphere and can generate off-spec products. When losses are added to wastewater streams, it is difficult to reuse water, such as cooling tower makeup. In Fig. 4.12, Ávila et al. (2004) presents the Water Balance in a chemical industry plant.

The balance of contaminants in organic effluents present the levels of solvents, raw materials, products and other reagents that are released to the organic effluent through voluntary or involuntary disposal, samples, equipment cleaning, and other procedures. The components are chosen and the balance is performed in certain control volumes, which facilitates the investigation on the generation of effluents at the source. This can be applied, for example, to the distillation top stream in case of a vacuum operation using ejectors and sending condensate to the organic effluent after passing through the barometric vessel.

The solid waste balance makes it easy to verify the chemical and physicochemical principles that promote precipitation of materials, thus generating solid waste. The solid waste can be found on: (1) solid phase [inorganic (clean or dirty), organic (clean or dirty)]; (2) in the pasty phase, mixed with organic effluent requiring concentration to avoid excess volume generation. Studying this mass balance and principles allows mathematical handling of precipitation efficiency by simulating situations.

The process balance includes internal (per unit) and external (effluent generation) processes, promoting a more complete exercise regarding the influences of each conversion and separation efficiency. This balance is represented by simple mathematical exercise and it is possible to include more accurate and nonlinear estimates. Another way to prepare a

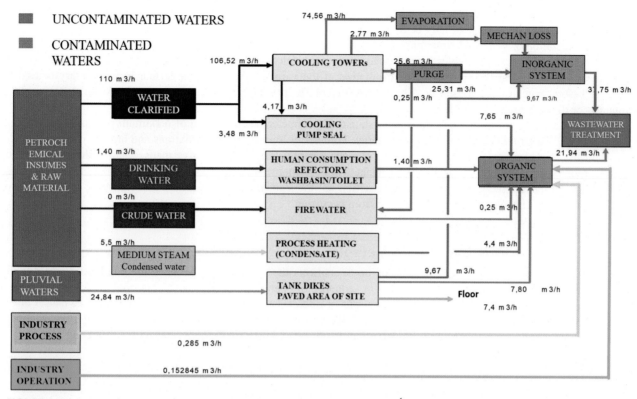

FIGURE 4.12 Water balance—reduction of water consumption through water balance. *Ávila, S.F., Alcântara, C.M.D., Kalid, R.A., Fiscina, F., Cunha, E., Siqueira, R., 2004. Reduction of water consumption through water balance. In: Congress of Excellence in Management, 2004, Niteroí. Anais of the Congress of Excellence in Management.*

process balance is through the Solid Material, which, aided by the data from control panel and process reports, becomes efficient in identifying processes with high material loss.

The integration between mass balances gives an insight into waste generation from the source located in the process. This overview makes it easy to study the source of process losses. It is important that both water balance and process balance include intermittent process currents as well as currents generated from the operational routine. To be able to include aspects of operation in the mass balance, it is necessary to know in depth the operational routines and the habits or technical culture of the operation.

Some tools are needed to maximize the accuracy of mass balance: pre-existing calibrated measuring instruments, field measuring instruments for the balance, confirmatory measurements for statistical validation, direct (laboratory) and indirect (rapid methods) stream analysis, specific mass balance packages (computer programs), estimates and considerations, pluviometry table, criteria for how to operate and use water/sample contamination and mathematical tool for reconciliation.

4.4.5.4 Statistical analysis of process and effluent variables

Statistical process analysis is part of the methodology to detect scenarios before losses occur, thus enabling preventive actions to avoid losses. To find the root cause of problems, when not immediately recognized, a robust data investigation is required. By understanding the relationship of variables, themselves and with events or occurrences, the intention is d to reduce the number of variables to be investigated.

The choice of process variables depends on the abnormal event map (MEA) and the construction of the causal nexus with the suggestion from a set of abnormalities that can cause process losses. The variables indicate the appearance or existence of problems. Variables are considered signs of failure when they exceed certain limits or when they occur sequentially, such as alarms that sound in sequential order.

Changing process variables can cause process failure or loss when it generates a state that causes system leakage or shutdown. The change of variable may be a consequence of other actions, or it may be corrective or preventive action in the normalization of generated situations.

Statistical monitoring of variables is performed using graphs/plots and statistical parameters such as: standard deviation, mean value and variability factor. The oscillation of the variable may be so large that it is necessary to find an average before generating graphs/plots. The monitoring of the process or quality variable should be done monthly, annually or during specific periods.

Some variables are represented by secondary statistical parameters such as: sum of maximum pressures at compressor discharge, Napierian Logarithm of hazard energy. For energy: sum, standard deviations and others.

The MEA technique investigates losses using statistical parameters and qualitative maps, which indicate the loss of process quality, product, raw material and effluents. These phenomena, when identified and analyzed through graphs/plots and tables enable the preventive-corrective action on: the standard, equipment and process setup; process control and design/project changes to prevent the failure from occurring.

4.4.5.5 Statistical analysis of abnormal events

Information extracted from the shift occurrence book confirms the relationships between the continuous and discrete process and production variables with the events or sequence of operational factors. It is important to indicate the statistical behavior of abnormal events. Certain assumptions or inferences are confirmed through statistics. The causal nexus of the events and their operational factors, that cause the discussed process losses, and validated through the statistical analysis of abnormal events (EVA), indicate statistical treatment of abnormalities considering the following aspects:

- Sector for occurrence process;
- Maintenance specialty; time of occurrence and events before and after; shift class;
- Product campaign; in the case of polycarbonate process, it influences a lot;
- Events with static equipment (obstruction, hole, rupture, failures in internals);
- Events with dynamic equipment (cavitation, disarming, sealing, and lubrication);
- Process and safety instruments (control valve, alarms, PSV, relief); and
- Information about people and culture (attention, memory, behavior, commitment, knowledge, writing, communication).

The Process Sector, in which process failure or loss is detected, should be recorded as well as the equipment and maintenance expertise involved. This information will compose the causal analysis. Although the failure is detected in a sector, it can originate in another sector of the process. The problem involves equipment where products pass, hence the need to check which area of maintenance is responsible for correcting the damage caused by operational failure. The most common classification involves: boiler, mechanical, electrical/instrumentation.

Knowing the shift class and time facilitates the interpretation of what probably happened and the construction of the mind map about events. Some aspects regarding leadership or the way to communicate in the shift class can be evidenced by this investigation.

Each process campaign has different characteristics in both procedures and process conditions. Thus some campaigns are more critical than others are. Another important factor is the time it takes to achieve stability.

Several events involve problems with vessels, (heat) exchangers, distillation towers and other equipment that do not have movement of their parts. In these cases, the movement occurs with the fluid passing through them and the failures can involve their internals, so it is extremely important to design the phenomenon. The failures that may be occurring in static equipment are: holes, obstructions, ruptures, and tampering.

Other events occur in fluid transfer equipment or in the use of centrifugal motion principles. In this way, some equipment such as pumps, compressors, extractors, face damage to their main function, which causes disarming and maybe reaching cavitation, as well as damage to auxiliary sealing or lubrication systems, which causes leaks.

Instrument failures can occur during equipment operation and industrial processes related to false indications and errors in control valve operation, such as improper use of bypass. Other instruments and accessories help control the safety of processes and equipment, such as safety valves, relief valves and rupture disks. Abnormalities may involve failures in these safety systems. It is also important to check when there are faults in the plant interlock logic and what needs to be changed to increase the safety level.

The operator, supervisor and engineer can be actors or authors of process losses, making events complex, involving human intrinsic variables and cultural influence on the resulting technical systems. Accordingly, if the social factor in the failure that causes process loss is identified, it is important to assess whether it involves these aspects:

- memory loss/replacement;
- Inattention (then skipping task steps);
- Non-acceptance of the rules/organizational pattern;
- Lack of theoretical and practical knowledge;
- Communication problems and lack of interpersonal relationship due to cultural preferences (not favoring cooperative behavior).

4.4.5.6 Production audits

Audits help confirm the state of the process that has been flagged by statistics and mapping of abnormalities. To prepare the diagnosis for process losses, it is necessary to interpret the trends for the variables, verify the causal link presented by the process mapping and confirm important aspects of production in the field. This field confirmation is done through production audits (Ávila et al., 2004).

The audits used for diagnosis, as indicated in Fig. 4.13, are for: (1) procedures and standards; (2) effluent; (3) cooling tower performance; and (4) control valve.

In general, audits for procedures and standards present the difference between the expected state and the actual state of the process after the procedure is performed. In this type of audit, the following are evaluated: the procedure writing, the necessary tools, following requirements and the accomplishment of the steps and the final process state after the procedure.

The audit for effluents presents the resulting process state, mainly in the area of organic effluents. By following the logic of the process and collecting samples, estimating flows, it is possible to know what happens in the process of integration with the effluent, where the contamination points to be reviewed at the source of the process and responsible equipment are.

The audit for cooling tower shows the performance in terms of energy, water consumption and quality of the cooling water produced. Here, important factors are the following: tower operation, maintenance, production system operation, recycling, water treatment, and the procedures involved.

The audit for control valve indicates the process state regarding control and field operation condition in case of over or under load design and the staff's level of knowledge.

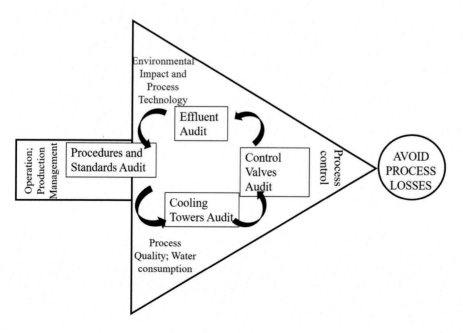

FIGURE 4.13 Audits to confirm hypotheses. *Author's original.*

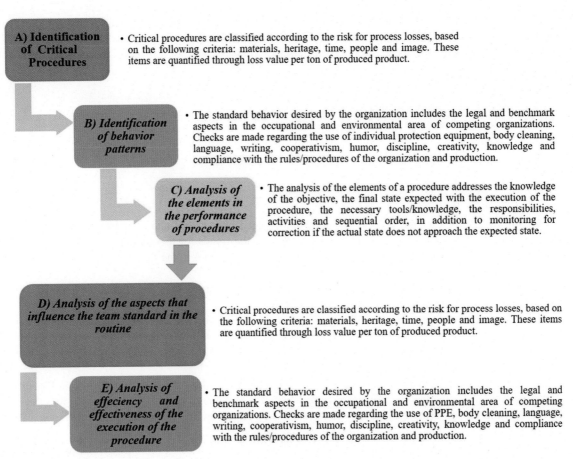

FIGURE 4.14 Audit for PADOP procedures and standards. *Author's original.*

Thus in a systemic manner, in order to diagnose process losses, opportunities should be identified in the areas of operation, production management, process technology and its environmental impact, process control, water consumption and process quality.

The elements that can cause human error are raised after reading the critical procedures (product quality, plant continuity, safety, team motivation and image) and after the field visit at overtime to perform the audit for procedures and standards. This audit methodology is summarized chronologically and functionally through Fig. 4.14, and is part of the PADOP Tool, Operating Procedures and Standards, developed by (Ávila, 2010; Ávila and Costa, 2015) and applied in the case of aromatic polymers and amines. The PADOP tool was described in Chapter 3 of this book after the discussion on human and social typology. It comprises three steps: Identification of Environment Task (IAT), Processing Cognition based on Task Execution (PCTE), Assessment of Task: Effectiveness and Efficiency (ATEE). In Fig. 4.14, Step A and B of the audit for procedures is part of the PADOP Methodology, Environment Identification Phase IAT. Audit Step C is in the Cognitive Processing and Task Execution Phase PCTE. Audit Steps D and E are inserted into the Task Control Phase, ATEE.

4.5 Cases: diagnostics with quantitative and qualitative analysis

The cases that apply the OD tools to a polymer plant and MD tools to a metallurgical plant are presented in the block diagram of Fig. 4.15. The steps related to Event Map, Statistical Event and Variable Analysis, and Audit Analysis were applied to the case of Polymers and Amines. The Process Mapping and Mass Balance steps were applied to the case of metallurgy. The first case was part of the discussion of the master's degree by Ávila (2004) and the second case of the master's degree discussion of Santino (2018). In the master's work of Ávila (2004) and Santino (2018), the results were discussed, showing the potentialities considering the OD and the MD.

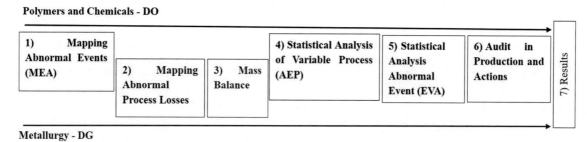

FIGURE 4.15 Steps of process loss methodology. *Author's original.*

4.5.1 Chemical and polymer case: diagnosis based on operator's discourse

Company Z is a medium-sized industry capable of producing 50,000 tons per year of polymer products. The industrial plant is of continuous process and is subdivided into reaction section (SR), separation section (SS), purification section (SP) and effluent treatment section (STE). This factory had process variation problems lowering its production capacity and impairing the quality of the final product. The technology of this process indicated a situation of greater comfort regarding the continuity of the plant, and there was a potential to reduce production costs, including the reduction of MC (solvent) consumption. For this activity, an analog type control panel was used for the polymer site. The period in which the methodology was applied in this polymer factory was of 2 years.

The final product of the polymerization reaction—P has specific physical characteristics depending on the amount of T, reaction-finishing element. Each type of P produced has different physical characteristics and implications in the production chain. The resin P produced in this plant is a condensed polymer of L and E, which have physical strength.

L is the limiting reagent from a gas phase reaction between An and B. To use An in reaction 1 it is necessary to promote its purification through physical processes. The preparation of P involves two-step condensation reaction (reaction 2 and 3). Initially E reacts with L in basic medium. In this reaction L is fully consumed. After these reactions, the separation processes take place only in the liquid phase.

The organic phase separated and sent to polymerization, third reaction, the inorganic phase sent to separators returning with P for reaction 3, and sending process wastewater for purification 2. In purification 3 solvent MC is recovered for the process. In reaction 3 the viscosity and P concentrations are monitored to avoid problems in treatment and drying—Purification 4. Viscosity adjustments are required by injecting MC causing P dilution.

Resin P when leaving reaction 3 has the chain formed but has quality affecting contaminants and needs to be removed through purification step 4. At this time contaminant ions, water and MC solvent accompanying the organic phase of the process are removed. Then, to the treatment section for removal of ions in the P molecule through rotating equipment, that mixes the aqueous phase and the organic phase. Also, in this section excess water and P are separated, after treatment, releasing the ions attached to the molecule to reduce moisture, leaving MC that will be removed on drying. For this operation, rotating equipment operating with heating is used to remove H (humidity) and MC (solvent). See P Flowchart—Fig. 4.16.

The product P is initially flashed on drying 1 promoting its concentration to the pasty state. P is then transferred to rotating equipment, which, with heating and vacuum, promotes almost total removal of MC and H, leaving a dry powder form.

The solvent (MC) and water (H) from drying are condensed and redistributed: MC is returned to the storage tank and H is reused in the treatment section and sent to the wastewater separator. Uncondensed MC goes to the chilling column using the cryogenics principle. P is transported to the storage silos and then sent for physical transformation. In this step are added masters with specific additives to meet commercial characteristics.

The methodology that uses the Operator's Discourse, recorded in the shift book, that will form the MEA and the various chains of abnormalities, needs statistical validation and audit actions to prepare for operation interventions and modifications in the project.

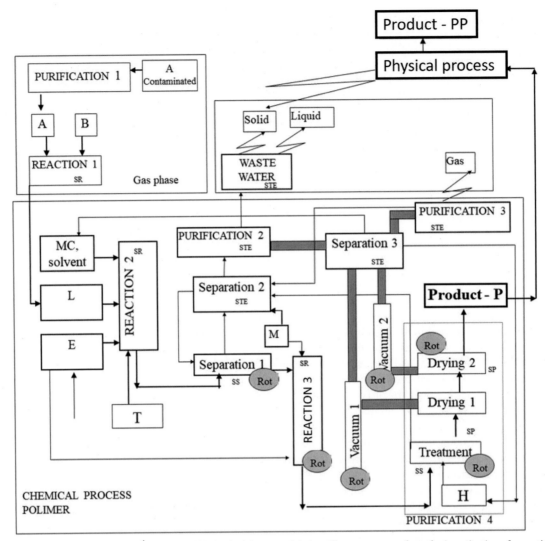

FIGURE 4.16 Polymerization process. Ávila, S.F., 2004. Methodology to minimize effluents at source from the investigation of operational abnormalities: case of the chemical industry. 2004. 115 p. Dissertation (Professional Master Management and Environmental Technologies in the Technological Production Process)—Federal University of Bahia.

4.5.1.1 Phase 1—Abnormal event map and chain of abnormalities

A 9-day period extracted from the MEA shown in Fig. 4.10, depicts the events described in the shift book during a critical 4 to 6-month production period at a Polymer Plant. The complete MEA is described by 23 complex abnormality chains, in Table 4.5. The abnormality chain of item 4, oscillations in reaction 2, is described in Fig. 4.17. Note the complexity and recycle of the failures.

TABLE 4.5 Chain of abnormalities in the polymers case.

Item	String name	Main function	Process
1	Purification disorders 2	Remove organic from effluent	Controlled vaporization
2	Purification operation 3	Retain plant off-gas MC	Adsorption
3	Polymer affects separation	Separate organic/inorganic	Separation
4	Reactions 2	Polymerization reaction	Reaction
5	E feed	Feeding flow	E Mix and Feed

(Continued)

TABLE 4.5 (Continued)

Item	String name	Main function	Process
6	Cooling oil missing	Cooling down	Cold for the plant
7	Purification Abnormalities 1	Purify A	Separation
8	Purification Abnormalities 3	Purify off-gas	Reaction abatement
9	S power	Basic medium	Feed
10	Ac power	Neutralize effluent	Reaction
11	Treatment lubrication	Remove ions	Extraction
12	Straps in treatment	Remove ions	Extraction
13	Treatment efficiency	Remove ions	Extraction
14	MC pressure	Solvent and sealing	Pumps
15	Airline obstruction	Fluid transport	Pumps
16	Dosing interns	Fluid transport	Pumps
17	Drying problems 1	P drying	Flash Vessel
18	Vacuum abnormalities	Dust transport	Pneumatic conveying
19	Water temperature	Cooling	Utilities
20	Feeder disturbances	Feeding Flow	E feed (solid)
21	Vacuum deficiency M	Vacuum	Off-gas
22	B in L	Reaction 1	Reaction
23	T abnormalities	T feed	Pump

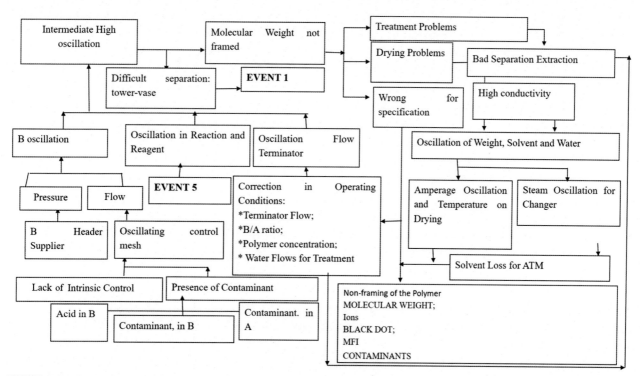

FIGURE 4.17 Chain of abnormality 4, oscillations in reaction 2, polymerization. Ávila, S.F., 2004. Methodology to minimize effluents at source from the investigation of operational abnormalities: case of the chemical industry. 2004. 115 p. Dissertation (Professional Master Management and Environmental Technologies in the Technological Production Process)—Federal University of Bahia.

4.5.1.2 Phase 4—Statistical analysis of process variable and effluent

Phase 4 is the result of Phase 1, where after Mapping Abnormalities it is possible to identify critical process, effluent and operation variables to meet cost, environmental, quality, quantity produced and safety goals. In the case of the aromatic amine plant, Table 4.6, shows the main process and product variables for controlling aromatic amine recirculation for the raw material tank. The quantity of input consumed gives an idea of the amount of waste to the effluent and the production indicators show whether the operating routine follows best practices, whether the production team has cooperation and commitment characteristics.

Fig. 4.18 presents effluent parameters to estimate process stabilization through loss reduction and reduction of solid material accumulation, also indicating loss reduction. From July to May the effluent profile was balanced in terms of flow and amount of accumulated solid material, in the case of the aromatic amines chemical plant.

4.5.1.3 Phase 5—Statistical analysis of abnormal event

Abnormal events were ranked by maintenance as priority, investigating what were the connections, seeming to be complex in routine analysis. Table 4.7 indicates the number of times per year the event occurred after which a correction for criticality effect is made depending on the presence or absence of chlorinated solvent MC. Thus the event to be statistically studied is ordered. Statistical indicators of events by specialty, shift class and time are cited in Table 4.8 and the discussion of the number of events for rotating equipment is in Table 4.9.

The graphs/plots in Fig. 4.19 indicate the EVA by shift class, time, and classes without abnormal events. Fig. 4.19 shows the grouping of events indicating the existence of a pattern with the respective events in the operational routine.

In Fig. 4.19, we analyze the data groupings that can bring important information if they are properly treated by statistical tools. Each group of information is identified by number, indicating situations likely to happen, and is analyzed through repetition pattern and various relations, then showing the existence of chronology/order in the events.

TABLE 4.6 Process and product variables at chemical case.

(a) Analytical results: raw material, process and product		
Raw material	Process	Product
A purity	AC/A ratio; % S in SN	AG in Pl; An in Pl
Pl at A	Pl at R1 (recycle 1)	viscosity
AC purity	Salt in Pl to SP	2Ø in Pl; Pl color
(b) **Process parameters:** Temperature in reactions R1–R3; SP temperatures		
(c) **Production:** Quantity produced; Stop hours and reasons		
(d) **Index:** Consumption A/production; Consumption B/production; AC consumption/production		

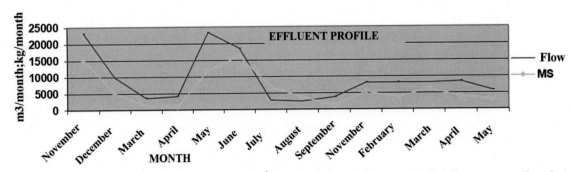

FIGURE 4.18 Monitoring volume, ms effluent chemical industry. *Ávila, S.F., 2004. Methodology to minimize effluents at source from the investigation of operational abnormalities: case of the chemical industry. 2004. 115 p. Dissertation (Professional Master Management and Environmental Technologies in the Technological Production Process)—Federal University of Bahia.*

TABLE 4.7 Priorities in chain of abnormalities at polymer case.

Event	Abnormality	Frequency	MC consumption	Result	Color %	Order	Event
01	Purification polymers 2	3.26	S	3.90	3.42	16.14	18
02	Purification disorders 3	4.35	S	5.22	4.57	14.29	17
03	Polymer affecting V-3/P-3	2.17	S	2.60	2.28	8.57	13
04	Oscillations in reaction 2, intermediary	4.35	S	5.22	4.57	6.29	5th
05	Improper feeding of E and S	5.98	S	7.18	6.29	6.29	14th
06	Lack of compressor oil	5.43	S	6.52	5.71	5.71	06
07	Purification abnormalities 1	3.26	N	3.26	2.86	5.71	23
08	Washer abnormalities	2.72	N	2.72	2.38	4.57	02
09	S power	4.89	N	4.89	4.28	4.57	04
10	HC feed	1.63	N	1.63	1.43	4.28	09
11	Lubrication problems	2.72	S	3.26	2.86	3.42	01
12	Extractor belt	1.09	N	1.09	0.96	2.86	07
13	Low treatment efficiency	8.15	S	9.78	8.57	2.86	11
14	M pressure for pumps	5.98	S	7.18	6.29	2.38	08
15	Airline obstruction	2.17	N	2.17	1.90	2.38	16
16	Internal pump problems	2.72	N	2.72	2.38	2.38	20
17	Drying problems	13.59	S	16.31	14.29	2.28	03
18	Vacuum abnormalities	12.5	S	15.00	16.14	2.28	21
19	Water temperature	1.09	N	1.09	0.96	1.90	15
20	Q-1 disorders	2.72	N	2.72	2.38	1.43	10
21	Vacuum deficiency header M	2.17	S	2.60	2.28	0.96	12
22	B no L	0.54	N	0.54	0.47	0.96	19
23	P-04 abnormalities	6.52	N	6.52	5.71	0.47	22

Total = 114.1.

Data groupings are gathered through rectangles that circle the set of events indicated in Fig. 4.18. Group 1 involves spheres (EF), pumps (B0303), filters (FDCORR) in LPG feed, and corrosion events (CORRL). In group 2, there is a pattern repetition, but with some dispersion involving the air (KAR) and refrigeration (KR) compressors, which would be indirectly related by the operation of the control valves. In group 3, various events are listed as high pressure in the systems (high Pressure): flash (FLASH), flash pump (condensate), refrigerated tank, and pump (TQ72), indicating that pressure fluctuations in the plant may cause oscillations in the flash circuit and refrigerated tank. In group 4, where general (GENERAL) and electrical problem (ELDUTO) events are repeated.

The blue arrow in Fig. 4.20 allows noticing that for each 12-day period, some events are repeated with a tendency to decrease between January and February 2006. The data groups on 5, 6 and 8 indicate the occurrence of events in clusters but not repeated for certain frequency. On 5, it is represented events with a fire system (SISINC). On 6, events involving cooling (REFRI) are indicated. On 8, events on the pier. Due to the higher data density and due to indicators related to plant performance, this analysis concludes the best is to choose the cluster that involves filters, or filtering including events with the sphere and the pumps.

TABLE 4.8 Statistical indicators of maintenance, classes and time.

Period	N of event	Shift without E	Load change	Maintenance				Class					Schedule		
				Mechanics	Instrument	Boiler shop	THE	B	W	D	AND	0–8	8–16	16–0
(1) 2nd May	60	13	13	13	5th	8th	5th	7th	8th	11	6th	13	12	13
(2) 1st June	61	11	22	16	14th	16	10	7th	16	12	4	8th	22	15
(3) 2nd June	45	22	30	16	3	14th	9th	4	16	2	6th	6th	25	6th
(4) 1st July	49	17	18	16	8th	10	2	4	4	9th	13	15	16	7th
(5) 2nd July	59	13	35	24	10	10	3	13	16	10	12	7th	33	13
(6) 1st August	57	8th	22	43	3	12	10	9th	10	10	6th	15	18	12
(7) 2nd August	64	11	22	30	5th	13	7th	10	16	8th	9th	10	19	21
(8) 1st September	84	9th	10	31	19	15	16	18	14th	5th	14th	20	30	17

TABLE 4.9 Unit of intermediary product statistical data (rotating with abnormalities per month).

Equipment	Total	May	June	July	August
BSR1	12S 1D	2 s	4 s	2 S 1 D	4S
BSR2	1S	–	–	1S	–
BSN1	1S	–	1S	–	–
BSN2	2S	1S	–	–	1S
BSP1	2S	1S	–	–	1S
Tot. pump Transf	18S 1D	4S 0D	5S 0D	3S 1D	6S 0D
BVSP1	2S	–	–	2S	–
BVSP2	12D	1D	9D	2D	–
BVSP2	3D	1D	1D	1D	–
Tot pump vacuum	2S 15D	0S 2D	0S 10D	2S 3D	0S 0D
BA	1S	–	–	–	1S
BAC	1D	–	–	–	1D
COLD	1D	–	–	–	1D
Total	21S 18D	4S 2D	5S 10D	5S 4D	7S 2D

D, trip; *S*, sealing; *BSR1*, reaction section pump 1; *BVSP1*, purification section vacuum pump 1; *BA*, pump transferring A to SR; *COLD*, ice water system.

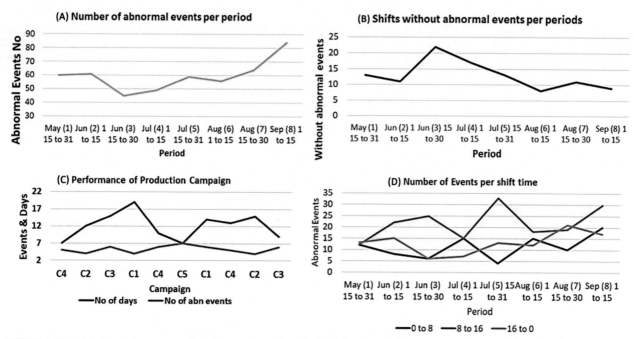

FIGURE 4.19 total abnormal events, number of events, campaign, time. (A) Number of abnormal events per time. (B) Number of shifts without abnormal events in the period. (C) Performance of production campaigns (Event A). (D) Number of events by period. Ávila, S.F., 2004. Methodology to minimize effluents at source from the investigation of operational abnormalities: case of the chemical industry. 2004. 115 p. Dissertation (Professional Master Management and Environmental Technologies in the Technological Production Process)—Federal University of Bahia.

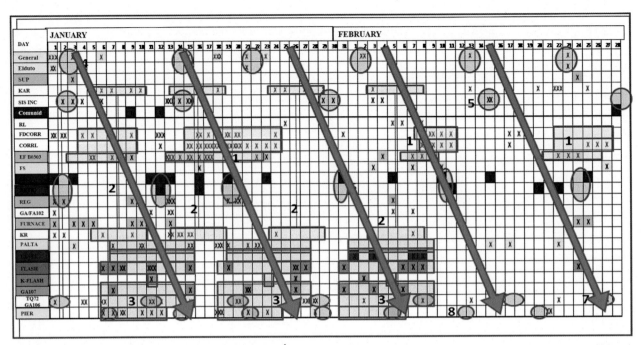

FIGURE 4.20 Analysis of grouping on process abnormalities. Ávila, S.F., 2010. Etiology of operational abnormalities in the industry: modeling for learning. 2010. 296 f. Thesis (PhD in Technology of Chemical and Biochemical Processes)—Federal University of Rio de Janeiro, School of Chemistry, Rio de Janeiro.

4.5.1.4 Phase 6—Audit for production and actions

The audit for production is performed based on the diagnosis from the identification of losses due to operational abnormalities. This diagnosis is initiated by mapping the chains of abnormality raised by analyzing the operator's discourse and after identifying the region where the hazard is being released. Preliminary assumptions and audits confirm what the necessary changes are for the task, personal, process, or equipment. Interventions begin and new audits are performed. Once the benefit of certain design changes is verified, the required investment is made.

In terms of diagnostics, process and environmental audits, it was found through investigation of abnormalities, that the loss of MC (methylene chlorine) to the atmosphere is a problem that involves virtually the entire plant and all activities. Thus after the establishment of statistical process monitoring, the focus was on the activities for reducing MC losses. If the work was done in isolation, as the investigation of abnormalities from the shift report, the actions would be based on hypotheses and would lose strength with the application of the proposed changes in the area. That is, the work performed was based on validation by the operation team, with formal leaders, informal leaders and those interested in the improvements.

Table 4.10 describes the activities performed, the period, the expected, and the achieved effect in the inventory, the expected and the achieved effect in the effluent. Codes were used to define whether the actions taken were Project (P), Operation (O), Process (Pr), Maintenance (M), Management (G), and Vendor (F).

Other activities performed to reduce and maintain minimum MC losses in the process are process monitoring (product quality vs operation), monitoring the quality of maintenance services, adoption of new standards regarding the area cleanliness and organization, and plan the plant stops; and daily monitoring of production indexes.

As presented in Fig. 4.21, the difficulties of correlating reductions or increases of the index in relation to the actions taken, in each day, were noted, thus making the necessary changes in the process, design, management, supplier, maintenance and, only four days later, it was changed to a new operating condition. In the beginning, very high results (above 2000 ppm) were obtained in the effluent A. Perhaps carelessness would lead to greater contamination. The fact that people being involved to reduce contamination provided more accurate care in operating the system. The entry of new equipment into operation (B) contributed to a greater reduction. The operation of purification section 2 (C) made it possible to achieve this reduction during two months.

There were disturbances that caused this behavior in D. Probably some out of control system, plant shutdown, equipment overflow with material in the effluent (lack of equalization).

TABLE 4.10 Activities, period, expected effect and effect achieved on inventory and effluent.

CODE	Performed activities	Period	M loss ratio		Effluent MC content	
			Expected effect (kg MC/ton P)	Effect achieved (kg MC/ton P)	Expected effect (ppm STE)	Effect achieved (ppm STE)
1Pr	Preliminary diagnosis	April		550		2400
		May		430		970
2P	Purification 2, new equipment	June	390	280	650	854
3P	Purification profile settings 2	July	347	285	575	830
4P	Temperature control	August		225		430
5 M	Reduce pressure pumps	September	320	350	520	468
6F	Supplier meetings	October	246	210	380	330
7M	Services: pipe will, vacuum pump purification equipment 4	November	236	200	318	305
8O	Review procedures p W	December	206	152	300	160
9O	E and S Power	January	200	100	300	215
10O			200	90	300	110
11M	Plan shutdown—Improvements	February	180	110	300	120
12G	Operator training	March	180	173	300	400
13P	New equipment for purification 3	April	100	94	250	513
14Pr	Process tracking					
15M	Cold system maintenance					
16O	Preventive work support O-RING	May	100	58	250	150

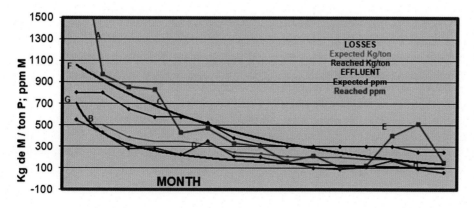

FIGURE 4.21 MC losses and effluent contamination. Ávila, S.F., 2004. Methodology to minimize effluents at source from the investigation of operational abnormalities: case of the chemical industry. 2004. 115 p. Dissertation (Professional Master Management and Environmental Technologies in the Technological Production Process)—Federal University of Bahia.

It was noted that the reduction tendency at the beginning of the process was more drastic (E). In the effluent, there was a 500-ppm drop from April to September. On the intermediate dates (September to January), this drop was of 300 ppm. On the final phase (January to May), the drop was reduced to 120 ppm. Drop reduction over time. In order to make big changes, it is necessary to invest in new technologies and/or culture change in the company.

Between April and September, there was a drop of 500 kg/ton of P. Between September and January, the drop was of 70 kg/ton. When considered from January to May, the drop was of 30 kg/ton of P (F). It was noted that the biggest achievement of this loss reduction process was the operators' desire to keep this item under control (G). It was reached, as indicated by the trend curve, 140 ppm for the effluent MC content (expected 250 ppm) and 100 kg/ton P for the inventory loss (expected 100 kg/ton), thus achieving this goal.

4.5.1.5 Phase 7—Discussion of results

In the research work involving the polymer plant, the terminology in Fig. 4.21 is used as the tools applied to achieve the results. The techniques used were more detailed than those presented herein, which were abnormality mapping, variable statistics, abnormal event statistics, and audits for systems and respective actions. The tools applied in polymer and chemical cases to achieve the results were:

1. DIPEA—Process dissection and abnormal events;
2. EVA—Statistical analysis of abnormal events;
3. AEP—Statistical process analysis;
4. EVOP—Test in normal operation and operational events;
5. PADOP—Standards and operating procedures;
6. ROTLIM—Clean routines;
7. MCA—Environmental control map with audits;
8. AUDIP—Channeled audit;
9. AUDIL Cleaning audit;
10. Fugitive emissions;
11. ALO—Localized activities;
12. Project physical improvement;
13. MULTI—Multiplier agents;
14. TREINOP—Training for operation;
15. EDUC—Educational programs;
16. Self-management for the operation and measurement of results.

As an example, process variable stabilization was achieved by applying the methods DIPEA, AEP, EVOP, PADOP, ALO, physical improvement and measurement.

According to Ávila (2004), during the period from 1992 to 1998, works were developed on factory operation in industrial plants that produce aliphatic, aromatic amines and polymers. In these works, the main goal to be achieved was the stabilization of the processes to avoid losses through the use of technology.

In the case of aromatic amines, the investigation of abnormal occurrences was performed, and a report was generated that was validated by the shift team. This validation was done through interviews, teamwork, tests for variables or changes in procedure. The identification of procedures, equipment and variables to be monitored and changed came from mapping abnormalities in the process. Results were measured after changes in patterns, parameters and variables comparing with the ideal state. This analysis of results in relation to the project is described below. The main result achieved in implementing this methodology was to reach the state of self-management, in which leaders were formed to implement training and solve problems.

In the case of polymers, the methodology was completely carried out except for the formation of leaders for self-management. In this case, the results measurement and task monitoring were done more concisely.

For each desired result, a technique for abnormalities mapping and diagnostics in the areas of production, process and environment are used. Part of the continuous variables were measured to achieve the 6 Sigma goal or 99% certainty within the standard range. Other items were measured in the aspect of time, compared to the standard time, such as equipment availability, overtime and reworked time.

Also, in relation to results, the following can be counted: the number of spare parts for process equipment, the number of accidents for a certain period of time, or the number of lost customers. Described below are the results, the techniques used and the expected and achieved results for each case (Ávila, 2004). Stabilizing process variables to reduce process losses. The expected result was that 80% of the process variables, around 15 variables, were within the pattern defined by Six Sigma, that is, the range of six standard deviations.

In the case of aromatic amines (C2), 80% fit the pattern defined by four sigma and in the case of polymer (C3), 13 variables, which account for 70%, fit into six sigma. Process mapping (MEA), statistics (AEP), standards review (PADOP), and testing (EVOP/ALO) were important to achieve these results. Correlation of certain variables reduced the amount of data to be controlled by the operation.

4.5.1.5.1 Improving liquid effluent quality

A very ambitious result was sought because 100% of the variables (12) were in the four-sigma range, that is, 95%. In addition, the procedures (PADOP) and processes (AEP/EVOP) were revised, which allowed to correct the effluent state in 12 hours. The result in case 3 was better than in case 2. There was an improvement from 3 to 4 sigma by 70% (8 variables) for Case 2, and by 90% (11 variables) for Case 3. The polymer industry (C3) reduced the time to solve effluent problems (from 48 hours to 12 hours), the same happened to amines although the situation in this case (C2) was less stable (from 72 hours to 24 hours).

4.5.1.5.2 Increasing the availability of critical equipment

Availability is considered the inverse of unavailability. It was expected to reduce the unavailability of critical equipment by 50%. In case 2, a drop was intended from 20% downtime to 10%, that is, increasing 80% availability to 90%. In case 3, the desired was a drop of downtime from 5% to around 3%. A reduction of 60% was achieved in case 2, which meant availability of 92%. A reduction of 70% was achieved in case 3, thus availability from 95% went to 98.5%. To make this analysis, critical equipment was chosen. In case 2, the cargo pumps and vacuum pumps; in case 3, the compressors.

4.5.1.5.3 Decreasing spare part consumption

The reduction in the use of spare parts in pumps depends mainly on the knowledge of operating conditions, maintenance procedures and equipment conditions that returns to the production process. The projected goal is to reduce the consumption of these parts by 30%, then indicating a reduction in purchases to maintain the industrial plant. Because of the work load on the pump and the control reaction in Case 2, a reduction of 20%, below the expected was achieved, while the work on alpha-Laval pumps (Case 3) and solvent control for the process reached a 40% reduction, thus 10% above the goal.

4.5.1.5.4 Decreasing overtime expenses

The goal set to be reached was a maximum of 1 overtime (HE) per employee. Considering that each employee works 40 hours per week, then 180 hours per month, this means 0.5% of the goal. On average in case 2, 5% HE (10 hours/month/operator) was already used and in case 2, 2% HE (4 hours/month/operator). The review of procedures and the knowledge of the causal nexus for operational problems led to a reduction in process oscillations, then leading to less overtime labor requirements, reaching 1.5% in case 2 (10−3), and 1% in case 3 (4−2), thus not reaching the goal but getting close.

4.5.1.5.5 Decreasing reworked hours

The intended was to reduce by 70% the number of reworked hours, which means that from all services performed with estimated time, those that are repeated within the period of two months, are considered rework. Considering that the rework in case 3 was accompanied by 5% of the hours of service performed, the plan was to reduce to 1.5% of rework. 50% was reached, which means to be above the goal (2.5%).

4.5.1.5.6 Improving motivation

As a result of the actions described above and some to be described below, mainly with the involvement of the operator, through the shift book and the revision of standards and procedures, the estimated participation in the volunteer events was 85% of the possibilities, thus showing that operator's interest in feeling included in the organization. In case 2, 75% was reached; and in case 3, the participation was lower, 65%, since the self-management state was not reached.

4.5.1.5.7 Decreasing accidents and incidents

Talking about motivation can also mean being more awareness of the risks of accidents and incidents. Thus the same previous reasons (motivation) are causing a decrease in the number of operational/environmental incidents and lost-time accidents. It was expected to reach 1 lost-time accident for each 2-year period and 2 operational incidents (involving downtime) each year.

In case 2 (aromatic amines), we started with two accidents per year to one accident per year (reaching the target). In terms of operational incidents, the change occurred from six to just four incidents per year (not meeting the target but reducing and improving the process state in terms of downtime without involvement with Man).

4.5.1.5.8 Minimizing waste at source

The industry's biggest challenge is to minimize wastewater generation to zero, making the plant with closed mass balance without losses, that is, with optimum conversion and separation efficiency and routines that meet the expectation of no waste or practices that cause waste considered as residue. A 70% reduction is desired in waste/effluent generation at the source, avoiding the application of "consecutive" methods for its treatment, wasting energy and generating more waste. To meet the goal of reducing 70% of waste/effluent at source, it is necessary to act on the operating routine, process, standard, and make physical improvements to the industrial facility. In the case of the described methodology (Ávila, 2004), the final state in case 2 and 3 was a 50% reduction. In the aromatic amines plant (case 2) 50% means effluent volume, COD, contaminants, and leftovers. In case 3, leftovers, effluent volume, solvent, non-condensed gases, unspecified product and packaging management.

4.5.1.5.9 Improving product quality

The industry's final product has two types of quality: intrinsic and apparent quality (Deming, 1990; Juran, 1992). In the case of chemical, petrochemical, pulp, and oil industries, the intrinsic quality is assessed, such as packaging and auxiliary services, are not as important as the use of the product in the downstream process economic chain. The improvement of processes with the reduction of oscillations in results means lower energy expenditure and better use of production inputs, leading to improvements in quality control items of the final product. The product can be reclassified, thus bringing better intrinsic quality and leading to increased sales, loyalty and profits. The goal of this work, which was carried out in three chemical industries, was to reach 95% of the lots that met the criteria used in the product market. This consideration has been used for more exigent Type A Customers, taking into account all defined specifications. In Case 2, it went from 75% to 85%. In Case 3, it reached 90% when originally it was at around 85%. Therefore, in both cases, there was an improvement.

4.5.1.5.10 Increasing load and continuity

Both load and continuity, in isolation, do not indicate that there was an improvement in process, as there may be losses that are indicated by reduced production and increased reprocessing, but still there is increased load and continuity. Both are a consequence of the process stability variables, optimized operating routines, and knowledge of the causal nexus to process and continuity problems. The goal in the implementation of the process stabilization method was the for chemical and polymer plants: increase in load by 20% and reduction in discontinuity by 50%. In case 2, the discontinuity was 30% (continuity 70%) and in case 3, the discontinuity was 6% (continuity 94%). Therefore, in case 2, 85% continuity was sought: and in case 3, 97% continuity. After the implementation of the program, in case 2 a goal was reached and in case 3, a reduction of 30% instead of 50%, means continuity of 96%. Load. The expected load with the improvements would increase by 20%, mainly due to the reduction of fluctuations for the best separation and reaction operation. In case 2, aromatic amine reached a 15% neck increase in the reaction pumps and their stoichiometry. And in case 3, only 10% increment, with difficulty of swinging light composition in rotary purification equipment.

4.5.1.5.11 Improving consumption rate

There is a difference between the reaction stoichiometry and the real situation of the reaction. The greater the difference between the ideal and the real state, the more possibilities may exist to optimize material consumption, thereby reducing this index. For each ton of finished product, there is the optimal situation where the loss is reduced compared with the actual situation. The difference between the optimal situations (which should be approximated to the ideal situation) is the potential (Δ) for process loss reduction. The intended goal in the industrial plants that were studied was to decrease 50% of losses, that is, $0.5*\Delta$. In the case of aromatic amines, a reduction of 20% in consumption was achieved, and in the case of the polymer plant, it was 15%. Both cases failed to reach the target.

4.5.1.5.12 Improving the company's image

As mentioned earlier, this intangible item is difficult to measure. The company's image depends on local actions that reflect positively on the external environment, understanding local actions as operational routines, operator's behaviors, communication with the community, affective involvement of the worker and his family near and around the company. To measure the improvement of the company's image in society, two indexes are adopted: (1) the number of awards that were gained annually and (2) the number of employees and scientific, technical and social works that were published in newspapers, magazines and books. The intended goal for case 3 (polymers) was 1 award per year and 3% participation in published papers (100 employees), thus meaning 3 papers per year involving the company in positive communication to the community. After the implementation of work to stabilize processes, the target was reached in the awards, but only 1 work per year published (below the target).

4.5.1.5.13 Increasing customer satisfaction

Customer satisfaction is measured in a number of ways as it has been presented. After applying a survey and interview, in which the following items were evaluated: compliance with contractual clauses, service, after-sales quality, product differentiation and image of the supplier company; the customer satisfaction rating was made in the last one-year period (depending on purchases intensity). In addition to the technical items of process stabilization, to achieve this goal, the preparation of the team to defend the company out there is considered. The goal was 95% of satisfied customers. Originally, there was only 60% satisfied in case 2 (many quality issues) and 80% satisfied customers in case 3. Following the implementation of the Process Stabilization Program, the following results were achieved: in case 2, 80% of satisfied customers (against 60% original and 95% target), in case 3, 90% of satisfied customers (against 80% original and 95% target). (Fig. 4.22)

4.5.2 Case of metallurgy based on manager discourse and technical issues

The metallurgical industry used for this case was an industry producing iron alloys. The objective was to produce a diagnostic report on process losses identified through the MD. In this case, the coordinator, supervisor and production manager represented the MD. In addition to the MD, a process report and a production management and HSE report were also used. It is noteworthy that this case was explained in Dissertation of the Federal University of Bahia (Santino, 2018), as well as validated by the company through a technical cooperation agreement.

To carry out this investigation, it was necessary to study the technological aspects in the subjects: process, procedures and operation standards, quality systems, equipment and process control with respective instruments.

For this case, interviews, data statistics, loss mapping, and mass balance were performed. With this type of methodology, it was possible to indicate actions to minimize the loss, as well as improvements in the operational area. After the methodology applied, the manager validated the results, and an action plan was presented to the production manager and the HSE manager.

4.5.2.1 Phase 1—Loss mapping through manager's speech

The manager has important information about how the process and staff work. One of the great advantages of the MD information source is that the manager has free access to production, environmental and costs reports as well as reports on the teams' profile per shift. This access allows the expansion of knowledge about operation routine, which favors management during decision-making.

RESULT \ TECHNIQUES	DIPEA	AEP EVOP	PADOP	ROTLIM	MCA+AUDIP	AUDIL+EMIFU	ALO	Physical improvement	MULTI	TREINOP	EDUC	Self management	Measurement	Projected result	Case 2 Result – Amines	Case 3 Result – Polymers
a) Stabilize process variables														80% of variables 99% = M+/-3 DP	80% of variables = 95%	70% of variables 99%
b) Improve liquid effluent quality														100% var 95% =M+/-3DP;12 h	70% var 95% within; 24 h.	90% var 95% 12 h
c) Increase equipment availability														50% reduction in unavailability	Reduction of 60%	Reduction of 70%
d) Decrease replacement parts consumption														30% reduction in parts	Reduction of 20%	Reduction of 40%
e) Decrease Hourly spending														MAX of 1 HE p/ employee (0.5%)	1,5%	1%
f) Decrease Re-worked hours														70% of the hour reduction retgrated		50%
g) Improve motivation														85% share	75%	65%
h) Decrease accidents and incidents														1 accident every 2 years and 2 incidents in the year	1 accident in the year, 4 incident in the year	
a) Minimize effluents at the source of process														Reduction of 70%	Reduction of 50%	Reduction of 50%
b) Improve product quality														95% of OK lots for Customer A	85%	90%
c) Increase load and continuity														Load + 20% and discontinuity - 50%	Charge +15% Descont.-50%	Charge +10% Desc. 30%
d) Improve consumption rates														50% loss reduction	20%	15%
e) Improve the company's image														1 annual award and 3% annual job		1 prize 1% jobs
f) Increase Customer satisfaction														95% satisfied	80% satisfied	90% satisfied

FIGURE 4.22 Results and statistics of abnormality event mapping. Ávila, S.F., 2004. Methodology to minimize effluents at source from the investigation of operational abnormalities: case of the chemical industry. 2004. 115 p. Dissertation (Professional Master Management and Environmental Technologies in the Technological Production Process)—Federal University of Bahia.

The interviews took place at the headquarters, besides some technical visits in the area and monitoring of the operational routine. The interviews allowed getting to know the general area, workplace and the main losses that would occur in the operational area. Initially, the losses by type of Material (M) and Energy (E) were listed. Material losses were divided into three types of waste: solid, liquid and gaseous, as shown in Table 4.11.

TABLE 4.11 Loss mapping in metallurgy.

Loss mapping—industry X

Loss classification	Waste type	Losses	Related problem	Communication/ program action	Location (Area)	Frequency (F)	IMPACT, I = (s + en + e)/3			Urgency of action (U)	Total (T) $T = F \times I \times U$	Priority
							Social (s)	Environmental (en)	Economy (e)			
(A) Material losses	Solid waste	Alloy "residue"	Opened Transport; wrong position of material in the trucks	NO/nonexistent	Production; Crushing; Segregation; Logistics; Expedition	2	1	2	2	2	6.67	5°
		Maintenance Waste	Maintenance Material—no spare parts—residue	NO/nonexistent	Production	1	1	3	1	1	1.67	7°
		Particulate Material	Materials logistics; segregation and crushing of materials in the open	NO/nonexistent	Production; Crushing; Segregation	3	2	3	1	2	12.00	3°
		Slag Class 2-B	Mixture of materials in the race: alloy in the slag and vice versa	There is no legislation or standard. But it is proportional to the production	Production	3	3	3	3	3	27	1°
	Liquid waste	Industrial waste water	Waste of water; sites without return of water for treatment	YES/Own treatment plant; Cooling towers	Production, auxiliary cooling area	3	3	3	2	2	16.00	2°
		Sanitary Effluent	Not filtering correctly	YES/Own treatment plant	Administrative area	2	3	3	1	1	4.67	6°
		Oil leakage	Materials logistics	NO/nonexistent	Logistic	1	1	2	1	1	1.33	8°
(B) LOSS OF ENERGY	Consumption	Energy	Unnecessary factory consumption	YES/specific sector for consumption control	Control panel of X factory	3	1	1	3	2	10.00	4°

The identification of process losses was possible by scoring factors (frequency, impact and urgency). Critical Loss is the aspect that has the highest score in the multiplication of factors, being it the slag. The problem identified in slag is related to "mixtures of materials during running: alloy in slag and vice versa." That is, the loss is caused when there is this mixture of materials, that is, a failure of the task.

4.5.2.2 Phase 2—Mass balance for materials

For the analysis of Mass Balance application for Material (Fig. 4.23), the STAN program was used. This balance provided the knowledge of the general flow of the case inputs and outputs, as well as the losses generated by the iron-alloy material processing. It was found that the critical loss, slag, really is a remark in the process, represented in the mass balance. Slag until some part of the processing is considered natural. However, the environmental and production impact of mixing alloys and slag occurs when the task mechanism fails. These failures result in errors or malfunctions in the operation, or management negligence or even technology failure. Thus these failures, when they occur, generate tons of slag, which affects the entire process sustainability chain and reduces productivity rates due to rework.

For this mass balance, values stipulated by an average were used, referring to the process inputs and outputs and converted to tons/day. These amounts were recorded through the interview with the production coordinator. The data collection work for validation via E. SANKEY was made with adjusted proportions of production and the correlated areas.

The following acronyms can be identified in this mass balance:

- OVEN: oven equipment;
- VAZ_LIGA: area for alloy receiving after exiting the furnace;
- VAZ_ESCORIA: area for slag receiving after exiting the oven;
- SEG: area for slag alloy recovering, also processing slag;
- BRIT_LIGA: area responsible for alloy processing, according to grain size requested by customers;
- STOCK: reserved area to store the alloy material until shipment (dispatch);
- FINO_ESCÓRIA: refined slag material ("slag dust");
- FINO_LIGA: refined alloy material ("alloy dust").

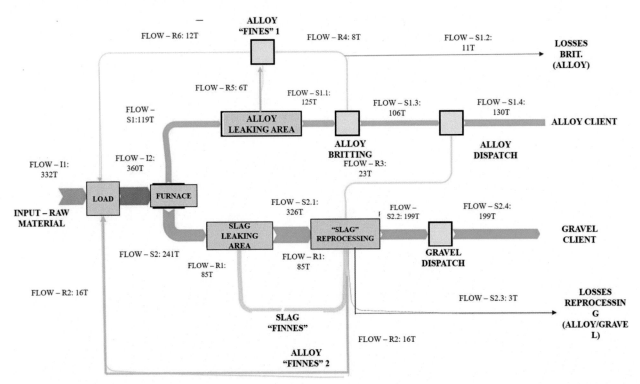

FIGURE 4.23 Material mass balance. *Santino, C.N., 2018. Methodology for mapping losses in a casting process, with application of the Likert Scale and Fuzzy Logic. 2018. 111 p. Dissertation (Master of Industrial Engineering)—Polytechnic School, Industrial Engineering Graduate Program—Federal University of Bahia—E. Sankey.*

The mass balance enabled the identification of the processes and tasks that cause possible losses. The furnace is supplied in inconstant batches, with variable accumulation and output. The furnace is also supplied as its load decreases. Each load contains an average of 60 tons of raw materials, inputs and others. For a common day, the value of 360 tons of material input was considered, with alloy recovery of around 24 tons, and unjustified losses (particulate material) of around 3 tons. In the mass balance, three main types of losses involving the production process are identified: recovery slag, alloy fines and alloy fine particulate material.

It is observed in the mass balance that after the "oven leak" the materials follow different destinations. The alloy goes to the resizing molds step in the crushing area and then to the shipping area. Losses in the process are presented as fine alloys. The LAN in the alloy are components in a powder form that are generated during crushing. It is noteworthy that this process of material fragmentation can occur, and consequently the loss of iron alloy during the slag reprocessing or in the process of iron alloys granulometry.

In this occasion, the loss is concentrated in the particulate material that is generated, in the slag reprocessing, because it is lost in the air, while the alloy fine is redirected to cover the mold area in the production and is also sent to compose the furnace initial load. Thus although the presence of a well-designed flow to receive and treat losses, there are inherent and unrepairable process losses. Waste disposal and reuse are important for the process, but it is also understood that a prevention plan in the production area could avoid a good part of this alloy recovery (326 TON). This prevention plan can reduce costs with energy, manhour (ph), and especially the loss of the product, as often during recovery the metal becomes alloy fin material due to the reprocessing. Therefore, a plan to prevent losses must be created.

4.5.3 Discussion

The operator's discourse brings the general perception of the events described through the shift book to indicate which variables to study and what cyclical behavior of failure or normality that occurs through the events.

The operator's discourse also presents (through omissions, commissions, and event patterns with or without human error) some aspects and movements called "social emergencies" that may hinder the accomplishment of the task, thus weakening the safety and reliability culture.

The manager's discourse presents the organization's formal leaders' perception of the failure causes that lead to losses, indicating then the best solutions. If the managerial and engineering biases are strong to the point of modifying the reality of practice to another wrongly perceived scenario or situation. Unfortunately, the manager's discourse may be incorrectly interpreted. Thus the most appropriate is the application of both discourses for the diagnosis of process losses.

References

Ahidar, I., Sarsri, D., Sefiani, N., 2019. Approach to integrating management systems. The TQM Journal. 2019, Mar 4.

Ávila, S.F., 2004. Methodology to minimize effluents at source from the investigation of operational abnormalities: case of the chemical industry, Dissertation (Professional Master Management and Environmental Technologies in the Technological Production Process), 2004. Federal University of Bahia, p. 115.

Ávila, S.F., 2010. Etiology of operational abnormalities in the industry: modeling for learning. 2010. 296 f. Thesis (PhD in Technology of Chemical and Biochemical Processes)—Federal University of Rio de Janeiro, School of Chemistry, Rio de Janeiro.

Ávila, S.F., Costa, C., 2015. Analysis of cognitive deficit in routine task, as a strategy to reduce accidents and increase industrial production. In: Proceedings of the 25th European Safety and Reliability Conference, ESREL 2015. Zurich. pp. 412.

Ávila, S.F., Alcântara, C.M.D., Kalid, R.A., Fiscina, F., Cunha, E., Siqueira, R., 2004. Reduction of water consumption through water balance. In: Congress of Excellence in Management, 2004, Niterói. Anais of the Congress of Excellence in Management.

Ávila, S.F., Santos, A.C.N., Pessoa, F.L.P., Andrade, J., 2011. Risk management and prevention program to avoid loss of process (LP3) based on human reliability, practical application and measurement. In: 7th Global Congress on Process Safety, Chicago.

Cerezo-Narvaez, A., Otero-Mateo, M., Pastor-Fernandez, A., 2017. Development of professional skills for project management of industry 4.0. In: 7th IESM Annual Conference. International Conference of Industrial Engineering and Systems Management.

Deming, W.E., 1990. Qualidade, a Revolução da Administração—Out of the crisis. Marques-saraiva.

Jarvis, R., Goddard, A., 2017. An analysis of the common causes of major losses in the oil, gas and petrochemical industries on land. Loss Prevention Bulletin, n.255.

Juran, J.M., 1988. Handbook of Quality. McGraw-Hill Book Co.

Juran, J.M., 1992. Quality from the Project. Enio Matheus Guazzelli & CIA Ltda, São Paulo, p. 551.

Kipper, L.M., Ellwanger, M.C., Jacobs, G., Nara, E.O.B., Frozza, R., 2011. Process management: Comparison and analysis between methodologies for implementation of process-oriented management and its main concepts. Techno-logic, Santa Cruz do Sul 15 (2), 89–99.

Kurdve, M., Shahbazi, S., Wendin, M., Bengtsson, C., Wiktorsson, M., et al., 2015. Waste flow mapping to improve sustainability of waste management: a case study approach. Journal of Cleaner Production 98, 304–315.

Mosier, K.L., Skitka, L.J., 2018. Human decision makers and automated decision aids: made for each other? Automation and Human Performance. Routledge, pp. 201–220.

Muchinsky, P.M., 2004. Organizational Psychology, 7th ed. Pioneira Thomson Learning, São Paulo, p. 508.

Neto, J.B., Ribeiro, M., Tavares, J.C., Hoffmann, S.C., 2019. Integrated Management Systems: Quality, Environment, Social Responsibility, Safety and Health at Work. Editora Senac São Paulo.

Pearson, K., 1896. Mathematical contributions to the theory of evolution. III. Regression, heredity, and panmixia. In: Philosophical Transactions of the Royal Society of London. Series A, 1896. The Royal Society, 187, 253–318.

Poveda-Orjuela, P.P., García-Díaz, J.C., Pulido-Rojano, A., Cañón-Zabala, G., 2019. An integral, risk and energy approach in management systems. Analysis of ISO 50001, 2018.

Rasmussen, J., 1999. The concept of human error: is it useful for the design of safe systems in health care? In Risk and Safety in Medicine. Elsevier.

Santino, C.N., 2018. Methodology for mapping losses in a casting process, with application of the Likert Scale and Fuzzy Logic, Dissertation (Master of Industrial Engineering)—Polytechnic School, Industrial Engineering Graduate Program, 2018. Federal University of Bahia, p. 111.

Sartal, A., Llach, J., Vazquez, X.H., Vila, R.C., 2017. How much does lean manufacturing need environmental and information technologies. Journal of Manufacturing Systems 45 (1).

Tavares, C.L. 2019. Integrated systems for quality management, environment and safety and health at work (SIGQSST): contribution to corporate sustainability. Doctor's degree thesis, Higher Engineering Institute of Lisbon.

Chapter 5

Learned lessons: human factor assessment in task

The concepts presented in the previous chapters are confirmed with the discussion of routine industrial cases, in a critical manner, and interpretation of the lessons related to human reliability. The method for diagnosing technical culture (culture resulting from processes, products and equipment) and of the operational culture (culture of the workplace in the factory floor routine) are exemplified in this chapter, which includes the diagnosis of human and social typology and the classification of human errors in each of the cases and articles studied in Chapter 5 (Fig. 5.1).

The discussion with analysis of the complete task is provided in Chapter 7 based on cases from oil and gas industries where practical exercises are presented.

Chapter 5 deals with routine analysis of common points to indicate management guidelines, detailed discussion of some cases extracted from reference articles and a presentation of the demands for human reliability from the various types of industry and public agencies. Chapter 6 presents the human reliability calculation in different types of industry, using SPARH method. Discussions about industrial routines with impact to the environment are also described in the survey on the perception of operators in clean industrial routines (Ávila and Santos, 1998).

5.1 Routine, environments, human types, and class of errors

The routine experiences in the industrial operation indicate how failures begin from technical and administrative decisions to production. Thus from the previous chapters, it is sought to develop the causal link of the accident and the failure and to indicate the appropriate barriers. The cases involve subjects such as knowledge, management, practices, diagnosis, culture, and behavior. Industry cases occurring in the routine are discussed at various levels of hierarchy and functionality. The activities of the operation that are discussed, role of shift supervisor, interpersonal relationships for the start of operations, technical investigation for statistical process control preventing or initiating the occurrence of events, and the elaboration-realization of practical tests to validate hypotheses in the area.

In the operation, personal scheduling for activities and how to measure productivity are also discussed. The company's internal cases that may impact the environment are discussed with the respective operational actions. These activities affect the quality-of-life company's internal community as well as the external community. Other cases discuss the role of the manager in managing the production team in a small company that faces the challenge of economic survival with professional-environmental ethics. The manager is a representative in the supply chains, where a macro view of environmental difficulties (economic and social) leads to decisions that contradict what is seen when involved with production (who try maintains the equilibrium in the universal ethics).

In each problem situation, the difference in linguistic and their interpretations are discussed: in the shift during administrative hours (8 a.m. to 5 p.m.); on the shift time, during hours without the administrative presence; as a technical engineering team at staff; as a manager who represents departments or sectors of the company; as a leader that represents the company in society in strategic issues such as, new energy matrix control or to achieve better productivity.

There is also a language or experience that is unique, not resembling any other previous situation. The language used in an emergency situation, where roles are confused, where planning loses value, where a clear vision of the whole is a factor of utmost importance, and where the initiative/leadership enables the mobilization of people.

In general, these 21 routine lessons learned (Ávila, 2011a) help the process safety manager (PSM) and professionals to discuss important gaps in behavior and production administration.

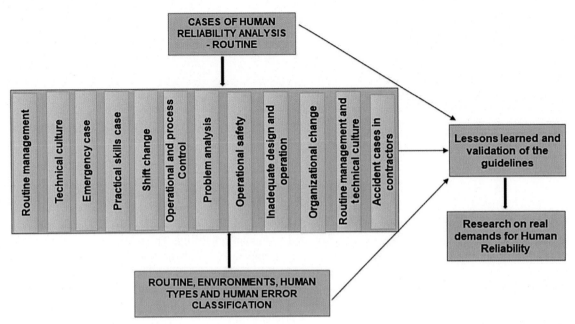

FIGURE 5.1 Gold map of the learned lessons.

This knowledge of the shop floor influences organizational policies and can allows business objectives to be achieved despite workforce restrictions. The active observation of the field including people, equipment and processes through the systematization the discourse of the operator/manager allows formulating appropriate ways to carry out the task efficiently and effectively.

The bottom-up technique or methodology intends to carry out this research work respecting the reading of the operator's and manager's discourse, in addition to identifying conflicting issues between organizational policies and the practices. The analysis of process losses and risk management goes through the study of routine situations in a neutral way and based on bottom-up techniques. The solutions to the biggest problems in the industry emerge from the routine and must be internalized organizationally to return as an action program for routine management.

In these cases of operational routine, several levels of decision making are analyzed, showing:

- When this commitment is not strong and gives way to attitudes of the "law of the least effort";
- What are the necessary and unrecognized skills in routine or emergency situations;
- What are the possible human errors at the levels: strategic, managerial and during the execution of tasks in normal or emergency operation mode;
- What are the mistakes made in decision making that have an impact on the loss of profit, image and credibility of the company vis-à-vis its work group and external institutions?

5.1.1 Routine management case GR: director's behavior (511—GR1)

A polymer company with low staff turnover (no layoffs), the case of the director who was an operator, engineer, chief operating officer is analyzed until he reaches his current role. This case refers to the plant startup activities, where the shift group has part of its initiatives inhibited by centralization of this director. The best form of operation for each situation is always indicated. Excessive centralization intimidates the shift team that would take operational decisions made from the start, and in some emergency situations. The disconnection of shift decisions by the team that is transferred to the director creates a function dependency. The shift team's decisions are dependent on the director.

The standard change was possible through the preventive action that took four months with reeducation, "shoulder to shoulder," and the acceptance of important requests by this professional, that is, action that requires review and development of procedures, new standards of behavior, confidence and delegation. Frame 5.1 presents in topics, an analysis of the routine administration (GR) in relation to the behavior of the company director.

FRAME 5.1 GR1 assessment—director's behavior.

Environment	Organizational climate, classified as an event involving the director's power with excessive centralization of decisions.
Leadership	This action inhibited formal and informal operation leadership by transferring this process to the director.
Skills	There was no development of technical or managerial skills, without practice.
Human error	A "full plate" of human error, not following rules, not believing in the company and, therefore, not having enough attention to carry out the task.
Decision	Anomalous due to excessive centralization of the board and no training and reaffirms the responsibility of the shift team for the normality of operations.

FRAME 5.2 CTO2 assessment—solution to the residual reaction problem.

Environment	Organizational environment with concentration of power and excessive centralization. The search for lower costs overrides environmental issues for only economic viability.
Commitment	The team demonstrated its commitment to the company due to the vision of being inserted in a larger environment and was not faithful to the latent and flawed policy suggested by the company.
Leadership	Engineering motivated the operation, which, when feeling committed and also responsible for the problem, suggested the method to minimize impacts.
Skills	The knowledge about the residue and the operational controls to make the dosing and monitoring viable for a long time indicate operational and environmental skills.
Human error	If the team and in particular the leader of this process did not cause a change in behavior in the shift, from an established latent culture, human error would happen.
Decision	The company's suggestion with a centralized decision would have a high environmental impact, to bury several drum. The internal discussion indicated a new solution, with delegation of responsibilities and ideas emerged that formed the solution, a methodology with reduced environmental impact.

5.1.2 Technical and operational culture: solution for waste in the reaction (512—CTO2)

In a chemical industry, management suggested an inadequate way to dispose of the waste generated in the reaction with impacts on nature, especially soil and groundwater degradation. This inadequate culture was widespread and practiced in an invisible way, but at times it seemed like an action approved by the company's management.

Ethical issues mobilized the shop floor team who, through dosages and monitoring, solved the problem for the operation without impacting nature, avoiding application of unethical solutions that would alter the natural balance. Thus it is confirmed that the popular saying "might is right, the sensible obey" does not need to be followed to the letter. The second analysis refers to the CTO2 technical-operational culture adopted in the operation, this case discussed the solution for reaction residue shown in Frame 5.2.

5.1.3 Emergency case: situation in reaction stoichiometry (513—ER3)

In another chemical industry, the reaction mass is unstable, losing fluidity and obstructing vitrified reactors, at risk of impact damage during cleaning. High risk of accident and loss of assets due to cleaning the reactor and handling utilities and tools not used in routine. This problem had an impact on loss of profit, as a result of low operational reliability.

The non-investment of top management in the purchase of flow controllers that would make the reaction more reliable, indicating the lack of vision in the analysis between production and safety. A leader, an operation engineer, with a global vision and who understood the difficulties of carrying out a local action, suggested a solution, which was carried out in the field. This leader suggested performing tests to visualize the color and fluidity of the reaction mass. By detecting the tendency to lose control, it avoided most of the risks of prolonged downtime in the plant. These input additive tests identified which chemical was in excess or was missing. This idea brought good results, even without formal management knowledge. Perhaps, the suggestion would not be made if the request for testing had been made official. In this case, formalization would be difficult due to the great centralization of decisions by management.

A procedure was written, a test was structured, and operators were trained, bringing a good field result, motivating the operation team, on the shift, to improve processes. The team was the author of its own improvement in the quality of work. This solution was bottom-up, and after the formalization and disclosure of the gains achieved, the board

FRAME 5.3 ER3 assessment—reaction stoichiometry.	
Environment	Senior management initially with a view to costs, brought losses. With the suggestion and effectiveness of the operation tests, management understood the importance of changing the instrument and procedure to increase operational continuity.
Commitment	The operation was involved, which corresponded to the review and application of the procedure to avoid major environmental and downtime impacts.
Leadership	The vision of self-management without depending on the influence of organizational culture and management guidelines led to bottom-up solutions appearing and being applied.
Skills	Chemistry, safety, and operations skills and knowledge were applied to make this action possible.
Human error	The low standard due to overflow of reactor and the dirt in the area generates dissatisfaction that generates fatigue, inattention, lack of commitment or even discouragement of the team, lowering the motivation for the work. The success in bottom-up actions led to the reversal of these trends.
Decision	Decision making was facilitated by the leader and shared with the team bringing the shared authorship of the work.

recognized and decided to invest in an appropriate flow instrument for the reaction. Frame 5.3 shows the analysis of case 3, reaction stoichiometry (ER3). Products were formaldehyde and aniline in reaction of methyl diphenylamine (MDA).

5.1.4 Practical skills case: perception and monitoring (514—HP4, HP5)

5.1.4.1 Acid plant (HP4)

At the start of this plant there was a loss of vacuum and unacceptable delays with a high loss of profit. Although everyone on the shift was involved for days to find this loss of vacuum, isolating the equipment and processes with rackets, finally, on the arrival of a specific operator, he managed to hear the vacuum, and this worker was dubbed as, tuberculous ear. We conclude that in the team there are perception and search skills, which help in solving problems, we do not always know these skills. In this case, it was necessary to know the skills of the shift that were not properly mapped, neither formally or informally. The correction of the problem was occasional and there was a lot of time loss and high cost in the startup of the plant, shown in Frame 5.4.

5.1.4.2 Chemical plant (HP5)

The oscillations in the distillation indicated in Fig. 5.2, cause a gradual loss of vacuum in distillation (with effect indicated in the statistical behavior with a decreasing "saw" appearance graph), generating contamination of dangerous chemical for the effluent and causing environmental fines. The interpretation of the event was published on the company's intranet through a technical report. This interpretation generated questions from the operation team, democratizing knowledge and allowing someone on the shift to investigate and find the REAL root cause (the investigator aimed to one side and the solution appeared on the other side). The owner of the idea and the respective action on the factory floor, triggered the solution of this old problem, without identification, having preferred to remain anonymous. The fact is that the solution to the problem appeared, although not in the equipment and in the process indicated in the report (the vacuum broke due to a hole in the exchanger hull and not due to obstruction in the bottom outlet of the barometric vessel). Frame 5.5 shows the HP5 Analysis.

5.1.5 Routine management case: meeting to change time in shift group

Operators and supervisors pass the shift quickly, not allowing to understand if the quality of the process is appropriate, that is, being able to pass the "square ball to the colleague" with a possible failure process in progress. When problems start on your shift, maybe the cost/benefit ratio indicate, that often worth to keep the worker, to "solve the problem," together with the colleague of the next shift, in this case the campaign was suggested, pass the ball to your colleague. The operator returns home only after directing the solution to the problem. It is observed in the analysis of GR6 the construction of a sense of responsibility of teamwork and the decision process in Frame 5.6.

FRAME 5.4 HP4 assessment—perception in acid plant.

Environment	Operational routine case without involvement with organizational environments in technical and operational routine issues. In fact, a mapping of skills was missing in the shift group.
Commitment	The evidence indicated the team's involvement, but certainly an individual made the additional effort to find the problem indicating greater commitment. He realized the problem and went to the field to try to contribute until he found the cause of the problem.
Leadership	If the team did not trust the leadership, the solution could certainly take longer to appear. Fear of being a volunteer could indicate weakness in the operational culture.
Skills	Those skills that depend on the five senses, in this case hearing, are less and less in practice with the overvaluation of instrumented systems. In this case, the classic and important skills of the operator appeared.
Human error	The non-functioning of the industrial plant would imply pressures that would cause stress that, probably, would create a fertile environment for the appearance of human errors, worsening the situation and making the search for solutions more difficult.
Decision	The decision could be faster if the team with the operator had been present at the beginning of the problem, or if the skills had been identified in advance.

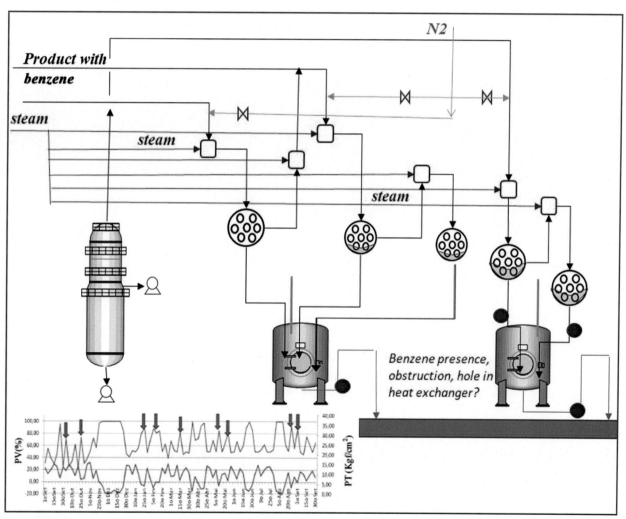

FIGURE 5.2 Perception of process failure.

FRAME 5.5 HP5 assessment—chemical plant monitoring

Environment	The organizational environment was conducive to the dissemination of technical works on the company's online platform. We noticed that people consulted the information, but did not want to formalize their participation in volunteer work, perhaps for fear of some penalty.
Commitment	Apparently, there was the concern of operators to solve problems indicating commitment to the company and to the manager-researcher-leader.
Leadership	In this case, the initiative of a leader who was able to prepare a map or causal link about the problem moved the critical sense of the team that found the root cause and not the secondary cause to solve the problem and kill the BUG, interrupting the cycle of problems in distillation.
Skills	The ability served to change the problem hypothesis and test solutions in the field. The disclosure of the initial hypothesis motivated the study of the root cause, the causal link. During the routine, the cause of the problem was investigated and finally, the problem was corrected.
Human error	This problem had been occurring for eight months and caused the loss of production, overwork and high risks of contaminant for the effluent (benzene). Human error would be caused if the team, both the one who issued the report and the one who discovered the real problem, had not acted.
Decision	The whole action was carried out without formalization, demonstrating that a large part of the solution to problems is in the informal routine of the operation. It is necessary to register these informal studies to standardize the best way to operate, even without the knowledge of the team.

FRAME 5.6 GR6 assessment—meeting to change time in shift group.

Environment	In the company there is flexibility to delegate operators to support the investigation—decide when to support the team to deal with non-conformities right from the start. The environment is conducive to the initiatives of the shift team, but the decision regarding the need for overtime requires knowledge of the situation.
Commitment	The operator with a sense of responsibility, during the arising of a problem, feels responsible and uncomfortable about not supporting the group, because the problem was generated in his shift. Solving the problem reduces uneasiness and increases commitment to the group and the company.
Leadership	The performance of the leader with the workers who stay at work to decide and to act in solving their problems, is healthy, when the cost-benefit is worth it. The vision of this leader and the assignment of authorizing overtime or transport indicate that the most suitable person is the engineer or technician of the administrative staff. It cannot be the manager to enable the bottom-up technique—avoid possible inhibitions in the shift.
Skills	The application of skills to correct technical problems and human errors, that were initiated in a previous shift motivates teamwork and promotes knowledge sharing.
Human error	By inhibiting the chain of problems at the beginning, the operator avoids the growth of the failure, and there is no unfolding of human error with process losses. The risk is not to discuss the problem and get lost in multiple omissions becoming a problem without an owner, in no man's land.
Decision	The decision-making process is considered a time of managerial learning in terms of taking responsibility for cost control in favor of benefits at the shop floor level. The leader of this process should report at the end of the month what the cost is in relation to the reduction of estimated losses.

POST-DISCUSSION

Unfortunately there are "moves" (of a low standard) made by the shift group that can compromise the next shift and must be cut at the root, such as, increasing the levels of process tanks to avoid closing problematic batches on their shift. This causes discomfort and undoes the sense of cooperativeness in the shift. This campaign, pass the ball to your colleague, making with these (inadequate) patterns decrease or even disappear.

5.1.6 Operational—process control: investigation of the process and wastewater (516—COP7)

The presence of solvent in the effluent is detected through a laboratory analytical bulletin with a time-consuming result; thus without viable corrective actions in a short time, removing the control process worsens the situation because of three hours of waiting. The tests on the purification column (stripper with steam) are not possible to be performed due to the lengthy time for the analytical results. The placement of sacrificial materials (plastic part prepared by product)

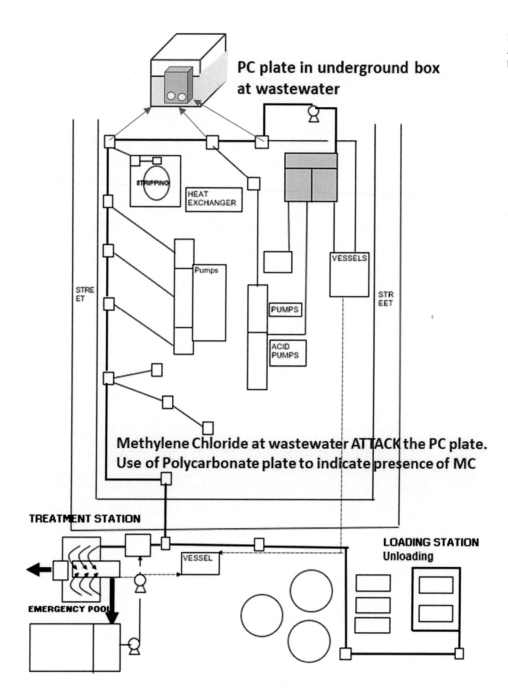

FIGURE 5.3 Indirect perception about wastewater quality, polycarbonate plate.

attacked by the solvent indicates the presence of solvent, calling for immediate action on the stripping column as indicated in Fig. 5.3. This idea "was born" during the shift with the installation of a finished product plate (plastic) in the effluent channel. There was creativity in the search for a sign of process by the shift team where corrective action was feasible. This action prevented uncontrolled process and loss of chlorinated solvent to the effluent, as shown in Frame 5.7.

5.1.7 Problem analysis: diagnosis and process mapping (517—AP8)

Mapping the abnormalities (MEA) based, mainly, on the shift book, is to value the operator's discourse (OD), to design the routine mapping (Frame 5.8), identifying the causality of normalities for the maintenance of patterns and causality abnormalities to avoid lack of control, that is, seek to stabilize processes (Ávila, 2004, 2010; Ávila and Santos, 1998).

FRAME 5.7 COP7 assessment—wastewater control by process, polycarbonate plate.

Environment	The organizational environment motivated the operation in search of creative solutions. Authorizations were not necessary, as a result of self-management, initiatives were taken to resolve the problem before it was too late.
Commitment	The birth of creative ideas to reduce losses or contamination to the effluent is a sign that there is a commitment from some people in the shift group.
Leadership	It is necessary to break cultural inertia to implement ideas born in the shift in search of increased competitiveness. We are used to wait for what the boss orders to put into practice. Thus the leadership that it delegates is fundamental in this process of change.
Skills	Basic knowledge and specific knowledge worked in conjunction with field skills and developed an innovative idea of reduced cost to reduce environmental impacts.
Human error	If the process control depended on a laboratory response that takes three (which is not always available on the shift), possible control problems in the stripper column could increase and spread due to the interconnection of the processes through the top condensation.
Decision	Process that serves as a learning experience for bottom-up activities and the development of a self-management culture on the shift and operational staff in the administration.

FRAME 5.8 AP8 assessment—diagnosis and mapping of processes.

Environment	Investigating the process to identify causalities indicates that the organization provides time for operational planning, indicating the concern to act preventively and not to rely on fragments of failure information, in isolation.
Commitment	When mapping the abnormalities based on the shift book and indications of process statistics, it means the authorization to use the techniques that arise on the factory floor, indicating a great possibility of the team's commitment to the result.
Leadership	To change operating standards and reach consensus on procedures, a leadership role is required to convince the best practices or the best standards or the best process variables.
Skills	The knowledge base to identify problems in complex systems combined with practicality in the implementation of changes in standards after consensus leads to positive results, owned by the operation team.
Human error	When the team is motivated and sees the result of their work (bottom-up technique), the risks of occurring human error are reduced. Operators are more focused, as their signature is on this beautiful work of art.
Decision	The learning processes for leadership, the application-development of technical skills, the increase in interpersonal relationships has the consequence of increasing human reliability. When the environment is conducive to the implementation of techniques on the factory floor, during the routine, the commitment occurs and everyone wants to enter this trip. Thus decision making happens for small changes in the shift itself and for the changes, it ends up involving leaders who activate the questioning and the search for new ways of operating.

The events are analyzed qualitatively and quantitatively, graphs are prepared for statistical analysis on: process, machine defects, production, logistics and product quality. Changes in standards, procedures and changes to the task steps are proposed. Some equipment are also improved or instruments are changed to achieve more stable operating and process standards toward the six-sigma program. Fig. 5.4 breaks down the complexity in the investigation of the root cause.

5.1.8 Problem analysis: negotiation for preventive action (518–AP9)

The temperature in the purification needs to be modified to avoid dragging the product to the raw material recovery section, and from there, it returns to the raw material tanking contaminating the chemical reaction. The identification of the causal nexus was made through tests, calculations, survey of history and internal discussion. The review and implementation of new operational standards were performed. The planner's understanding of the actions is not always the same as that of the executor, and, depending on the commitment to the result, "hidden" resistances occur. Knowing what had happened and giving up on the attempt to convince the team using force, he set out to identify the resistance and its treatment, that is, the negotiation to change operational standards.

Panel operators who eventually showed commitment to the result were clearly asked why the resistance and, thus clarifying the fact based on historical skills, a new procedure for changing the temperature in the purification was defined, reducing the possibility of problems occurring. This negotiation transformed resistance into an ally in changing operational standards. There was an acceptance by the leadership to make the change in a period eight times higher than intended, shown in Frame 5.9.

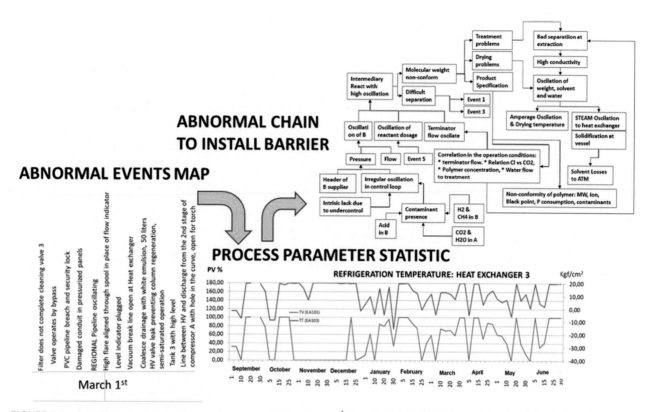

FIGURE 5.4 Complexity in the investigation of root cause. *Adapted from Ávila, S.F., 2004. Methodology to minimize effluents at source from the investigation of operational abnormalities: case of the chemical industry. Dissertation (Professional Master's degree in management and Environmental Technologies in the Productive Process)—Teclim, Federal University of Bahia. 115 p.; Ávila, S.F., 2010. Etiology of operational abnormalities in the industry: modeling for learning. Thesis (Doctor's degree in Chemical and Biochemical Process Technology)—Federal University of Rio de Janeiro, School of Chemistry, Rio de Janeiro. 296 p.*

	FRAME 5.9 AP9 assessment—negotiation to preventive action.
Environment	The openness to dialog about resistance regarding instructions demonstrates that the organizational environment allowed negotiating standards without blaming the operation for not immediately adopting the action, after all, one must respect the experiences.
Commitment	In this case, resistance was understood as a commitment to avoid events that worsen the plant's operational continuity.
Leadership	The leader's role in opening communication with panel operators and lowering the authority role in the process by sharing the decision, motivated the team in search of safer solutions.
Skills	The rescue of the operation history and the application of phase balance knowledge, specific knowledge of alignments and vacuum systems, skills in performing the task and interpersonal relationships to discuss technical issues and standards maintenance brought positive results to the adoption of procedures appropriate.
Human error	If the "might is right, the sensible obey" policy were adopted, the sudden change in temperature would certainly lead to the dragging of materials, maybe a plant shutdown, and future problems of omissions in the operation's tasks.
Decision	The process was democratized in the shift and served as a basis for other improvements that took less time for the joint decision.

5.1.9 Operational safety case: about safety culture (519—SO10, SO11, SO12)

5.1.9.1 *Failure of personal protective equipment, occupational health risk: new standard (SO10)*

Operators are not concerned with their own health and do not notice that their feet are being contaminated through cracks in the safety boot. An occupational health audit role is necessary with the intention of noticing yellow feet ("foot audit in the locker room.") and failed boots to indicate a "dark" future that this operator awaits if he continues with this low standard. If workers followed the personal protective equipment (PPE) exchange rules, there would be no yellow feet. See the discussion on Frame 5.10.

FRAME 5.10 SO10 assessment—new standard: personal protective equipment failure and health risk.

Environment	Area contamination, unfortunately, can be accepted as normal by the operation for a long period, without noticing the impacts on health in the routine. It is necessary to develop (or recover) in the operator, self-esteem, and make tools available so that protection against chemical and physical agents at work is possible.
Commitment	Not taking care of one's health is a low self-esteem or a loss of hope in organizational policies when it comes to occupational health and the environment. The commitment can be recovered with hand-to-hand educational campaigns like "foot audit" in the locker rooms.
Leadership	To develop campaigns of this type on the shop floor, it is necessary to detach certain values and invade the personal space of each operator. This campaign is carried out in the locker room. This means the need for courage, good humor and detachment as well as a sense of humanity.
Skills	In this case there are no apparent deficiencies in technical skills, although, to interpret contamination, knowledge of the area and toxicology of products is required. It requires interpersonal skills, as it involves gaining trust and invading one's own space to raise awareness about the health of each one.
Human error	The unhealthy situation generates low motivation for work, causing an inappropriate and conducive environment for the occurrence of human errors.
Decision	This case involves cultural changes indicating the need for more aggressive educational campaigns and a decision to change standards. Decisions are more individual on the basis of achievement and awareness of occupational health and less of guilt.

FRAME 5.11 SO11 assessment—new standards, requirements to the task.

Environment	The availability of paints, materials, time for organization and cleaning, in addition to authorization from top management, leads operators, with the leadership of the operation engineer, to be motivated to keep their areas clean. If the problems are complex, the campaign to assign owners to areas and equipment can become a reason for motivation-demotivation and cause human errors.
Commitment	If the attribution of responsibility by area does not depend on complex problems still outstanding, but on routine adjustments, there is a great possibility of increasing the commitment of the operation. Everyone wants (at first) a clean area, especially when the organization makes time and tools available.
Leadership	The role of the leader is fundamental when the previous culture or previous custom leads to disorganization. The transformation agents orchestrated by an informal leader can guarantee the change process.
Skills	To keep the areas clean, it is necessary to know the equipment and procedures related to that (your) area well.
Human error	The identification of product leakage risks and their signaling increase the motivation to act safely. Circling in the area under the care of operators is safer and facilitates the investigation of complex problems. There should be no clash between policies, guidelines and practices.
Decision	When owning the area, there is some autonomy. The leader of the informal operation interfere in the operators works to keep the areas organized and in normal state, that is, there is a control scheme but it is not absolute autonomy.

5.1.9.2 Tools and requirements for the task: new standards (SO11)

In API770 (Lorenzo, 2001), it is considered that one of the situations that cause human error is inadequate order and cleanliness in the production area. In a certain chemical industry, operators are given tools to keep their control areas clean and organized; new cleaning standards must be internalized. The identification of the areas by the operation and the painting of areas with the possibility of leaks make it the responsibility to keep it clean. If the problem is in the equipment or due an unstable process, it is useless to hold the operator responsible because the root cause has not yet been identified. The tying of sociability rituals increases the team's integration and confidence, leading to the possibility of adopting new standards of cleanliness in the area. Among the rituals, the owner of the area is mentioned, placing the operator's name on areas. Stoppage for maintenance of the facilities can be promoted with the direct participation of the operation with the maintenance, where motivating factors inserted in the local culture are installed to increase the productivity of the services. Planning is carried out when defining responsibilities for leaders (formal and informal) and in the physical deliveries of these services that are monitored by the leadership and team. Presentation in Frame 5.11 in Analysis SO11.

5.1.9.3 Guilt and information for Environmental, Health and Safety: new standards

Often the rules of the Environmental, Health and Safety (EHS) indicate an impeccable conduct to the team, but this is not possible, in the face of the technical-operational culture installed in the company. This culture in the factory floor is

FRAME 5.12 SO12 assessment—new standards to EHS: guilty and information.

Environment	Excessive registration gets in the way because it requires a lot of time from each employee to prepare documents in the management system and does not focus on studying the root cause of problems. The operator's discredit in relation to the organization and vice versa comes as a consequence of what has already been discussed as an unhealthy state of the organizational system for certification. Incomplete use of the glove.
Commitment	In the informal rules of the operation, small deviations cannot exist because they lead to human behavior accordingly not worrying about small flaws authorizes living with errors, or even skipping some steps in the procedure and so on.
Leadership	Thus the commitment is compromised when adopting SMS, excessive control by guilt and not by teamwork.
Skills	Discussing issues that only have to do with SMS culture today, is not a good deal, as they are top-down policies, acquired by senior management, and which are attempted to be implemented if they respect the informal rules of the operation. This fact leads to the inhibition of leaders regarding the circulation of information to increase productivity.
Human error	The culture of guilt leads to penalization and leadership inhibition for change. On the other hand, it keeps latent problems from emerging into the visible zone. In this way, it promotes the desire not to use its skills in full capacity in the work of the shift, during the operational routine,
Decision	Conflicts of HSE policies and routine practices lead to an environment conducive to human error, thus increasing the risk of loss of process.

a partial result of regional-global social culture that influence the operator's task. Some small slip-ups that happen accidentally, such as getting your hand dirty for not wearing a glove or breathing gas for not wearing a mask, should be recorded to monitor the change in health and safety standards in the company. In dealing with EHS issues, certain items adopt a culture of excessive centralization and official knowledge of data, transforming it into statistics. This happens up to the level of the deviation leading to the registration of the authors' names and generating an entire procedure for handling non-conformity in EHS. So, the dirty hand is reflected in the employee's record too, guilt can lead to unemployment. Indicating blame culture (Fragoso et al., 2017). This anomalous procedure generates behavior of avoiding the report of small events in order not to damage the operator history in the company and not blaming the operator by the deviation that could be registered in the individual file. Just as the operator gives himself the right not to record what happens in the routine, the same occurs for not to follow the procedures to the letter. Thus this safety culture is contaminated by the normalization of deviations. Analysis SO12 is shown in Frame 5.12.

5.1.10 Inappropriate design and operation: technological solutions (5110–PJ13)

When the plant started up ten years ago, it was decided to operate under different conditions in relation to the project for alternative volumetric pumps. The o-ring of the sealing of this pump had its diameter increased allowing the passage of a large amount of solvent into the process (without monitoring) and avoiding the exit of solid materials (product) out of the process, causing large leaks and dirt in the plant. These environmental aspects joined with other operational events, can cause plant shutdown.

Thus the contamination and operational continuity problem was apparently reduced in the first start of the plant, at the expense of non-standard operation in relation to the project. This operating profile, adopted since the start, did not allow the control of solvent flow, which was also the sealing fluid of alternative pumps. Thus oscillating solvent inputs without being accounted for altering the reaction balance and promoting oscillations in the return line feeding the fault to the beginning of the process, a true interconnected set of fluctuations (solvent flow, vessel pressure, content of solvent in off-gas). This situation presented in Frame 5.13, in Analysis PJ13, was analyzed and adjusted, and the operation pattern was revised in relation to various equipment, considering the possibilities of obstruction considering all the solvent that is used in the plant, thus reducing oscillations of process.

5.1.11 Organizational change: change in practice without consulting past ritual (5111-PR14)

The manager of a specific Plastic Manufacture worked for five years with sensitivity to the rituals of organizational integration and social inclusion. He insisted on maintaining the integration rituals of the operational and administrative staff, at Christmas and São João parties. In this way, the unspoken issues in the shift routine are presented once or twice a year in relaxed organizational events. In this ritual there is the breaking of hierarchical barriers and group castles are broken.

FRAME 5.13 PJ13 assessment—inappropriate design and operation.

Environment	The company allowed, after living for 10 years with problems of excessive solvent consumption, the analysis of problems using abnormality mapping techniques in the shift leading to believe that it is willing to adopt programs resulting from bottom-up techniques.
Commitment	Many times, one lives with old problems since the plant started due to lack of time or getting used to the current situation, so removing filters, changing o-rings, and changing temperatures are changes that are adopted in the matches and transfer a load of possible non-conformities for a good period in the operation. Organizational commitment may exist at all previous opportunities, but the technical basis for evaluating the best ways to operate is lacking.
Leadership	In order to return to old cases of coexistence in the technical culture and promote its rediscussing, it is necessary to lead groups of different specialties promoting questioning of standards. Technical or technological support, tests and statistical validation are essential.
Skills	In this case, mechanical, interpersonal, statistical and operational skills were applied to arrive at an operational standard closer to the original project and that did not repeat its mistakes, making it possible to decrease the fluctuation of the processes.
Human error	Living with inadequate standards and many process fluctuations lead to dissatisfaction with the low quality of the work and the consequent increased risk of making human errors due to inattention, loss of memory, lack of commitment to the rules and lack of common sense in the results of operations.
Decision	Basis of work in the shift book and specific technical discussions made it possible to improve mechanical systems. The decisions regarding tests to be carried out were made jointly, but the beginning of work had its inertia broken through the agent of change.

FRAME 5.14 PR14 assessment—changes to past good practices and rituals.

Environment	Before, the organizational environment promoted the resolution of conflicts without guilt, and when they occurred they were treated in the integration and insertion rituals. Then, with the change in leadership, occupational problems in this type of industry exploded into forms of accidents and social and legal occupational issues.
Commitment	There was a break in the commitment to the technical-operational culture in the company as a result of the change in the managerial profile and the lack of organizational integration and social inclusion, under the new leadership.
Leadership	The new leader in the routine of the operation treats the matter in a classic way, adopting the hierarchy in the decisions regarding the issues of the shift time and the administrative.
Skills	There was a general inhibition in the use of skills, leading to problems of accidents and absence due to occupational health. There was no more effort of subjective rituals for the formation or treatment of the team.
Human error	The accidents were the result of the operator's lack of commitment to the organization and occurred through warnings not heard by the management level.
Decision	Standard management following regulatory standards and laws leads to a lack of motivation for work. People like to be involved in an affective bond and participate in the decisions of managers, the lack of this ritual in daily life and in commemorative events caused the great split: worker and organization.

With the change in the managerial profile, the organization annulled the integration rituals, resulting in the loss of employee's bond in relation to business objectives, changing productivity, increasing labor complaints and occurrence of occupational diseases. Judicial issues multiplied and accidents that did not happen before appeared. Frame 5.14, demonstrates the influence of manager, who rather than be aggregating, adopted a managerial pattern of disaggregation. This issue is discussed in the Book dealing with the management for Organizational Change (Center for Chemical Process Safety (CCPS), 2013) methodology to avoid accidents.

5.1.12 Routine management, technical-operational culture: bias in execution (5112–VIO15/19)

5.1.12.1 Intentional violation: false sampling of effluent masking abnormality

During the visit of an environmental agency to the factory, to carry out the daily sampling of the effluent, the operator traveled to the area and performed unlawful sampling operations. The effluent was being shipped with non-standard Ph to ease the workload of the operation while the other side of the effluent basin was framed with correct Ph and at a low level. It was out of the standard and due to the visit of the environmental agency it returned to normal situation. There was an intentional violation whereby the motivation was the least effort for the operator and/or of the supervisor or even motivated by inadequate postures by the manager or staff in relation to the environmental agency. For this to

FRAME 5.15 VIO15 assessment—false sampling of effluent masking abnormality.

Environment	There was an inappropriate environment in the operation team or in the shift or with discredit in relation to organizational practices. The shift showed a behavior during the day, and apparently, the shift showed another behavior in other periods related to control of liquid effluent.
Commitment	Perhaps a work of rescuing commitments with leaders can result in positive results.
Leadership	Apparently some leaders work in opposite directions to the organization or to what the organization apparently defined in the period.
Skills	In this case, counter-skills (violations) or skills to commit slips were applied, such as sending the effluent out of the standard.
Human error	This human error is classified as a violation with the possibility of impacting the environment or financially with environmental fines.
Decision	Decision-making followed the law of least effort or the famous law dictates that some command and others obey without critical sense.

FRAME 5.16 VIO16 assessment—great loss of solvent in group omission.

Environment	Apparently the environment was not conducive to communication due to the inhibition caused by the company's director. Thus omissions in some events were common practice. The environment was undergoing organizational change and to technical standards, but shift leaders were afraid of change.
Commitment	There was no commitment from the 3 operating groups to clarify this event.
Leadership	The leaders organized themselves in favor of group omission.
Skills	The skills were applied to organize the group omission, without registration in the shift and no traces in the area. The fact is that there was a loss of more than five tons of solvent.
Human error	Serious real impacts, the risk of product loss has become a fact.
Decision	All the solvent sent to the effluent is lost and the decision making at the managerial level was a penalty for the shift leaders who participated in the group omission.

happen, it is a sign that at any time other violations may occur in the execution of the task. This analysis is found in Frame 5.15, which discusses the implications for identifying a false sample.

5.1.12.2 Violation: great loss of solvent with group omission (VIO16)

During the weekend, seven tons of chlorinated solvent are sent to the effluent. All this quantity ends up contaminating the atmosphere and promoting serious risks of fire and occupational disease. After investigating data and meetings with the shift teams, the probable cause of this problem was not reached. There was a group omission on the causes and it was only with statistical analysis and confirmation of events in the field that it became possible to identify the initiating cause.

From this cause, it is noted that there was an unintentional violation, but that could have been avoided with the findings, of the impact in the beginning of the drainage/overflow that caused the sending of this large amount of solvent to the effluent (Frame 5.16).

5.1.12.3 Intentional violation: forced vaporization of solvent in the area, occupational ethics or false environmental image? (VIO17)

Two actors in this case play inverse roles: (1) the engineer who detects the placement of a trace of steam into the effluent channel, insistently, causing the vaporization of excess methylene chloride into the atmosphere and, to correct the problem, this engineer removes that steam line from the effluent and return to feeds the trap; (2) the leader of the administration who, to meet the limits of solvent in the effluent, quality, requires maintenance of the alignment of the vapor trace to vaporize contaminant before reaching the effluent, and not accuse the results that go to the environmental agency. The two actors meet in one day and the clash of values is initiated. This finding can be understood step by step, in Frame 5.17.

FRAME 5.17 VIO17 assessment—forced solvent vaporization: ethics and image.

Environment	The organizational environment was not conducive to the participation of change leaders due to the centralization of management and practices not consistent with environmental and safety policy.
Commitment	The administrator, director of factory, without any commitment environmental and occupational aspects and, lack of commitment with the vertical hierarchies, downwards, give an undue example.
Leadership	There was an annulment of the leadership's actions to change standards, avoiding errors such as the diversion of vaporizing organic vapors in the area causing loss of quality in health. There was an overlap in the formal leadership of the administration, with an order to maintain steam for effluent.
Skills	There was no adequate application of operational skills and an attempt was made to apply the "might is right, the sensible obey" law.
Human error	If there were no self-management and self-motivation, patterns and environments would be conducive to human error.
Decision	There was no learning in this decision and action, a centralizing agent canceled the participation of the shift or staff in the decision making.

FRAME 5.18 VIO18 assessment—unaccompanied drainage, passage of oil to the sea.

Environment	Many tasks performed at the same time or lack of instrumentation may be the result of the organization's policy, which adopts low automation in the industrial plant and multiple tasks for each function without a cognitive capacity criteria.
Commitment	The law of the least effort may indicate the worker low commitment with the organizational objectives or, the operator does not see, in the administration rules, the appropriate policies that induce best practices, leading them to skip important steps.
Leadership	The leadership role does not play a role in this drainage, if the culture of the organization overlaps the managerial profile or the supervision of the shift.
Skills	There was a failure to monitor the operator by letting oil pass into the sea.
Human error	The impact of this human error caused environmental accidents in the bay of all saints.
Decision	The decision-making process was inhibited by low-standard culture.

5.1.12.4 Violation: unaccompanied drainage, aqueous phase and sea oil (VIO18)

Often the separation of oil from water or water from oil is done by direct decantation in buried equipment or process vessels/tanks. The interface of this separation is so good that drainage or suction of the aqueous phase is enough to remove the water and keep the oil stored for more suitable destinations. The problem is that without the proper instrumentation, this drainage will rely only on the observation, full time, of the material to avoid passing the oil to the destination of the aqueous phase. Drainage or extraction can be contaminated with oil to the sea or river, or even to equipment that does not have the capacity to retain this oil, also leading to the receiving body (rivers or seas). In Frame 5.18, it is clear that the procedure is not properly monitored, with the possibility of generating an environmental accident.

5.1.12.5 Risk analysis in critical systems, cost or security decisions? (VIO19)

In a certain chemical plant there was a reactor bore processing highly toxic products. The decision to stop it for correction involves high loss of profits, taking the pressure to resolve the problem with maximum agility to continue the operation. management demanded that the return of the operation should take place in record time; impossible due to the lack of specialized material inspection services at that time, at the risk of high impacts on the return of the operation. The reactor leak with toxic gas (phosgene) leads to incomplete inspection procedures carried out by the technical team and forced by pressure from the administration. Despite the risk claims, the reactor returned to operation. The staff engineer published a technical report for directors requesting the shutdown schedule in the near future for the most adequate inspection of the equipment. The publication of this Report presenting the correct routing regarding this reactor caused confrontation with the organization that demands a loyal stance in relation to the low environmental standard solution. In the description in Frame 5.19, risk analysis is questioned as a cost or a necessary procedure for safety.

FRAME 5.19 VIO19 assessment—risk analysis: cost or security?

Environment	Unfortunately costs and billing guide decisions in relation to environmental and occupational impacts. But professional ethics led the Engineer to prepare a Report on the risks of environmental accidents. This report was criticized by senior management.
Commitment	There was a commitment by the staff to common sense and ethics when sending a Report with high risks of maintaining a reactor without due material inspection, this caused organizational risks of dismissal.
Leadership	Leadership in the process of changing culture or patterns goes through confronting situations, in the team and in top management due to the use of bottom-up techniques.
Skills	The technical knowledge was applied and the inspection skills also, although it did not satisfy the need for safety.
Human error	If the staff's attitude had been to accept organizational practices, there could have been a rupture of the reactor with a large leak of toxic gas.
Decision	Studies, risk analysis, inspection techniques, team discussion, and organizational risks in search of higher safety standards.

FRAME 5.20 VIO19 assessment—risk analysis: cost or security?.

Environment	An organization that, despite placing priority on security among its policies, does not actually practice this, developing high-risk services on a primary basis. The designs of the instruments lead us to believe that this risk of accident (isolation of the tank and its pressurization) was not known, requiring a review of the practical concept of equipment.
Commitment	There was no commitment by the leaders to assign high-risk services such as releasing a tank that had a toxic product for maintenance, involving beginners. At certain times, vices also decrease commitment.
Leadership	There was no apparent leadership for processes and process safety indicating that what everyone's responsibility is may not be anyone's.
Skills	As already mentioned, there was little application of knowledge (calculation of inventories) and skills (checking drain or vent in loco) generating the serious accident.
Human error	There was a serious accident with material loss and human life due human error of diagnosis to release the tank.
Decision	The inadequate decision-making process, relying only on temperature, level and pressure instruments. No additional checks for local physical sensors, vision, touch, hearing and other auxiliary tools, pump curve, tank inventory and more.

5.1.13 Accident cases in contractors: inadequate standard for services (5113—AC20 a 21)

5.1.13.1 Accident with death during service in TDI tank (AC20)

In a certain chemical industry, a tank of product in process would be released for maintenance. Therefore, in previous shifts, the industrial section is stopped, the tank is completely drained, depressurized and isolated to provide the opening of the manhole and performance of the services.

When draining the product in the tank, mass displacement was not monitored to compare what dropped in level in one tank and what went up in another. Proper local inspection of temperature and pressure was not carried out by carrying out physical tests and fully relying on level, pressure and temperature indications. The instruments indicated level 0%, atmospheric pressure and appropriate temperature to release the tank. In reality, there was a reaction of the isocyanate product with moisture (due to the failure of the N2 mattress) that formed insulation with residue throughout the tank covering the instruments, falsely indicating the release position.

A Work Permit was issued and sent for follow-up by an inexperienced operator and to an inexperienced boilermaker. As a result, the operator went to the area and released PT to start service, the boilermaker started to open the manhole in a clockwise direction and the tank was pressurized with gases resulting from the reaction between moisture and the product. The tank exploded through the manhole, causing the boilermaker's loss of life. Frame 5.20, presents the whole context that led to the accident with fatality.

5.1.13.2 Accident with death when handling forklifts (AC21)

During the weekend, a waste transportation service was carried out, from one place to another, using forklifts and drums. The area through which the forklift passed to reach its destination is the patio in front of the organic product tank park. One of the tanks, of volatile and flammable organic matter, had a small hole that was not noticeable due to the rapid vaporization on sunny days.

FRAME 5.21	AC21 assessment—accident with death when handling forklifts.
Environment	Probably low value contracts leading to the use of forklift, and respective drivers, without sufficient level of preparation, despite having received certification. The inadequate inspection of tanks, the non-operability of the dike valves, are all indicative of an inadequate culture of safety and maintenance of equipment.
Commitment	It is in the simplest operations where the highest risks may be, so it happened and most likely there was no supervisor to monitor. At the weekend there are reduced people on the shift.
Leadership	No leadership influence, only the one who scheduled the services for the weekend and without due monitoring (at least someone from the operation would know about the existence of volatile organic matter on the floor—by the smell).
Skills	Skills for handling forklift failure.
Human error	Serious accident with loss of life and severe burn caused by human errors.
Decision	Unattended action leading to decision making without any routine risk analysis.

As the day was rainy and cold, what leaked was on the dike and as there was a lot of rain it ended up overflowing to the patio. The valve that transfers the dike to the rainwater system was stuck, something common. Thus the water that overflowed to the patio had the mentioned organic matter that was supernatant. The forklift faulted (electric battery) exactly where the pooled water was, it was stopped, demanding another forklift from maintenance service. There was a simple solution to put the systems to work, which was to charge the battery using the other. With the spark, it ended up causing a fire in the yard and the death of the forklift driver and serious burns in another. Some important aspects regarding knowledge and standards caused the accident. In Frame 5.21 the conditions that triggered the forklift accident are clear, the context and the irreparable consequence of the forklift driver's death.

5.2 Routine learning: guidelines for human reliability

Each case discussed has learning points that can make up diagnosis or training. The hazard energy flow that causes human error, technical failure, the incident or the accident will indicate the best management guidelines to develop appropriate Standards to be followed by the production Team. The conclusions on how to avoid an environment conducive to failure, even if they are managerial, cultural and organizational aspects, are the most subjective area of reliability, and which can culminate in occupational, process and public accidents.

5.2.1 Learning points

The reading of these experiences motivates the discussion about learning points in matters important to Human and Sociotechnical Reliability. These points are based on 20 years of practice on the factory floor, which includes production, Maintenance, Technology, management and Research activities. The subjects chosen based on this study are:

- Culture (cultural bias; policy & management; culture of safety and guilt; technical and operational culture);
- Knowledge, commitment and standards (investigation of causal connection; skills; alteration of procedures and standards; education, commitment and health risk);
- Social relations at work (group work and communication; organizational integration and social inclusion);
- Operation, project, maintenance and contractors (operation and project standard; task and shutdown planning; contractors and commitments).

5.2.1.1 Culture
5.2.1.1.1 Cultural vices

The popular saying "might is right, the sensible obey" deconstructs the strategies to improve the standard in the routine. The "law of least effort" and opportunism also undermines standards for process safety and may indicate low commitment by the operator or manager or supervisor with organizational objectives. In some cases, the worker do not believe in management due to the objectives not being respected in the strategic area.

When the worker does not see the policies at work in routine practices, he can react by not following the procedures, leading to skipping important steps in the task. The team's discredit in relation to the leaders and the organization is a reason for omissions that can cause an accident.

In addition to these factors of believing in the company, the influence of regional or global culture can hinder the establishment of management guidelines promoting a scenario of conflicting priorities.

What other cultural vices are there in our work environment?

_____
_____
_____
_____
_____
_____

5.2.1.1.2 Corporate policy, management profile, and conflicting priorities

The integrated management system is wrong, when it formalizes procedures excessively, making initiatives to change the pattern difficult. Excessive formalization at certain times makes initiatives difficult.

In certain organizations, costs still overlap with more global sustainability issues. It is necessary to provoke the breaking of paradigms in the face of aspects of an inadequately established culture.

In certain organizations, costs and billing lead decisions regarding environmental and occupational impacts. In companies and organizations, policies are established that announce the priority in safety, but, in fact, do not practice safety, developing activities of high risk and in a primary way. Excessive centralization intimidates shift teams that do not participate in operational decision making, at startup and in normal operation. The impacts of non-critical participation in the operation may be greater in some emergency situations. It all depends on the boss. While we wait for the boss, the accident is approaching and will happen.

These routine issues undermine the safety culture and allow, in certain task environments, the establishment of guilt for non-compliance with the procedure.

What are the other management issues that need to be discussed?

_____
_____
_____
_____
_____
_____

5.2.1.1.3 Culture of safety and guilt

Often, the rules in the area of corporate safety, health, and the environment indicate an impeccable conduct that, at the same time, is impossible to be met in view of the current technical-operational and regional culture. As much as we don't believe it. This culture directly influences the operator's task. The operator gives himself the right not to record what happens in the routine, he also gives himself the right not to follow the procedures to the letter.

It is important to know this scenario to make organizational changes, changes that will modify the Technical-operational culture and, consequently, the safety and organizational culture.

Is there evidence of just culture or blame culture in your company? Which are?

_____
_____
_____
_____
_____
_____

5.2.1.1.4 The new technical-operational culture

The behavior and attitude of self-management does not depend on the organizational culture installed at that time in the company and, also, it does not depend only on the current management guidelines, at the same time, it respects these demands.

Self-management facilitates the appearance of innovative bottom-up solutions, and tests are carried out in the field, to validate the actions and enable subsequent standardization. The organizational environment motivates the operation in search of creative solutions. No authorizations are necessary and due to the effect of self-management, initiatives are taken to solve the problem before it is too late and the failure is installed. If there were no self-management and self-motivation, patterns and environments would be conducive to human error.

When the team is motivated and sees the result of its work (bottom-up technique), the risks of occurring human error are reduced. Operators are more focused, as their signature is on this beautiful work of art z. Leadership due to a change in culture or pattern requires attitudes that, at certain times, are confronting, especially when bottom-up techniques are used.

How to identify which operational culture is installed on the factory floor?

_____
_____
_____
_____
_____
_____
_____
_____

5.2.1.2 Knowledge, commitment, and standards

5.2.1.2.1 Knowledge and investigation of causal link

The knowledge necessary to identify problems in complex systems combined with practicality, in the implementation of changes in standards after the consensus, leads to positive results in terms of transferring the sense of ownership of the results to the operation team.

The concern to act preventively in the company allows investigating the process to identify causalities and not to rely on fragments of information, in the sequential and failure chronology, in isolation.

This indicates that the organization must make time available for operational planning where it is located, and the areas under the care of the operators and the areas under investigation for complex problems are identified. The resolution of problems must include the team that acts in the respective processes.

Knowledge facilitates the construction of a conceptual basis for problem solving. But, without skill, we do not put the procedure to avoid the problem in practice. Thus we must respect the experience and the willingness to get it right by those who will listen to the process in the field and then carry out the necessary interventions.

Why training and reviewing the procedure alone do not resolve issues related to incidents and accidents?

_____
_____
_____
_____
_____
_____
_____

5.2.1.2.2 Skills

Sufficient knowledge to identify problems in complex systems combined with practicality, in the implementation of changes in standards after the consensus, leads to positive results in terms of transferring the sense of ownership of the results to the operation team.

The concern to act preventively in the company allows investigating the process to identify causalities and not to rely only on fragments of information, we need to understand the history behind the sequential failure chronology and events, together and in isolation.

This indicates that the organization must make time available for operational planning where it is geographic and cultural located. We need to identify the informal rules and respective areas under the care of the operators. We need to identify what are the areas that must be investigated to solve complex problems. The resolution of problems must include the team that acts in the respective processes.

Knowledge facilitates the construction of a conceptual basis for problem solving. But, without skills, we do not adjust the procedure to avoid the problems in daily practice. Thus we must respect the experience and the willingness of process operator. We need to listen carefully the opinion of panel operators and field, and then, carry out the necessary interventions.

How to register the skills that assist in the operational routine and how to disseminate it in the operation team?

5.2.1.2.3 Changing procedures and standards

To change operating standards and reach consensus on procedures, it is necessary to take a leadership role or to conquer or convince the best practices or standards, or the best process variables. The understanding of the action planner is not always the same as that of the executor, and, depending on the commitment to the result, silent resistances can occur that undermine the result of the task and the company.

The interpretation of routine events needs to be published to raise questions about the facts and investigations. In the routine it is possible for operators to disagree or complement the interpretations in reports with field tests. In this way, the shift can find the true root cause, which differs from the hypothesis built on the event. Sometimes, the staff interprets in one way, but through joint action, the solution appears in another way.

Some behaviors are required when performing the procedure in an operation group. Valuing principles of security and humanity, knowing how to cooperate in a team, being keenly aware to detect new signs of emerging failures, and having perseverance and motivation to react when the crisis is installed. Attitudes toward organizational resilience are welcome in the operation group. Many of these issues are of the cultural-educational dimension and increase the commitment to health and decrease trends in our risk aversion culture.

In the review of procedures, there are technical requirements to be considered, but there are also aspects of the failure process related to the interpretation of complex processes that need to be known, as, commitment (affectivity at work), organizational belief and intuition. How to discuss this type of cause or event in complex process?

5.2.1.2.4 Education, commitment, and health risks

It is necessary to develop self-love in the operator and provide tools, so that it is possible, protection against chemical and physical agents at work, leading the team to a condition of better occupational health.

The worker's commitment can be recovered with educational campaigns that invade the personal space, hand to hand, such as the audit of feet in the locker room, to cause the breaking of "crystallized" structures. To develop campaigns of this type on the shop floor, it is necessary to detach certain values and invade the personal space of each operator. The unhealthy situation generates low motivation for work, causing an inappropriate and conducive environment for the occurrence of human errors.

These customs and habits are passed through routine in society and at work. It is in the hands of the task planner and work organizer to establish appropriate social relationships that shape the team with the necessary integration through transparent communication.

Should educational programs respect local customs and operational culture in the unit under intervention? What would these campaigns look like in your industrial units?

5.2.1.3 Social relations at work

5.2.1.3.1 Group work and communication

Operators and supervisors pass-on the shift quickly (priority in economic value) without allowing the team to understand if there is conformity in the quality of the process. If the communication is incomplete, the classes pass noises, that is, they pass the square ball to the colleague with ongoing failure processes. Unfortunately, there are ways to shift the blame for problems to leaders of other incoming classes upon entry to work indicating low standard in group formation. The previous shifts transfer the blame along with the decision of problems that arose in your shift time, and compromise the performance of the next shift. This is an example of a lack of commitment and cooperation in group work. The root cause is culture and difficult to identify.

The work of the operation group in the industry is improved with the greater integration of performers and maintainers in their work function, during normal operation, during maintenance stops, or even stops for the assembly of new projects. Work depends on a good internal and external social relationship, which involves work and society, facilitated by social inclusion work. Communication is more transparent in a work environment with an established sense of justice and with people who reduce their resistance so that there is a good result of teamwork.

How can the production manager better understand the effectiveness of group work? How to measure feedback and formal communication tools? How to correct following the just culture and not the culture of guilt?

_____
_____
_____
_____
_____
_____
_____

5.2.1.3.2 Organizational integration and social inclusion

The bonds of sociability rituals increases the team's integration and confidence, leading to the possibility of adopting new standards of cleanliness in the area. It is important to maintain rituals of approaching the operational and administrative staff, where hierarchical barriers break down and bad groupal castles are destroyed. The redemption of commitments with the leaders can result in positive results.

The discussion on safety gains value when it involves the maintenance and assembly contractors. Accident indicators are of concern to maintenance companies working for the chemical process industry.

The lack of experienced people in the production and security area causes the company to hire outsourced workers, who, in fact, do not know the operational routine.

Are these people able to integrate new hires into the organization? What about the ability to include new generations of operators in the shift routine?

_____
_____
_____
_____
_____
_____
_____

5.2.1.4 Operation, project, maintenance, and contractors

5.2.1.4.1 Operation and design standard

Many times we live with ancient changes of project, solving old problems, changing the pattern since the plant started operation. We confirmed, in several cases, that no investigation was done about the ancient pattern change due to, either lack of time or due to getting used to the current situation.

Living with poor standards from this changed project is not appropriate because it causes process fluctuations and leads to dissatisfaction of the operation group. This group living with low quality of work generating human errors of inattention, memory loss, lack of commitment with the rules, and, lack of common sense in the results of operations.

This conflictual situation has serious consequences due to forgetfulness imposed by organization, inducing the sensation that the change represent the real project. Besides that, the organization can induce a loss of memory, accepted by the operation group, causing loss of critical sense in the operation, nothing can be done.

The operation is responsible for the activities of control of the processes of transformation of the raw material in product. The departments of maintenance, laboratory, and logistics are responsible for structural support for the production. It is important that there is transparency of communication and the same security between the departments in production, which includes operation, maintenance and other support services. In this way, an accusative relationship is avoided in routine and emergency situations between the areas.

We need to learn about these social-organizational-department phenomena.

The organizational model needs to act on specific activities, new programs and projects, as well as act on the routine of operations. How to maintain adequate and updated standards in the organization without conflict between departments to own the power of decision?

5.2.1.4.2 Task and shutdown planning

In the routine of the operation, there is often no due commitment from the leaders when assigning high-risk services, such as releasing tanks of toxic and flammable chemicals, to inexperienced or non-competent workers, both in maintenance and in operation. Why?

Is it due to lack of workers, for relying too much on instruments and not seeking experience?

Why do you think the event is rare and shouldn't happen at your facility? And why do they happen nowadays?

Is it because you believe that when you are hired for a certain service, it is obvious that safety values are accepted and practiced in your routine?

Anyway, not everything we believe happens the way we expect. It is necessary to renew the commitments of operation and maintenance on an ongoing basis and always create a critical environment for safety.

5.2.1.4.3 Contractors and commitments

Low value contracts lead to the use of forklift drivers without sufficient level of preparation. The inadequate inspection of tanks, the non-operability of the dike valves, all of which are indicatives of an inadequate culture of safety in the maintenance and operation of equipment.

The lack of attention or lack of sensitivity for contractors who participate in bidding at low prices, leads to inadequate safety standards. How to improve the relationship with contractors while maintaining an adequate performance in the number of incidents and accidents?

___

5.2.1.5 Summary

We started learning about culture to avoid biases that hinder routine management, weakening security and allowing human errors to occur. In fact, what is sought is to create an appropriate climate for carrying out the critical task.

A well-established culture facilitates a commitment to applying knowledge and skills in practice. The habit or the bad habit is only modified through educational campaigns but it is more complicate when it involves primary and/or strong archetypes. The values of health and safety must be remembered in these educative campaigns including primary symbols.

With a stable culture, knowledge and skills in application without filters, it is important to maintain cooperation and communication, quality social relationships in the work environment. Only in this way, the standards and procedures in the production functions will be revised to maintain best practices.

5.2.2 Route of human, group and organizational error

At this point, we show in Fig. 5.5, the conclusions of the 21 case analyses involving routine, human errors, and the possibility of actions to increase Human Reliability. The conflict between policies and practices reduces commitment and risk perception, bringing a low standard situation to the operation routine. Decisions meet more the pressure for production than in relation to environmental ethical demands. production management in this situation may be involved in a technical culture that solves problems from evident causes in active failures. A side effect of poorly established policies is the promotion of omission as protection against the possibility of fault for abnormal events. This situation of omission

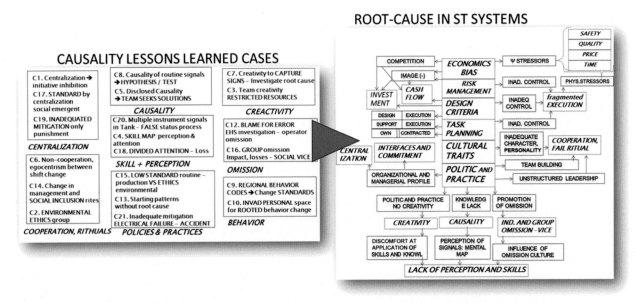

FIGURE 5.5 Causal link of technical failure/human error in the analysis of 50 deviations/accidents. Ávila, S.F., 2011a. Assessment of human elements to avoid accidents and failures in task perform, cognitive and intuitive schemes. In: 7th Global Congress on Process safety, Chicago. Reproduced with permission from Ávila. S.F.

can attain a greater degree reaching groups of shift and even production teams. Omission is a behavior that has traces of Brazilian culture, leading the organization to the construction of strategies to break stabilized vices.

Conflicts between priorities in the company's sustainability functions and inadequate management practices, impact the team's creativity, making it difficult to investigate the root cause of problems. Thus innovative solutions are not suggested because the signs of operational failure are not studied, and the actions are classic and have higher costs.

When the team is not interested in studying the signals, it can lose important skills for operational control. Human errors begin to happen, involved with perception and attention, including many parallel activities leading to divided attention and low quality in the operation and maintenance services. To keep routine activities under control, managers starts to act by centralizing decisions and inhibiting the team's initiative. The mitigation of impacts resulting from the task in the operation is inadequate because the root cause region is not identified.

The conflict between policies and practices and the tendency of human-type behavior in globalized society lead to low cooperation in the team, impairing social and technical relations in the tasks of the shift. Communication during the shift and the feeling of not belonging to the group can lead to various human errors.

These occurrences or cases of routine operation are discussed in the Risk Analysis indicated in Table 5.1, in relation to aspects of leadership, influence of environments and the possibility of derived human error (Ávila, 2010; Ávila and Barroso, 2012). Some actions were taken by the organization, operator or engineer or were perceived by the actions of the organization, operator or engineer, all on the production line. We seek to examine the following aspects: environmental analysis, leadership, skill, human error and the decision task.

These situations, experiences and actions were classified according to the risk level for lack of control, failure and accident as indicated in Table 5.1: the actions are considered appropriate in green (value = 1); the actions are considered adequate but not effective in yellow (value = 2); the actions are considered inadequate in orange (value = 3); or the actions are considered inappropriate with a high impact including fatality, in red (value = 4).

Table 5.1 shows the adequacy of the actions in relation to costs and image. We call attention to the fact that events that involve: errors in the decision on critical tasks are in red (45); inadequate leadership (40) and human errors in performing the task (40). In terms of the sum of points, incorrect decision with 70 and human errors with 66.

Regarding the cases of experiences analyzed, violations 15, 16 and 17 that involve practices carried out by senior management had the worst impacts (25). Experiences 12, 13, 14 are related to practices resulting from the culture of safety (guilt) and also have results with an accident tendency (21).

5.3 Lessons learned and validation of the guidelines

Although item 5.2 deals with experiences to design recommendations in the form of management guidelines, in 5.3 it presents and discusses real cases investigated and reproduces in the form of articles that deal with the cognitive academy, the executive function, the archetypes in the organization, the operator discourse for industry, blame culture, operational control and safety-organizational culture. These investigations present methods, valid as guidelines, and reaffirm tools already included in this book.

A group of characteristics at work imposes different behaviors and reactions that can impair operational discipline (Center for Chemical Process Safety (CCPS), 2013) with negative consequences for safety and production. Many researches referred in this book have attempted to test and explain how these factors influence in the task, and in the final we validate products to avoid the consequences, as accidents and disasters.

The proposal made by Academia Cogntiva (Avila, 2015b; Ferreira, 2018) is discussed in the cases of the fertilizer industry with the educational campaign *Amigos da Lagoa de Emergência* and with the work carried out to reduce losses of methylene chloride in the chemical industry.

The techniques of Archetype Analysis (Ávila and Menezes, 2015) applied in the control room of the industrial gas industry and the technique of study of the executive function, applied at the start of the reaction section in the chemical industry, were previously presented. In this discussion, in Section 5.3.2, it is intended from these practices to learn more about the technique. We also added the discussion with the Petroleum Industry manager about differences of results in control rooms of the same technology and the same managerial learning, understanding that actually, the manager does not have prediction where the hazard will leak and where the fault in the barriers will occur.

As we investigate the flaws in the operation of industry, we begin to understand the need to disclose important conclusions, as presented by American Congresses and Conferences in the area of process safety and Human Factors. The lessons learned that involve technological aspects are part of the subjects of this chapter. We will deal with an environmental accident with HCl overflow in the chemical industry (Souza et al., 2018a), with equipment breaks in the sulfuric acid plant (Ávila and Costa, 2015; Santino et al., 2016), presented in the Behavior and culture on Chapter 2. We will

TABLE 5.1 Operational routine cases and risk of human and organizational elements.

Case discussed	Failure Agent / Element	Environment	Leadership	Skills	Human Error	Task Decision	TOTAL
511 – GR1	Director at Start	3	5	5	3	5	21
512 – CTO2	Waste Disposal	3	1	2	2	2	10
513 – ER3	Control Stoichiometry	3	1	1	2	1	8
514 – HP4	Tuberculous Ear	3	2	1	2	3	11
514 – HP5	Root Cause Hypothesis	1	2	1	1	1	6
515 – GR6	Shift change	2	5	3	3	5	18
516 – COP7	Solution by Operation	2	1	1	2	1	7
517 – AP8	Routine Signs and Warnings	1	1	1	1	2	6
518 – AP9	Resistance and Behavior	2	1	1	1	2	7
519 – SO10	Foot Inspection	1	1	2	2	3	9
519 – SO11	Organization and cleaning	2	2	1	2	2	9
519 – SO12	Fear Punishment PPE Glove	5	5	3	3	5	21
5110 – PJ13	Startup Pattern Plant o-ring	5	5	3	5	3	21
5111 - PR14	Change of Manager, ritual	5	5	3	5	3	21
5112-VIO15	Improper alignments	5	5	5	5	5	25
5112-VIO16	Social Vices	5	5	5	5	5	25
5112-VIO17	Health: Staff vs Director	5	5	5	5	5	25
5112-VIO18	Transfer Control	3	3	3	5	5	19
5112-VIO19	Ethical Conflict	5	2	2	2	2	13
5113 – AC20	Improper service release	3	3	5	5	5	21
5113 – AC21	Forklift Failure	3	3	5	5	5	21
TOTAL		65/35	63/40	58/30	66/40	70/45	

also work on the case of alarm management for the Hydrogen compressor at the Oil Refinery (Souza et al., 2018b) and the case of operation control for the caprolactam plant that deals with the application of the FMEAH (Ávila et al., 2012).

These discussions on lessons learned are complemented with the calculation of complexity in the forklift accident in the Refining industry, and with the search for solutions that involve the Just culture of Communication in the Metallurgical and Oil production Offshore Industry (Ávila, 2011a; Ávila and Drigo, 2017; Ávila et al., 2018b).

5.3.1 Cognitive and behavioral academy: routine and program friends of emergency pool

Chapter 4 proposed the development of a Cognitive Academy to treat organizational disease with vaccines, which also involves aspects of Behavior and Culture. In this Chapter 5 we will address each social failure tree that impacts the task. This discussion presents Case 1 of the fertilizer industry that involves the cognitive-organizational campaign named as, "friends of the emergency pool to avoid environmental consequences," held during the 8-month period; and presents Case 2, with the recovery work of methylene chloride lost to the effluent and to the atmosphere, for a period of 24 months (Ávila, 2004, 2010).

The anatomies of the failures shown in Fig. 5.6 have results in low motivation, bad decisions by leaders, conflict of priorities between policies and practices, wrong patterns, procedures that do not represent reality, and bad habits.

In *A*, Fig. 5.6, unresolved technological issues that cause failures and stoppages to the industrial processes cause the operation to adopt inadequate standards, as discussed in Case 1, where an operational group adopts the dilution of effluents as a common practice. The control of the effluent becomes complex when it is not possible to avoid contamination from ammonia nitrogen and from other different sources. This complexity hinders the control of the process and opens the door to bad habits. In reality, in this case, environmental and safety issues are put in second place in relation to production and billing.

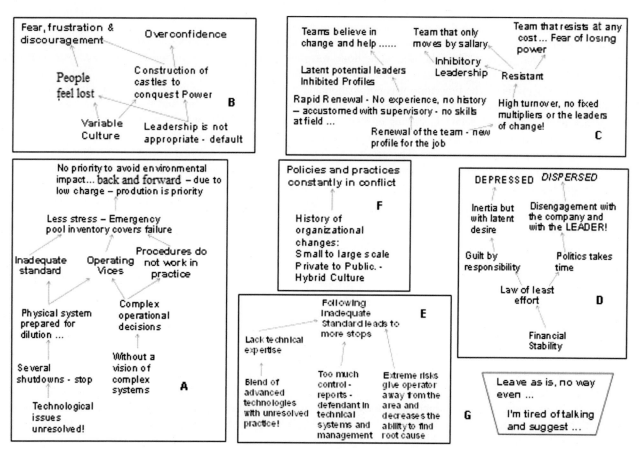

FIGURE 5.6 Sociohuman anatomy of failure on the factory floor. *Adapted from Ávila, S.F., Souza, C.R.O., Silva, C.C., 2013. worker behavior by changing dynamic culture. In: 9th Global Congress on Process safety, San Antonio.*

In the Methylene Diphenyl Amine and Isocyanate Plants, both Chemicals Plant, this complexity in the control of effluents is noted due to the failure to detect the point of contamination with organic matter, with acid and the subsequent lack of control causing chemical reactions and decantation of solids in the effluent (Ávila, 2004). On the other hand, at the Polycarbonates Plant, it is noted that modified patterns since the project alter the process condition, mainly in solvent recirculation currents, causing difficulties for the monitoring and control of effluent, it is not possible to control the quality of the effluent through of the steam stripping column (Ávila, 2010).

In B, Fig. 5.6, the lack of a "genuine" leadership that at the same time represents the group and the organization, causes the team to be discredited and often distrustful in the organization. People in the work group feel confused. The constant conflict between production and security is confirmed by underreporting. Part of the team is afraid to speak or record deviations, another part is frustrated by the absence of positive results for the team and the environment. And there is still a part that feels unmotivated. This confusing situation creates protections, which can be the shift group or the administrative group, to prevent damage to the team. This situation aggravates the efficiency of the operation and creates for the leaders the role of "FATHER" of the group, where, it also generates the behavior of overconfidence in itself.

At the Isocyanate Plant (MDI, MDA, Sulfuric Acid) points out Ávila (2004), the old leaders of the local professional career, lived with problems in the plant that were not resolved since the project, and try to respond to the Organization, without knowing how and what to do. The search for new solutions are suggested by engineers and operators. They tested technological barriers possible to be adjusted due to the informality, with more humble leaders, without excessive confidence in absolute truths about the causes of failures and incidents.

Meanwhile, at the Polycarbonate Plant, the leaders knew better about the technology, but did not share with the engineering and thus hindering the solution of problems. The group omission showed that certain supervisors defined the truths to be told. This behavior had environmental impacts. At the time the Operator's discourse technique was used, the group omission was identified and undone.

In the Oil and Gas Plant (Ávila, 2010), the politicization achieved by the production team established owners of areas, with own power, in relation to the power of Coordinators and managers. Thus the only way to convince the importance of reviewing procedures and designing cognitive barriers to avoid human error was by adopting transparent communication to the team regarding the human reliability program.

At the fertilizer unit, the shift coordinator represented the most appropriate standard of operation, since there were no procedures. What the boss "thought" should be done was done. The question discussed and analyzed was related to the appropriate moment and how to stop production in the synthesis of urea, without impacting the environment with ammoniacal nitrogen load. The procedure was not written and the stop left large reaction mass liabilities leading to the effluent. Routine operators were unable to correct or mitigate the complex situation, and they took the problem to the emergency pool, resulting in delays in starting of plant operation and the generation of environmental fines.

In C, Fig. 5.6, cases of "quick" hiring of new operators to fill vacancies for retired workers and "quick" selection of workers for new projects are discussed. The passing of skills happened in a disorganized way. Another discussion is about management who spends little time leading their team. This situation causes: low competence in carrying out complex and high-risk activities; fear of loss of position and power promoting inhibition of leadership. The petroleum industry managers remain in the production activity for only 2 years. When the manager begins to understand the problems of the team, and the team understands the managerial style, it is time for a new change of managers.

In D, Fig. 5.6, the lack of political and financial stability, in Brazil, brings the behavior of the law of the least effort, where, workers do what brings immediate economic result (without taking care of environmental priorities) and apparently does not need modifications. This position induce deviance normalization and hinders the innovation process. The behavior does not change although there is a desire for change, the team becomes discouraged, depressed. The lack of changes causes the group's lack of unity, bringing about dispersion. Both dispersion and depression favor human error.

Management must promote an environment of continuous learning and with projects, challenges that motivate the team, avoiding the inertial effect of normality. Or worse, inertia, due to a change that is impossible to happen despite the need for innovation.

These sensations were seen in human reliability experiments, in chemical and petrochemical plants, where normality indicated no need for innovation. In the isocyanate industry, on the other hand, there were so many accumulated problems that a change was impossible to happen, according to the team's feeling, generating a state of fear and depression.

The fertilizer public industry, which had the characteristics, job stability, constant change of managers, low standardization of good practices, low credibility for not solving accumulated problems, kept the team's motivation for innovation and for cooperative and committed work low.

In E, Fig. 5.6, there is a self-explanation: when improper behavior occurs in production, it can direct the team to follow inadequate standards that cause high operational discontinuity and accident. Among these behaviors are: (1)

excessive control, without flexibility in performing procedures; (2) low availability to investigate problems in complex production systems that involve systemic failure; (3) much demand by the organization for carrying out bureaucratic work (writing reports and reviewing procedures) by the production team; (4) high risks in the routine that take the operator out of the field for fear of a fatal accident.

In F, Fig. 5.6, where organizations make profound changes without the concern of adapting tasks, leaders, and rituals to maintain operational safety. The metallurgical industry that was family enterprise became a private society and changed its functional body, to meet technological issues, but did not change to understand about social-human issues included in the organizational culture. The chemical industry of isocyanate allowed the hiring of its team, with the establishment of different rules for different audiences bringing internal revolts into and inter groups, from different regions of Brazil, from Bahia, Rio de Janeiro, Rio Grande do Sul and São Paulo.

In G, Fig. 5.6, younger people or innovative people have the habit of suggesting improvements, always indicating a new way of acting. The inactive listening of the leaders causes the loss of motivation for innovation. Tiredness promotes loss of interest, many industrial problems are not solved, once there is no active listening by the leaders and also within the team. Operational groups need to wake up from this sleep stage for lack of motivation. In the fertilizer industry, we created the friends of emergency pool educational program and in the polycarbonate industry, the human reliability program. These programs can start informally so that they are not just linked to top management and seeking agreement from the factory floor.

5.3.1.1 Diseases and vaccines in the study of behavioral aspects

The fault trees in human and social behavior mentioned in Fig. 5.6 from A to G cause, along with organizational, technological and social culture aspects, cognitive and behavioral diseases. In Table 5.2 we list the diseases and vaccines proposed to be treated in the Programs of the Cognitive Academy under discussion in Fig. 5.7.

One of the diseases is the low perception of risk, which is considered the main cause in the investigation of accidents in the chemical and oil industry. It is related to aversion and risk and the normalization of deviations. It may also be due to new characteristics of attention in the workplace due to the conflict of generation and also due to automation. It is necessary to list the causes, measure and develop preventive actions (Ávila et al., 2018a). Still in the area of perception, we often do not diagnose correctly when the failure is systemic and multidimensional. For this, it is necessary to exercise technical-human and organizational causality with hypotheses (Ávila, 2015b; Ávila and Dias, 2017; Ávila and Drigo, 2017).

The feeling of not believing in the future, in people and even in teams in the work environment, makes it difficult to implement innovations discussed in Table 5.2. When developing interesting technical works, there can be a change in the motivation status of individuals and teams. In social terms, we can also classify as social and behavioral illness, an example is unwillingness to cooperate. In this case, demanding rituals and educational campaigns to activate this desire to groupal work. It is noted that the lack of motivation affects society and work in general, impacting on the perception and interest in research for the solution of problems.

The lack of memory activation due to the lack of commitment to work is different from that caused by the high level of complexity, and also different, resulting from the new generation occupying the job. All may be interrelated and require preventive actions to control the task. Selfishness-egocentrism is characteristic of the mass culture that isolates the worker from cooperative work. There is an excessive protection of subjects and personal space. Sharing is avoided and makes routine work as well as improvement work difficult. Meanwhile, characteristics of regional culture and also of selfishness, which bring about the formation of tribes, do not allow groups or people to signal vices and bad habits, they are only characteristics of tribes, which must be respected. Thus many times, specialists who are foreign to the environment and who are part of the just culture, are neutral social and politically, these multidisciplinary professional are accepted to signal vices and bad habits. But these strangers must observe the behavior in the routine respecting the establishment of the informal rules of the group, without judgment of values. Only after assessing the risks to critical activities that suggest modifications in the routine.

Commitment, an essential feature of the job, can be corrupted or very corrupted. (Sennett, 1999) People resist when they present goals and policies that are impossible to execute successfully, highlighting the differences between explicit policies, informal management guidelines and what is done in practice. It is important to identify the technical-operational culture to correct behavioral illnesses.

In this context of cognitive and behavioral diseases, it is questioned which is the best profile for leadership in the high-risk industry, and which is the best leader to motivate current teams. We have to quote Collins (2007) when discussing the level 5 leader who is different from the leader who values power in the pyramid of organizational responsibilities.

TABLE 5.2 Diseases versus vaccines.

Diseases	Vaccines	Modus operandis
1 Low risk perception (LRP)	Waking up to the need for change: training, measuring and intervening	In LRP training and in Measurements for C4t—SARS Program to propose the best interventions
2 Low failure perception, wrong diagnosis about failure mode	Exercise technical and behavioral causality with hypotheses	Prepare summary reports, newsletters and disseminate interpretations to the routine group—see reaction and results in practice
3 Not believing in the future, no innovations and attempts to be better in work	Evidence from good statistical work, persistence & appropriate method	Prepare interpretations that can be presented in texts or training for shift group
4 Little desire to cooperate	Neutralize political forces against Cooperation	Perform rituals and educational Campaigns to motivate the group
5 Lack of motivation to find the causes of complex problems as well on simple ones	Being involved in a structured team to seek results, motivate persistence and the search, also, on mitigation	The Group needs to be willing to investigate routine solutions, but also be comfortable of having practical and concept specialists searching for solutions. Leadership is the key.
6 Without activated memory, believing in guilt, without enough responsibility and commitment	Review activity responsibility in practice, activate memory	In the Control and review of the Task and in its Failure Analysis—Review
7 Excessive protections of personal space	Share more spaces, develop assets owners, share goals and ideas	Perform task force and training on shared modules
8 Refusal to work with strange actors who signal vices	Experts and Visitors help break vicious cycles	Improve understanding: vices and health, open the company to criticism
9 Compromise corrupted by policy and goals that are impossible to achieve, big differences with practices	Review policies and goals based on common sense including practice and Technical-operational culture (TOC) evaluation	Identify the Technical-operational culture and confirm through TOC Exams
10 Lack of true leadership, leader 5 discussed by Collins	Identify and develop leaders	Humility, fierce determination, active listening, knowledge, training and communicating. Develop clear tasks.
11 Inappropriate formal and/or informal communication	Investigate the quality of communications in company departments, hierarchies and areas	Conduct research on formal communication and open feedback, and directed through the operator and manager discourse
12 Very simple mind map	Investigate the complexity of the task, process and social relationships to define new actions for understanding and control	Develop people, carry out task forces, study and present cases and lessons learned
13 Inappropriate behavior for difficulties, lack of resilience	Develop activities to promote emotional balance	review the profiles of leaders and groups, anticipating dynamic threats and the need for resilience
14 Resistance to technical and organizational changes	Prepare informal leaders, give the feeling of ownership of change and the direction of the result	Prepare a cognitive and behavioral academy

In the routine of the operation, there is a lot of low-quality communication that is part of the organizational procedures or that is informal. Communication failures cause accidents and loss of productivity. It is important to do research what are the types of formal communication and what is the quality of open feedback for production. This research begins with the discourse analysis of the operator and manager (Drigo, 2016; Drigo and Ávila, 2016).

The management of complex and redundant processes must be handled by leaders and teams that are able to prepare the respective mental maps for investigation and decision. Thus a culture shock occurs when people are placed with a simplified mind map to solve complex problems.

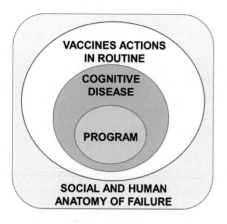

FIGURE 5.7 Cognitive academy and behavior. *Adapted from Ávila, S.F., Souza, C.R.O., Silva, C.C., 2013. Worker behavior by changing dynamic culture. In: 9th Global Congress on Process safety, San Antonio.*

What to do to combat this cognitive illness? What skills are needed and how to develop leadership for complex decisions? What kind of tool do we use to measure and correct behaviors (Ávila and Drigo, 2017; Ávila et al., 2018a)? How to maintain resilient teams in an environment of unknown threats? How to deal with resistance to technical and organizational changes without economic, social and environmental impacts?

_____
_____
_____
_____
_____

5.3.1.2 Cognitive and behavioral academy program (Avila et al., 2016)

The suggestion is to maintain continuous and integrated activities involving the areas of: technology, education and culture to shape behavior in the routine and in emergency.

Technological research is related to the application of statistical tools, routine analysis through the operator's discourse and the manager's discourse. When interpreting the facts that have an impact on heritage, billing and image, it is possible to provoke the team through operative group tools in search of the truth. Technological, human, and social criteria and restrictions are included in the routine analysis. The educational and training programs are part of the cognitive academy, shown in Fig. 5.7. A major concern is to keep formal and informal leaders in line with the intention of increasing human reliability.

This motivation and knowledge will guarantee the sustainability of the cognitive academy. The identification of bad habits, human-social typology and technological typology, in addition to the classification of human errors, allows to reduce the vices at work and to propose new forms of intervention to guarantee the increase of human reliability. The analysis of culture, its conflicts and the consequences on behavior for work are of high importance.

5.3.1.3 Fertilizer case, public industry: PROGRAM friends of emergency pool

The cognitive and behavioral academy had the purpose of acting on the team's behavior controlling the organic industrial effluent, mainly at the source. With this motivation, this academy created the educational program friends of emergency pool that receive industrial wastewater.

Preliminary knowledge about the events that contaminated the emergency lagoon of the fertilizer plant caused discomfort and generated discussions about the possible truth and the need for change, both due to technical and to social causality. A change in the behavior of the operating team was then the goal.

This research work involved the university and industry with technical activities such as water balance, statistical analysis of effluents, and study of process and equipment performance. These investigations intended to indicate how to reduce water consumption and the generation of organic effluent. The high concentration of contaminants (nitrogen) exceeded the negotiated standards for treatment, thus bringing the negative image to the company because of environmental fines.

The researchers focused on reducing effluent contamination at the source. This educational-technical work, entitled friends of emergency pool, promoted the sensitization of the company's employees, encouraging them to seek solutions to change the pattern of effluent release.

The educational program involved all employees, with greater emphasis on the participants of the production process in the areas that most contribute to generation of effluents or contaminants. The problems identified and the goals established required actions carried out to create leaders, monitor effluents, correct poor operational practices and evaluate results. These problems and breakthroughs were discussed by a group of investigators and subsequently reported to employees through booklets, lectures and reports. A technical bulletin generated for each event involving the plant and consequences on the effluent helped toward an organizational change to avoid recurrence of events.

The educational program was successful, because as the reduction of the level of the emergency pool by 50%, it was possible to identify the real problem of effluents, the poor management of passive inventories of plant shutdowns. There was a reduction in effluent flow, contaminant loads and in value of effluent treatment bills. Reducing waste improved the company's image for employees and environmental agencies.

The methodology of this environmental education program consisted of: identification of critical problems; setting goals; training, integration, monitoring and correction actions; and evaluation of the results. These steps form a continuous feedback cycle similar to the PDCA presented in Fig. 5.8.

5.3.1.3.1 Step 1: Problem definition

The environmental education program begins by studying the company's processes and monitoring its effluents. The survey of the effluent profile indicates possible disturbances in the operation and its sources of contamination. With this, it is possible to understand the oscillations of temperature, flow and contaminants, and determine the effluent release patterns, and the critical procedures that contribute to ammoniacal nitrogen contamination. At this stage, the reduction of load and the effluent flow were the principal goals.

5.3.1.3.2 Step 2: Formation of training multipliers agents

With the identification of the problem is possible identify the employees that are allocated in the areas, that, probably, mostly contribute with the contaminant load to the effluent. These workers were trained as agents of transformation through lectures, manuals and meetings. The objective of this training was to guide the workers about the project and its stages, inform the results of the preliminary study of the effluent profile and study the identified problems, thus finding possible causes and solutions for them. Leaders of program presented the standard behavior of the multipliers on the front line. This program, friends of the emergency pool, was applied when demands or opportunities occurred.

These are the additional requirements to apply the program:

1. Accept contributions to the program and respective guidelines;
2. Have booklets to distribute to the concerned parties;
3. Create forms to record ideas of opportunities for effluent reduction and contamination;
4. Inform about the feasibility of the solutions suggested by the company's technical group;

FIGURE 5.8 Method for the PROGRAM friends of emergency pool.

5. Guide the use of the portal for registration and monitoring the development of ideas;
6. Always have a printed technical bulletin so that the operator can read;
7. Discuss the results indicated in the bulletin with operators;
8. Be courteous;
9. Listen carefully;
10. Make sure that proposed ideas are understood correctly;
11. Win over new employees, volunteers, for the Program Friends of the Lagoon;
12. Moreover, create multipliers agents of training.

5.3.1.3.3 Step 3: Preparation material

Documents with technical and environmental aspects are prepared for dissemination. The group prepared booklets, technical bulletins and presentations based on case studies on the various points and critical procedures of contamination.
The documents included:

1. Booklets. It is the first material distributed in the company, as a way to introduce the project. It should contain information about the toxicity of the product manufactured by the company, the concentration of contaminant in the effluent and what is allowed by legislation, the difficulties for effluent adequacy, in addition to presenting the scope of the project and the work team. The booklet is an invitation for the entire company to work as a team participating of the project.
2. Technical bulletins: the purpose of these bulletins is to provoke discussions about problems, possible causes and solutions. They are derived from monitoring, case studies, simulations, feasibility ideas study of proposals for effluent improvement as well as information on changes in operating and management patterns.
3. Presentations: these are materials produced for lectures and training, made available electronically. It aims to keep the contacts network informed about the project stages. The designed materials are delivered in technical, formal and informal visits (step 4). If there is, during some period restriction of human resources, the information should be passed from leaders of educational program to company Engineers by electronic means. Thus keeping the interested parties always aware of the operational and environmental conditions. The flow of information must never be cut off, otherwise the company's employees would lose confidence in the project.

5.3.1.3.4 Step 4: Technical visits

The group of training agents from the company and the University presented the technical bulletins and gave these during visits to the units. The objective of this dissemination is to provoke further discussions in the operation team about the situation of effluents and releasing standards adopted. This is the "shoulder-to-shoulder" awareness, which encourages those involved with productive activity to participate in problem solving. For the level of integration with the production team to be higher, it is important that the physical headquarters of Friends of Emergency Pool to be in the control room of one of the industrial units, preferably to contribute, more effectively, with flow or contamination to the effluent.

5.3.1.3.5 Step 5: Maintenance and evaluation

The evaluation of changes in the pattern of the production routine is carried out on the level of execution and management where, tools are needed to make these changes in operational actions and decision-making. The changing of standards must be homogeneous for both areas, management and operation, and disclosed in the form of technical bulletins, both to help the decision moments and actions. Thus with clear references, it is easy to introduce to the team the new standards that meet the legal requirements and, mainly, the internal demands of much of the workforce. This stage of the program is carried out through inspections, monitoring of critical maneuvers and technical meetings.

5.3.1.3.6 Step 6: Results

Effluent monitoring should be maintained so that changes in its profile can be evaluated. These data are analyzed in technical meetings between multipliers. If necessary, new opportunities for improvement of the system are defined until the predetermined reuse potential is reached during step 1. The outcome is reducing environmental fines as a result of reducing contamination.

5.3.1.4 PROGRAM friends of emergency pool's activities description

Before the friends of emergency pool actions, it was possible to relate the increased loss in the process of fertilizer production with the contamination of the effluent increases, indicating when the effluent become inappropriate. According to the previous study, the non-continuity of the production process is responsible for almost 43% of the contamination of the plant's organic effluent system. This contribution is the result of draining the reactor at stops lasting more than 24 hours, because the highly corrosive reaction solution can cause damage to the equipment.

To prevent this reaction solution from being sent directly to the stabilization tank, lung tanks were installed, designed to receive the inventory of stops for future storage and reprocessing. However, due to poor management of this inventory and several attempts to startup and stop successively, these tanks reached risky operating levels, being necessary to send the inventory to the stabilization pond, generating a serious environmental problem and waste of raw material. Thus the lagoon designed to receive the stop inventory sporadically became process equipment and the inventory disposal maneuver became routine.

By identifying the causes of contamination, the friends of the pool began to sensitize the operators, also through technical visits and bulletins publication focused on the impacts caused by their operational practices.

During sensitization, a case study was conducted on the stopping patterns of urea production, before and after the beginning of the Program Friends of the Pool. The two stops chosen had the same cause—lack of raw material—and took place soon after washing the granulation tower.

For the weekly washing of the tower, the sending of semi-finished product to the tower is interrupted. The production of reactors is diverted to storage tanks that are the same that, in case of stop over 24 hours, receive inventory, which prevents damage to reactors.

At the first stop of this study, shortly after washing the granulation tower, there was an order to wait to reprocess the tanks only the next day. However, early the next morning the ammonia synthesis area stopped, thus cutting off raw material supply, forcing the urea area to stop. With the lung tanks filled with the wash inventory, it was necessary to send its contents to the pond before emptying the reactors, which caused ammoniacal nitrogen overload in the effluent and consequently high environmental impact and fine issued by the effluent-handling company.

Management and the urea operation team changed their stop patterns after becoming aware of the financial and environmental damage caused by the delay in decision-making and in reprocessing the tanks. Thus when the second stop occurred due to lack of raw material from the ammonia area, the lung tanks of the urea area were at reduced levels, even after washing the tower. There was also a quick management decision, which changed the patterns of stops and starts of the fertilizer-producing unit. It was established that during the stops, the entire inventory of the lung tanks would be conducted for the completion of the process, contrary to the previous orientation that was to discard it to the lung tanks and, later, to the stabilization pond. With such a decision the environmental and financial impact was reduced. This solution was possible only after opening a communication channel between engineers and operators using the intermediation of researchers.

5.3.1.5 Polycarbonate case: reduction of methylene chloride losses (Ávila, 2004)

The company that produces polycarbonate is medium-sized, with the capacity to produce 50,000 tons per year of products in the area of polymers. The industrial plant is continuous process and is subdivided into reaction section (SR), separation section (SS), purification section (SP) and effluent treatment section (STE).

Despite the technology of this process indicating a situation of greater comfort regarding the continuity of the plant, with a potential for the reduction of production costs, including the reduction of M (solvent) consumption; the factory had problems of process variation by lowering its production capacity and impairing the quality of the final product.

The final product of the polymerization reaction (P) changes its physical characteristics depending on the amount of reaction terminator (T). Each type of P produced has different physical characteristics and implications in the production processes. P is produced in this plant as a condensate of two reagents, here, L and E. L is the limiting reagent from a reaction in the gas phase between A (CO) and B (Cl2). To use A in reaction 1, it is necessary to promote its purification through physical processes.

P preparation involves two-step condensation reaction (reaction 2 and 3). Initially E reacts with L in basic medium. In this reaction, L is totally consumed. After these reactions, the separation processes start to occur only in the liquid phase. The organic phase is separated and sent to polymerization reaction 3. The inorganic phase will be sent to separators that return with P and M for reaction 3, and send process residual water for purification 2. In purification 3, the solvent M is recovered for the process. In reaction 3, viscosity and P concentrations are monitored to avoid problems in treatment and drying—Purification 4. Viscosity adjustments are required by injecting M, then causing P dilution.

The P resin resulted from reaction 3 has the chain formed but also has contaminants that affect quality and need to be removed through the purification step 4. At this moment, contaminant ions, water and solvent M that accompany the process organic phase are removed, later addressing the treatment section for ion removal in the P molecule, through rotating equipment that mixes the aqueous phase and the organic phase. It is made, in this section, the separation of excess water and P. After treatment, the ions attached to the molecule are released to reduce humidity, leaving M that will be removed in drying. For this operation, rotating equipment that operates with heating is used to remove H (humidity) and M (solvent).

Product P is initially flashed in drying 1, which promotes its concentration to the pasty state. P is then transferred to rotating equipment that, with heating and vacuum, promotes almost total removal of M and H, leaving after drying, in powder form.

The solvent (M) and water (H) from drying are condensed and redistributed: M returns to the storage tank and H is reused in the treatment section and sent to the wastewater separator. The amount of M that is not condensed goes to the abatement column using the cryogenic principle. P is transported to the storage silos and then sent to physical transformation. In this step, master with specific additives are added to suit commercial characteristics.

The factory operation is composed of people who produce polymers from raw materials and, to do it, they control systems and equipment with pre-established standards and procedures.

In the cognitive processing analysis of the polycarbonate plant operation group, some different behaviors that can result in team's speed, sharpness and cooperation in decision-making were observed. Different types of behavior can help or impair teamwork in production:

- The team has good visibility, which allows the causal link construction, and inadequate characteristics are compensated by qualities on the shift operational group;
- The team does not have good visibility. There are situations of failure, and people have porous memory, leading to situations in which the group cannot reduce the possibility of individuals' failure. Inadequate characteristics are covered by qualities on the shift;
- The team does not have good visibility. There are situations of failure, and people have porous memory, leading to situations in which the group cannot reduce the possibility of individuals' failure.

Knowledge of the process is essential to enable its control. Thus the availability of technical information, the experience of routine in the plant operation and the overall view of the equipment and systems, which are being operated, are important to achieve the result of the finished product with desired quality and quantity. Fig. 5.9 illustrates the polycarbonate plant.

A case of reduction of methylene chloride contamination is discussed in this industrial plant. In terms of environmental diagnosis, it was detected, through the investigation of abnormalities, that the loss of M to the atmosphere is a problem that involves practically the entire plant and activities.

The statistical process assessment (SPA) and the analysis of abnormalities from the shift report directed the focus of this research to the reduction of M losses. The hypothesis is that M goes to effluent and atmosphere due to incorrect process variables and procedures. The abnormality chains identified from the fragments of information at shift book and the tendencies of process variables help to identify the best barriers, including, review of procedures, process parameters and new standard in the use of technology (MEA, EVA, AEP). The barriers are validated after tests in the operation routine and the opinion of the leaders (formal and informal).

Table 5.3 describes the activities performed, the period, the expected effect and the effect reached on the inventory, the expected effect and the effect reached on the effluent. The following codes are used to describe in which area the intervention was performed: project (P), operation (O), process (Pr), maintenance (M), management (G), vendor (F). It is important to affirm that each intervention or action on equipment, people and processes respected at least 15 days, to ensure the renewal of the material under processing in the industrial plant. The average time to renew the inventory of this polycarbonate plant was of seven days, plus five days, when the subject involves standardization and communication.

Preliminary diagnosis (1Pr) is the application of the operator's discourse tool to map the chains of abnormality. Then, with the participation of the operation and maintenance technicians, these chains of abnormality are validated and priority is established, in our case, reduction of the methylene chloride consumption. This stage remained for 4 months due to the work of investigating the shift book and process variables. At the end of the diagnosis we had the variables, procedures, equipment and hypotheses about the group's behavior.

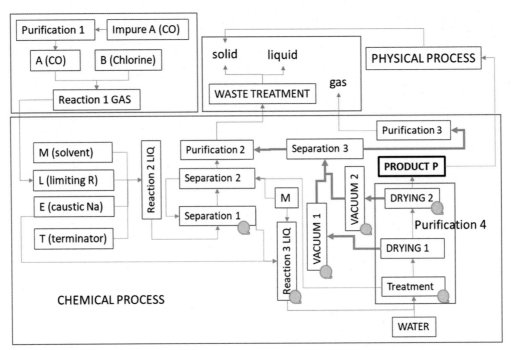

FIGURE 5.9 P plant, polycarbonate. *Based on Ávila, S.F., 2004. Methodology to minimize effluents at source from the investigation of operational abnormalities: case of the chemical industry. Dissertation (Professional Master's degree in management and Environmental Technologies in the Productive Process)—Teclim, Federal University of Bahia. 115 p.*

TABLE 5.3 Activities performed, period, expected effect, and effect achieved on effluent inventory.

Identifier	Activities for M loss investigation	Month	M loss index; M in the product			
			Effluent M content (ppm)		M loss (kg of M/tonP)	
			Expected	Result	Expected	Result
1Pr	Basic preliminary diagnosis OD, MD	M1,2	800	2400	550	
2Pr	Preliminary diagnosis construction MEA REA	M3,4	800	970	550	430
2P	Equipment in purification 2—Pf2	M5	650	854	390	280
3P	Adjustments in purification Profile 2	M6	575	830	347	285
4P	Temperature control	M7	600	430	360	225
5M	Reduce pressure for pumps	M8,9	520	468	320	350
6F	Meetings with supplier	M6	380	330	246	210
7M	Vacuum service and Pf4 equipment	M7	318	305	236	200
8O	Revision of procedures for W	M8	300	160	206	152
9O	E and S feeding	M9	300	215	200	100
10O	Revision of Extractor Procedures	M10	300	110	200	90
11M	Plant shutdown—improvements	M11	300	120	180	110
12G	Operator training	M12	300	400	180	173
13P	New equipment in purification 3	M13	250	513	100	94
14Pr	Process follow-up	M14				
15M	Maintenance of the cold system	M15				
16O	Preventive work support O	M16	280	150	100	58

We noticed customer complaints about quality in the finished product, black dots, and saw that, in the purification, the critical equipment was the Kneader. We worked together with the maintenance team to revise equipment loading procedures and vacuum control. It was necessary to carry out inspections until it was understood that the oscillation problem came from the feeding of the plant. This stage lasted a month. At this moment, we investigated the concentration of M at the Kneader process in purification 2 (2P). The complaints from the environmental agency led to the end of the process and we concentrated for a month on the temperature control in the stripping column (4P), which removes M from the effluent to send to the effluent treatment plant. In fact, this stage of the case under discussion was presented in the experience, in which the temperature pattern needed immediate performance analysis, but the analysis was time-consuming, which made it difficult to decide on the goals. Thus it was proposed the use of polycarbonate plate to adjust the temperature of this unit operation (Fig. 5.3, Frame 5.7—COP 7–516).

In the discussion about the actions, it was verified that the M content continued to vary in the various process stages. It was noticed that the alternative Alfa-Laval pumps were always dirty in the area with the presence of polymer and with methylene chloride odors. When investigating the sealing of these pumps, we saw that it is external and uses M recirculated from the process. It was noted that since the start of the plant occurs, out of specification O-ring is used. We did tests involving several pumps to reduce the pressure for these Alfa-Laval (5M) pumps, with the modification of the O-ring diameter. That change took a couple of months (Frame 5.13—PJ13–510).

The investigation focused on purification and reaction. In the reaction, we investigated the oscillations in CO pressure, reagent for phosgene production. These oscillations influenced the stability of the polymerization reaction. We had meetings with the CO supplier (6F) to improve supply standards.

The investigation that carried out interventions on the Kneader continued working on the stability of the vacuum pump (7M), in revising the procedures for feeding W reagent, bisphenol for reaction (8O), and it was also necessary to investigate the events of obstruction of the Soda feed pump (9O) for the reaction. A review of the process parameters was necessary after all these interventions related to the extractor starting procedure (10O).

To achieve the improvement in the reduction of effluent contamination and reduction of methylene chloride (M) consumption, it was necessary to make changes in the plant through its shutdown, when the leaders were delegated by shift group (11M). Then, the operation teams were trained: (A) training for operators on new standards in equipment and process (12G) and training for operators on safe behavior (12G).

By detecting the need to reduce M losses, the board had authorized the purchase and installation of cold box, a new equipment in plant purification (13P). In a meeting with the director, he said that by this time, only changes standards, maintenance interventions and process changes (14Pr), reaching 200 ppm in the effluent were achieved. To complete, the maintenance of the SABROE cold system (15 M) was performed and preventive work with operational support (16O) carried out.

The other activities that acted to reduce and maintain M losses in the process at the minimum values were: (1) process monitoring (product quality vs operation); (2) monitoring quality of maintenance services; (3) adoption of new standards regarding the cleanliness and area organization; (4) plan plant stops; and (5) daily monitoring of indexes (Fig. 5.10).

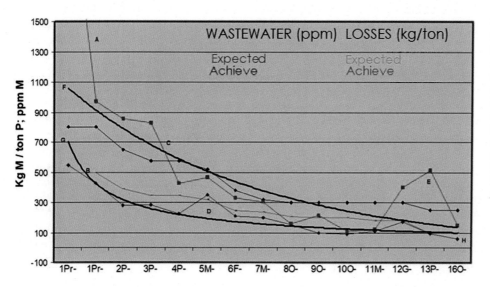

FIGURE 5.10 Graph of losses of methylene chloride to effluent, index consumption in P product. Ávila, S.F., 2004. Methodology to minimize effluents at source from the investigation of operational abnormalities: case of the chemical industry. Dissertation (Professional Master's degree in management and Environmental Technologies in the Productive Process)—Teclim, Federal University of Bahia. 115 p. Reproduced with permission from Ávila, S. F.

5.3.1.6 Comments about the chart in Fig. 5.10

General—There were difficulties in correlating reductions or increases in the indexes in relation to the actions taken, day by day, thus the necessary changes were made in the process, project, management, supplier and maintenance, and only four days later, changed to a new operating condition.

In A—At the beginning, there were very high results (above 2000 ppm) in the effluent. Perhaps the lack of care would promote higher contamination levels. The fact that people were involved to reduce contamination provided more accurate care in the system operation.

In B—The entry of new equipment for operation contributes to a greater reduction.

In C—The way to operate the purification Section 2 made it possible to achieve this reduction for two months.

In D—There were disturbances that provoked this behavior. Probably some system out of control, plant shutdown or equipment overflow with material in the effluent (lack of equalization).

In E—It was noticed that the trend of reduction at the beginning of the process was more drastic. In the effluent drop of 500 ppm between April and September and September to January, this drop was 300 ppm. In the final phase (January to May) the drop was decreased to 120 ppm. Reduction of the fall overtime. In order to make major changes, new technologies and/or culture change must be invested in the company.

In F—Similar to F, between April and September a drop of 500 kg/ton of P, between September and January the drop was 70 kg/ton. Between January and May the drop was of 30 kg/ton of P.

In G—It was noted that the greatest achievement of this loss reduction process was achieved after operator involvement—keeping this item under control. 140 ppm for the M content was achieved in the effluent, as indicated by the trend curve (but 250 ppm was expected). For the loss recorded by the inventory 100 kg/ton of P (expected 100 kg/ton) was obtained, thus reaching the target.

5.3.2 Application of tools for archetype analysis and executive function in the industry

The analysis of human behavior considers only "cold" and simple cognitive models, probably, this is the cause that causes the activities to be unpredictable. Probably, if the current cognitive models included intuitive, historical and affective aspects, with the evaluation of emotional balance, with the analysis of cultural biases and inheritances on the action of routine, the activities would be more predictable.

Many of the human errors that occur in the construction of major accidents are not expected, and may arise from apparently unexplained factors. The existence of latent, non-visible faults is proven when identifying specific characteristics (resident pathogens) that can lead individuals and equipment to start a chain of abnormalities.

The analysis of tasks sequentially, as the actions were performed in the procedure sequence, does not allow the visualization of the chain of operational factors that lead to failure, and it is necessary to understand the operator's discourse through a tool for abnormal event mapping (Ávila, 2010, 2012).

Regionalisms lead to rules on behavior and trends for specific decision-making models. There is no universal worker that has a single behavior in the face of high level of stress or of high complexity. Thus the combination of organizational and socioeconomic environments produces specific stressors that can change the behavior of different types of workers. At the same time, the multiple data, linguistic aspect and external influences on tasks become a challenge not to cause human error

The available data on human error presented in the models are limited for application due to these regionalisms, thus requiring correction in this analysis. The current model search the approximation tendency through signs of failure for the large event or accident.

Marais et al. (2006) propose a dynamic model to assist in the interpretation of human behavior in group activity. This model considers the existence of group behavior archetypes that are common, repetitive, and not explicit. This makes it clear why security-related decisions do not always result in expected behavior.

Thus it is necessary to study human error and know how to discern when a repetition of error occurs, as if it were a cultural habit, as if something triggered this repetition, as if it were an "obsessive disorder" but this time motivated by social fantasies coming from the collective unconscious (Jung, 2002), which are built as an actor in the task. This construction process is named an "archetype." Some very old and others new, which comes from group vices or even vices of leaders transferred to the group.

In accident analysis, it is important to know the plant history to identify which erroneous actions may have been the result of bad habits and that are ingrained in organizational or social archetypes.

This subject was commented in Chapter 3 but was not discussed as lessons learned. Here, there will be a discussion considering cases from the oil and chemical industry.

The scenario of a control room is set up as in the article about the influence of social characteristics on the cognitive ability of solving problems in the connivance of the group, which variables and controls are performed through the man-machine interface. This case discussed in the AHFE (Ávila and Pessoa, 2015) involves the adaptation of the Liquefied Natural Gas project with reverse operation, from plant that received the gas, to a new function, a plant that sends the gas. In this case, local behavioral characteristics were studied to make adaptations in the control room.

The criteria and standards established in the critical relations in blue, indicated in Fig. 5.10, can cause human error. Thus the archetype of hero can affect the decision in emergency situations, with a suggested improvement, a new safety barrier, in the visual appearance and in the logic of the emergency screen, as a barrier to the cultural situation.

The critical control variables chosen in the project for the LNG (Liquefied Natural Gas) plant primarily meet the need for quality control and safety, but do not see the possibility of a Crisis Scenario and what is the typical behavior in that region. Then some new important variables for cases of great release of hazard, such as furnace explosion and release of fireball after breaking the sphere would be suggested (Ávila, 2010).

The excessive flexibility identified through sociohuman risk analysis (Ávila, 2015a) and confirmed in research on human reliability (Ávila, 2011a, 2011b) applied in chemical and LPG plants can affect perception and attention. These essential functions for cognitive processing lower the quality of decisions in emergency situations or in the case of complex operations. This cultural characteristic, when it occurs repetitively, can generate a bad habit with the "birth of an archetype" and its related actions. Because of this, computer screens in chemical and petroleum units should be revised to increase the operability of control systems.

The revision of the screens layout and functionality aims to trigger attention in the operation of emergency screens and facilitate the operability of control systems, like facilitate the interpretation of events by giving hints of probable calculations or providing ways of rapid calculation, or rapid memory retrieve for the interpretation of events.

Bophal's accident showed inadequate characteristic-attitudes of operators, supervisors and staff. With the chain of occurrences, the release of progressive hazard energy, and the respective barriers that become failures due to economic biases, it is noticed that the team had no action and the arms folded, while they waited for the safeguards to work. As a result, the chances of avoiding the major accident were decreasing until the release of toxic gases into the city became reality.

The group's behavior indicates an inertia in the face of the release of hazard and a lack of perception regarding the consequent events and the need to interrupt industrial processes. This isn't happening to us.

Is this inertia caused by cultural issues? As the phenomena of risk aversion?

Is there any way to stop this freezing behavior?

Why was the group virtually hypnotized and believing that the barriers would work despite the reduction of investment in process safety?

What actor promoted this state?

Or, was the organizational culture that promoted this state?

Please, answer these questions.

_____
_____
_____
_____
_____
_____

These are questions to be discussed through techniques and tools that consider the failure as a systemic event and that bad habits can cause great damage. Until that moment, then, we talked about different tools on the same subject: archetype analysis, human error analysis through executive function (ef), preliminary sociohuman risk analysis inserted in local cultural influences such as risk aversion, spectator posture, paternalism, and others.

The human-process interface project must be accurate, at the same time, must include social flexibility. This interface needs to include this cultural-social-human discussion about the risks to the occurrence of accidents resulted from chain events. To reduce the possibility of human error, Ávila (Ávila, 2010; Ávila and Costa, 2015; Ávila and Menezes, 2015) proposes new criteria for the design of control systems by suggesting changes in cognitive limits in alarm systems, the establishment of priorities based on operator's discourse analyses, and the maximum level of automation to be monitored by the operation and the need for discrete authorizations among these automatic actions.

At the moment we discuss human error and technical failure in the operational routine, let's talk about a discussion held with a manager of a public industry oil well during negotiation on the research project. He managed 11 FPSO ships and said he did not know what ship, where in the ship and what function is related with the next occurrence regarding occupational or process safety.

This manager indicated that the technology is the same in all oil production ships, as well as the contracted firms management, the same investment in safety, and the same tools learned in the school of managers, thus indicating appropriate criteria for production. He asked me, since the standards are the same, why he couldn't anticipate effective actions for the next big and impacting occurrence to avoid time waste or an accident.

Initially, it is known that the choice of leadership is a fundamental part of enabling actions that avoid the accident. The leadership in production and safety must have high listening capacity and must admit that the influence of regional-global culture on behavior occurs. Leadership should also practice justice in the group by avoiding the formation of fiefdoms with preferences and avoid the separation of the group from the contracted companies in relation to the group that operates. Formation of social-economic castes and feuds.

The repetition of sentences, already discussed in Chapter 3, such as "might is right, the sensible obey," should be avoided because it provokes a religious mantra against safety. Actors assume this behavior and radiate this phrase and have as consequence archetypes against safety. The immediacy in the actions and the lack of group discussion, where the leader assumes that he/she does not fully know the problem, and has the humility to accept other opinions and doesn't rest while there's no evidence indicating that the region of the accident was removed.

The leadership style affects the development of team skills, the level of knowledge and also communication between the components. The centralizing style is not suitable for process safety and adopts many biased behaviors of protecting smaller groups and not caring for the result of the whole team. These differences between leaders hinder the standardization process and the relationship with actions in the control room. An indicator on the interfaces, computer screen, could help the measurement of this approximation of the practices to the written procedure by counting performed alarms per operator in relation to the respective standard.

Ávila and Costa (2015) stated that if there is no cognitive-intuitive-affective analysis on the task in progress, it will not be possible to correct the barrier flaws in the areas of behavior, management and technology. The analysis of the operator's discourse allows for mapping the main process abnormalities and suggesting how the group relates socially and what the effect of socioeconomic-affective utility is on the execution of the critical task.

We are talking about human errors of omission and also incorrect decisions in the strategic, emergency, tactical and operational areas, and the relationship with the bad habits resulted from failures in hiring and developing teams as well as the incorrect rituals for maintaining the safety culture. These rituals, such as the daily safety dialog, intend to act by isolating, canceling and deviating from behaviors resulted from the archetypes already present, that promote accident process. On the other hand, to understand the group's collective unconscious and the resulting situation of progressive stress, Ávila (2011b) suggested the LODA that discusses the quality of decisions in crisis situations with the fireball scenario in industry with a unit in Brazil.

The type of commitment depends on the affective bond between the company and the employee at all levels of the organization. In this case, the oil industry is discussed and the social-cognitive matrix for modifications in the human-process interface at the control room. The oil and gas industry has a high risk for fire and explosion. The fear of these events versus the safety quality of barriers and organizational behavior will reflect on the operator's performance in routine and emergency.

Human error depends on cognitive functions and the social environment where the worker is historically involved, or at the present moment in the workplace. If low cooperation is a regional characteristic and is not properly treated by the organization, in this case of LNG facility discussion, it will be necessary to design better communication tools in the control of operations and in the work social environment. Avila (2015b) argues that poor feedback, especially in Brazilian culture, causes underreporting, mainly due to the acceptance of the blame culture, and because, also, there is no effort to seek a fair culture that privileges communication in the critical task. The supervisory screen has to be adapted for the improvement of feedback through control systems.

The workers' impulsive style in performing tasks affects emergency decisions as well as general decisions during operation, which requires time delay adjustments in some controls of the production supervisory. This criterion can be adopted in the alarm systems and in the operability, in the article, the case of LNG plant. The characteristic of multiculturalism requires special attention in the choice of linguistic signs to enable interpretation and monitoring of critical tasks. Communication failure should be avoided through corrections to the criteria adopted for operational safety. There should be clarity and consistency in standardization.

The question from the Oil Well manager that managed oil production in 11 FPSO (Floating production Storage and Offloading) units in Brazil indicates the unsafe condition in the programming of resources for maintenance and operation in production planning. After the discussion on tools for reducing human errors and incorrect decisions, the following conclusions are indicated: (1) the platform manager and well director influence or create bad habits that can be part of heuristics in production; (2) the installed team can adopt an informal leader who has a better sense of justice in working with the team; (3) the collective unconscious may induce regionally or globally addictive behaviors that harm production; (4) natural and economic environmental threats; (5) material and human fatigue; and (6) the leadership and team behavior features can cause failures and accidents. We need to analyze all these phenomena, together, to predict the group behavior and the probable production in future time.

Interfaces should therefore help to decrease the likelihood of human error. Normally, interface projects do not include the most important cognitive and social limitations to be incorporated into patterns. Among ISO and EEMUA criteria, the following can be summarized in Fig. 5.11 from the Cross Matrix designed by Avila (2015b): (1) configuration of the control device for several variables in an emergency situation; (2) trend configuration in the release of hazard in chain reactions; (3) screen and training that integrates complex factors in the industry that, when are connected, can cause the accident; (4) improve the operability of systems through cognitive verification of automatic actions; (5) increase robustness in feedback systems for large and complex plants; (6) increase the clarity of the written text for the multicultural team; (7) avoid the characteristics of excessive flexibility; (8) avoid spectator posture; (9) avoid the low-affective bond with work; and (10) avoid inadequate communication style and low-speed response. Environmental factors affect the following cognitive aspects: perception, attention, and emergency decision, decisions to control complex operation, skill, commitment and diagnosis.

5.3.3 Communication in routine—environmental accident with HCl (Souza et al., 2018b)

The chemical process of the chlorinated product hydrolysis generates a by-product, HCl. Part of this HCl is condensed, sent to storage tanks in the production area and transferred to 20-ton-capacity trucks. The system operates with two production units and the need for communication between the respective panels. It is a simple operation that needs an interface in the activity with effective communication from field operator and panel and between two production areas. In the last week of September, at 12:50 p.m., in the case of an HCl truck loading operation, the panel operator began mapping the operation screens to begin to send the team for lunch, and noticed the start of HCl transfer. The field operator identified that the truck was loaded but it was indicated by measuring instruments as empty and that the loading station containment system was flooded with HCl.

Although previous communications between the panel operator and the field operator were indicating that the system was normal, the transfer valve was in the closed position, and the panel operator didn't notice that there was no HCl flow.

Social → Cognit.	Hero, anti-h	Flexible	Paternalism	Central ization	viewer	Leadership	Region alism	Affectiv link	Cooperation	Communication style	speed	Multicultural.
Perception		B		D			EF		EF	B		
Atention		B			OD					B		
Memmory				D		D	D				D	
Routine decision				IJQ					IJQ			
Emergency decision	K	K				K					D	
Simple operation											C	
Complex operation			C		C	L	CL					
Skills						GHI						
Knowledgement										K		L
Commitmment			GH	GH				MNO				
Diagnosys								CIN				
Communication noise						DGHI				G		G
Labor charge						DGHI						
Physical eforcement											P	
Displacement											A	

(A) Design and layout of stations
(B) Operation screen information
(C) Operability of the control device
(D) Alarm system organization
(E) Number and layout of the screens
(F) Ease of navigation on the monitor
(G) Clarity, Consistency and standardization
(H) Flexibility level operations
(I) Feedback information for the operation
(J) Presentation of information and screen layout
(K) Emergency Screen: provision and Information
(L) Automation and controls
(M) Size of the plant - product inventory
(N) Complexity of operations (recycle)
(O) Redundancy: critical system "cognitive laziness"
(P) Segregated stations
(Q) Identification of the hierarchy of screens

FIGURE 5.11 Cross matrix: social, cognitive, and interface. *Reproduced with permission of Ávila, S.F., Pessoa, F.L.P., 2015. Proposition of review in EEMUA 201 & ISO Standard 11064 based on cultural aspects in labor team, LNG case. In: 6th International Conference on Applied Human Factors and Ergonomics (AHFE 2015) and the Affiliated Conferences, Jul 26–30, 2015, Las Vegas, USA.*

The start of loading with the valve in the closed position caused increased pressure on the line, thus resulting in the opening of the PSV (pressure safety valve) that directs the flow of HCl to the secondary containment of the loading station. A sequence of operations were performed until the occurrence of the containment loss event, in which HCl was sent to the effluent system as the plant stopped.

5.3.3.1 Normality status (human factors and operational control)

The perception of the operator's discourse indicates the level of normality of the events in the maximum abnormality scale until complete normality. Automatic or discrete control systems are triggered when abnormal events occur in these high-risk activities; safeguards are triggered. The normality states can be classified as follows:

1. Normal State (NS—between 91% and 100%, indicated by green color). The process and product state meets the standard desired by task planners and also field and panel operators' perception. Normality indicates values close to 100%. Small deviations are not enough to affect this state of normality.
2. Intermediate State of Normality (ISN—scale from 71% to 90%, indicated by blue color). In this state, part of the events reported in the shift book and analyzed in the operator's discourse tool, have the occurrence of 20% deviances in the operational routine. These deviances may be related to critical variable alarms, product quality, equipment failures, non-standard effluent, and other events. These deviances require historical analysis in the shift book for the last month.
3. Intermediate State of Abnormality (ISA—scale of 41% to 70%, indicated by orange color). In this state there are process losses with consequences on the cost of the company—loss of quality, production, safety or environmental impact, which leads to reduced load. It is necessary to trigger a multidisciplinary task force to raise hypotheses about the causes and how to avoid chain reactions. At this stage, tests are performed in operation.
4. State of abnormality (SA—scale from 0% to 40%, indicated by red color). Minimally one of the actions can be the plant stop to avoid the hazard circulation. If the hazard keeps being released, accidents or even major environmental impacts can happen. This situation requires specific tests to identify causes of operational problems.

Based on the series of events and their chronology, the normality-abnormality scale graph can be constructed.

5.3.3.2 Chronological description of occurrences with HCl in chemical installations

(08:30–09:00) Field operator establishes a connection between the truck and the HCl tank. A test is performed involving the flexible tube, ventilation, and e a contact established between panel and field operator until finalizing the truck connection. The HCl tank had not yet been released for loading. 80%.

(09:00) Pneumatic load valve remained almost completely closed. 95%.

(11:25) Panel operator in the area advised the panelist in this event area to release the field operator to start loading. The event area panel operator made the preparations for loading, collected information and the checklist from the field. The knowledge was provided by the field operator and released from the checklist. The field operator checked the information released by the panel. The field operator began loading through the valve and requested confirmation from the panel operator. 80%.

(11:34) Panel operator selected the HCl loading screen and observed the automatic opening of the valve. There was no verification of the HCl flow through the flow indicator according to the procedure, despite confirmation of panel and field loading. 72%.

(11:45) Area field operator inspected leak-free hoses and flanges. There was no verification of fluid flow through the indicator in the field, procedure. Panel operators changed the truck loading operation screen. The loading remained without continuous monitoring, not following the procedure. 80%.

(12:00) Field operator initiated another truck disconnection activity in another product near the HCl. 72% loading station.

(12:16) Charging station operator initiates the change-over for lunch time. The operator now goes to lunch and returns 40 minutes later, to another area of the process to take tank sampling. 65%.

(12:55) Panel operator begins mapping the monitoring screens and realizes that the HCl throughput and the totalizer are zeroed. The panelist requests immediate confirmation from the field operator regarding the loading station. There was no HCl flow to truck with the tank level changed and confirmed by the other panel operator. 60%.

(12:58) Field operator arrives at the loading station and, when climbing the loading walkway, finds the charging valve closed. The operator of the loading station gets off the truck and realizes that the ramp was leveled and there was a smell of HCL. 40%.

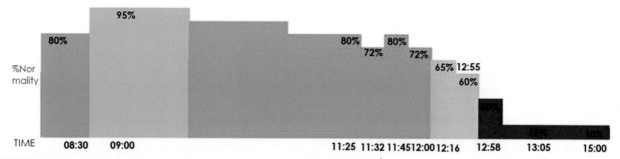

FIGURE 5.12 Level of normality from the occurrence of HCl. Based on Souza, M.L., Ávila, S.F., Cerqueira, I., Santino, C.N., Ramos, A., 2018a. Failure Analysis Cases in Operational Control and Recommendations for Task Criteria in the Man-Process Interface Design. In International Conference on Applied Human Factors and Ergonomics. 21–25 July 2018. Springer, Cham, pp. 317–329.

(13:05) Field and panel operator discuss the event and then report that the pneumatic load valve remained closed during the event. The panelist informs about the loss of containment when operating the pressure safety valve (PSV) of the loading line, confirmed. 15%.

(15:00) Cleaning and organization of the area, and transport of the material to effluent treatment that, due to the acid characteristic, without pH control, the effluent caused the plant to stop. 10% (Fig. 5.12).

The discussion of these events that caused the environmental impact of HCl shedding requires the identification of the task environment and the risk analysis of cognitive failure through the analysis of the cross matrix already discussed in this chapter in the cases of LNG and LPG plant.

The external environment, culture and natural phenomena can threaten behaviors with the possible loss of discipline in the execution of procedures, triggering human errors with incorrect decisions. During the event investigation, a series of anomalies were observed in the process of system reliability and operations, thus highlighting aspects of human performance and procedure execution. In the case of human performance, the flaws are in the following gaps: knowledge, perception, memory, communication and decision in the routine, in procedure execution, when HCl loading was not followed by the field operator and the panel operator who did not check the HCl flow.

The organizational failure (safety Culture) of not performing simulated training for leakage or overflow situations with HCl make the team weak in risk perception by not calibrating attention through an improved perception of unusual situations.

The simulated training for the perception of unusual situations activate the flow of information to the working memory involving the supervisory system, the central screen and the transition of operators in the control room.

Training, flexibility, leadership style and cooperation should be part of the certification process for field and panel operators. This certification process should be improved for HCl loading events as evidenced by environmental impact failures in this chemical industry. Thus certification should involve the hazardous activities of the chemical industry including acid handling and truck loading.

Performing the procedures requires special attention regarding the level of flexibility in accomplishing the steps and the level of redundancy in the communication of these critical operations. There is no evidence that operational instructions were being read and understood by operators both the original procedures and their last revision. Redundancy in written and spoken communication can highlight the need to verify the actions of the field or panel. Thus confirming and reconfirming the mind map is necessary.

The human-machine or man-process interface must meet quality-production and safety goals. Thus it is important to make a diagnosis indicating possible control device failures and also memory retrieval for the decision. The control needs to be improved, since it has been verified that the manual valve used to load the truck is pneumatically actuated, but without indication of position (open or closed) on the panel.

The installation of an appropriate risk perception for these chemical activities including involving HCl depends on the auxiliary memory or auxiliary processing of the inventory and that can be provided by the computer. This is a necessary review to assist the panel operator who is in the decision-making process regarding the procedure.

The protection safety and process control system is also an important topic for discussion. Apparently, in the event with HCl, no appropriate diagnosis was made regarding the risk of activities in this chemical industry. This lack of diagnosis brings false demands of devices and auxiliary memory or does not even consider these demands. This system in the present case should be improved by the prevention of new scenarios involving PSV, the logical control, inserting automation in the opening and closing of the HCl valve that must be activated in a viable time to avoid the accident.

The lack of cognitive field conference due to the comfort of automatic systems can introduce low risk perception and inertia of actions. Thus the characteristics of persistence and perseverance in field activities are important to maintain a standard of excellence in operational safety.

HCL loading tracking charts were not used by panel operators to interpret the events that were occurring. This situation indicates cultural characteristics present in the production team such as excessive flexibility in the criteria for decision, paternalism in operational group protections, leadership difficulties for process safety, lack of knowledge, communication noise, lack of compromise and tools in auxiliary memory and auxiliary information processing.

Thus there are repetitive failures that require tools to prevent the recurrence of this accident: training, leisure, organizational integration, social inclusion, simulated for chemical emergence, adaptation of rituals and review in communication tools.

The panel operator was in the process of fatigue because, during three consecutive days, he did not have an adequate rest in the shift regime. This situation has implications in increasing workload through both increased physical and cognitive efforts. The lack of diagnostic, practice and the failure to cooperate indicate the existence of an inappropriate leadership profile.

The risk analysis of the system was failed because it didn't consider important aspects of the interface between the worker and the panel, the procedure and the equipment. The non-perception of the scenario in a timely manner caused the accident, which was connected with flaws in the protection system and in the decision for the operation in a complex process, although apparently the activity is simple.

The large amount of HCl that went to the effluent caused the plant to stop in an emergency regime which could affect the production and the process with other damage. This means that there has been a lack of training for operators and leaders in emergency communication

It was found that there was no risk analysis of the system (not the human interface) for possible accident scenarios, protection systems and their effectiveness in decision-making (complex operation). Due to the accidental production interruption during the disposal of the waste generated, there was a need for training of operators and shift leaders in communications and emergency actions. In this case, the trainings including emergency decisions, avoidance of the HERO archetype, the centralization in the decision, and the failed feedback in the operation can be mentioned.

5.3.3.3 Some final considerations

5.3.3.3.1 Cognition

The matrix study indicated that the main cognitive characteristic to be worked on was the perception of risky situations and redundancy in communications. Another fundamental element is routine risk analysis, always considering the possibility of hidden factors are causing the failure.

5.3.3.3.2 Culture and behavior

Is there flexibility at work? Because they are people who know each other and have some comfort and confidence, they get used to it and lose the "intense" desire for process safety. It can include listening, perseverance, and determination through task forces and the development of informal leaders.

5.3.3.3.3 Level 5 of leadership and operational discipline

No fault. Is there any regionalism or leader's specific characteristic that may have created the communication noise?

5.3.3.3.4 Human-machine interface

review the amount of information on the loading-process screen. Revision of the emergency screen in cases of effluent overload. Revision of the effluent system related to the loss of pH control.

Your comments...

5.3.4 Investigation of technical failure and human error in the sulfuric acid plant

5.3.4.1 Process description and main events

The reconcentrated acid production unit aims to provide sulfuric acid above 95% for amine preparation and isocyanate as final product. This unit removes water from dilute acid. For this removal, it is necessary to operate with special equipment that resists high temperature, acid and vacuum.

Diluted acid with high dinitrotoluene content is stored and sent to the purification section, SP1, where gases and the organic matters are removed (at least it's the intention). The gas current is directed to WS (Water Separation), where gases are liquefied and form diluted acid solution. The liquid current goes to SP2, where both high temperature and vacuum promote the removal of water and thus reconcentrate the acid. The Reconcentrated Acid (RA) is stored in tanks and transferred to the production of isocyanate. RA production is in Table 5.4, where the parts of the reconcentration process are detailed.

5.3.4.2 Critical discussion on the culture of operational safety (Ávila, 2004; Avila et al., 2016)

The sulfuric acid reconcentration plant (84% to 96%) was sized to meet the content of organic, dinitrotoluene (DNT) and nitrous gases (NOx) in raw material, much lower than in operational practice. Probably, the increase charge of the TDI (toluene diisocyanate) plant brought low efficiency of the nitration reaction with inadequate separation and high content of these contaminants due to the high presence of organic matter. It seems that the investment in the expansion of the TDI plant was not adequate, taking into account the increased events of failure in the process quality and reduced in RAM (principally reliability and availability).

Horizontal Budget cuts are common following the bank culture, repeating various situations of major accidents happening in the chemical industry, such as what occurred in the Bophal Accident, where the competition forced cost reduction with cuts in the maintenance area of systems of Process safety. Since the project, the operation lives with the situation of excess DNT and NOx. The project in this bank organizational culture imposes low standards in the operation of sulfuric acid reconcentration industry.

The quality and quantity failures can have penalties such as commercial contractual fines with political advantages in the supply chain. In the case of the sulfuric acid plant, where the spent acid leaves the dinitrotoluene preparation plant and goes to be reconcentrated in the SARU (sulfuric acid reconcentration unit). The supplier is the maintainer of the sulfuric acid plant, so it "pushes" contamination problems (organic matter and NOx) that return with product availability problems (lack of specified acid). This situation forces the purchase of pure acid from another plant and the external sale of this unspecified co-product.

The raw material contaminated with organic matter is directed to the plant where the process consists mainly of glass and vitrified equipment. At first it is stocked in tanks where the decanting of this organic matter occurs, so getting immediately the benefit of solid retention. But, with the saturation of the tanks, this accumulation of organic matter can deflate and cause equipment failures. These failures, due to mechanical and thermal impact, can cause many breakdowns of the plant glass parts.

The reconcentration plant has two production units, named side A and side B. When stopping the plant on one side due to the breaking of a glass part, the phenomenon of "cannibalism" occurs, in which one side is "fed" by the other.

TABLE 5.4 Elements of the acid unit process—AU.

Input	Systems					Output	Effluent
	Physical	Laboratory	Mechanical	Instruments	Pipeline		
Acid Spend AC1	Separation SP1 Gutting Purification—Flash SP2; Vacuum; Absorption	Content of organic matter in the acid	*Purification Pump *Vacuum Pump	Flow of: AC1 Temperature SP1; SP2 Vacuum: SP	Obstruction of pipelines and equipment with Organic matter. Hole in pipelines and equipment.	RA Reconcentrated Acid	Organic; RA gases, AC2; Acidic liquid effluent

Thus to maintain operational continuity, one side is adopted as a storehouse for the other, thus decreasing production capacity and delaying the investigation of problem cause.

Repetitive replacement of glass parts brings situations of acid leakage and cracking of flanges. Glass or glazed tube flanges can easily be damaged if the maintenance team does not adopt appropriate standards with specific torque by pipeline type. Due to the repetition of the events at the Sulfuric Acid Plant, the SCHOTT of Brazil and Germany were invited several times for technical assistance. Also several times, an opinion was issued of breakage in the vacuum system (large pipeline), which shows that the processes are complex and require investigation of cause, taking into account a chronology of events, with periods of above one week.

This unstable situation of pressure oscillation in the feeding of the reconcentration plant due to unexpected blockages with DNT of the acid, requires a revision of procedures for new process states. This decision seeks to increase operational reliability. Thus the research in this plant suggested and revised: the procedure for vacuum testing and for interpretation of problems; the start of the plant, considering the condition of the raw material tanks; the starting of the organic removal column through stripper steam, the first larger equipment that receives the load of organic matter; the feeding of raw material to avoid obstructions with organic matter; plant stop to avoid sudden dilutions of acidic water, which would generate thermal shock; and other procedures for maintenance and for safety.

The acquaintance of operators for 10 years in a high-risk industrial plant, where the problems are repeated consecutively and insistently, causes the loss of motivation for the work and the low credibility in relation to the company. The relationship between professionals in the operational routine was irritating, part of the information was not communicated and the commitment to the organization did not occur. The safety culture was weak, and the affection relationships too.

The fear of dying in an accident is repeated in industrial plants where these types of events occur and when one does not trust the company, low credibility. This fear has already been discussed in the case of the ammonia pump operator (120 kgf/cm^2) in the fertilizer plant and in the case of the LPG plant operator during the sphere drainage procedure. Here the fear was related to hot acid burn and affected the priority of micro-decisions in the operational routine.

The reduction of profitability due to inconstancy of production and the reconcentrated acid quality causes, at certain times, this 96% acid to be sold as residue (presence of organic matter). Unfortunately, the price of residue is so low that the cost of acid drops to zero and the amount paid is only for transportation, this situation unfortunately generated environmental impact. The carrier discarded sulfuric acid in the middle of the trip between states of Bahia and Sergipe causing environmental impact.

5.3.4.3 Work and people

The visitation of the plant at random time on the shift time, without previous warning, and during a minimum period of a month was adopted to enable the identification of root cause. Contact with field operators, without the communication filter of supervision and panel operators, facilitated the detection of failures and the interpretation of processes, on a daily basis, and in abnormal incidents.

After this consultation, it is noted that operators had access to basic information of plant startup operating procedures. But, there was, then, no global knowledge of the process, harming the interpretation of the problems. Since the technological packages do not provide an answer to all questions about process abnormalities, the answers have to be obtained from the commitment of the local team.

A training program was developed for operators, with the following steps: preparation of material adapted to the reality of the company, including how failures occur; preparation of instructors; programming and execution of training. The following aspects were noted in the transfer of knowledge about the technology: (1) many data were monitored and did not contribute to the decision; (2) important information was not available; (3) supply chain information was not analyzed for problem solving; (4) few people analyzed the process or the operation in period above more than one month.

Due to the accumulation of functions, the production Engineer and, in some cases, the consultant should sensitize the management/board of directors as to the need to know the chains of abnormality in the process. For the generation of this knowledge, it is necessary to: develop research during a certain period; seek results of increase in productivity, cost reduction; and improvement of product and environmental quality.

The research team sensitized the four levels of functions in the company: management and Board of Directors (allowed the use of resources to start research and conduct tests in the area); technicians and shift supervisors (facilitated information for the execution of the investigation and to assist in the execution of tests); panel operators (gained their trust, invested at the time of modifying certain operational parameters, and also keeping the requested tests under control); Field operators—to report abnormalities that affect the operational factors of the facility, being called "process investigation eyes," for their attributions.

5.3.4.4 Investigation of abnormalities

Investigation of abnormalities requires recording of abnormal events, day by day, and the effort to connect the cause and relate the consequences. The identification of the problems involved investigation during a 4-month period, when it was possible to build possible chains of abnormality in the process, in a complex format, which required confirmation by field tests. In order to interpret these abnormalities, it was necessary to interconnect process flowcharts with utilities and new process flows. After mapping the abnormalities and the discussion for validation, a report of abnormal events (REA) was written. This report indicated actions to be taken in the field in equipment and revision of procedures.

Much of the operation problem is related to failures in process technology and a detailed investigation of critical variables is required, which will indicate improvements, operation values and trends for more severe consequences.

The diagnosis entitled mapping of event abnormalities (MEA) indicates which regions and chains of abnormality, and consequently which variables, are critical for process control and the maintenance of normality from operational procedures. This diagnosis indicates the need to revise and test certain procedures as indicated in Table 5.5.

An anonymous communication channel for suggestions coming from the team brought a sense of justice and cooperation to facilitate the work and maintain a program improvement credit from operational and human reliability. This bottom-up communication about changes in the routine usually had feedback regarding the total or partial approval and the speed of responding to the suggestions. All incidents that occurred at the factory were documented and later transformed into a manual about events that occurred and the probable causes evaluated.

Corrective and preventive actions are indicated after the application of the OD (operator's discourse) and after an abnormal events mapping, MEA. The revision of some procedures and the change of some process variables require the formation of work groups. The tests in normal operation adapted to the company's culture required registration and correlation with process variables. The form of communication and discussion about the interventions must respect a sense of justice regardless of where the components of the production team come from (Rio de Janeiro, Rio Grande do Sul, São Paulo or Bahia). The actions were directed to equipment, control mesh, process system, utilities system, or on the process that involves the recycle line. The goal when setting up a working group for operational reliability is to direct actions to solve abnormal events. The chains of abnormalities in the case of sulfuric acid are described in Table 5.6.

5.3.4.5 Activities carried out on the team to achieve self-management

A program to improve operations requires the team's continued motivation and the dissemination of results on industrial processes. The operators who participate of this program have different behaviors that depend on different impulses and needs. It is intended to reach the respect for the company's values. If this is not possible, it is better not to start a motivational program. Deming (1990) points out that the constancy of purpose authorizes the self-motivation of the process main leader and also his team.

The MEA diagnosis, tool for abnormal event mapping, recommended activities for barriers against failures (procedure review, technological change, and training). These barriers were scheduled to be carried out at the sulfuric acid plant. The team with voluntary participation brings the highest probability of sustainable and resilient projects. The fear

TABLE 5.5 Procedures reviewed in the sulfuric acid unit—AU.

Sulfuric acid unit	1	Bubble test	Operate	Avoid loss of vacuum
	2	Abnormalities table and action	Operate	Agility in solving problems
	3	Starting the plant	Operate	Avoid obstructions
	4	Departure from SP1	Operate	Avoid problems in the stripper
	5	AC1 supply	Operate	Avoid obstructions
	6	Plant stop	Operate	Avoid thermal shocks
	7	Critical equipment release	Operate	Accident risks
	8	Safety procedures	Operate	Avoid accidents
	9	Cautions in handling	Operate	Avoid thermal shocks, ruptures

TABLE 5.6 Abnormalities assessment.

Abnormality	Consequence	Probable causes	Actions	Recommendations
Tube break in glass heat exchanger—SP1	Elevation in temperature	Water Header pressurization; Thermal shock, water with AC1	Evacuate area; Cut water and lower AC1 flow	Evaluate changer head to avoid the increasing of leaks at the stop
Pressurization of tanks—AC1	Deformation in ceiling	Obstructions by organic matter false indication	Interrupt blowing, equalization, loading	Attention to operational procedures
Missing or falling flow rate for SP1	Acid plant stop	Disarm the pump AC1; Obstructions in flow valve	Cut AC1 to SP1; control column level	Heating for tanks Transfer—direct supplier
Loss of level in the column—SP1	Acid Unit stop loss of vacuum,	False level indication; Failure—control loop flow rate; AC1 not specified	Lock AC1 to column—panel and field	Measure level mesh check panel indication versus field
Very high level in the column—SP1	Smash in filling; Failure of procedure	Uncontrolled flow and level meshes	Lock AC1 to column, panel and field	Check alignments and meshes control—start
High pressure in column—AS	Break of rupture disk in column	Obstruction in condenser WS—column jammed; hole	Inform about pressurization	Keep condenser available
Level in the bottom tube—SP2	Tube breakage column break	Air intake through vacuum sealing, solid	Reduce load; Stop plant; reduce vacuum	Avoid solid in AC1 avoid organic matter in tube
AC1 leak in SP1	Corrosion, losses, risk of accidents	Thermal shock, increased flow, inadequate flange tightening	Evaluate extension, stop plant and fix problem	Proper torque in flange tightening
Acid leakage in SP2	Corrosion, losses, risk of accidents	Improper tightening on line and equipment flanges	Evaluate extension, stop plant and fix problem	Proper torque in flange tightening
Tanks shedding	Product loss, risk of accidents	False level indication, incorrect transfer, alignment	Stop operation, inform, isolate area, wash	Increased transfer care, routine level check
Hole in SP2 exchangers	Corrosion of condensate system	Use of overheated steam, hydraulic hammer	Inform panel, stop, cut steam (panel and field)	Always operate with saturated steam
SP1 glass column crack	Plant stop, risk accidents	Thermal shock, tensioned flange, mechanical impact	Inform panel, stop, drain column, isolate area	Slowly warm the column
Water leakage to AU	Thermal shock	Corrosion in water lines, protection for rain	Place repairs to prevent splashing	
Hole in RA line	Losses, corrosion; Contamination	High temperature, acid concentration	Stop transfer and change lines	Keep plant output temperature at 45°C
Crack-break in serpentine Changer	Acid concentration, loss, hole	Increased water pressure Thermal shock	Inform, stop, block water and drain equipment	Check water alignment before start.

of participating in these projects is due to the maintenance of jobs. The guilt culture was very strong in this scenario. It is noted that the production staff has the following profile during organizational changes:

- Passive worker (to do what was defined without asking or even participating);
- Marginal passive or latent active—individuals who said they did it, but in fact did not do it and did not let others do it due to ignorance or lack of negotiation;

- Resistant passive—individuals who claimed to be carrying out the activity, but in fact did not do it and did not let others do it, due to real resistance;
- The real active—individuals who accepted the challenge of trying to improve factory processes, and;
- The faltering active—individuals who were easily overcome by difficulties in changes that improve the process.

The preparation of knowledge multiplying agents is necessary to maintain the dissemination of the successful practices carried out, here known as lessons learned. Initially, the aim is to prepare the training materials and choose the agents who can be informal leaders in the transformation process.

The change alters the behavior of the team, and former leaders feel threatened by informal leaders. Probably the resistance to change was due to these reactions and appeared in meetings to solve problems. Although, as the practice-based courses were taught, there were different behaviors, some individuals detected the flaws and corrected them in the routine while other individuals resisted and, after attempts with process leaders, were finally invited—then leaving the company.

Every dismissal of an employee is uncomfortable for both the company and the employees. Thus despite previous communication and attempts to improve employee performance, changes were made to the team. This action caused a delay in the team's work. Depending on the case, the delay was shorter (15 days) or longer (1.5 months). These delays are considered to be manageable in a change process. This issue must be taken seriously by the leaders of transformation to avoid the risks of delay that could be fatal and block the change.

The lessons learned in correcting routine problems required changes in production planning and management in the shift. The search for problem solving was intended to reduce the "manual" work to enable the planning of the teams in the administrative and the shift. The assessment and discussion of the various problems and solutions were part of daily meetings that took place in front of the control panel. Team leaders were encouraged to plan their shift growth through scheduled or spontaneous training. The most advanced people in certain subjects were oriented to assist those who were weaker in these subjects. A manual was prepared with guidelines on the planning and administration of production in the shift, and, daily, these matters indicated in Table 5.7 were divided into the following classes: organizational matters; team motivation; training and change; and production routines.

The Manual highlights the role of leadership in the activity on the shift, then promoting an environment marked by responsibility and commitment, and in the context of production, the organization's credibility is affirmed. In the guideline that deals with the motivation and involvement of the team, motivational programs are suggested as a way of holding back resentments or any disagreement in the daily coexistence between team members. The training comes in the

TABLE 5.7 Guidelines for production planning and administration in the shift.

Organizational matters		
Values in productive system	Motivational obstacle-tool	Managerial myths
Credibility recovery and maintenance	Discipline versus responsibility versus professionalism	Leader's role
Team motivation		
Motivation at work & motivational programs	Participation and involvement	Resentment—legacy from the past
Training and change		
Choice of key people	Team knowledge	Need for communication
Study group formation	Studies and improvements	Formation of multipliers
Training plan and program	Implementation of institutional programs	
Production routines		
Production management	Planning for production	Work area and interfaces
Routine work	Shift supervisor's role	Standard operator characteristics
Interface with the Laboratory	Interface with maintenance	

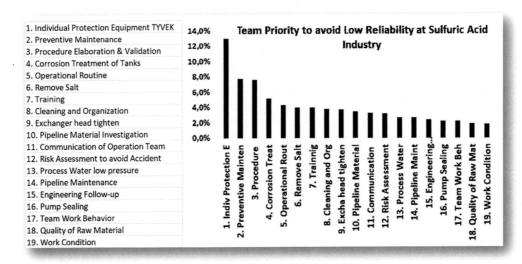

FIGURE 5.13 Team priority.

context of being able to reach the largest number of employees through the formation of multipliers, informal leaders. For the development of training, it is important to get to know the team and its demands, and the form of communication that develops knowledge and contributes toward better results. Finally, the production routines will include the design and structured procedures based on the guidelines previously developed, trained and applied in the production phase in the shift.

In the sulfuric acid unit, training was developed for the operation regarding new procedures. A lecture to raise awareness about the 14 principles of Deming (1990) applied to production and a tool for the definition of team priorities, with the intention of minimizing operational problems such as low continuity and low product quality. After the awareness stage, the operational procedures were presented.

At that moment, we presented the experience in the acid unit regarding the definition of team priorities. There are several types of actions that were indicated in this work to achieve the objective of improving quality or productivity. Productivity can be translated into the case of the acid unit such as, operational continuity, lower production cost, process-occupational safety due to the product feature and the process.

The preventive or corrective actions that have been suggested are in these areas and functions: Managerial actions, Technical actions (Maintenance, Operation, Process, Laboratory), project, safety and Environment, Product Market and Supplier.

It is common for the priority setting to be done by a limited number of people which does not equate to global opinion—sometimes we are surprised at what people think as a priority. In order to bring about a change in the production system, it is important for the group to participate in the definition of priorities, as this makes it easier to implement the change in a short term.

A non-uniform behavior was observed of the statistics obtained per shift class, which indicates the existence of specific problems for each class that may not be reflected in the overall result. As an example of what was done, Fig. 5.13 shows a Pareto Chart with the priorities defined by the shift team. We only put the 80% most impacting in the program.

In this chart, the most important planning items for the sulfuric acid plant are observed. The part of the activity planning discussed by the group is mainly related to the use of PPE and the big discussion is about the fear of using Tyvek because it is fragile, prone to rip and therefore, does not protect the operator.

The development of a critical sense in the operating group regardless of the company's motivation was the objective for the work on improving operational continuity led by Ávila. The keen critical sense will maintain an excellent Operational safety despite the possibilities of organizational or technological change.

The ideal professional to act in an environmentally correct way is the one who can see the global despite performing specific tasks. The professional understands that his/her role is part of a larger system and that his/her influence in relation to environmental issues is important in the air that his/her children breathe.

Provoking self-motivation for environmental issues or the self-management of activities in the industry is a difficult job that apparently does not pay and has to do with voluntary non-profit activities. It does not pay because the result does not appear explicitly, but implicitly as long-term results. Seeking self-motivation and self-management pays off. In the operation group, resistance indicate following questions:

Why should I do this? Or do this way? Does anyone watch my performance? Am I wasting time?

But after discovering that someone noticed the team's work and that it produced an improvement in mood, it is motivating.

It can be said that companies' policies change overtime and that people also change their way of thinking.

Does offering security and balance lead to an increase or decrease in participation in the company's controversial issues? Which company do I belong to? Does the company make me, or do I make the company? What are the proposals to improve the environmental issues?

Really, being critical is difficult due to the corporate culture, the managerial style, and the fear of launching himself/herself based on critical and constructive thoughts?

Comments:
_____
_____
_____
_____
_____

Other questions to be asked regarding the appreciation of critical sense:
Does the company need professionals with a critical sense?
Why do the challenges for survival depend more on teamwork than on the technology itself?

Multifunction is present in all sectors of the company, and usually only has the capacity to carry out several activities at the same time has a critical sense.

Comments:
_____
_____
_____
_____
_____

5.3.5 Industry alarms and shutdown (Ammonia, HDT, Cyclohexane): H_2 and CO compressors

The continuous processes that have gases as raw material or as a by-product from chemical reactions always demand pressure control to enable their conversion and selectivity. In the chemical, petrochemical and oil industry, the use of compressors to increase the pressure of these gases is common. In the book, these compressors are of the centrifugal or alternating types and are large equipment. As the risk is high for working with flammable and toxic gases, it is intended to maintain the compression operating continuously, with constant speed, without axial-radial displacement, and without leakage to the toxic-flammable gas environment.

The cost of this equipment is high, the flow must be constant as well as the pressure. The auxiliary control systems in the equipment indicate the high complexity, and in case of failures or human error, they indicate the consequent high impacts. So, let's discuss two cases of hydrogen compressor and one case of CO compressor.

The Hydrogen Compressor transforms benzene into cyclohexane. It is centrifugal and has low operational reliability resulting from low human reliability. The application and creation of the FMEAH (analysis of human factors in equipment failure) that discusses technical failures and human errors in maintenance and operation has already been discussed in this book and now in this chapter.

The refinery's hydrogen compressor works with high flow and the discussion is around Alarm management and process failures. The CO compressor at the fertilizer unit was initially used to compress methane from gas production on an oil rig. After adjustments, this compressor left the offshore production (hydrocarbons) to work on ammonia synthesis at the fertilizer unit.

5.3.5.1 Hydrogen compressor in benzene hydrogenation unit (FMEAH Ávila et al., 2012)

The process for the production of caprolactam (monomer used in the production of nylon 66) includes the production of cyclohexane, through the hydrogenation of benzene in addition to the production of cyclohexanone from the catalytic oxidation of cyclohexane and thus the oxidation of cyclohexanone with toluene. The cyclohexanone oxime is rearranged in the presence of "sulfuric acid," which produces caprolactam.

The catalytic hydrogenation of benzene to cyclohexane (Fig. 5.14) occurs in the gas phase after vaporization. For this purpose, a vertical heat exchanger is used, which works as benzene/hydrogen mixer and vaporizes benzene by exchanging heat with the reactor's pre-heated thermal oil. The gas that feeds the benzene vaporizer is made up of "fresh" hydrogen and circulating hydrogen from the compressor. The compressor has the purpose of increasing the pressure of H_2, then causing it to recirculate to the vaporizer and to lower the partial pressure of the benzene vapors, thus achieving a better conversion to cyclohexane in the reactor.

The drop in this pressure impairs the efficiency of the reaction in the first reactor, resulting in a higher fraction of the reaction occurring in the post-reactor. The post-reactor was designed to react the benzene residue of the first reactor and remove impurities that cause the catalyst poisoning, it does not have a jacket for cooling. The benzene hydrogenation reaction is exothermic, therefore, an increase in the fraction of benzene will result in an increase in temperature, which may cause undesired reactions such as an isomerization in methyl cyclopentane and, in a critical state, to cause a plant interruption.

The continuous pressure drop will not only hinder the process, but below a certain value it will hinder the compressor's operation, causing the surge phenomenon. The surge occurs in an area where the compressor reduces its inlet flow to keep the system pressure high until a critical value where the flow is reversed and the pressure drops to adjust the flow in the normal direction. This cycle is repeated with the new drop in flow and inversion of flow.

For safety reasons, the compressor is connected to a safety system, which when a limit pressure is reached, a hydraulic relay activates the solenoid valve and causes the turbine to stop. This safety system is connected to an interlock switch that stops the plant.

The compressor increases the pressure from 29–29.5 to 30.4–31 kgf/cm^2 ($\Delta = 1.4$ kgf/cm^2) at a total flow rate of 19.9 tons per hour of hydrogen and with a temperature of 50°C, as shown in Fig. 5.14. Its operation, in addition to promoting an increase in pressure to meet the chemical reaction, also causes an increase in temperature until it reaches

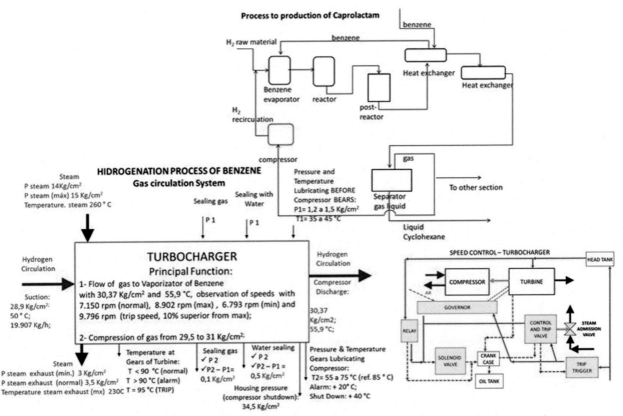

FIGURE 5.14 FMEAH in cyclohexane production, compressor and turbine speed. *Reproduced with permission from Ávila et al. Ávila, S.F., Mendes, P.C.F., Carvalho, V., Amaral, J., 2012. Analysis of human factors in failure of turbocompressor FMEAH equipment in chemical plant. In: 27th Brazilian Maintenance Congress (CBM), September 10–14, 2012, Rio de Janeiro, Brazil.*

56°C. The turbine moves with medium pressure steam between 14–15 kgf/cm^2 (maximum) with a temperature of 269°C. This turbine delivers steam of at least 3 kgf/cm^2 and can reach 3.5 kgf/cm^2. The exhaust temperature is 230°C. It is expected that there will be no condensation in this turbine.

The sealing of the compressor promotes a pressure difference of 0.1 kgf/cm^2, which prevents hydrogen from going to the atmosphere, and gas may get into the equipment. Due to the risk of hydrogen, this compressor also has a sealing system with water in a pressure difference of 0.5 kgf/cm^2. When the pressure in the frame exceeds 34.5 kgf/cm^2, the compressor will stop, for safety. The lubrication system sends oil under the following conditions at the inlet: $P = 1.2-1.5$ kgf/cm^2 and temperature from 35°C–45°C, and the temperature at the alarm output with a positive change of 20°C. The turbine is stopped when there is a positive change of 40°C. A maximum reference temperature is 85°C.

The possibility of low pressure during normal operation or at plant shutdown are failures to be analyzed by the FMEAH. Another study under investigation is the speed control of the turbine, which is therefore a subject of systems maintenance area.

As shown in Fig. 5.14, the components of the turbine speed control are: the Trip Actuator (TA), the Regulating and Trip Valve (RTV), the Steam Intake Valve (SIV), the Hydraulic Relay (HR) and the Solenoid Valve (SV).

The turbine speed control comprises 3 important elements: sensor, transmission and correction. The sensor element is made up of the Governor part, which, to keep the turbine speed constant, controls the opening of the Steam Inlet Valve—SIV, compatible with the power required by the load. The transmitting element is the hydraulic pressure of the lubricant in the control oil circuit and the air pressure over the Governor Head, as well as the electrical signals over the elements of the Interlock Component. The intake element is the Steam Intake Valve.

The focus of this work was to discuss human causes in operational problems with pressure changes in the process and compressor failures that involve speed control in the turbine. Among the causes the inversions in the sequence of the task and rework or repetitions can be mentioned. The mistakes can be related to non-complying with rules or lack of the necessary knowledge.

The human errors characterize human actions that did not reach the intended goal. This may be caused by lack of attention in the execution of a task. Failures or omissions occur when a step of the task is not performed, ranging from the simple fact of not transmitting information to the non-execution of any task. The omission is a serious failure, which can lead the process to the lack of control and activate emergency procedures, or the lack of knowledge of new operating conditions in the next shifts, causing the process to become unstable. Mistakes consist of failures in the decision-making related to the operation, regardless the source. These effects are more present in cases in which employees have no experience and are in new situations, where a standard procedure has not been developed for detailing tasks.

The organizational environment affects the human work quality in the processes control explained in the equipment technology. Without the necessary knowledge by the engineers in the area of turbine speed and in the area of operational control, failures occur, which cause stops of this important compressor. Another additional possibility is the poor disposition of the resulting work environment and influenced by centralized management.

The application of FMEAH (Ávila et al., 2012) indicates the failure mode of the stops of the cyclohexane unit due to the interlocking of the compressor or the low reaction efficiency. In the case of control, turbine operation and compressor as a failure mode for plant shutdowns, the human failure mode includes subjects such as lack of experience, lack of attention, omissions of information, error in setting priorities, team demotivation, lack of training, and inadequate procedure.

5.3.5.2 Hydrogen compressor failure analysis in HDT unit, refinery (Souza et al., 2018b)

The oil industry has human factor criteria for the design of the control room, and also in process safety barriers such as alarm management, nevertheless there are still plant shutdown problems.

The refinery units located in Brazil have the following general characteristics: high risk of fatalities by accident; high severity of impacts from possible accidents; high pressures and temperatures; and circulation of flammable-toxic products. In a simplified way, the production of oil products in this refinery involves the following: (1) distillation with the respective cuts of LPG, diesel, kerosene, naphtha and diesel; (2) conversion of heavier compounds into lighter ones to increase sales and; (3) treatments (impurities, sulfur).

This work has as scenario a medium hydrocarbon treatment unit to produce diesel with low sulfur content. Hydrotreatment units (HDT) are designed to treat oil fractions in the presence of hydrogen to remove sulfurous and nitrogenous compounds that cause fuel instability.

The HDT unit (Fig. 5.15) receives high sulfur diesel from the distillation in the Refinery, in addition to others through the cargo ship. These streams are drained through a pump, where pure hydrogen is added, destined to the heating furnace. The output is of around 350°C.

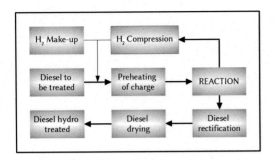

FIGURE 5.15 HDT process. *Araújo, R.B., 2018. Evaluation of the performance metrics of the alarm system of the industrial refining units. Course Completion Work (Mechanical Engineering)—Federal University of Bahia, Salvador.*

Once leaving the furnace, the liquid load added to the hydrogen enters the reactor, where hydrodesulfurization reactions take place. To control the temperature of this exothermic reaction, a second injection of hydrogen is made.

The liquid part of the high-pressure separator vessel is sent to the low-pressure separator vessel and to the rectifying tower, where, in counterflow with steam, more H_2S is removed to the ERU. Leaving the rectification tower, the diesel is sent to a vacuum tower. It is responsible for removing the steam condensate from the previous tower and separating it in a container.

There is also an alternative compressor that powers the process, responsible for replacing the hydrogen "lost" in the formation of H_2S, from the hydrogen generating unit (UGH). The energization and de-energization of the solenoid valves are controlled by the compressor's capacity control system, pressurizing or depressurizing the actuator to open or close the valves, allowing the compression or return of the gas through the inlet valve without compacting it. The outlet valves of the compression chambers are opened as the piston movement increases the pressure inside the chambers, surpassing the pressure downstream of the outlet valves and increasing the pressure at the compressor discharge.

A higher demand for hydrogen in the reactors is due to the receipt of hydrocarbons containing a higher sulfur content and, as the reactions are facilitated by the increase in the reaction temperature. The formation of H_2S in the reactors decreases the partial pressure of hydrogen in the unit, which leads to an increase in the demand for hydrogen with a high degree of purity and an increase in the capacity of the compressor. At the end of the process, a high-quality diesel with very low sulfur content is produced.

The unit has about 5000 sensors and actuators that monitor and control all sensors, including pressure, flow, temperature, vibration, displacement and speed sensors, control and interlock valves, flame, smoke, hydrocarbon and hydrogen detectors. The variables are controlled through the DCS (distributed control system) using pre-established parameters in the project.

The pressure control of the hydrotreating unit is carried out by changing the capacity of the compressor with the higher demand for hydrogen in the unit; by relieving excess gas for compressor suction through a pressure valve automatic control; and by opening a second valve for the flare. The compressor capacity control system consists of a hydraulic unit, solenoid, and actuator and suction valves, in addition to reservoir level and temperature transmitters and oil pressure at the inlet valves of the compression chambers.

The failure identified in this work is related to the low-pressure alarm in the compressor's capacity control system. It is an electrical problem that causes the hydraulic unit to stop and the loss of oil pressure in the actuators, thus causing a lack of control in the operation of the hydrogen inlet valves.

The extension of this scenario leads to a stop of the UGH that does not have a "lung" for production and, usually, an increase in the gas supply demands more time than the speed with which the makeup compressor sucks, leading to UGH low pressure in the last stage of the process and causing the plant to stop.

From the evaluation of the standards for Alarm Management, we arrived at the following proposal of methodology, Fig. 5.16:

The failure process occurs, therefore, when there is no action by the operator to overcome the loss of control command for the compressor capacity. This lack of action can occur for three reasons:

1. There is no clear identification of the alarm, that is, it does not draw attention for the information it is intended to transmit, a low-pressure alarm;
2. It is not classified as a high priority alarm, the availability for response time is short (less than three minutes) and the severity of the consequences is medium;
3. There are several alarms that far exceed the operator's cognitive limit.

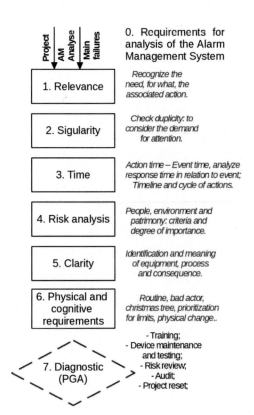

FIGURE 5.16 Methodology for the alarm management system. *Souza, M.L., Ávila S.F., Cerqueira, I., Brito, R., 2018b. Discussion about Criteria for the management of alarms and cognitive limits for the chemical industry. In: International Conference on Applied Human Factors and Ergonomics. Springer, Cham, pp. 330–342.*

The alarm identification and rationalization process can be understood as a single step in which all alarms are documented, containing at least this information:

- Info & tag (process report, alarm identification, alarm description);
- Root cause (factors that trigger the abnormal situation);
- Action (operator response to the alarm);
- Available time (time between the alarm warning and that in which the operator initiates the action);
- Impact (consequence of the plant in case of failure to execute the alarm response);
- Alarm priority (relative importance given to the alarm indicating the urgency of the response);
- Suppression strategy (condition in which the alarm can be suppressed);
- Adjustment valve (limit value of the variable in which alarms should be announced (setpoint)).

Rationalization is a continuous process, especially considering that the plant is in operation and any type of change, could be made in relation to the initial project, such as increased load, change in the inputs quality or in the processed load. Such changes need to undergo a change management procedure to assess, mechanical, ergonomic, environmental, impacts and gains related to the change. In this way, new alarms can be designed for the system, as others can lose their function and must be removed. This entire process must be documented and this information can be tracked.

The next phase is the detailing and implementation related to the way the alarms are presented to the operator, mainly with regard to the human-machine interface (HMI), the system's ability to segregate alarms by priorities to identify alarms by different aspects, such as color, symbol and sounds, and monitor and evaluate the performance of alarms according to what was defined at the beginning of the project.

The implementation of the alarm system is the transition from the project to the operation, through functional validation tests, training in operation, alarm performance measurements and final adjustments of the system. This phase is very important, as instruments that may require repair or replacement will be identified. Also, the operation of alarm systems depends on the reliability of the measuring instruments. Proper maintenance is also important to maintain a good management of the performance goals established in the philosophy, such as the number of alarms treated by operator per hour. The treatment of "bad actors," alarms that occur in a much larger number than the others in the same period of time, is a very effective practice in the management of metrics, as it greatly reduces the occurrence of alarms not associated with an operator action, and helps to maintain an adequate response condition when other alarms sounds occur.

Periodic audits should occur throughout the lifecycle of alarms, since the philosophy stage to the monitoring, management change and maintenance stages. Performance audits and alarm management can reveal gaps and develop best practices to make work more efficient. These audits include human factors and organizational as criteria.

5.3.5.3 Failure analysis of CO compressor in ammonia unit, fertilizers

The Fertilizer Industry case study (Ferreira, 2018) with complex chemical processes and high operational risk. In this case study, the reliability of rotating equipment and the influence of human factors on the performance of the machine's operation were analyzed. The consequences of the failure are easy to identify due to the visibility resulting from the impact. The failure mode, the failure history, the cause-effect relationships, depend on the understanding of the process under normal conditions and on abnormalities.

This work in the fertilizer industry was carried out in 4 stages: preliminary analysis; task analysis; ishikawa diagram and equipment reliability; human reliability.

5.3.5.3.1 Step 1: Preliminary analysis

The first stage of the investigation is the identification of the operating condition with the respective process variables and critical equipment for the preparation of a preliminary diagnosis that indicates the system's complexity. The delimitation of the object of study includes the identification of critical equipment inserted in the systemic failure mode. With this intention that the losses of the unit were studied, in trying to relate to the causative equipment involved and its characteristics.

The unit under study is expected to produce more than a thousand tons of product per day. All production below this target is classified as loss and a record is made by management for report monitoring.

This report, which presents the management discourse of all events that took place at the unit, was the main source of data for the quantification and classification of production losses. The collection of information was obtained through shift reports, the unit's operating manual and visits to the area.

Thus loss mapping followed the steps below to delimit the object of study:

1. Loss of global of production—losses due to the following factors: equipment problems (EQ); scheduled stops (SS); lack of utility (UTL); strike (STR); other internal events (IS); clearance (CL); problems in the supply of cargo (SC); and other external events (EE);
2. Loss of production due to rotating equipment—comparison between rotating equipment and all other equipment in the unit;
3. Interruption of operation, participation of pumps, compressors, fans and blowers in the loss of production.

When identifying critical equipment, its function, the impact in the event of downtime and/or loss of efficiency, the critical variables that have the potential to affect the process/production and the performance of the equipment, also with the description of the main system and auxiliary, we can initiate an investigation.

5.3.5.3.2 Step 2: Operational context and task analysis

In step 2, an analysis of the operational context was carried out to identify the main elements that affect the performance of the operation, identify the critical task and classify the level of complexity of the operation of the CO_2 compression system. To achieve these objectives, the following aspects are analyzed: the existence of standard operating procedures and their accomplishment in carrying out the operation activities, the level of automation, the impact of monitoring auxiliary systems on the operational routine of the main system and the level of interconnection of the system under analysis.

The analysis of the task performed in this work was based on the methodology proposed by Ávila (2013a), which aims to identify elements that maximize or minimize human performance in the operational context and, thus provide a basis for the application of the other two reliability tools.

The 2nd stage was carried out from the application of the following actions and tools: analysis of the workplace, analysis of health and safety rules, analysis of the complexity of the task taking into account the compliance with procedures, level of process automation, auxiliary systems of compressor control and analysis of the recycling lines.

The interviews carried out for the second, third and fourth stages followed forms structured in face-to-face questions applied to engineers, managers and operators, including questions related to human factors in the application of SPARH.

5.3.5.3.3 Step 3: Ishikawa diagram and components reliability

The third stage consists of the elaboration of the Ishikawa diagram and the modeling of compressor failure through the application of reliability tools. The construction of the Ishikawa diagram aimed to present the relationship between the loss of compressor production and efficiency and the respective causal factors. The result of this analysis of the Ishikawa diagram was compared with the behavior obtained in modeling the compressor failure rate function (Ferreira et al., 2016).

The construction of the fishbone (Ishikawa) diagram involved the following activities: description of the problem, identification of the causal factors, brainstorming about these factors and developments and analysis of the priority causes.

Once the causes of failure modes have been identified in the diagram, the next step is to estimate parameters and select the model that best describes the system's behavior. The selection of the model takes into account statistical tests, graphical analysis and the characterization of the failure performed in the Ishikawa diagram as shown in Fig. 5.17.

5.3.5.3.4 Step 4: Human reliability

The information obtained in the previous steps was a basis for the application of the human reliability analysis (SPAR-H) in the 4th stage. This method consists of a qualitative and quantitative analysis of eight human performance factors (HPFs) that result in the quantification of the human error rate for a given task.

Throughout the process, visits were made to the company with the application of quality and reliability tools alongside the different hierarchical levels (operators, supervisor, engineers and managers).

Due to the existence of several factors that act on individuals and that influence the failure of the technical system, the studies on human reliability were started only after the identification and study of critical equipment—which

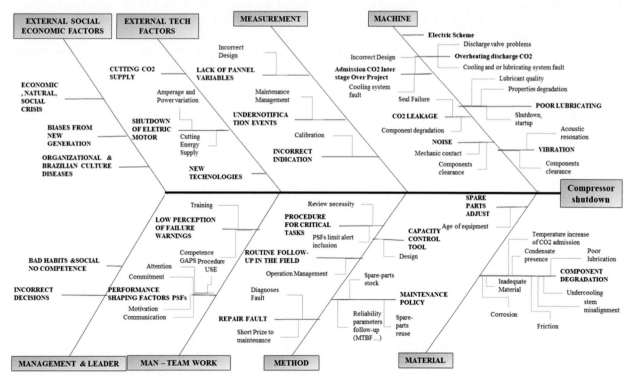

FIGURE 5.17 Ishikawa diagram of the Fertilizer Industry, CO_2 compressor. *Ferreira, J.F.M.G., 2018. Analysis of the reliability of the CO compression system in a petrochemical plant considering the technical-operational and human factors. Master Dissertation from Post-Graduation Program of Industrial Engineering, Federal University of Bahia, Salvador, 128 p.*

includes the analysis of equipment reliability operational reliability. The latter following the methodology of Ávila (2015a).

The analysis of human reliability used in this work was done using the SPAR-H method based on the manual prepared by the North American Nuclear Regulatory Commission. This method allows quantifying the human error rate through assessments of HPFs based on the observation of the task in study and interviews with those who perform and/or supervise the activity. This method takes into account eight human factors: available time, stress and stressors, complexity, experience and training, procedures, ergonomics (human-machine interface), aptitude for service and work processes.

The application of this method was assisted by a field operator with over ten years of experience in the profession, but who had only been in the study unit for two years. In this stage, all the necessary steps for applying the SPAR-H method were followed, namely:

1. Identification of the task and the context; 2. Analysis if the basic event involves diagnosis, action; 3. Assigning quantitative values to each HPF to calculate the human error probability (HEP); 4. Determination of HEP without dependence; 5. Calculation of total HEP; 6. Analysis of the results. The results were validated in conjunction with the operator and the unit's operation manager.

5.3.5.3.5 Case study

This case study was applied to the fertilizer industry. The urea manufacturing process can be separated into three stages: (1) synthesis, (2) decomposition and recovery, and (3) concentration and granulation. The first stage refers to the process of manufacturing ammonia carbamate through the reaction between carbon dioxide and ammonia. The final product (urea) comes from a process of dehydration of carbamate. At the outlet of the reactor, urea and non-converted reagents (CO_2 and ammonia) are present, as well as amounts of biuret (undesirable product). Thus in the second stage, the unconverted products are separated and recovered. After that, in the last stage, the urea concentration process is carried out, which becomes possible to carry out the granulation that goes to commercialization. Figure 6.12 shows the simplified form of the urea production process.

The CO_2 compatibility and synchronization system consists of four compressors, two identical alternative compressors powered by an electric motor and two turbocharged compressors, one being centrifugal and one alternative (Fig. 5.18). The compression process is carried out in two units of the plant: part of the CO_2 is compressed in the urea plant (about 56% in the total volume) and the remainder (44%) will be compressed in the hydrogen plant by another alternative compressor.

In the synthesis section, however, excess ammonia helps to keep the low content of this product. The components recovered in the recovery section are recycled to the synthesis stage and the formed urea proceeds to the concentration and granulation stage.

Stopping the compressors does not necessarily mean stopping the plant. However, the unavailability of these compressors directly affects production, that is, it reduces the capacity of the plant by 28%, if one of the alternative compressors stops; by 56%, if the booster compressor fails; and 44% if the turbocharged alternative compressor becomes unavailable. In these scenarios of equipment unavailability, it is assumed that the event did not cause any major disruption to the process to the point of stopping the plant.

The fact that the compressor in study is a process equipment, without another one to replace it in case of failure, makes it essential for production. Therefore, early diagnosis and maintenance planning are important to ensure a quick return to operation.

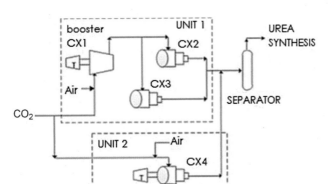

FIGURE 5.18 Diagram of interconnection of compressors in the process. *Ferreira, J.F.M.G., 2018. Analysis of the reliability of the CO compression system in a petrochemical plant considering the technical-operational and human factors. Master Dissertation from Post-Graduation Program of Industrial Engineering, Federal University of Bahia, Salvador, 128 p.*

The quantification of the probability of human error took into account all eight factors of HPF in the execution of the task, whether of cognitive or action characteristics. The levels of the HPFs' delimiting factors were evaluated in two scenarios. In the milder scenario, the level of the delimiting factor was considered as nominal. In the most negative scenario, the work relationship is perceived as bad, because there are enough elements to negatively influence activities. The results of the HPFs are shown in Fig. 5.19 and the final value of the probabilities of human error are: soft scenario, HEPaction = 0.01%, HEPdiagnosis = 3.2%; negative scenario, HEPaction = 0.2%, diagnostic HEP = 13.9%.

5.3.6 Task complexity, low efficiency, and accident investigation

The oil industry is increasingly concerned about risks of disorganized growth, due to this motive, they increased complexity through automation where leaders have the illusion that technical systems can completely prevent human error. The accident anatomy in this article is simple, without many events and apparently linear. Through the use of the example of a forklift truck, during placement of sleepers, with the presence of a worker who supported the activity near the forklift, the operator was hit in the right hand or foot and had a fractured fingers.

5.3.6.1 Critical task complexity, forklift accident

The design of the task in new installations is defined by its criticality, impacts on the economy, nature and image. Some factors influence the criticality of the following: frequency of task execution, the possibility of failure, and severity of failure events related to the task and the complexity involved in the production (process, procedure and social relations). The frequency refers to the number of times that the execution of the procedure to maintain production during a period of time is scheduled. Severity is classified by the intensity of the negative impact on society, the environment and the economy during the execution of the task. Therefore, knowing the severity of the events helps to identify the level of risk. Understanding the line of action of the hazard released for the product and process quality and for the physical and social environments helps in the development of barriers, here it is considered that the risk is sociotechnical and that the failure is systemic.

The complexity is related to the type of communication on the social network (macro—in the community) and on the social node (micro—in the workstation); the level of automation of critical operations by type of technology; the

SOFT SCENARIO			
ACTION TASK		DIAGNOSES TASK	
PSF	Multiplier	PSF	Multiplier
Time	0,1	Time	0,1
Stress, Stressor	1,0	Stress, Stressor	1,0
Complexity	2,0	Complexity	2,0
Experience/ Training	1,0	Experience/ Training	1,0
Procedure	1,0	Procedure	20,0
Ergonomics/ Man Machine I	1,0	Ergonomics/ Man Machine I	1,0
Fitness	1,0	Fitness	1,0
Job relations	0,5	Job relations	0,8
HARD SCENARIO			
ACTION TASK		DIAGNOSES TASK	
PSF	Multiplier	PSF	Multiplier
Time	0,1	Time	0,1
Stress, Stressor	2,0	Stress, Stressor	2,0
Complexity	2,0	Complexity	2,0
Experience/ Training	1,0	Experience/ Training	1,0
Procedure	1,0	Procedure	20,0
Ergonomics/ Man Machine I	1,0	Ergonomics/ Man Machine I	1,0
Fitness	1,0	Fitness	1,0
Job relations	0,5	Job relations	2,0

FIGURE 5.19 Soft and Negative Scenario of Organizational Environment. *Based on Ferreira, J.F.M.G., 2018. Analysis of the reliability of the CO compression system in a petrochemical plant considering the technical-operational and human factors. Master Dissertation from Post-Graduation Program of Industrial Engineering, Federal University of Bahia, Salvador, 128 p.*

type of equipment and control systems; the type of control installed in the return currents—recycling of intermediate products (mass), energy and information.

In addition, in public and private services, complexity is related to the various interconnections between events in the task and the level of attention after the calibration required by cognitive processing.

Embrey (2005) states that complexity is related to the number of steps that the procedure has and that influences the level of attention and, consequently, compliance. This complexity can also be associated with aspects that cause noise in the mind map, which impairs decision-making in the task. So, in summary, we will approach (1) accomplishment of the task; (2) type of technology and automation level; (3) auxiliary equipment systems; (4) level of attention; and (5) linearity in the process.

1. Accomplishment of the task

    The accomplishment of the task is related to the operational discipline and the operations' behavior in the work routine. On the other hand, the result of the task may be related to indiscipline resulting from vices, bad habits and archetypes installed in the shift operational group. Often, production management does not care about the influence of regional and global behaviors, which is a serious mistake, because conflicts can exist, which cause indiscipline when they are not resolved. The calculation is based on the percentage (%) of non-fulfillment of the task, for example, 10% is not carried out according to the written procedure. Five of the 50 steps are not accomplished according to what is written, or other things are done, or simply nothing.

2. Type of technology and automation level

    The Technology indicates the type of process control and process safety and the type of equipment control. Technology also indicates the level of automation that decreases human physical effort but increases the need for knowledge as it has a higher complexity. The operator must think about how to control the processes in a discreet and automatic way, facing the difficulties of transforming the current abnormal state, back to normal. Higher complexity requires higher competence and more transparent communication. The calculation is performed in a simplified way, in which, for example, it is detected that 30% of the main controls are in closed loop compared to 70% of the controls that are in open loop. Closed loop because a certain measured variable is automatically controlled through another manipulated variable, with the correction of errors in relation to a desired parameter and considered normal. Thus the complexity in this technological system is of 30%.

3. Auxiliary systems of equipment control

    The auxiliary systems necessary to attend secondary functions related to quality, comfort, environmental impact and safety require attention and connectivity in operations and controls to compose the actions of the equipment's main function. Thus more auxiliary systems increase the complexity of the system, a hydrogen compressor, in addition to the function of increasing the pressure, has to maintain vibration, sealing, axial and radial displacement under normal conditions. The number of auxiliary systems that require the operation's attention may indicate a higher level of complexity. The calculation is performed considering the percentage of auxiliary systems in the equipment in relation to the total. Considering that five auxiliary systems represent 50% of the attention and no auxiliary is 0% of the attention. In this example for calculation, two auxiliary systems will be considered indicating 20% complexity versus 80% in activity focused on the main function of the equipment.

4. Level of attention

    The level of attention on the task depends on the types of workers selected and trained. The worker may have more scattered or focused attention. Complexity requires more focused attention to problem solving. This operator characteristic must be compatible with the task, in which, operation tasks performed in a non-exclusive way by the same person that develops only one activity, indicates less complexity. Thus the calculation regarding the task considers for example that there are 4 out of 12 tasks performed in a non-exclusive way, 33%, with divided attention, thus increasing its complexity level to 33%.

5. Linearity in the process

    The linear flow of the process is indicated when the currents of energy and mass go in the direction of inputs for the final product. If the process has a linear flow, it increases the predictability, probably indicating less dispersion. When the process involves several return streams, recycling and reprocessing, then forming a skein, the forecast for future results is reduced. The calculation is given by the number of recycling streams, of types, product, inputs and energy per number of main streams, with the weigh 5. Thus as an example, the percentage resulting from the streams that recycle, return to the beginning (inputs, products and energy) divided by the number of main product streams, leaving the plant with weight 5. Thus 2 input streams, 3 product streams and 5 energy streams recirculate in relation to 5 main streams. Calculation: $((5 + 2 + 3)/(5*5)) = 40\%$.

In an attempt to calculate the complexity, the calculation is performed in which the individual complexities are multiplied together in decimal format (not percentage) and then by 100,000 (non-dimensional correction factor). In an average condition:

For the case designed in the 5 complexities exercised above, we calculated the average level of complexity = (1) * (2) * (3) * (4) * (5) * 100000 = 0.1 * 0.3 * 0.2 * 0.33 * 0.4 * 100000 = 79. This number is compared with the product and process risk and then analyzed, which may suggest safeguards to prevent accidents.

A highly complex situation becomes, 0.2 * 0.4 * 0.5 * 0.5 * 0.5 * 100000 = 0.01 * 100000 = 1000, this number compared with the product and process risk and analyzed through safeguards to prevent accidents. A low-complexity situation becomes 0.05 * 0.2 * 0.1 * 0.1 * 0.3 * 0.01 * 100000 = 0.000003 * 100000 = 3, this number compared to product and process risk and analyzed for safeguards to prevent accidents.

5.3.6.2 Qualitative approach for risk and task complexity

The risks of products and processes depend on the type of hazard and barriers designed/installed to block the causal event or mitigate consequences. These risks are expressed in the classic risk matrix (after analyzing the frequency and severity of specific events). The estimated amount of leakage and the impact on the population depends on the critical physic-chemical properties (toxicity, flammability and explosion) and the way this product leaks from the process equipment.

In some cases, the data is used to define the frequency and severity and to indicate the real need for auxiliary memory—Embrey. In other cases, proportional barriers to the resulting level of attention are defined by the number of activities per person and per 8-hour shift. In other cases, the risk level is defined considering the amount of double shifts (%) and the number of signs of abnormalities (Discussed in Figure 3.36).

Therefore, it is necessary to establish criteria for risk classification based on the impacts of products, processes, technologies, risks (frequencies and impacts) and respective complexities. The establishment of risk levels requires knowledge about social reality (communication and culture), human and technological types such as risk and complexity (Perrow, 1984).

Studying the task is a priority for industry and support services. The task risk is analyzed at each stage of the task and in its overall aspects. This risk analysis is carried out from the design phase to the operation and plant shutdown. It intends to avoid accidents through barriers in systems with a high level of automation or discrete systems with complexities related to Business Economics.

Some difficulties increase the need to study the task and create controls to avoid errors and failures: (1) the lack of people with managerial and control competence, and for the execution of services with resilience; (2) the search for insurance to cover the growing risks caused by larger plants (more hazardous materials) and with communities that are even closer; (3) managerial difficulties in dealing with the human types available for the task; (4) the need to revise communication tools and rituals to strengthen the safety culture.

A current issue for task planners and executors is over-information as a task requirement regardless of whether they are simple (low risk) or critical tasks (Ávila et al., 2019). Embrey (2000) recommends in CARMAN Methodology, the use of a checklist for the risk analysis of critical tasks, in which, depending on the risk, the level of training and the type of auxiliary memory are chosen.

Non-critical tasks, when performed with the same rigor as the critical task, for example, in the accomplishment of a large checklist, without major occupational safety problems and without higher demand for attention, cause the operator to commit the violation both in critical and non-critical tasks by considering the procedure as simply trivial. Thus individual and group omissions are part important in the operational routine.

This practice is not adequate, but it is common to happen, situations in which the checklists are completed without confirming the requirements, subject and partial steps. It is very dangerous when this practice reaches critical tasks and can cause accidents and incidents.

5.3.6.3 Case study of a forklift in the oil industry

The forklift is used to support the operation, and to transport waste in barrels and materials, such as sleepers. In the oil industry there is a high frequency of forklift accidents with damage to the worker, in particular their hands or feet (Fig. 5.20).

In the current scenario, the company that transports materials performs many services during the weekends to move waste, sleepers and parts. The rules for this service contract are of a minimum price, requiring the company to meet safety requirements and integrated management. Although the contractor company's manager tries to include safety values, it is not successful, when it is necessary to save on hiring the team. This contracted company offers its group of employees very different conditions from the contractor company. The training tries to meet the requirements, but not

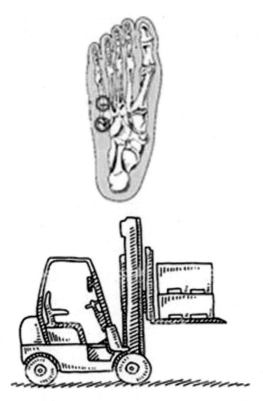

FIGURE 5.20 Forklift accident, fractured finger.

exactly the service quality in the routine, and the benefits and salary are not enough to guarantee the commitment to avoid incidents and accidents, despite the effort.

We are here dealing with an accident that occurred in the handling of the forklift, in which, during the placement of the sleepers, the worker who was close to the forklift, supporting the activity, received an impact on the right hand or foot and fractured fingers.

According to accident investigators, the causes identified by the accident analysis were: failure in supervision (inadequate operation in the process of unloading cargo); non-conformity of the operational procedure for unloading the machine part (fall); failure in the risk perception of the forklift operator; failure to comply with the procedure (cargo handling and Preliminary Risk Analysis); supervision failure (people close to a heavy part); and failure in risk perception.

Despite the high frequency of this accident, corrective actions have always been to review the procedure and train the team. Until now, these actions have not been efficient and effective.

What to do?

_____
_____

The complexity calculation in the case of the forklift is simple because it is a logistics activity and not an industrial process. Therefore, due to the lack of commitment mainly resulting from low value contracts, 40% of the steps are not carried out according to the standard procedure. The technology and level of automation are considered low, so we apply 10%. Automation can be related to closed loop communication, for example, low value.

There are no auxiliary systems but concerns with gas control (prevent leakage) and battery control (operational reliability), thus a minimum value of 10%. Again, considering the possible amount to be paid to supervisors and drivers, unfortunately, it is expected to hire some workers with scattered attention. Thus the complexity rises to 50%. And finally, low complexity due to the characteristic of the non-recycled or reprocessed forklift service.

Complexity = $0.4 * 0.1 * 0.1 * 0.5 * 0.1 = 20$, low complexity, below 70, but it could be lower if it were possible to hire people more committed to the service. Accidents happen frequently considering the discussion about procedures not performed properly and the impact on increasing the frequency.

When knowing the high frequency of this type of accident, we consider that this simple failure is systemic, but, only the technical variable of logistics is analyzed. When considering the technical variable with the lack of attention and non-compliance with the procedures (items of high complexity), we reviewed the accident investigation procedure to design a better diagnosis. Thus we propose that the accident investigation follow the method in Fig. 5.21.

After identifying the underlying causes, it is necessary to recommend the various areas of the organization. These recommendations include the following: contract management, image, organizational aspects, and human factors in different action level (strategic and routine). Other important factors are measurement (monitoring and control of contracts), staff training, education to maintain motivation with the safety, assumptions and criteria, appropriate leadership, harmonization between expectative and reality after high level of stress and imbalance between the cost and the investment.

5.3.7 Just culture in metallurgy and oil industries

5.3.7.1 General aspects about Brazilian culture

Preventive and corrective actions in the treatment of accidents, both social and technical, depend on knowledge about the bad habits and human errors promoted by the biases of Brazilian culture. Blind loyalty and informality diminish the critical sense in relation to the scenarios, which makes that in the risk analysis and investigation meetings, conflict is avoided in advance, believing that reaching a consensus is good for safety. The centralized power always personalizes actions, and power is maintained if relationships are strong in that way. In this direction, Group protection may be a necessity. Teams are protected by the influence owners; conflict is avoided so that there are no losses in the decision process. Paternalism is a need in the network of operational and strategic decisions. This is a problem for real safety, for culture.

Now, to maintain the strength of the group, we seek to protect them as the leader protects the tribe. However, this practice requires centralization of decisions that goes in the opposite direction of the delegation and brings inappropriate characteristics to avoid the accident.

Unfortunately, when people in the operation group are not observant and critical, they have no initiative and wait for someone to take action (spectator behavior). This situation transfers many operational micro-decisions to managers that can in this way, induce abnormal events, chain reaction and major process losses.

Although these negative traits of Brazilian culture occur, they are mixed with challenges for the company and also vices of Global culture such as, attempting to achieve goals within a short time, selfishness, self-centeredness, and lack of connection with work.

Avila (2015b) states that there are positive points of Brazilian culture that fight against some negative points of Global culture as well. Innovation is an adaptable movement in Brazil, especially in the new generations, where, when admitting group formation, cooperation is quick and a strong sense of group is established. For the most part, there is a lot of desire to do things the right way and the high creativity of the part of the group, "contaminating" and facilitating investigations in the routine operation, this when the group sense is established.

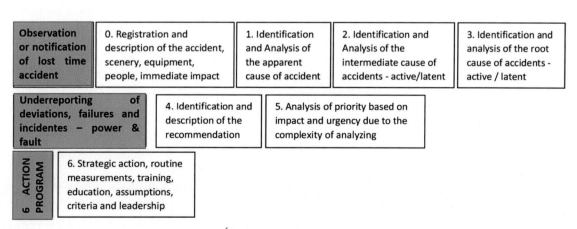

FIGURE 5.21 Accident investigation method. *Based on Ávila, S.F., 2013a. Review of risk analysis and accident on the routine operations in the oil industry. In: 5th CCPS Latin America Conference on Process safety, Cartagena das Indias, Colombia.*

When Brazilian people face new things, they are very willing to learn. Here and in other discussions, it depends on the generation and the understanding of what the working group is.

We commented in other texts that multiculturalism from migrations (due to work, climate, or geopolitics), in addition to virtualism (resulting from media culture and the internet) strengthens negative traits such as the immediatism habit and the lack of tolerance in admitting different groups in the work or living environment in the city.

In investigations carried out based on the Brazilian civil code and environmental laws, we noted that the government blame in advance the entrepreneur who operates risky industrial projects. This situation requires care from business owners to prevent abnormal events, such as accidents or inappropriate organizational practices, from becoming a fine. Thus procedures are written more on the basis of legal fundamentals to avoid anticipated blame than for technical reasons of maintaining the normality of the technology.

Unfortunately, when writing procedures and rules for organizational control, there is an excess of documents to be monitored and reviewed, which become a disease of the information system, or even a disease of integrated management. This disease reach the point that the registrations show conformity in case of accident investigations. Eventually, the worker signed records and reports that confirm training, regardless of its quality, largely because of the formal signature record. The protection of the employee is the omission in the face of deviances made, which can lead to group omission and even violation when the comparison between individual values and organizational present an impossible negotiation.

In parallel, it is necessary to defend fair culture, which represents the culture of communication in the task and without underreporting which comes from the blame culture. In blame culture the anticipated transference of responsibility and guilt occurs, from the government to the entrepreneur, from the entrepreneur to the worker, and finally the omission, a protection measure (Ávila et al., 2018b).

The culture acquires traces from the technology applied in the production system and at the same time, the operation of the technology depends on the culture installed in the manufacturing operation. It is important to identify which are the technology restrictions and which are the main risk scenarios for the elaboration of cultural transformation. The analysis of the highest-impact risks already carried out are reviewed including human and organizational factors.

The task is classified as critical from the analysis of objective functions regarding social and economic risk. Thus in the case of the metallurgical industry, the objective function is defined by the number of accidents with leave (weight 50%), interruption of operations (30%) and quality of life (20%). Critical tasks and jobs are analyzed (Ávila and Costa, 2015; Ávila et al., 2018b) taking into account risk and complexity in the environment. The constraints of perception, interpretation and cognitive processing in addition to the restrictions for controlling the task are also analyzed through the investigation of the executive function.

Although we work with the control of the task that has objective indicators, it is important to discuss the worker behavior and the group that includes subjective aspects.

Behavior analysis is based on the C4t tool, which is a title in another article. The human type is identified from the research on human error. The commitment is defined based on data from human resources and complemented with the supervision of the shift. Cooperation depends on the level of aggregation, disaggregation and resolution of small emergencies. Competence, as already discussed, depends on the identification of gaps in knowledge and skills. These gaps are traced from the operator's discourse.

Finally, the quality of communication is analyzed at three hierarchical levels: bottom—located at the floor factory; medium—staff and manager activities; and top—strategic decisions. This quality is assessed in two situations: standard or feedback. The level of stress directly influences the result of the 4 Cs. Leaders must identify ways to minimize these consequences (Ávila and Drigo, 2017).

The surveys are prepared and applied to identify the behavior, the level of fair culture and the level of the culture of guilt in the organization. These are steps to measure the installed culture. It is important to define the dominance of antagonistic behaviors and also to establish the focus of culture investigation, which in this case, is in the accident and in the energy loss.

A series of actions is applied in an organization and the respective results is analyzed step by step, through measurements that involve risk, the task, behavior and culture. These measures indicate the success or the need for revision in interventions for technical-human-organizational change, or still, for cultural change.

This diagnostic research collected data from operator's discourse of production in critical periods of 3 to 6 months to enable the mapping of regions of organizational efficiency. It enables the future projection of this efficiency with the culture transformation. The control of offshore activities or in the metallurgical industry (exercised application cases) requires preventive or corrective actions, which are triggered by means of conformity indicators. The priority in decisions and investment levels depends on the risk of loss. Thus the proposal is that after the investigation is carried out, key indicators (KPIs) are developed to control the task and behavior, thus authorizing the organization's resilience within the operational culture.

The analysis of visits to production units in the industrial oil segment and the discussion of characteristics of the metallurgical industry, mining and the involved human factors made it possible to build the application exercises and analysis of projected results. The methodology is applied considering the specific characteristics of the Metallurgical Industry, Mining and Offshore Oil Exploration, in general. The current discussion is based on proposals for the metallurgical industry (ferroalloy) and for the onshore and offshore oil production industry.

5.3.7.2 Discussion on fair, organizational and safety culture in metallurgy

Most of the production workers in the mining and metallurgy industry are from cities close to the factory, far from the capital. Engineers and some supervisors live in the capital or in larger cities and travel by centralized transport to the company. Often, small cars are used by managers from logistic area or from strategic area, forgetting safety principles and remembering time necessities. The route between home and factory, or between mining—metallurgical—and energy facility can be a problem that causes frequent accident.

The simplicity of the mind map is opposed to more complex controls to avoid losses with the increase in production, brings points to be adjusted in the organizational and safety culture. Teams live in antagonistic situations of following old leaders, who have a strong bond, or meet the demand for leaders who work with indicators and goals and seek a cold relationship in meeting organizational demands. Part of this cold behavior are from outside of Bahia.

The history of losses in the routine and accidents with lost time indicates the potential for improvement in the production area, safety and energy. The workplace has advanced systems such as the digital control system and, at the same time, has classic systems such as transportation, grinding and furnaces. Few instructions are written, so the pattern is considered flexible. It is difficult to guarantee Operational Discipline if there is no reference to the standard guide of activities.

The metallurgy production process involves high temperatures, noise, ore transport by large equipment and transportation belts, generation of particulates, 6-hour shift regimes and high-capacity process due to the low metal content in ores. The furnace uses the arc-flash principle with high electricity consumption.

The risks of occupational accidents may be high in addition to the risks of rework or reprocessing. Therefore, some tasks are considered critical, mainly the separation between waste and metallic alloy. Confidence in the results of the operators' work requires a commitment to the organization that can reflect a behavior of blind or critical loyalty. Blind loyalty can be the result of centralized management and critical loyalty can be the result of delegated management.

Communication involves omissions mainly on feedback in an organizational environment that adopts a blame culture. Sometimes omission is a protection to avoid guilt due to failure to perform the procedure. But, sometimes, omission is due to informality in the operating instructions. In the case of metallurgy where the number of procedures is smaller, omission is probably due to the non-officialization of routine instructions.

Low quality in the feedback of routine abnormal occurrences harms revised standards. Low quality in the managerial feedback harms revised organizational standards. Both impact on the possibility of the occurrence of accidents with absence and in the routine events with high energy losses.

Undue safety behavior has to do with the loss of perception of risks in the work environment involved with the mechanical impact of parts, machines, cables and others. The improper behavior of worker in the use of energy has to do with the absence of a mind map for the use, transformation, losses and cost of energy. Often the operator does not realize the impact of energy consumption when reworking and reprocessing materials in production.

The competency project depends on the type of technology and on the task to identify the experience level and what are the special requirements. It is important to understand the specific skills and the knowledge necessary to choose the best workers and to develop them in the industry. It is realized that the activities in the metallurgy industry, in part, despite being unhealthy, do not have as high complexities as those seen in chemical processes. Thus recognition through human errors or technical failures indicate the competence gaps for the task. Apparently, handling high temperature products is the biggest difficulty.

A cooperative climate in the workstation is a signal that the leaders rotate the roles of key workers in the organization, from time by time. To achieve goals the operational mass need a better perception about the hazard products and activities. In this case, Leaders must adopt an attitude of listening and negotiating resources and achievements. The result of a cohesive group indicates that the movement of aggregation is higher than the movement of disaggregation in the routine work. In the case of the metallurgical industry, cooperation is a necessity to reduce routine stress caused by high temperatures and handling of material with risks of mechanical impact.

If we consider a ranking between the culture of Guilt and the Fair Culture, in which, the culture of guilt has a relative minimum value of 40%; when the value goes up, the influence of the culture of guilt decreases and superimposes characteristics of the fair culture. This growth is possible with learning and communication processes in strengthening groups and in the clarity of organizational purposes. The maximum value of 140% is reached when one considers that the organizational culture is fair and does not suffer the influence of the protections or reactions resulting from a culture of guilt. The 140% value for the organizational culture indicates a result of resilience. The metallurgical industry is 90% based on visits and interviews.

The planned interventions for the metallurgical industry involve raising awareness of the front line and preparing the staff to investigate the task and human errors. The rituals must be changed to an intermediate configuration between the economic result and the "team union."

5.3.7.3 Discussion on fair, organizational and safety culture on the OFFSHORE oil platform

The geographical distance from offshore oil wells to the continent, to the capital of state, and to the workers' hometown can be very large. This fact and the time of on-board work (14 days) bring specific characteristics to the environment of the task. Stress is a real possibility and if the manager adopts a centralized and autocratic stance, the possibilities of human errors are diverse, involving the contractors and the inspection team.

Drilling and oil exploration activities are complex because they involve a flammable fluid, activity in the sea with possible impacts to the environment and the possibility of serious events such as fire and contamination that can eventually cause fatalities (Ávila, 2013b).

The history of fire events with recurrence suggests that the investigation of the causes is not in the right direction. Some platforms operate remotely through automatic controls of control rooms on the continent, thus increasing their complexity, and if there is no correct integration of operations, there will certainly be losses.

The regional characteristics of Brazilian culture and the characteristics resulting from globalization in culture, compose a multiplicity of characteristics that, if not treated, affect the workplace. In the case of the offshore oil production industry located at sea, regionalities may indicate the formation of groups or "fiefdoms," which makes interpersonal relationships more difficult.

The behavior of excessive self-confidence at work is remarkable, which brings conflicts between the contracted people/firm and contractors, or even continent and sea, operation and safety, operation and maintenance. Thus problems of social exclusion occur, that is, there is a lack of group belonging, in addition to organizational disintegration, lack of organizational belonging.

In the organizational and safety culture, care must be taken with regard to historical beliefs and the migration to current beliefs. There were major changes in the regulation of the oil market and many people were motivated to leave the sector, especially the older ones. What are the consequences of these changes?

What is asked in the current model, in which people believed and believe now? These beliefs are determined by events, discourses and confirmations. The rituals must be studied to cancel the move in the direction of the accident, or in the direction of the hazard route being released. Some archetypes deserve attention for their effect in favor of safety and energy or against them. The Hero archetype must be avoided.

The oil production process involves low temperatures, high pressures, flammable liquid and gas, transportation of oil and gas to the continent including operations under the sea. The 12-hour shift regime during 14 days, in the middle of the sea, demands care to avoid stress and cognitive loss. Work with pumping, compressors, tanks, rig and other equipment is constant. Water processing and power generation are part of the activities of a platform (Ávila, 2013b).

The risks of occurrence of process accidents are high as indicated by the accidents in history. The risks of reprocessing are lower. Some tasks are considered critical such as storing hydrocarbon condensate or drilling and starting oil production.

Confidence in the results of the operators' work requires a commitment to the organization that can reflect on behavior resulting from centralized management and/or management by delegation. In the case of platforms, this depends on the "luck" of working with a GPLAT (platform manager) who delegates the activities and has a strong sense of justice.

Behavior is marked by a culture of guilt with many omissions. In the case of the oil industry where the number of procedures is large, omission is probably due to the defense regarding the fault of the victim for the accident. Low quality in the feedback of abnormal routine occurrences for revising standards and low quality in the managerial feedback to revise organizational goals are aspects that impact the possibility of accidents occurring with absence and, in the routine, with high losses of energy to the atmosphere, and product losses, which go to the sea (Ávila, 2013b).

The improper unsafety behavior has to do with the perception (inverted or failed) of the risks in the work environment involved with the transfer of flammable fluids, high pressure of the systems, and high costs for the production.

In a similar manner to what was presented about cooperative environment in Metallurgy Industry, in the case of the oil industry, cooperation is a necessity due to the fact that the routine is carried out in a stressful environment (extended shift) and there are fire and explosion risks allocated in the ocean, where access is difficult.

The criteria adopted to classify the fair culture in oil production are the same as those used for the metallurgical industry. The oil industry is situated at 70% of the fair culture scale, based on visits and interviews.

The interventions involve sensitizing the front line and preparing the staff to investigate the task and human errors. The rituals must be changed to an intermediate configuration between the economic result and the "team union." Failed processes must be documented in the form of training modules. Measurements of task, behavior and culture should indicate movements for or against fair culture. The knowledge gaps are detected through the perception of the operation, the operator's discourse—human errors in the occurrences, and quality of the decision through the examination of the technical culture.

5.3.7.4 Final considerations

Process risks that include human factors are identified and are related to routine procedures and specific operating instructions. Tasks are controlled based on performance indicators in the various areas: production, quality, asset management, safety & environment, productivity, time and others. A good performance is considered when it exceeds 80% on average of the task indicators, except for safety and environment cases, which indicators must be close to 100%. The risks that impact the workers through their occupation are included in safety. The risks that impact the population, the neighboring facility and nature include the risks of process accidents and are also discussed in the area of environmental impacts.

The first metric to be monitored and used to control cultural transformation is related to the task indicators. The analysis of infrequent events, or anomalies, also indicates the result in the form of indicators mainly when it involves process losses, fatalities and, occupational and environmental fines. But, no less important, are the deviances that are often considered with normal situations, but they can be indicators of accumulation of danger energy. Deviance monitoring is important to trigger interventions before it becomes an anomaly or failure and incident.

The second metric to be followed is in the comparison between the behavioral gaps related to competence and communication, where communication quality is measured and the difference between supply and demand for competence is verified. These gaps indicate that the basis for safe behavior and prevention of energy losses is fragile.

Communication creates the environment for the task and the competence enables its successful execution. In the C4t analysis, the missing commitment ensures the full application of competence and improves the clarity of the standard and the feedback. Finally, it is important to analyze the team's level of aggregation, which creates an appropriate cooperation climate for the good result in production activities.

The third metric relates to the culture, in which the intention is to evaluate beliefs and values regarding the level of acceptance through the analysis of archetypes, their stories, characters that help and hinder organizational performance. The final result are indicators such as the number of accidents with lost time and or product leaks to effluents, which are certainly not interesting for the company. The culture metric is divided into final results, intermediate consequences, signs of behavior.

The Industry and Services Segment adopt an organizational resilience posture when they capture changes or environmental requirements and adapt to new restrictions. These dynamic corrections are only possible in a place with Fair culture where the flow of communications occurs without high distances between the factory floor and top management.

In Metallurgy and the Offshore Oil Industry there are differences and coincidences in behaviors that require corrections in the organizational culture project, or even in the cultural result on the factory floor. The intensity of thermal stress and the isolation regime of the team in the production of offshore oil are characteristics, which must be treated with seriousness to minimize bad practices and negative results in the company. This treatment is possible with learning, transparent communication and with a Fair culture installed.

Cultural transformation for avoiding energy losses and accident losses requires actions, measurements and goals to be achieved in the field of Task, Behavior (C4t) and Culture.

5.4 Human reliability, sociotechnical reliability, culture of safety demands

In the context of the research and services on dynamic risk, we were searched by companies and by society that did not understand the complex relations (processes-tasks-social) and the occurrence of unexpected accidents, and in more serious cases, the arrival of the Black Swan, or, the occurrence of disasters.

Initially, our research was based on the Chemical Industry (isocyanate, sulfuric acid, alkylamines, polycarbonate, caprolactam), in which there was a discovery of a method that identifies the "anatomy of the failure" from the organic causal nexus (high multidimensional connectivity), here called as operator's discourse—OD. From this method, the warning-signs that indicate the presence of systemic failures were valued. Then, this OD method, considered as the integration between Task and culture by the University of Manchester Institute of Science and Technology (Paul Sharratt) was incorporated by the area of Risk, Reliability, Energy & Water efficiency, and Human Factors. These methods were applied as learning model by the Petrochemical Industry, Fertilizers, Oil & Gas, and, currently Social Defense Public Agencies.

The experiences reported here will be coded by type of Industry and Public Agency to preserve confidentiality, regarding social and political aspects in each Organization. The serial number is the register of the institutions where there were demands. In this way we have:

CI means Chemical Industry with products and, DEE means Distribution of Electric Energy

CI 1—Caprolactam (BR)
CI 3—Aromatic Aroma (BR)
CI 5—Sulfuric Acid Reconcentration (BR)
CI 7—Explosive (USA)
DEE 9—Electricity Distribution (BR)

CI 2—Alkylamine (BR)
CI 4—Isocyanate (BR)
CI 6—Polycarbonate (BR)
CI 8—Acrylate (BR German headquarters)

PI represents the Petrochemical Industry that produces the following products,

PI 9—Basic Petrochemical Input (BR)
PI 11—DWC Petrochemicals (USA)
PI 13—EaTM Petrochemicals (USA)

PI 10 BTX (Benzene, Toluene and Xylene) (BR)
PI 12—Petrochemical DUPT (USA)
PI 14—Ammonia and Urea (BR)

OGI represents the Oil and Gas Industry, onshore and offshore, producing or distributing products,

OGI 15—LPG Compression and Distribution of Gas, Oil and Derivatives (BR)
OGI 16—Offshore Oil Exploration, production and Transport (BR)
OGI 17—Exploration, production and Transportation of onshore Oil (BR)
OGI 18—Offshore Oil Exploration, production and Transportation (USA)

ReFeI represents the Refining and Fertilizer Industry which produces the following products,

ReFeI 19—General Oil Refinery (BR)
ReFeI 21- CO boiler (BA and RJ—BR)
ReFeI 23—Ammonia and Urea (BR)

ReFeI 20—Diesel Hydrotreating (BR)
ReFeI 22—General Oil Refinery (USA)

MiMeI represents the Mining and Metallurgy Industry which produces the following products,

MiMeI 24—Metallurgy BR

MiMeI 25—Article demanded in Brazil Mining

FI represents the Food Industry,

FI 26—Food Industry—Chicken Processing

PuA represents Public Agencies that carry out the following activities,

PuA 28—World Cup Risk Analysis by CNPq

PuA 30—Cooperation: Public Security Secretariat

PuA 29—Carnival Megaevent Risk Analysis by PMS (the city hall of Salvador, Bahia)
PuA 31—Master's Degree for Firefighters and PRF (the federal police for highways)

5.4.1 Chemical industry and electricity distribution

5.4.1.1 General cases

In the Chemical Industry, the study required planning and control of the task and methods were developed to investigate the sociotechnical failure mode (FMEAH). Environmental impacts, operational continuity and production costs were failure and human errors consequences, handled in the period. Aspects of communication in routine between departments and in the level of cooperation between people included intrinsic behavior variables, in which fear, motivation, application of competence and communication are intrinsically related.

5.4.1.1.1 Caprolactam industry

Initially, it was noted that the relationship between powers in departments impairs communication and responsibility in the routine of support services. The training material for the routine of the shift period in the laboratory and operation was also improved, with a view of the service provider and its respective importance in the speed and quality of the data. The Hydrogen Compressor and the hydrogenation of benzene were the focus of the study to increase the operational reliability of the plant. This study motivated the operation to develop training on complex processes in the case of operation of hydrogenation of benzene.

5.4.1.1.2 Alkylamines

The operator's discourse (OD) in the Alkylamine Industry brought a statistical survey of the production campaigns differentiating cycles and bottlenecks, thus indicating which shift groups have the best performance based on the number of operational deviances. This stage was published in a specialization monograph in the area of organizational culture consulting (CESEC, UCSal). This case was in a chemical little industry with good operational reliability.

5.4.1.1.3 Industry of aromatic amine, isocyanate and sulfuric acid

The complexity of the process indicates the need for specific studies on the task, equipment and the role of the worker in the operations routine. At this moment, when understanding the risks, actions that use visual communication were suggested to increase the team's commitment. Design errors had to be recognized to reduce the environmental impacts of the units' operation. The fear of damage to equipment and occupational accidents hinders the relationship between management and the group.

The recognition of informal leadership to make changes possible requires respect for the operation history. Finally, the formation of groups or "castles" with their own informal rules significantly reduces the company's results. These experienced sociotechnical processes indicated the existence of systemic failure and the need to identify social contingencies to perform assertive interventions in the group and in the equipment. The omissions were studied and the development of educational programs showed good results when understanding the limitations and characteristics of the population. The main gain in this experience was the decrease in accidents and the increase in operational continuity.

5.4.1.1.4 Polycarbonate industry

The experience of working in the operational routine, in equipment improvement and human factors was intense in the polycarbonate plant where the anatomy of the failure was studied from OD with goals of reducing solvent losses to the atmosphere and reducing the cost of production. The actions and corrections were carried out in the following areas: prioritization of abnormalities based on the study of equipment components in operation, maintenance and design criteria; routine activities relationships between maintenance and operation; process testing for changes in the procedure; training and task forces to establish responsibility and commitment; in addition to continuous investigation of operation events. The lived experiences were transformed into lessons learned and training, which, through preventive-corrective actions and also cases of punishment for violations (behavioral and consequent antecedents), the pattern was gradually changed. Unfortunately, this pattern was not kept due to a lack of investment in self-management. The director finalized this important work before establishing self-motivated leaders.

5.4.1.1.5 Electricity distribution (DEE) in Brazil

In the network of contacts in human reliability, happened a demand from the electricity distribution sector appeared to reduce losses in the function of maintenance and operation and to avoid accidents. The Brazilian Association for Awareness of the Hazards of Electricity (ABRACOPEL, 2010) indicates a high number of deaths from electrocution and the increase of fire events from short circuits. In addition, in chemical process industries, the main cause of death is the operation in substations with drawers to release equipment in the industry. In the case of losses, 7% is calculated for the indicator of equivalent total failure in the supply of Electricity in the State of Bahia. Investigations always involve failures in equipment, safety control systems and natural environment results. Little is discussed about the possibility of an improper decision causing incorrect actions to solve problems and about human factors resulting from organizational issues.

5.4.1.2 Real cases of chemical industry demand and electricity distribution demand

In the Isocyanate Industry, the sense of ill-established justice and the formation of fiefdoms caused a break in cooperation and communication between workers who were born in Bahia, Rio Grande do Sul and Rio de Janeiro, with the CEO from São Paulo, which causes low operational continuity and accident. In the American chemical explosives industry, the Corporate Security manager stated that changes in the social environment probably changed the Human and Social Typology discussion carried out during the process safety conference in Chicago. Another fact detected apparently is the change of informal work rules, causing lost-time accidents in two physically and culturally distant units, thus breaking no-accidents records. management stated that it lost something and did not adjust the safety culture (SC). A diagnosis was requested on Human and Social Typology based on the operator's discourse. The demanded proposal is detailed below.

At the same time, another American Chemical Industry (also for explosive product) affirms that, unfortunately, it did not understand the reasons for the operation to have been considered normal, when in fact, it was living with routine deviances; in conclusion, the accident came back to facilities.

Already in 2019, we discussed Low Risk Perception (LRP), an actual case of the chemical industry, at the process safety symposium in Texas AM University. When discussing with peers the causes of LRP and Dynamic Risk, we discovered about a network of relationships established between Human factors in the operation and human elements in the project. They can cause the failure and the accident.

A clear question was asked: are Center for Chemical Process Safety (CCPS) (2007) guidelines able to view LRP in a timely manner to avoid an accident? The author of the article (same of this book) answered the question by demonstrating that the leaders' lack of preparation in terms of human (intrinsic variability) and social issues (change in informal work rules) allow the passage of hazard energy, which causes repeated pattern of accidents and even disaster. Then, CCPS do not avoid the accident, totally, with their guidelines.

On the other hand, in this same Symposium, TEES-TAMU offered training services for the industry, with the objective of presenting a concept or practice of how the issue of generation conflict in training should be treated and thus the work of adjusting the team at the workplace. In fact, TEES-TAMU had no method to deal with the matter and is expecting guidance from the contracting industry. There is a lack of knowledge about the influence of social phenomena on human error.

In this scenario, the chemical, petrochemical and petroleum industry continue with the occurrence of accidents with the respective production losses. Probably it happens due to not understanding how to intervene in social phenomena and how to organize adequate training methods to promote cooperation, commitment and operational discipline. Without these intangible and subjective barriers, we cannot stop the deviance normalization, we continue having the chance for accidents.

Understanding how human behavior works in the work environment, the performance factors, the modeling, both regionally and in the timeline, is a technology frontier. In managerial terms, the intention is to find out which models are appropriate for the decision and which are the most effective barriers or safeguards for each situation.

The justifications are tangible and intangible, which makes this understanding difficult. It is intended to establish ways of immediate measurement in the areas of production costs, safety, and image to demonstrate the feasibility of working on research projects in Human and Operational Reliability. These projects involve dynamic risk management and the operational routine. These projects try to develop an understanding about cognitive schemes for decision-making both in routine and emergency.

Learning to design projects for the chemical process industry helped us to build and propose a research project and services for the energy industry.

A 24-month project applied to the electricity distribution area will result in the validation of Methodologies and Tools in the area of Human Reliability. The return on investment is guaranteed, considering the cost of the methodologies and the reduction of losses due to greater reliability. The suggested methods are validated in a pilot area that should involve around 100 employees of the company and include other workers from outsourced company. The procedures throughout the job such as pre-dispatch, service, efficiency analysis, costs control and others will be reviewed.

Human errors will be avoided mainly in the maintenance team with the help of monitoring team (operation) which services always impact on safety and productivity. The hypothesis is that losses due to blackouts in the distribution of energy will be significantly reduced. It should be noted that the research approved by the National Electric Energy Agency involving the Brazilian Electricity Generation and Distribution Concessionaire focuses mainly on the development of devices or equipment, and does not carry out financing in the area of Human Factors management for production and safety. This is a contradiction in an industry that has so many accidents.

Thus to make the research project viable, the preparation of a strategic intelligence software is offered to enable the analysis of human factors avoiding future accidents.

5.4.1.3 Proposal for the blast-explosive industry (USA)

Consultancy on human and operational reliability in 2012. The health safety environment corporate director of an American Chemical Industry that produces explosives needs a revision of the management barriers to prevent the increase in hazard energy that cause failure and accidents. They need actions to be carried out at the base of the "BIRD" pyramid, near the root cause. The history of this company demonstrates excellent safety performance, without accidents, but due to recent events, in which accidents with high severity have occurred, diagnoses of the human, these multidisciplinary professionals, social and organizational factors are required.

The methodology presented in Fig. 5.22 is divided into (1) identification of the company's image and structure, (2) mode of technical and social failure, (3) identification of typologies (social, human and technical), (4) critical task evaluation, (5) dynamic risk assessment about human-social performance factors; and (6) program to avoid human error and Human-Operational Reliability guidelines.

5.4.1.4 Human reliability research proposal for electricity distribution (BR), private industry

In 2011, a Research-Service proposal was presented to the Energy Concessionaire in Northeast Brazil with the title: methodologies and tools for reducing human error in maintenance services of the electricity distribution network in Bahia

The development of management methodologies in pilot areas of the company; studies of the cognitive, intuitive and affective models for the decision and execution in the task by the supervisor and by the activity executor of these pilot areas, thus making it possible to change the task and the work environments.

As shown in Fig. 5.23, a program for data management, research and interventions indicate the flow of information on reliability and human factors from the various areas of the company: health, safety, maintenance, operation, people, management, technology and others. This database allows a systematic and efficient management decision, reducing task uncertainties. By classifying errors and identifying aspects that can cause their appearance, using techniques for sociotechnical systems, comparing ideal and real competences in the task, it is possible to intervene on tasks, environments and management systems.

A program for measuring results makes it possible to analyze whether changes in standards, procedures and programs have achieved their objectives in the human, social and technical areas. Thus it is proposed to adapt, prepare and validate the following programs and procedures in the managerial area.

The results or numerical goals of the research will be established in percentage values in the areas of safety and operation/maintenance of electrical systems. The non-numerical goals included in the products to be delivered involve the installation of management Methodologies and Programs to strengthen the Technical culture and the safety culture in the local facilities. It is understood that Technical and safety culture are the result of consistent and transparent policies applied to the dynamic team with a self-management profile. The Team and Leaders seek the lowest risk of accidents and of the interruptions in electricity supply.

The results are divided for the area of safety and operations, although the human and social factor is the same. Some aspects that influence the results are:

Level of commitment (COM); physical distance (DIS); competence level (KOM); type of leadership (LID); knowledge of the company's goals (ME); social rituals (SOC); type of established technology (TEC); physical environment (CHU); relationship between policies and practices (PXP); type of risk management (RISK); task writing, records, pattern and complexity (TK); regional characteristics (REG); characteristic human type (HUM); type of contract (CONT) such as revenue, personal rotation, leadership rotation, caste formation; compliance with legislation (LEG).

The projected results in the Operation and Maintenance area are:

With regard to revenue: Loss of revenue due to interruption (FAT); Business loss in the company (LUE); Frequency of unexpected cuts (FEQ). With regard to Cost: Time (standard vs realized: programming, execution and output) (TIME); Overtime (EXT); Re-Service (RES). Regarding the Technical Area: Accumulated annual cut-off time (DEQ); Region (classes); Impact of the geographic area of the stop; Area where services are performed. With regard to Image: Loss of Profit on the Client (LUC); Affected population (POP); Commitment and contracted company; Type of company (class); Service productivity (TIME); Quality in service (QUAL).

Increase Reliability in task, decrease possibility of accident			
CULTURE	STEPS	PROCEDURES	DIAGNOSES
Elements of Culture, Opinion of STAFF, Practices	(1) Identification of Company Image & Structure	Preliminary Diagnoses API770; Perception of Manager and STAFF; Trainning Introd Human Reliability	Current Situations between polices and practices and Organizational Factors
Data from routine, reports and observation	(2) Failure Mode of Social and Technical System	Anatomy of failures ST Systems at blast indust; Barriers Program in Routine deviation; Training about Root-cause investigation	Immediate review of critical tasks, variables and Organizational Efficiencies
Company database from Human, Healthy, Safety, Ergonomics	(3) Identification of Typologies: Social, Tech, Human	Diagnoses of typologies; Failure Database; Organizational Procedures; Task Assessment; Typology Procedures	Competence; Work; Workstation; Stress; Rules of Behavior; Decision Models; Procedures elaboration
Current procedures and tasks; Failure Anatomy; Aspects of Culture	(4) Critical Task Assessment	Training about Task Assessment; Diagnoses and Review of Critical Tasks	Increase of efficiency at task; Decrease Task Cost and Time; Decrease Accident; Increase Reliability; Task requirement and Communication
Incident and Accident assessment; Aspects about Dynamic Risk	(5) Dynamic Risk Assessment	Training Recommendation and Implementation; Human and Social Factors; Cognition; STPRA; Social Hazop; LODA; Culture and Behavior Assesment	Deviations, Incidents, Accident reduction; Intangible benefits: communication, quality; Programs about psycho social and educational aspects
Structure and Company Image; Failure Modes ant ST Systems; Task Dynamic Risk	(6) Guideline in Human and Operational Reliability	Validation of Procedure to HR Program; Global Diagnoses of Human and Operational Reliability; Guideline to Task, Information and Training	Organizational Strategic Procedures/ Specific Competences; Database System; Calculation Method

FIGURE 5.22 Method for human-operational reliability in the explosive industry in the USA.

The relationships between factors and expected results in operation and maintenance are shown in Fig. 5.24. The result of this analysis indicates that the Task (TK) and the Competence (KOM) influence the results, and in addition, the reduction of Losses is in re-service (RES), service productivity (TIME) and Quality (QUAL).

The projected results analyzed in the safety area are:

With regard to Accident: Number of accidents in the community, own and contracted (NA); Number of fatal accidents in the community, own and contracted (NAF); Number of non-fatal accidents in the community, own and

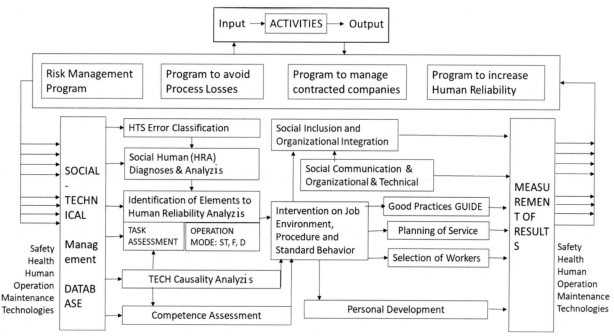

FIGURE 5.23 Management of databases, techniques and organizational programs in human-operational reliability energy BR.

Company	FAT	LUE	FEQ	TIME	EXT	RES	DEQ	LUC	POP	QUAL	Total
ME											7
COM											6
PXP											5
RISK											5
LEG											2
IMAGE											6
Physical											
DIS											1
CHU											1
Social											
HUM											4
LID											9
SOC											5
REG											2
Task											
TEC											5
KOM											9
TK											8
Loss Reduction	10%	5%	5%	15%	10%	20%	10%	10%	10%	15%	

FIGURE 5.24 Relationships between factors and indices with expected results in operations/maintenance.

contracted (NANF); Frequency rate of accidents in the community, own and contracted (TXF); Severity rate of accidents in the community, own and contracted (TXG); Security level of own and contracted work (NS). With regard to Image: Legal actions of the population against the company (CPE); Legal actions of employees against the company (CEE). With regard to Cost: Cost per accident (CAC).

The relationships between factors and expected safety results are shown in Fig. 5.25. The result of this analysis indicates that the Task (TK) and the Risk (RISK) greatly influence the results, in addition the highest reduction in

	NA	NAF	NANF	TXF	TXG	NS	CPE	CEE	CAC	Total
Company										
ME										3
COM										7
PXP										6
RISK										8
LEG										5
Physics										
DIS										3
CHU										2
Social										
HUM										6
LID										7
SOC										5
REG										3
Task										
TEC										2
KOM										2
TK										8
Loss Reduction	10%	10%	20%	25%	10%	20%	5%	5%	10%	

FIGURE 5.25 Relationships between factors and indexes with expected results in safety.

Losses are in the Frequency Rate of accidents (TXF), Number of non-fatal accidents (NANF) and Level occupational safety (NS).

5.4.1.5 Proposal for chemical industry of acrylates (2019–20), private industry

The research and consultancy work proposed and carried out in the human, operational reliability, loss of process, culture and task areas motivated the preparation of international and national Knowledge and Services NETWORK in these areas. The new discoveries are released and then the demands arise. A Chemical Industry of Acrylates requested training to discuss the low perception of risk from the point of view of human factors.

The indication of the Corporate safety manager and conversation with the local safety staff suggested the application of training for the entire operational mass, 180 operators and engineers. This demand resulted in risk consultancy, presented as follows: 1st work proposal; diagnosis and training execution; 2nd work proposal.

The objectives of the 1st and 2nd assignments, the main training topics and the topics that compose the preliminary diagnosis are discussed.

5.4.1.5.1 1st Task—training and preliminary diagnosis—risk perception: cases and practices

Objective: Develop, through informative discussion of concepts and critical analysis of real cases with perception exercises on scenarios, investigation about causes and consequences in routine and emergency situations in risky activities.

Method: Relate the concepts with real situations that caused an accident, with the failure triggering element, the lack of perception. Conduct quick questions at each stage of training to indicate the level of the team's risk perception after the discussion.

Diagnosis: Establish the possible causes for the poor perception of risk and the possible interventions. Indicate differences and leadership traits.

5.4.1.5.2 Main topics in training

(1) Principles: Understand the Hazard, Mechanisms and Risk in Sociotechnical Systems; (2) Cognition and Practices: Concepts and Practices: Cognition, Routine Decision and Emergency. Lack of Perception: Causes and Consequences; (3) Workplace Criteria: Cognitive (Task, Competence); Organizational (Commitment); Social (Cooperation) and

Physical; and (4) Human Factors Management: Failure and Safeguard for Accident; Human Factors, c4t, Leadership and Stress. How to intervene?

5.4.1.5.3 organization of the preliminary diagnosis

Report Topics: management Summary; Objective; Phases; Concept (Chemical Industry and Requirements, Plant Design, Operations Control, Routine Risk, Training Concepts); Training Comments; Preliminary Diagnosis (Sensitizing in favor of safety; Grouping of Questions into Classes, Question—Sense—Statistics, General and Class Behavior, General Behavior—Average and Standard Deviation, Analysis of operational culture); Hypotheses to be validated; Conclusion.

Multinational company that adapts to local cultures, allowing social differences to exist and seeking to improve their safety standards. It works on behavioral aspects in line with technological aspects, taking into account the current characteristics of the hazardous energy generating environment and the complexity of technologies. The current challenge of CCPS is to discuss the causes of the deviance normalization and how to avoid the accident after organizational changes.

Unfortunately, without knowing, different regions, leaders, technologies and organizations are driving hazard energy toward the accident. The work of adjusting: the safety and organizational culture; worker interfaces with production; the quality of communication (standard and feedback); the level of cooperation; and the level of commitment needs to be confirmed through routine (operator's discourse) or by survey during times of awareness regarding the position of the operational culture and points of concentration of hazard energy.

Training on risk perception based on behavior brings an approximation to this "operational truth." In this management summary, it is concluded that the preliminary diagnosis on general and specific aspects that affect the team's behavior was carried out and interventions are needed. An exercise of a Bayesian network considering the hypotheses concluded that the company has positive and negative aspects that indicate the presence of deviances that can be avoided.

For the diagnosis of the causes of LRP behavior, questionnaires were interpreted, which after being processed resulted in Fig. 5.26, which presents, according to operators, in six training classes with the entire group of company operations, the causes and consequences of LRP.

5.4.1.5.4 2nd Task—diagnosis for changes in safety culture: phase 2 of risk perception diagnosis

In the preliminary diagnosis, the need for interventions was presented in order for the company to leave the region of incidents it is in today: between failure and accident control.

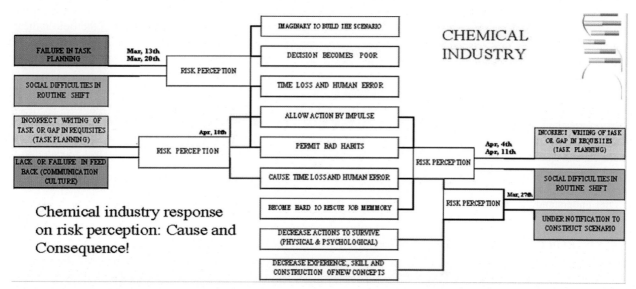

FIGURE 5.26 Causes of low-risk perception in the chemical industry. *Reproduced with permission from Ávila, S.F., Ahumada, C.B., Pereira, L.M., 2019b. Analysis of the low perception of risk: causes, consequences, and barriers. In: 22nd Annual International Symposium-Mary Kay O'Connor Process Safety Center, College Station, Texas.*

In the preliminary diagnosis, it was discussed, from the point of view of the 180 operators, technicians and engineers of the industry (nicknamed as operational mass) what the main causes and consequences for the low-risk perception, indicating, as causes: social difficulties in the shift operational routine (social phenomena); and miswriting of the task or requirements gap. As consequences for low-risk perception, there are: the group act on impulse, repetition of bad habits, waste of time and human error.

The proposal is related to the improvement of risk perception and the safety culture in the industry. The preliminary diagnosis that dealt with safety principles should be validated with additional behavioral data and with the statistical processing of complementary questions of cognition, workplace and human factors. After the statistical analysis of the complementary questions, investigations will be carried out on the operational routine, group dynamics on industrial accidents and interviews with production and safety leaders. This activity, focused on questions related to the principles, will indicate the appropriate interventions, objective, time, necessary resources, methodology, implementation systems and relationship with the safety and organizational culture.

The interventions depend on the validation of the hypotheses and it can be in the area of rituals (DSD, daily safety dialog), training, educational processes, operational groups, intellectual capital for the failure, adjustments of the HMI, alarm system, risk review and tasks including human factors and other tools, such as indicators in the areas of human reliability and human factors. It may be necessary to set a team of researchers in human reliability with knowledge of Human Factors.

The construction of safeguards to prevent the hazards present in the area of safety Culture, Behavior, Leaders' Role and Work organization from being carried out depend on actions that are part of this proposal.

These actions follow methodology in 5 phases: Phase 1—Validate Operational and safety Context; Phase 2—Interpret Critical Events to identify Hypotheses: Cognition, Workplace and Human Factors; Phase 3—Test and Validate Hypotheses; Phase 4—Accident Construction and Risk Analysis Training; and Phase 5—Final Diagnosis and recommendation of interventions for the main aspects defined in the Preliminary Diagnosis.

5.4.2 Petrochemical industry

The work in the petrochemical industry was performed during a project to reduce water and energy losses in basic input plants. The study of the procedures indicates non-conformities regarding the loss of water and energy due to the non-synchronization between the work from the operation (practices) and the staff (concepts and standards). High water losses were noted in the cooling towers and improper production schedule at the BTX (Benzene, Toluene and Xylene) plant.

The Research project with a Brazilian Petrochemical company and the resulting publications recorded the results of this stage, thus indicating that the higher HR will guarantee the reduction in energy consumption. The technical problem was revealed in a Workshop and it was noted that the operations team made decision mistakes when controlling the flow monitoring of the tower with a deviation, considering a normal situation. The appearance in the field indicated a high level of losses that was calculated from the sum of the evaporation flows. This situation shows that diagnostic and management errors have high impacts on the environment and perhaps higher impacts than individual human errors.

A PhD Research in the HR area would be initiated at the Basic Inputs Unit in Petrochemical industry in Bahia, but after meetings to deal with the flow of information from the operational routine, the medical sector signaled to the production area that, studying HR would be like "shooting themselves in the foot" for production managers, the wrong decisions would appear and there would be a risk of unemployment. Promptly, the research was suspended because there was no agenda for the meetings, and the beginning of fieldwork was suspended. This behavior was repeated in the Public Refining Industry in Rio de Janeiro, São Paulo and Bahia.

Organizational resistance to work with human factors is evident, due to the discomfort of Classical managers, who do not understand the importance of working with behavioral uncertainties.

5.4.2.1 Real cases of petrochemical industry, demand for services

The integration of contracted companies in the petrochemical industry is performed through training, with the maximum of 3 hours, similar to examples of selling the company's image, they try to show an image of safety culture using positive marketing in the company. This marketing is inappropriate for organizational integration, in which gaps must be presented in the communication system for critical services with real examples. These trainings do not demonstrate the availability of listening to enable the dynamics of knowledge to migrate from the company's leaders to those hired and from those hired back to the company's Leadership. This symbiosis between contracted companies to maintenance services and the petrochemical industry is necessary to enable preventive measures that avoid the occupational accidents and of the processes.

This demand and discussion, in particular, was observed at an American enterprise, which indicated that, at different times in its history, to be aligned with the challenges of dynamic risk, they really worked with Human Reliability. In the period of 1995–2005, internal seminars were promoted for the shift groups on the "Human Element" Program, which brings together Local-Regional culture and Organizational Culture. Recently, at the TAMU Symposium, discussions were about the cons of assigning extreme responsibilities to a BIG DOG, the great supervisor, which coincides with our discussion of developing informal leaders to break the resistance to changes and innovate the production system.

On the other hand, another American Multinational Industry in the Chemical-Petrochemical segment, in successive moments, demonstrates loss of competence in the area of Process safety, as it cannot explain why fatalities occur in its units in Brazil and in the USA. The answer from the Organizational managers is that they lost something with the "decentralization" of the safety Culture. In their words, it was probably because deviances became normal. It is believed that this problem has not yet been solved.

During a Global Conference, a chemical industry leader in operational discipline, which leads studies in this area, promotes lectures at GCPS asking what are deviances and how do they normalize. They say that they adopt CCPS "guidelines," that show the existence of the event, deviance normalization, but fail to advance in explaining why it occurs and the necessity of multidisciplinary team for the investigation.

These American companies and associations try to develop solutions based on their experiences and from their internal consultants in Process safety. But because they don't understand exactly what Human and Sociotechnical Reliability is, these institutions adopt exclusively Technical and often Political postures with Economic Biases, which they defend with broad explanations, but which in fact they do not know how to define, the influence of culture on unsafe Behavior that causes human error and accident. Therefore, these institutions do not know how to design safeguards, because they do not admit that the system is sociotechnical.

These American Institutions are multinationals and positively influence the growth of the Industry, but due to the lack of knowledge about measurement of subjectivity in human behavior, or due to overconfidence affirming that they know what is Sociotechnical Reliability and Human Factor, they end up bringing results that do not prevent abnormal events. Accidents will keep happening in the Global Chemical Process Industry.

5.4.3 Onshore offshore oil and gas industry

Research work on the production of Oil & Gas (Ávila, 2010, 2013a; Ávila et al., 2008; Carvalho, 2016; Huang et al., 2014) also started by analyzing the OD and identifying the Technical Typology (Process, safety, Maintenance, Logistics) and the Human-Social Typology (Behaviors, Sociodemographic and Rules of Communication) indicate the trends of the operational culture regarding human errors and technical failures.

This culture is recognized through exams and analysis of specific cases in the operation. A study of the main cognitive and behavioral gaps and the correlation with process losses (gas volume, energy cost, failure warnings) indicates what the old patterns are and proposes new patterns to be adopted by the team, and indicates how execute do this transformation.

Technical failures and human errors occur during the execution of tasks, thus raising questions about the quality of operational standards and procedures. An important analysis is suggested with the Standards and Operational Procedures tool (PADOP) with the investigation of the environment, cognitive processing, execution and control of the task. At the same time that PADOP is developed to be applied in the Chemical Industry, this tool is validated for Complex Processes in the Oil Industry.

The rituals for entering and leaving the oil production fields are seen from the social point of view as affecting the quality of the services and which can cause accidents (Ávila, 2013b). The demands of the Petroleum Industry involve aspects of HR inserted in the sociotechnical system that has a high impact in the safety area, in a similar manner as to what occurs in the fertilizer-refining industry.

5.4.3.1 Real cases of demand for research and services

These cases are the result of the discussion during training on human reliability for 20 groups, with 24 hours for each, during the period from 2009 to 2013, for the Brazilian Oil and Gas Industry. The classes comprising Engineers,

Psychologists, Administrators, supervisors and Operators, had, on average, 15 students, accounting for up to 300 professionals in the technical and sociohuman area in Oil & Gas Industry, including the organizational standards area and assessment of human performance shaping factors. The main complaint from the team of technical-engineers was regarding the absence of managers, who should learn about Human Factors and how to work on the Task and Deviances in the Routine. These experiences and the development of research in the area of communication and leadership for the crisis brought important information about the gaps in the company's culture.

In offshore oil production, after research with contracted firms that were taking a course for shipment & salvage, critical findings occurred: managers divide the work group into classes that are proven through the priority of transportation and benefits; managers are not aware of the effect of stress levels at work; managers are unaware of human factors and the importance of human design elements; entry and exit rituals are poorly planned and performed, with economic biases. Health professionals commented that leaders do not notice when their employees are depressed, in psychological isolation, and do not know how to analyze the risks of this situation for an accident.

One case of accident with leave and coma was caused by late shipment of the maintenance worker and due to a delay experienced by a boilermaker professional that had no time to rest, thus going straight to the service. Due to haste, probably related to tiredness, the wrong valve documents were gathered and the safeguards for the existence of the spring were not prepared. At the opening of this device, the professional suffered a mechanical impact and fell into coma.

This situation was confirmed in the accident on the São Mateus FPSO ship (ANP—Agência Nacional de Petróleo, 2015), considering that in the investigation, the main reasons were technical and management of the contracted company. In actual fact, only after inquiring about communication through the presence of UFBA specialist (always present in workshops about oil accidents), it was confirmed that all operating instructions were written in English and the vast majority of operators did not speak the language.

A Public Oil & Gas company manager demanded research on Human Factors, and during 9 months, meetings were held with the staff. As at other times, the meetings begin and do not converge to the jobs demanded. This company does not end investigations on Human Factors and Operational safety Culture. The staff's self-confidence and economic biases interrupt this process.

On that occasion, an Oil production manager in southeast states was unable to establish a causal link to the accident. The manager could not find the social-human origin because the type of subjective relationship (often organizational) was not understood. That makes man a tool for incidents and fires. Therefore, this manager clearly states that it was not possible to locate where to install a more effective barrier to failure.

The manager's question was whether this situation was possible. I confirmed that yes, where leadership profiles, installed archetypes resulting from regionalism, inadequate rituals and a low sense of justice promote the uncertainty of where the safety event will take place and the level of release of hazard. It was stated that the control room is the center of communications and the main responsible for the generation of communication noise, which provokes bad mental maps for decisions.

These problems in oil production occur at sea and on land. On onshore production it is noted that the quality of communications is low due to the lack of understanding of regionalism and due to the lack of credit in relation to the organization, thus making the feedback communication full of gaps, underreporting, which promotes an unsafe environment.

These examples demonstrate that a fair culture, based on communication and task is not well established in the oil industry, while the culture of guilt generates an environment with a low sense of justice and omission at all levels.

In the LPG industry, where large drain valves in the spheres bring fear and failure to remove moisture, then cause plant shutdowns due to the low capacity of the other barriers such as the coalescer.

At two different times, a proposal was developed for the Oil production Industry, the first for the onshore industry in the northeast at different times, for training and research, and the second for offshore production in the southeast, both regions in Brazil.

5.4.3.2 *Proposal for onshore, Brazil, 2008, public industry*

In 2008 a proposal was presented for the human resources area with the title "sensitizing the operation to issues related to human error and the impact on safety and productivity." A Competency analysis method is initiated through awareness, process mapping with diagnosis based on operator's discourse, classification of human error, loss analysis (for then to perform the task analysis), and skill analysis to implement human reliability programs. It was offered to a production offshore oil in southeast.

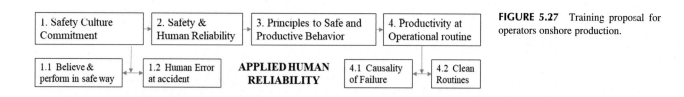

FIGURE 5.27 Training proposal for operators onshore production.

The work of sensitizing the operation team to human aspects of behavior that affect safety (in terms of accidents and incidents) and affect productivity (in terms of costs and production schedule) will be carried out in lectures to the shifts.

Lectures that aim at preparing the terrain (environment) to discuss (or investigate) aspects of process losses and human reliability in a second moment. Thus it seeks to present important themes with the discussion of the influence of human behavior on safety and productivity controls. The topics in Fig. 5.27 could be developed in a 12-hour lecture based on the operations practice.

5.4.3.3 Proposal for offshore, Brazil, 2010, public industry

In 2010 a demand from the oil industry to prepare training for a specific operation on oil wells was presented. They intended that the safety culture of this new production field with new operators, started with concepts of human reliability to avoid contamination with the old culture, which experienced deviances and problems of operational discipline. The title of this proposal was "Operational safety based on Human Reliability," thus repeating the 2008 proposal. The stages presented: Training Program for Operator Awareness (similar topics), Course for staff as multipliers and investigators on accidents based on human factors, diagnostics for risk perception and human reliability. In this proposal, we draw attention to the tools used in the diagnosis as indicated in Fig. 5.28. A series of Guidelines (Fig. 5.29) in addition to Principles and Methods (Fig. 5.30) are proposed to be implemented in the Oil Industry built after the interview with production and safety managers. The main question is how to ritualize these guidelines?

It is important to affirm that this production unit after conclusion of the meetings, decided to contract Nordic university confirming the archetype that the product from outside Brazil is always better than inside. Actually, we noted a great demand of American companies in this product, even though they are new products and being launched, formatted in Brazil, they propose to do a pilot and after validation, apply throughout company.

5.4.3.4 Proposal for onshore, Brazil, 2012, private industry

The work for the onshore oil production industry is a preliminary diagnosis for the Analysis of Human Factors in the Task and, based on the conclusions, recommendations are made. This work intended to analyze, in a preliminary manner, the human factors in the production and maintenance tasks of rigs in a mature field managed in the State of Bahia, Brazil. This analysis deals with the possibility of accidents and loss of productivity. This work requires a pilot study group that represents the company's workers, a team of 25 workers from different functions.

With this diagnosis it is intended to recommend priority actions to be carried out in the following areas: safety culture, preparation of leaders, analysis of requirements for the task, and allocation of human types.

The activity is proposed to be done in three stages: (1) Survey of sociotechnical data, (2) Analysis of causality, and (3) Diagnosis.

In the survey of sociotechnical data, questions are asked about company characteristics to identify the stage of culture, task, and people behavior through a remote questionnaire, and then, interviews with leaders, and maintenance and production staff. Then, the Causal Analysis will be carried out, in which, from the visit to the workplace, the installation and the worker's role will be observed.

This observation of routine activities will be carried out during administrative hours, when the behavior in the routine, the feeling of group inclusion, control, leadership and cooperation will be seen. Routine communication modes will also be included in causality. Together with the staff/manager, causality for the incident/accident impact will be designed to analyze the dynamic team behavior in psychological stress.

The last step will be the elaboration of a Diagnosis with the following analysis: technical adequacy through training, educational programs that unsafe practices may occur, a program for leadership development, adjustments in the safety culture seeking effectiveness in the results of non-accidents, and criteria for planning and analysis of critical tasks.

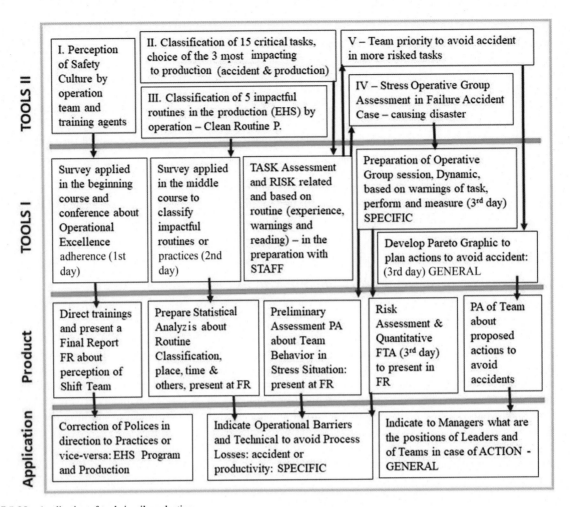

FIGURE 5.28 Application of tools in oil production.

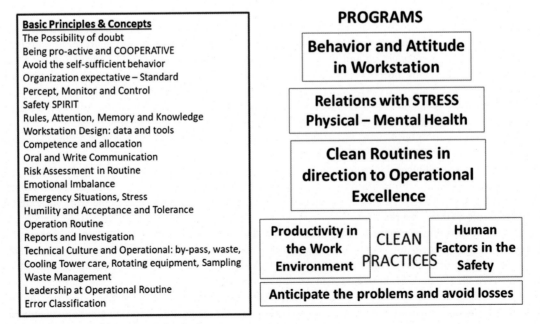

FIGURE 5.29 Principles and methods for operational safety.

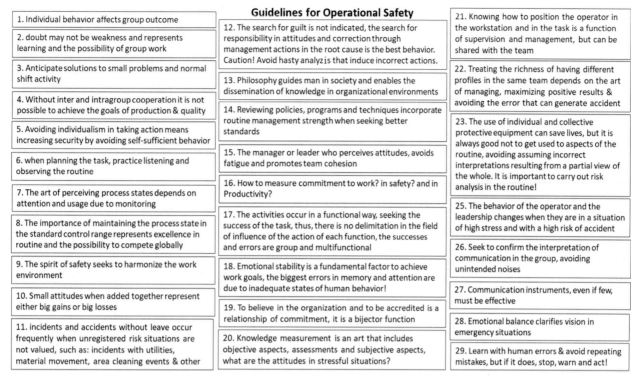

FIGURE 5.30 Guidelines for operational safety.

5.4.3.5 Proposal for offshore, Brazil, 2016–17, public O&G industry

This project was demanded by other manager from an offshore oil field, for a public oil and gas company also in southeast of Brazil. The project title was: "Analysis of human factors to reduce unscheduled stops in oil production with a focus on offshore oil and gas production." The Methods to increase competitiveness by increasing the productivity and availability of processes and equipment were tested, based on Services and Consulting carried out for the Chemical and Petroleum Industry. The top-down practices and the limitations of the specialists in understanding the subjective dimension in the calculation of the sociotechnical systems reliability, causes the industry to make the classic mistakes in solving problems around the cause indicated by the deterministic view (immediate cause). managers still do not understand the importance of analyzing human factors and, consequently, they are not willing to break the classic paradigms yet.

The research team is multidisciplinary, composed of 40% of non-engineers and social, humanity and health sciences professionals. Engineers account for 60% in the discussion about the task, processes and equipment.

This project, based on bottom-up techniques, and knowing the top-down practices uses the operator's discourse and the observation of the workplace to understand the origin of the tacit rules in working on a production platform. The methods used follow the next main themes: (1) Operational and Dynamic Risk Management, limitations, criteria and control modes; (2) Diagnosis of sociotechnical system failures; (3) Analysis of Accidents and Operational Risks based on history, similar cases and projection of future accidents; (4) Analysis of the critical task in complex processes; (5) Analysis of prioritization and adjustment of Human and Organizational Factors; (6) Adjustments and Programs.

The suggested techniques were developed at GROD in (Research Group on Operational and Dynamic Risk Management) during the last 20 years and applied to the energy and chemical process industries in addition to applications in public management in big events. It is intended to map, prioritize and design (revise) safeguards for critical systems through the analysis of approximation to the root cause (technical, human or organizational). An analysis of the plant's history (management and routine reports) intends to indicate how much the human error influences the causes.

The research stages (Fig. 5.31, 5.32A and B) have the following dimensions: Start with the beginning of the research; Strategy with the alignment of concepts and the systematization of a dynamic risk management model; Diagnosis, in which critical areas of reliability are identified, root causes are analyzed and barriers are identified to reduce the probability of failure events considering the sociotechnical system; Operations with the application of risk and failure analysis techniques including human factors, based on the events already identified and analysis of the planning and execution of critical procedures with the possibility of failure; Analysis of Human Factors of the hypotheses

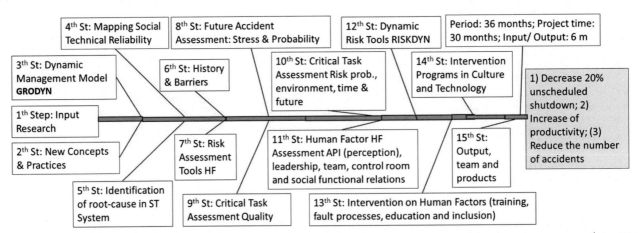

FIGURE 5.31 Stages of the research project (3 years) and objectives—HR in onshore production. *Reproduced with permission from Ávila, S.F., Bittencourt, E., 2017. Operation risk assessment in the oil and gas industry—dynamic tools to treat human and organizational factors. In: K.F. Olson (Org.). Petroleum Refining and Oil Well Drilling, Vol. 1, 1st (ed.) Nova Science Publishers, New York, pp. 1–125.*

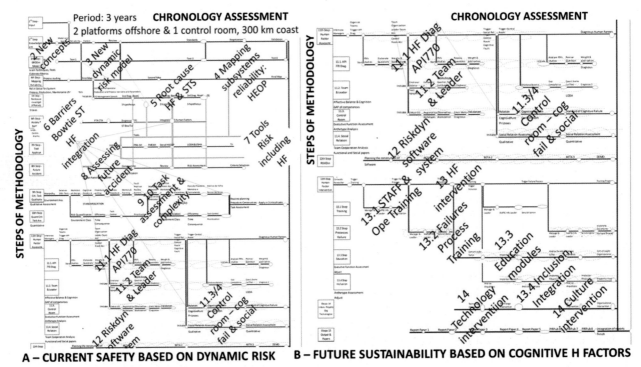

FIGURE 5.32 (A, B) Application of bottom-up tools in offshore research. *Reproduced with permission from Ávila, S.F., Bittencourt, E., 2017. Operation Risk assessment in the Oil and Gas Industry—Dynamic Tools to treat Human and Organizational Factors. In: K.F. Olson (Org.). Petroleum Refining and Oil Well Drilling, Vol. 1, 1st (ed.) Nova Science Publishers, New York, pp. 1–125.*

raised in the stages of diagnosis, analysis of operations and workplace for the verification of the critical human factors, which are involved with the stop events; The development of a beta-stage software for application at the end of the project period that includes subroutines and calculation modules for the reliability of sociotechnical systems, dynamic risks and human factors; Interventions in the area of human factors, technological and organizational projects for decision-making; and elaboration of scientific and technological production.

The products to be made and delivered are:

1. Models and techniques validated for risk reduction based on human factors;
2. Development of an OSH (occupational safety and health) Reliability and Human Factors Analysis for interventions;

3. Identification of appropriate standards based on bottom-up techniques;
4. Tested Model for operational and dynamic risk management, adapted and validated for the oil producing company;
5. Reliability mapping to prioritize interventions;
6. Root cause diagnostics and barrier identification;
7. Bowtie-sociotechnical diagnosis with review of human factors;
8. Recommendations based on the application of risk analysis and sociotechnical failure techniques;
9. Standardized recommendations in procedures;
10. Report: Future accident risk analysis for behavior and decision, resulted from the workshop of stress in future accident;
11. Critical task analysis reports by the qualitative dimension and application standards;
12. Task analysis reports by the quantitative dimension and standards;
13. Training, education and inclusive ritual programs;
14. API770 diagnosis, perception of human-organizational factors with classification;
15. Diagnosis regarding the level of leadership, team aggregation and competence gap;
16. Analysis of human errors in the control room and characteristics for the executive function;
17. Program for adapting social and functional relationships to a production platform;
18. Preparation of beta-stage software version for operational and dynamic risk management;
19. Training program and incorporation of intellectual capital on the failure process;
20. Examination of the technical culture based on the GAP of competence;
21. Intellectual capital documentation on failure processes;
22. Implement a multidisciplinary educational program;
23. Analysis of archetypes to establish integration and inclusion rituals;
24. Report on the inclusion of bottom-up elements in the organizational and safety culture;
25. Report: HMI and equipment for necessary adjustments in the task due to failure occurrence.

5.4.3.6 Proposal for onshore, Brazil, 2017–18, public industry, research

Research to define Technological Solutions to improve Human, Operational and Equipment Reliability, decrease the Risk of Operations and decrease process losses for onshore Oil production units in Bahia. That is, the purpose of this Cooperation between the University and the Petroleum Industry is the development of technological solutions in the areas of: (1) analysis of the production task, (2) reliability of sociotechnical systems, (3) analysis of human factors, (4) maintenance, RCM (reliability-centered maintenance) and maintainability management, (5) material and corrosion analysis, (6) quantitative risk analysis.

The University's proposal was to divide the work plan into two parts for a total period of 24 months, which can be extended for another 24 months, incorporating other subjects. Part 1: production and maintenance management systems, influence of both human factor and risk analysis. Between 2018 and 2019. Part 2: Equipment and systems in onshore production. Between 2020 and 2021.

To facilitate the organization of the groups, working groups and leaders are established for each subject and a specific work plan is developed. Thus the subjects are divided into specific projects: TASK (production and logistics task analysis); ASSET (systems reliability, losses, asset management and equipment reliability and their criticality); RECHA (human reliability, human factors and safety culture); and CORR (structures and materials reliability).

The research would start with process loss mapping, and then, would go to work on human reliability in the areas of communication and leadership, as described in Fig. 5.33. The results of the research through postdoctoral theses and undergraduate theses were presented every 3 months at the University or at the company.

5.4.3.7 Proposal for offshore, USA, 2018–19

The main objectives suggested for a year of operation in the Oil Industry: (1) Develop procedures to prioritize human and organizational factors that can cause process losses, including accidents and disasters in the occupational, environmental and process safety areas; (2) Develop a data and information system to monitor and measure human, technical and organizational factors, by indicating the approach of the main or sinister event; (3) Map the reliability of Sociotechnical Systems using the criteria, parameters and tools developed in (1) and (2); (4) To test interventions, validate methods and tools, publish methodologies (preserving intellectual property) to improve reliability in new and operational projects.

In parallel, we operate in the Sociotechnical Reliability area, applying the following article with data from the American Petroleum Industry with a branch in Latin America. Summary of the work done and published on AHFE

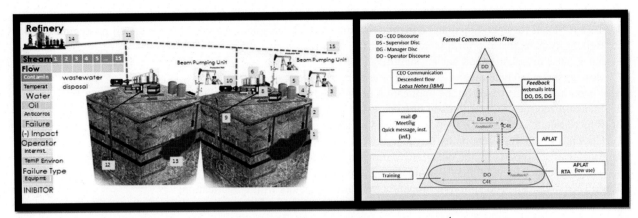

FIGURE 5.33 Process mapping and pyramid of organizational communication. *Based on Drigo, E.S., Ávila, S.F., 2016. Organizational communication: discussion of pyramid model application in shift records. In: Katola, T.B., Salman N., Andre, T. (Org.). Advances in Human Factors, Business Management, Training and Education. 1st (ed.), Springer International Publish, Springer, Vol. 498, pp. 739–750.*

2019 (Ávila et al., 2019) in Washington. This research work was attended by a Senior Reliability Engineer in the Oil Industry. Research Paper Title: "Management Tool for Reliability Analysis in Sociotechnical Systems—a Case Study."

Abstract: The objective of this work is the development of a management model for system reliability analysis influenced by technical and human factors (Ávila et al., 2019a) This analysis is based on an understanding of technology, operational and organizational contexts and the determination of human reliability, equipment reliability, operational reliability and a process known as integrated reliability or sociotechnical reliability. With this management tool, equipment and process failures, as well as human failures, will be reduced and the result will be higher reliability and operational availability levels. This work was applied to the Oil Industry whose headquarters are located in the USA, Texas.

5.4.4 Fertilizer industry and refining units

The Government Fertilizer industry, linked to oil&gas public company, has characteristics that increase the process risk. There are: high pressures, above 120 kgf/cm^2; high temperatures, above 800°C; toxic products as ammonia; and complex chemical processes. The operation has a high level of automation, but due to the non-standardization of tasks, problems of stoichiometry and contamination in the processes may occur. This situation indicates low operational reliability and fluctuations in product quality.

On the other hand, with similar aspects, the oil refining units are large-scale, highly politicized, able to handle flammable products and depend on high competence when the type of raw material is changed, thus changing operational parameters. There may be high emissions of CO and CO_2, explosions caused by the modification of product properties, for example, viscosity, presence of salt or sulfur, causing operational discontinuity and the risk of environmental-process-occupational accident.

The Group that produces oil products is the same that produces fertilizers, and is immersed in a conflictive environment, with policies and powers. In addition to assuming an egocentric and vain posture, it consequently brings excessive self-confidence and the game of positions and powers. In this context, the opportunity arises to research the operational context in the two environments, Refinery and Fertilizer Industry.

In the Fertilizer Industry, during a 4-year survey to reduce water consumption, unaccomplished or unplanned standards led to variations in the quality of the effluent and in parallel, excessive plant shutdowns. The under-specification (or over-specification) of the project is one of the items that is repeated in relation to the chemical process industry and the gas industry, and that impacts on the low operational reliability by changing the capacity of the control systems. This means that to keep the plant operating, many discrete, one-off decisions are made. To avoid contamination in the effluent, decisions are delegated to operators who are not prepared.

When noticing the non-standardization and the decision biases, it was decided to analyze the routine, interpret the events, and return interpretations to the operation, in the form of lessons learned through the Friends of the Emergency Lagoon Educational Program. It was realized that a major factor that promotes this process oscillation is the decision at the moment of the plant shutdown. Thus decision errors cause contamination of the effluent, which results in the violation of dilution, both high water consumption and environmental fines. These topics were previously discussed in this chapter.

The discovery of incorrect decisions and bad habits demanded new methods in the area of behavior and team reeducation in addition to the standardization of critical procedures. In the Fertilizer Industry, this low continuity demands corrective maintenance stops, thus increasing the production cost. This situation becomes even worse due to the fact that the integration of contracted companies is done through training with instructors who have no experience in production, as they give examples for impossible risks (possible electric shock when changing lamp in a pool filled with water) concealing the real risks in the plant.

During 3 years we had the opportunity to work with the Public Refinery (Scientific-Academic Cooperation) to reduce energy consumption through Reliability Management, Good Practices and Tools to avoid losses of electrical and thermal energy.

In the Refining Industry, we had the opportunity to participate of safety integration training in the public oil and gas industry. The integration of the contracted firms is unfortunately performed by means of "packaged" training by instructors without the necessary experience in operating conditions. The company is obliged to do the training, but, in fact, it does not reach the expected result for a production stoppage. Virtually, the instructor is not heard, or the student does not believe in the presentations.

5.4.4.1 Real cases of demand

In the Fertilizer industry, there is a fear of operating close to the alternative ammonia pump, where accidents have previously occurred with the launching of a cylinder into the atmosphere, because people did not put the pump to recirculate at the time it was demanded (protection against synthesis urea back-pressure). This situation caused uncontrolled reaction, loss of assets, plant shutdowns and discontinuity of operations. Fear is possible to occur in the operation environment when the safety pattern appears to be good but, in practice, is of low standard.

Similarly, "Fear" occurred at the Sulfuric Acid Plant, where the Tyvek coverall, which did not protect against acid because it was physically fragile, influenced operators not to go to the area to perform field services, as well as drainage of spheres at the LPG plant. Fear is a primary feeling, which occurs in the routine when operators, who do not trust PPE or devices to safeguard the process, do not carry out the control because of their physical survival instinct.

At different times at the process safety conference in the world and in Brazil, due to demand from the Oil Industry, it was repeated patterns were verified in accidents, with finding that reveal that the real causes were not adequately treated.

Failure in solving the causes of accidents leads to public pressure in the case of consecutive accidents in the Refining Industry. Fatalities in consecutive periods raise questions in the Public Ministry regarding preventive actions. Unfortunately, the Oil Refinery, through meeting the demands from government and population, in Brazil, believes that an efficient way to solve problems, is only training and the review of certain procedures. Well, this means they will have the same problem again soon because it is a "rooted" organizational problem.

For these cases, it is necessary to recognize the operational culture present in the Industry and analyze the conflict between policies and practices. The operator's discourse Method would help a lot.

This demand was not attended due to offer of a diagnosis (based on operator discourse assessment) before training but the company only wanted training and a review of procedures of the operation team.

The Petroleum Industry demands research to avoid explosion in the control of processes in CO boilers. The question is about the warning of failure before the occurrence of uncontrolled process and the boiler explosion. Was it just a case of lack of attention? Communication failure? Or lack of knowledge regarding the analysis of routine occurrences?

This CO boiler explosion accident situation is attributed to cognitive gaps in the routine, but what is the organizational implication and the relationship with the tasks complexity? And what about the organizational belief feeling in the operational mass?

The startup of a steam generating boiler is simpler than startup of a CO boiler connected with previous processes that are complex, by indicating that the poorly planned or the poorly performed task can contribute to the start of the event. Unfortunately, this was another research project that aborted and did not go forward. The fear of discovering latent truths causes management to stop projects.

In general, the oil refining industry has a great concern for standardization and consultation of the procedure in the routine. This theme was presented by David Embrey's CARMAN (Embrey, 2005) Method in the British Petroleum (BP) industry.

Due to the fact that the workers' group is highly politicized and unionized in Brazil, the term "standard operation" applied by the union refers to the written procedure, impossible to be done, then if fulfilling all the steps, the plant will be stopped. This was verified during a human reliability course, when half of the group was in favor of making the procedure more flexible (Rasmussen) and the others, in full compliance with the written procedures.

So, the following question is raised, is it possible to keep the plant operating with quality and quantity following procedures of more than 50 stages being accomplished to the letter?

This questioning joins the event of underreporting in the oil industry, which, for the occurrence of a fatality (1), there would be a certain number of deviances (30,000). Unfortunately, all of us live in the culture of Risk Aversion and low feedback. This situation creates the anomaly for the accident to appear without warning.

Our methodology applies the analysis of the operator's discourse (OD) and does not rely only on the technical variables of the process or on the managerial reports of routine control, or only in alarms log.

It is important to draw attention to human reliability research initiated at a refinery in southeastern Brazil by a local university. Once completed the diagnostic phase on Human Typology on the factory floor, the manager interrupted the work disagreeing with the results because it did not coincide with the production Manager's expectations. Both may be wrong for this case: the researcher may not have an appropriate tool to deal with subjectivities or the manager may not be structured to lead with the reality that, my team is an ET (extraterrestrial) or, in an inverse thought, I am ET to my team.

Maybe the manager is not be structured to lead a human reliability research.

This fact was repeated when in the research for doctor's degree (Ávila, 2010) the theme developed presented the human typology and respective errors in the Oil & Gas operational routine in Brazilian Industry, but, unlike the previous case, which was interrupted, in the current case, the research was concluded with the defense of the Thesis. The manager was structured for discuss HR.

The manager validated the result of this research, and showed a diagnosis from International Consultancy that indicated a team's behavior that did not coincide with the practices of their routine. The Method that analyzes the operator's discourse (Ávila, 2010; Drigo, 2016) indicates the best way to represent the chain of abnormalities and which are the informal rules of the shift operations.

The Oil & Gas area manager agreed that the team's behavior followed the rules indicated in the diagnosis based on the operator's discourse. Why does it happen? Could it be that International Consulting interviews are biased due to guilt and power issues? That is, through answers with vices to satisfy economic biases?

And in the case of research-based diagnosis, in which the routine discourse was analyzed, the O&G Plant manager recognized the team's standards from the discourse's long-term conclusions. We know that mistakes, incorrect decisions and bad habits and defense mechanisms are in the discourse.

In international trips to GCPS congresses in the USA, effects of Dynamic culture (Ávila et al., 2013) were presented regarding the workers' behavior performing critical tasks. In this discussion, the character Johnny Big Head is presented, an Operator-Executor, with his various psychic-psychological functions, performing the task and living performance cycles, which indicate maturity at work.

The mental fragments from the facts and the cognitive-intuitive extrapolation to routine and stress solutions demonstrate the importance of investigations on dynamic risk.

The American Oil Industry has praised the theoretical-conceptual models as being true for what happens on the factory floor, but has requested that we develop application tools with steps and objectives in the areas of safety, energy and production. They affirmed that they are obliged to invest in safety after the occurrence of accidents and know that unfortunately, the consultants bring poor solutions for the behavior prediction and the dynamic risk. They are organizational hostages of flawed methods.

5.4.5 Metallurgical industry and chicken manufacturing

The regionalization of the industry facilitates the transportation of employees and related costs, but brings biased relations in the workplace. This characteristic is perceived in the cases of mining, metallurgical, forestry and food industries.

When discussing the characteristics of Brazilian culture, it is said that there is a tendency toward impunity, as indicated by Motta and Caldas (1997). The excess of flexibility, the spectator attitude and paternalism are mentioned, but there are other derivative behaviors for the case which people with a family relationship in regions of small cities are workers in large industries.

We will discuss the case of a chicken Industry, research of a professional master's degree, where most of the workers of this factory are relatives and live in a city, in the Northeast of Brazil. A statistical study carried out indicated that the deviances in production increase when expected or unexpected events occur in the city and in families, part of the operation group had kinship ties that influenced in the work quality.

As a result of this research, the production leader (and master's researcher) established a communication system as preventive measure to, after detecting events in the city or in families, organize preventive actions for each type of production campaign, increasing or optimizing the redundant barriers when the risk of problems for certain campaigns is higher.

The metallurgical and mining industry is also located cities in the interior of the Northeast of Brazil, thus requiring the same type of preventive measure. In this case, care must be greater due to higher physical stress (thermal and mechanical). The pace of production involves a lot of material handling and requires PPE to avoid occupational or process accidents.

The need for both recovering materials in production and increasing the pace, or accelerate the cycle, increasing the quantity produced of alloys, motivated technological change, which allowed bringing new automatic systems to the metallurgical industry. The organizational changes, including this company in the stock market, which is sensitive to the reflected image of production impacts on local and global society, then changing the positions about the occurrence of an accident. Thus social, environmental and occupational responsibility becomes the organization's value in the stock market.

The Brazil Mineral Magazine from *Editora Signus* (2013c) in São Paulo, Brazil, suggested that we prepared an article for the Mining Industry at the request of Entrepreneurs, and so, we strove and compared the type of technology and relationships at work. Later, we presented the theme: "Human Factors Engineering in the Mineral Industry, a discussion to increase competitiveness." In this work the activities of: Extraction of Ore and Sand; Iron Mining by Classical Technology; Mining and Iron Concentration; and Copper Extraction and Concentration in Metallurgy were described. This demand for work by the Magazine presented guidelines for human reliability in mining and metallurgy with their conclusions.

In the organizational environment, economic bias can activate human error and change the priorities between safety and production generating conflicts among policies and practices. In the management of Industry and of Services, critical human factors can cause human errors related to:

- worker behavior, human errors, intentional (violation) or unconscious;
- Design and production resultant from feature of risky activities;
- Failure in mitigating actions that can cause serious environmental impacts;
- Wrong period of time applied for management and execution in maintenance service;
- Type of technology, complexities;
- Types and characteristics of materials,
- Level of automation;
- Type of people from internal or external community;
- Task characteristics.
- Type of communication, by building a database and measuring the impact of human error on routine, which feedback helps to avoid human errors in the decision-making process;
- Quality of the control panel communication that depends on good relations with the community throughout the pipeline, for example.

Human performance seeks to avoid wasting time in operations, through improvements of the indicators and good quality of services in maintenance, production and logistics. Maintenance service must be fast and efficient with suitable spare parts and leaders with attitude to avoid wasting time.

The Contracted company management must assess the rules and guides of the contractor to execute the services. They must understand how to Plan Stops to avoid re-services that indicate low standard of the contracted parties.

In Change Management, the company must pay attention to the Positions Plan, commitment and team skills. Increasing the workload or scale of operations requires harmonization of teams and preparation of new leaders.

For the task, it is intended to: work with the task's risk analysis; Analyze the physical and cognitive effort to define safeguards against human error. Task Complexity can be incremental in logistics and production operations. This situation requires a mental model for complex controls due to the level of automation. An important discussion is to avoid excessive signs or warnings that hinder the control of operations.

In the Selection and Development of Teams it is intended to: meet the demand for specific knowledge and executors' skill; Update knowledge of New Technologies; Understand the factors that affect equipment, operations and processes efficiency, available through courses in the operational routine. It is important to maintain a good perception of risks in the field and develop skills to avoid rework and delay in ship-train-truck operations.

Another important concern is to avoid Technical Idleness and lack of commitment to avoid low productivity. The equipment maintenance plan must be carried out with commitment resulted from the organizational climate. Leadership commitment is fundamental in expanding business. Check the possibility of individual and team fatigue, which would cause loss of commitment and low-quality accomplishment of tasks.

On Leadership in Mining and Metallurgy, the preparation of leaders must be constant to meet the demands of new production lines. It is necessary to define and develop the practical leader in ship-train-truck operations. Good sense must be installed in the leaders to enable the good quality of mind maps and the correct decision-making in the routine or in emergency situations.

5.4.5.1 Real cases of demand

A Ferroalloy company located in the interior of Bahia, Brazil, undergoing a technological update due to organizational changes, now complies with new requirements from the society regarding the safety and Environment indicators. Part of the company's leaders and part of the operators still have a family business culture when, in fact, the industry became a joint-stock company, therefore it is in the stock market.

The production and safety managers were not comfortable with applying tools that change behavior and perform measurements of appropriate control items. In other words, they demanded tools to measure, make behavioral changes in the team and in the leadership, meeting the pressure of shareholders for better results in the safety area.

managers, with an engineering background, did not know how to change human behavior. They did not have the capacity to understand subjectivity in decision moments and what are the appropriate tools. Non-lost-time accidents, lost-time accidents and unfortunate fatalities, some not directly related to the production activity, keep occurring, despite using International safety Managerial systems and CCPS guidelines Human Factor (Center for Chemical Process Safety (CCPS), 2007), Deviance Normalization. Why?

GRODIN (Research Group on Operational and Dynamic Risks) met the demand of the Metallurgical Industry in the formulation methodologies for Diagnosis and adaptation of the safety culture and the Organizational culture through changing the behavior of the team. Thus the organizational actions intended to achieve the safety goals, in an indicators ROADMAP, going through intermediate phases for 3 years, until reaching indicators similar to the external benchmark in culture and Behavior.

Upon receiving this demand, the metrics of Cooperation, Commitment, Communication and Competence were unified together with the level of stress, the characteristics of the Level 5 leader (Collins, 2007) and the indicators for complexity, transforming into a long-term program of interventions in cognitive-intuitive-affective behavior.

Unfortunately, the economic world and fear also interrupted the work of Human Reliability, due to the discomfort of entering into the subjectivity of this work, a region unknown to engineering, and we can say, unknown even to the current ergonomics. The applications of current ergonomics are in the physical and cognitive area for information flow in the Control Panel.

The applications demanded by the Metallurgical Industry are in the area of behavior and culture, and not related to anthropometry nor to the philosophical discussion about suffering at work, French Ergonomics.

Therefore, tools were sought for behavior change to improve micro-decisions and reduce the likelihood of human error. Tools that work in learning cycles and cognitive academy. The company stopped this initiative and preferred the classic means, that is, acting with interventions after the accident, relying on the events history for the decision, or neither analyzing the Operator Discourse nor the causes of low-risk perception, or measurement indicators of the 4 Cs, worker behavior.

5.4.5.2 Metallurgy, mining and cellulose industry proposal, 2018–19

During the years from 2018 to 2019, there were demands to improve safety culture considering management and Behavioral tools. The Service Title was "Tool Intervention for the Evolution of safety culture in the Metallurgy Industry, a reference for mining and cellulose."

The purpose of this service was to support the publicly traded Brazilian Mining Industry to design an action plan aligned with the safety culture Internship research, to influence the variables of its formation. The planning-analysis and training-awareness actions will enable the evolution of the maturity stages and the achievement of excellence in preventive practices, the basis of change is behavioral with management tools.

The project presents interventions in the tools, training of multipliers and intervention of identified processes for the elaboration of the Road Map Graphic with indicators for transforming the Culture. A series of interventions were proposed to be done over the three-year period: reviewing the daily ritual of talking about safety, implementing tools to investigate accidents and a tool to manage consequences in addition to training the team in risk perception.

In the execution of the safety tool, daily safety dialog (DSD), a review was proposed on the importance, scope, function, and application methodology, assistance in opening the communication channel and in analyzing the risks of the tasks. Develop and apply a tool to investigate and analyze abnormal occurrences. In this case, it is intended to enhance the methodologies used to awaken managers about the importance and participation in the investigation of accidents. It is necessary to improve the identification of the root causes and what the activators for risk behaviors are.

The ABC (antecedent, behavior, consequent) tool for management by consequences should support the company in the development and dissemination of a clear and comprehensive policy-making undesirable behaviors known to all (Nery, 2009; Fitriani and Nawawiwetu, 2017). The improvement in risk perception contributes to production and safety. In the development of risk perception, it is intended to enable the development of the employees' sense of risk and to provoke an early reflection on the execution of activities, which will make possible the execution or not of potentially risky tasks.

Intervention in specific tools and processes will allow treating the activators and motivators of an improper behavior and triggering the failures in the processes. The following subjects and tools are the focus of research for improvement: Work permit and accident; Care in the management of Technological and Organizational changes; Routine training; Accident investigation; Approach to behavioral change and the new safety Culture.

The project's actions are monitored monthly through factors and characteristics of the Task, Behavior and culture areas. Monthly the interpretation of these results indicates the evolution of the safety culture. At the end of 2 years, a new diagnosis is made to verify whether the evolution of the safety culture has reached the expected objectives. The result of this final diagnosis is presented in a workshop with the presentation of a new ROADMAP for continuous improvement and the evolution of culture.

The Approach is Behavioral, based on methodologies resulting from research on human factors. Operator Discourse Analysis (OD) and tools that involve human factors will support the identification of latent behavior, and inadequate rituals toward the accident. The leadership must adjust to the cultural challenges that move the team, such as the shock of generations, the cultural scheme related to a regional linguistics, and behavioral changes based on the moments in the organization. Identifying the appropriate characteristics for the Leader will facilitate cultural transformation.

Behavior, leadership and stress will be measured using new indicators of the results in the task and culture suggested in the diagnosis. The construction of behavior indicators depends on the identification of gaps in Competence, Communication, Cooperation and Commitment.

> Excellence in Operational safety is only achieved if the operator's discourse at his workplace opens the door to the convergence between Task and Culture, according to Paul Sharrat at a meeting at UMIST in 2003 …

Behavior is the result of psychological activation in organizational moments that will be analyzed in this work. Deficiencies in rituals and organizational procedures that imply safety will be discussed in the diagnosis.

Thus the analysis of cultural influence, the characteristics of the workplace where accidents occur, the respective cognitive functions, the type of behavior and what the critical HPEs are will be part of the methodology.

Human error and process losses are intertwined with the likelihood of accidents. Thus deviances and incidents are earlier stages of the accident, which, with the chaining of the sociotechnical failure, keep releasing hazard energy.

Large corporations in the Petroleum Industry and the Chemical Industry have questioned classic safety methods based only on pre-formatted methodologies in "foreign" cultures. This has motivated these companies toward the safety culture based on Behavior resulting from the management of Human Factors. In this case, the conflicting priorities must become a symbiotic process involving safety and production. The behavior in the operational routine is latent, in which a forgetfulness that caused a fatality has a whole fault tree below the slip.

The steps of this intervention consist of: 1. Data collection and Interviews for Preliminary Analysis of the Routine from the Operator's discourse; 2. Risk & Reliability Analysis of specific Events based on History and risk review including Human Factors: scenario, deviances, past accident and future accident; 3. Classify Tasks (complexity and risk) for the investigation, analyze critical tasks that reflect the operational routine of maintenance and or of the operation; 4. Identify the most common Human Error and the relationship with the accident, typology, analyze the leadership style before and after the transformation from Familiar Business to Open Market; 5. Identify the quality of communication, competence, cooperation and commitment and the consequences for safe behavior based on bottom-up techniques; 6. Analyze the stress level, the focus group's perception of the impact of stress and the consequences for safety; 7. Analyze the gaps between the desired standard for excellence in operational safety with the measured and perceived state in the operational routine; and 8. Propose interventions for the transformation of culture.

5.4.6 Public security agencies: security, mobility and health, firefighter—PuA

The National company of oil&gas, involved with many political crises, had its research work on human reliability in industry affected. On the other hand, public accidents intensified in this period, raising questions about risk management and crisis in public services. In view of the trends for the industry, we started, in 2013, researches on risk analysis in sports megaevents (world cup) and of the entertainment (salvador carnival)

We see that there are many cognitive flaws during contingency actions in the great public accidents in history.

During this period, the Rasmussen's model was applied to investigate accidents involving shipwrecks on transport boats in Bay of Todos os Santos (Ávila et al., 2018d) in addition to the collision of ships with piers and other facilities, through the Logistic companies that work with Ships, using practical simulation describing the different aspects in publication of UFBA Congress and AHFE (Ávila et al., 2019c). The question is,

How does the chain of events that cause accidents with vessels occur? How to analyze the cognitive, natural, leadership and stress dimensions, maritime safety devices, legislation, supervision, economic and behavioral aspects of the local population or resulting from the culture on the vessel-ship?

_____
_____
_____
_____
_____
_____
_____

In view of the investigation history on industrial process accidents, it was intended to apply the emergency team's behavior models to identify the cognitive and behavioral gaps of the leadership and the team. These behavioral models were created in the master's degree in HR with supervisor that have 30 years of experience in the oil & gas area. And in this scenario of demands, there are also participations in simulations promoted by the Firefighters Department and Brazilian institute for environment and renewable natural resources protection (IBAMA). Thus the understanding for crisis management and the failures resulting from the teams that deal with the emergency are being prepared.

The presence of GRODIN in the Disaster Congress that took place in Feira de Santana in 2018 was remarkable. In this Congress' schedule, we formulated and presented the following works carried out during the congress:

- Opening of CIDEM, with the presentation of the emergency team cognitive gaps in the treatment after accidents in history, published by LACPS in 2018 (Ávila et al., 2018c);
- Discussion about the accident in Pojuca, Bahia: the technical, human and cultural causes, the current legislation, a comparison with the Mexico Accident, where gasoline theft occurred. A debate on aspects of Latin American culture—risk aversion and opportunism;
- Progressive Stress Dynamics—LODA (Ávila, 2011b) based on Pichon Riviere Operating Groups for the accident caused by a gasoline spill in the USA, adapted to our situation, close to the Continent region beside a northeast bay in the state of Bahia. This technique was applied 3 times in tests. The first involving inflammable gas at Oil & Gas Public company in the routine investigation; the second discussing fire at the Polytechnic School of UFBA; and the third opportunity with 165 participants in an American case study;
- Critical Observer of the Social Defense Agencies' role during the tipping of a truck with benzene load, impact on the bus transporting people, terrorism and logistics to assist people involved in this disaster;
- Coordination of scientific works presented in the Disaster area.

5.4.6.1 Real cases of demand

The Analysis of the Risk Industry with chemical and social hazards located in Bahia, Refining and onshore oil production, led the research group and its leader to develop research cooperation work to avoid plant shutdowns, leaks of chemicals and oil products and their consequences, such as fire, explosions, environmental impacts to the effluent and contamination at sea, as well as accidents involving ships.

The Public Ministry of the State of Bahia (MPBA) was invited to visit GRODIN's research programs based on multidisciplinary tools involving the task technological aspects, human behavior at work, and a culture of organizational safety. Meetings were held to address CO emissions at the local Refinery.

Policies and powers hinder the work of human reliability based on the rules and the blame culture model, a theme published in ABRISCO and which was studied by chemical engineers and lawyers who currently assume a role in the state in the areas of research financing and professors at the federal universities and private ones.

GRODIN developed a procedure for cooperation involving UFBA and the Public Security Secretariat, public bodies with limited budgets. Expectations from SSP (a Brazilian department for public security) regarding teaching and new tools offered by the University were high in relation to reality. The University does not currently have the capacity to offer these tools to public agencies considering its current Modus Operandis in Research, Teaching and Extension. There was no financing for the activities. Even so, we worked on preparing groups for social and civil defense at the Carnival megaevent in Salvador, Bahia. The theme on the analysis of progressive stress (Ávila, 2011b) and the respective impacts on operational decisions (LODA) was considered as an opportunity for joint activity between the Police and the University, and as product, training for the police execution groups in the state of Bahia.

Knowledge of the decision under a situation of high stress brings different types of responses according to the level of preparation of emergency leaders (Ávila et al., 2018c):

- Not recognizing the release of the hazard, not studying the hazard line, and neither the probability-impact of the accident-disaster. People resist or prefer forgetting the responsibility that, in comparative terms, has to do with the reptiles' reaction to hazard. The part of the human brain that induces the FREEZE, RUN and ESCAPE behavior is the spinal cord that includes the cerebellum and neuronal network in the spine.
- Avoid excessive affective bonding, in which leaders are concerned with protecting their own skin in the first place. Thus it loses the ability to mitigate situations for the large population. This is the brain developed in mammals and has to do with the Hero archetype.
- The rational brain indicates risk analysis as the best option, where statistics and the study of scenarios will indicate the ways to release hazard and where to install safeguards. The frequency and severity as well as how to cause damage should be analyzed. Here it is discussed the main barriers in preventing and correcting for preventing or treating the hazard.
- The pineal gland represents intuition. In the complexity of environments with threats that include: biological invasion that intersects with negative cultural influences with characteristics of the local-global human typology and with health protection and human coexistence technologies; we have to use a little intuition as well. The emergency leader often does not have time for risk analysis.

Any comments on that description?

_____
_____
_____
_____
_____
_____
_____

The theme discussed in meetings on skills available to act in the Crisis, motivated Public Agencies (Security Agency of Bahia, Engineer and Architect of Salvador City Hall, Firefighter company of Bahia, Federal Police) to request from UFBA, preparation of Professional Master's degree with themes registered in the proposal, under the coordination of Professor Salvador Ávila.

The discussion of disaster and how to design barriers includes assessing how events called Black Swans appear, and the difficulty of developing solutions in an environment with limited resources of all kinds. It should be noted that, in the case of accidents during structured megaevents (Carnival and World Cup), of accidents promoted by multiple threatening environments (biological vectors, cultural and economic aspects, human typology), it is necessary to discuss which are the best measures to avoid the biggest impacts.

Thus considering preventive and corrective actions, the following is listed:

- Fast and transparent data communication to identify the hazard;
- Know how to communicate with the impacted public, avoid crowds and lack of resources, avoid chaos in the crisis;
- Realistic but Optimistic Communication to avoid people's mass depression or population revolt;
- Integrated management of public social and civil defense agencies to avoid chaos, gather and distribute resources;
- Plan tasks to avoid the lack of resources and the loss of the action moment;
- Analyze common solutions, and for specific groups, measure resources, apply resources overtime;
- Innovate in the development of barriers, once resources are restricted, and perform the risk analysis of barriers;

- Study behavioral changes to avoid cultural bias resulted from archetypes and bad habits.
- Preparation of appropriate leaders to meet the challenges.
- Know how to contain the population at various crisis moments.
- Know how to get the population to help in the various crisis moments.

The cases of ship collision, terrorism events in the carnival and World Cup, terrorism with a chemical leak, rupture of a mining dam, in addition to the COVID-19 + economy + improper behavior, can all together become the Black Swan and then need to be treated with respect and clarity. In our case, while we finished writing this book, we prepared the following technical bulletins TB to clarify the network of contacts on human factors in health and economic crisis management.

TB1—Starting the subject of Crisis and BT Management; TB2—Concepts on Hazard, Risk, Safeguards and Crisis; TB3—Geopolitical-Economic and Environmental Scenario in the COVID-19 Crisis; TB4—Natural Environment, Demography and Social Culture; TB5—Contamination and Maximum Curve; TB6—management Strategy and Contamination Curve; TB7—Preventive measures and bad habits in isolation; TB8—Cultural Threats to Treat Crisis and Plan Resilience; TB9—Black Swan and the Treatment of Death; TB10—COVID-19 Contamination Update; TB11—Contamination and death data scheme and Decision COVID-19; TB12—Introducing the COVID-19 database; TB13—Interpretation of contaminated data in Brazil and the World; TB14—Interpretation of data on Contaminates 2; TB15—Speed and acceleration of people contaminated in Brazil; TB16—Who is Salvador Ávila; TB17—COVID-19 indicator; TB18—Average Standard Deviation and Criteria per infected; and TB19—Integrated management at the Academy, and others.

Presently we are at 45 Bulletins.

5.4.6.2 Master's degree proposal for the Bahia firefighters department, 2018–19

When comparing the demands for competence and the management of crises, emergencies and accidents, it can be suggested that, in general, skills are missing for emergency teams, specifically for the Fire Department. So, we started by checking which gaps would complete this low competency reality. What are the skills and knowledge that we intend to transfer and develop for graduated students in this Master's Course?

The basis in the area of intelligence to plan resources for crisis control is part of the content of the master's program. Defense agency actions lose quality in crisis management based on critical scenarios built for cities, industries, fields and megaevents (COVID-19).

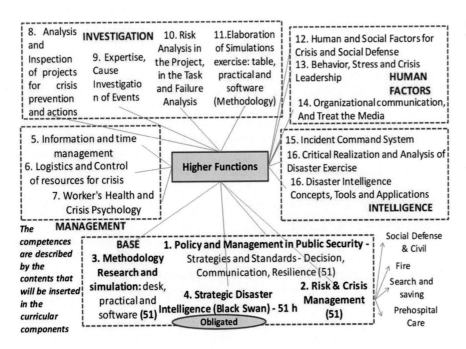

FIGURE 5.34 Map of missing skills and the subjects for the professional master's.

TABLE 5.8 Appropriate and inappropriate behavior in human factors.

Feature	Appropriate behavior	Inappropriate behavior
Research continuity	Perform priority analysis and maintain continuity and number of resources to finance research in the critical issues	Low funding for all research—political action with little recourse, does not prioritize, has no continuity
After event, 8X80	Analyze impacts and decide how much to invest and where to invest	Investing nothing or investing everything, depending on the damage
Knowledge migration in the search	Dual-knowledge migration, development of solutions for research and migration of financial support for research	Knowledge migration to enterprise and migration of solutions, without any exchange; only the company wins.
Risk aversion	Respect history, lessons learned, statistical data, scenarios; perform risk analysis and design recommendations	Not respecting history, not observing what happens with the neighbor, not studying scenarios, not performing risk analysis; waiting for lucky situations.
Opportunism	Have cooperative actions respecting humanity; respect the neighbors and help where it is possible	Use opportunities to get along, the good is the one who takes advantage on everything over the community, government and society.
Budget cut in project	Conduct risk analysis and plan resources; if any need to cut resources, study where this event is located in a situation of long-term unfeasibility and where there is resilience	Make cuts of 20% to 30% whenever work delay or exchange rate changes occur. Horizontal cut without risk analysis.
Budget Cut on production	Reduce production costs without losing materials quality and the spirit of cooperation	Cut the budget horizontally without analyzing the effects on other people.
Task planning	Have minimal competence to analyze the task and recommend minimum requirements that respect the legislation. If flexible, use new criteria	Reduce requirements to reduce costs and allow low standards also to reduce costs. Flexibility of the procedure steps without criteria.
Deviances normalization in operational Discipline	Awareness of the existence of hazard, from the deviance to the accident; disaster indicates that the team treat deviances to maintain the values of safety and discipline	By adopting the normalization of the deviance, indicating acceptance of risk aversion and also the lack of awareness about the hazard liberation process. Also, bad actions are flexibility, paternalism and waiting for the solution.
Competence and decisions	The matrix of responsibilities must be planned from the project, in which the necessary knowledge and skill to revise procedures is the one that is actually used; thus the success rate of decisions is high.	Planning and assignment of critical tasks by people without sufficient competence will increase the likelihood of incorrect decisions.
Operational routine versus strategic decisions	Establish policies, open the vision of the future; Instrumentalize management is as important as conducting operational routine knowing the scenarios, the hazards, and the lines of action and risk	The operational routine is the result of awareness of the action over the risk by the operating team and their supervisors. The missing strategic decision hinders the implementation of good practices, worsening the scenarios with the respective risks
Strategic intelligence	Data and information circulating in real time. Respect cooperative and integrated roles of public management. Active listening, valuing each role.	Delay, poor quality, and inadequate systems for circulation of data and information. Disrespect for its role in integrated management. Strategy versus operation. Not listening to routine warnings.
Logistics in disaster and management Resources	The identification of the necessary resource and the displacement for use are activities that can prevent damage in cases of accidents and disasters.	Not understanding the logistics services and not developing resource management will interrupt the supply by not carrying out preventive actions, correction and mitigation properly.
Stress effects progressive	In the emergency team, it is needed energization for quick and calm decisions to check the line of expert release—roles must be respected.	If each party performs the work inappropriately, the effect will be of increased damage.

In the subject of Management, the administration of resources, equipment, information and people are indicated. We also miss the ability to implement the investigation process regarding the birth of the crisis, using risk analysis techniques or constructing an event scenario to enable the prediction of the future disaster. The barriers or safeguards will be improved through risk management, in which the analysis of accident scenarios is applied.

Security and social defense agencies must be structured to know how to act, recognize the hazard line, and deal with high stress situations and how to communicate to the team and the community, what the risks while acting in incidents. Once these skills are obtained and the knowledge on how to act in contingencies, it is also important to know how to communicate events through the social leaders, media and organizations involved.

The competence to identify behavioral changes that can increase the crisis and the losses to society is worked on during this master's degree through the elements: Human performance factors; Cultural and social aspects in society and in labor activity; Behavior and leadership for emergencies and urgencies; and Communication (content and means) for the respective scenarios.

The Map of Skills Gap indicates the presentation of concepts and skills (Fig. 5.34) to be worked on in the master's degree.

5.4.7 Context conclusion: research versus society's demand

Brazilian companies and institutions are immersed in the relationship between powers and guilt, an inappropriate situation when joint actions are required when managing deviances and accidents of all proportions. The appropriate leader to manage, design, and apply barriers must know how to listen and, at the same time, have fierce determination. The leader is accompanied by the team that will provide comfort in strategic intelligence, identify the real scenario (data, spatial location and hazard action) of the past (learning), present (reduce impacts) and future (prevention) accident. Leader 5 is humble and recognizes the lack of resources and prioritization, then requires help from the team to define the best solutions. Pride and vanity hinder decisions.

Salvador Ávila's research on human reliability, risk analysis, and crisis management had dropouts at various times, which unfortunately will be demanded and necessary after major events occur. The appropriate and inappropriate characteristics for research projects in the area of industrial and public accidents prevention are mentioned in Table 5.8.

References

ABRACOPEL—Brazilian Association for Awareness of the Hazards of Electricity, 2010. Statistic data about accidents with electricity. Available from: <https://abracopel.org/a-abracopel/> (accessed July 2010).

ANP—Agência Nacional de Petróleo, 2015. Investigation report of the explosion incident that occurred in 11/02/2015 no FPSO Cidade de São Mateus, Rio de Janeiro. Available from: <http://www.anp.gov.br/wwwanp/?dw = 78834> (accessed 27.07.19).

Araújo, R.B., 2018. Evaluation of the performance metrics of the alarm system of the industrial refining units. Course Completion Work (Mechanical Engineering)—Federal University of Bahia, Salvador.

Ávila, S.F., 2004. Methodology to minimize effluents at source from the investigation of operational abnormalities: case of the chemical industry. Dissertation (Professional Master's degree in management and Environmental Technologies in the Productive Process)—Teclim, Federal University of Bahia. 115 p.

Ávila, S.F., 2010. Etiology of operational abnormalities in the industry: modeling for learning. Thesis (Doctor's degree in Chemical and Biochemical Process Technology)—Federal University of Rio de Janeiro, School of Chemistry, Rio de Janeiro. 296 p.

Ávila, S.F., 2011a. Assessment of human elements to avoid accidents and failures in task perform, cognitive and intuitive schemes. In: 7th Global Congress on Process Safety, Chicago.

Ávila, S.F., 2011b. Dependent layer of operation decision analyzes (LODA) to calculate human factor, a simulated case with PLG event. In: Proceedings of 7th Global Congress on Process Safety—GCPS, Chicago.

Ávila, S.F., 2012. Failure analysis in complex processes. In: Proceedings of 19th Brazilian Chemical Engineering Congress—COBEQ. Búzios, Rio de Janeiro.

Ávila, S.F., 2013a. Review of risk analysis and accident on the routine operations in the oil industry. In: 5th CCPS Latin America Conference on Process Safety, Cartagena das Indias, Colombia.

Ávila, S.F., 2013b. Risks in the Platform: Social and Human Guidelines Avoid Accidents and Alter Standards in Oil Production. Revista Proteção.

Ávila, S.F., 2013c. A Discussion to Increase Competitiveness, Human Factors Engineering in the Mineral Industry. Signus, São Paulo, Brasil Mineral. 30 years edition, n 329.

Ávila, S.F., 2015a. Reliability analysis for socio-technical system, case of propene pumping. Journal of Engineering Failure Analysis 56, 177–184.

Avila, S.F., 2015b. Prevention of operational and dynamic risks in the process industry, an exploratory discussion. XX Congresso Brasileiro de Engenharia Química. Blucher Chemical Engineering Proceedings 1 (2), 778–779.

Ávila, S.F., Barroso, M.P., 2012. Preliminary analysis of socio-human risk: an application exercise in a chemical industry. In: Global Congress on Process Safety—GCPS, Houston, Texas—USA.

Ávila, S.F., Bittencourt, E., 2017. Operation risk assessment in the oil and gas industry—dynamic tools to treat human and organizational factors. In: 1st (ed.) Olson, K.F. (Ed.), Petroleum Refining and Oil Well Drilling, Vol. 1. Nova Science Publishers, New York, pp. 1−125.

Ávila, S.F., Costa, C., 2015. Analysis of Cognitive Deficit in Routine Task, as a Strategy to Reduce Accidents and Industrial Increase Production. Safety and Reliability of Complex Engineered Systems, London, pp. 2837−2844.

Ávila, S.F., Dias, C. 2017. Reliability research to design barriers of sociotechnical failure. In: 27th European safety and Reliability Conference on safety and Reliability. Theory and Applications, ESREL, 18−22 June 2017, Portoroz-Islovenia.

Ávila, S.F., Drigo, E.S., 2017. C4t tool to assess human factors. In: Brazilian Association of Risk Analysis, Process Safety and Reliability Congress— ABRISCO, 27−29 November 2017, Rio de Janeiro.

Ávila, S.F., Menezes, M.L., 2015. An influence of local archetypes on the operability & usability of instruments in control rooms. In: Proceedings of European Safety and Reliability Conference−ESREL, 7−10 September 2015, Zurich, Switzerland.

Ávila, S.F., Pessoa, F.L.P., 2015. Proposition of review in EEMUA 201 & ISO Standard 11064 based on cultural aspects in labor team, LNG case. In: 6th International Conference on Applied Human Factors and Ergonomics (AHFE 2015) and the Affiliated Conferences, 26−30 July 2015, Las Vegas, USA.

Ávila, S.F., Santos, L., 1998. Clean routines in the industry. Course Conclusion Paper (Specialization) —Teclim. Federal University of Bahia, Salvador, Brazil, p. 107.

Ávila, S.F., Pessoa, F.P., Andrade J.C., Figueiroa, C., 2008. Worker character/personality classification and human error possibilities in procedures execution. In: Proceedings of CISAP3−3rd International Conference on Safety and Environment in Process Industry, Rome, pp. 279−286.

Ávila, S.F., Mendes, P.C.F., Carvalho, V., Amaral, J., 2012. Analysis of human factors in failure of turbocompressor FMEAH equipment in chemical plant. In: 27th Brazilian Maintenance Congress (CBM), September 10−14 2012, Rio de Janeiro, Brazil.

Ávila, S.F., Souza, C.R.O., Silva, C.C., 2013. Worker behavior by changing dynamic culture. In: 9th Global Congress on Process Safety, San Antonio.

Avila, S.F., Santino, C.N., Santos, A.L.A., 2016. Analysis of cognitive gaps: training program in the sulfuric acid plant. In: Proceedings of ESREL 2016 European Safety & Reliability Conference, Glasgow.

Ávila, S.F., Cerqueira, I., Drigo, E., 2018a. Cognitive, intuitive and educational intervention strategies for behavior change in high-risk activities— SARS. In: International Conference on Applied Human Factors and Ergonomics. Springer, Cham. pp. 367−377.

Ávila, S.F., Cerqueira, I., Ferreira, J.F.M.G., Drigo, E., 2018b. Task & communication in just culture, oil offshore exercise. In: 2018 Spring Meeting and 14th Global Congress on Process Safety, 22−25 April 2018, Orlando, Florida.

Ávila, S.F., Ávila J.S., Drigo, E., Cerqueira, I., Nascimento, J.N., 2018c. Intuitive Schemes, procedures and validations for the decision and action of the emergency leader, real and simulated case analysis. In: 8th Latin American Conference on Process Safety, 11−13 September, 2018, Buenos Aires, Argentina.

Ávila, S.F., Cerqueira, I., Nascimento, J., Ávila, J., 2018d. Application of tool to evaluate the causes of accidents according to the technical partner model of Rasmussen: a case of study of the shipment in Baía de Todos os Santos. International Conference on Applied Human Factors and Ergonomics. Springer, Cham, pp. 367−377, AHFE 2018—July 21−25 2018, Orlando, Florida.

Ávila, S.F., Ahumada, C.B., Cayres, E., Lima, A.P., Malpica, C., Peres, C., et al., 2019a. Management tool for reliability analysis in socio-technical systems—a case study. International Conference on Applied Human Factors and Ergonomics. Springer, Cham, pp. 13−25.

Ávila, S.F., Ahumada, C.B., Pereira, L.M., 2019b. Analysis of the low perception of risk: causes, consequences, and barriers. In: 22nd Annual International Symposium-Mary Kay O'Connor Process Safety Center, College Station, Texas.

Ávila, S.F., Cerqueira, I., Junior, V., Nascimento, J., Peres, C., Ahumada, C.B., et al., 2019c. Integrated management in disaster: a discussion of competences in a real simulation. International Conference on Applied Human Factors and Ergonomics. Springer, Cham, pp. 33−45.

Carvalho, A.C.F., 2016. Diagnosis of gaps in the skills of the team to cope with situations outside of routine. Graduate Program in Industrial Engineering—*UFBA*.

Center for Chemical Process Safety (CCPS), 2007. Guidelines for Human Factors Methods for improving Performance. Center for Chemical Process Safety, American Institute of Chemical Engineers, New York, NY.

Center for Chemical Process Safety (CCPS), 2013. Guidelines for Managing Process Safety Risks during Organizational Change. Center for Chemical Process Safety, American Institute of Chemical Engineers, New York, NY.

Collins, J., 2007. Level 5 leadership. The Jossey-Bass Reader on Educational Leadership 2, 27−50.

Deming, W.E., 1990. Quality: The Management Revolution. Marques Saraiva, Rio de Janeiro.

Drigo, E.S., 2016. Analysis of the Operator's Discourse and His Communication Tool between Shifts as a Tool for the Management System. Federal University of Bahia, Salvador, p. 105, Master's degree dissertation.

Drigo, E.S., Ávila, S.F., 2016. Organizational Communication: Discussion of Pyramid Model Application in Shift Records. In: first ed. Katola, T.B., Salman, N., Andre, T. (Eds.), Advances in Human Factors, Business Management, Training and Education, Vol. 498. Springer International Publish, Springer, pp. 739−750.

Embrey, D., 2000. Preventing Human Error: Developing a Consensus Led Safety Culture Based on Best Practice. Human Reliability Associates Ltd, London, p. 14.

Embrey, D., 2005. Understanding Human Behavior and Error, Vol. 1. Human Reliability Associates, pp. 1−10.

Ferreira, J.F.M.G., 2018. Analysis of the Reliability of the CO Compression System in a Petrochemical Plant Considering the Technical-Operational and Human Factors. Federal University of Bahia, Salvador, p. 128, Master Dissertation from Post-Graduation Program of Industrial Engineering.

Ferreira, J.F.M.G., Ávila, S.F., Fontes, C.H.O., Melo, S.A.B.V., 2016. Failure analysis of an alternative CO_2 compressor using the Ishikawa diagram and the Weibull distribution. In: 6th Brazilian Congress of production Engineering (CONBREPRO), November 30–02 December 2016, Ponta Grossa, Paraná, Brazil.

Fitriani, A., Nawawiwetu, E.D., 2017. The relationship between antecedent and consequence factors with safety behavior in pt. x. Journal of Vocational Health Studies 1 (2), 50–57.

Fragoso, C., Ávila, S.F., Sousa, R., Massolino, C., Pimentel, R., Cerqueira, I., 2017. Blaming culture in the investigation of accidents at work: current discussion of the model and requirements for changing to a collaborative model. International Conference on Applied Human Factors and Ergonomics. Springer, Cham, pp. 318–326.

Huang, J.L., Ryan, A.M., Zabel, K.L., Palmer, A., 2014. Personality and adaptive performance at work: A *meta*-analytic investigation. Journal of Applied Psychology 99 (1), 162.

Lorenzo, D.K., 2001. API770—A Manager's Guide to Reducing Human Errors, Improving Human in the Process Industries. API Publishing Services, Washington.

Marais, K., Saller, H.J., Leveson, N.G., 2006. Archetypes for organizational safety. Safety Science 44 (7), 565–582.

Motta, F.C.P., Caldas, M.P., 1997. Organizational Culture and Brazilian Culture. Atlas, São Paulo.

Nery, S.S., 2009. Performance management. Revista de Ciências Gerenciais 13 (17), 131–140.

Perrow, C., 1984. Normal Accidents: Living with High-Risk Technologies. Basic Books, NY, p. 453.

Santino, C.N., Magalhães, R.S., Ávila, S.F., 2016. Process loss analysis considering the environmental impacts-case study in a metallurgical industry. Technical-Scientific Congress of Engineering and Agronomy. Foz do Iguaçu, Paraná, Brazil, 29 August—1 September 2016.

Sennett, R., 1999. Corrosion of Character. Record, Rio de Janeiro, p. 204.

Souza, M.L., Ávila, S.F., Cerqueira, I., Santino, C.N., Ramos, A., 2018a. Failure analysis cases in operational control and recommendations for task criteria in the man-process interface design. International Conference on Applied Human Factors and Ergonomics. Springer, Cham, pp. 317–329, 21–25 July 2018.

Souza, M.L., Ávila, S.F., Cerqueira, I., Brito, R., 2018b. Discussion about criteria for the management of alarms and cognitive limits for the chemical industry. International Conference on Applied Human Factors and Ergonomics. Springer, Cham, pp. 330–342.

Chapter 6

Human reliability: SPAR-H cases

This chapter presents the cases of application of simplified plant analysis risk-human reliability analysis (SPAR-H) during human reliability research in master's dissertations and publications in national and international congresses, conducted by the operational and dynamic risks group (GRODIN-UFBA).The themes of the chapters of this book with the calculated indicators of human reliability and the relationship with the respective human performance factors (Fig. 6.1) are provided.

We then work on the importance of analyzing the operational context for the calibration of factors when the company's results and the manager's perception do not match with the results of this methodology. The actual cases in which SPAR-H was applied include industry and critical services. A comparative analysis of these cases is performed for learning about the improvements in understanding and application of the SPAR-H Method.

Finally, the theme socio-technical reliability is introduced, which has been already mentioned internationally and in publications in ESREL, JEFA, ICEFA, AHFE, ABRISCO, and TAMU that adds human reliability with reliability of operations, process, and equipment. It also describes the social attractiveness of the result of culture and management profile.

6.1 Introduction

6.1.1 Reviewing the chapters

We start with the topics addressed in the chapters of this book and provide a discussion of the relationship with the SPAR-H Tool. The human performance factors analyzed in the human reliability analysis (HRA) tool are part of different multidisciplinary dimensions that relate the organization, the task, and human in the elaboration of indicators and in the design of safeguards to avoid future accidents. It is not enough to discuss situations that induce human error, but it is also necessary to measure human factors. The big challenge is to understand how cognitive processing happens, and what social influences alter the decision according to global, regional, or local rules of behavior.

The combined social and technical variables indicate how to formulate actions respecting limits of the operational culture that can induce human error. Thus an analysis of context and complexity is made so as not to rely only on the discrete attribution of values in the SPAR-H and to calibrate the method for different situations.

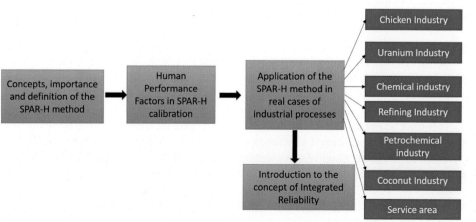

FIGURE 6.1 The gold map in the application of SPAR-H. *Author's original.*

Cognitive models should be related to task-related shop floor rules, and through the level of complexity, it is possible to define the probability of systemic failure. Currently, we propose the tool analysis of the operator's discourse as an option of the 3rd generation in case of unestablished leaders, unstudied social phenomena and rules of non-apparent decisions.

By deepening the discussion of human and social typology, we analyzed the psychology and administration to explain where to intervene, to achieve the best results respecting more subjective issues. Educational programs, changes in social relationships, behavioral analysis, and implications in decision and research are discussed.

The task is the focus of research, regarding its complexity, the relationships in the workplace, possible cognitive failures, and the social phenomena resulting from regional and global culture. Human reliability depends on human error databases, incorrect decisions, and bad habits. This scenario is treated differently when the operating mode is routine or emergency. Therefore, SPAR-H separates the diagnostic phase from the execution phase and connects these two steps (Gertman et al., 2005).

Operational abnormalities resulting from the operator's discourse unite the human and social characteristics to explain the best intervention tools for human behavior, for operational safety and in production programming. In this analysis of tools, the API770 (American Petroleum Institute) is adopted to enumerate interpretations and guidelines (Lorenzo, 2001). In this way, there is a discussion about management guidelines for the control of human, operational and organizational factors.

The discussion of process losses is presented in Chapter 4, Process Loss Assessment, which concludes with the presentation of human errors related to the main causes of low quality, low reliability, and increased consumption of inputs. In this context, the main difficulty is the low visibility of the problems' root cause, which occurs due to dynamic operational risks.

Regarding process losses related to the control of operations, these losses require methods and techniques for identifying root causes. Each technology has its peculiarities, identified from the need to process data and information, usually not available in the description of the technology.

In Chapter 4, Process Loss Assessment, techniques to identify, measure, and treat process losses, and discuss the types of losses, specific characteristics of each industrial segment, skills for analysis of process losses, methods and techniques of data collection, identification of losses and respective diagnoses, application of tools.

In Chapter 5, Learned Lessons: Human Factor Assessment in Task, the lessons learned and routine issues in the chemical process industry are clarified, thus indicating aspects of human reliability regarding the cultural environment, leadership profile, decision model, possible human errors and successes, construction of mental maps, and organizational culture. The routines are related to the operational and safety culture, such as the auditory perception at the plant start-up, the quality of communication in the shift change, in addition to cases involving technology.

In Chapter 5, Learned Lessons: Human Factor Assessment in Task, the authors present cases in the oil industry, fertilizers industry, and other cases from other industrial areas, including suggestions for improving human and technological factors. Additionally, presenting the history of service and research demands involving human errors in high-risk activities, in the industry and in public agencies.

SPAR-H is significantly used in calculating the probability of human error because of the simplicity of applying workplace problems to the team of executors and managers. The application of SPAR-H does not require database, and the method needs expert opinion (Di Nardo et al., 2015). This discussion generates a calculation compared to economic losses in operations and can recommend the calibration of the method. It is important to discuss SPAR-H through management indicators and the best regions and functions for investment in Risk Management (Rasmussen et al., 2015).

We highlight the introduction of indicators that include human reliability of processes, operations and equipment, called sociotechnical reliability (Ávila, 2015a; Ávila et al., 2019). This chapter presents a simplified script for the SPAR-H application and its relationship to process loss assessments and cognitive gaps.

In Chapter 7 (Ch7), Human Reliability: Chemicals and Oil & Gas Cases, the methods and techniques for calculating human reliability, human errors, human factors, process loss auditing and operational culture diagnosis were applied in the LPG industry in oil & gas industrial segment. In this investigation, concepts, techniques and procedures were designed and confirmed to obtain maximum organizational efficiency in production. Management programs and tools indicated for application in the prediction of failure situations, presenting the steps for the implementation of the methodology developed in an industrial plant, as well as the roadmap of activities to investigate the technical culture in the workplace of the oil and gas industry.

The sociotechnical reliability study (Ch6 and Ch7) based on the operational culture is presented from the investigation of the analysis of the operator's discourse in intersection with technological and social aspects of the work environment. This discussion results in the definition of new criteria for the design of human-machine interfaces, methods to measure human reliability, considering the theory of nebulous mechanics and with primary and secondary indicators for adjustment of human factors.

This nondeterministic method (Ch7), that starts with the identification of signals to measure emerging risk energy, can point and evaluate parallel chains of abnormalities to better install safeguards. Transforming large numbers of data into smaller quantities with principal component analysis enables this graphical visualization of the optimal approach of organizational efficiency points in the scatter plot.

In this chapter, paradigms are being broken by new conceptual models, and calculation algorithms are revised. Risk management is punctuated with practical exercises and with the result of discussion of these tool applications. The discussions are presented on the measurement of human factors, Bayesian networks, concept of elements that enable hazardous energy, and calculations for integrated or sociotechnical reliability.

6.1.2 Human errors in the context of critical activities

The current situation of sustainable economy imposes the need to produce with efficiency, quality, and safety. Organizations invest in equipment, quality programs, and infrastructure with the perspective of obtaining products that exceed the expectations of their customers and provide greater visibility of their brand and image in the market. In addition to these product concerns, companies are increasingly working on reducing losses: in the quality of processes; in the employee and process safety; and in the environmental concerns, such as energy and water. Accordingly, they plan, create procedures and set goals to avoid process and task failures that may be due to human error in the practices (Di Nardo et al., 2015). According to Reason (1990), human error happens when planned actions do not occur, and pre-established objectives are not achieved.

From this perspective, it is understood that the studies of human factors are related to the analysis of the individual under several aspects. In the relationship that employees have with the company and its organizational culture, reflecting on how the worker validates and understands the mission, vision and values of the company, affecting its performance and satisfaction. Experienced leaders aligned with organizational culture are able to make assertive decisions in a short time; and a mobilized and committed team is able to establish self-management and achieve the established goals. The organizational climate can lead to the strengthening or weakening of the performance of employees and their bond with the environment and co-workers. The stress caused by an unstable environment generates emotional changes, which cause oscillations in judgment, attention, productivity and decision-making, as well as environments prone to the occurrence of human errors (Ávila, 2010; Souza et al., 2002).

Reason (1990) mentions that human errors can occur from two different perspectives: failure to execute and failure in intent. The first brings adequate planning, however it is not consolidated as expected due to the non-execution of the tasks as planned. The second is the planning that does not conform to the goal to be achieved. These errors can be classified as: lapses, slips, mistakes or violations, or even as latent or active failures, as already seen in Chapter 2, Human Reliability and Cognitive Processing.

In this scenario of changes in priorities in companies, human error should be considered of extreme importance in the management of losses. This event can transform environments and make them complex (Perrow, 1993), with a need for changes in values, change in the concepts, and necessity of paradigm breaking (Pires and Macêdo, 2006). Human errors have a direct connection to the success of the task and on the success of product quality. For Ávila (2013b) there is a need to reduce human errors in the task execution in the industry, because this would reduce loss of time, materials, the problems of quality in product, failure in assets, earnings decrease, and absent employees, as well as problems regarding deadlines compliance and image of the company.

According to several studies in the context of human reliability, human error is considered as the main cause of many accidents in various sectors (Helmreich, 2000; Kariuki and Lowe, 2007; Ren et al., 2008; Wu et al., 2009). For example, over 80% of failures in the chemical and petrochemical industries (Kariuki and Lowe, 2007) were caused by human error. Based on these reports, it is increasingly necessary to develop an effective analysis of human reliability, to reduce the probability of error in the process, increase productivity, and reduce risks.

French et al. (2011) state that human behavior becomes evident in many causes of system failures, and includes it as an important part in risk assessments of plant operations. In this manner, it is important to identify factors that can influence operator performance in the work environment.

Modeling Man's behavior in the work environment is not a simple task. Several methods of analysis have been developed to generate a reliable estimate of human error, and are based on the evaluation of one or more experts, that assign error probability values, using direct and indirect numerical techniques (Boring and Gertman, 2004; Di Nardo et al., 2015).

Amaro (2014) points out the elements that induces human errors as follows: in procedure elaboration and execution; in equipment operation at work; in the operators' lack of knowledge; in the conflict of priorities affecting decision; in the lack of signaling to review procedure; in poor asset management; in the availability of financial and human resources; in the communication processes causing problems; in technology failures; in the operational behavior and respective cognition; and in maintenance without planning.

HRA is performed to identify human errors in operating systems. At present, there are several methods to verify the human error probability (HEP) in different systems, such as Technique for human error-rate prediction (THERP) (Swain and Guttmann, 1983), Accident Sequence Evaluation Program (ASEP) (Swain, 1987; Kirwan, 1997), A Technique for Human Event Analysis (ATHEANA) (Cooper et al., 1996), SPAR-H (Gertman et al., 2005), and new methods that are being developed. In order to assist in human error investigation, Lorenzo (2001) presents the API 770 Guide to help managers in performing diagnoses to avoid failures.

The SPAR-H method is a simplified method used to analyze the level of human reliability in the process. This method was created by the American Nuclear Industry Regulatory Commission (NRC), together with the Idaho National Laboratory (INL), in 1994, to quantify HEP in a given task or task steps, through the multiplication of a basic error probability, and the coefficients applied for each PSF (performance shaping factor) listed in the API 770. The SPAR-H method was developed to succeed the ASP precursor accident sequence program (Gertman et al., 2005; Rasmussen et al., 2015). In addition, SPAR-H is also considered, among HRA techniques, the most practical technique that can be used by professionals in any industry and this method calculates the probability of risks in post-initiator scenarios.

The human reliability analyst, responsible for applying the SPAR-H method, should know how to consider various aspects of diagnosis and planning, as well as the chance of the operator's success in performing actions correctly. The distinction between diagnosis (information processing) and actions (response) is what results in two forms, one for action and one for diagnosis, to fill out with different probabilities of error (Gertman et al., 2005). It is noteworthy that external factors can influence the operator's ability to perform a diagnosis, or action, correctly, for this reason they are taken into account in the SPAR-H method.

For analyzing human reliability in several industrial environments, this chapter presents different cases of investigation using the SPAR-H method. It is noteworthy that the demonstrations of the resolutions of the cases were analyzed in different types of industrial processes: Chicken industry; uranium production industry; propylene industry and fertilizer industry and in the logistic area.

6.2 Concepts and SPAR-H calibration

The calculation of human reliability is defined based on the analysis of eight human performance factors in tasks and activities that are discussed in the API 770 document. Remembering that in the API770 performance factors, physical-psychological stressors and situations that cause human error were discussed. These factors are discussed and classified based on operational, managerial experience and perception, in addition to the perception-experience of technicians who plan the activities of operation and production control.

The factors that compose the human reliability calculation will be identified and discussed in Section 6.2.1 and deal with the operational context:

- Available time,
- Stress/stress-causing agents,
- Task complexity,
- Experience/training,
- Standards and procedures,
- Ergonomics/human-machine relation,
- Aptitude for work, and
- Labor relations

The type of task performed indicates the process state, operational aspects, and human limits to elaborate specific values for each factor. The calculation performed seeks to define the HEP, and from the HEP, calculate the human reliability. This issue is found in the IDAHO National Laboratory Report describing the SPAR-H methodology (Boring & Gertman, 2004; Gertman et al., 2005).

6.2.1 Operational context

The company project includes the definition of the product, process technology, and organizational standards to establish the organizational and safety culture. Project feasibility requires business risk analysis with the projection of future scenarios and with the confirmation of economic and financial results in compliance with the demands from society.

Once this is discussed and formatted, we seek to physically implement it in the company, where the concept of the product and the process becomes reality from the development and assembly of the equipment and the interconnection of processes. In this phase of the investment, feasibility analysis is also required through the projections of operational and economic-financial scenarios. As the future operability is measured, work begins for the transformation of matter, that is, the operation in processes and equipment. Man will make decisions, carry out actions, investigate these actions and plan productions. Thus to make possible the start of operations, the hiring and development phase of the operators-engineers-technicians, professionals, who will form the teams for the beginning of production activities is necessary.

The team chosen to work in this process and product technology must have experience, work and personal motivation, leadership potential, good interpersonal relationships, sincerity in the work environment, ability to develop the work, competence to achieve solution to problems, and to be interested in group activities.

The technology and organization project influences and defines the workplace routine. When the original project was assembled, from the concept to the detailing and assembly, it must be understood that the job will be operated by people. Routine and strategic decisions will be made by people in this work environment. In this technology and organizational project, the main criteria to prepare the job must be defined, which will come into operation in the future, and which barriers must be predicted to avoid production shutdown.

Contrary to the classical way of dividing functions in the operating teams, that weakens team facing new challenges, we indicate the organizational model that allow adaptation to face organizational threats, a model presented as organizational Jam in Fig. 6.2.

In this case, mobility of competencies ensures organizational resilience. The organizational norms, part in informal format, are gradually modified to a new behavior, in a new list of rules that define the operational culture. Thus it seeks to adapt the workplace, its organizational, physical and cognitive requirements, so that the operational culture becomes close to the safety culture and organizational culture.

Analyzing the operational context requires identifying in an apparent way the operation of the team in the routine and then confirming through the recognition of informal rules, the way that, in fact, the team works at the workplace, the way that, in fact, the operational culture is established.

Two different realities are discussed: the continuous process industry design, using the chemical process industry; and the manufacture design with discrete product, as examples, the chicken industry and the sports materials industry, where we discussed the issuance of packing list to release the product trucks. The subjects addressed here will compose the operational context and establish operational safety. Each of them will be discussed to correlate with performance factors in SPAR-H.

Subjects:

1. Work organization
   a. Work time, shift, administrative area, and task force;
   b. Flow of information and routine decisions in the shift and administrative area;
   c. Policy on the formulation and control of procedures;
   d. Responsibilities on activities and decisions, cognitive models adopted;

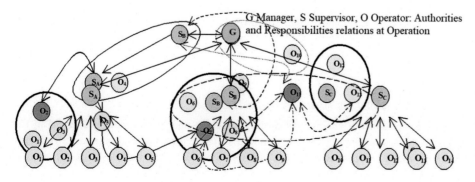

FIGURE 6.2 Organizational Jam. Ávila, S.F., 2010. Etiology of operational abnormalities in the industry: modeling for learning. Thesis (Doctor's degree in Chemical and Biochemical Process Technology)—Federal University of Rio de Janeiro, School of Chemistry, Rio de Janeiro. 296 p.

e. Process chronology analysis and expected results;
   f. Professional functions with cognitive and physical limitations;
2. Knowledge and skills
   a. Knowledge of technology limitations and necessary and available documentation;
   b. Level of established and applied competence, competence, and commitment;
   c. Level of perception of signals that increase risks until the failure and accident occur;
3. Behavior
   a. Expected team behavior in routine operation mode, operational problems, and emergency situations;
   b. Quality in social and labor communications, standards, and feedback;
   c. Responsibilities over activities and decisions, level of cultural guilt, and consequences;
4. Leadership and team
   a. Sense of justice in shift and administrative areas;
   b. Manager's discourse in the establishment of strategic, managerial, and operational guidelines;
   c. Market rules that influence in the logistics and production—purchasing, sales, and maintenance;
   d. Level of physical and cognitive stress established in the team;
   e. Guidelines for organizational integration and social inclusion;
   f. Treatment of social conflicts in the workplace.

6.2.1.1 Operational context in the manufacturing or intermittent process industry

The organization of work in the intermittent manufacturing process has specific features to be assessed. For example, the process of production of plastic bottles for filling with mineral water have these human resource aspects:

- Two work systems, administrative time and shift;
- The administrative team works 44 hours during the week, including every Saturday morning;
- Three fixed work shifts, with a leader for each one, that operate from Monday (8 a.m.) to Saturday (4 p.m.);
- A group of operators work in the administrative area for replacement of vacation and other operational activities in task force;
- Sunday is a break where important maintenance works are performed and once a month there is a change in shift time.

In the manufacturing industry, the flow of information is simplified, where the main objective is the flow of materials for physical transformation and respective functional features. Some activities are performed every month (weekly and daily): (1) monthly demand survey; (2) weekly production schedule; (3) daily production order; (4) report of occurrences in the administrative hours; and, (5) report of occurrence regarding the critical process variables. This information enables the analysis of history and authorizes routine decisions.

The simplicity of manufacturing processes requires less complexity in decisions, so fewer people with specialized skills maintain process stability. As most abnormalities involve noise, vibration and defects, the perception of signals is clear in the process stages. The operation shift team and administrative team have simple functions, requiring less specialized staff where the risks are lower and the decisions to avoid accidents are more visible. Barriers are designed and easily applied when compared to continuous processes.

The working group is more intimate, due to less hierarchical levels, which at first improves communications. This statement can be reversed when manufactures employ people with affective relationships in communities near the factory.

The sense of justice established in the routine is independent of technology. The Manager of a large industry or small manufacturing industry can have appropriate or inappropriate solutions in the same way. The risks of events and their consequences are higher for the continuous process industry. Problem investigation can be more complex in this industry as well. The lack of record of shift occurrences or excessive automation can cause information flow failures, low risk perception and decisions.

6.2.1.2 Operational context in the continuous process industry

In the case of the continuous process, we mention the chemical industry, which is one of the cases presented in this chapter. The monitoring of production processes is more critical and requires higher levels of competencies to control these technological complexities, task and social relations in the work environment due to the high level of automation, production rate, risky events involving the toxicity and flammability of inputs and products.

The work organization involves five 8-hour shifts in addition to the administrative regime. Owing to distance issues and access to the production area, the shifts can be twelve hours in a period of 14 days, as in the case of offshore oil production.

The control of procedures is formalized through the ISO system for quality and, due to the complexity and high risks of the processes, requires the choice of critical procedures for review regarding practices and its consensus. The implementation of these procedures requires information from different sources for decisions that, if incorrect, can cause high-impact accidents such as Bhopal in 1984.

Despite the need to delegate micro-decisions and critical actions, it is necessary to evaluate the responsibility for the acts for the consequence analysis investigation in accidents. Thus it is important in the chemical and nuclear industries to consider what decisions, who should make them, and what is the appropriate model to ensure their quality. Knowledge of the timeline, function, risk, complexity of each critical procedure or task, allows anticipating barriers and changing the way that process state scenarios are perceived to classify modes of operation. The time of performance of each activity should be referenced in the best practices of this activity. The analysis of the failure and the accident in performing the activity should take into account the time of each event, where it is located in the action of the procedure and the amount of hazard energy released.

The greater the knowledge about the technology and the greater the skill to perform tasks in complex systems allows identifying where memory reinforcement is needed, new alarms or forms of management and other barriers in the interface between the operator and the procedure or between the operator and the process-equipment. Knowledge affects most performance factors affecting work organization and cognitive stress level.

Attention and perception of the level of normality requires comfort regarding the sense of justice and decisions in the organization at all levels. Behaviors of high attention, valuing the principles of safety and warning for last minute changes are expected, which characteristics lead to a Safe, Alert and resilient behavior (Ávila et al., 2018). Organizational comfort comes from greater commitment of the operator who feels comfortable in the composition of the work group. Thus due to the high risks and the intensity of circulation of the danger, high quality is required in the social and work communication tools.

In the continuous process industry, greater agility is required in corrective and prevention actions than in the manufacturing industry. The team should feel comfortable in giving feedback to improve operational routines. If this does not happen, hazard energy will easily flow through barrier faults, also in projects with safety redundancy.

If the manager's discourse is carried out in a vicious manner, it can build a situation of difficult commitment to safety making it complicated to implement a relationship of cooperation necessary for a shift regime. Micro-decisions are built from strong strategic, managerial, and operational guidelines.

The logistics in the production system can complicate the design and control of human performance factors. The non-inclusion of sensitive competencies towards levels of stress and its consequences can cause biases to cognitive decisions without considering cultural influence in decisions. The organizational comfort that permeates the entire organization of work for the operational routine depends on good guidelines for organizational integration and social inclusion.

6.2.1.3 Culture and operational safety

The standards established in the organizational project are designed based on the company history in its headquarters or other subsidiaries, the wishes of shareholders and owners, and also based on the external references of organizational consulting. These standards represent the desire of the owners respecting the way the future company operates in the region and country where it will be installed. Also, in relation to the standards, these must be compatible with the product technology design, process, and safety control.

The organizational standards resulting from the project represent the desire of the board of directors, which are not necessarily the same standards performed in practice during the operational routine. In other words, the guidelines resulting from organizational and safety culture may not be carried out on the shop floor, thus bringing a different operational culture.

There is a difference between the investigation of the SPAR-H procedure and the procedure-method-tool that involves routine task analysis in the identification of operational abnormalities, where in the former the study is made based on the manager's discourse (technical staff), and in the latter, on the operator's discourse analysis. In this analysis, the investigation into systemic failure may indicate the rules accepted by the operation group in relation to the formal rules demanded in the organizational standards.

The job as it really is should have the perception record of a group of operators who do not feel included in a culture of guilt and that indicates what the performance factors in the operation routine are. This result should be compared to interviews with managers and engineers. The application of the chemical industry and publication in the

CONEM—National Congress of Mechanical Engineering (Souza et al., 2018) confirm that the engineer of Process Safety was frightened by operator perception in the application of the SPAR-H questionnaires (NUREG 6883, 2005).

The rules formally established by the organization culture project and process safety project may not reflect the informal (heuristic) rules installed in practice.

The risk of technology and manufacturing units Operation may change over time due to social conflicts that alter human typologies, due to the wear of equipment materials, and due to climate changes that modify operational constraints.

6.2.2 Calculation of human error probability

The HEP calculation is used to identify probability of operator/team failure in the analyzed task. From this calculation, the human reliability of such operator/team can be calculated.

To perform the calculation of the HEP and the calculation of human reliability (R_{hu}), it is necessary to understand the parts that make up these calculations. The equations presented below are the results from our entire study until this topic, involving differentiations between action tasks and diagnostic tasks.

The calculation of the HEP is done by multiplying the nominal probability of failure by the coefficients found in the analysis. If, in the activity, the operator needs to perform actions (action task) together with cognition activities (diagnostic task), the probabilities found using the coefficients for each of the cases should be summed.

Human reliability was calculated from the methodology described by SPAR-H, which proposes equations for calculating human failure probability for diagnosis (HEP Diag.) or action (HEP action), which can be calculated using Eqs. (6.1) and (6.2):

$$HEP_{Diag.} = NHEP_{Diag.} \times \prod PSF \qquad (6.1)$$

$$HEP_{Action} = NHEP_{Action} \times \prod PSF \qquad (6.2)$$

The direct multiplication of the PSF coefficients, for the calculation of the final HEP, is only approximately correct. The fact of using a scalar (PSF coefficient) to multiply a probability, depending on the number of negative PSF, may generate a HEP greater than 1 (Boring & Gertman, 2004). Eqs. (6.3) and (6.4) represent the adjustment made for the calculation of HEP when three or more PSF negatively influence operator performance.

$$PSF_{composed} = \sum PSFr \qquad (6.3)$$

$$HEP = \frac{NHEP \cdot PSF_{composed}}{NHEP \cdot (PSF_{composed} - 1) + 1} \qquad (6.4)$$

Being,

$NHEP$ = Nominal human error probability

$PSF_{composed}$ = multiplication of all the coefficients established by the HRA analyst.

Once the errors are calculated for the assessed aspects of the activity (action and diagnosis), Eq. (6.5) is used to assess the activity error. Equations are used to determine the diagnosed human error probabilities = HEP_{Diag} and related to the action = HEP_{Action}. The total human error probability (HEP_{Total}) is defined as the sum of the two factors ($HEP_{Diag} + HEP_{Action}$), according to Eq. (6.5):

$$HEP_{Total} = HEP_{Diag.} + HEP_{action} \qquad (6.5)$$

And finally, to assess the probability of the team not committing failures, human reliability ($R_{hu.}$), the Eq. (6.6) is indicated:

$$R_{hu} = 1 - HEP \qquad (6.6)$$

6.2.3 SPAR-H calibration

In HRA, operator tasks are classified into action and diagnostic tasks. The action tasks, for example, are to operate an equipment, perform alignments, calibrate instruments and others. An example of the diagnostic task is to assess operating conditions to take decisions, including availability of people, equipment and processes, planning of production and prioritizing activities.

Research indicates that tasks that depend on cognitive abilities commonly present a failure rate of 0.01, that is, the operator has a 1 out of 100 chance of performing a misdiagnosis. For action activities, the chance of failure is 0.001 or 1 out of 1000 (Boring & Gertman, 2004). These values are also called nominal human error probability (NHEP). The nominal value depends on the type of technology involved, the operational discipline and commitment of the operation team. Thus there are questions about this nominal value. Therefore, it may be necessary to calibrate the SPAR-H with modifications in this nominal value and in the values regarding the criteria of human factors, that is, validation along with process management, essential to validate the existing NHEP or even to adjust them. To fit the calculation of SPAR-H, the values adopted in NHEP should be evaluated, due to the direct influence on the HEP and R_{hu} results. After this definition, the HRA specialist has the responsibility to fill out a form or worksheet with the multipliers of each PSF, according to their observation in the field, and according to interview answers with the operators, and the checking of the plant state in operation.

For the cases that will be presented, the following criteria will be adopted for the application of NHEP calculations:

- NHEP in the Diagnosis were equal to 0.01, that is, assumes a value of 10^{-2} for diagnostic tasks.
- NHEP in the Action = 0.001, that is, assumes a value of 10^{-3} for action tasks.

However, it is also important to emphasize that the SPAR-H method was developed for environment of the nuclear industry (Gertman et al., 2005). In this sense, in cases where this technique is applied outside the nuclear industrial environment and the results are not compatible with the manager's perception, there may be a need to calibrate the NHEP values of the PSFs. Thus the operation management must be aligned with the response of the operators that indicate the calculation of human reliability in the operational routine, with the success and failures of equipment, people and processes.

The criteria of the nuclear industry to perform the calculations of HEP indicated by the SPAR-H procedure are applied to the chemical process industry considered as complex processes, or even applied to the linear manufacturing industry. Although the results of the application of SPAR-H are satisfactory for the level of human error in these types of industry, they may not be the truth expected by the Managers or Engineers of the facilities, which then may be true or not. Thus it is important to test with new calibrations of the criteria in the calculation of human reliability. These new calibrations can be performed by changing the original NHEP values of the nuclear industry (0.01 and 0.001) as well as to the values considered in the Factors for formatting human performance criteria (PSFr).

6.2.4 Discussion of performance shaping factors and calibration

This topic presents the PSF that make up the SPAR-H analysis method. PSFs deal with human, environmental, organizational or task characteristics that can improve or worsen the operator's performance and, consequently, decrease or increase the HEP in a given task (Boring & Blackman, 2007). The probability of failure increases with the negative influence of PSFs and decreases when the influence of PSF is positive. To understand the analysis of the PSF's in the SPAR-H method it is necessary to know how the task is assessed in practice.

In 1994 the accident sequence precursor HRA methodology divided the tasks developed into two components, the first called the processing component and the second, the response. The comments received by those who tried to implement the method indicated that this name "processing and response" were well understood by the professionals of HRA and Human Factors, who worked with the method. However, training the operators and inspectors who were collaborating in the implementation of the method became difficult. In 1999 these components were renamed in the SPAR-H method with the name of diagnosis and action. By using this assessment in SPAR-H, the team of analysts make decisions regarding the definition of a particular activity: pre- or post-initiator, for any diagnostic or action task (NUREG 6883, 2005).

Guidance for diagnostic tasks has to do with identifying the most likely causes of an abnormal event by checking systems or components whose status can be changed to reduce or eliminate the problem. The analysis of the diagnostic task includes interpretation and, when necessary, decision-making. Diagnostic tasks usually require knowledge and experience to understand existing conditions, to plan and prioritize activities, and, therefore, to determine the course of the appropriate actions possible.

To better understand, the question is: Does this task contain a significant number of diagnostic activities? To respond, it must be considered whether the operator or team has to expend mental energy to observe and interpret the information that is present, or hidden, determine what is happening, think about possible causes and decide what to do. The greater the number of observations, interpretations, thoughts and decisions that the operator or team performs, it will result in a significant number of diagnostic activities (NUREG 6883, 2005). The orientation to the action has to do with the accomplishment of one or more activities, indicated by the diagnosis, rules of operation, or by written procedures. Examples of action tasks include operational equipment, pump commissioning, calibration testing and others.

When performing HRA, it is sometimes practical and reasonable to model the tasks or sub-tasks of an event at the basic level that is, based on the diagnostic and action aspects (NUREG 6883, 2005). The diagnostic and action tasks are quantified as part of the HEP determination process in the SPAR-H. For this calculation of the HEP, among the more than fifty PSFs cited for the various techniques of HRA, only eight of them are used: (1) Available Time, (2) Stress/Stress-Causing Agents, (3) Task Complexity, (4) Experience/Training, (5) Patterns and Procedures, (6) Ergonomics/Human-Machine Relation, (7) Aptitude for work and (8) Labor Relations. Following the report of the Idaho National Laboratory describing the SPAR-H methodology (Gertman et al., 2005), the PSFs and its coefficients will be discussed individually, and verified for practice in the forms indicated in this chapter.

1. Available time

    The "available time" factor expresses the idea of the time that the operator has to perform a diagnosis or action due to an abnormal event in the task (Boring & Gertman, 2004). Performing tasks in a hasty manner can increase operator stress, which can affect cognitive and motor skills. This combination of pressure in activities and also the high stress generated increases the probability of the action being performed incorrectly. Reducing the available time also increases the Task Complexity, since the actions that must be taken by the operator, such as opening a valve and checking the level of a tank, should be taken at a shorter interval than the nominal one.

    Other factors, such as experience/training, ergonomics, quality of standards, and methods of operation, influence the operator reaction time. Experience makes the worker to need less time to perform actions correctly. A confusing or little-accessed layout can decrease reaction time, as well as poorly designed or insufficient procedures.

    In their report, Gertman et al. (2005) suggest a minimum time of 30 minutes for the decrease in failure rate when performing a diagnostic activity. This value, however, was determined for the application of the SPAR-H method in the analysis of nuclear plants. Therefore, it is always necessary to verify the need for an adaptation or recalibration of the NHEP, according to the expert's evaluation after the HEP calculation result.

    For the calculation of HEP, the multipliers, for action or diagnostic activities, are defined accordingly:

    - *Diagnosis*

        Time is considered inappropriate if the operator or executor can't perform the whole critical task, in which, considering the possibility of skipping steps, the failure is considered certain with a probability of failure equal to 1. When the time is considered slightly below the nominal or the standard, for example, if the availability is two-thirds of the nominal execution time of the task, the PSFr applied is 10.

        The time is nominal when it is sufficient to diagnose the problem, PSFr is equal to 1. When there is extra time to perform the task, that is, the available time is about one to two times the nominal time required then the PSFr applied in the equation is 0.1. Here we consider a lot of extra time to perform the task when the available time is greater than twice the nominal time, that is, the PSFr is 0.01. When the information is insufficient, the SPAR-H indicates that the PSFr is equal to 1, or if it is necessary to adjust to the reality of the industrial routine.

    - *Action*

        The time for action is considered inadequate when the operator can't perform the whole task in the available time, the failure is a certainty, the probability of failure is equal to 1. When the available time is equal to that required then the time for execution is the minimum required and the PSFr is equal to 10.

        In the execution of the task, time is considered nominal when a little more time occurs than the minimum time required, therefore, the PSFr is equal to 1. And considering that the PSFr is equal to 0.1 when there is extra time, that is, the available time is equal to or greater than five times the nominal time required.

        The SPAR-H Method considers a lot of extra time when the available time is equal to or greater than fifty times the nominal time, then PSFr is considered equal to 0.01. When the information is insufficient the PSFr is equal to 1.

    - *Calibration of available time*

        The PSFs are related to each other when they have high complexity in the task and the level of risk of the product or activity promotes a state of stress that induces human error, or a slip, with rework that can hinder reaching the objective, in the expected time. The production team needs more time to complete the critical task. This phenomenon may have different probabilities according to the type of product and the automation of processes, in which occurrences have different characteristics when comparing the nuclear, mining, petrochemical and food industry.

        The way to manage also causes differences in the available time criterion. When working with teams and individuals who have different reactions and behaviors at the time of demand for the task, it is observed that flexibility in the steps can increase success. When tasks are automated at a high level, and depend more on decisions than actions, the fixing of steps with specific times and respective requirements can be a necessity, as is the case of the catalytic cracking process in the refining industry. In these cases, it is necessary to revise the values of multiplication of the available time, both for the action and for the diagnosis.

2. Stress/stress-causing agents

   Stress can be defined as "a reaction of the organism, with physical or psychological components, caused by psychological changes that occur when a person is confronted with a situation that irritates, frightens or confuses, or even makes him/her happy."

   Upon receiving a stimulus, the subconscious activates emotions such as fear, anger, and guilt, among others, then triggering a reaction in the nervous and glandular system, with consequences, especially at the physical level (Murphy, 1984). The exposure to high levels of stress (Holman et al., 2018) can trigger a number of complications such as increased blood pressure, palpitations, myocardial infarction, chronic bronchial dilation, hand tremor, and increased predisposition to diseases caused by viruses, fungi, bacteria, and microorganisms.

   In the analysis made through the SPAR-H, the level of situations and circumstances that can lead to decreased operator performance is observed, considering stress-causing agents: the excessive contingent of work, noise and excessive heat, in addition to poor ventilation and other events that may cause apprehension, anxiety or fear. It is important to emphasize that stress does not always act negatively. At moderate levels it can cause worker productivity to become high, being considered "nominal stress," although higher levels impair performance in the execution of the task. The Yerkes and Dodson (1908) indicates the relationship between stimulus levels and performance, graphically represented by Fig. 6.3.

   Yerkes and Dodson (1908) report that once considered the same research subject, tasks considered "simpler" may present a higher level of performance, while tasks considered "more complex" may present a lower level of performance. Yerkes and Dodson concluded that performance reaches its highest levels when the stimulus presents moderate levels, that is, when the stimulus level becomes too high, the level of performance tends to decrease appreciably, and very low or very high stress stimuli tend to impair the level of performance, emphasizing the need for balance and moderation in the task.

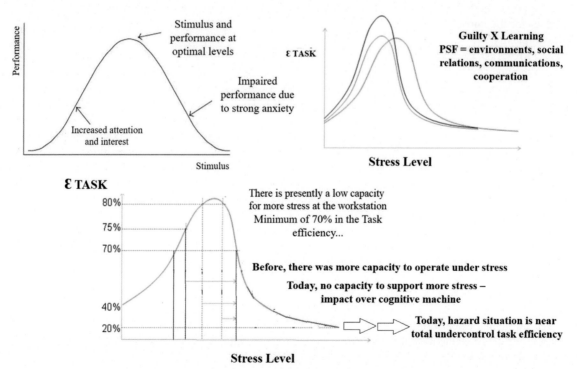

FIGURE 6.3 Yerkes-Dodson's law and interpretation. *Based on Yerkes, R.M., Dodson, J.D., 1908. The relation of strength of stimulus to rapidity of habit-formation. Journal of Comparative Neurology and Psychology, 18 (5), 459–482 by Ávila S.F., 2013c. LESHA—multi-Layer progressive stress & impact assessment on health & behavior. Global Congress on Process Safety, San Antonio.*

When classifying the level of stress that the operator is subjected to, the HRA analyst should perform an interpretation based on operational knowledge and the influence of stress on human behavior.

- *Diagnosis and action*

    Extreme stress (Souza et al., 2002) deteriorates operator performance and depending on regional characteristics, type of task and some specific characteristics, can have different consequences in performing the procedure. It usually occurs when stress-causing events occur unexpectedly and remain for a large period of time. It is also associated with the threat to physical integrity, self-esteem or position; for these cases, the PSFr is 5.

    The stress level is considered high or above the nominal, when, for example, several alarms trigger at the same time, decreasing the cognitive abilities of the operator, or when the task to be performed involves serious risk to the plant operation. In such cases or in similar cases, PSFr is 2.

    The SPAR-H Procedure classifies stress as optimal or nominal level when the incentive for work keeps the operator motivated and the stressing agents are controlled. This causes an increase in the productivity of this or these operators. In this case the PSFr is equal to 1. Finally, when the information is insufficient, the PSFr = 1 is also considered.

- *Stress calibration*

    Stressing agents generate an environment that overloads cognitive processing or contributes to physical fatigue. Fatigue and mental overload feed the environment with more stress, thus bringing a growth that needs to be stopped. The big concern is that teams and people don't value leisure and turn themselves into bombs ready to blow up. This lack of control, according to Fig. 6.3, impacts on: health, decisions, diagnostic investigations and actions.

    The great difficulty in calibrating the multiplier values of this "Stress" PSF, is the prediction of the behavior of the group or individuals who are led in a specific managerial style during a certain emergency situation for a specific technology. The subjectivities of the behavior should be analyzed before assigning discrete values to the PSF "Stress."

    The individual characteristics and the respective consequences of high stress for the body and mind may be different due to: phase of the enterprise cycle, emotional stability of critical functions at work, availability of memory aid and cognitive processing, and quality of simulated situations for high-risk conditions and possibility of accident.

    Some cultural characteristics related to guilt and communication can also change the criteria of multiplier values for the impact of stress on the HEP.

3. Task complexity

    The PSF, "Task Complexity," refers to how difficult it is to perform the procedure, considering the environment in which the operator is. The fact that the operator needs to perform complicated calculations, memorize long procedures, know how complex control equipment works, or have to pay attention to several indicators at the same time, make the Task Complexity higher (Ávila, 2012; 2013a,b,c). The consequence of this increase in complexity is the increased HEP (Rasmussen et al., 2015).

    High physical exertion with multi-step sequences and compliance analysis can greatly increase stress, thus necessitating reduced task complexity through cognitive, technological, and physical barriers. In the SPAR-H Procedure, the individual specific skills of plant operators are not evaluated, such as a person's natural ability to perform calculations or memorize a large amounts of information with ease. The possibility that the task has a high level of complexity for all operators can cause a deviance between the actual probability and the calculated probability of the execution failure rate.

    According to Boring and Gertman (2004) the main factors that contribute to the increased task complexity are:
    - High degree of memorization required;
    - High number of actions;
    - Low fault tolerance;
    - Equipment unavailable;
    - Simultaneous activities;
    - Need for mental calculations;
    - Need for a lot of communication, task requires coordination with activities in control room.

    The level of complexity for SPAR-H is divided as follows:
    - *Diagnosis*

        The very difficult task to be performed due to the lack of clarity about failure events and their causes and consequence can make it difficult to diagnose to design the appropriate safeguards, named as a highly complex task. Usually, for this type, many variables are involved and activities or actions are carried out simultaneously, as Ávila indicates when discussing the parallelism of the task (Ávila, 2015a; Ávila et al., 2015c). In these cases, the PSFr is considered equal to 5. The procedure or task is moderately complex when the number of variables is high, but lower than the previous case. There are difficulties in making diagnoses. In addition, there may be ambiguity in what is being evaluated. SPAR-H adopts PSFr equal to 2.

The Task with Nominal Complexity is not difficult to perform. Few control and check variables are involved, as well as little ambiguity about what should be done, avoiding double interpretation. The normal level of complexity has PSFr equal to 1.

On the other hand, when the diagnosis is obvious, bringing the pitfalls of not consulting procedure or still relying on the experience and not asking, human errors and failures in the accomplishment of the task can occur. The diagnosis becomes quite simplified and the problem becomes so obvious that it would be difficult for an operator to diagnose it. The most common and usual reason for this is that validation and/or convergent information is available to the operator. This information may include auto-actuation indicators or additional sensory information, such as odors, sounds, or vibrations. In this case the PSFr is equal to 0.1. The PSFr is equal to 1. when the information is insufficient.
- *Action*

    A very complex task is very difficult to perform, as there is no clarity about what should be diagnosed, in case of failure, or performed in an action. Many variables are involved with activities that should be performed simultaneously, for these cases, the PSFr is equal to 5. For the case of moderately complex activities, the number of control variables is smaller than the very complex one. There may be ambiguity resulting from complexity, and PSFr is equal to 2.

    Complexity at the nominal level is not difficult to perform. There are few variables involved and there is not much ambiguity about what should be done, thus considered a normal level of complexity, so the PSFr is equal to 1. In the case of insufficient information, the PSFr is equal to 1.
- *Complexity calibration*

    In the case of complexity, the ability to analyze, decide and act with the complexity of the situation, the existence of auxiliary systems, the level of risk for accidents, the number of requirements in the task and the respective number of steps must be compared. It is also important to analyze how to operate it, and also, what is the kind of product technology and process safety. In this way, we can reduce the HEP if we apply HRA in manufacturing. In more complex cases than nuclear power plant, such as high-risk surgery, there is a need to review the multiplier values of the Task Complexity and then review workstation.

4. Training and experience

    The PSF, of the type "Training and Experience," refers to the training time or experience of the operator involved in a task. Although the amount of training or experience indicates an increase in competence and lower probability of occurrence of human error, unfortunately, the acquired competence is not always applied. Thus it is necessary to raise the number of hours and the level of commitment.

    The Organization that has a stable psychological contract with the worker, begins to consider the level of commitment guaranteed. Thus the higher the experience, the shorter it takes to act correctly. The operator's memory is activated through signals and will be transformed into action in minimum time, after reasoning and the necessary actions.

    An experienced operator exposed to a greater number of adverse situations reacts in a calmer and more correct way compared to an operator who has never experienced the same situation.

    Insufficient knowledge about the process involved in the task increases the likelihood of accidents. Operating with an emery board (sander) on a flammable product line during racket insulation activities requires a certain skill and knowledge of the risks in practice. Recycling training is intended to reduce the likelihood of accidents in routine or service release, work permit.

    The effective training time is calculated based on the total time in the training course, in addition to the recognition of the application of these techniques and the commitment to perform them in the appropriate time in the work activity. The choice of the multiplier value for the PSF in the HRA is related to the hours of training, ability in function and indicator of impairment.

    For the calculation of the probability of failure, the classification of the degree of training is done as follows:
    - *Diagnosis/action*

        Less than 6 months of routine training and experience in the activity indicates low PSF level. This level of experience does not provide the knowledge to perform the task, does not allow sufficient practice in the task or does not expose the operator to sufficient abnormal situations for it to act correctly. In the case of diagnosis, the PSFr is equal to 10 and in the case of action the PSFr is equal to 3.

It is considered that the PSF training and experience is nominal when the time is above 6 months. This level allows the operator to acquire sufficient technical and practical knowledge for the execution of day-to-day tasks, for this case the PSFr is equal to 1.

The PSF is considered high when the operator has a high level of knowledge about the process related to the task to be performed. The operator has extensive practical experience because he/she has already been exposed to a large number of abnormal situations, in this case, the PSFr is 0.5. For insufficient information, PSFr equals 1 is considered.

- *Training calibration*

    The calibration to be done in the case of training is related to the level of commitment and the need for specific training of skill and knowledge. The type of training is directly related to the Task Complexity and the risk of accidents occurrence with high impact. Therefore, an efficient and robust safety culture with appropriate decisions for threatening environments in a resilient organization will bring adequate training to the operational routine.

    The leadership that has the ability to replicate knowledge in the routine will help at the level of training. The reverse is true and can cause incidents and accidents. In a nuclear power plant, with core fission reactions and redundancies in temperature control, it requires a much greater commitment than in the food industry. In case of accidents, the impacts can be disastrous. Therefore, there would be a reduction in multiplier values in the food industry.

5. Standard/procedures

    The PSF, "Working Pattern and Procedures," indicates the level of standardization of critical activities. Tasks should be planned following the Good Practice guidelines presented by David Embrey (2000) in the CARMAN Tool. The commitment to the accomplishment of the procedure steps in high-risk processes indicates the need for operators to be critical and the need to review the task.

    The Success of the Task in industrial and service activities depend on the standardization and respective application of the procedures. The application of these standards results in positive indicators in the area of safety and production, certainly the reverse causes human errors and eventual accidents.

    Procedures are memory activation codes for performing working practices. This is the most effective way to perform an activity, seeking to increase productivity and reduce risks (Embrey, 2000). The procedures must include the steps of performing the task with necessary cognitive resources and aids, safety guide documents for execution, limits of process variables and list of those responsible for the execution and supervision of the activity. According to Embrey (2000), the procedures describe what would be the "best practice," but they are not always the practices preferred by the operator. This is mainly due to two factors:

    1. The procedure is usually written by specialists who have no practical experience in performing the task, which may cause him not to be attentive to factors that make it impossible to perform the task satisfactorily.
    2. There is no system that ensures that procedures are changed according to changes in working practices, constant updating.

    In his work, Embrey (2000) conducted a survey with 400 operators on the reasons for not using the standard procedures developed for the activities. For tasks where safety or quality are considered critical, the use of procedures is 75% and 80% respectively. For problems involving diagnosis, only 30% of operators use written procedures and, for routine tasks, only 10% use them. Among the justifications given in the research, four stood out:
    - If the procedure was followed to the letter, the work would not be performed in time.
    - People are not aware that there is a procedure for the task.
    - Operators prefer to rely on their experience and skills.
    - People say they know what is in the procedure.

    It is concluded that most activities are performed without the use of patterns. In addition, the use of personal notes on the execution of the activity was observed, which leads to disparities in the accomplishment of the task from one professional to another and that, sometimes, the notes are very different from the written procedure.

    People will not follow procedures if they understand that they are not practical or that they have sufficient experience to perform the work. In order to develop a culture in which best practices are also preferred by operators, it is necessary to re-evaluate the procedures so that they are appropriate to the conditions of operation, to present up-to-date and clear information to realize of the work.

SPAR-H indicates for the PSF "Standard and Procedures" different multipliers for diagnostic or action activities:
- *Diagnosis*

    The non-existent procedure does not actually exist or is not available for a given critical task and by the SPAR-H Tool has PSFr equal to 50. The incomplete procedure, on the other hand, does not have all the information necessary to perform the activity, with PSFr equal to 20.

    The procedure may be available but has been poorly designed due to formatting or content problems. The consequent ambiguity and lack of consistency can decrease the performance of the task performer, in which case the PSFr is 5. The procedure or standard that achieves its objective when it effectively assists the operator in performing the activity, classified as Nominal, has PSFr equal to 1.

    The specific procedures for performing diagnosis are based on symptoms characteristic of a given abnormal situation, in this case, the PSFr is equal to 0.5. When there is insufficient information, the PSFr is considered equal to 1.
- *Action*

    The procedure that does not exist or is not available for the execution of a given task or activity has PSFr equal to 50. The Incomplete Procedure may not have all the information necessary for the execution of the task or activity, in which case the PSFr is equal to 20. The procedure or standard available, but poorly designed due to formatting or content issues, can cause ambiguity or inconsistency by decreasing operator performance during the task routine. In this case, the PSFr is equal to 5.

    The nominal procedure was clearly elaborated, without double interpretation, and will bring the success of the task as expected. In this case the PSFr is equal to 1. This value is also the same when there is not enough information about the procedure.
- *Calibration of the standard or procedure*

    The type of process, product and safety technology indicates the need for standard and procedure to maintain the quality and expected quantity of production. On the other hand, when the production process is bureaucratic, human errors occur. Hence the question, what is the appropriate level of standardization for production processes?

    This question also depends on organizational culture and organizational resilience. If the company's culture allows leaders to adapt to new situations and protect against environmental threats, procedures are allowed to be revised without extreme centralization, with some flexibility. The calibration of multiplier values for standard and procedure in the application of SPAR-H therefore depends on culture and technology.

    The human and social typology of the execution, planning and diagnosis teams may also indicate the need to correct the values attributed to the PSF procedure. Thus the numbers available according to the original SPAR-H certainly need to be modified to reflect the reality of technology, the organization and the people who work or transit in it.

6. Ergonomics and human-machine interaction

    Physical ergonomics, anthropometry, is the study of worker comfort at workstation. The project of industrial facilities need to adapt the system to the action of Man, and Man has to try a better form of work, as best as possible, to avoid diseases and discomfort.

    The "Ergonomics/HMI" Factor is related to the quality and quantity of information available on instrumentation and control systems of equipment and processes. This information and knowledge makes it possible to perform complex tasks. Some of this information is aided by the appropriate layout of the control room and control panel.

    Process and equipment control actions are reliable when dashboard operators perceive the signals emitted by the system clearly and accurately. Reading instruments in non-ergonomic positions (poorly located-positioned) may cause operator reading to be performed incorrectly. This failure occurs because of the operator's improper reading position, which results in a parallax error.

    A Human-Machine system is an operative combination between Man and machine that complement each other to perform a certain function, starting from input stimuli subject to the conditions of a given environment. Thus in the operation of a machine, Man receives information (input stimuli), processes and transforms them into command actions.

    In this factor the compatibility of the instruments offered with the activity to be performed is considered. Therefore, after processing the received signal, the operator must perform actions through the operation of the machine. It may be that these alarms or indications are not available to the operator. In the case of physical layout of the poorly elaborated plant, with pipes in the way, few or tight accesses, lack of identification of equipment and instruments, there is an increase the time of action of the operator, with less time to perform other actions, thus increasing the probability of failure in the process.

The contribution to Ergonomics/HMI for the calculation of HEP is:

- *Diagnosis/action*

    Ergonomics or Instrument for Control may not exist or indicate process states that do not match the reality. The instrumentation available to control processes and equipment may not help in diagnosis, or instruments may not be accurate, causing the diagnosis to be made incorrectly. For this case, the PSFr is equal to 50.

    The design of the plant can negatively affect the operator performance, in this case, this factor is classified by the SPAR-H Method as bad. The measuring instruments for this situation are not being visualized in the position where the operator and the machine are located. The PSFr is equal to 10. When operator performance is neither improved nor worsened due to plant design, it is considered as a nominal factor in ergonomics and the PSFr is equal to 1.

    The Ergonomics and Human-Machine Interaction Performance Factor is considered good when the instruments are easy to read, easy to use and with appropriate units of measure for the task. In addition, when a good identification of the equipment occurs and when the access and escape routes are well signaled, the PSFr is equal to 0.5. When the information on Ergonomics and adequacy of the instruments is insufficient, the PSFr is equal to 1.

- *Ergonomics calibration*

    In a Nuclear Power Plant, there may be control systems and instrumentation with redundancy of signals to avoid instrument failure. This redundancy philosophy is also used in petrochemicals and refineries to prevent explosion of tanks, vessels and reactors. But redundancy is rarely used in the food industry. Use is reserved to guarantee quality. In these cases, it is simpler, and may be related to Ph confirmation through three different methods. Thus depending on process control technology, there may be a need to calibrate the PSF Instrumentation and Ergonomics. Gross work, which involves physical effort and high logistics may also require change in multiplier values for the calculation of SPAR-H.

7. Aptitude for work

    The PSF, "Aptitude for work," is related to both physical skills and mental abilities of the operator that allows the adaptation necessary to perform the activity. There are several factors that can affect the physical conditioning of the operator to properly exercise their activities, such as: fatigue, illnesses, drug use (legal or illegal), excessive self-confidence, problems and personal distractions. However, in this PSF, factors related to the level of experience or stress are not considered. Boring et al. (2011) argue that many of these aspects are not reported internally in companies, such as issues on drug or alcohol impairment, distraction due to family problems, or whether a person is mentally or physically able to perform a task. These authors highlight, in an organization, fatigue as the most commonly recorded, and the one that has the most influence for non-compliance with work. Fatigue is a symptom that is characterized by the feeling of weariness, tiredness and lack of energy, and when present in critical industrial environments can compromise the safety of the environment and people.

    According to Williamson et al. (2011) tiredness is defined as a "biological impulse for recovery rest." Fatigue may include drowsiness, but also mental, physical and muscle fatigue, depending on the nature of its cause. These authors also developed a model of the relationship between fatigue and safety. In the model, there are three factors that cause fatigue and drowsiness; time of day, time since the worker woke-up, and task-related factors, such as a task that requires constant attention or monotony for a longer period. The time of day plays an important role in "Aptitude for work," for example: it is not uncommon for people to become drowsy in the early afternoon after lunch. For individuals accustomed to a night shift, cognitive functioning in the early hours of the morning is worse than during the day. It is important for the company to always check and document about the poor skills or people difficulties in the environment that the work is performed; mainly in tasks essential for site, process and people's safety, such as: prolonged surveillance or monitoring tasks, which usually drop performance after about 30 minutes of continuous monitoring (Boring et al., 2011).

    Given the mechanization of functions, the organization has a more decisive role to ensure the reliability of the entire system. It is worth remembering that of the eight PSFs used in the SPAR-H method, only one of them (the aptitude for work) does not have a direct relationship with the organization. Therefore, a change in the work environment has the potential to have a major impact on the performance in the activities.

    The classification for the PSF "Aptitude for work" for SPAR-H is performed as follows:

- *Diagnosis/action*

    The unfit operator cannot perform the task due to illness or some motor or mental disability, in which case the failure is certain, the probability of failure is equal to 1. Aptitude can be considered degraded when the ability to perform the task is diminished by the following factors: intake of sleep-inducing medications, tiredness, and intake of alcohol or drugs. In the case of degraded factor, the PSFr is equal to 5. It is considered that the aptitude for work is nominal when there is no degradation in the performance of the operational routine and the operator is able to perform the tasks normally, for this case, the PSFr is equal to 1. And when the information is insufficient, the PSFr is considered equal to 1.

- *Work fitness calibration*

   Operational discipline, commitment, human and social typology are directly related to physical and cognitive aptitude for work. At the moment when the worker is aware of his physical and cognitive limitations in addition to understanding the importance of achieving organizational goals, he will certainly develop the activities, adapt to the environment and the task or suggest management changes in the environment and of the task. This will be done to avoid future human errors.

   On the other hand, centralization in management, excessive self-confidence, budget cuts make aptitude to work become static and non-dynamic. Thus depending on the understanding of limitations and adaptations, it is necessary to perform calibrations in this PSF. Not always the multiplier values originally suggested by the Nuclear Industry are applicable to the Refining or Petrochemical Industry or the Manufacturing Industry.

8. Labor relations

   The PSF, "Labor Relations," refers to the social aspects for the accomplishment of the work, including the belief in the safety culture, the quality in the planning of work, the quality of communications in the organization and the belief in policies and practices. These Labor Relations provide the necessary support for the management of processes in the operation routine. Also, the way in which work is planned, executed and transmitted to track team performance for achieving organizational objectives. For this reason, it is essential that there is good communication between those involved, in addition to monitoring and controlling the execution of the work.

   A good articulation and coexistence of the team is transformed into a higher performance of the procedures through the full execution of the work with the highest reliability of the process, equipment and operations (Drigo and Ávila, 2017). If the registration in the shift register is not filled out correctly, misunderstandings about the procedures can happen, thus affecting, for example, the quality of maintenance service for critical equipment, or even causing the current shift team to need to verify what has already been done, by the previous shift group, in the field. The rework in the shift can causes a loss of precious time that could be used for other activities (Drigo et al., 2015).

   For calculations, the HRA analyst must choose between the following levels:

- *Diagnosis/action*

   Performance is negatively affected when Labor Relations are bad. It can be affirmed that the lack of communication between shift change and the low commitment of supervisors lower the efficiency in the task or even, when the procedures are poorly planned due to the difficulty of understanding and linguistics. The PSFr Labor Relations, when diagnosed is equal to 2. The PSF action or performance of procedures is equal to 5.

   Labor Relations are appropriate and classified as nominal when team performance is not significantly affected by these Labor Relations. The necessary information is available, but is not transmitted proactively, in which case the PSFr is equal to 1. On the other hand, Labor Relations are classified as good when teams are formed by people who have a good level of cohesion among themselves, supervisors who provide support and at the same time supervise work properly. Communication in team at all hierarchical levels is good and involves the team around organizational goals. The PSFr is equal to 0.8 for diagnostic activities and the PSFr is equal to 0.5 for execution activities.

- *Calibration for labor relations*

   In this performance factor, the interpersonal relationship in the work environment is also considered, especially between the leaders and the operation team. Conflicts between colleagues can cause the team to have its decision-making power affected, and the group to decrease the results of the routine.

   Despite the difficulty of evaluating the way people interact with each other, SPAR-H tries to aggregate the contributions of interpersonal relationships with Labor Relations. This is done to avoid deviations in probability calculation in real situations. The great intrinsic variability of human behavior in the shift regime can result in different attitudes from supervisors for each situation. The coefficient assigned to relationships at work should be treated as an average in evaluation behavior of teams, but in some critical functions, behavior of these operators should be considered. A production supervisor does not need to demand so much from a class when, for the most part, people are naturally disciplined and interested in work. Relying on the team's result can decrease excessive centralization, and improve the quality of supervision at other critical points in the process.

TABLE 6.1 Summary of values for performance factors based on SPAR-H.

PSF	Rating by suitability level	PSFr Activity Diagnosis	PSFr Activity Action	PSF	Rating by suitability level	Activity and PSFr Diagnosis	Activity and PSFr Action
Available time				Stress	Extreme.	5	5
	Nominal time (~2/3)	10			High.	2	2
	Nominal time	1			Nominal.	1	1
	Extra time (between 1 and 2x nominal time and > 30 min).	0.1			Insufficient information.	1	1
	Expansive time (> 2x time required and > 30 min).	0.01		Training/ Experience	Low	10	3
					Nominal	1	1
					High.	0.5	0.5
	Nonexistent	1.0			Insufficient Info.	1	1
	Available T = T required.		10	Procedure	Nonexistent	50	
	Nominal time		1		Incomplete	20	
	Time ≥ 5x T required.		0.1		Available and Poorly elaborated.	5	
	Time ≥ 50x the required T.		0.01		Nominal	1	
	Nonexistent		1.0		Diagnosis/symptoms	0.5	
Complexity	Highly complex.	5			Insufficient Info	1	
	Moderately complex.	2			Nonexistent		50
	Nominal	1			Incomplete		20
	Obvious diagnosis	0.1			Available, poorly elaborated		5
	Insufficient information.	1			Nominal		1
	Highly complex.		5		Insufficient Information		1
	Moderately complex		2	Aptitude for work or service	Unable	P (failure) = 1	
	Nominal		1		Degraded aptitude	5	5
	Insufficient information.		1		Nominal	1	1
Ergonomics	Non-existent / poorly designed	50	50		Insufficient Information	1	1
	Bad	10	10	Labor Relations	Bad	2	5
	Nominal	1	1		Nominal	1	1
	Good	0.5	0.5		Good	0.8	0.5
	Insufficient information.	1	1		Insufficient Information	1	1

9. Summary of PSF criteria and levels by SPAR-H

Table 6.1 was designed to show the values applied to the PSFs of the SPAR-H method. It presents a summary of the values corresponding to the performance factors. This table was prepared based on the Manuscript "The SPAR-H HRA method" by Gertman et al. (2005).

6.3 Case studies

All cases investigated were derived from real applications of the GRODIN research group led by research Professor Salvador Ávila Filho. These cases were presented in the form of articles, theses, dissertations, monographs and course completion papers. The researches were carried out in the areas of production engineering and human reliability. The cases investigated include the following industrial segments: *food and beverages* (1 chicken and 6 coconut water);

chemical (3 chemical and 5 fertilizer industries); *oil* (4 refineries); *nuclear* (2 uranium mining industries); and *control services in the logistics* (7 finished product from the manufacture of sports materials).

We consider that the cases (3), (4), (5) are cases of high complexity, we consider cases of average complexity case (2) and low complexity cases (1), (6) and (7).

1. Chicken industry
2. Uranium industry
3. Chemical industry
4. Refining industry
5. Fertilizer industry
6. Coconut industry
7. Packing list services, logistic control, in manufacturing

In the cases listed, the eight PSFs were analyzed to calculate the HEPs, both for diagnostic and for action, with values corresponding to 0.01 and 0.001. The scenarios of the industries will be presented and their processes will be detailed. HEP equations were applied through SPAR-H applications in real situations in the various process industries.

6.3.1 Chicken industry

The descriptions of the case were based on the Dissertation presented at the Federal University of Bahia (2017) for the master's degree of Dórea, (2017) and in an article for the ABRISCO Congress. He participated in GRODIN under the leadership and guidance of Professor Salvador Ávila Filho, Dsc. (Dórea, 2017; Dórea and Ávila, 2015).

6.3.1.1 Industrial context

The first case of application of SPAR-H refers to HRA in a chicken slaughter and processing (Fig. 6.4) unit located in the interior of the state of Bahia, Brazil. This unit accounts with an adequate location for the flow of the product in the international markets and for regions of Brazil.

The distancing of traditional poultry production centers in the South and Southeast makes it difficult to integrate corporate suppliers, making their products more expensive due to the high freight cost. Another difficulty is regarding the technical assistance and maintenance of equipment due to the distance from spare parts suppliers.

The chicken industry was built on a rural property with the possibility of expansion in the slaughter plant as in the industrialized plant. This industry has procedures to change the production mix, thus varying the productive capacity between whole chicken units and parts (with low added value), embedded product units, with high added value.

The consumption of water and energy is higher in the Bahia Site and the fixed cost, payroll of personnel, is lower in relation to the industries of the South and Southeast. The employees of this chicken industry reside close to the industry, thus reducing turnover and absenteeism. The distance is short when comparing with the southern industries that are located 200 km away from their employees' homes.

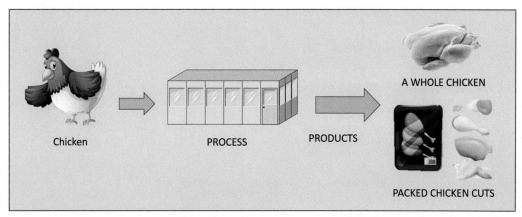

FIGURE 6.4 Representation of the chicken industry. *Courtesy: Freepik.com.*

The company meets standards required by Occupational Medicine and accident prevention policies, as well as various local actions, to ensure the safety of its employees, where uniforms and safety equipment due to refrigerator environment with extremely cold and or hot areas are distributed.

Ergonomically, their equipment complies with the labor Regulatory standards regarding the design and operation in an environment appropriate to the development of tasks, preventing the event of body injuries and/or occupational diseases by errors in the structuring of the workstation.

The main raw material in chicken growth is the chicken food, composed of: corn grains (70%), soybean grains (20%) and other grains and components (10%). Stocks are managed following the demands of the market that accompanies the seasonality of religious festivals. The periods of low consumption are during religious festivals (Lent), when people avoid eating beef and chicken. In this period the food is based on fish and seafood. This occurs for 40 days, annually.

In the last 15 years, since its inauguration, this industrial unit had 5 organizational changes. During this period, changes were made to the procedure and replacement of leaders. These issues of organizational uncertainty probably caused an inadequate organizational climate with human errors and technical failures.

The rules from the Federal Inspection Service (FIS) regulate that the slaughter of chickens from a producer can only be started after the end of the previous one. The production restrictions indicate the design of criteria for the scheduling of transporting chickens in the farms supplying this industrial unit. Other important factors are observed: distance from the farms to the slaughterhouse; charging time; waiting in the refrigerator to meet the best practices indicated by Hazard Analysis Critical Control Point; and ensuring animal well-being.

6.3.1.2 Process analysis

The process of the chicken unit takes place in two steps: slaughter and processing. The first starts after the arrival of the trucks with the poultries at the refrigerator unit, and follow the flow presented in Fig. 6.5. In the phase of chicken processing, at the evisceration step, the chickens are cooled in the chiller, equipment specific to this operation. The poultries remain for an hour, time divided between pre-chiller and chiller. The operating principle is heat exchange to keep chickens at low temperature.

The chickens still hot are placed, due to the scalding process, and eviscerated at a temperature between 0°C and 4°C. To perform this heat exchange, the chickens are transferred by a transport thread, in the counter water flow at 0°C. This flow is sufficient to cover the standard described in the procedure for controlling good production practices.

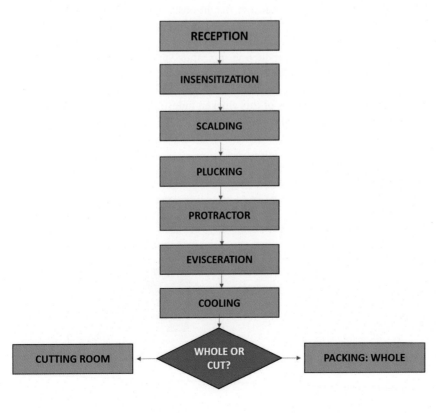

FIGURE 6.5 Process flowchart of the chicken unit. *Reproduced with permission from Dórea, S.S., 2017. Investigation in the operations of a poultry refrigerator to define product mix and understanding of sociotechnical failures. Master's degree dissertation in Industrial Engineering—Federal University of Bahia.*

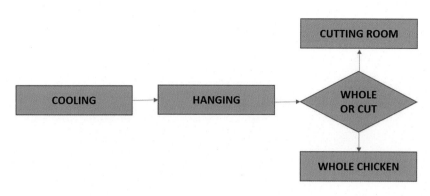

FIGURE 6.6 Poultry processing flowchart. *Reproduced with permission from Dórea, S.S., 2017. Investigation in the operations of a poultry refrigerator to define product mix and understanding of socio-technical failures. Master's degree dissertation in Industrial Engineering—Federal University of Bahia.*

Following the flow, the poultry processing will take place in two steps: the whole chicken packaging, and cutting room, according to what is represented in Fig. 6.6.

The Production Planning and Control Sector of this plant receives the production needs of the corporate sector and organizes the production plan. The sales sector puts on the market this monthly offer on the delivery date.

Process management aims to control production through procedures for performing the task, analyzing and treating anomalies, and also through quality control. The unit underwent controller changes more than three times, and each controller had its own preference for execution mode. Local supervision is difficult to work with the new model, since there was no sharing and acceptance by employees. Despite being a unit that has undergone changes, a high degree of adaptability is observed.

6.3.1.3 Equipment

For the analysis of human reliability, through the SPAR-H, a survey is carried out in the process/task scenario of the equipment considered critical in the slaughter process. One of the main tasks the "triggering of the tray packing machine."

The tray packing machine, Fig. 6.7, is a machine consisting of a camcorder and a thermal tunnel, equipment with an average level of automation and complexity. The part responsible for placing the film is composed of an electronic and another mechanical part. It has a control panel, as well as sensors and electrical controls.

The steps of this critical task involving the packer are the following:

1. Turn on the equipment fifteen minutes before the start of the process in the cutting room;
2. Check the operation of the elevators and the temperature in the heating tunnel;
3. Adjust the positioning of the plastic film;
4. Test the process on an example tray.

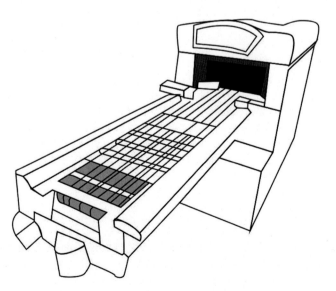

FIGURE 6.7 Projection: tray packing machine. *Author's original.*

TABLE 6.2 Performance shaping factor values for the chicken packing activity.

Performance shaping factor (PSF)	Diagnosis	PSF level	Action	PSF level
Available time	1	Nominal	1	Nominal
Stress	2	High	2	High
Task complexity	1	Normal	1	Normal
Training/experience	10	Low	3	Low
Procedures	1	Nominal	1	Nominal
Ergonomics	1	Nominal	1	Nominal
Aptitude for work	1	Nominal	1	Nominal
Working relations	2	Bad	5	Bad

6.3.1.4 Application of the SPAR-H method

The application of the SPAR-H method in this case intends to calculate the HEP in the diagnosis of this activity of triggering the packer, which is considered as the critical equipment, but human reliability should be considered in the accomplishment of the task. To collect the data, it was necessary to conduct interviews with the application of questionnaires to the operators who perform the task in the tray packing machine. For the effectiveness of the values, the form was applied to the operators and supervisors of the process. According to Table 6.2, the values for PSFs in the action task and in the diagnosis task for the operation of the tray packer equipment were described.

The HEP in the task "Triggering of the tray packing machine," using Eqs. (6.3), (6.4) and (6.5) as the calculation base.

$$\text{HEP}_{\text{Diag}} = \frac{0.01 \times (1 \times 2 \times 1 \times 10 \times 1 \times 1 \times 1 \times 2)}{0.01 \times ((1 \times 2 \times 1 \times 10 \times 1 \times 1 \times 1 \times 2) - 1) + 1} = 0.29$$

$$\text{HEP}_{\text{Action}} = \frac{0.001x \, (1x2x1x3x1x1x1x5)}{0.001x \, ((1x2x1x3x1x1x1x5) - 1) + 1} = 0.029$$

The total probability of error is calculated using Eq. (6.5):

$$\text{HEP}_{\text{Total}} = 0.29 + 0.029 = 0.319$$

Eq. (6.6) is used to calculate human reliability. The human reliability obtained using the SPAR-H method is:

$$R_{\text{hu}} = 1 - 0.319 = 0.681 \times 100\% \rightarrow R_{\text{hu}} = 68.1\%$$

6.3.1.5 Considerations of the case of the chicken industry

It is concluded that the task "trigger the tray packing machine" presents a high probability of human error of 31.9%. Some aspects should be taken into considerations in the activity "Trigger the tray packing machine," such as: stress involved in the activity; the level of training of operators and issues involved in work (labor relations). Considering the result of human reliability (68.1%), it is demonstrated as a fragility in the execution of this task, signaling the lack of qualification of the operators, high level of stress and noises in communication or lack of motivation regarding the team or organization.

This task/process has productions focused on the items of higher added value and, therefore, the importance of failures rises. Given the condition where the unit of poultry slaughter and processing is located, the error in the task needs to be treated with management, so that it is possible to reduce the effects of stops and consequently improve the operating condition.

6.3.2 Uranium industry

The case descriptions were based on the article published in the ABRISCO Congress (2015) by Ana Caroline Alcoforado and Salvador Ávila Filho. Ana Caroline participated in GRODIN (Research Group on Operational and Dynamic Risks) under the leadership and guidance of Professor Salvador Ávila Filho, Dsc. (Alcoforado & Ávila, 2015)

6.3.2.1 Industrial context

This second case is related to the uranium production chain industry (Fig. 6.8). This chain includes mining processes and nuclear industry, which generates electricity from fission. Uranium is a distributed element in the Earth's crust whose main use is as fuel for nuclear energy reactors from the fission of the atom. It is through fission that the division of a chemical element nucleus caused by the bombardment of neutrons is understood. This split releases other neutrons that divide with other nuclei in chain reaction.

Because of the power of this isotope, to generate the same energy contained in 10 g of U-235, it is required 700 kg of oil or 1200 kg of coal. The use of radioactivity and nuclear energy is increasing in the world and despite the risks, it is considered an alternative to the use of oil and coal.

The company's mission is to ensure the supply of nuclear fuel for the generation of electricity with safety, quality, transparency and social and environmental responsibility, through integrated management, diversification of the product line and technological autonomy in the fuel production.

Nuclear energy accounts for 17% of the world's electricity generation. Although it does not generate greenhouse gases, the hazard lies in high radioactivity residues and the possibility of accidents in the plants, which can be devastating. The largest nuclear disaster in history occurred in Chernobyl, Ukraine, in 1986, when a reactor at the plant presented technical problems, releasing a radioactive cloud with 70 tons of uranium and 900 tons of graphite into the atmosphere. The accident caused the deaths of more than 2.4 million people nearby and reached level 7, the highest of the International Nuclear Event Scale (INES).

There are, then, numerous lessons, including reactor safety and crisis management, intervention criteria, emergency procedures, communication, medical treatment of irradiated people, monitoring methods, radio-ecological processes, region supervision, and public information.

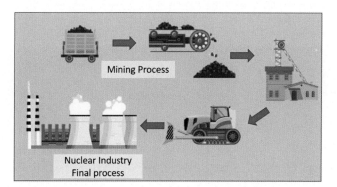

FIGURE 6.8 Representation of the uranium production chain. *Courtesy: Freepik.com.*

6.3.2.2 Process analysis

Uranium ore is mined by the open-pit mining technique. In the development of the mining, it also has sterile material, with low concentration of uranium, and its recovery is economically unfeasible. This material is stored in hillside locations, called solid tailings deposit of the mine and plant, located near the mine.

The ore, transported from the mine through tipper trucks, is discharged into a silo (or hopper) and a vibratory feeder. The ore goes through four crushing phases, in which its granulometry is reduced to 13 mm for the formation of piles. The uranium liqueur is extracted in the leaching process.

The separation/concentration of uranium is carried by extraction process through organic solvents. Initially uranium is "extracted" from the liquor, separating it from other chemicals that were also solubilized (leached) from the ore, in the leaching stage with sulfuric acid solution.

At this stage, the uranium contained in the liquor is transferred to the organic solvent, separating it from impurities. The liquor, now without uranium, becomes a liquid tailing, called "raffinate," being transferred to the treatment area of liquid effluents for treatment.

The final product of the company's production process, called Ammonium Diuranate (or DUA) is placed in 200-L metal barrels, with simultaneous sampling, thus generating a batch of the product and a set of representative samples of the batch. The samples are analyzed in the company's laboratories, for issuance of the corresponding Certificate of Analysis.

The control room is strategically located in the ore processing plant, from where it is possible to visualize the entire crushing area and the leaching yard, except the tunnel of resumption of the ore of the lung pile and the hopper, where the cameras are installed to monitor these activities of the operation.

In this room, there is a microcomputer-based operation station that controls the entire primary, secondary, tertiary and quaternary crushing process. In addition, the Programmable Logical Controller and power distribution racks are also installed, equipped with an uninterrupted power system (no break).

The control is done using a software which runs in a Windows NT 4 operating environment from which it is possible to turn on and off all electromechanical equipment in the area of crushing and leaching yard. The system is composed of eight screens with representation of the area flowchart for remote operation of the process, through charts or trend curves, historical records, index of alarms, reports and others.

The primary crushing operation works independently of the re-crushing. In other words, the operation of the re-crushing is interlocked and the shutdown of any equipment or machine causes production interruption. The re-crushing operates continuously, while the primary crushing operates only when there is transport of the ore mined or deposited in the pre-stocking yard of uranium ore.

6.3.2.3 Application of SPAR-H

For the analysis of human reliability in the uranium processing process, forms (Annex A and B) were applied to the operators, supervisors and coordinators involved in the process control room. In the control room there is all the monitoring of the processes, from the entry of the raw material to the crushing and the re-crushing. Thus the participation of the whole team (including all shifts) was important in the application of the method. The results from the forms for the action task and diagnostic task are summarized in Table 6.3.

Following the methodology, it is important to evaluate the number of PSFs that has a negative influence, that is, when the multiplier value is greater than one. This check is necessary to evaluate whether it is needed to perform the adjustment in calculating the HEP. In this scenario, none of the tasks presented three or more PSFs with negative influence, which means that the calculation can be performed using Eqs. (6.1) and (6.2).

Thus for the tasks of action, diagnosis of the control room, the probability of human error is given:

$$HEP_{Diag} = 0.01 \times (1 \times 2 \times 2 \times 1 \times 1 \times 1 \times 1 \times 0.8) = 0.032$$

$$HEP_{Action} = 0.001 \times (1 \times 2 \times 5 \times 0.5 \times 1 \times 1 \times 1 \times 0.5) = 0.0025$$

The total error probability is calculated using Eq. (6.5):

$$HEP_{Total} = 0.032 + 0.0025 = 0.03$$

Eq. (6.6) is used to calculate human reliability. The human reliability obtained using the SPAR-H method for the control room of uranium processing was:

$$R_{hu} = 1 - 0.03 = 0.97 \times 100\% \rightarrow R_{hu} = 97\%$$

TABLE 6.3 Performance shaping factor values for the activity in the uranium industry.

Performance shaping factor (PSF)	Diagnosis	PSF level	Action	PSF level
Available time	1	Nominal	1	Nominal
Stress	2	High	2	High
Task complexity	2	Medium complexity	5	High complexity
Training/experience	1	Nominal	0.5	High
Procedures	1	Nominal	1	Nominal
Ergonomics	1	Nominal	1	Nominal
Aptitude for work	1	Nominal	1	Nominal
Labor relations	0.8	Good	0.5	Good

6.3.2.4 Considerations for the uranium ore extraction industry

The study of uranium ore extraction industry demonstrated how environmental conditions and the importance of the activity and its complexity can influence performance. With this data, Human reliability (97%) was within the typical range for this type of critical process, according to the validation of an area specialist. However, it is important to emphasize that the construction of an action plan to improve the "stress" factor is necessary and adequate for positive maintenance of this reliability value. The company's intention is to increasingly elevate the system reliability in general to reduce the HEP to the value of 1%.

6.3.3 Chemical industry

The descriptions of the case were based on the article published in the National Congress of Mechanical Engineering-Souza et al. (2018) by Salvador Ávila and partner of the chemical industry in Camaçari) who participated in GRODIN (Research Group on Operational and Dynamic Risks) under the leadership and guidance of Professor Salvador Ávila Filho, Dsc.

6.3.3.1 Industrial context

In this third case, a chemical industry was analyzed (Fig. 6.9). In this industry, one of its main by-products is hydrochloric acid (HCl). This by-product is neither treated nor used by the company, being sold as a secondary way for obtaining profit. To do this, there are HCl loading docks, where trucks are coupled, through hoses and pipes, for the transfer of this product. The loading is carried out through the supervision of a local operator with assistance of the operators of the control room. This alignment between operational and control panel is very important for the correct performance of maneuvers to enable the loading of vehicles.

FIGURE 6.9 Chemical industry. *Courtesy: Freepik.com.*

6.3.3.2 Process analysis

In this industry, there are several critical and complex processes with handling of toxic products. This article investigates the task of loading HCl, co-product of chemical reactions. This process of transferring acid to trucks is critical due to the possibility of an accident that impacts people, the plant and neighboring communities. It is important to evidence the high number of unplanned maintenances in the acid area equipment, thus indicating low reliability of equipment, operation and safety systems.

The Ergonomic Improvement of the HCl loading system aims to increase availability and reliability of related equipment, and thus promote a higher level of process safety to avoid possible incidents and accidents.

6.3.3.3 Application of SPAR-H

The original calibration and questionnaire were used to apply the SPAR-H, which can be applied with the data obtained in interviews with workers who perform or are involved in the activities, as well as those who manage and understand the operation.

TABLE 6.4 Performance shaping factor values for the activity in the chemical industry.

Performance shaping factor (PSF)	Diagnosis	PSF level	Action	PSF level
Available time	0.1	Extra Time	1	Nominal
Stress	2	High	2	High
Task complexity	2	Medium complexity	2	Medium complexity
Training/experience	0.5	High	0.5	High
Procedures	20	Incomplete	1	Nominal
Ergonomics	10	Bad	10	Bad
Aptitude for work	1	Nominal	1	Nominal
Labor relations	0.8	Good	0.5	Good

Visits were made to the site of the activities to understand the environment culture, as well as the search and evaluation of procedures and checklist related to the tasks. From the collected perceptions, the method was applied to find the value of human reliability.

Thus the results were compiled and analyzed with the assignment of levels for each of the eight criteria (Table 6.4). These levels are related to multiplier values for calculating the HEP.

According to the method, the HCl loading task can be classified into diagnosis and/or action, and the following analyses of the items are considered:

Available time criterion: It is related to the time the operator has to diagnose or act in response to an adverse situation. This time influences how the operator can react and suggest alternatives, that is, how it will affect his/her performance in a situation that demands decision. This time is considered nominal for operator action, even if the operator in this task needs to act quickly in response to an adverse situation with demands such as: activation of button-ups, communication to the panel by radio to call the firefighter team. The time diagnosis was considered extra in situations of abnormalities, the safety devices will act and lead the plant to a safe condition. A diagnosis can be made after returning to a safe condition such as containment and pressure control.

Stress criterion: Excessive heat, noise, lack of ventilation, and uncomfortable positions to perform the task trigger stress. Thus it was possible to evaluate through the interviews that the stress level is high. This is mainly because the task is performed with PPE throughout the unloading, causing discomfort, mainly during the summer days when the heat is excessive. Another factor observed is the variation of the truck dome—in some situations it is small, which makes the task of connecting the pipes ergonomically uncomfortable for the accomplishment of the procedure.

Complexity criterion: The evaluation of this criterion allowed us to conclude that the complexity related to the action task and the diagnosis task is moderate, since the HCl loading operation can be performed simultaneously with unloading of two raw materials in case the plant is operating near its maximum capacity, thus generating elevation in the HCl tank level and high demand of raw materials for production. In addition, HCl is a product considered dangerous, which due to its high volatility, releases toxic vapors that when inhaled is corrosive to the mucosa, eyes and skin.

Experience and training criterion: As more than 90% of the operators have extensive experience in performing plant activities, this was considered high, according to the evaluation of the SPAR-H method criterion.

Procedure criterion: It was possible to identify that the procedures have adequate guidelines for carrying out the activity. Sometimes, it was indicated in the interviews that it would be possible to perform the task unless there was a lack of experience for execution. However, with regard to the procedure for diagnosis, it was considered incomplete, since the lack of specific information about the process necessary in decision-making is considered.

Ergonomics/HMI criterion: Ergonomics/HMI was considered negative, considering that the responsibility for the loading dock is from area "1," however, operation of the truck loading is attributed to the operator and panelist who belong to area "2," which requires extra communication between the areas and hinder visualization and the action in the control panel.

Aptitude for work criterion: The analysis determined that the level of adequacy for the activity is nominal, since there is no loss of operator performance.

Labor relations criterion: Individual or group performance may be affected depending on how work is planned, communicated and executed. In the case under study, it was verified that the work process is positive, since there is good communication, well understood and supportive policies, and the work team behaves well.

To determine the HEP, factors determined in Table 6.1 were used considering values of the evaluation from the diagnosis and the action, through equations of 6.3, 6.4 and 6.5.

$$HEP_{Diag} = \frac{0.01 \times (0.1 \times 2 \times 2 \times 0.5 \times 20 \times 10 \times 1 \times 0.8)}{0.01 \times ((0.1 \times 2 \times 2 \times 0.5 \times 20 \times 10 \times 1 \times 0.8) - 1) + 1} = 0.2442$$

$$HEP_{Action} = \frac{0.001 \times (1 \times 2 \times 2 \times 0.5 \times 1 \times 10 \times 1 \times 0.5)}{0.001 \times ((1 \times 2 \times 2 \times 0.5 \times 1 \times 10 \times 1 \times 0.5) - 1) + 1} = 0.0099$$

The total error probability is calculated using Eq. (6.5):

$$HEP_{Total} = 0.2442 + 0.0099 = 0.254$$

Eq. (6.6) is used to calculate human reliability. The human reliability obtained using the SPAR-H method for *HCL* transfer process was:

$$R_{hu} = 1 - 0.254 = 0.746 \times 100\% \rightarrow R_{hu} = 74.6\%$$

6.3.3.4 Considerations for the case of the chemical industry

The influence of factors (PSFs) related to the task of loading, such as: stress, task complexity, procedures and ergonomics negatively impacted value of human reliability (74.6%). Considering the high probability of human error (25.4%), efforts should be directed to improve these factors. Errors in this loading area, which is highly critical, can cause unwanted events in the company, such as incidents, failures and accidents. The state of the operator can generate distraction, disinterest and lack of attention to the tasks performed. Thus the operator can act slowly when an alarm is triggered, then causing an overload of the equipment, which cumulatively can contribute to the failure. Once the causes that are negative for critical PSFs are mapped and addressed, failures and the probability of error can be significantly reduced. Knowing the critical factors in the main tasks, thus knowing potential for correct answers and errors, helps in decision-making to direct investments for improving the organization, which leads to assertiveness at the management level.

6.3.4 Refining industry

The descriptions of the case were based on the Work for Course Completion of Diego Rocha dos Santos for Graduation in Santos (2014) at the Federal University of Bahia, guided by Salvador Ávila. Ávila and Diego published at ICEFA in 2015, International Conference on Engineering Failure Analysis. Finally, this work was revised and published in JEFA (Journal of Engineering Failure Analysis). Diego participated in GRODIN (Research Group on Operational and Dynamic Risks) under the leadership and guidance of Professor Salvador Ávila Filho, Dsc. (Avila & Santos, 2014; Santos, 2014).

6.3.4.1 Industrial context

In this case, a routine activity is analyzed in the Refining Industry. The analysis of human reliability uses the SPAR-H applied in this company to analyze the factors that modify the operator performance in executing the propane pumping activity. The task is logistics and deals with vertical centrifugal pump starting procedure. This equipment, the pump, is used in the transfer of the propylene product. Propylene is an unsaturated hydrocarbon (alkene) of formula C_3H_6, which usually presents as a colorless and highly flammable gas. In pumping, with high pressure, it is in the liquid state.

6.3.4.2 Process analysis

The transfer of propylene is carried out through a set of centrifugal and vertical pumps, without recirculation with the plant and in batches or campaigns. It is a continuous process and depends on customer demand and refinery production schedule. This transfer is represented in Fig. 6.10.

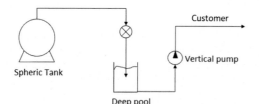

FIGURE 6.10 Flow of the propene transfer process. *Reproduced with permission from Santos, D.R., 2014. Human reliability analysis in the process of pumping propane in a refinery using the SPAR-H method. Course Completion Work- Mechanical Engineering—Federal University of Bahia.*

Propylene is an unsaturated, flammable and colorless hydrocarbon (C3H6) that, when subjected to pressure and cold, liquefies. It is found in the composition of LPG, derived from process of pyrolysis or catalytic cracking of naphtha, mixed with other gaseous compounds such as propane, n-butane and isobutane. Propylene is obtained by separation of remaining components by distillation of LPG and, depending on its use, it can be produced within a large range of purity degrees. During storage, propylene is in its saturated state at an operating temperature of about 28°C. There are fluctuations due to climate change. LPG temperature and pressure are considered stable, with a high probability of being inserted into the 5 SIGMA (about 99% probability of process compliance and product reliability).

6.3.4.3 Task

The pumps transfer propylene under the control of two operators per shift group. There is the total of five groups (ABCDE) and 10 operators. Repeated operation of this rotating equipment follows standards and procedures. According to the response from the operators of the activity, the available procedure does not guarantee a good operation of this activity, pumping propene. Responsible operators and supervisors should know the limits and nominal conditions of the process variables for these pumps, including to avoid phase change and cavitation in the pump.

The flowchart in Fig. 6.11 shows the activities performed for the start of this set of centrifugal pumps. For the proper functioning of this equipment, it is necessary that the lubrication and sealing system work properly. At this stage of the process, the operator must observe the kerosene level in the cup in the sealing system. If the level is below minimum, the pressure of the sealing fluid (kerosene) will drop and the seal will leak. Lubrication is done by fog, thus generating good distribution. For proper lubrication, the operator should check the tank level, the oil level in the collector, and the process variables.

The process consists of partially opening the discharge valve before the start to reduce frictional forces at the time the fluid is transferred out of the pump and pressurizes the valve. The correct start of pump operation includes fully filling the carcass and filling the suction line. This procedure, when performed correctly, ensures that there are no unfilled spaces inside the enclosure, reducing the probability of cavitation and increased vibration of the equipment. Cavitation is a critical event that should be avoided through procedure control.

Online monitoring (vibration, flow and pressure) and manual monitoring at regular intervals (bearing temperature) ensure quality control in pump operation. This monitoring prevents the appearance of peak vibration, triggering preventive actions that ensure the stabilization of the pumping process. Vibration readings are done in real time using the PI Process Book software, which receives data from the bearing-mounted sensor and plots the graph of absolute vibration versus time values.

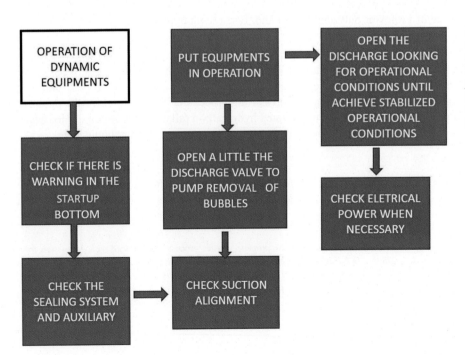

FIGURE 6.11 Activities for starting the centrifugal pump. *Reproduced with permission from Santos, D.R., 2014. Human reliability analysis in the process of pumping propane in a refinery using the SPAR-H method. Course Completion Work- Mechanical Engineering— Federal University of Bahia.*

6.3.4.4 Equipment

This investigation is focused on a device manufactured by Flowserve Worthington, model 8L-12 CAN, extracting from the liquefied propylene tank through the use of a pump 3.86 m deep, Fig. 6.12. The pumping operation is not constant; therefore, operators need to perform the procedures to start and stop several times during the day.

Regarding the failure analysis history (4 years), 17 interventions were indicated in pump 1 and ten interventions in pump 2. The main defects were vibration and leakage of the mechanical seal. Data collection for the analysis of human's influence in the process was performed through interviews with the shift supervisor, the operation technical coordinator and the workers who operate the pump. Discussions on performance factors (PSFs) indicate the values of the coefficients for the calculation of the failure probability at pump startup.

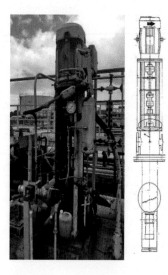

Operation Characteristics	
Numbers of stages	14
Power of Motor driver	75 HP
Power of Eixo	60,6 HP
Minimum flow continous	21,2 m³/h
Average flow in operation	44,3 m³/h
Efficiency	66%
RPM	3500
Discharge pressure	36,7 kgf/cm²
Suction pressure	11,9 kgf/cm²
Sealing Plan API	13/52

FIGURE 6.12 Equipment and its operating variables. *Reproduced with permission from Santos, D.R., 2014. Human reliability analysis in the process of pumping propane in a refinery using the SPAR-H method. Course Completion Work-Mechanical Engineering— Federal University of Bahia.*

6.3.4.5 Application of SPAR-H

In the analysis of performance factors (PSFs), the following observations were reported about the task of "centrifugal pump starting":

Available time criterion: The main cause of increased vibration in the pump is cavitation, which is caused by poor starting. According to the responding operators, it is common to wait 10 minutes to stabilize the pump. Once it is a vertical pump, large displacements can occur at its free end if there is cavitation. To avoid more serious damage to the bearing or shaft, the operator should interrupt the process as vibration increases, detecting noise or increasing engine temperature. The shutdown is done by button without the need for additional action. Once adopted a nominal time of 5 minutes, the operator has enough time to act in case of any abnormality (e.g., increased vibration due to cavitation), being assigned a PSFr equal to 1 for action and a PSFr equal to 0.1 in diagnosis.

Stress criterion: The operator's stress level is higher when some problems occur in the start step or if the redundant pump is defective. The pressure that the operator oversees requires speed in restoring operations. This may cause the operator to not perform the procedure properly because of the short time available for the completion of a new start. Starts are made, especially in the late afternoon or evening. When performed in the late afternoon, that is, near the end of the shift, fatigue can negatively influence, lowering the quality of the procedure implementation. Failure to start the pump increases operator stress due to supervisor pressure and the need to meet transfer schedule. In view of these aspects, we consider the PSFr equal to 2, high.

Task complexity criterion: The equipment park consists of four centrifugal pumps, six spherical tanks, which must be monitored by both the control room and field workers. The starting procedure requires a large number of radio-controlled communications to ensure correct alignment between the pump and tank, as well as the control of process variables, sphere level, vibration, and flow. While activities are performed, other steps of other procedures are performed. A noise in communication not transferring important information, or deviation of behavior regarding the non-use of PPE, for example, can induce the operator to perform an action in the wrong way. Errors in reading and interpreting two suction temperature meters and unloading of lubricating oil and the kerosene vessel line level. The operator must have prior knowledge about the equipment, inspection, proper operation to treat the faults. Because it is a moderately complex activity, a PSFr equal to 2 was assigned.

Experience/training criterion: Operators are trained in 6 months in various areas, such as instrumentation, inspection, and operation of dynamic and electrical equipment. Recycling training is conducted every 2 years with a time of 8 hours. Company procedures and new techniques can be used in the process. The study reserved for the principles of operation of dynamic equipment takes 40 hours. During the interviews, it was observed that the operator's training should be performed with the task. The people responsible for the bombs have at least 5 years of experience, enough to assimilate the practice and theory for the successful completion of getting to know the task, and is therefore not subject to possible deviances of execution, so a PSFr of 0.5 (High) was assigned.

Procedure criterions: A standard procedure is available for the initialization of dynamic equipment, but it does not specify important requirements in the case of propylene pump. This specific information for the pumping process is provided in the Operating Instructions (OIs) and is therefore not treated as operating patterns; there may be variability in the way to operate. Through field observations and interviews with operators, it seems that neither formal nor OIs are used during the execution of the task. According to the operators, the standard is not used because it contains all the information necessary for the proper accomplishment of the activity or because they believe that they will lack time if it is strictly followed, as indicated by the results of a search by Embrey (2000). The lack of supervision and guidance of supervisors contributes to the non-use of standards in the field. Due to the failures of the procedure, a PSFr 20 (Incomplete) was assigned.

Ergonomics/HMI criterion: In this PSF, the pumping operation was evaluated and compared with the results of the interviews. The park pump has enough space in the layout, with good accessibility to the equipment, except for some pipes that can hinder the operator to perform a procedure, firefighting equipment easily accessible and escape routes with easy access and signaling. The operator has no visual indicator if the beginning was done correctly. The indication of the kerosene level in the vessel is unclear. The motorized valve in the pump unloading was not working, which impeded its closure with consequent increase in vibration. No alarm or safety system if the flow in the pump is below the recommended minimum. There are factors that contribute negatively to perform cognitive tasks attributed to a PSFr 10 (Bad).

Aptitude for work criterion: All refinery employees undergo periodic medical examinations to certify physical and mental capacity. During the interviews and field visits, no operator was physically unable to perform the work, then PSFr 1 (Normal) was assigned.

Labor relations criterion: Information is transferred from shift to shift through operational instructions and is discussed in weekly meetings. Shift change is a factor that can cause errors in the next task group. In the shift register, information about procedures, scheduling transfer of available equipment out of operation and maintenance instructions for the next class are transmitted. If this register is not properly completed, the following group will not be able to know the specific procedures for the operation. A PSFr of 1 is assigned (Normal).

From these observations, the values were given for the PSFs, both for the multiplier referring to the action task, and for the multiplier of the diagnostic task (Table 6.5).

The set of procedures for starting and monitoring equipment requires the operator to have motor skills, such as opening valves, and cognitive skills, such as monitoring the pressure at unloading. For the PSFr, we established that: considering the diagnoses of NHEP = 0.01 and the action of NHEP = 0.001. For the determination of human error, the

TABLE 6.5 Performance shaping factor values for the activity in the refining industry.

Performance shaping factor (PSF)	Diagnosis	PSF level	Action	PSF level
Available time	0.1	Extra time	1	Nominal
Stress	2	High	2	High
Task complexity	2	Medium complexity	2	Medium complexity
Training/experience	0.5	High	0.5	High
Procedures	20	Incomplete	20	Incomplete
Ergonomics	10	Bad	10	Bad
Aptitude for work	1	Nominal	1	Nominal
Labor relations	1	Nominal	1	Nominal

factors determined in Table 6.4 were used considering the evaluation of the diagnosis and action for the calculation of the HEP, in each of these activity aspects, through Eqs. (6.3) to (6.4).

$$HEP_{Diag.} = \frac{0.01 \times (0.1 \times 2 \times 2 \times 0.5 \times 20 \times 10 \times 1 \times 1)}{0.01 \times ((0.1 \times 2 \times 2 \times 0.5 \times 20 \times 10 \times 1 \times 1) - 1) + 1} = 0.2877$$

$$HEP_{Action} = \frac{0.001 \times (1 \times 2 \times 2 \times 0.5 \times 20 \times 10 \times 1 \times 1)}{0.001 \times ((1 \times 2 \times 2 \times 0.5 \times 20 \times 10 \times 1 \times 1) - 1) + 1} = 0.2859$$

$$HEP_{Total} = 0.2877 + 0.2859 = 0.57$$

The total error probability is calculated using Eq. (6.5):

$$HEP_{Total} = 0.2877 + 0.2859 = 0.57$$

And finally, the application of the Eq. (6.6) to identify value corresponding to human reliability yields:

$$R_{hu} = 1 - 0.57 = 0.43 \times 100\% \rightarrow R_{hu} = 43\%$$

6.3.4.6 Considerations for the case of the refining industry

For specialists in the field and considering the use of the SPAR-H formula, the human reliability presented (43%) was a value much below the ideal perspective for a process so critical as that of propylene pumping. There are several factors that are negatively influencing the correct execution of the task, and among them the following can be highlighted: stress, procedures and the ergonomic issue. The performance of these factors should draw the special attention of managers for immediate treatment, regarding the reformulation of procedures, evaluation of the scenario that composes the task and analysis of the main stressors of the task. For a HEP of 57%, it is very important to invest in an action plan for improvement and identification of the main causes that are compromising these performance factors.

The need to calibrate the multiplier values for the HEP could also be considered. It may be that procedures and stress have values originally higher because SPAR-H has its origin in the nuclear industry.

6.3.5 Fertilizer industry

The descriptions of the case were based on the dissertation of José Filipe Michel Gagliano Ferreira, Master of Production Engineering at the Federal University of Bahia in 2018, guided by Salvador Ávila. Michel participated in GRODIN (Research Group on Operational and Dynamic Risks) under the leadership and guidance of Professor Salvador Ávila Filho, Dsc. (Ferreira, 2018).

6.3.5.1 General context

In Chapter 5, Learned Lessons: Human Factor Assessment in Task, we worked on the fertilizer case, both discussing the compressor and environmental education programs, like friends of the emergency pool. In this fifth case of application and calculation of human reliability, we deal with the fertilizer industry, which resembles the high complexity of a petrochemical plant.

The application of SPAR-H occurs in the startup task of the horizontal piston alternative compressor, a critical procedure in the production of urea. The final product of this industry can result in three types of urea, each with specific quality standards: premium, fertilizer and industrial. In addition to these, there is also the production of high purity liquid urea. Manufacturing processes are carried out in several steps, with activities on the control panel and on the factory floor.

The plant is classified as high product and process risk, also with high complexity in communications. This industrial installation has the following characteristics: high automation, equipment with complex control systems (main and auxiliary) and with high interconnection between units, in energy and mass; large process equipment; high operating conditions (temperature and pressure)—at some points on the site, for example, the pressure reaches levels of 260 kgf/cm^2; in addition to the fact that the company has a history of accidents incurred in rotating equipment (compressor). Several investments have been made to automate critical processes as well as investment in instrumentation of these critical systems. With automation, field actions were replaced by monitoring and action in the control room in the operation routines.

The criticality of the equipment for the process, the product, production and the safety, makes it imperative to have a greater monitoring of the parameters and process variables in the main and auxiliary systems.

Operators (panel and field) monitor and operate all unit systems related to compressor equipment, one which mostly causes plant shutdown. This means that there is no unique team or a unique operator of a given system that can perform multiple controls. The investigation unit is complex and numerous activities are carried out for ensuring safety, quality, and the quantity produced in the plant.

6.3.5.2 Process analysis

The urea manufacturing process can be separated into three steps: (1) synthesis, (2) decomposition and recovery, (3) concentration and granulation, shown in Fig. 6.13. The first step refers to the manufacturing process of ammonia carbamate by the reaction between carbon dioxide and ammonia. The final product (urea) is the result of the process of dehydration of carbamate. At the reactor outlet there is the presence of urea and unconverted reagents (CO_2 and ammoniac), in addition to a small amount of biurate (undesirable product). In the second step, unconverted products are separated and recovered. After that, in the last step, the urea concentration process is carried out, through which it is possible to granulate, dry and subsequently commercialize. Fig. 6.13 outlines in a simplified way the urea production process.

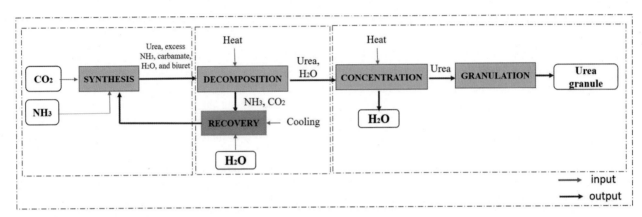

FIGURE 6.13 Simplified industrial process of urea manufacturing. *Modified from NZIC (2008) by Ferreira, J.F.M.G., 2018. Reliability analysis of CO_2 compression system in a petrochemical plant considering technical-operational and human factors. Master's degree dissertation for Industrial Engineering—Federal University of Bahia.*

6.3.5.3 Task

The level of automation and advanced instrumentation facilitate the start and stop of equipment operations, especially the compressor, thus minimizing the risk of failure. This characteristic modified the activities performed by the operators, who began to perform more monitoring and verification tasks than intervention in the field. Even so, there are situations that require the presence of the operator near the equipment, especially in this operation of starting and stopping the plant.

Operators perform inspection, monitoring, and maneuvering at the start and stop of equipment, with the aid of procedures and respective checklists. The task "compressor start" was chosen because it is considered by the staff and the operation team as the most critical and high-risk task, with a lot of human intervention during its execution. The compressor start-up procedure has approximately 36 steps, which must be followed by the staff, thus demonstrating complexity in its execution. The "compressor start" task involves action and diagnostic activities. Compressor start-up maneuvers are performed via control room and field. Therefore, efficient communication is essential.

One of the premises of compressor start-up is that operators carefully follow the guidelines of the procedure. The issue of safety is a priority for the organization and, considering the level of importance of the activity, it is recorded that the written procedure is followed. However, a deviance margin of 10% (non-compliance) is allowed, which can be attributed to changes in the design or unavailability of devices.

Despite the involvement of panel operators in the compressor start-up, the application of the SPAR-H method is based on the activities developed by the field operator. The task of start compressor can be as follows:

- *Cognitive characteristic:* performing the reading and interpretation of the variables (pressure, temperature, flow rate, oil level, etc.) and identifying abnormalities (leakage, abnormal noise, abnormal vibration, etc.).
- *Action characteristic:* opening/closing valves, testing auxiliary systems and softening components (e.g., procedure of gaskets joining).

6.3.5.4 Equipment

The CO_2 compression system consists of four compressors, two identical alternative compressors powered by electric motor and two turbined compressors, one centrifuge and one alternative. The compression process is carried out in two units of the plant: part of the CO_2 is compressed in the urea plant (about 56% of the total volume) with the participation of three compressors (one booster compressor C-01, and two alternative C-02/03) and the remainder (44%) will be compressed into the hydrogen plant by another alternative compressor (C-04).

The shutdown of the compressors does not necessarily imply the plant stopping. However, the unavailability of these compressors directly affects production, that is, reduces plant capacity by 28%, if one of the alternative compressors C-02/03 stops; by 56%, if the booster compressor fails; and by 44% if turbined reciprocating compressor becomes unavailable. In these scenarios of unavailability of the equipment it is assumed that the event did not cause further disturbances to the process to the point of stopping the plant.

The CO_2 compressor studied in the task is alternative, consisting of five horizontal cylinders with electric motor trigger. The gas is received in the first stage at a pressure of 4.5 kgf/cm^2 and compressed sequentially until the last stage. This compressor uses a connecting rod-crank system to convert the rotating motion of a shaft into the translational motion of a piston and then raise the gas pressure. Thus at each rotation of the trigger, the piston makes an one-way go and return, establishing an operation cycle.

6.3.5.5 Application of SPAR-H

The quantification of the HEP considered all eight human performance factors (PSF) in the execution of the task, whether cognitive characteristic or action.

The levels of the delimiting factors of the PSFs were evaluated according to the following criteria:

Available time criterion: The equipment start-up task is not limited by the time factor. Despite the extreme importance of the equipment, the longer the time unavailable, the higher the lacking profit will be, the need for the rapid return of equipment operation is not a limiting factor of the task. The procedure and work instructions are designed to ensure that there is time off for all activities to be carried out safely and even allow response actions to be carried out in an abnormal event. Therefore, in action activities, the available time level corresponds to a time greater than five times the required time (nominal). For diagnostic activities, the level indicated is the extra time—between one and two times the nominal time.

Stress criterion: This PSF can be influenced both by physical environmental factors (heat, noise, humidity, radiation, ventilation, etc.) and by mental factors (excessive workload, apprehension when facing hazardous tasks, etc.). Given the possibility of combining these factors, the evaluation of the PSF did not result in the selection of only one level. The starting activities performed by the operator in the field are subject to nature issues (excessive heat at some times of the year, rain, etc.) and noise. Operating conditions offer a potential risk of causing irreparable damage to the equipment in the event of an accident occurrence and, consequently, cause injury and even death of people close to the equipment. So, there is a certain level of tension about this risk. On the other hand, considering the experience of the operator in detecting the abnormalities that can trigger such a disaster and the safety offered by the protection systems, which force the equipment to stop in case of abnormality, it is considered that the stress level is nominal for the diagnostic and action activities. Regarding the workload, it is common for employees to perform a double working day, especially in cases which the presence of the most experienced professional is required for the accomplishment of more complex activities. Another issue has been changing the work environment in the organization: the company has been undergoing an internal restructuring and had some staff reductions, which means fewer people to carry out same activities. Thus in a scenario in which there may be an accumulation of such factors (reduction of staff, excessive workload per doubled shifts, environmental factors and the risk of the activity itself), the stress level can also be classified as high.

Complexity criterion: Automation and experience aid in the operation of the equipment, but do not make the activity trivial. The start of the compressor depends on the proper functioning of all its components, including its auxiliary systems. This means that more variables will have to be monitored and auxiliary systems need to be initialized. The activities do not require great physical effort, although they are influenced by environmental and ergonomic conditions. A mature knowledge on principles of operation of the equipment is also required, mainly to assist in diagnosing abnormalities that may occur. Therefore, there is a need for experienced professionals. Considering these aspects, it is understood that, despite automation and instrumentation, the level of diagnostic and action activities corresponds to the moderately complex.

Experience/training criterion: Although there may be some operator with little experience in the unit, the starting of the compressor always accounts with more experienced professionals. In addition, regardless of experience, all employees, when joining the company, go through a period of one year of training. After this entrance, employees are continuously submitted to training recycling. The interviewed operator has more than ten years of experience in the role, however, he has only two years in the unit under study. Thus the nominal level for both activities was selected.

Procedure criterion: The starting task of the CO_2 compressor has a specific standard procedure, which is followed by the operation. This document is reviewed continuously whenever necessary, but it is not seen as a 100% complete document. For example, there is no evidence of human factor alerts (need for greater physical fitness, greater concentration/attention, communication, etc.) in the execution of the task. The instructions are oriented to the aid of cognitive and action activities in a single document—there is no specific operational procedure for diagnosing abnormalities. Thus for the purpose of quantifying the HEP in the action task, the classification of the procedure was considered as nominal and, in the diagnostic task, the existence of an incomplete procedure was taken into account.

Ergonomics/HMI criterion: The compression system design (Clegg, 2000) supports the correct performance, but it does not maximize performance. The computer interface is adequate, however, not all control variables are available on the dashboard. Some readings are performed in the field in non-ergonomically favorable regions. Therefore, it was considered that the delimiting factor is nominal.

Aptitude for work criterion: From the point of view of physical conditions, the activities performed at the start of the compressor do not require special skills. Although the compressor is above ground level, only two flights of stairs are required. Auxiliary systems also do not require efforts in performing the maneuvers. Due to the risks involved in the task, an operator will hardly perform the activity without the necessary focus. This same reasoning eliminates the possibility of the individual being under the influence of drugs (legal or illegal) or distracted. Safety is always put first. If there is something that could endanger the one's life or co-workers' life, some action is taken, even if it results in the stop of the activity. Therefore, in this work it is considered that the level of the delimiting factor is nominal.

Labor relations criterion: Work processes include any management, organization, or supervision factors that may affect activity performance. In general, there is a good interpersonal relationship between the members of the organization. On the other hand, some factors negatively affect the performance of activities:

1. Communication between the operation, maintenance and auxiliary teams present obstacles related to the investigation of events—culture of guilt;
2. There is a lack of policies more directed to the investigation of incidents and abnormalities that occurred in the operation daily life with the involvement of professionals from different areas;
3. Inadequate completion of the shift register (lack of record of failure signs, outdated information);
4. Improper reporting of failure events—incomplete information;
5. Behavioral issues affect the activities accomplishment, such as the delay in releasing work permits;
6. The change in the staff—change of managers, and transfer of team members.

Therefore, when evaluating the whole set of characteristics raised, two scenarios are proposed. In the most appropriate scenario, the delimiting factor level was considered as nominal. In the most negative scenario, the labor relations are understood as bad, because there are enough elements to negatively influence activities, and stress was considered high. Tables 6.6 and 6.7 show the levels of the eight selected PSFs, as well as the value of each multiplier, considering the scenarios.

TABLE 6.6 Performance shaping factor multipliers—softer scenario—petrochemical industry.

Performance shaping factor (PSF)	Diagnosis	PSF level	Action	PSF level
Available time	0.1	Extra time	0.1	$T \geq 5$ times the required
Stress	1	Nominal	1	Nominal
Task complexity	2	Medium complexity	2	Medium complexity
Training/experience	1	Nominal	1	Nominal
Procedures	20	Incomplete	1	Nominal
Ergonomics	1	Nominal	1	Nominal
Aptitude for work	1	Nominal	1	Nominal
Labor relations	0.8	Good	0.5	Good

TABLE 6.7 Performance shaping factor multipliers—most negative scenario—petrochemical industry.

Performance shaping factor (PSF)	Diagnosis	PSF level	Action	PSF level
Available time	0.1	Extra time	0.1	Time ≥ 5 times required
Stress	2	High	2	High
Task complexity	2	Medium complexity	2	Medium complexity
Training/experience	1	Nominal	1	Nominal
Procedures	20	Incomplete	1	Nominal
Ergonomics	1	Nominal	1	Nominal
Aptitude for work	1	Nominal	1	Nominal
Labor relations	2	Bad	5	Bad

In the first scenario (Table 6.6), none of the tasks presented three or more PSFs with negative influence, which means that the calculation can be performed using Eqs. (6.1) and (6.2), according to the proposed calculation methodology.

Thus for the task of action and diagnosis, the HEP is given by:

$$HEP_{Diagnóstico} = 0.01 \times (0.1 \times 1 \times 2 \times 1 \times 20 \times 1 \times 1 \times 0.8) = 0.032$$

$$HEP_{Ação} = 0.001 \times (0.1 \times 1 \times 2 \times 1 \times 1 \times 1 \times 1 \times 0.5) = 0.0001$$

The total error probability is calculated using Eq. (6.5):

$$HEP_{Total} = 0.0320 + 0.0001 = 0.0321$$

And finally, the application of total HEP in Eq. (6.6) to identify the value corresponding to human reliability:

$$R_{hu} = 1 - 0.032 = 0.968 \times 100\% \rightarrow R_{hu} = 96.8\%$$

Considering the most negative scenario (Table 6.7), the presence of three negative PSFs was verified for the action task and four negative PSFs in the diagnostic task. Thus HEP is calculated from Eqs. (6.3) and (6.4).

$$HEP_{Diagnosis} = \frac{0.01 \times (0.1 \times 2 \times 2 \times 1 \times 20 \times 1 \times 1 \times 2)}{0.01 \times (0.1 \times 2 \times 2 \times 1 \times 20 \times 1 \times 1 \times 2 - 1) + 1} = 0.139$$

$$HEP_{Action} = \frac{0.001 \times (0.1 \times 2 \times 2 \times 1 \times 1 \times 1 \times 1 \times 5)}{0.001 \times (0.1 \times 2 \times 2 \times 1 \times 1 \times 1 \times 1 \times 5 - 1) + 1} = 0.002$$

The total HEP is also calculated using Eq. (6.5):

$$HEP_{Total} = 0.139 + 0.002 = 0.141$$

And finally, total HEP in Eq. (6.6) is used to identify the value corresponding to human reliability:

$$R_{hu} = 1 - 0.141 = 0.859 \times 100\% \rightarrow R_{hu} = 85.9\%$$

6.3.5.6 Considerations for the fertilizer industry case

Compressor-related activities are supported by the level of automation and instrumentation of the system. The task of starting the equipment is not different, although it requires more physical effort than usual and a higher level of attention, but a very high probability of human failure was not expected. The application of the method resulted in a value of 3.2% for HEP. The positive evaluation (value less than or equal to 1) of the PSFs contributed to a final result of high human reliability—96.8%. The team ensures that the procedures are followed in their entirety, and are also accompanied by a checklist. Although the plant is quite old, the equipment has a modern control and protection system in which the human-machine interface provides an advantage for operation.

In this scenario, two observations are contrary to this result. The first is that currently the company has been going through a series of changes and this, directly or indirectly, caused a change in the work environment. This possibility of a more negative scenario was also modeled in SPAR-H. In this scenario, there was a change of two PSFs—"stress/stressing agent" and "Labor Relations"—which resulted in a change in the failure probability—an increase from 3.2% to 14.1% of the HEP was observed, that is, almost five times more.

With the mechanization of functions, the organization has a more decisive role to ensure the reliability of the entire system. It is worth remembering that of the eight PSFs used in the SPAR-H method, only one of them (the aptitude for the service) does not have a direct relationship with the organization. Therefore, a change in the desktop has the potential to have a major impact on the performance of activities. The second observation was raised by the operation coordination engineer. The engineer considers that, based on experience, daily operation routine is strongly influenced by sociocultural aspects. This thought confronts results found in the first scenario. Although it is not explicit in which activity this phenomenon has the most influence, the HEP of 3.4% may not represent the reality of the current scenario. Therefore, the engineer's view corroborates the result of the second scenario.

6.3.6 Coconut industry

The descriptions of this case were based on the completion of course work of Ghabriel Anton Gomes de Sá for the graduation in Chemical Engineering (2019) at the Federal University of Bahia, guided by Salvador Ávila. Ghabriel participated in GRODIN (Research Group on Operational and Dynamic Risks) under the leadership and guidance of Professor Salvador Ávila Filho, Dsc. (Sá, 2018).

6.3.6.1 Operational context

In this sixth case, the application of the SPAR-H method in an industrial coconut processing unit, located in state of Bahia, will be discussed. This industry is an activity that is integrated in agriculture through the raw material, coconut, as shown in Fig. 6.14.

Coconut presents advantages because it is ecological and renewable, belonging to the family of hard fibers, has cellulose and wood as the main fibrous components, which give it high levels of stiffness and hardness, thus becoming a versatile material, given its strength, durability, and resilience.

The process of plant extraction and the logistics of transporting coconut to the processing demands care in the harvest, transport and storage of these fruits. In general, for water and pulp processing, coconuts should be uniform in their characteristics. They should be harvested when they reach their maximum size, and, depending on the variety, this can occur between five and seven months after fertilization.

At this age the coconuts present the water with highest quantity and quality, have high solids content (cream in the form of film), and have the highest composition of mineral salts and albumin dissolved in water.

The quality of the fruit is affected by the time between harvest and processing, hence the need for the fruits to be handled with care and that transport needs to be carried out as soon as possible, in covered vehicles and at times of milder temperature. Care in logistics from the farm to the industrial plant unit is extreme, and it is recommended that the transport be done up to three days after harvest.

The storage of raw material in the industry aims to regulate the flow of processing and avoid discontinuities in the operation of the production line. The sheds to store the fruits must also meet specific requirements: screen protection to avoid small animals; good ventilation; protection to prevent sunlight; and to avoid temperature rising.

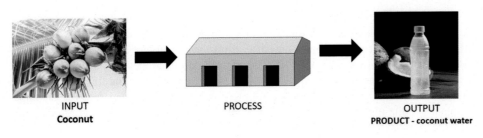

FIGURE 6.14 Coconut processing industry. *Courtesy Freepik.com.*

6.3.6.2 Task

The task chosen for the application of the SPAR-H method is the monitoring of coconut processing processes, which takes place in the control room that is the operators' living environment. In this environment the process control interfaces and the maintenance and engineering support team are installed. It is essential that there is good communication between teams and people in the various roles. It is also important to know how to interpret the data and everyday situations.

It was found in the initial interview about the team profile that all employees, including the supervisor, have worked for 2 to 5 years in the company. In addition, it was possible to observe that no employee is older than or equal to the supervisor. This fact may be due to a possible management strategy of the company, where it is believed that the chronological difference of ages between the leader and operators results in greater respect towards the leader, who is the most experienced.

6.3.6.3 Application of SPAR-H

For the application of SPAR-H in this monitoring team, a form was applied to the entire control room team: the operation technicians and the Supervisor. After applying the forms, the results reached a final value based on the most operators' opinion. Table 6.8 showed the presence of three negative PSFs for the diagnostic task. Thus for three or more negative PSFs, HEP for the diagnostic task is calculated from Eqs. (6.3) and (6.4).

$$\text{HEP}_{\text{Diagnosis}} = \frac{0.01 \times (10 \times 2 \times 2 \times 1 \times 1 \times 1 \times 1 \times 1)}{0.01 \times (10 \times 2 \times 2 \times 1 \times 1 \times 1 \times 1 \times 1 - 1) + 1} = 0.288$$

For the action task, Table 6.8 does not present three or more PSFs with negative influence, which means that calculation can be performed using Eqs. (6.1) and (6.2), according to the proposed calculation methodology.

$$\text{HEP}_{\text{action}} = 0.001 \times (10 \times 1 \times 2 \times 1 \times 1 \times 1 \times 1 \times 1) = 0.02$$

The probability of total human error is also calculated using Eq. (6.5):

$$\text{HEP}_{\text{Total}} = 0.288 + 0.02 = 0.308$$

And finally, the application of total HEP in Eq. (6.6) to identify the value corresponding to human reliability:

$$R_{\text{hu}} = 1 - 0.308 = 0.692 \times 100\% \rightarrow R_{\text{hu}} = 69.2\%$$

TABLE 6.8 Performance shaping factor values for the activity in the coconut processing industry.

Performance shaping factor (PSF)	Diagnosis	PSF level	Action	PSF level
Available time	10	2/3 of T required	10	Available $T = T$ required
Stress	2	High	1	Nominal
Task complexity	2	Average complexity	2	Medium complexity
Training/experience	1	Nominal	1	Nominal
Procedures	1	Nominal	1	Nominal
Ergonomics	1	Nominal	1	Nominal
Aptitude for work	1	Nominal	1	Nominal
Labor relations	1	Nominal	1	Nominal

6.3.6.4 Considerations for the coconut processing industry case

From the application of SPAR-H it was possible to perceive, unanimously among the team's responses, that the main factor that is negatively influencing the error occurrence is the "Available Time." The time available in this monitoring activity is very close to the required time, which causes discomfort, pressure regarding the accomplishment of the activity and, consequently, the increase of stress to perform it in a timely manner. A readjustment or adaptation in the process is necessary, revision of the procedure and, if required, verification of the technology used. The "Available Time" factor is one of the main factors in the analysis of SPAR-H, due to its assertiveness in being able to perform the task, or not to perform it, as well as in meaning the calculation of the HEP directly at 1, that is, 100% probability of failure, represented by: the available time is less than the time required. This is an extremely important item for applying an action plan.

6.3.7 Packing list services in the manufacture of sports products

The descriptions of the case were based on the Article published in ESREL (2016) by Priscila Pereira Suzart de Carvalho and other authors, which includes Salvador Ávila Filho. Priscila participated in GRODIN (Research Group on Operational and Dynamic Risks) under the leadership and guidance of Professor Salvador Ávila Filho, Dsc. (Carvalho et al., 2016).

6.3.7.1 General context

In this seventh and final case study, a service process in the industry is evaluated, not involving the area of industrial operation, but involving the area of materials control in logistics. The SPAR-H method will be applied to the process involving material manufacturing. Technical or human failures occur in the operational activities of material control in the sports industry. The company works with the manufacture of sports products, has updated technology and accounts with high insertion in market. It is a multinational company with 100% national capital and is present in 14 countries. The company has been in the market for more than 70 years and has factories located in the northeastern states of Brazil, in addition to a unit in Paraguay. There is the total of 3000 direct employees.

This industry began its activities in 2000 with the manufacture of balls produced with synthetic material for the practice of sports. This production consists of 5 final product lines and 6 semi-finished material lines, being supplied by a range of approximately 450 production inputs. The production is composed of an ERP System (Enterprise Resource Planning) comprising several modules.

The management profile is centralizing and the company owners came from banking activities (banks financial culture). The company does not have a career plan for its employees, which negatively affects the team's motivation, increasing its turnover. These organizational characteristics impact on the company's human capital that has low standardization of administrative tasks, does not perform investigation of failures and does not collect data. The training of new employees takes place from practical training based on experience.

6.3.7.2 Process analysis

The service process is related to controlling the movement of the product. The sports balls elaborated in two stages, we will focus on the first, which is the processing of synthetic and natural rubbers and synthetic laminates with fabric-based structure and resin coating (Jesus et al., 2013) and which is highlighted in blue color in Fig. 6.15.

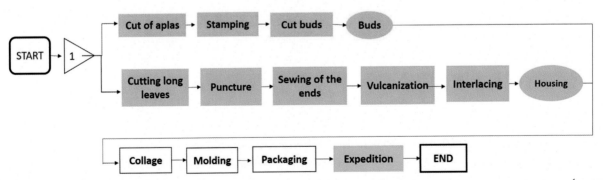

FIGURE 6.15 Flowchart of the production process. *Reproduced with permission from Carvalho, P.P.S., Fonseca, M.E., Kalid, R.A., Ávila, S.F., 2016. Integrated reliability analysis in task of emission production orders in ERP system: a case study on a company's sports segment. In: Proceedings of ESREL—European Safety & Reliability Conference, Glasgow.*

6.3.7.3 Task

The finalization of each product batch is recorded in inventory control through the logistics system. This batch is made available for sale and earnings. Product quantities manufactured in the system are entered and, in parallel, the reduction of the components and raw materials used for the production of this product is posted. This decrease in materials occurs according to the structure of the product informed by the company's engineering sector and prepared through a bill of materials. This reduction updates the inventories in the warehouse for the control of the production lines. This set of activities is equivalent to the output of the raw materials from its inventory and the entry of the final product into the finished product stock/inventory through the system. This task is performed by the Production Planning and Control (PPC) team, which has 4 employees composed of 2 production engineers, 1 administrator and 1 intern. The responsibility for the logistics and warehouse area is assigned to the industrial manager. Employees receive a set of guidelines to follow and must work with the management hierarchy to make any changes to work roles. The employee who controls the stock/inventory is an engineer, male, aged 25–30 years and responsible for the activity control report. This task requires information about inventory management, bill of sale, and issuance control. This task requires skill in the use of computers and in handling software such as enterprise management system.

Despite being part of a team, the employee who carries out the posting of production orders does not work on a group basis, because the divisions of tasks in the PPC sector are specific to each member. The operator has a working day of 8 hours, and when necessary, works overtime. The operator is subjected to constant pressure, since the company's earnings depends directly on the correct execution of its task.

6.3.7.4 SPAR-H application

The methods for calculating human reliability were initially developed for the nuclear industry. In the case of application in other industrial segments and services it may be necessary to review the multiplier values for human PSF and values for the initial NHEP.

In this case, the SPAR-H method was applied in the production order posting activity for the sports industry to reduce the probability of errors in the process, increase productivity and reduce financial risks. Data was collected to analyze the influence of Man in the process through interviews with the collaborators who perform the activities.

The PSFs that influence the task were collected, listed and evaluated according to the original method. The questionnaires were applied to the company's workers in the material handling control service. These values are in Table 6.9.

6.3.7.5 Scenario with original calibration

For the determination of human error (HEP) the factors determined in Table 6.9 were used, considering NHEP = 0.01 for diagnostic task and NHEP = 0.001 for action task. The presence of three negative PSFs was verified for the action task and also for the diagnostic task. Thus the HEP is calculated from Eqs. (6.3) and (6.4).

$$HEP_{Diag.} = \frac{0.01 \times (1 \times 2 \times 1 \times 10 \times 50 \times 1 \times 1 \times 1)}{0.01 \times ((1 \times 2 \times 1 \times 10 \times 50 \times 1 \times 1 \times 1) - 1) + 1} = 0.91$$

TABLE 6.9 Performance shaping factor for product control service in manufacturing, original SPAR-H method.

Performance shaping factor (PSF)	Diagnosis	PSF level	Action	PSF level
Available time	1	Nominal	1	Nominal
Stress	2	High	2	High
Task complexity	1	Nominal	1	Nominal
Training/experience	10	Low	3	Low
Procedures	50	Nonexistent	50	Nonexistent
Ergonomics	1	Nominal	1	Nominal
Aptitude for work	1	Nominal	1	Nominal
Labor relations	1	Nominal	1	Nominal

$$\text{HEP}_{\text{action}} = \frac{0.001 \times (1 \times 2 \times 1 \times 3 \times 50 \times 1 \times 1 \times 1)}{0.001 \times ((1 \times 2 \times 1 \times 3 \times 50 \times 1 \times 1 \times 1) - 1) + 1} = 0.23$$

Eq. (6.5) was used to calculate HEP considering aspects of diagnosis and action in the activity evaluated. The human reliability was calculated using Eq. (6.6):

$$\text{HEP}_{\text{Total}} = 0.91 + 0.23 = 1.14 (114\%)$$

High values were perceived in the HEP results for both the probability of error for the action task and for the diagnostic task. The sum of these probabilities resulted in a $\text{HEP}_{\text{total}}$ of 1.14, that is, a probability of 114% of the error to happen in this task. This probability exceeded the maximum value, usually limited to 100%.

There is the possibility of being true, so, to do the task correctly, it is necessary to repeat the same task twice or more, or by two people or more. This situation is quite abnormal, thus indicating a high fixed cost for keeping a team and constant reviews in the emission of packing list.

The other possibility may be due to differences between operation activity and nuclear power plant diagnostics compared to the manufacturing industry.

This possibility is closer to the reality found in the company. It is concluded that this analysis of Human reliability by the original calibration of SPAR-H is inadequate for the reality of the task, as well as if we admit a probability of 114% in the task, it would be to affirm that the occurrence of the failure is certain, and its frequency is high.

This probability value is an example of a poorly calibrated NHEP or PSF for the task/industry profile. As discussed in the operational context and in the job description, this is not an activity of the industrial area, but an activity of registering product entries and exits in a system.

Therefore, in an attempt to adapt these values to the task reality, the recalibration of NHEP and PSF was suggested to the PPC responsible. To facilitate for the company, we performed an exercise with the change, considering the following task characteristics.

- The occurrence of human error in the activity has a higher frequency than in an industrial environment, considered as reference the original calibration in the case of the nuclear industry;
- The risk in the activity is lower than in an industrial environment, also considering the reference of the original calibration in the nuclear industry;
- The high frequency of releases of entry and exit of materials, mainly finished products in the system, makes it repetitive and induces greater error, more frequent, amenable to correction without major damage;
- Cognitive pressure and the consequent stress caused by management seek to avoid the occurrence of bottlenecks in the system (inventory, production, financial), and cause high stress situation, in repetitive and non-complex activities;
- The lack of procedures in low-risk activity, brings non-standardization, requiring training and experience for the task.

The PSF Procedure was recalibrated, reducing the score from non-existent to 20. The PSF Training/Experience was also recalibrated for diagnostic task, due to alignment of action task values and diagnostic tasks (reduction of errors severity), leaving 5 for diagnosis and keeping 3 for action. Based on management observations, the NHEP was also adjusted for the severity and frequency of the analyzed activity. For the new HEP calculation, the NHEP values were reversed due to the fact that it is a posting system, where when an error occurs, it automatically already presents where, when, what failed, besides not requiring a pre-analysis before performing the action. The system automation and the practicality in the diagnostic task caused the occurrence of diagnostic failure to become less than the occurrence of the failure in the action task, so we established that:

- NHEP for diagnosis is equal to 0.001, meaning the operator has a 1 out of 1000 chance of performing a misdiagnosis.
- NHEP for task action is equal to 0.01, that is, the operator has a 1 out of 100 chance of performing a wrong action.

For the new determination of the HEP the same factors (PSFs) were used but with some different values (Table 6.10). The presence of three negative PSFs for the action task and for the diagnostic task was again verified. Thus the HEP is calculated from Eqs. (6.3) and (6.4).

$$\text{HEP}_{\text{Diag.}} = \frac{0.001 \times (1 \times 2 \times 1 \times 5 \times 20 \times 1 \times 1 \times 1)}{0.001 \times ((1 \times 2 \times 1 \times 5 \times 20 \times 1 \times 1 \times 1) - 1) + 1} = 0.167$$

TABLE 6.10 Performance shaping factor for product control service in manufacturing, SPAR-H method changed.

Performance shaping factor (PSF)	Diagnosis	PSF level	Action	PSF level
Available time	1	Nominal	1	Nominal
Stress	2	High	2	High
Task complexity	1	Nominal	1	Nominal
Training/experience	5	Low	3	Low
Procedures	20	Nonexistent	20	Nonexistent
Ergonomics	1	Nominal	1	Nominal
Aptitude for work	1	Nominal	1	Nominal
Labor relations	1	Nominal	1	Nominal

$$\text{HEP}_{\text{action}} = \frac{0.01 \times (1 \times 2 \times 1 \times 3 \times 20 \times 1 \times 1 \times 1)}{0.01 \times ((1 \times 2 \times 1 \times 3 \times 20 \times 1 \times 1 \times 1) - 1) + 1} = 0.548$$

Eq. (6.5) was used to calculate the HEP considering the aspects of diagnosis and action in the activity evaluated.

$$\text{HEP}_{\text{Total}} = 0.167 + 0.548 = 0.71$$

And finally, the application of total HEP in Eq. (6.6) to identify the value corresponding to human reliability:

$$R_{\text{hu}} = 1 - 0.71 = 0.29 \times 100\% \rightarrow R_{\text{hu}} = 29\%$$

6.3.7.6 Considerations on the new calibration of performance shaping factor and NHEP

As the PSF and NHEP are adjusted, the values were closer to the reality of the analyzed task. The new values: HEP = 71% and R_{hu} = 29%, were validated by management, which exposes the constant frequency of failures due to the critical factors involved in the task, such as stress, low task training, and the lack of a procedure to guide, or signal observations or frequent system problems. The company needs an action plan that involves: intensive training and the strengthening of the experienced operators' knowledge; the hiring of more workers to reduce stress and pressure in the execution of the activity; technology analysis; and preparation of procedures with best practices, observations about the systems, sequencing of steps and observations on critical items/ practices of the process.

6.4 Comparative analysis

The variation of social culture, discredits of management, degradation of equipment materials, different types of procedures and processes influence the team performance, the leaders and the production staff. The cases of industry and services presented in this book have different results of HEP.

The lack of commitment in the operational routine and related to good organizational practices induces the degradation of human labor in industrial activities. This situation can bring negative results that can lead to discontinuity of business, by fines, high costs or negative image of the organization for society.

Critical analysis using the SPAR-H method can prevent accidents and increase productivity. For this reason a summary of the applications that facilitate through this comparative analysis was prepared, and suggestion of general recommendations to be implemented for the various types of activities.

Table 6.11 gives the application descriptions, industry and product type, process type, equipment, people and tasks as well as the presentation of the HEP, and the main PSFs. Finally, recommendations are presented.

TABLE 6.11 Comparative analysis for cases of SPAR-H application.

Description	Industry product	Process	Equipment	People	Task	Critical performance shaping factor (PSF)	Recommendation
(C1) Regional industry with organizational change	Chicken (Food)	Manufacturing	Packer	Low rotation employees and with kinship	Trigger the packer	HEP = 32%; Labor Relations, stress, and training	High HEP and loss of time. Work in the area of culture and organization credibility
(C2) Regional industry with environmental risk	Nuclear Uranium Processing—Fuel	Continuous in Mining	Gyratory crusher	Regional employees from Salvador	Control Panel Activities	HEP = 3%; stress and complexity	This probability is compatible with the risk of working with uranium. To reduce, it is important to improve stress.
(C3) Chemical process industry, environmental impact	Hydrochloric Acid	Continuous and Complex	Panel and Field, Tanks and Pumps	Trained operators from Salvador	Load HCl	HEP = 25.4%, Procedure, Ergonomics, Complexity and Stress	HEP is high for the risk of HCl, it is important to have time to analyze the task and simulate situations.
(C4) Chemical process industry, safety and environment	Propylene	Batches in continuous process	Vertical pump	Trained operators from salvador, organizational climate	Pumping of propane	HEP = 57%, procedure, ergonomics, stress and complexity	The HEP is very high and was not clearly felt because it is a batch operation. High production cost. A specific procedure is required.
(C5) Chemical process industry, safety and environment	Urea fertilizers	Continuous process involving CO_2	CO_2 compressor	Trained operators from salvador, organizational negative climate	Control of urea synthesis and production stops	HEP = 14.1%, procedure complexity and stress	Although the standard HEP value is 10% and has been calculated 14%, it is recommended to improve the procedure and automation and training.
(C6) Coconut manufacturing industry	Coconut water (Food)	Manufacturing and continuous process	Control room	Regional employees and manual labor	Monitoring and process control	HEP = 30.8%, available time, stress and complexity	High HEP for Industry, thus indicating high rework level and perhaps quality problems. Inserting automation into processes can be an option.
(C7) Control service on product release	Sports balls (Manufacturing) Issuance of packing list	Manufacturing and Logistics	Computer and software	PPC engineers and technicians	Registering inputs and outputs in the PPC control system	HEP = 71%, stress, training and procedures	Although it calibrated NHEP and PSFs, HEP remained high; however, it is compatible with the company reality. Inserting procedures into critical transactions, increasing workers number, and investing in training are good options to reduce HEP.

HEP, human error probability; *NHEP*, nominal human error probability; *PPC*, Production Planning and Control; *PSF*, performance shaping factor.

6.4.1 Recommendations

(C1 and C2) Equipment failures caused by human errors cause production delay, requiring, for its recovery, the execution of activities in a shorter time. The cycle of non-quality at work is then formed, thus generating stress due to lack of time. But the question is, what is the root cause of this problem? It's probably in the area of culture, where one thing is said and another is done, they do not match. This situation can create the lack of credibility of the organization. This situation becomes even more difficult and stressful if the product hazard is very high-risk.

(C3) The repetition of the task, the boring routine, often decrease the level of commitment despite the high toxicity of the product handled. Routine is dangerous when working memory and commitment to organizational goals are not valued. Therefore, human errors can occur and incidents and accidents may happen, resulting from the procedure.

(C4) Batch operations or administrative services are apparently more repetitive than field procedures, thus leading to the possibility of human error. It is important to create or use tools to draw attention to the procedures steps. Therefore, specific tools are required.

(C5) The analysis of the task and its requirements can increase the understanding of the operational procedure realization and what are the desired goals. In addition, it is recommended to implement automation and training to avoid accidents resulting from human error.

(C6) The "fair" time or lack of it to perform the tasks is the critical factor of this analysis, so it is recommended to insert automation in the processes, as well as the adjustment of the task procedure along with reevaluation of the technology. These options are suggestions that aim to re-adequate the critical PSF "Available Time" with the time required to perform the activity in a timely manner.

(C7) The absence of procedure, high stress derived from pressure by management, and the low level of training cause errors, financial damage, and bottlenecks in the systems and in the production line. As it is not a critical process in a matter of threat to life, the error occurs very often, thus causing rework and unnecessary costs. It is recommended to insert procedure for critical transactions, increase the operator's team to reduce pressure and stress, and offer practical and online training for critical systems/transactions, with update cycles.

6.5 Integrated reliability: the beginning

In recent decades, industries have sought to identify the "best" strategies and practices that provide superior performance in their processes because the pressure of competitors leads organizations to increasingly idealize competitive and sustainable positions. Factors such as market globalization, mergers and acquisitions and emerging new technologies are the biggest reasons for increased competitiveness. In this intention of business success, both the internal and external environments must be understood, as they have variables that influence the companies' development over time. With this, many rules and restrictions arise, which are imposed on the industry operation, requiring integrated management.

The low integration of activities or functions is the result of inadequate communication tools that affect process efficiency in industry and services. Work shifts are often concerned with isolated results, not integrating events more broadly, thus leading to hasty decisions and adopting actions that can maximize failure as a result of human errors.

The integration demanded by globalization makes the organization integrate its information systems, improving the quality of the decision, in which it becomes possible to clearly identify the requirements for activity in the production routine.

Despite the integration of control systems in Quality, Safety, Environment and Health, serious accidents still occur in industries caused mainly by incorrect decisions of production leaders. This is the case of the Fukushima Nuclear Power Plant accident (Aoki and Rothwell, 2011) and the accident on the Horizon Platform (Beyer et al., 2016) in the Gulf of Mexico. These major accidents raise questions about the effectiveness of the integrated management system and the production leaders' capacity to solve complex problems, which in practice compromises industrial operability (Mohaghegh et al., 2009).

Successful operations, on the other hand, must be identified and repeated to seek better results. Not always the operations written in the report are, in fact, what is presented on the shop floor. So, what to do when the omission of supervision or staff occurs?

In fact, the operational leader is not structured for quick and complex decisions. In addition, operators are not part of the decision process, due to management centralization error, lack of training and perception, or because most important actions are automatic in process and safety controls.

Increased complexity can increase the risk of accident, therefore, to control the situation, it is important to recognize the signs of failure in the operation routine and in the natural environment. The operation requires the strengthening of structures and control systems to carry out field activities. The lack of integration, team preparation and tools for deciding over complex situations unfortunately cause crisis situations to occur, thus the importance attributed to the integration of operations, with guarantees of mutual commitment between companies and employees (Ávila et al., 2019).

The best operational performance results from successful management processes to meet the current socioeconomic and environmental challenges, which have a dynamic format. The control of operational reliability in risk activities is achieved with new tools or tools adapted for task analysis and training focused on the acquisition of skill in operation (Rognin et al., 2000).

Operational activities achieve excellence in safety and production when they are planned considering social, technical, organizational and natural environment requirements. Thus they achieve a result of high reliability of the system, which brings as a consequence good indicators in production and safety. Of course, social requirements are subjective and depend on intangible aspects that are difficult to measure and control. Managers should learn to work and control with human performance factors.

The operations team integration with technology and support areas requires changes in procedures, adjustments in operating systems and design of new systems (equipment and processes) to reduce operational costs and environmental impacts. Knowledge about the characteristics of processes, people and culture, allows to adapt the operational routine to the best practices. The definition of best practices should be carried out from the beginning of the plant, allowing investments to reduce the cost and size of the new facilities (Laidoune et al., 2016; Pence et al., 2019).

Chidambaram (2016) analyzed the influence of human and organizational factors on equipment and process failure in petrochemical companies. This analysis indicated that the operation was unprepared to control the automated production system and the possibilities of human error in the design, operation and maintenance of production systems.

Avila and Santos (2014) state that reliability programs for the demand of sustainability must be implemented through the integration of operations. This integration aims to reduce costs by reducing process losses through the adequacy of the job (Ávila et al., 2019).

According to Borges and Menegon (2009), operating systems are formed by two subsystems: (1) Technical Subsystem that consists of mechanisms, machines, equipment and components; and (2) Social Subsystem consisting of human beings.

Ávila (2013b) agrees that the analysis of reliability in the industry is not only about equipment, but also, includes, requirements of operation, process and human factors. Therefore, it is necessary to assume the need for a new management style to deploy integrated reliability.

This new manager will set controls in a trusted organization that recognizes changes in natural and social environments. These settings influence the performance of processes, equipment, and people, which corresponds to new resource needs and cultural changes that affect safety. Ávila (2013b) suggests a conceptual model of Integrated Reliability described in Fig. 6.16. The objective is to facilitate the development of integrated techniques to analyze failures, accidents, tasks and hazards, including, simultaneously, the aspects of operation, process, maintenance and human factors.

FIGURE 6.16 Integrated reliability model. Ávila, S.F., Santos, D., 2014. Integrated reliability: failure assessment to process, operation, equipment and human integration (POEHint), a case of propene pump operation. In: Sixth International Conference on Engineering Failure Analysis, October 2015, Lisbon.; Ávila, S.F., 2015a. Reliability analysis for socio-technical system, case propene pumping. Engineering Failure Analysis Journal, 56, pp: 177–184.

The integration between Man and machine is facilitated by a good human-machine interface design or human-process. Clearly controlled processes help in this integration between human reliability and Process Reliability. The business high sustainability is the result of a measurement program with the establishment of operational limits for the process, product and critical equipment.

Finally, successful integration of System Reliability is achieved with adoption of new organizational standards and new management styles. The analysis of sociotechnical indicators of the routine allows the calculation of Integrated Reliability, thus showing the areas of discomfort for equipment, process, operation and human factor that format performance.

6.5.1 Control and metrics to achieve integrated or sociotechnical reliability

The industry always states in reports that the factors involved as root cause of operational problems are technical. However, political, organizational or functional decisions are forgotten, which, due to discomfort with the loss of power, causes human errors in diagnosis and in activities related to procedures.

In this way, organizational policies and practices can impair actions, controls and degrade organizational resilience. To discuss the Integrated Reliability in the organization, we should consider the effect of cognitive load on the work team, which occurs from inadequate decisions and policies of leadership, impoverishing the time for the functional study of failure.

The organizational environment can raise the level of stress at work. In addition, the natural climate, management profile and group leaders also influence level of stress and provoke different situations and task effectiveness levels.

Other factors that influence the appearance of failure in the production system are: (1) qualitative or quantitative aspects in materials, equipment and products; (2) change in legislation with different local and global environmental requirements; (3) specific demands of labor legislation; (4) corporate social responsibility requirements in relation to society; and (5) emerging social processes that induce new behaviors at work and in the market.

The task complexity depends on the characteristics of the worker, the machine, and the process. This complexity affects the level of investment in new projects and depends on the level of dynamic risk from the calculation of indicators.

Ávila & Pessoa (2015b) developed a system of equations to calculate the integrated reliability based on the algorithm in Fig. 6.16, in which the final Integrated Reliability (IR) is the result of the correct adjustment of social attractiveness, elaborated from Eq. (6.7). Attractiveness depends on the analysis of leadership, the team's level of aggregation, the number of abnormalities warnings in the routine and the dynamic indices of stress on the task and occupational health. Increased attractiveness decreases the dynamic risk of incidents occurring.

Integrated Reliability also depends on the level of complexity (high, medium and low) of production systems with different Eqs. (6.8), (6.9) and (6.10). For each complexity level, we suggest using equations for calculation based on expert opinions on operating failures in the chemical process, manufacturing, and service industries. According to Embrey (2000), complexity is related to the number of steps that have the procedure, which affects attention and commitment. We understand that this complexity may also be associated with aspects that cause noise in the mind map, thus impairing decision-making in the task, such as:

1. The accomplishment of the task;
2. The type of process control and safety technology indicates the level of automation;
3. The presence of auxiliary systems that require operations and controls for the main function of the equipment, may indicate complexity;
4. The level of attention in the task depends on the types of workers selected and trained;
5. If the process has a linear flow with higher prediction or a complex flow with reduced forecast.

$$IR_{final} = Attractiveness \; Ж \; IR \qquad (6.7)$$

Ж Subjective function with discrete number (social aspects at organization)

$$High\ complexity \rightarrow IR = (R_{pr})^2 * R_{eq} * (R_{hu} * R_{op})^{1/2} \qquad (6.8)$$

$$Medium\ complexity \rightarrow IR = R_{pr} * R_{eq} * (1.5 * R_{hu} * R_{op}) \qquad (6.9)$$

$$Low\ complexity \rightarrow IR = (R_{eq})^{1/3} * R_{pr} * (1.7 * R_{hu} * R_{op})^2 \qquad (6.10)$$

Being:
$R_{pr} = f$ (capacity of critical process variables and intermediate purity)
$R_{eq} = f$ (critical equipment failure rate)
$R_{hu} = f$ (task, work organization, human type)
$R_{op} = f$ (task complexity, time effectiveness, relative cost)

In the method we suggest the use of Eq. (6.11) to calculate the individual complexities, and Eq. (6.12) to calculate the reliability of the task or operation:

$$\text{Complexity} = (1) * (2) * (3) * (4) * (5) * 100.000 \tag{6.11}$$

$$R_{op}(\%) = \left(100 - \left(\frac{\text{LOG(complexity)}}{4}\right)\right) * 100 \tag{6.12}$$

As shown in Fig. 6.17, reliability is calculated based on the level of complexity identified by project information (techniques such as HAZOP and FMEA, social, cognitive or technical), following green or red lines. After the operation of the process, a revision is made to change the premises for the calculation of reliability.

Each term is a function of factors defined by the method, where: "R_{pr}" is defined as the process reliability, depending on the criticality of the critical variables and how they interfere in the outputs; the term "R_{eq}" refers to the equipment criticality, being a function of the failure rates and mean time between failures of the critical machinery; "R_{op}" is defined as operational reliability, being a function of the tasks complexity and relative cost; finally, "R_{hu}" refers to Human reliability, being a function of the task, work environment and organizational culture (Ávila & Pessoa, 2015b).

The analysis of the task provides opportunities for the appreciation of the operation-man integration, discussing the elements involved in its execution, especially with regard to risks and their impacts. Human reliability is the product of a favorable organizational environment with routine management and problem solving in teams, practiced and measured in this chapter through SPAR-H. The reliability and maintainability of equipment depend on the degree of automation and reliability in the design process, and may fail due to time of use or human errors.

It is noteworthy that human reliability (R_{hu}) was this chapter's object of study, being calculated through the SPAR-H method, and presented in several cases of industrial processes, such as chicken industry, uranium production industry, chemical industry, propylene production industry (refinery), fertilizer industry, and even in a service process.

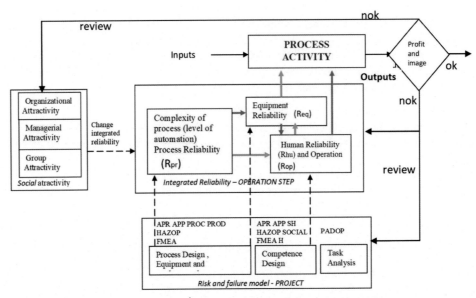

FIGURE 6.17 Algorithm to calculate integrated reliability. Ávila, S.F., Santos, D., 2014. Integrated reliability: failure assessment to process, operation, equipment and human integration (POEHint), a case of propene pump operation. In: Sixth International Conference on Engineering Failure Analysis, October 2015, Lisbon.; Ávila, S.F., 2015a. Reliability analysis for socio-technical system, case propene pumping. Engineering Failure Analysis Journal, 56, pp: 177–184.

6.5.2 Simulation of the application of the integrated reliability method

For this simulation of the integrated reliability method we will use the chemical industry case—hydrochloric acid loading process (HCl). This case has already been studied in item 6.3.3 for application of the SPAR-H method. This is one of the Company's critical processes, due to the criticality of the product, which is highly aggressive, and can affect the safety of people, plants and neighboring communities. In addition, it is a process in which the equipment undergoes various unplanned maintenances. This detailed study seeks to increase the availability of related equipment, a higher degree of ergonomics and comfort for people performing this procedure and a higher level of process safety to avoid possible incidents that are undesirable for any chemical industry. In general, there are HCl loading docks in this process, where trucks are coupled through hoses and pipes. Truck loading operations are supervised by a local operator, in addition to the assistance of control panel room operators.

To assist in composition of the operational context when applying this method, some key questions have been elaborated:

6.5.2.1 Illustrative questions from experts
- What kind of production, capacity and automation?
- What is the product and the raw material?
- What are the main characteristics of culture and management?
- What technology is used?
- How do you evaluate the communication on the shift?
- In your opinion, in this plan, are variables and actions visible or hidden?

The simulation will be based on the calculation for Integrated Reliability, involving evaluation: process reliability, equipment reliability, task reliability and human reliability based on human factors. A guide to the respective reliabilities was elaborated (Table 6.12).

TABLE 6.12 Reliability, risk, tools, complexity and measurement (Ávila, 2015a; Avila & Santos, 2014).

Reliability	Risk	Tools	Complexity	Measurement
Process	Quality of product and process	Statistical process and control, HAZOP	Costumer relation correlation	Capability—Cpk process reliability R_{pr}
	Automation and Control	Task analysis, human-machine interface, alarm management	High automation	Standard deviation % (automatic/manual)
Equipment	Failure, maintenance	FMEA, FTA, failure analysis	Human factor and hidden failure	Failure rate and probability—R_{eq} Equipment Reliability
	Technology	PERT, GANTT, CPM	Late events	Spare part substitution maintenance cost
Operation/task	Accident control	Task analysis	Multiculturalism	Efficacy, complexity—R_{op}
	Environmental impact	Communication, feedback	Age conflict	Quality
	Operational target	Risk analysis	Leadership	Cost of production
Man	Competence	Competence analysis	Behavior	Knowledge—R_{hu}
	Commitment	Social HAZOP, PRA-SH.	Organizational management change	% Organizational climate
	Communication	Organizational climate, LODA, interpretation of procedure, SILH, LOPAH.	Management of change	% Change acceptance

The operational context, process, task and equipment have already been studied in item 6.3.3, so we will align the content to the required evaluations, and calculate the integrated reliability. But first. we will read some key tips and questions for any investigation and evaluation of the integrated reliability in industrial processes.

Tips and questions by reliability type in integrated reliability calculation:

6.5.2.2 Complexity
- What is the level of automation?
- How much does equipment influence complexity?
- What is the level of attention in tasks?
- What is the influence of recycling lines?
- What is the level of non-compliance with the procedure?

6.5.2.3 Process reliability
- What are the three variables that cause product quality loss, performance reduction, partial safety stops, or environmental impact?

6.5.2.4 Human reliability
- Is the time available for the task adequate?
- What causes human stress?
- What is the level of complexity?
- What is the employees' level of experience and training?
- Are the procedures available and consistent with the task?
- Is ergonomics suitable for the job?
- Are the employees suitable for the job?
- Is the level of the work process adequate?

6.5.2.5 Equipment reliability
- List the equipment that causes severe shutdown or loss of performance in industrial plants. What statistical model is applied?
- Estimate the reliability through an exponential function.

6.5.2.6 Operational reliability
- Use complexity definition assumptions to calculate operational reliability.

6.5.3 Results

This topic presents the values found for the Reliability of the hydrochloric acid (HCl) loading process. With regard to the process involving HCl, the temperature and concentration conditions ensure the process control and are maintained through it, in addition to redundancies and system safeguards, being classified in the 5 SIGMA. Thus it is premised that the Process Reliability (R_{pr}) is fixed, resulting in 99%. However, the human reliability (R_{hu}) already calculated and discussed in item 6.3.3 resulted in 74.6%. Other reliability such as equipment (1) and operational (2) will be discussed below:

1. Equipment reliability (R_{eq})

    The calculation of Equipment Reliability (R_{eq}) is directly related to the capacity that the studied system has to remain operational during its use. Following the methodology proposed by Lafraia (2001), the exponential distribution was used, which considers that the equipment has a constant failure rate and its only dependent variable is the Failure Rate (λ).

    As a configuration for the system, it was considered that the two pumps participating in HCl pumping are in parallel, and that elements such as Level Switches, pipes and pressure safety valves are in series with the pumps as they are commonly used for both pumping systems, thus creating a mixed system.

    The history of failures was raised to the system, and the failures were identified, thus signaling causes to be investigated, such as: if there was a loss of equipment availability when they were requested by the production team, or if there was product leakage to out of the control volume studied.

After filtering the history data, selecting only the interventions relevant to the analysis, the reliability values of the equipment were collected and the value of 72% was found for the HCl pumping system.

2. Operational reliability (R_{op})

    The calculation proposed by Ávila (2013c) for operational reliability was performed, first identifying the complexity of the task studied. Values were identified, by considering consideringthe five factors for determining complexity:
    - Accomplishment of the task: It is related to operational discipline, considering bad habits, regional traits and behavioral traits. The value defined was 5%, considering the robustness of the company's organizational culture in which the process was studied.
    - Type of technology adopted in process control: The pumping process has a whole structure of instrumentation and control systems that ensures that the variables (fluid flow, pressure, tank level, etc.) are always being monitored through the operators located in the control room, thus the value of 30% was defined due to the high degree of automation that the company has and the concern with the hazardousness of the product.
    - Type of equipment with control instruments installed: The analysis of the process showed that it does not have many auxiliary systems linked to it that require operations and controls for the primary function of the system to be met, so the value of 10% was defined for this criterion.
    - Number of process steps: This item refers to the number of process steps that are performed with an interface between more than one operator; this characteristic of the process increases the process complexity, as there may be a loss in reliability due to the different levels of training that exists between the operators of the same company and the loss of information that can exist in the direct communication of both in the work area. After detailing the steps of the HCl loading process, it was found that 20% of the tasks are performed with different workers.
    - Degree of attention required to perform the task by the operator: by analogy to the propane pumping process studied by Ávila (2013c), a value of 20% was defined for this criterion.

    Thus multiplying (Eq. 6.11) the defined values of the five criteria studied, we have that the Complexity Level is equivalent to 6, and, applying Eq. (6.12), the value for R_{op} of 80.5% was found.

3. Integrated reliability

    Although the HCl loading process presents high risk related to the risks of the product, the complexity of the loading task was taken into account with regard to operation, pumping system and experience of the operators involved in the field.

    Based on reliability assessments and using data based on suggested practice to perform the exercise of Integrated Reliability analyses, the results of individual reliability are presented:
    - $R_{pr} = 99\%$
    - $R_{eq} = 72\%$
    - $R_{op} = 80.5\%$
    - $R_{hu} = 74.6\%$

    Thus Eq. (6.10) was applied to the calculation, using the individual reliability of R_{pr}, R_{eq}, R_{op} e R_{hu}.

$$\text{Low complexity} \rightarrow IR = (0.72)^{1/3} * 0.99 * (1.7 * 0.746 * 0.805)^2 \rightarrow IR = 0.926$$

Finally, the activity social attractiveness was defined. For Ávila (2015), in this case social attractiveness can be classified as:

$$\text{Very good} \rightarrow 1.2$$

$$\text{Medio} \rightarrow 1$$

$$\text{Low} \rightarrow 0.9$$

Considering the average social attractiveness (value = 1), and the low complexity adopted for Integrated Reliability, we will have the following value for IR_{final}, according to Eq. (6.7):

Then IR_{final} yields:

$$IR_{final} = 1 \text{ Ж } 0.926 \rightarrow \times 100\%$$

$$IR_{final} = 92.6\%$$

4. Final considerations on integrated reliability in HCL charging case

    The development of the integrated reliability methodology has provided satisfactory results that corresponds to the reality of this industrial system. It is possible to verify the individual reliabilities and understand the interactions between the calculated reliabilities. It is also possible to validate the algorithm suggested by Ávila, showing the potential of the reliability analysis method for standardization of evaluations.

The influence of factors and aspects related to the task and human factors can be observed because, although the value found for system equipment reliability is 72%, and human reliability is 74.6%, the positive influence is perceived of the other reliability on the final result, reaching 92.6%. These factors can be explained by organization solidity and consolidation of the operation procedures. The results of the different reliabilities are effective in decision making for the direction of investments to improve the organization, leading to assertiveness at management level.

References

Alcoforado, A.C., Ávila, S.F., 2015. Application of SPAR-H in the uranium ore crushing operation. In: ABRISCO Congress, 23–25 November 2015.

Amaro, R., 2014. The influence of human factors in the use of operational procedures in a petrochemical company. Master's degree Dissertation. Post-Graduation Program in Industrial Engineering, Federal University of Bahia. Salvador.

Aoki, M., Rothwell, G., 2011. Coordination under uncertain conditions: an analysis of the Fukushima Catastrophe. ADBI Working Paper Series, n. 316, Tokyo, October.

Ávila S.F., 2013a. LESHA—multi-layer progressive stress & impact assessment on health & behavior. In: Global Congress on Process Safety, San Antonio.

Ávila, S.F., 2013b. Operational excellence through integrated reliability. In: 1st Brazilian Conference in Reliability, Risk and Safety of Process, ABRISCO, 23–25 November 2013.

Ávila, S.F., 2010. Etiology of operational abnormalities in the industry: modeling for learning. Thesis (Doctor's degree in Chemical and Biochemical Process Technology)—Federal University of Rio de Janeiro, School of Chemistry, Rio de Janeiro. 296 p.

Ávila, S.F., 2013c. Review of risk analysis and accident on the routine operations in the oil industry. In: 5th CCPS Latin America Conference on Process Safety, Cartagena das Indias, Colombia.

Ávila, S.F., 2015. Reliability analysis for socio-technical system, case propene pumping. Engineering Failure Analysis Journal 56, 177–184.

Ávila, S.F., Ahumada, C.B., Cayres, E., Lima, A.P., Malpica, C., Peres, C., Drigo, E., 2019. Management tool for reliability analysis in socio-technical systems—a case study. In: Advances in Human Error, Reliability, Resilience, and Performance: Proceedings of the AHFE 2019 International Conference on Human Error, Reliability, Resilience, and Performance, 24–28 July 2019, Springer, Washington DC, USA, p. 13.

Ávila, S.F., Pessoa, F.L.P., 2015. Proposition of review in EEMUA 201 & ISO Standard 11064 based on cultural aspects in labor team, LNG case. In: 6th International Conference on Applied Human Factors and Ergonomics (AHFE 2015) and the Affiliated Conferences, 26–30 July 2015, Las Vegas-USA.

Ávila, S.F., Santos, D., 2014. Integrated reliability: failure assessment to process, operation, equipment and human integration (POEHint), a case of propene pump operation. In: Sixth International Conference on Engineering Failure Analysis, October 2015, Lisbon.

Ávila, S.F., Cerqueira I., Santino C.N., Drigo E. SARS, Safe, alert & resilient behavior in social technical system. Spring Meeting and 14th Global Congress on Process Safety Orlando, Florida April 2018.

Ávila, S.F., Sousa, C.R., Carvalho, A.C., 2015. Assessment of complexity in the task to define safeguards against dynamic risks. Procedia Manufacturing 3, 1772–1779.

Beyer, J., Trannum, H., Bakke, T., Hodson, P., Collier, T.K., 2016. Environmental effects of the Deepwater Horizon oil spill: a review. Marine Pollution Bulletin 110 (1), 28–51.

Borges, F., Menegon, N., 2009. Human factor: reliability to the instabilities of the production system. GEPROS Magazine, 2009. Production, Operations and Services Management 4 (4), 37–48.

Boring, R.L., Blackman, H.S., 2007. The origins of the SPAR-H method's performance shaping factor multipliers. In: 2007 IEEE 8th Human Factors and Power Plants and HPRCT 13th Annual Meeting. IEEE, pp. 177–184.

Boring, R.L., Gertman, D.I., 2004. Human error and available time in SPAR-H. In: Workshop on Temporal Aspects of Work for HCI, CHI, 24–29 April, 2004, Vienna, Austria.

Boring, R.L., Whaley, A.M., Kelly, D.L., Galyean, W.J., 2011. The SPAR-H step-by-step guidance. INL/EXT-10-18533. Rev 2.

Carvalho, P.P.S., Fonseca, M.E., Kalid, R.A., Ávila, S.F., 2016. Integrated reliability analysis in task of emission production orders in ERP system: a case study on a company's sports segment. In: Proceedings of ESREL—European Safety & Reliability Conference, Glasgow.

Chidambaram, P., 2016. Perspectives on human factors in a shifting operational environment. Journal of Loss Prevention in the Process Industries 44, 112–118.

Clegg, C.W., 2000. Sociotechnical principles for system design. Applied Ergonomics 31, 463–477.

Cooper, S.E., Ramey-Smith, A.M., Wreathall J., Parry, G.W., 1996. A technique for human error analysis (ATHEANA), United States Department of Energy, United States.

Di Nardo, M., Gallo, M., Madonna, M., Santillo, L.C., 2015. A conceptual model of human behavior in socio-technical systems. In: International Conference on Smart Software Methodologies, Tools and Techniques. Springer, Cham, pp. 598–609.

Dórea, S.S., 2017. Investigation in the operations of a poultry refrigerator to define product mix and understanding of socio-technical failures. Master's degree dissertation in Industrial Engineering—Federal University of Bahia.

Dórea, S.S., Ávila, S.F., 2015. Investigation of failure in chicken industry operations based on human and technical factors. In: Proceedings of PSAM 2015, Rio de Janeiro, ABRISCO.

Drigo, E.S., Ávila, S.F., 2017. Communicative skills and the training of the collaborative operator. In: 8th International Conference on Applied Human Factors and Ergonomics (AHFE2017), 2017, Los Angeles. Proceedings AHFE.

Drigo, E.S., Ávila, S.F., Souza, C.R.O., 2015. Operator discourse analysis as a tool for risk management. In: European Safety and Reliability Conference, 2015, Zurich. Proceedings of ESREL 2015, v. 1.

Embrey, D., 2000. Preventing Human Error: Developing a Consensus Led Safety Culture Based on Best Practice. Human Reliability Associates Ltd, London, p. 14.

Ferreira, J.F.M.G., 2018. Reliability analysis of CO_2 compression system in a petrochemical plant considering technical-operational and human factors. Master's degree dissertation for Industrial Engineering — Federal University of Bahia.

French, S., Bedford, T., Pollard, S.J.T., Soane, E., 2011. Human reliability analysis: A critique and review for managers. Safety Science 49 (6), 753—763.

Gertman, D., Blackman, H., Marble, J., Byers, J., Smith, C., 2005. The SPAR-H human reliability analysis method. United States Nuclear Regulatory Commission 230, 35.

Helmreich, R.L., 2000. On error management: lessons from aviation. BMJ (Clinical Research ed.) 320 (7237), 781—785. Available from: https://doi.org/10.1136/bmj.320.7237.781.

Holman, D., Johnson, S., O'Connor, E., 2018. Stress Management Interventions: Improving Subjective Psychological Well-Being in the Workplace. Handbook of Well-Being. DEF Publishers, Salt Lake City, UT.

Jesus, A.R., Neto, A.O.G., Cerqueira F.C., Bulos L., Souza L.G., 2013. Manufacture of sports balls. Technical dossier. Brazilian Service of Technical Answer.

Kariuki, S.G., Lowe, K., 2007. Integrating human factors into process hazard analysis. Reliability Engineering and System Safety 92 (12), 1764—1773. 2007.

Kirwan, B., 1997. Validation of human reliability assessment techniques: part 1—validation issues. Safety Science 27 (1), 25—41. Available from: https://doi.org/10.1016/s0925-7535(97)00049-0.

Lafraia, J.R.B., 2001. Reliability Guide, Maintainability and Availability. Qualitymark, Rio de Janeiro, 388 p.

Laidoune, A., Gharbi, M., El, H.R., 2016. Analysis testing of sociocultural factors influence on human reliability within sociotechnical systems: the Algerian oil companies. Safety and health at work 7 (3), 194—200.

Lorenzo, D.K., 2001. API770—A Manager's Guide to Reducing Human Errors, Improving Human in the Process Industries. API Publishing Services, Washington.

Mohaghegh, Z., Kazemi, R., Mosleh, A., 2009. Incorporating organizational factors into Probabilistic Risk Assessment (PRA) of complex sociotechnical systems: A hybrid technique formalization. Reliability Engineering & System Safety 94 (5), 1000—1018.

Murphy, L.R., 1984. Occupational stress management: a review and appraisal. Journal of Occupational Psychology 57 (1), 1—15.

NUREG 6883, 2005. The SPAR-H Human Reliability Analysis Method. United States Nuclear Regulatory Commission. Washington, DC.

Pence, J., Sakurahara, T., Zhu, X., Mohaghegh, Z., Ertem, M., Ostroff, C., et al., 2019. Data-theoretic methodology and computational platform to quantify organizational factors in socio-technical risk analysis. Reliability Engineering & System Safety 185, 240—260.

Perrow, C., 1993. Small Firm Networks. Institutional change: Theory and Empirical Findings, 111—138.

Pires, J.C.S., Macêdo, K.B., 2006. Organizational culture in public organizations in Brazil. Public Administration Magazine 40 (1), 81—104.

Rasmussen, M., Standal, M.I., Laumann, K., 2015. Task complexity as a performance shaping factor: a review and recommendations in standardized plant analysis risk-human reliability analysis (SPAR-H) adaption. Safety Science 76, 228—238. Available from: https://doi.org/10.1016/j.ssci.2015.03.005.

Reason, J., 1990. Human Error. Cambridge University Press, UK.

Ren, J., Jenkinson, I., Wang, J., Xu, D.L., Yang, J.B., 2008. A methodology to model causal relationships on offshore safety assessment focusing on human and organizational factors. Journal of Safety Research 39 (1), 87—100. Available from: https://doi.org/10.1016/j.jsr.2007.09.009.

Rognin, L., Salembier, P., Zouinar, M., 2000. Cooperation, reliability of socio-technical systems and allocation of function. International Journal of Human-Computer Studies 52 (2), 357—379.

Sá, G.A.G., 2018. Analysis of Human Reliability in Sociotechnical Systems: A Case of Agroindustry. Course Completion Work—Chemical Engineering—Federal University of Bahia.

Santos, D.R., 2014. Human Reliability Analysis in the Process of Pumping Propane in a Refinery Using the SPAR-H method. Course Completion Work—Mechanical Engineering—Federal University of Bahia.

Souza, AD, Souza, J.O., Campos, S.S., Silva E.C., 2002. Stress and Work. Course Completion Work (Specialization in Occupational Medicine). Estácio de Sá, Campo Grande, 68 p.

Souza, M.L., Ávila, S.F., Ramalho, G., Passos, I.C., Oliveira, A.R., 2018. Analysis of the reliability model for interdependent failure mode, a discussion about human error, organizational and environmental factors. In: National Congress of Mechanical Engineering (CONEM), 20—24 May 2018, Salvador, Bahia.

Swain, A.D., 1987. Accident Sequence Evaluation Program: Human Reliability Analysis Procedure. United States. Sandia National Labs.

Swain, AD, Guttmann H.E., 1983. Handbook of Human-Reliability Analysis with Emphasis on Nuclear Power Plant Applications. Final report. United States Department of Energy, United States.

Williamson, A., Lombardi, D.A., Folkard, S., Stutts, J., Courtney, T.K., Connor, J.L., 2011. The link between fatigue and safety. Accident Analysis & Prevention 43, 498—515.

Wu, S., Hrudey, S., French, S., Bedford, T., Soane, E., Pollard, S., 2009. A role for human reliability analysis (HRA) in preventing drinking water incidents and securing safe drinking water. Water Research 43 (13), 3227—3238. Available from: https://doi.org/10.1016/j.watres.2009.04.040.

Yerkes, R.M., Dodson, J.D., 1908. The relation of strength of stimulus to rapidity of habit-formation. Journal of Comparative Neurology and Psychology 18 (5), 459—482.

Chapter 7

Human reliability: chemicals and oil and gas cases

Methods for the Analysis of Human Reliability based on errors in the execution of the task on equipment and processes cannot infer about the organizational environments and moments, limiting the conclusions that follow the expected trend when the processes are stable and under management control, without surprises. This book, which was concerned with presenting new concepts about operational culture and technical methods for its evaluation, demonstrates through real cases from the chemical and oil industry the possibility of avoiding emergent accidents, but not totally visible. In Fig 7.1 is the Gold Mine that will be worked on during Chapter 7 including investigations already carried out and third generation methods that go beyond the reliance on single expert discourse to the measurement of human performance factors influenced by technical, human and social typology. This method intends to indicate preventive programs resulting from the analysis of the operational safety culture and organizational culture variables, correcting the intrinsic, managerial, and technological interface human factors. The roadmap for investigations is outlined in Fig. 7.1.

7.1 Methodology description

7.1.1 Guiding the algorithms to apply technological tools and social environment

Organizations fail to learn about the past (Leveson, 2011) and do inappropriate changes to avoid process losses. In fact, these organizations do not recognize the causality of the failure that can cause the accident. Retrospective analysis can predict future events from the chain of human errors and failures during the task. Currently, the accident analysis methods do not discover the root cause of events and experiments do not indicate better ways to operate, and then trap in the immediate evidence about the factors involved in the top event. Unfortunately, the investigation of accidents indicates functions, locations, and structures that are not related to the main event.

Research on human reliability in the industry aims to reduce process losses by studying the task in the operation routine. Process loss events happen with accidents, incidents, equipment shutdown, out-of-control variables, and low team productivity. All these events cause direct financial losses.

Research on external, socioeconomic, and natural environments, and its influence on human behavior at work is mandatory and can be done through human reliability analysis (HRA) techniques that include investigation of cultural aspects.

The action plans defined from this analysis (HRA) will reduce the risks of activities in industry, reduce process losses, and increase the value of the company in the production chain.

HRA should consider the following:

1. Intrinsic aspects of the individual that format behavior in the work function: authority, responsibility, centralization, cognitive processing, memory, attention, and acceptance of rules.
2. Aspects of complex social relation, leading to the formation of specific structures in the organization, with their fields of influence.
3. Study of abnormalities in factory operation based on human, group, and technical errors.
4. Data collection for investigation from the shift register, process monitoring reports and management reports.
5. Data collection on group performance from simulated exercises and tests for the decision.
6. Adequacy of organizational and safety culture.
7. Efficiency of the interface or the man-equipment and man-process relationship for physical and cognitive comfort.
8. Human Error Analysis in technical systems.

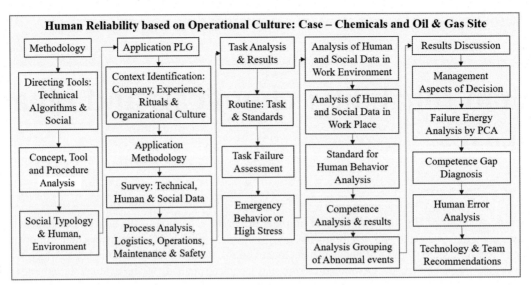

FIGURE 7.1 Gold map of oil and gas application.

Human Error Analysis requires the following:

1. Understanding of the productive system in the dynamic form.
2. Prior analysis of human behavior in the presence of leaders with their resistances and vices.
3. Projection of man's action in the control of equipment (technical and human structures), in the elaboration of the product and in the process control (control in transformation).
4. Asset and routine management with low and high variability of control parameters.
5. Difficulties in returning to normal state (resilience).
6. Critical analysis of organizational stability (gains and image).
7. Models for management and training available with the best learning processes for stability in management processes.

Environment analysis is a requirement for reducing error and increasing reliability:

1. Verification of stage of change in economy, nature, society, and organization.
2. Analysis of change impact on structural and procedural changes.
3. Identify the technical typology (product, process, operations, and events).
4. Identify the social and human typology (behaviors, management, and rules of behavior).

In the identification phase, investigation of the technical and social types is a tentative part to avoid process losses. The type of technology and mapping of abnormal events in the shift indicate an important tool explained in this book. Together with the identification of the type of human error, the type of social behavior will facilitate the analysis of the origin in the life cycle of the failure, thus allowing intervention in the process.

The actions located from the studies and mentioned mappings will comprise a program to avoid process losses composed of other programs in human reliability and process stabilization. In this research function objectives, gross profit (cost and billing) and image (safety) are analyzed, where dynamic aspects of operational failure are incorporated in the organization culture. Success in performing routine tasks makes it possible to achieve maximum gross profit with the best image of the company, that is, maximum productivity with minimal incidents and accidents. These programs are listed in Fig. 7.2, as well as data collection and processing activities to identify the anatomy of the failure. The methodology in the stage of identification of technical, social and human types is divided into three parts:

1. Process stabilization and abnormality mapping related to technical or technology behavior as shown in Fig. 7.3.
2. Human reliability and types of human behavior individually or in groups, in addition to competency analysis (using personality rule technique or competency analysis).
3. Analysis of the influence environment through the identification of social typology or types of stressful environments, as shown in Fig. 7.4.

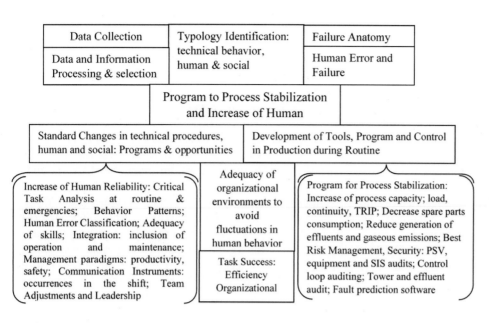

FIGURE 7.2 Avoid losses: process stabilization and human reliability. *From Ávila, S.F., 2010. Etiology of Operational Abnormalities in the Industry: Modeling for Learning (Thesis—Doctor's degree in Chemical and Biochemical Process Technology). Federal University of Rio de Janeiro, School of Chemistry, Rio de Janeiro, 296 p.*

Knowledge about the impact and causal nexus of organizational and social environment and related processes corrects inadequate environments and promote human reliability. The intention is to act preventively and not act after the environment impact on the indicators of human reliability, processes and operations. The analysis of causality, information vectors, communication noises and scenarios where the failure occurs, together with the impact analysis on the objective function, will result in alternatives with criteria and weights for action on organizational environments.

In the case of the chemical industry, ABIQUIM (Brazilian chemical industry association) includes the strategic map of the responsible care, a program to control the industry performance (ABIQUIM, 2021) with aspects such as process control; company image; organizational culture focused on environmental sustainability; accident risks; and production control. The human reliability and process stabilization work are activities compatible with ABIQUIM's vision, to promote the reduction of process losses. Despite the intention, the ABIQUIM values mainly incorporate actions generated through *TOP-DOWN techniques*, following the same direction of ISO certifications. These programs do not present the necessary mobility for organizational oscillations originated with globalization and crises, thus not reaching neither the state of self-management, resilience nor sustainability.

7.1.2 Concept, tool, and procedure for technical, social, environment, and human typologies

The ideas of Rasmussen (1997) on multidisciplinarity in the investigation of accident anatomy, the analyses of Dodsworth et al. (2007) related to the safety culture and organizational culture, and the concerns of Shönbeck (2007) about quantification of the safety culture influence on the occurrence of accidents gave the base to propose conceptual models for tools implementation and procedures. Furthermore, considerations of Lees (1996), Embrey (2000) and Reason (1990) on cognition and task analysis formed the basis to construct techniques and procedures including heuristics, algorithms and use of principal component analysis (PCA) for the failure prediction in a sociotechnical environment. The new conceptual models will present steps for the implementation of process loss programs in the areas of process stabilization and human reliability, which begin with the identification of typologies or types of human, social and technological behavior indicated in Figs. 7.3 and 7.4.

The concepts about interaction between human and technical structures, also the new methods, and tools and procedures explain the behavior of technology in operation and try to avoid failure with task assessment, as indicated in the Fig. 7.3. The investigation of technical—operational culture is possible from the reading of the operator's discourse (OD) in the routine, and through mapping of abnormalities in the shift operation or mapping of events in abnormal state (MEA). The statistics follow-up on process variables (AEP) and EVents in abnormal state (EVA) allow to validate some hypotheses and exclude others, enabling the construction of chains of abnormalities, called anatomy of failure, and the Identification of Culture mode, Technical—operational —ICT.

Tools and methods to identify social and human typologies discussed in Fig. 7.4 explain features of the social-technical system in investigations. Some important aspects such as cooperation, leadership style, best pattern of behavior, skill development, intervention to adjust the leaders, planning and administration of production on the shift, development of critical sense regardless of the company's motivation, and return to normality of the company through organizational practices that value critical sense.

The presentation of the concepts, techniques, and procedures for analysis applied in the identification of technical, social, and human behaviors have two steps. The first step, related to technology and the production tasks (Fig. 7.3), followed by the 2nd step related to environments of influence on failure and human type, and their group relationship in the execution of the task (Fig. 7.4).

The understanding of this dissertation about the method of investigation uses codes to discuss each form of elaboration of hypotheses and their validation. The new Concepts designed in this research are represented by (C), the Techniques or Tools for data collection for statistical analysis, in addition to others, are represented by (T) and the Procedures in Application (PA) of task analysis and analysis of human behavior and their interpersonal relationship.

The knowledge of technology through the reading of equipment manuals, processes, and procedures allows the analysis of the OD (C3) observed through the shift register and the experience in the workplace. This knowledge of the technology also allows starting the analysis of the standard task (PA1). Data collection from the shift register uses the principles of the dynamic model of productive system (C5), thus enabling the mapping of abnormalities (MEA) or abnormal events (T2). The MEA, on the other hand, indicates the use of techniques and analysis procedure: statistical monitoring of processes (T3—AEP) is done based on variables defined by the MEA that also defines the statistics of abnormal events (T4—EVA) that are operational factors of routine failure. After the MEA, it becomes possible to prepare the chain of abnormalities (T5) and it is possible to evaluate the failure of the task (PA2).

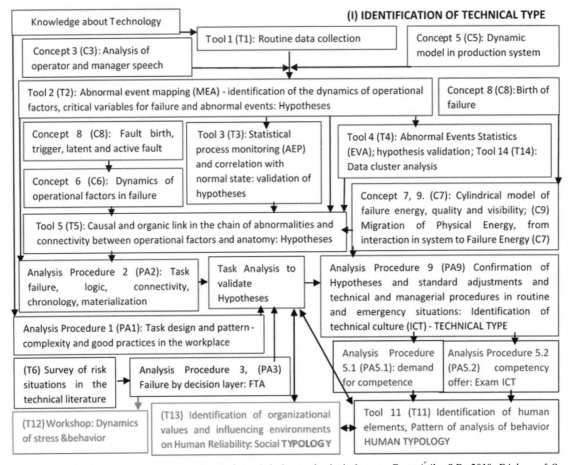

FIGURE 7.3 Sequence of concepts, procedures, and tools for technical or technological types. *From Ávila, S.F., 2010. Etiology of Operational Abnormalities in the Industry: Modeling for Learning (Thesis—Doctor's degree in Chemical and Biochemical Process Technology). Federal University of Rio de Janeiro, School of Chemistry, Rio de Janeiro, 296 p.*

The concepts about the birth of failure (C8), the dynamics of operational factors (C6), the model of the hazard energy released or failure energy, and a learning model (C7) are used to support the construction of the causal nexus and draw the chain of abnormalities.

The validation of some hypotheses from the abnormality chains depends on the knowledge and identification of the failure in the task, with the analysis of the standard task and with the help of statistical techniques (T3 and T4). From the choice of dankest tasks, the risk analysis (PA3) based on signs of abnormalities (MEA) and organized in the form of a causal nexus begins.

After confirming the hypotheses raised, new standards and technical procedures for routine and emergency situations are proposed. At this moment, we can affirm what are the features of the technical culture, what is the type of technology behavior, and what are the critical tasks that can cause process losses.

The technical typology is related to social typology (social, economic, and organizational environments) and human typology (behavior of man and group in the workplace). The technical or technological typology indicates the best procedures for competency analysis (PA5, 6) and assists in the investigation of the group's behavior in an emergency environment during workshop preparation (T12) for data collection.

The information migrates from the stage of identification of the technical typology to the failure analysis procedures (PA7) to make up the workshop, thus enabling a dynamic with psychological stress application (T12). Through this tool, it is possible to analyze individual and group behavior following the more instinctive than cognitive decision in the treatment of emergencies. The technical typology also includes information for the analysis of competency demand (PA5) and the preparation of the Technical—Operational Culture exam to verify the supply by competence (PA5). In the identification of human typology as indicated in Fig. 7.4, new concepts are used about cognitive processing (C2) and for the formation of conjugated memory (C1) with the inclusion of typologies, then enabling competency analysis and other analyses of human behavior. In the workshop held to collect social and human information (T7, T8, T9), stress dynamics (T12) and the technical culture exam (ICT—PA9) are also applied. By collecting human and social data, it is possible to analyze the emotional balance of the team (PA10), team cooperation (PA11) and leadership (PA12). From the results and observations of dynamics with psychological stress, it is possible to analyze behavior of the group in emergency scenery. After collecting Human Resources

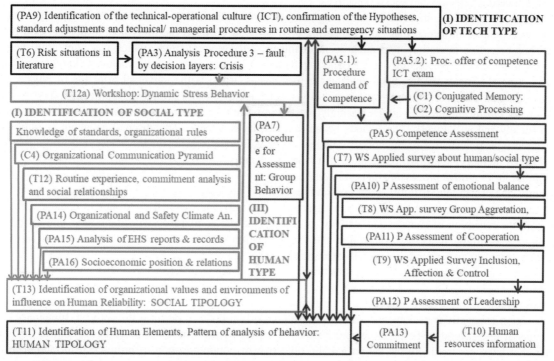

FIGURE 7.4 Sequence of concept, procedure, and tool for social and human type. *From Ávila, S.F., 2010. Etiology of Operational Abnormalities in the Industry: Modeling for Learning (Thesis—Doctor's degree in Chemical and Biochemical Process Technology). Federal University of Rio de Janeiro, School of Chemistry, Rio de Janeiro, 296 p.*

(HR) data (T10) it is possible to infer about commitment (PA13) of the operators in relation to organization. Almost at the same time, the identification of the social typology can be made (T13) through techniques and procedures of analysis. Previous knowledge of organizational norms and rules facilitates the understanding of the company's policies, so on, in a chronological way, where experience in the routine to analyze commitment and social relations (T12) is necessary. It is also essential to interpret the company's organizational and safety climate reports (PA14), analyze the records and HSE reports (PA15), confirm socioeconomic positions and identify the main internal rituals in the work environment (PA16).

7.1.2.1 Concepts—C1 to C9

The new concepts, discussed in the previous chapters, detailed here, define tools for analysis and enable interventions through procedures of application task. These procedures aim to adapt human performance factors to achieve better human reliability results.

C1—Construction of conjugated memory

The principle of memorizing seems to be simple. Every psychic or social fact that has affective value is delimited as "history" and linked to the conscious center to be rescued when requested (Jung, 2002). However, it is known that, in addition to the conscious facts experienced by the individual in interaction with the family, at work and in society, there are traces of history present in individual memory, the collective unconscious (Jung, 2002).

Some of these traces of history make up memory and also form the daily routine. These traits may also have come from previous periods, but recent ones are current archetypes. They can also be old memories, related to physical survival and meeting the needs of pleasure.

Archetypes (Jung, 2000) are social memories that are considered as a pattern of behavior absorbed by personality. This mnemonic material is present in the individual in the form of a trace or psychic fact, also a trait when it is lost in the middle of the universe of information circulating in the mind, in the instance of personality named as ID by Freud (1989). This mnemonic material is a psychic fact when the affective bond establishes a historical continuity that is retrievable when requested by the conscious mind of the individual in society.

The perception of the facts that occurred in the external environment requires sensors that perceive the signal and send it for processing, rescuing from memory the references, and in view of the usefulness for society, are used in decisions and in meeting personal and social objectives.

The affective experience, in addition to uniting the perceived and similar sensations, relating the present with the past, creates a field of influence that can change the type of bond between the individual's memory and behavior in society. Thus cognitive operations and memory mechanisms, cited several times, which can alter both the perceived current signal and the recent memory itself (working memory), or can alter the old memory (knowledge base). This change, in principle, meets the universal demand for life, that is, it meets the universal principle of physical (survival and preservation of the species) and psychic stability (social inclusion and affective stability). Today, with the new social diseases, *borderline profiles* are increasingly present in society, indicating a change in priorities for human motivation in society and in the work. The esthetic becomes a priority in society. This discussions about cognition concerns indicate three typical scenarios of memory in relation to the individual's behavior in society and in the job (Fig. 7.5): (1) Fragmented memory; (2) Modified memory; (3) Normal memory.

C2: Model for cognitive processing, decision, and action

Currently, the processing of information considers aspects of memory to facilitate the preparation of algorithms for decision and action at work and in society, as indicated in Fig. 7.5. The external environment changes decision, both passively, in the reception of signals; and actively, in the processing, comparison with memories, action and transmission of signals. Learning processes involved in the transformation information include intuitive (environmental), primary (psychosomatic) and secondary (cognitive) learning.

The processing of information, cognitive functions, intuitive changes, and psychosomatic phenomena are important parts to usefulness in the task. The information processed is useful for taking action that has its own operation, in terms of beginning and development of the action, in unfolding, in the motor performance, either verbal or imaginary.

Motor action is usually conscious and uses the movement of the body to perform the activity. Imaginary action can be conscious or unconscious, where, in the conscious case, it may be prior to motor or verbal action, or thoughts about memories that have occurred. Verbal action has the objective of communication to promote team action to meet the usefulness of the operational activity.

Memory is an important psychological function and is related to all other functions. Memory function is the result of the influence of internal and external environments. After the influence, we consider as conjugated memory, which

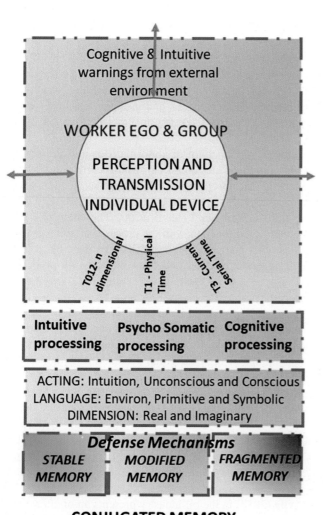

FIGURE 7.5 Conjugated memory and cognitive processing. *Adapted from Ávila, S.F., 2010. Etiology of Operational Abnormalities in the Industry: Modeling for Learning (Thesis—Doctor's degree in Chemical and Biochemical Process Technology). Federal University of Rio de Janeiro, School of Chemistry, Rio de Janeiro, 296 p.*

depends on defense mechanisms in personality. An exercise is done in the application of the oil and gas Case, in the description of human errors, including a tendency to fragment memory (Dalgalarrondo, 2000). Each experience in society or in the work, each psychosomatic learning in personality development, and influence of older instincts, impose changes in memory and priorities in decision and action.

C3: The importance of the operator's discourse

The operator's discourse discussed in Chapters 1–4 and applied by Ávila in different studies (Ávila, 2010, 2012; Ávila et al., 2018) is the transcription of the occurrences of the shift recorded, in writing, by discourse and by the way of routine behavior in the face of situations in the work environment. Similarly, the manager's discourse (MD) is the transcription of the management guidelines, written and spoken, formal and informal, in the form of instructions and recommendations for the team, or also, reported through the monthly production reports.

The operator discourse (OD) analysis uses the operational factors collected in the occurrence register in the shift. This investigation indicates possible chains of events with signs of abnormalities, failures and related behaviors. The experience in the job and performance of routine tests confirm the hypotheses elaborated from this analysis. It is important to make passive observation and diagnosis based on the understanding of the information present between the lines of the routine.

The operator's discourse observed from group reaction of operators on scenarios of future accidents, based on the experience of operators during history and experience of abnormal events help to understand operator's behavior in the use of technology. These scenarios in stages can help to devise future accidents for risk analysis. The construction of these scenarios is based on the: (1) history of events in the routine of the operation, shift or administrative; (2) technology specifications and limits; (3) study of similar accidents that have occurred in history; and (4) insertion of possible events to interconnect the local routine with the accident of history after meetings with experienced operators.

This risk analysis allows to infer about the behavior of leadership and the team from the OD tool in different cases. We tested the effect of stress on operators and leaders in several simulations. These cases indicate the human behavior in liquefied gas, an industrial Company in Brazil (Ávila, 2010, 2011); fire in public building, case of the Polytechnic School in UFBA (Cerqueira et al., 2017), presented in the United States; and gasoline leak in pipe near an American city, presented at the International Congress of Mass Disasters (CIDEM), in Feira de Santana, Bahia, Brazil, in 2018.

The type of language used in the shift allows the interpretation of the chain of abnormalities and the realization of inferences about group relationships, mechanisms of human error and technical failure control, and cooperation between groups and operators, a subject discussed in the proposition of cognitive academy (see Fig. 3.15B).

The frequency and function of operational factors from the shift register build causal nexus. The chain of events, signs of operational abnormality and equipment failures are interconnected together with the process variables and production. Also identifying the rare behaviors and frequent behaviors that can cause the accident. The experience in the workplace can help to confirm some hypotheses raised from the operator register and from the monthly production reports. The OD was presented in Section 1.6.

C4: Pyramid of organizational communication (Fig. 7.6)

The corporate guidelines of some multinationals, DuPont, for example, are used as benchmarks for safety solutions in different regions of the planet, as models of successful management systems. After the Brazilian oil and gas industry adopted the method and procedures for process safety elements, we would like to ask if this acquisition of this safety management system was a correct business attitude. Without adaptations or regional group adaptations, as is normally done, in the implementation of the total quality control program (Falconi, 2014).

Unfortunately, many managerial or operational errors occur due to the resistance of senior management to understand the local technical—operational culture, installed on the shop floor. It can impede or hinder the success of the imported safety culture.

When discussing organizational communication, we avoid the impacts of strategic actions that run in the TOP-DOWN direction without disrespecting movements in the BOTTOM-UP direction. An outdated organizational culture is born out of top-down management, with low sustainability and little resilience, mainly due to the MIX of local and global cultural resistances.

The research work carried out and published by Drigo and Ávila (2017) in the oil industry presents an organizational communication model that runs through the organization chart of the company in top-down and bottom-up cyclical movements to enable the incorporation of advanced technologies in diverse management environments and regional cultures.

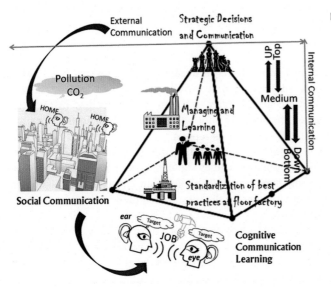

FIGURE 7.6 Pyramid of organizational communication.

The conceptual model of the pyramid of organizational communication in Fig. 7.6 follows the concepts of OD analysis (written, spoken or practiced). The communication assessment tool is the result of this model-method that contributes to achieve operational control, risk control and crisis control.

This model analyzes the quality of internal organizational communication (IOC), external organizational communication (EOC), and quality of cognitive communication. The organizational communication tool can treat situations in five-year life periods of operational plant with enough history to review the strategies and management systems originally adopted.

In IOC, the bottom-up and top-down information movements happen. The OD method can set the standards and schedule the new projects. Also considered are the movements in the middle of the organizational hierarchy that includes the manager's discourse (MD), both in planning and standardization that should be robust and based on the inputs of the OD. Then, the manager's discourse (MD) strengthened by the standardization process and the OD reviews the policy that guides EOC and returns reinforcing standardization and authorizing control through the OD.

The OD at the base of the pyramid of organizational communication offers hypotheses about events that are validated by the interaction between tendencies of process variables and alarms, indicating the paths for operational control. Then, the discourse analysis contributes for process control. Going up this pyramid, this control is incorporated at the managerial level, when the plant manager accepts the already validated hypothesis, and speaks to the team in the form of management guidelines. Thus the manager's discourse is aligned with the OD.

The actions resulting from the unfolding of the management guidelines result in criteria for the standardization of procedures, here named as control resulting from the manager's discourse. This standardization, already accepted by the operation team, is now inserted into the routine through the execution of the procedures, here named as standard. This acceptance is a process of interpretation, analysis of requirements, corrections, until they enter a consensus of the best practice for critical operations.

The movement in the *bottom-medium/medium-down* directions strengthens the technical−operational culture and increases the chance of achieving operational excellence but depends on the resulting complementary communication cycle in the organizational pyramid. By maturing the management guidelines, there is the return of common modes, rules and behaviors, previously not seen by the strategic area, and that needs to enter, in a transformative way, into the policies, adapting the old view of shareholders with a mixture of contribution of the technical−operational culture.

Thus the *bottom-medium-up movement* is complete and returns enriched, having the external benchmark of corporate views in the organizational areas, added to contribution offered by the OD, manager's discourse, standards, and business policies.

The conclusion of this work is the strengthening of a culture with unique language and without stratification filters between discourses. Finally, the policy reflecting the shareholder's view, is represented here by strategic discourse and its controls (ED). ED has the same meaning as the manager's discourse (MD) and equates definitively with the OD and its controls (OD).

Two dimensions in the organizational communication field are important: worker cognitive structure and external communication (EC). The cognitive structure shaped in local society composes technical−operational culture. The EC is important because it allows a good interaction between the organization and the society. The EC based on environmental limitations direct the company's operation. EC must have appropriate format to treat the cases of accident and crisis that affect the neighborhood and the public.

C5: Dynamic model in production system

According to Alvarenga et al. (2009), human factors can be evaluated at three hierarchical levels: level 1, cognitive behavior, and human−machine interface; level 2, cognitive behavior when performing work in groups (interaction—PSF); and level 3, influence of organizational culture on human beings (PSF). This dynamism demands a way to analyze the technical and human structures, as well as the management processes of leadership and learning in addition to the organizational and social environments on work.

The model of dynamic productive system C5, indicated in Fig. 7.7, facilitates the interpretation of factors that influence the appearance of human error and the operational failure. In this model, the figure that has the central cylinder in the horizontal position represents technology structure and human-social structures (little cylinders beside the big). Technology includes equipment and processes, that is, they are technical structures. The human-social structures are lateral cylinders grouped around the central cylinder and representing the shift teams and the administrative team. The details of dynamic productive system model are explained in Section 3.2.4.

Each individual cylinder represents a shift team operator with specific characteristics that can initiate human error and equipment failure. Stressor environments, external-colored circles, influence management processes and involve the cylinders (or technical, human and social structures). This way, the figure represents sociotechnical system in operational and dynamic Risk.

Management and educational processes are dynamic vectors that bring smaller cylinder clusters (people and groups) closer to/ away from the larger cylinder (technical structures), to allow the task to be accomplished. Social, economic, and organizational environments influence differently on equipment structures, processes, groups, individuals and on management and learning processes.

The operator acts on process, performing procedures delegated by production manager and those who work as a team have life and history experiences that are transformed into a social function. This history of the individual is the basis for acting in the productive system and is constructed from individuality, social coexistence and in a labor group.

The relation between the team and the equipment described by the management processes include standardization (procedures) and learning (incorporation of knowledge). The management and learning processes are influenced by the physical (natural temperature and rainfall), organizational (culture) and occupational environments (risks of accident and legal aspects of man at work). Failure occurs dynamically in structures that exchange energies with environments, for this reason, there is the statement that operational factors feed or drain energy for the increase or decrease of failure (corrective and preventive actions).

Human and social structures have links with physical structures (equipment and process) through leadership processes in production teams and relationships that promote learning about the task on the shop floor. Along with the dynamic model in the production system, these new concepts deal with integration of physical energy and failure energy by development of operational factors marked in the anatomy of failure. These system characteristics that trigger a latent failure are named "pathogens."

C6—Dynamics of operational factors in the failure process

The worker or team participate in the construction of the operational failure during the task. Failure state is compared with the actual state of the process and equipment to understand the possible gaps with the expected standards (technical, economic, social, and environmental). Both states recorded as occurrence in performing the task serve to study the quality and availability of operational and organizational procedures.

The mapping of abnormalities in the shift use, as database, the record of occurrences and the process and production monitoring reports in the industry. The task in failure and respective written operational procedures and records are important materials in the investigation, but after the knowledge of the principles of technology. In reading the routine, it is possible to extract aspects of abnormality with the identification of signs or warnings that indicate the existence of operational failure in progress. The signals released and encoded on the map of abnormal events with long-term records (3–5 months) will allow the analysis of relationships that occur in the industrial plant and that, due to low visibility, do not allow the correct decisions and the respective assertive actions.

The operational factor that changes the failure energy geometrically increases the power of failure (elevating the consequences) or, on the other hand, decreases this failure power, it depends on the resultant direction of vector from that failure energy. You can and must differentiate, causes, from warnings of events, and from the consequence of the failure. It can be a cyclic event and recorded differently from the chronological order. This understanding makes it possible to find out the root cause of the problem.

The study of operational factors allows differentiating the signal or warning of failure in relation to its consequence, which is the energy that performs the work of failure. In addition, it allows finding that the failure energy has returned to begin from a point, as already discussed in Chapter 2, Human Reliability and Cognitive Processing. This failure energy feedback refers to the recirculation of materials, energy and information loaded with pathogens. It is possible to evaluate these phenomena by analyzing the logic, connectivity, and materialization of the failure.

Operational factors related to unrecorded events due to omissions in writing or in the discourse are also inherent to vices because of the social culture or poor safety culture. Omissions in shift written reports are difficult to identify due to use of the chronological rule of analysis. Certain factors happen in advance and others are false because they have no logic and causality in the anatomy of the operational failure.

C7—Conceptual failure energy model

Ávila et al. (2006) discuss a new dynamic format for failure analysis in industry operations. He states that the influence of the personality structure and the worker's learning processes, the management profile and organizational culture that translate into the possibility of Man in committing failure. The influence of aspects of the organizational and physical environment determines the possibility of failure in the process. The characteristics of process fluids, operational parameters, equipment and process capability are temporal and linked to operating patterns. It is important to analyze the failure from its various dimensions and quantify its hazard energy to avoid the big event, that is, the accident.

Vertical cylinder in Fig. 7.7 represents the power model of the failure, C7, discussed earlier in this book (Chapter 2: Human Reliability and Cognitive Processing). The construction of this cylinder is made from facts recorded in the occurrences register, where the failure warnings are inserted (Lees, 1996) as well as the various operational factors (causes, consequences, and actions that modify the failure energy). These operational factors help in the reconstruction of history, making it possible to infer about future events. Thus they enable the development of programs and tests to prevent production loss.

The root cause of the failure adds energy, directing the vectors that move to the consequence of the failure. This ascending energy moves toward the risk zone at the elevated part of the cylinder. The consequences of the failure are the operational factors that perform work, thus affecting the cost and image of the company. Corrective and preventive actions drain the failure energy and impede the existence of the consequences, then directing the energy resulting vector to the base of the cylinder.

This model, indicated in Fig. 7.7 as C7, uses cylindrical coordinates for qualitative and multidimensional failure analysis. The radial coordinate represents the measurement of the level of visibility of the failure, the axial axis that indicates trajectory and accumulated energy of the failure that can perform work or not. This depends on the types of operational factors participating in the composed failure, or multidimensional failure. Quality of failure classes in this model are: (1) occupational environment or environmental impact, (2) natural physical environment, (3) organizational environment, (4) managerial profile, (5) social factors, (6) human factors, (7) procedures and (8) technology. When the failure has mixed formation, it will be characterized by the dominant classification between the established angles attributing weight and calculating the weighted average. The vector resulting from the failure will be the sum of the previous vectors, positioning the failure at a composed angle of the failure quality in an updated position.

Operational factors with low visibility refer to causes, which may be the root or secondary cause. Normally, the operational factors with high visibility are the consequences, which perform work, and classified up to the fourth level or quaternary. Failure warnings have high visibility or not, preventative actions have high visibility. Corrective actions have variable visibility, depending on the operator's level of knowledge about the role of this action or the operational factor.

Depending on the knowledge about the anatomy of the failure, its chronology, the strength and the ability of systems to withstand high risks without accidents, operational factors have several movements described in the learning model about operational failure. The type of product and process technology, the type of people's behavior in the team, and the type of social relationship directly affect the failure energy behavior and the appearance of operational factors.

The maximum height of the vertical cylinder that is part of Fig. 7.7 depends on the distance of the accumulated failure energy from the state of normality, without latent failures, at the base of the same cylinder. Failure energy is the result of technical aspects and the influence of social aspects.

Informal rules at different moments at (liquefied gases) oil and gas plants influence human behavior and contribute to the team, mainly in case of the oil segment. These rules and movements: union political movement, which affect the organizational climate and influence of leaders committed to correct, or to keep vices. It is important to analyze the economic, image and time factors available that influence the appearance and increase of failure energy.

C8—Birth and development of operational failure

The birth of the failure is the moment when "the defects" of the physical (equipment) and social (personalities) structures are connected. The resident pathogens (Hollnagel, 1993; Reason, 1990) are interconnected by pulsating energy from organizational factors, causing the stress of these structures and the beginning of life of the failure with a position at the base of cylinder C7 in Fig. 7.7.

The characteristics of the equipment structure, the individual structure and the group have defects or "pathogens" that increase the possibility of committing individual, group, technical or system failure.

Some characteristics of human or group structures can cause individual or group failure. The "pathogens" cause behavior conflicts with the organization and begin in the steps of personality formation (Fadiman and Frager, 2002), for example, the nonacceptance of authorities in the working environment. Structures may have "pathogen" linking failures or chains of abnormalities, favoring the network format, and making it difficult to diagnose and to identify the root cause in operational failure.

Operational failure can happen individually or when it involves a group of people and affects equipment. Individual failure, which has low immediate impact on the group, can cause fatigue and stress when its effects happen during a long period of time.

In the cylindrical model of learning about the operational failure, at the time when the failure has zero energy, means that this failure has not yet entered the inertial zone, that is, the base of the cylinder. With the existence of a "pathogen" in the structures and process group and after the pulsation of environmental energy (e.g., nonacceptance of changes in the corporate composition), the birth of the failure occurs and the entry into the failure cylinder with energy greater than zero. The beginning of the failure is in an inertial zone with low visibility, considered, therefore, as latent failure. From then on, the failure is receiving energy, produces work with consequences, and if nothing is done, it can even achieve maximum goal of failure, that is loss of the company utility. The size of the cylinder indicates the risk tolerance of the system.

C9—Integration of concepts—exergy of operational failure and relationship with productive system

The multidimensional and complex failure in the sociotechnical system (Ávila, 2010) goes through high and low visibility areas, which indicates the difficulties of detecting emerging and latent problems. The loss of physical energy of technical systems is indicated by loss of equipment performance in the industry. This loss of physical energy of the system causes its least utility, increases the power of the operational failure. At the same time, the loss of physical energy caused by noise in the communication, is a latent failure in the inertial zone of the failure cylinder.

Yantovski (2004) tries to discuss the disorganized socioeconomic state relating to energy but does not present models for operational failure. The low technical performance, that includes machinery, unit operations and management systems migrate energy to the cylinder, which has maximum capacity of energy accumulation. This cylinder saturates over time until it reaches the maximum level, causing the production system to lose its usefulness in the local and global economy.

Integration of physical energy with failure energy

The physical energy of the equipment and processes represented by the boiler is integrated in the failure energy (Fig. 7.7) and is represented by the increase of energy in the cylindrical model of the failure. Thus it creates the view that the noise in the communication of the routine, where the procedures performed can cause human error and the consequent increase in energy losses, increasing the position of the failure in the cylindric model and, in physical terms, reducing energy efficiency in the production process (Ávila, 2012). The initial design of the equipment is calculated based on ideal models considering the absence of failures in the operation, therefore, the amount E of energy is expended. From this ideal project, part of the energy is removed considering that it will be operated by people and that human error is a possibility despite the control systems included in this project, so we reduce the input energy to E-Δ. Finally, when there are high losses due to failure and incidents in the process, with high impacts on the economy of the business, we consider the loss of energy equivalent to Λ, that is, to produce products, we use the equivalent of E—Δ—Λ. We then consider that the proportion of energy lost in the equipment is equivalent to the loss of power per failure. We do not consider serious accidents in this simplification.

Migration of energy of failure in the productive system, cylindrical model

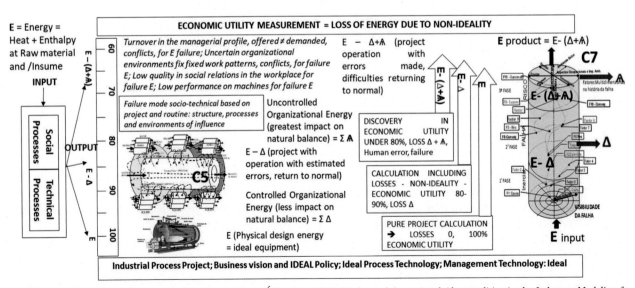

FIGURE 7.7 Operational failure energy C9. *Adapted from Ávila, S.F., 2010. Etiology of Operational Abnormalities in the Industry: Modeling for Learning (Thesis—Doctor's degree in Chemical and Biochemical Process Technology). Federal University of Rio de Janeiro, School of Chemistry, Rio de Janeiro, 296 p.*

As mentioned in Section 2.1.5, failure energy is cumulative, with entry points, the causes, and exit points (consequences that sum, or corrective-preventive action that decreases). The failure migrates between equipment and social systems through functional connections that may not follow the logic of physical or temporal proximity (Perrow, 1984). The failure starts when the first energy input (initiator event) occurs in the cylinder. This start is not visible, indicating that it is in the inertial phase. The failure energy is zero at the base of the cylinder, minimal and barely visible in the inertial zone, average in the development zone where the complexity of the failure is constructed, and high in the risk zone where the major event may occur.

The failure can branch and move up or down the Y, of the vertical failure cylinder, with its energy level. The failure can move radially outward/inward in the cylinder, depending on the visibility. If the structures (equipment or people) have defects without impact on the economic utility, the failure does not already begin. There was no accumulation of energy to overcome inertia. The failure energy works causing consequences of primary, secondary and tertiary levels, which migrate in the form of economic loss to the environments. Operational failure has an entropic tendency to lose energy.

Failure energy is under control when the system power losses at the design level (loss of Δ) are considered. The failure energy is uncontrolled, and at that moment represents energy, when its measurement begins, energy that loses economic value. By understanding about the anatomies, the failure quality, its energy level, and its visibility, it is possible to indicate strategies to avoid the risk zone or to eliminate the beginning of the failure at the start of the project or energy of operational failure.

Principles of operational failure energy (Ávila, 2012)

By establishing business policies that reduce failure, accidents and energy loss in the industry, we present important principles for their operation and risk activities. These principles follow the concepts in this chapter. The industry that understands the importance of implementing these rules from the project and throughout the productive period will be a sustainable and resilient industry.

Principle I—Currently, the use of nature's energy has low yield in a continuous state of abnormality. At this moment, it is important to read the signs until understanding the causality of the social, technical and sociotechnical events. There are difficulties in aggregating the teams: inadequate leaders and isolated groups.

Principle II—Technical-social project directs events to incidents and accidents. Internal knowledge is not valued, and managers and leaders believe that technical skill is enough, causing group and individual omission. Process controls are not compatible with product risks. The unmeasured signals toward failure indicate only corrective actions after the top event occur.

Principle III—The technical (control of materials in processing, control of leaks of hazardous products) and human (omission and excessive self-confidence) structures were defective. organizational environments inspired mistrust of power games (groups and leaders). The managers worked on impulse generating stress in the team (accident).

Principle IV—Many of the failure processes were already installed emitting signals and giving chances for decision-making, but the way of investigating did not value routine information. If we know how to gather fragmented information, in other words, when knowing the human type and possibilities of communication failure, it becomes easier to investigate abnormal events. An important point was the oscillating quality of the raw material, it should be valued to the point of distrust of the sampling and the quality of the recycle current.

Principle V—Continuous loss of power occurs from equipment by failure and from the failure cylinder to the outer universe. The more diversions and incidents, according to the pyramid of the accident, the more likely a fatal accident will occur. Corrective maintenance of the systems does not have positive result due to not knowing where to install the safety barriers.

Principle VI—Conflicts between policies and practices in addition to inadequate social coexistence of individuals and groups are the propitious environment for system entropy and promote operational failure. The Board was focused on solving the problem but was not knowing the complexity of the processes involved.

Principle VII—The sociotechnical systems were unbalanced, and it was impossible to maintain the failure energy under control. Incoherent policies.

Principle VIII—The failure energy conservation is not reached, and the failure cylinder is not managed. There is no awareness of its existence. The industry was stuck in accumulated management vices with technical problems, and low competence over the risks of the process.

Principle IX—The treatment of consecutive fires does not allow analyzing the causality of the failure, then causing corrective maintenance actions in regions that do not solve the problem. The lack of knowledge of the dynamics of the failure ends up generating a more stressful environment, causing more losses. There is no focus the right direction.

Principle X—The failure occurs in the operating systems by the production teams. Maintenance, that can cause the correct drainage of the failure energy, is not being well performed. Knowledge about the dynamics of the operational failure's power draining, allows choosing the most appropriate point for each situation.

Principle XI—The nonperception of operational risk and the habit of not valuing signs from routine and only valuing signals from instruments (which because they were in excess, hindered the analysis) cause the system to be very close to being partially and totally uncontrolled. The maximum point of entropy occurs with great impact on nature.

7.1.2.2 TOOLS—T1 to T11

The tools described in this subchapter are part of the steps to identify the technical, human, and social typologies indicated in Figs. 7.3 and 7.4 that present the methodology for HRA, based on the technical and operational culture of the risk industry.

The behavior resulting from the technology in the operational routine is identified from different information and interpreted continuously and with multidisciplinarity. Thus shift and administrative data are interpreted indicating aspects of logistics, maintenance and problems related to production, occupational safety, process safety and environmental control in the case of these oil and gas industries. The process data are statistically processed and interpreted for pumping and compression, in the temperature process variables, and pressure in the drying of liquefied gases. Productivity issues in logistics are addressed to understand the reasons that promote human errors.

T1—Technical and socio-human data collection

The data of the shift are collected in the register of occurrence, which, despite having an omission content regarding the events, presents high robustness due to the collection being always exhaustive in the period studied. The omission of occurrences in writing exists and depends on local cultural rules, interpersonal relationships and leadership aspects in each shift or administrative hours. OD previously registered in the shift register, as well as its resistances, clashes, safety and fears. The competence of the shift team is presented from the level of quality, safety and production achieved in the routine of the operation.

The main database used in this research on technical–operational culture is the shift register, where the occurrences of the routine are described and which has, depending on the habits of the operation or the social behaviors of the shift, errors by omission and commission, and many errors in the basis of competence.

This type of behavior is confirmed by Reason (1990) in concerns applied in the factory operation. The omission refers to the lack of registration of the facts as they happened, impairing identification of the causal nexus. The committee refers to records indicating the change of events involuntarily.

In the shift register, information is extracted from logistics operations (type, number, volume and number of operators, as well as identification of tanks and ships), and maintenance specialties related to the event (mechanical, electrical, instrumentation and boiler services, steels).

We can identify the impacts of events on productivity, environmental problems, and safety events. How can we identify these events? Investigating the shift register, where data collected from operations and process parameters, incident/accident reports, as well as human socio-human information from HR and operators' perception or discourse.

Process report data comes from supervisory information and interface for process control. The type of technology used to confirm which knowledge profiles are necessary for the jobs in the company researched. Identifying process variations, in operations, attempts to establish a less unstable pattern of operation.

Raw material quality data, such as contaminant content and density, directly influence process capacity. Together with the data collected from the shift register, it is possible to prepare the causal nexus of the events, relating the shift class and the work hour.

Data collection is performed using database software such as work or access, and spreadsheets are used in the excel tabulator for statistical processing (mean, standard deviation and graphic). In the case of qualitative data processing of the shift, the abnormal event mapping (MEA) methodology is used, which results in the cause-consequence chain of events considered abnormal. An example of the items that are part of this tab is in Fig. 7.9.

T2—Abnormal event mapping/MEA

Failure, not common frequency in deviations, is classified into different new events that can indicate increase of future failure as abnormal events or abnormalities—this definition is based on a time choice to assess, it can be a shift, a day, a week, a month, or a year—depends on the type of investigation. According to Ávila (2004), the MEA, as indicated in Fig. 7.8, with the routine events, facilitates the vision of a long period of production. This map allows checking several events that occurred using the handling of large amounts of data, in an easily identified scenario. This technique facilitates the verification of the analysis of consecutive events and their causation relationships. The method evaluates

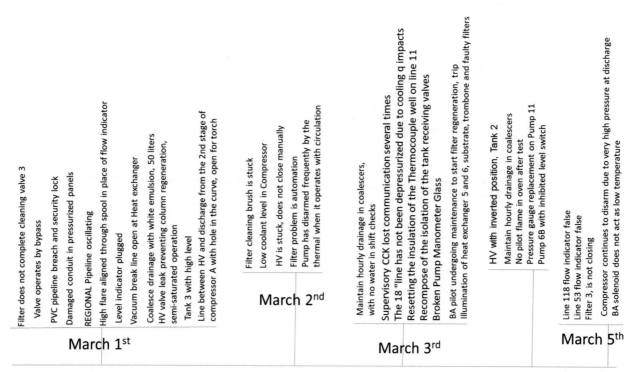

FIGURE 7.8 Abnormal events mapping before the relationships between events in an industry. *Adapted from Ávila, S.F., 2010. Etiology of Operational Abnormalities in the Industry: Modeling for Learning (Thesis—Doctor's degree in Chemical and Biochemical Process Technology). Federal University of Rio de Janeiro, School of Chemistry, Rio de Janeiro, 296 p.*

what is cause and what is consequence in complex events and what levels of causality and effect are. Knowing the causes and consequences of the events, besides distinguishing which synergistic, false and anticipated factors, tables are prepared to avoid the repetition of the same abnormality. In summary, the MEA identifies what are the cyclical events, their relationships and the periods of these cycles. The MEA also tries to correlate sporadic events not expected with the dimension of culture and environmental aspects external to the process. There is a report focused on the discussion with the shift teams, the so-called REA (report about events in abnormal state). Failure-related hypotheses discussed in the report, after tested, are used to validate or nullify instructions for the projected event.

T3—Statistical process control/AEP (Fig. 7.9)

Process data reproduce the technology, processed materials, and type of people who make up shift groups. Industrial logistics activity depends on processing time and inventory availability. These are operations that, eventually, can stop the plant, and certain characteristics are important in terms of process during routine, start-up and shutdown.

The operator is constantly involved with numbers and standards that indicate condition of operation of the system, process sector or equipment (Ávila, 2004); thus it is necessary to handle data and information through statistical methods. The operator has resistance in interpreting charts and numbers with distant time trends, his vision is busy with data and information for immediate use in performing the task.

Although there is the commitment of the operator to know what is happening and the continuous search for solutions to the problems, part of the abnormalities require significant attention because they are not visible within the limitations of diagnosis of the operation team. The lack of knowledge about the whole hinders the administration of procedural, technical and social conflicts that have been damaging the process.

The handling of data in an automated way is a trap that hinders the interpretation of the process from statistics of events and variables. The excess of recorded data is indicative of a lack of organization of production management in relation to process control. In addition, excess data impairs the establishment of the sense of ownership over this collection result. The team in operation needs to be adapted to the intense use of statistics at all steps of the process, but selectively. Without this training it is difficult to decide critically.

The interpretation of the data presented in the form of graphs and numbers, which summarize the whole, depends on the training and leadership over the team. Leaders should be committed to interpreting abnormal occurrence statistics (EVA, Statistics of Abnormal Events) and relating it to constructed cause-and-effect diagrams or abnormal event map

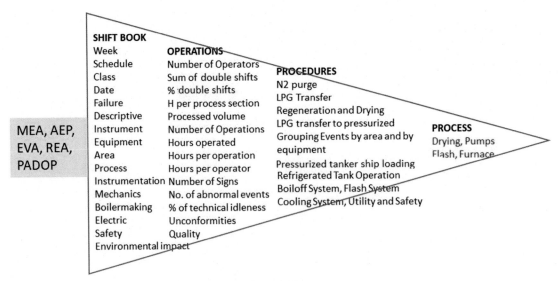

FIGURE 7.9 Data for typology: shift register, process, and production variables.

(MEA). The interpretation of the statistical data of process control and production (AEP) should be presented-interpreted continuously in the shift; with the quality and quantity of data selected to indicate the paths to control and anticipation of future problems.

The statistics on the shop floor should be understood as a work of conquering knowledge in the team, so that it becomes a habit of interpreting. This practice should be disseminated uniformly and the lack of knowledge in statistics should be gradually met by the production manager and by the formal and informal leaders in the factory.

T4—Statistical analysis of abnormal events /EVA (Fig. 7.9)

The information from the statistics of abnormal events is related to the classification of the signs and operational factors of these events, regardless of its functional position in failure as cause, signal, consequence, false factor, anticipated factor, or even factor that potentiates failure. The frequency of these signals is measured indicating approximation of the failure event and that can cause larger events in the area of loss of productivity in logistics, loss quality, or events with impact on safety (occupational or process safety) or impacts to the environment.

From the statistics of abnormal events, aspects are raised by maintenance specialty. The interpretations of abnormal event analysis represented in chain format connected with other chains of abnormality may clarify aspects regarding beginning of failure; pathogens of technical and human structures; and root cause or secondary problems. The lack of knowledge of how to operate and maintain rotating equipment such as pumps and compressors causes failures in operation or maintenance, which depending on other human errors and system failure can cause accident.

T5, T6—Process data, safety, logistics, operation, and maintenance (Fig. 7.9)

T5—Information from process mapping (MEA) indicates the causality of events involving logistics and safety. The causal nexus resulting from the events is compared with the items or variables of the process, effluent, logistics and safety. From this causal nexus and its interpretations, a diagnostic report is issued to be validated by the formal leadership and the operational team. This mapping methodology indicates the possibility of low quality in the data collected because of omission in writing during the recording of occurrences. The guarantee of effectiveness in use of shift register is its high robustness, a long term with all work groups. As the method is based on the OD, it ends up suggesting the existence of failures with the production of an Industry.

T6—The information of incidents and accidents from the reports of the safety, quality, health, and environment area statistically analyzed, can feed the interpretation of technical, human, and social types of behavior. These technical types or technology behavior indicate interpretations about process variables, abnormal events, logistic control, and technological demands. The interpretation methods in T6 are: follow-up of process variable (AEP); discussion of parameters to patterns (PADOP); abnormal events statistics (EVA); mapping analysis indicating causes and signs (MEA); analysis of logistics activities and operations; and technological demands for the job indicating which profile (demanded) is appropriate for the operator and team to perform the operations.

T7—Individual and social data about the worker (Fig. 7.10)

The identification of human types is made from the analysis of behavior and human and group structures. Some questionnaires are directed to operators and supervisors through scenarios, to assess whether characteristics and actions are appropriate to the work environment. At the same time, information from the HR and Company Safety (HSE) sectors helps identify the group's performance in the routine.

Other questionnaires are used to measure emotional balance, human characteristics under stress, and the groups' level of cooperation. From the identification of the behavior type, in routine, in situations of failure and in emergency, we can conclude whether the worker has sufficient knowledge and ability to perform the task.

The data collected from the departments and those collected from the self-assessment questionnaires or scenarios indicate the level of competence through the confirmation of the technical and operational culture. Personal characteristics are interpreted from the information collected regarding memory, attention, and behavior in the routine, such as rationality, profile in changes and technical causal nexus.

T8, T9—Questionnaire on individual and group behavior (Fig. 7.10)

The questionnaires applied to estimate the level of cooperation, the relationship of leadership, and emotional balance of individuals and groups can indicate the behavior at work. These questionnaires are addressed to operators and supervisors to infer about functionality (application of competence) and fidelity (commitment to the organization), performance with stress in the work environment, level of cooperation in social relations (rituals), style of group communication, roles assumed at work, and informal leadership analysis.

T10—Collection workshop (Fig. 7.11)

The workshop is promoted to collect data from the operation. Its participation is voluntary with 20% of the operation group, randomly, in the case of oil and gas. The workshop presents the importance of research on human reliability for process safety and productivity. This activity includes the application of a questionnaire on individual and group behavior, on cooperation, leadership, and group negotiation.

A complementary poll collects social and human information, obtained voluntarily by the worker, to explain aspects of emotional balance that influence commitment and application of competence when performing the task. Some technical questions evidenced concerns about routine of the operation and the interpretation regarding the possibilities of operational failure, increasing the risk of accidents. With the analysis of the group's response, the examination or test on operational routines and behavior when applying the LODA technique (multilayer operational decision assessment),

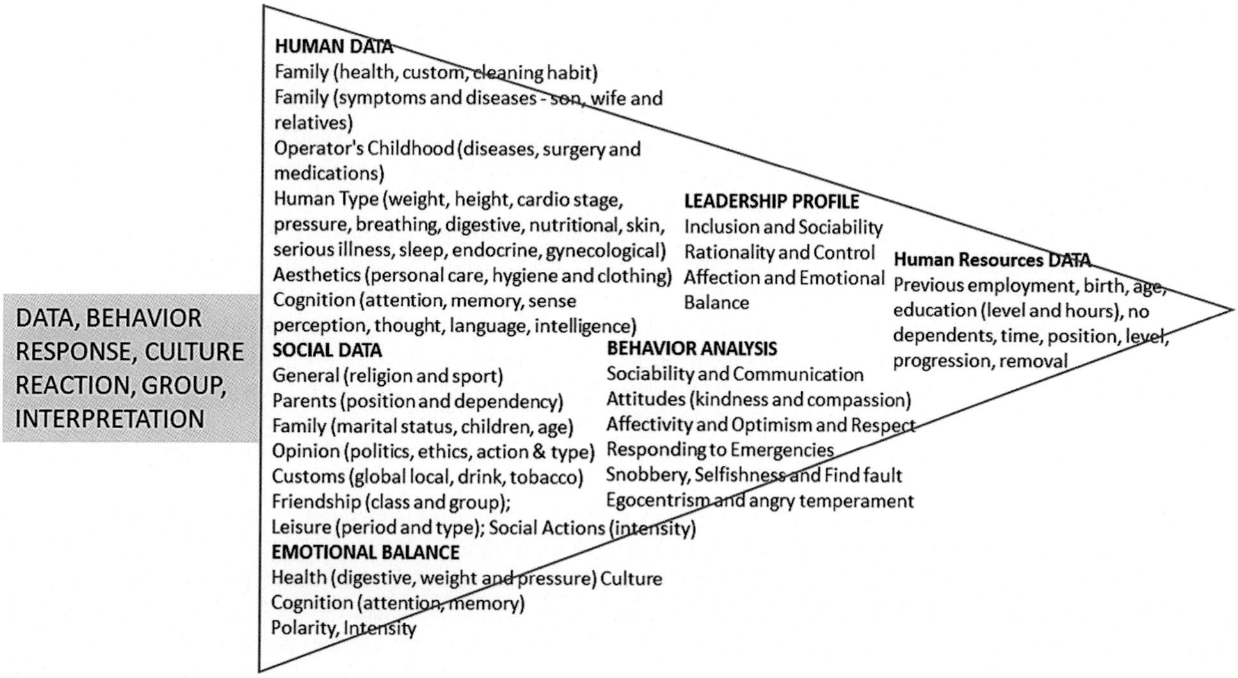

FIGURE 7.10 Data for human and social typology.

Awareness Lecture: Environment and Technical-Operational Culture
    Importance of Human Reliability
    Results so far, Methodology

Survey of Human Social Aspects
    Behavior and sociability - 160 questions (yes / no)

Survey: Human Behavior and Cooperation
    Group relationship in the workplace
    Based on FIRO-B, 54 questions (from 1 - 6)

Data Collection - Human and social information
    Responses based on social and human characteristics

Technical Culture and Operational Identification
    Risk, communication, case decision with LPG
    17 questions from the abnormalities identified from the OD

LODA-based Stress Dynamics - LPG industry fireball
    5-level decisions, Observation, behavior, risk analysis
    Levels: failure warning, Process uncontrolled, task failure, fire and fireball (accident)

FIGURE 7.11 Data collection and human typology for human reliability.

the results are compared with the OD analysis and with the process variables, permitting to identify the technical—operational culture (ICT).

In the application of the LODA, a group dynamic with five levels of decision-making takes place, in which the shift teams, after each situation presented, are requested to build a program of actions aimed at avoiding or mitigating the proposed situation.

T11—Human Resources department data (Fig. 7.10)

Based on the company's HR database and from questionnaires applied to supervisors and operators, it builds the social and human type, presenting the profiles available for the jobs and the demanding profile of logistics technology. The data collected involve previous employment, birth, age, education (level and hours), number of dependents, time, job position, level, progression, and leave of absence to calculate the level of commitment. The investigation includes the calculation of the following intermediate indicators, age per company year, time of company per dependent, time off per years of work, and others. Another survey is regarding the number of training hours in each area, to estimate, through operator commitment level how much will be applied in the task.

7.1.2.3 Assessment procedures—PA1 to PA16

The implementation of the human reliability program through actions that also avoid Process Losses depends on procedures analysis organized to apply in the industry. These procedures were tested and validated in the chemical industry (Ávila et al., 2006, 2008b) and during the doctoral research (Ávila, 2010), they were systematically applied in the oil and gas industry.

The PADOP method includes three types of Task Review: Task Planning assessment (PA1 with standard task); task failure assessment (PA2); and Assessment of Emergency Task (PA3). Another Assessment procedure is the classification of Human Error with its derivative (PA6). The investigation of human behavior depends on routine (PA4) and data-information of human and social relations (PA7). The Competence Analysis (PA5) can indicate aspects about the level of knowledge, skills, attitude and cognitive gaps to perform tasks in the Operational routine and emergencies.

On the other hand, we need procedures to indicate when the group work in the correct direction. Some mathematic and statistic tools, heuristics (informal rules), algorithm for directing the procedures to predict future production campaign, and finally, PCA—Quantitative analysis of failure or accident approximation (PA8.1, PA8.2 and PA8.3) help the construction of diagnosis.

The introducing of some procedures in the social area related to human characteristic ensure the success in avoid process losses. Initially with the identification of Technical—Operational Culture—ICT (PA9), after with Emotional Balance at Work (PA10) indicating a good environment for Cooperation and Commitment. With measurements from Cooperation Analysis (PA11), Leadership Analysis (PA12) and Commitment analysis (PA13), we can adjust some social aspects. The organizational Climate and safety culture Analysis we define procedures adapted to local Operational Culture (PA14) that use together the data and information from analysis of reports and records in Health, Safety and Environment Area—HSE (PA15). Other aspects from contracted Company is treated after an assessment about socioeconomic position and analysis of internal rituals (PA16). These procedures, discussed briefly in this book, are initiated through the investigation of critical procedures.

PA1—Routine task standard

The task analysis tool is important because it investigates the efficiencies of teams in the workstation, in mass production. This tool is discussed in Chapter 3, Worker Performing Task. This discussion about the failures caused by low motivation and attention in the routine of production is subjective (Louart, Maslow and Herzberg, 2002) and does not detail aspects of culture that affect cognitive functions and the teams' productivity in ensuring quality and production on the shop floor (Deming, 1990). The subjective analysis of the standard task is important for the design step and for the step in which the degradation of industrial facilities occurs.

The interaction between the worker and the production process (rotative machines, critical equipment, complex process, and control supervisory) includes the physical issue of accessibility and the cognitive perception of the indications of the process states. In fact, the Human—Machine Interface considered as a human factor, requires attention for the success of process control and depends on the local management style and the human type installed in that factory at a given time.

The behavior resulting from human grouping formats in an operational culture defines differentiated mind maps and styles for decision-making. The aspects that influence the construction of a local mind map must be identified before labeling situations as abnormality or as normality, when, in fact, it is the inverse that is happening. Repeating ways of drawing up the mind map and preferences for the decision indicate that informal rules are ruling production. Work behavior responds to a typical social behavior that responds to formal rules that is only a part of the local culture, the other part is from subjectivities, such as archetypes derived from leaders and local history. In this case, we need psychological knowledge to avoid bad habits and social vices.

The task analysis must fulfill characteristics that can be complex in two ways, project and control the Standard Tasks and adjust the critical tasks in Failure Mode.

The definition of design patterns or even its revision are based on what is intended to be achieved as a goal for normality, that is a state resulting from process control. On the other hand, the analysis of the task using routine information can indicate trends for abnormality. This identification helps the manager and staff to prepare preventive actions before shutdown or before the accident. We have already dealt with the concepts and tools of task analysis in Section 3.2.

In the case of task planning, the following techniques are used: task architecture; task project based on requirements, steps and goals; risk dominance and parallelism analysis in steps of task; analysis of organizational factors and cognitive academy; analysis of physical and cognitive effort and complexity task; and finally, analysis of task steps by type and location.

PA2—Task failure assessment

Dynamic analysis of failure in task, for the specific case (Ávila, 2010) of Liquefied Gas drying, used routine information that indicates trends for abnormality, failure or even accident. The investigation of failure in critical tasks needs to fulfill some requirements: understand how phenomenon of failure occurs during task performing; set the control volume; identify the physical body that circulates promoting this failure, danger-bearing body.

The tools of failure analysis in the task, discussed in Chapter 5, Learned Lessons: Human Factor Assessment in Task, here performed in a case of an oil and gas site. In addition to the knowledge about the phenomenon, we intend to analyze the logic, connectivity, chronology and materialization of the failure. These tools include sociotechnical dimension of analysis. We try to confirm how failure materializes to establish the main barriers.

PA3—Behavior in emergency task

The behavior of the operating team can be different between routine times compared to emergency times, in which extreme stress occurs for workers and leaders. In our case, we worked on the furnace fire in the drying section of the liquefied gas plant, considered an emergency stop and which can get even worse if it hits vessels and launches a fire. Application of techniques for analyzing the behavior of operators in crisis conditions aims to identify cognitive and leadership gaps. Crisis Control requires specific knowledge, skills, emotional control, attitude and mutual trust; therefore, it seeks to avoid flaws in the team's work behavior emergency situations. The proposed methodology called LODA, Dependent layers of operation decision assessment (Ávila, 2011), promotes the state of stress with the approximation of the accident in a simulated way. The operation team was tested during a simulation about cognitive and affective features' gap. A diagnosis performed after this simulation, confirms whether the team has enough knowledge to solve the emergency. The tools applied to diagnose the team are the dynamics of the team to investigate if behaviors and management mechanisms are enough; subsequent risk analysis with the calculation of the frequency of accidents; and discussion and choice of defense mechanisms for humans, groups, equipment, and processes with the intention of

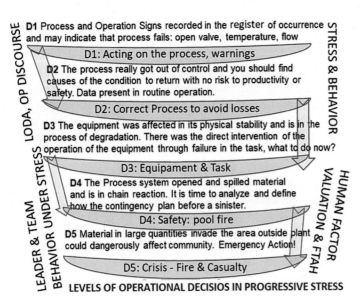

FIGURE 7.12 Dynamics-based risk analysis. *Adapted from Ávila, S.F., 2010. Etiology of Operational Abnormalities in the Industry: Modeling for Learning (Thesis—Doctor's degree in Chemical and Biochemical Process Technology). Federal University of Rio de Janeiro, School of Chemistry, Rio de Janeiro, 296 p.*

reducing the risk of accidents. Fig. 7.12 indicates the progressive level of stress suffered by the team up until the disaster scenario.

A series of events sequenced at different levels with questions about what to do, in turn, at each level, generates discussion with assessment of behaviors and quality of decision. In each series of events, the decisions made by the operation, regarding the signs of failure, indicate the best safety barriers. LODA recommends process corrections, new ways to operate equipment or perform tasks and keeping the equipment in a safe state.

The first tool aims to measure the reaction of the team and the quality of the proposed actions, to avoid the continuity of the accident process. The second tool calculates frequency of the accident based on the frequency of the events estimated in the first technique. The second tool also analyzes the consequences of the accident for the company and for the public, proposes the installation of barriers to avoid the accident at the management level and with safety devices for the installations and recalculation of the frequency of accidents. We intend to detail this case in the next subchapter that will describe this tool applied in a liquefied gases facility.

1stTool: Steps of risk analysis from team decision in simulated situation.

1. Identification of signs, warnings and failures to construct the abnormality chain (MEA).
2. Coding and chaining in causal nexus (REA).
3. Choice of critical task for team risk analysis (HSE, Production and human reliability).
4. Data collection with HSE and Operation on the chosen event, case of the furnace.
5. Meetings *with staff* and operation to correct causal nexus prepared from the MEA.
6. Construction of the situation divided into critical decisions:
   D1: Process/operation signs.
   D2: Out-of-control process.
   D3: Abnormalities in equipment and tasks.
   D4: Unsafe situations with fire and explosion.
   D5: Out-of-control situations affecting the neighboring public.
7. In a team dynamic, interpret what decisions to avoid the event from continuing.
8. Analyze stress caused by situation in teams and classify by means of scores behavior.
   Clarifications requested.
   Communication between operators (1, 2, 3)
   Established leadership (yes/ no)
   Rotary leadership (yes/ no)
   Stress level (1, 2, 3)
   Commitment to the accomplishment of the task (1, 2, 3)
   New ideas that emerged in the group (yes/ no)
   Excessive physical movements (yes/ no)
9. Data processing and consensus decisions, resulting in behavior analysis with stress.

2nd Tool: Construction of failure tree analysis following decisions in the operation, steps.

It also uses the signs of abnormalities mentioned in the dynamics with the team for risk analysis repeating the same steps from 1 to 5 before. Steps:

6. Define the anatomy of failure by decision level.
7. Identify for each failure anatomy factor the appropriate logical operators: AND or OR.
8. Estimate the probabilities of each event.
9. Calculate the probability of the final failure.
10. Check the impact on the public.
11. Prepare a management program, technical actions and recalculate the probability of failure.

The tool presented herein aims to reduce problems of divergence between the profile of the operator and the culture of the organization, then promoting a greater psychological commitment and resulting in a lower possibility of Man and team to commit the failure. This technique aims to increase operational reliability, that is, the reliability of Man and the team in the actions on the process. At the same time, it aims to increase the reliability of the process, equipment, and environmental and safety impacts, promoting greater comfort for man. This methodology allows that the influencing environments can change over time, changing balance of processes, structures, and other environments in the industry.

Influencer environments are mainly the social values and rules, the economic scenario that affects the way society, groups and families function. Another factor, or influencing environment, is the nature that circulates society and the economic activities in which the industry is inserted. Climate change and the imbalance of natural ecosystems will alter the environment where the manufacturing activity is inserted and, consequently, cause changes in technical and social systems.

All the necessary infrastructure to produce products or the execution of services are understood as a technical system. Thus equipment, systems, procedures, human-process interface are factors or processes, environments and structures of the technical system. As social systems, structural aspects are understood, such as the personality of the individual and the team; procedural aspects, such as forms of learning, management, leadership and, consequently, team building; and finally, environmental aspects, such as the organizational environment.

PA4—Socio-functional relationship at the workplace and behavior diagnosis (Fig. 7.13)

This procedure aims to indicate the steps to achieve the best efficiency of the team and leadership in the job. Initially we visualize the relationships in the operational routine and which influence task, leadership roles, decision,

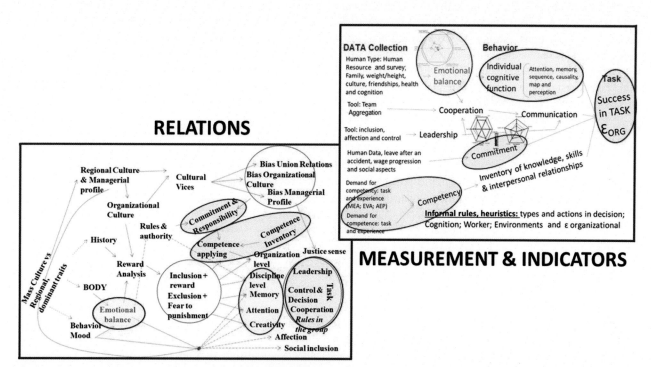

FIGURE 7.13 Socio-functional relationship in the job and diagnosis to intervene. *Adapted from Ávila, S.F., 2010. Etiology of Operational Abnormalities in the Industry: Modeling for Learning (Thesis—Doctor's degree in Chemical and Biochemical Process Technology). Federal University of Rio de Janeiro, School of Chemistry, Rio de Janeiro, 296 p.*

cooperation, functional division, and execution of the task. This view includes analyses of organizational and cultural bias, cognitive and functional characteristics, desired competence, social inclusion, reward analysis and emotional balance aspects. Remembering that the emotional balance or lack of it affects all cognitive functions for the task and can promote human errors.

This discussion was held in the doctoral thesis (Ávila, 2010) detailed in this book. The emotional balance, level of cooperation and leadership to evaluate the quality of cognitive function and communication in performing the routine task was defined. In addition, we calculated the level of commitment of shift groups using functional data and compared the supply with the demand for skills to structure the shift team as to the possibility of cognitive gaps and human errors in the operational routine.

By understanding the cold or hot, objective or subjective factor that weighs most on the operation team, we can act preventively to avoid failures in future campaigns. And knowledge about predicting task efficiency is an indicator for a fair and resilient safety and organizational culture.

PA5—Competence assessment

Selection and training of personnel are important functions around people management, also called HR management. The selection activity aims to reduce the risk of organization's efficiency loss, resulting from lack of knowledge, lack of skill or even lack of employee understanding about the company's rules represented by organizational policies and norms.

Type of competencies required for routine activities depends on process and product technology. Often, this competence is not ready, and development is necessary through training. It is important to identify whether there is ability to form new concepts (Vygotsky, 1987) to solve problems of new subjects, where experience and skills are completed with ability of innovation in accomplishment of tasks.

The identification of technical–operational culture is confirmed through interpretation of the OD that indicate cognitive gaps and confirmed by conducting an examination on the practices of the industrial plant. This test will assess the operators' behavior pattern regarding communication and the type of mind map for the decision. Items constructed from the history of performing the task by the team.

Operator's role in the routine discussed by Ávila (2006a) did indicate that the execution of procedures using the inventories of knowledge and skill installed allows operational control. Avila affirms that the task, despite being influenced by external environments, depends on the success of the team learning process.

The learning process is efficient if the knowledge installed and its application promotes success of the task. At first, the installed knowledge must meet the following demands: concepts for the creation of new solutions that meet complex problems; presence of skills measured by experience and problem-solving skills; knowledge about specific processes and operations; ability to seek additional information about the systems; and behavioral aspects in the workplace. This knowledge allows adequate decisions for routine problems and for new situations.

Demand for technology compared with available supply of technological knowledge, shows gaps in the competence and respective corrections in training and behavior. Identification of technical and operational culture is an examination of knowledge and behaviors to correct installed competencies as indicated Fig. 7.14.

PA6—Human error classification

Ávila et al. (2008a) discusses the types of human errors based on the workers' personalities with different traits of behaviors and makes this analysis in the research of the oil and gas and Chemical sectors (Ávila, 2010). He states that, by knowing the types of human behavior, type of technology behavior during the operation routine and, the influence of the environments on the work of the group, it is possible to identify the root cause of latent failures. It is also important that, for competency programs and adjustments of organizational environment, to classify human errors even if the action performed does not act on the root cause, but on secondary causes, thus decreasing failure energy and inhibiting future accidents.

Several authors have reported human errors while trying to facilitate management decision-making and rationalizing resources in preventive and corrective actions. Based on the PSFs (performance shaping factors) cited by API770 (Lorenzo, 2001) and Hollnagel (1993), it delimits actions based on history of error. In this case, there are possibilities of failure in routine, but it does not try to analyze whether the source these events are interrelated between these classes of errors. The matter of "How to resolve, how not to repeat it" was not discussed in the document about PSFs.

Lees (1996) also presents, in the chemical industry, the influence of human performance factors indicating regions of possible action in the case of error. Reason (1990) deepened this in discussions about human error on cognitive side, where he states that errors are based on skill, knowledge and rules.

The skill is part of a behavioral discussion and includes aspects such as omission or commission at work. The errors in not performing the action according to the rule is at the contextual level, and has to do with language, word association and memory. At the conceptual level, the knowledge installed and practiced relates to cognitive mechanisms and the possibilities of errors.

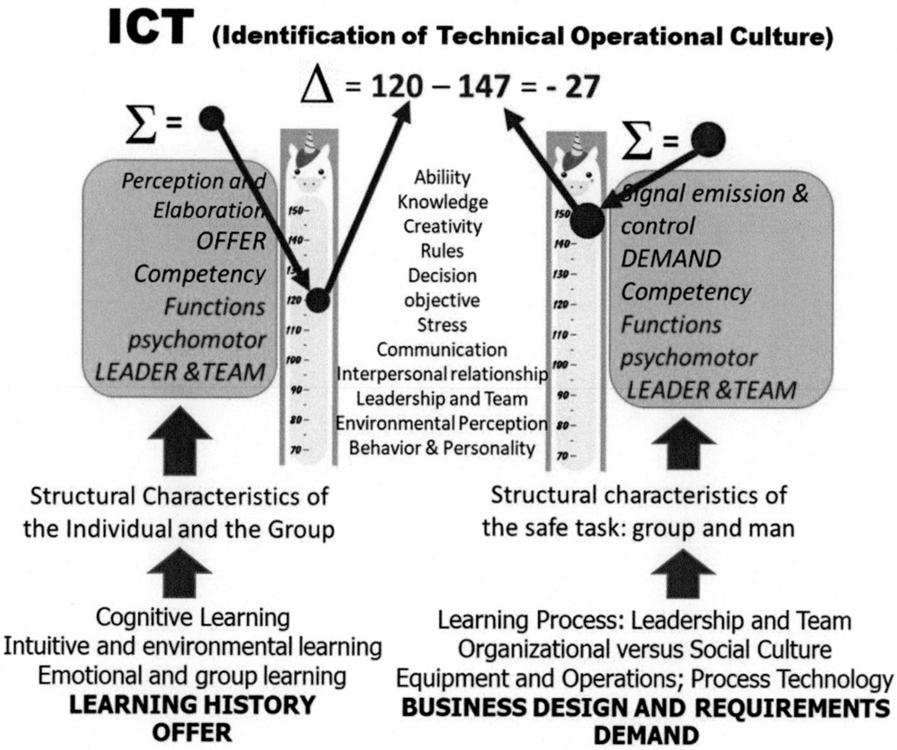

FIGURE 7.14 Competency analysis: supply versus demand or rule of personality. *Adapted from Ávila, S.F., 2010. Etiology of Operational Abnormalities in the Industry: Modeling for Learning (Thesis—Doctor's degree in Chemical and Biochemical Process Technology). Federal University of Rio de Janeiro, School of Chemistry, Rio de Janeiro, 296 p.*

Hollnagel (1993) also presents a classification for sub-specification of errors in tasks related to human behavior: word identification, repetition of a list of verbs, generalization of category, language and action slips, failures, prospective memory, and others. Sternberg (2008) presents psychological explanation for human error citing the following types: capture error, omission, repetition, correct actions on the wrong object, incorrect data from the sensory, associative activation and loss of activation to finalize the task.

Based on the various models of human error classification, errors have been derived into 52 types, depending on aspects of knowledge, skill, technology, personality, decision-making, tools and rules. Table 7.1 indicates some of the main human errors derived for the chemical industry and oil and gas.

In summary, human error classification procedure follows what is presented Fig. 3.28:

1. Perceive and recognize the signs of abnormality.
2. Identify the latent failure that can turn into active failure.
3. Discuss whether the failure is technical, human or social.
4. Relate the signs, errors and failures with deviances in the formal or informal task and relationship with goals.
5. Establish an objective function for calculating improvement in changes in the cause of the failure.
6. Relate constraints or rules to behaviors through the task type.
7. Prepare exercises to extrapolate behaviors by changing the cause.
8. Analyze impact and frequency to estimate the level of risk and effort in the investigation of the failure.
9. Classify the risk: (1) noncritical and without potential for impact (disregard); (2) Noncritical but with high potential (not to analyze but to file and monitor); (3) Critical with low potential (delegate a person to investigate); (4) Very critical, when losses occur (delegate task force).
10. Abnormalities Map listing signs and failures, recommending treatment, identification of roles on the map, itemizing functions and locations.
11. Identify which levels are involved in the failure process or in the chain abnormalities: Operation (routine or emergency); (2) Management; (3) Strategic.
12. Classify the causative human error and its factors and failures involved, in which 60% will define a specific type to be considered, or human error is considered MIX.
    a. Cognition (K knowledge; S skills; R rules; CP cognitive processing).
    b. Design (E specification, I interface, SS safety system).

TABLE 7.1 Classification of derived human error.

No.	Derived human error—subtype	Cognition (K, S, R, CP)	Project (E, I, SS)	Behavior (Mem, AT, Ge, E2)	Task (Gr, Coo, T)	Environment (N, O, S, MP)
1	Inadequacy of task planning (function)	CP			T, Gr	MP
2	Failed task control	CP			T, Gr	MP
3	Originality in the task, fear of the unknown	K, S, R, CP		AT, Ge, E2	Gr	MP, O
4	Failure to perform the Task under high stress	K, S, R, CP		AT, Ge, E2	T	O
5	Self-confidence leads to taking risks and skipping steps			E2		MP
6	Accomplish the task without knowing the rules	R				O
7	Inadequate environment to perform the task for decision	CP		Ge, E2	Coo	S, O, MP
8	Forgetfulness and slip errors when performing the task			Mem	T	MP
9	Inattention caused by too much activity/ lack of affective bond	S, CP		AT		O
10	Missing/replacing memory to avoid repeating past event			Mem, E2		O
11	Risk/Mental load high guilt anticipated for incorrect decision			Ge	T	O
12	Error for not noticing change in routine procedure	CP		AT	T	
13	Error monitoring and sampling by failure tool S, K	S	E	AT		
14	Error in control panel, process supervisory	CP	I		T	
15	Measurement instrument failure, tool failure, S, K	S, K	I			
16	Incorrect definition of management guideline: lack of clarity, dubious			Ge	Gr	O, MP
17	Incorrect data, lack of reliability for the task		E	Ge	T	O
18	Loss of time in the execution of the task, missing SK			Ge	T	MP
19	Incorrect associative capacity makes memory difficult SK	S, K		Mem		
20	Lack of knowledge about technology for decision TECH	K	E			MP
21	Lack of skill due to low practice and or experience	S		Ge	Gr	
22	Error due to inaccuracy in performing the work, lack K	K	E			
23	Human error by improper spatial orientation and TAG from area and equipment	S	E	AT		

(Continued)

TABLE 7.1 (Continued)

No.	Derived human error—subtype	Cognition (K, S, R, CP)	Project (E, I, SS)	Behavior (Mem, AT, Ge, E2)	Task (Gr, Coo, T)	Environment (N, O, S, MP)
24	Task plan error for priority and originality	K			T	MP, O
25	Error by function omission, manager indicates leader in task force	CP		Ge		MP
26	Excessive interference error in family business routines	CP		AT		S, O
27	Error by omission of communication on the shift or administrative				Gr, T	S, O
28	Incorrect choice of alternative in decision by organizational rules	CP			Gr	S, O, MP
29	Error by side effects of the task, safety or environment, K	K	SS		T	N, S, O
30	Misdiagnosis: causal link, complex relationships	K, CP	I	Ge	Gr, T	MP
31	Error in the construction of the causal link, lack concepts or vision	K, CP	I		T	
32	Incorrect re-call error—false alarm	K		AT	T	
33	Violation, individual profile, difference: Organization and team				Gr	S, MP, O
34	K, S not properly applied (discredit in the company)	K, S			Gr	S, MP, O
35	Nonacceptance of responsibility-authority over the task				Gr	S, MP, O
36	Lack of effective training, K review for Routine	K, S		Mem, AT	Gr, T	MP, O
37	Management errors when centralizing or delaying in decision				Coo, Gr	MP, O
38	Error in work pattern, safety and health, accept and fear		SS	E2		MP, O
39	Failure to interpret-perform written requirement procedure	CP	I	Ge	T	MP
40	Error in the use of equipment by the individual		E, I	AT	T	MP
41	Error in maintenance of equipment by the individual		E	Mem, AT	Gr, T	
42	Error caused by stress in emergency procedure		SS	AT, E2	T	MP
43	Error in tests for safety system validation	CP	SS		T	MP, O
44	Error in complex system analysis—difficult diagnosis	K	E, I	AT, Mem	T	MP
45	Operator's difficulty in learning new concepts	K		Mem	T	MP
46	Failures in the analysis of logical functions (relate, compare...)	K, CP	I, SS		T	

(Continued)

TABLE 7.1 (Continued)

No.	Derived human error—subtype	Cognition (K, S, R, CP)	Project (E, I, SS)	Behavior (Mem, AT, Ge, E2)	Task (Gr, Coo, T)	Environment (N, O, S, MP)
47	Error in linguistic sign, cultural symbol in communication	CP	I	Ge	Gr, Coo	S, MP, O
48	Errors due to difficulty in teamwork, tolerance			AT	Gr, Coo	S, MP, O
49	Error in estimating goals above the team load project	CP	E, I, SS	AT, Mem		MP
50	Rule-based errors—nonacceptance	R, CP			Gr	O
51	Human errors of unsuitable knowledge for the task	K			Gr, T	MP
52	Communication errors of the computer man interface		I	AT, Mem		MP

  c. Behavior, human types (MEM memory, AT attention, General G—emotional balance).
  d. Task (Gr group; roles—communication, COO level of cooperation, Task requirements, step and goal).
  e. Environment (natural N, organizational O, Social S and MP managerial profile—level of commitment).
13. List in the diagnosis of human error that causes the greater failure the items from (1) to (12).

Make a critical analysis by creating context in the chemical process industry for the 52 situations of derived human error classified in Table 7.1.

PA7—Procedure for data collection, processing, and information

The objective of this procedure is to collect data with the intention to identify elements that influence operator performance while performing tasks, that is, related to potential human error.

Forms of data collection can be a questionnaire in workshops, interviews with leaders or survey applied to the operation team, also collecting data from database of the Human Resource area. Questionnaires are applied to supervisors and operators. Interviews applied to coordinator of operation, collection of individual or group data in system of people or Human Resource of the company, where data related to formation of competencies, behavior and occupational health are concentrated.

These questionnaires or data collection schemes will be the second step of the methodology in oil and gas. The subjects treated are (1) general question, (2) competencies module, (3) social information, and (4) personal information. There is no need for personalized identification by the respondent of the questionnaires, or the reference by the institution of personal data by operator, but it is necessary to identify in which shift group or class these operators are a part of.

It is important to say that people's names will not be available due to ethical questions. The ethics of secrecy is kept by not performing individual discussion, only group behavior will be part of this research. The Health and Safety areas will give information in new projects after data group statistic processing. The groups organized by age, by shift, by origin, in short, by the factors that may be part of the human error statistical treatment.

The general data includes time of experience in previous companies; experience with the product; income in the company. Knowledge Module includes Classification of the employee in selection; Basic operator training; Specific training for functions; Knowledge recycling in the operation routine; Results of tests and exams; Measurement of knowledge and skills. Module of positions and functions in the workforce includes Position; Occupation; General Service level; and knowledge application rate. Routine work model includes the following aspects: job type; accidents/incidents; apparent commitment; functionality and related efficiency analysis.

Interpersonal relationship module deals with the following subjects: level of cooperation; clarity of verbal communication; clarity in writing and logic thinking; social participation and external training. The routine of the leader function module depends on the leadership style and acceptance by the team. Social aspects module deals with individual, family, and sociocultural information (social aspects, public, habits, social relationships, leisure). Health module and human typology discuss selection, classification; basic individual information; health/disease information (body, behavior, cognition, and self-image).

PA8—Quantitative analysis of the failure approximation (heuristic to human reliability)

Heuristic rules prepared in the routine, construction of algorithm or procedure with steps analysis, database management and calculation to estimate approximation of occurrence of main failure or top event, in production system are activities in research to control or to increase human reliability.

Statistical tools include historical analysis of indicators used for the mining of related data and the analysis of key components of these grouped items, according to their functional connection in sociotechnical systems. These items, apparently, does not have relations between each other, because they are from different areas, technological, human behavior, and safety culture. However, we know that the performance of operator depends on multidisciplinary factors.

Once we have identified the items in the PCA tool that affect organizational efficiency (billing, safety, stress level and cost), we can adjust production campaigns toward the best results of this hybrid objective function. Thus we will have classified what are environmental conditions and the factors that significantly alter the behavior at work, leading to human errors and persistent failures that can cause accidents.

Therefore we demonstrated that the prospecting of variables and indicators through research on technical, human and social typologies, enables the development of formal rules on the work in the routine. With the chosen variables we can map organizational efficiency and predict the future in production programming. If we do not make previous interventions, certainly the abnormalities will increase, and failure energy, and may reach the state of crisis, indicating in the cylindrical model of the failure, an approximation to the top of the cylinder. This level of approximation can be calculated and seen through the graphs where we release the variables derived from the PCA.

The application of concepts for dynamic failure analysis has as practical objective to predict failure and facilitate production programming. The surveys of the technical, social and human type enable the writing of the formal rules that govern the behavior of the sociotechnical system.

The analysis of the routine based on the shift register indicates causality of the problems, the grouping of structures and functioning of management processes around problems. We intend to identify the relationship of discrete variables with continuous variables to build an analysis of main components and path of organizational efficiency. These summarized indicators tell the story of the indicators like a bee swarm toward the collection of honey and returning to hive with its work product (particle swarm).

The procedures for analysis and control of human factors have four steps described in Fig. 7.15: heuristics in operational—technical culture; elaborate and test of algorithm to predict failure or treat environment, with less probability of operational failure and human error; apply main component analysis using technical, human, and social data-information; and plot data produced from PCA in dispersion diagram. This Includes the use of graphs to calculate failure energy according to occurrence of the events, including environmental factors.

PA8.1—Heuristic rules for human reliability

This procedure aims to record heuristic rules that will lead to a more comfortable way of operating toward lower risk environments and higher productivity. Heuristics indicated in Table 7.2 is applicable to any industrial productive environment, and it is divided into research activities of the type of group behaviors in decision-making, depending on the environment, people, and technology.

The mapping of abnormalities (I) indicates consequent behaviors that cause production failures. It is important to register event variables and discuss appropriate behaviors and biases classes resulting from the analysis of competences (II) that result from the comparison between supply and demand and result in cognitive gaps and human error. It is also important to relate human errors to technical abnormalities, indicating the type of cognitive gap and personality structure involved. Current knowledge inventory is a starting point for identifying through the ICT Tool operational—technical culture examination, the need to correct profiles at the competency level. In the analysis of the environment (III), we identified organizational pressures and climate issues, indicating the rules. In the analysis of environments, there are factors such as workload, stress, company's economic positioning or specific issues that will be discussed and classified.

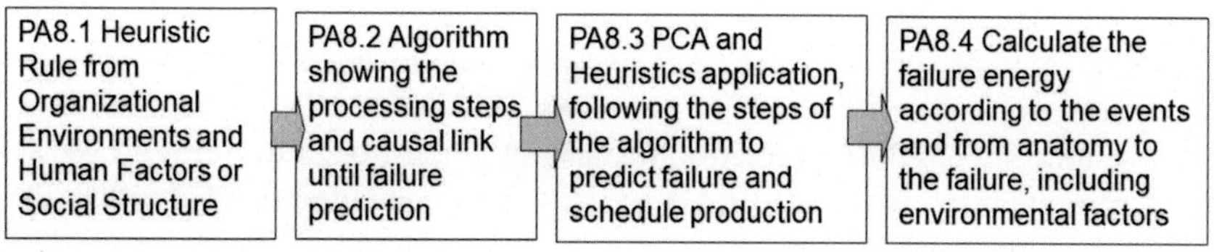

PA8.1 Heuristic Rule from Organizational Environments and Human Factors or Social Structure	PA8.2 Algorithm showing the processing steps and causal link until failure prediction	PA8.3 PCA and Heuristics application, following the steps of the algorithm to predict failure and schedule production	PA8.4 Calculate the failure energy according to the events and from anatomy to the failure, including environmental factors

FIGURE 7.15 Procedures: quantitative analysis and intervention for human reliability.

TABLE 7.2 Generic heuristic with human reliability.

Heuristics = human reliability	(IV) Human type
(I) Abnormality Mapping = Technology, Task Analysis and Operator Discourse	(B) Types of behavior Profile 1 = Applied competency, Profile 2 = Scatter and slip, and Profile 3 = Unruly. (C) Cognitive processing Profile 4 = good visibility allowing the construction of the causal nexus; Profile 5 = low visibility and planning failure. HEURISTIC If Profile 4- satisfactory efficacy in mind maps If Profile 5—unsatisfactory efficacy and F increase (D) Decision and action process Profile 6 = Broad view of alternatives, consequences. Profile 7 = Restricted view of alternatives, because it has low long-term memory redemption. HEURISTIC If Profile 6—Promotes agility/assertiveness in the decision. If Profile 7—Faltering and incorrect decisions, high possibility of promoting or incrementing operational failure. Environment 3 = Workload by intensity of operations, low occupancy, average occupancy, high occupancy.
Causal Nexus with Validation of process statistics, events and logistics operations. Identification of signs of abnormalities and investigation of risk scenarios. 1. Identification of independent variable = Tech Typology. 2. Identification of team behaviors, way of creating mind map, processing information, deciding and acting = Human and Social Type. 3. Identify technical and social parameters that directly influence the hazard energy in release, that is, failure energy. 4. Survey of inventory of knowledge and skills.	
(II) At the competency level	
(A) Knowledge inventory Profile 1 = Good knowledge, good skill and good interpersonal skills; Profile 2 = Little knowledge, good skill and good interpersonal relationship; Profile 3 = Any level of knowledge, skill and little interpersonal relationship; Profile 4 = Good knowledge, poor skill and good interpersonal relationship.	(V) Technical type
	(E) Dynamic analysis of the task and operational failure: Complex system: Architecture, dominance, complexity, Logic, chronology, sequence, connectivity and materialization of the body of operational failure and others. (F) Principal component analysis Group 1 = Critical system 1, continuous and intermittent Group 2 = Critical system 2, continuous and procedure Group 3 = Critical system 3, Equipment and processes
(III) In environmental analysis	
Time of year regarding the wage campaign; resulting from bending or economic positioning. Environment1 = Period of the year: stressful and calm; Environment2 = Labor load per double shift.	

Classified among human types of behavior (IV) are the following: type of behavior identified through polls; resulting from analysis of technical culture; commitment information based on company historical data; dynamics with stress for risk scenarios; and observation of experience in company during shift hours. Thus the first type of behavior occurs regarding the application of competencies (B), which can also be considered a high level of commitment; the second profile is due to people who are dispersed and who cause slip errors, the third is classified as unruly behaviors, having a continuous need to establish focus and not keep during much time in one activity. Cognitive processing (C) divides the behavior regarding visibility about the production process and clarity in planning. In the rules for the human type are analyzed: behavior, cognitive processing, decision, action, and how they work per human type.

Decision process (D) depends on leadership profiles and personal characteristics related to knowledge base (alternatives), level of abstraction (knowing how to project future, consequence analysis and long-term memory activation). At this moment, inductions or constructs in the human types appear, where the effectiveness in the task is analyzed from its planning and clarity. These characteristics define profiles 6 and 7, which promote agility or oscillations of behavior in decision-making.

In the technical type (IV), the complex characteristics of the task, process, and social relations are enumerated. The technical variables identified from the MEA are mined and posted to the task failure analysis tools to prepare the release in the PCA tool.

PA8.2—Algorithm for correction of human factors

The algorithm is divided into five group of activities and decisions (Fig. 7.16): Group I, time request by society and economy for production or service; Group II construction map to assist in decision-making; Group III, failure rate prediction, or calculation failure energy, indicating trend of success when changing *the setup* of operations, machines and processes.

Group IV, builds maps used for decisions, new simulations, prediction of new situations, or to perform a task and after its completion, where the data is stored in a memory that will feed Group V. After completing the task, some data are stored in the memory of Group V, where failure occurs. The processing helps algorithms to learn and corrects

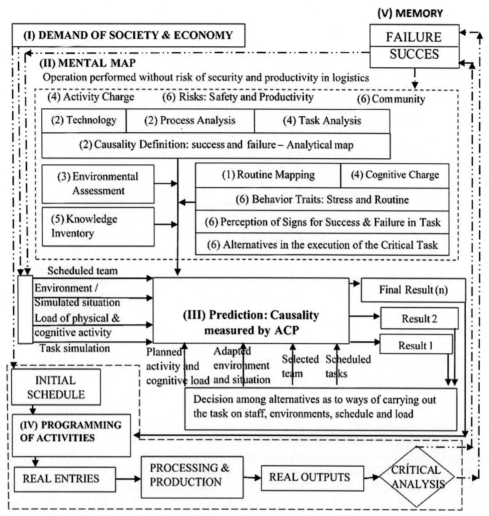

FIGURE 7.16 Generic algorithm with steps in predicting operational failure. *From Ávila, S.F., 2010. Etiology of Operational Abnormalities in the Industry: Modeling for Learning (Thesis—Doctor's degree in Chemical and Biochemical Process Technology). Federal University of Rio de Janeiro, School of Chemistry, Rio de Janeiro, 296 p.*

preexisting causal nexus; in the event of success, right working environment and best way to operate are marked for repetition next time.

Group II deals with information that builds mind maps, being an activity for processing, but that needs to be updated with key information for the success of this algorithm. The activity of Group III involves processing information for simulation alternatives, that is, mathematics for decision-making on the speech of the operator analyzed in Group II. The activities of Groups I, IV and V are carried out at the time of the task, that is, during execution.

GROUP I. Analyze the demand of society and the economy that triggers the task, it is an important moment due to the times involved in planning and executing the task. If performed late there will be implications as to the load of activities and the elements chosen for the production schedule.

GROUP II. For the construction of processing in this algorithm, the first step is the construction of the types of technical, human and social behavior at work, that is, to prepare block II of the algorithm. The mental map (with heuristics) depending on research based on (1) mapping abnormalities and influencing environments in the routine activities of this logistical work. It is important to build the model by identifying the tendency of the signals that lead sociotechnical systems to critical environments (major events such as accidents) in relation to failure, to form the knowledge base that will allow us to continue with complementary research (2) aspects of technology, and processes that allow definition of mapping of abnormal events with respective chains of abnormalities or causal link of failure are required.

Environmental analysis (3) indicates which are the influencing factors that cause changes in behavior of the operator and the team, which should constitute rules, actions to correct the technical performance of the operator and, consequently, the process. For the design of a heuristic or decision-making map, it is necessary, (4) to analyze the task, its complexity and the load of activities involved, cognitive and physical. The knowledge inventory (5), with skill analysis, will allow mapping the shift deficiencies and the knowledge required to perform the task. This additional knowledge allows you to analyze the team's deficiencies for an effective production schedule.

To define work heuristics, it is important to study (6) risks involved and proximity to the public, which are technological and social aspects, respectively. In terms of human action, it is important to assess (6) behavioral traits in times of stress and routine, and level of perception of signs of normality or abnormality for the task. In production schedule, it is important to check the decision points of the critical tasks, the possible alternatives for each situation and their impacts or consequences, thus completing the knowledge base to form the decision map.

GROUP III. From the characteristics raised to establish the mind map, simulating the decision in the work, heuristics are constructed, which are the basis for the prediction of behaviors of the sociotechnical systems of the company in which this methodology is applied. Decision-making on the scheduling of activities is the result of the alternative of the elements involved regarding the load, the environment, the team and the schedule of the task. For each new situation of each possible element, heuristics constructed in Block I of the mind map are tested and results for analysis in production schedule are obtained. Results measured in terms of resulting organizational efficiency: productivity, processed volume, quality, accidents and environmental impact, profit or cost, quality of life, satisfaction of public and employees, many others that compete, depending on the strength of the stakeholders in this economic activity.

GROUP IV. Initial schedule, with its simulated result of organizational efficiency, allows adjustments in new production schedule to reapply tools resulting from the respective corrections obtained through heuristics. Thus it is possible to compare alternatives in relation to organizational efficiency, from first to the last, allowing decision-making in best option (closest to the maximum organizational efficiency), as well as actual entries for the accomplishment of the task.

GROUP V. From new result obtained from simulations, activities are then programmed, carried out with great chances of success, and compared with expectation. When carrying out real activities, results are memorized for future use, when company and market request a new load of activities. In this way, bias in decision is avoided and routine risk situation is simulated, which can become an emergency for the surrounding public.

PA8.3—Application of principal component analysis and heuristics

Technical behavior is the result of discrete or continuous variables that indicate, in a causal connection, factors that favor the achievement of operational failure. The choice of the type of problem to be analyzed depends on the objectives in relation to organizational efficiency. Thus if the company's objective is profit, most likely abnormal event map proposed for analysis differs from the case in which company's objective is to avoid accidents and environmental impacts. Thus construction of the map based on technical aspects allows, together with the technological information collected (process, event, event, and technology statistics), to choose discrete and continuous variables that formed PCA groupings. According to Minitab, PCA is data reduction technique used to identify small series of variables that account for large proportions of total variances in original variables.

Human behavior based on the socio-human typology varies according to the influence of personality traits during the application of task competency. Slip on the task is related to the way in which individual or group process information and connect by causal nexus of factors. Knowledge of this phenomena helps problem solving and indicates the best way in which individuals or groups can take decisions and best alternative choices for action.

These human behaviors, in addition to environmental and social variables, are part of the decision heuristics to change the values resulting from plotting the PCA variables.

After identifying types of technical, human and social behavior in the work environment, it is possible to suggest individual and group actions, bringing the human response closer to the best execution of the task. In parameters of operator and process behavior, those that will integrate groups of variables for PCA are chosen.

This resulting calculation or estimate indicates a tendency to control or lose control, depending on the direction of the artificial variables in relation to the goal or standard optimal organizational efficiency. Regardless of the results, approximation or distance from efficiency, records are transferred from organization's memory, which comes back influencing behavior of process. Past memory influences resulting control (PCA) in production systems.

Process yield (organizational efficiency) in relation to failure functions as concentric level curves in relation to target cloud (maximum yield), moving away when failure energy increases, which is dimensional (example: greater loss of profit). The main component is changed through changes from interventions: (1) analysis of environments, (2) analysis of socio-human typology and (3) competency analysis. Thus resulting PCA is adjusted that can direct value to goal or its opposite direction.

Simulation of a situation: from choosing the measurement results to reducing the frequency of accidents that can affect public, chains of abnormalities involved with this possibility are chosen, and from discrete and continuous variables the indicators (measures) are extracted, which are indicators of the system that is out of control, leading to accidents. These technical variables are grouped by similarity in variability, and artificial PCA variables are calculated. Two chosen groups of variables indicate the position if close to or outside accidents regions, group 1 (specific pressure and temperature and time of execution of the critical task), group 2 (contaminant in the raw material and number of times the filter is cleaned in the shift). When joining these variables, a scatterplot is constructed. It is noted that some socio-human aspects are inscribed in variables chosen from the groups and presented in the graph of artificial variables of PCA. Knowing the results of organizational efficiency for this variable, in the past, grouping of data derived from PCA artificial data is classified as to point of accident (maximum failure) or without accident (maximum efficiency) in the organization.

History-based mapping serves to indicate characteristic regions of hot and cold clouds in relation to accident. Using the chart as reference, it is possible to make actual measurements. Real situation refers to new points resulting from artificial PCA variables, measuring the distance from optimal point and *changes are made in preprogrammed setups* leading to corrections through heuristics, regarding the team, tools, environments, and tasks.

Therefore it verifies the level of risk found and change the situations, or PSFs, in the search of lower risks. The simulation permits not to perform real tests in industrial plants, and improves with the experience of heuristics, abnormality maps, choice of discrete continuous variables, and allows the development of a preventive program to increase human reliability.

PA8.4—Construction of the failure cylinder, a quantitative format

When reaching the full height of the vertical cylinder of the failure, H (Figs. 7.17 and 7.18), from the base to the top, this indicates 100% of the hazard energy released until the loss of economic and social utility, which means, until the accident or the crisis occurs. If the failure energy is too low, it always returns to the normal state point (with little noise or errors without enough power to go in the upward direction always returning to normality) as shown in Figs. 7.18 and 7.19. If instead of the failure cylinder we have clouds to plot the variables derived from the reduction in PCA, the failure cylinder will be replaced by point clouds as shown in Figs. 7.17 to 7.19. The distance between the

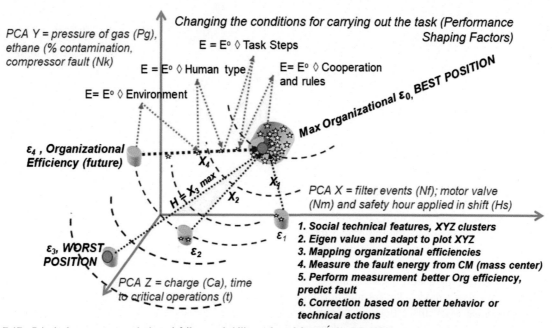

FIGURE 7.17 Principal component analysis and failure probability. *Adapted from Ávila, S.F., 2010. Etiology of Operational Abnormalities in the Industry: Modeling for Learning (Thesis—Doctor's degree in Chemical and Biochemical Process Technology). Federal University of Rio de Janeiro, School of Chemistry, Rio de Janeiro, 296 p.*

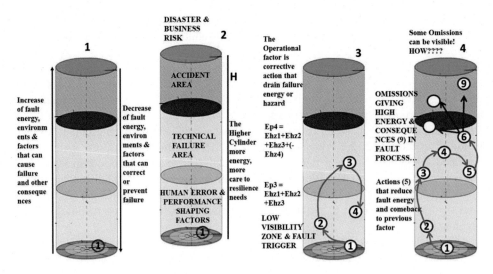

FIGURE 7.18 Analysis of situations in the failure energy cylinder (hazard release1).

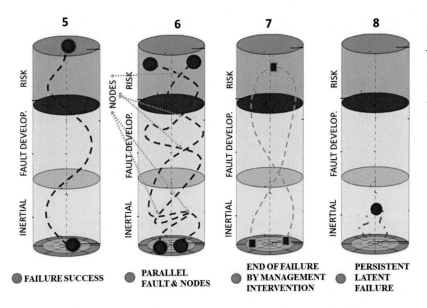

FIGURE 7.19 Analysis of situations in the failure energy cylinder (hazard release2). *Adapted from Ávila, S.F., Dias, C., 2017. Reliability research to design barriers of sociotechnical failure. In: 27th European Safety and Reliability Conference on Safety and Reliability. Theory and Applications, ESREL, Portoroz, Slovenia, June 18–20, 2017.*

cloud mass center of maximum organizational efficiency and the cloud mass center of minimal organizational efficiency (Fig. 7.17) will indicate proximity to the accident. This is the full height of the cylinder, in a graph of corrected variable grouping to estimate the point of the main failure and the accident.

The measurement of the failure is made from the *actual setup conditions* chosen for the task, considering the environments of the moment, and indicating, through the relative height of the cylinder, the frequency of occurrence of the accident. It is important to update the mind map that allows decision-making (using as tools the heuristics and PCA of discrete and continuous variables), because processes change with time and the chains of abnormality undergo modifications. For critical systems, it is recommended to measure critical environmental and human variables, at least every 6 months, and a restudy of the chains of abnormalities at least every 2 years.

To represent the above model, the formulation indicated in Table 3.1 is suggested, failure F^o depends on the efficiency of the machine (ε_M), the efficiency of the *continuous process* (ε_{CP}) and the efficiency of the activities and operations developed (ε_{IP}). Failure energy, F^o does not have the influence of organizational and socioeconomic environments. The F failure energy with environmental influences incorporates environmental (P_{AF}, $£_{AF}$), management (P_G, $£_G$), organization (P_O, $£_O$) and individual structure (P_H, $£_H$) influences.

The maximum cylinder height of Figs. 7.18 and 19 depends on the distance of Failure Energy (resulting from the failure energy) from the state of normality without latent failures, base of the same cylinder. E_F is result of technical aspects as presented in equation (E_F^o) and influence of social aspects, both related in the equations that symbolizes Failure Energy.

Some factors influence the calculation of failure energy in the human area to a sociotechnical environment (E). These factors are vices of being an absolute owner of the plant, working in a public company, periods which political movements occur that affect organizational climate and influence the leaders too. We are imbued with correcting the vices. If the company is private, other factors influence failure energy such as: the economic result, image, and time available for troubleshooting.

$$E_F^o = f(\varepsilon_{Machine}; \varepsilon_{Continuous\ process}; \varepsilon_{Intermittent\ process})$$

$\varepsilon = f$ (lifetime, mode and failure rate, Technology, Cycle, Capability, Failure Body, Frequency)
ε_M—Machine efficiency; ε_{CP}—Continuous process efficiency; ε_{IP}—Intermittent process efficiency
Cycle = activity repetition by time and load
Body = type of materialization of the equipment failure, that have hazard energy inside

$$E_F = E_F^o + E_F^o P\diamond(AF) + E_F^o P\diamond(H) + E_F^o P\diamond(G) + E_F^o P\diamond(O) = E_F^o + \pounds(AF) + \pounds(H) + \pounds(G) + \pounds(O)$$

In 1, Fig. 7.18, an increase and decrease in failure energy shown, respectively, energy input with factors that cause failures and energy output in case of corrective and preventive actions. In 2, Fig. 7.18, the regions of human error and PSFs (performance shaping factors), technical failure, accident, and disaster are indicated, thus bringing high risk to business. In 3, Fig. 7.18, the region that initiates low visibility failure (trigger) is indicated, in addition to the indication of a power outage for preventive action. In 4, Fig. 7.18, the possibility of omissions that make actions in risk area unfeasible are demonstrated, besides indicating fluctuations in level of hazard energy when there is rework.

In 5, Fig. 7.19, the time when "fault success" occurs is indicated, when accidents occur, triggering a crisis environment. Fig. 7.19 shows existence of parallel chains of abnormalities that intersect in common factors. In point 7, Fig. 7.19, shows that when Management act, at the heart of the problem, failure energy drops to zero. Fig. 7.19 shows that poorly treated failure, without action on the root cause, continues to feed energy to the system insistently.

7.2 Chemical industry case application

The application of the abnormal event mapping methodology (MEA) using the shift register as a basis is described, listing the abnormalities with discussion of common points, failure logic, and some specificities. The types of plants where the methodology was applied are (1) fine chemistry plant, (2) chemical intermediates plant and (3) polymer plant. The methodology was applied to different plants for different purposes:

- In case (1), the methodology was applied to diagnose possibilities for improvement, already knowing that the continuity in this plant was high—diagnostic study time: 5 months.
- In case (2) there were serious problems with the load, continuity and quality of the finished product: the work was done despite the manager, who did not believe in improvements in procedures but in new technologies—time of implementation until the phase of self- management: 5 years.
- In case (3), the aim was to reduce solvent consumption and reduce environmental impact—time to implement the reduction in chlorinated solvent consumption: 2 years.

In the implementation of these 3 cases, in the period from 1992 to 1998, the goal was to process the data and information acquired to form a methodology on abnormalities on the factory floor. To minimize losses at the source, initially the aim was to improve production costs and increase the productivity of the industry.

For each implementation process, some of the topics of the methodology for stabilizing processes are presented (Ávila, 2004). The abnormalities were studied using the abnormal event mapping methodology and to facilitate the studies, information was gathered that are similar in the investigation of the chains of operational abnormalities in the three cases. Thus events were classified into the following classes: (A) Pumps and rotary extractors; (B) Compressors; (C) Process in general; (D) Auxiliary systems; (E) Reaction and Feeding of Raw Material; (F) Purification and (G) Production Schedule.

7.2.1 Abnormalities inventory

The 45 chains of abnormalities are described with the respective functionality of the operational factors in Table 7.3. The abnormalities are classified according to the types presented above and distributed with the identification of the case or the industrial plant where this investigation was carried out.

In general terms, the following comments can be made by type of process abnormality: In the abnormalities related to (1) pumps and rotary extractors: (a) there are more events with the alternative and gear pumps, the centrifugal pumps present less problems; (b) alternative pumps have many events that relate to design and operating conditions and pulse problems; (c) the gear pump, with constant spindle changes (flow adjustment) to correct the grade, has malfunctions.

In the abnormalities related to (2) reaction and feeding of raw material (MP): (a) cargo pumps have a great influence on the quality of the chemical reaction; any oscillations generate stoichiometric imbalance; (b) events involving a catalyst are rarer; in the case of the fine chemical plant, these events probably occur due to the mechanical impact of the gas phase on the reactor; (c) Gaseous input also causes problems when the supply (load, quality and continuity) fluctuates; (d) Some important inputs cause abnormalities in the process, among which, caustic soda for reaction medium, hydrochloric acid for treatment (ion removal) and others; (e) the operational parameters directly influence the reaction

TABLE 7.3 45 Abnormality chains by type.

Pump, rotative extractors			Auxiliary systems		
(1)	Extractor lubricating	C3	(10)	Cooling water	C1
(2)	Extractors belt	C3	(11)	Contaminated flare	C1
(3)	Alternative pump—bottle	C3	(12)	Tanks and vessels	C1
(4)	Alternative pump—interns	C3	(13)	Cold box	C3
(5)	Abnormalities in terminator pump	C4	(14)	N2 Header contamination with NaOH	C3
(6)	Performance compressor raw material	C1	(15)	Abnormalities at caustic scrubber	C3
(7)	Coolant compressor	C3	(16)	Acid feed to treatment	C3
Process in general			(17)	Sealing pressure/solvent	C3
(8)	Batch process variation	C1	(18)	Vacuum low-performance header off-g	C3
(9)	Raw material composition	C2	(19)	Vacuum abnormalities	C3
Reaction and feed or raw material			(20)	Cooling water abnormalities CW Brine	C3
(22)	Catalyst pass	C1	(21)	Heating coil hole	C2
(23)	Oscillation of charge Pump—batch	C1	**Purification**		
(24)	Performance of charge pump	C1	(37)	Polymer to stripping	C3
(25)	Oscillation of gas input	C3	(38)	Adsorption with carbon	C3
(26)	Undue input feed	C3	(39)	Polymer to pump and separator	C3
(27)	Undue soda feed	C3	(40)	Phase inversion in water separation	C2
(28)	Problems in solid feeder	C3	(41)	Extractor efficiency	C2
(29)	Acid in gas input	C3	(42)	Solid to flash vessel	C2
(30)	High temperature at reaction	C2	(43)	Purification recirculation	C2
(31)	6th reactor overflow	C2	(44)	Heat-exchanger plugging	C2
(32)	Reaction shutdown	C2	**Production schedule**		
(33)	Viscosity high variability	C2	(45)	Production indexes versus campaign	C1
(34)	Reactors heating	C2	**Observation**		
(35)	Reactor draining	C2			
(36)	Wrong measurement of B	C2			

condition: thus in the studied cases, temperature, flow of raw material and operational procedures that involve reactions are determinants in performance; (f) some chemical reactions have viscosity as a control parameter, to avoid problems of low fluidity.

In the abnormalities related to (3) compressors: (a) this rotating equipment is very important for processes that work with gases; thus the transport of gaseous raw material or its recirculation can cause disturbances in the reaction when instability in the gas composition occurs (presence of liquids or solids); (b) cold systems also have variables and loops to keep the process stable, but, on the contrary, instability can happen mainly due to the load and the low quality of maintenance.

In the abnormalities related to (4) auxiliary systems: (a) water has an important function in the process, being very important the type of treatment used; (b) sealing of rotating equipment is an important part; if contamination occurs, the effect caused is potentiating; (c) various contaminations occur through the gas phase, either in excess of vaporization in vacuum systems, causing deficiencies, or by the contamination of the inertization header; (d) there are services that are directly involved with the chemical reaction and involve other services; thus acids and bases used in the process can cause various types of contamination by auxiliary systems.

In the abnormalities related to (5) processes in general: (a) The fluctuation in the batch, mainly due to the composition of the gases or liquids of recycle and oscillations in the feeding (stoichiometry), causing imbalance of the chemical reaction; (b) stoichiometry can also be affected if the quality of raw material varies with time.

In the abnormalities related to (6) purification: (a) in the final phase of the process, in unitary operations, such as gutting, adsorption, flashing and separation, variations in the process can have a negative effect, amplifying the negative result; (b) when there is no stability in the production process, the tendency to reduce product losses is to reprocess the off-spec product; this recirculation, when it occurs punctually, is healthy, but when it becomes a procedure and exceeds 5% of the total amount, it is a sign of accumulated problems; (c) problems of high viscosity, both due to excessive vaporization and due to the presence of salt in the organic stream, cause low fluidity and clogging of the filters, which can even obstruct exchangers.

In the abnormalities related to (7) production schedule: (a) Production depends on sales which, in turn, depend on the availability of resources (people and materials) for the maintenance of equipment and instruments in operation. If there is no organization in the schedule, the resources can be used at the wrong time and causing recirculation.

7.2.2 Abnormality logic in complex processes

The abnormalities were divided into sections or process functions. With this division, visibility increases and apparently the identification of the causes for the problem is easier. In fact, the root causes are hidden in the network of abnormalities where the rotating equipment is related to the reaction, which, on the other hand, is influenced by the utilities, and which affect the quality of the final product through the purification section (SP). Anyway, there is a complexity to be investigated with several connection loops called failure connectivity.

Thus the events involved in reaction, in the form of a network, are procedures, feedstock, reactor and column. Thus the following events occur in the MP supply: dirt in the filters, unregulated PSV (pressure safety valve), constant corrections in the stroke (flow regulation device), inadequate flow control, contamination in the raw material, packaging with solid and improper feeding of inputs. Abnormal events can occur in the following procedures: at stop, in the load table instruction, reaction control, instrument calibration, maintenance of solids feeder and others. In the reactor and the columns connected to it, events occur false level, obstructions, fluctuations, sealing pressure, temperature, and viscosity control.

In auxiliary systems, interconnected in all process areas, abnormal events can occur in water, steam, inert, vacuum, caustic scrubber, and other systems. Thus the main parameters that are normally affected in the abnormality are: process back pressure (in the case of pumps); temperature, level, and flow; various contaminations with salt, solid and moisture; valve obstructions; process corrosion; loss of vacuum; and others.

In purification, the main equipment cited in this work to investigate abnormalities are cryogenic separation (cold box), stripping column (effluents), adsorption of chlorinated solvents; liquid/liquid separation, extraction (centrifugal), flash separation and thermal exchange. From a brief description in the cold box, it was found that contamination of the gas with methane and water greatly affects performance, changing pressures and levels of process vessels. The presence of a solid body in the flash vessel and an unstable composition in the load of this vessel impairs the drying performance, including with excessive vaporization in the vessel and oscillation of the kneader's amperage (equipment that vaporizes solvent and breaks formed polymer blocks).

The liquid/liquid separations form a third phase with emulsion due to the campaign to produce a random grid (with cross links between polymer chains). The purification-related events originate in the chemical reaction and among these events there is the use of excess solvents, filter clogging, leakage in gaskets, oscillating concentration of inputs after mixing for the chemical reaction, oscillation of reaction intermediates, oscillation, and high conductivity values.

In the investigation of events in the process it includes oscillations in the composition, from the reaction to the purification and consequent variation in pressure, temperature, loads and recycles of the chemical reaction. Procedures are also changed to correct process fluctuations.

In the investigation of rotating equipment, the equipment involved is alternative and centrifugal pumps (reaction recirculation), gear pump (polymerization terminator), centrifugal extractor, reciprocating and centrifugal compressors, in addition to a cold compressor using ammonia (SABROE). In this equipment, problems are found in operational parameters such as pressure, temperature and flow, changing its performance according to the molecular weight being produced. Process facilities that are in an altered state are also detected, such as unregulated PSV, slack or slipping belt, inadequate lubrication, defective seals, leaking gaskets, defective pump internals, excessive or inconsistent vibration, dirty filters, and many others (use of bypass, for example).

In the investigation of problems in the production schedule, the analysis of stock of raw material, inputs, materials and finished product, the analysis of sales for the "coming" period and the analysis of resources necessary to make productive capacity feasible are activities that can bring production problems.

Still on the study of the fault nexus, as indicated in Fig. 7.20, the formation of a network is present in practically all chains of abnormalities. The following comments can be made about the network format of operational abnormalities in the industry (based on the plants already mentioned): 1. Auxiliary systems that use cooling water, chilled water and steam directly affect the performance of the chemical reaction; 2. The auxiliary systems in (1) and also the inertization and vacuum systems affect the purification performance, and cannot be separated in the investigation; 3. Problems with deficient inertization in equipment, such as tanks, vases and silos, generate solids that accumulate and impair the transfer of fluid, as well as the quality of the product. In the case of silos, there may be a tendency to pack, also causing a solid supply to stop the reaction.

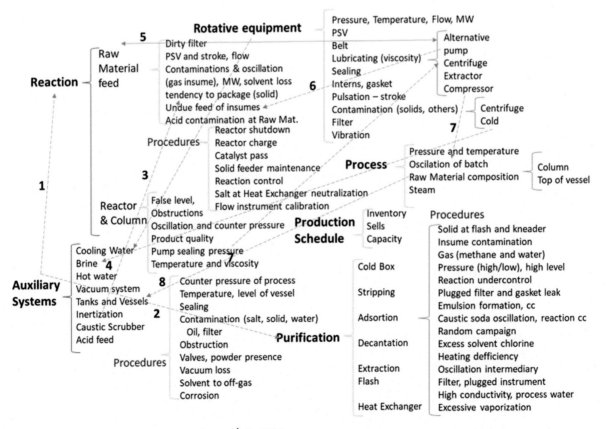

FIGURE 7.20 Failure logic and relation between factors (Ávila, 2004).

4. When operating with low efficiency, CW (cooling water) system pumps and cold compressors (BRINE) cause temperature changes in CW and BRINE consumers; 5. The feed of raw material can oscillate due to low efficiency of the pumps, or the pumps can cause fluctuations in the supply of MP (raw material), or even, both events can happen at the same time, requiring action in both directions; 6. The performance of alternative or gear pumps affects the supply of inputs (terminator, caustic soda, acid), causing problems in the process; 7. Oscillations of the MP feed compressor and the recycling of excess reagents causes oscillations in the process as a whole, both continuous and batch; and 8. Problems with PM composition generate waste at the bottom of vases and tanks.

7.2.3 Aliphatic amines

7.2.3.1 Technology

A small industrial company with the capacity to produce 13 thousand tons per year of products that are raw material for the fine chemistry area. This industrial plant is multi-purpose, where depending on the product, equipment, instruments (ranges) and specific instructions are used. Different products are processed from different setups (processing information), named as campaigns. The coincidence in all campaigns is the use of hydrogen as a gaseous raw material, and the use of liquid raw material. This company has no problems with specifying the final product and has no critical problems of operational continuity. The work of stabilizing and correcting abnormalities focuses on increasing campaign productivity. The production process is accompanied by Fig. 7.21, flow diagram of the production process.

7.2.3.2 Operational context

A plant that produces aliphatic amines (alkyl amines) in batch process, with several campaigns in the same month, sometimes even in the same week. This plant has a continuous process in the reaction stage, mainly hydrogenation, with a raw material compressor located at the supplier. Few employees and high continuity. When the Production Manager was asked about the development of process stabilization work, he showed interest, but stated that it would have no application in this plant, as in his understanding the process was fully resolved: after the diagnosis, he agreed that there were operational improvements with possibilities to further increase the plant load.

The chains of abnormalities in the fine chemical plant are listed, as follows: (1) Loading of raw material for the reaction—refers to excessive adjustments of flow control and stroke of the cargo pumps for raw material; (2) Fluctuations in batch—uncontrolled reaction, loss of performance of cargo pumps, pressure fluctuations in the compressor cause fluctuations; (3) Fluctuations in the purification—oscillating load for the columns due to the fluctuations described in (2); (4) Cargo pumps—failure in the oil replacement activity; (5) Compressor—compressor sensitivity to the presence of a foreign body; (6) Negative effects of cooling—low quality of cooling water causes excessive heating; (7) Contamination of the system—product for the flare, dirt in the cargo tanks and false level in the vessels; (8) Catalyst passage, causing dirt in the purification—filters do not retain the catalyst; (9) Influence of the production schedule on indexes achieved in the campaigns. The production rates were influenced by the schedule, as shown in Fig. 7.22.

7.2.3.3 EVA—Statistic abnormal events

The abnormality statistics in the shift as shown in Tables 7.4 and 7.5 indicate behaviors described in the graphs in Fig. 7.23.

There was a gradual increase in abnormal events from May 15 to September 15, as shown in Fig. 7.23. There was a reduction in the number of shifts without abnormal events indicating loss of pattern in the production area in Fig. 3.9. Note that the patterns are repeated for each campaign in relation to the number of abnormal events, except for campaign

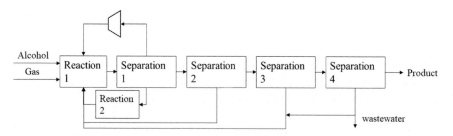

FIGURE 7.21 Production of alkyl amine (Ávila, 2004).

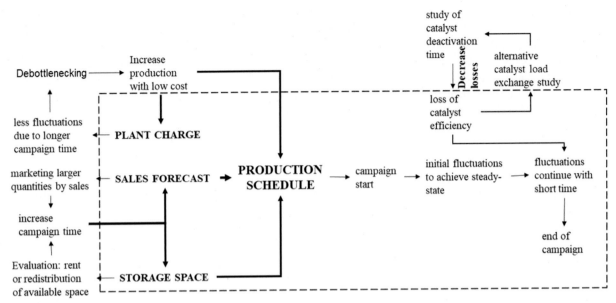

FIGURE 7.22 Abnormality chain of production schedule (Ávila, 2004).

1 (C1), which had a change in behavior, in this case there was a change in the number of days of operation in the campaign.

Analysis of the number of events per hour can indicate specific needs. Some corrective actions can be easily applied and related to the operator's job or ritual in the field or on the panel, during the task. The graph in Fig. 7.23 indicates that there are more abnormal occurrences during administrative hours, due to the greater presence of inspectors and controllers in this period.

7.2.4 Aromatic amines

7.2.4.1 Technology

To produce PF (or final product with isocyanate group) it was necessary to initially produce PI (intermediate product with amine group), or aromatic amine (methyl diphenyl amine). PI came because of the reaction between A (aniline) and B (formaldehyde), under hydrochloric acid (AC). This reaction occurs in a battery of mixing reactors and receives, in the supply of A, the recycle line (R1) from the purification. The production process flowchart is shown in Fig. 7.24.

After the reaction in mixing reactors (SR) is completed, neutralization of the aqueous and organic mixture (SN) is promoted, the neutralizing element being called S (caustic soda); in the next section, phase separation occurs—the aqueous phase goes to the water section (SW) and the organic phase goes to purification (SP). The raw material A (aniline), which did not participate in the reaction (excess), is extracted in the SP that works in two stages. Initially, by adding heat and negative pressure, A and water are removed and returned to the beginning of the process through the recycling line (R1). Then the PI proceeds to the second purification stage (SP2), where the drying and final purification of the PI occurs. All purification works under negative pressure.

The liquid phase originated in SP1 returns to the reaction, being called the recycle stream R1. The liquid phase of SA proceeds to purification in three stages, all under negative pressure. In SP2, distillation takes place with an intermediate cut return, becoming the second recycling stream consisting mainly of A and traces of water (R2). In SP2, the removal of A. is complemented. PI is obtained at the bottom of SP2, with high viscosity, requiring a heating system in the pipes. After purification, the PI is sent to stock the final product (Fig. 7.24).

7.2.4.2 Operational context

A plant that produces aromatic amines (methyl diphenyl amine) with critical reaction, where stoichiometry defines the fluidity of the reaction mass. It produces chemical polymer to produce polyurethanes and reaction between amine and aldehyde with homogeneous acid catalysis. The charge is also critical due to the reaction control. The generation of inorganic phase makes phase separation an important step in the process. Purification is simple, with a vacuum system

TABLE 7.4 Statistic indicators to maintenance/groups/shift time (Ávila, 2004).

Period	Number of events	Shift without event	Charge change	Maintenance			Groups					Shift time		
				Mec	Inst	Iron	A	B	C	D	E	0–8	8–16	16–24
May 2	60	13	13	13	5	8	5	7	8	11	6	13	12	13
June 1	61	11	22	16	14	16	10	7	16	12	4	8	22	15
June 2	45	22	30	16	3	14	9	4	16	2	6	6	25	6
July 1	49	17	18	16	8	10	2	4	4	9	13	15	16	7
July 2	59	13	35	24	10	10	3	13	16	10	12	7	33	13
August 1	57	8	22	43	3	12	10	9	10	10	6	15	18	12
August 2	64	11	22	30	5	13	7	10	16	8	9	10	19	21
September 1	84	9	10	31	19	15	16	18	14	5	14	20	30	17

TABLE 7.5 Statistic indicators of plant section/campaign (Ávila, 2004).

Period	Sector of plant							Campaign prod. (days)		
	ANC	REA	GAS	X	Y	Z	W	V	Type	Days	No events
May 2	5	17	8	3	1	2	6	0	C4	7	5
June 1	17	20	3	1	3	1	6	0	C2	12	4
June 2	9	15	3	5	4	3	1	1	C3	15	6
July 1	4	25	5	1	1	4	0	3	C1	19	4
July 2	6	25	13	7	7	6	1	2	C4	10	5
August 1	3	30	8	4	1	1	1	1	C5	7	7
August 2	4	38	1	10	2	1	0	5	C1	14	6
September 1	4	38	2	14	1	7	0	5	C4	13	5
X	X	X	X	X	X	X	X	X	C2	15	4
X	X	X	X	X	X	X	X	X	C3	9	6

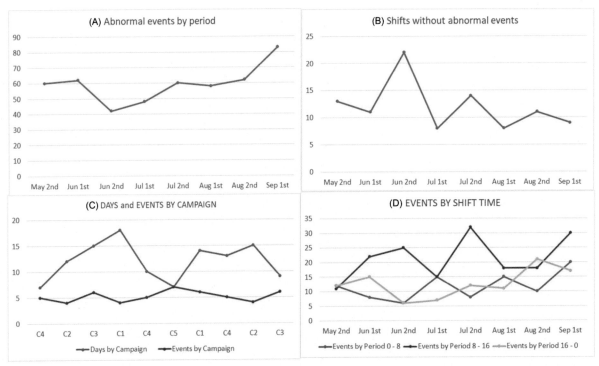

FIGURE 7.23 Statistic of Campaign: A—Number of abnormal events; B—Number of shifts without abnormal events; C—Number of days and events; D—Number of events by shift time (Ávila, 2004).

that uses a vacuum pump; however, with the previous variability in the process, it ends up causing fluctuations in composition and continuity problems with recirculation of off-spec product.

The unstable separation causes the product to be sent back to the recycling line, feeding the reaction, and increasing the process failures. Plant with low continuity and low average load. Processing in the liquid phase with increased viscosity in the final purification phase, using film evaporator to obtain fluidity. Effluent treatment hindered by the amount of phase inversion and the implications of currents from other acidic plants that cause an undesirable reaction in the effluent itself. In this case, in addition to technical work, work was done in team building with a self-management profile.

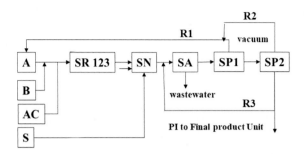

FIGURE 7.24 Production of aromatic amine (Ávila, 2004).

Knowledge of chemical process technology is necessary to analyze the technical typology and stabilization of processes, so important information is: type of inputs or inputs of the reaction with the physicochemical functions and characteristics; chemical reaction conditions and reactors in addition to unit separation operations; process controls from analytical and instrumentation data; types of pipes and events that occur with them; exits and difficulties in handling due to physical condition and toxicity; and control of effluents, waste and gas streams.

Owing to the accumulation of functions, the production engineer and, in some cases, the consultant, must make Management aware of the need to know the abnormality chains in the process. For the generation of this knowledge, it is necessary to develop research for a certain period, seeking results of increased productivity, cost reduction and improvement of product and environmental quality.

In this case, the team was sensitized from the four levels of functions in the company: Management and Board—to allow the use of resources to initiate research and perform tests in the area; Shift technicians and supervisors—to facilitate information for the execution of the investigation and to assist in the execution of the tests; Panel operators—to earn your trust, invest in the moment to modify certain operational parameters, and still keep the requested tests under control; Field operators—to report abnormalities that affect the operational factors of the installation, being called "eyes of investigation of the process," due to their attributions.

7.2.4.3 Statistical monitoring of processes

At the beginning of the application of the methodology, the efforts involved in monitoring processes were greater, because when the process is stabilized, the abnormal events report (REA) always indicates several parameters above the real need. Statistical process monitoring (AEP) as indicated in Table 7.6, has important functions, such as: (1) "indicator" of improvements in each section; (2) handling of operating values and limits, seeking optimization in the overall quality of the process and product (tests in normal operation); (3) evaluation of process trends at different times, in terms of mean and variability.

AEP is a continuous activity, being carried out in an average period of 1 year, in the most complete way possible. After the tests were carried out and some operational procedures were reviewed, it was found that the amount of data evaluated is reduced. The first diagnosis carried out on production confirmed some localized actions, working to solve the main causes of operational problems.

7.2.4.4 Operational diagnosis on standard

At this moment, certain procedures need to be changed to allow the process to stabilize and, as can be seen in the more detailed description of this work, the reaction mass showed constant instabilities that were reduced with the visual tests of the reaction mixture. The operation procedure of the reaction recirculation pump needed to be corrected to avoid innocuous interventions in this equipment. Temperature control in neutralization (SN) needed to be improved because of salt accumulation in the exchanger, depending on the capacity of the plant.

Drainage procedures for vacuum equipment needed to be revised; at the same time, it was noted that there were operators and supervisors of varying performances acting in the process and operational change aspects: (1) supervisors and operators who performed without questioning or participating; (2) supervisors and operators who claimed to execute, but who, in fact, did not do it and did not let others do it either—due to lack of knowledge or lack of negotiation or, still, due to real resistance; (3) supervisors and operators who have accepted the challenge of trying to improve plant processes; (4) supervisors and operators who were easily overcome by difficulties in changing to improve the process.

TABLE 7.6 Process follow-up to diagnosis (AEP) (Ávila, 2004).

(a) Analytical results		
Raw material	Process	Product
A purity	Proportion AC/A; % S in SN	AG in PI; A in PI
PI in A	PI in R1 (recycle 1)	Viscosity of PI
AC purity	Salt in PI to SP	2Ø in PI; Color in PI
(b) Process parameters		
Temperature in the reactors—R1, R2, R3	Temperature in SP	
(c) Production		
Quantity of production	Shutdown	causes
(d) Indicators		
Consumption A/production	Consumption B/production	Consumption AC/production

In this period, there were difficulties in choosing the most suitable methodology to review the operational procedures without interrupting activities and with minimal expenses in this process. Some groups of people were formed who accepted the challenge and, as the success happened, the operators became interested in participating in this improvement.

It was very worthwhile to create the alternative communication channel for the acceptance of suggestions coming from the team, and it was also essential to clarify the team as to the speed of service or noncompliance with the suggestions. All incidents that occurred at the factory were documented and, later, transformed into a manual on the facts that occurred and an assessment of the probable causes.

The formation of the working groups also depends on the profile required for the technical or behavioral activity and the time for its execution, which may be short or long term. Short-term activities were considered: tests in normal operation, review of the report of abnormal events, tests for review of operational procedures, investigation of incidents, specific training, disclosure of lectures and internal and external seminars for the production team, and others. And of long duration—process monitoring, continuous training, educational campaigns, environmental improvements, monitoring of production costs, suitability for carrying out small maintenance services and others.

Experience has shown that each group must clearly understand the motivation of their work and what are the simple and direct communication tools. The group leader was freer to create new paths without needing practical intervention from the company manager. This knowledge implied the definition of guidelines for the improvement process, which were based on: stabilization of the industrial plant process; decrease in production costs; improvement of the local environment, the neighborhood, personal and mental hygiene.

To maintain the ideal standard of speed in actions, quality and effectiveness in results and critical sense in variations, it was necessary to maintain training in the routine, leaders motivated by the result and self-management. The experience indicated the need to write a small manual on the improvement activities, the identification and maintenance of the standards reached and the self-management rules. To maintain the results, it is essential to multiply the knowledge acquired. For this, an instructor is hired who, based on the last data survey, will teach courses A and B during period X.

Question: Is this the most correct procedure?

Experience shows that during the period in which he was involved with the production of PI and PF, it was necessary to rescue the knowledge of the other people on the team to add to the investigation of abnormal occurrences, maintaining statistical monitoring of processes, testing, and reviewing many times to learn from mistakes. It was also necessary to organize the documentation and history built over time in the process stabilization Program. For each activity, an element was delegated to write the story, using words that are easy to understand in the operation. Among the various documents that were used to form the training manuals, we can list: (1) Investigation abnormal events mapping (MEA); (2) Report about abnormal events (REA); Reports about Statistical Process Follow-up; (3) Report about Wastewater Follow-up; (4) Register—Report of Test during Normal Operation; (5) Report about nonconformity of production; (6) Process Guide document; (7) Abnormal chains and relations; (8) Registers from Seminars and Conferences about the same topic; (9) organizational Programs; (10) Managerial Reports (indicators, charge, continuity); (11) Reviewed Operational Procedures; (12) Operational Incident Reports; (13) Information and Report about safety and environment; and (14) Equipment and Processes Guidelines.

Some basic rules were used to organize the training: (1) prepare manuals from internal staff, based on practical experience and absorption of new knowledge; (2) establish the goals to be achieved for training, time and performance; (3) prepare the logistics sector to carry out the training, that is, make available a room, photocopying services, and other materials; (4) preparing to release a person for low cost training—training in times of low sales, for example; (5) prepare the training tree, with multiplying agents at various levels, and monitor the quality of this training.

It is natural that in the course of this work, some former leaders felt discredited, for not being able to do naturally what other informal leaders were doing. Some resistance occurred in these old leaders who ended up being motivated through behavior-focused meetings where some of these leaders detected their flaws and corrected them, others resisted the change and, after trying together with the process leaders, were invited to leave the organization.

Every employee dismissal was uncomfortable, both for the company and for the employees themselves. Thus despite the previous communication and attempts to adapt and improve the employee's performance, the team ended up being modified. This action caused a delay in the team's work. Depending on the case, the delay was shorter (15 days) or longer (1.5 months). These delays are manageable in a change process. If there were not several attempts and team meetings, directly or indirectly, to address this issue, the delay could be fatal and block the process of change.

The search for problem solving aimed to reduce the "manual" work to make the planning of the teams feasible in the administrative time and in the shift. The assessment and discussion of the various problems and solutions were part of daily meetings that took place in front of the control panel. Team leaders were encouraged to plan their shift growth through scheduled or spontaneous training. The most advanced people in certain subjects were instructed to assist those who were weakest in such matters. A manual on production planning and administration was prepared on the shift, and the following subjects were dealt with daily: (1) values in the production system; (2) motivation at work; (3) motivational obstacles and tools; (4) communication need; (5) management myths; (6) participation and involvement; (7) discipline, responsibility and professionalism; (8) choice of key people; (9) recovery and maintenance of credibility; (10) resentment—inheritance from the past; (11) formation of study groups by subject; (12) production management; (13) production planning; (14) desktop and interfaces features; (15) team knowledge; (16) routine at work; (17) studies and improvements in process and production; (18) leader roles; (19) role of the shift supervisor; (20) standard operator characteristics; (21) motivational programs; (22) maintenance interfaces; (23) interface with the laboratory; (24) training plan and program; (25) training of multiplying agents; and (26) implementation of institutional programs.

7.2.5 Polycarbonates

7.2.5.1 Technology

This Process Technology was presented in Fig. 4.15.

Materials: Raw Material—CO (A)—gas; Chlorine (B)—gas; Phosgene (l)—gas; Bisphenol A (E)—solid. Inputs—Reaction Terminator (T); methylene chloride (M); sodium caustic; water (H).

Final Product—Polycarbonate (P). OPERATIONS: Reaction of Polymerization (SR); Separation (SS); Purification (SP); Wastewater treatment (STE).

The company is a medium-sized industry, with the capacity to produce 50,000 tons of products in the polymer area per year. The industrial plant is a continuous process and is subdivided into a reaction section (SR), a separation section (SS), a SP and an effluent treatment section (STE). This factory had problems with process variation, which reduced its production capacity and affected the quality of the final product. The technology of this process indicated a situation of comfort regarding the continuity of the plant, with a potential for reducing production costs, including reducing consumption of M (methylene chloride).

For this activity, an analogic control panel was used for the polymer plant. The methodology was applied in this polymer factory for 2 years. This plan is presented in other parts of the book and we will be here only remembering the application of the methodology as a whole.

The final product of the polymerization reaction (P—polycarbonate) changes its physical characteristics depending on the quantity of terminator (T), the finalizing element of the reaction. Each type of Polycarbonate produced has different physical characteristics and implications for the production chain. The polycarbonate resin produced in this plant is a condensed polymer of phosgene (L) and bisphenol A (E) which have excellent physical properties.

The limiting reagent (L) comes from a reaction in the gas phase between CO (A) and chlorine (B). To use A in reaction 1 it is necessary to promote its purification through physical processes. The preparation of P involves a condensation reaction in two steps (reaction 2 and 3). Initially E reacts with L in a basic medium. In this reaction L is completely consumed. After these reactions, the separation processes start to occur only in the liquid phase.

The organic phase is separated and sent to the polymerization reaction 3. The inorganic phase will be sent to separators that return with P and M (methylene chloride solvent) for reaction 3 and send residual process water for purification 2. In purification 3 the solvent M is recovered for the process. In reaction 3, the viscosity and P concentrations are monitored to avoid problems in the treatment and drying—Purification 4. Viscosity adjustments are necessary, injecting M to cause the dilution of P.

When resin P leaves reaction 3 with its chain already formed, it contains contaminants that are removed in the purification step 4. These contaminants (leaving, water and solvent M) that accompany the organic phase of the process go to the treatment section (removal of ions) through rotating equipment that mixes the aqueous phase and the organic phase. In purification 4, excess water is separated after treatment. The solvent M is removed on drying using rotating equipment that operates with heating to remove H (moisture) and M (solvent).

Product P is initially "flashed" on drying 1, promoting its concentration to a pasty state. After transferring P to rotating equipment, with heating and under vacuum, M and H are almost completely removed, leaving the final product after drying in powder form. The solvent (M) and water (H) from drying are condensed and redistributed: M is returned to the storage tank and the product H is reused in the treatment section and sent to the wastewater separator.

The M that is not condensed goes to the slaughter column using the principle of cryogenics. P is transported to the storage silos and then sent for physical transformation. In this stage, "master" with specific additives is added to meet the commercial characteristics.

7.2.5.2 Operational context

A plant that produces polymers (polycarbonates) in an emulsion polymerization reaction. In this plant there are processes in the gas, liquid and solid phase. It also works with a solvent that serves as a sealing liquid for alternative pumps. The solvent is chlorinated organic, which is a priority pollutant. It has several campaigns (several grades) although it is a continuous integral process. In this plant there are several problems that go in parallel and have a point of interconnection or influence. The chlorinated solvent has a high consumption rate, and it is necessary to recover what is lost as liquid effluent and to the environment by sealing rotating equipment. The reaction in the gas phase requires an earlier purification of CO and a subsequent chlorination reaction. The delivery of impure CO (A) and chlorine (B) is made directly from the supplier.

The application of this methodology, in Case 3, was carried out following steps already validated where, each timing and goals were part of schedules. The process of building the MEA was closely monitored by the team leader who was the operation technician. As the map was being prepared, it eventually started to work in the control room. To choose the most problematic months in terms of operation, the following factors were considered: (1) period of low continuity, (2) period of low load, (3) quality problems of the final product during the month following the events, (4) nonstandard effluent, (5) absence of consensus in the team, (6) record of occurrence of events different from what occurred. It was noted that the appearance of the map (MEA) was similar in cases 1 and 2. In other words, the map of abnormal events can be applied in a process of the chemical industry in general, of fine chemistry or polymers. The mass balance, the schematic flowcharts (recycling, reactions, inertization, transport of solids), the study on the internal constitution of equipment, and the operation manual enabled a better understanding of the relationship between the events. The chains of abnormalities were of different natures and are shown in Table 7.7 regarding the main function and type of process involved.

7.2.5.3 Statistical monitoring of processes

In the P production plant, there was already the practice of monitoring processes and recording operational parameters. With the investigation of the abnormalities, there was a review of the control items that continued to be monitored, with the inclusion of new ones. It is noted that, due to the number of abnormality chains detected, the process had to be followed intensively, reaching 22 control items. Table 7.8 presents a brief description of the variables. Process variables serve as indicators of the functioning of a system and the intention is to follow trends even if the system maintains its efficiency. It is important to review the system's normal standards to monitor trends and schedule corrective and preventive actions. There are several movements that indicate normality or abnormalities or simply indicate behaviors of one or the other equipment, or of a section. In addition to monitoring the 22 process items for 1 year and 6 months, items of the quality of the final product of the chemical process and the physical process were controlled, in addition to

TABLE 7.7 Abnormality chains at polycarbonate industry (Ávila, 2004).

Item	Chain abnormality	Principal function	Process and failure comments
1	Disturbance at purification 2	Remove organics from wastewater	Different content of M cause uncontrolled vaporization
2	Purification 3 operation	Retain M from off-gas from plant	Adsorption with carbon in cycles, do not operate well, stop every time
3	Polymers affect the separation	Separate organic from inorganic	In Blow molding campaign, we have a lot of problems
4	Oscillations in reaction 2	Polymerization reaction	Probably in pressure, caustic or terminator
5	E feed	Feed flow	Mixture and feed of E cause different composition of intermediary
6	Lack of oil to compressors of refrigeration	Refrigeration of M	Cold temperature to avoid M losses at the plant
7	Abnormality in purification 1	Purification of A	Separation in cryogenic temperature of CO from other gases
8	Abnormality in purification 3	Purification off-gas	The off-gas feed reaction to reduce reagent in recirculation and avoid losses
9	S feed	Promote a basic pH reactive input to reaction	Feed of E in sodium mix
10	AC feed	Neutralize wastewater in 2 sections	Reaction of effluent
11	Lubricating in treatment extractor	Ions removing	Rotative extraction
12	Skating straps from treatment	Ions removing	Rotative extraction
13	Treatment efficiency	Ions removing, water balance	Rotative extraction
14	M Pressure as external sealing	Solvent in all plant, without control of feed from sealing	Alfa Laval pumps
15	Obstruction of air pipe	Alternative pump with difficulties to transport fluid	Polymer come to sealing pipe
16	Interns in doser	Transport of fluid	Mechanical problems in pump
17	Problems in drying sector 1	P drying (powder)	Flashing
18	Vacuum abnormalities	P transportation (powder)	Pneumatic transportation
19	Water temperature	Refrigeration	Utility
20	Solid feeder problems	Feeding flow to reaction	Control by density is very uncertainty
21	Deficiency in vacuum M	Vacuum system pump	Off-gas
22	B in L	Bad reaction 1	Problems and different color of reaction sampling
23	Problems with T feeder	Feed of T	Pump

effluents. It is recommended to include consumption indexes that were updated daily through the plant's inventory (the assessment was daily based on cumulative data).

The trend graphs were presented to the operation team, who returned with many criticisms and difficulties to accept certain interpretations of the process. The biggest advantage of the open exhibition method is that, as the understanding of the subject gains space, many operators become leaders of the change process. Furthermore, it is important to state that, after a year of monitoring processes, the number of variables has been reduced; in addition, the complete review of the records brought the observation that the evaluation work was reduced, allowing a greater commitment to prevention.

TABLE 7.8 Process follow-up variables to investigation about abnormalities (Ávila, 2004).

No	Item	Area	Section	Follow-up
1	Production P	Production	General	Charge and continuity of Plant, quality of P
2	Consumption of M	Production	General	Losses of M to wastewater and off-gas, process
3	B pressure	Gas phase	Supplier	Quality of supply
4	Ratio B/A	Gas phase	Reaction 1	Control of reaction
5	Separator 1 level	Chemical	Separator 1	Operational control of reaction 2—after
6	Intermediary in reaction	Chemical	Reaction 2	Operational control of reaction 2—during
7	E in the reaction	Chemical	Reaction 2	Operational control of reaction 2—during
8	S in the reaction	Chemical	Reaction 2	Operational control of reaction 2—during
9	Ratio M/P	Chemical	Reaction 3	Operational control of reaction 3—during
10	Sum of flow deviations	Chemical	Reaction 3	Operational control of reaction 3—during
11	Molecular weight of P	Chemical	Reaction 2/3	Operational control of reaction 2/3—during
12	M to purification 4	Chemical	Treatment	Ions extraction
13	Viscosity of P	Chemical	Drying	Operational control of reaction 2/3—after
14	Level of treatment 1	Chemical	Treatment	Operational control of treatment and drying
15	Sum of Flow deviations	Chemical	Treatment	Operational Control of Treatment and Drying
16	High pressure	Utilities	Treatment	Operational control of treatment and drying
17	Level of treatment 2	Chemical	Treatment	Operational control of treatment and drying
18	Conductivity	Chemical	Treatment	Efficiency in remotion of ions at extractor
19	Steam to drying 2	Utilities	Drying 2	Vaporization control avoid solid appearance
20	Amperage drying 2. kneader	Physical	Drying 2	Operational control of drying
21	Temperature drying 2	Physical	Drying 2	Operational control of drying
22	Cold M temperature	Physical	Separation 3	Condensation and vacuum efficiency

The results of the work were: (1) reduction in the consumption rate of M per P; (2) reduction of contamination of the liquid effluent by M; (3) greater stability in the plant's process, increasing the maximum load; (4) reduction in maintenance interventions and consumption of spare parts for equipment; (5) a frictionless relationship with maintenance, due to the better quality of its services; (6) decrease in customer complaints about the quality of the final product; (7) better general aspect of the production area and the behavior of operators; (8) P production campaign with grid W, without operational problems.

7.3 Oil and gas case application

The present case study was conducted in a gas logistics company during more than 1 year, where the research on the type of technology behavior was carried out following the same methodology of work in process stabilization in the chemical industry (Ávila, 2004). We used the OD together with statistical analysis of critical process variables.

The innovation of the methodology is in the inclusion of the following techniques: task analysis; statistical analysis of security and logistics aspects; analysis of social and human behavior. Innovation is also in the development of procedures and application in: competence analysis; heuristics for social and human behavior; and the analysis of the main components of failure forecast in production schedule. One of the major difficulties in the engineering area is to migrate from objective aspects (equipment and product) to subjective aspects, such as communication, which depends on human behavior at work, then returning to objective aspects to find new indicators. This movement cannot be lost on production routine that includes safety, product quality and cost control requirements.

The subjectivity of theme under discussion is "strange" for strategic management, which is based only on the classic subjects of selection, training and social assistance to the worker. In fact, this board does not scientifically discuss human, social and organizational factors that affect process efficiency and task effectiveness.

Knowledge about psychological functions and health aspects, in conjunction with engineering competences, made it possible to develop this discussion, indicating a new tool for management decision-making.

7.3.1 Context identification: company, experience, rituals and organizational culture

7.3.1.1 Application of methodology at an oil and gas company

Concepts, techniques and procedures presented in Figs. 7.3 and 7.4 guide application of the tools and the measurement of the results indicated in Fig. 7.2. The intention is maximum organizational efficiency when achieving success of Task. Management tools based on these concepts, techniques and procedures after being tested in other industrial facilities, as described in this book, were applied in the Industrial oil and gas Segment following the specific adaptations indicated in Fig. 7.25.

Implementation work was divided into phases: (1) identification typology and analysis; (2) analysis of impact each type behavior (human, social and technical) on possibility of failure; (3) application of mathematical techniques with

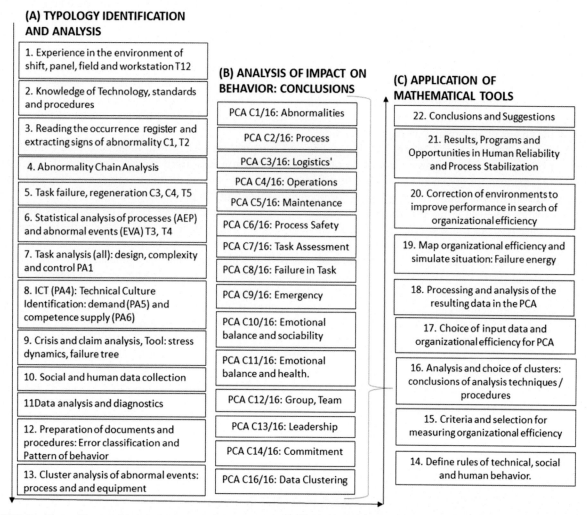

FIGURE 7.25 Steps of implementation of the program in oil and gas unit. *Adapted from Ávila, S.F., 2010. Etiology of Operational Abnormalities in the Industry: Modeling for Learning (Thesis—Doctor's degree in Chemical and Biochemical Process Technology). Federal University of Rio de Janeiro, School of Chemistry, Rio de Janeiro, 296 p.*

heuristic analysis and algorithm for failure energy calculation. That is specific practical methodology for Industry with risky activities, as indicated in Fig. 7.20.

The identification and analysis of typologies (A) depend on perceptions regarding the rules of Social and human behavior constructed by steps (7.3.1.2), (7.3.7), (7.1.2.3) and (7.3.8.2). Aspects related to signs of failure and identification of chain of abnormalities are described in steps (7.3.3), (7.3.5.2) and (7.3.7). Statistics for validation of hypotheses are performed in step (7.3.4) and from there; procedures and reports are created for diagnosis at each stage. The analysis of the impact of each type, behavior (human, social and technical) on possibility of failure (B) occurs from systematization of partial conclusions and that can be data entry for PCA. Application of mathematical techniques, heuristic analysis and construction and use of the algorithm (C) are performed in the cycle between design and test of algorithm with real data from the oil and gas plant. This algorithm will be validated using available mathematical and statistical tools and will be debugged through practice and results.

The stages of methodology divided into three Phases (Fig. 7.20) can indicate Social-Technical Reliability based on Operational Culture. 1st Phase—Typology identification (A), from step 1 to step 13. The 2nd Phase refers to studies and conclusion and have objective of data mining (priority conclusions and variables). This data was selected to apply in the tools, in 3rd Phase, stages 14 to 22, where we used statistical and mathematical tools such as PCA.

7.3.1.2 Knowledge of technology and experience in the process
Knowledge of technology

Knowledge of process and product technology is fundamental for the implementation of the methodology in industry, in the initial phase, as presented in Figs. 7.3 and 7.25. Then, experience in the workplace (T12 in Fig. 7.4 and item 1 Fig. 7.25) in shift times allows investigation of the production team's rules of behavior in the work environment.

This method was applied in the analysis of logistics, that is, in safety activities in the transport and storage of liquefied gases, composed of a mixture of gases. This analysis seeks the planning of an environment favorable to the nonevent of accidents, reducing the possibility of occurrence of Risk Scenarios. This logistics and storage activity is in Brazil, that is, in some cases, located near residential properties. Although the risks of accidents involving the public are rare, this event is possible and requires preventive analysis, including the nonidentification of latent failures, as signaled by the chains of abnormality in the investigation.

The Process shown in Fig. 7.26 includes the cooling and compression system with alignments and control meshes. Operation of drying cycles requires automation of operations that use motorized valves. Presence of light gas is a restriction on the load for drying and cooling of liquefied gases. Operations rely on alignments for cooling and compression systems with the possibility of using bypass for operational flexibility in performing tasks.

The activities involved in the process include transfer and storage of Inflammable Gas. The operations performed in the plant: transference of Liquefied material; filtration and drying; cooling through thermal cycles with gas; pressure control; drying and regeneration with the use of furnaces; sulfur contamination from the production unit; and gas quality (other gases affecting thermal cycle). Production work teams deal with these kinds of failures: motorized valves, leak detector, pressure safety valve (PSV) and torch. In addition, they work with low temperature or high-pressure operations, transfer operation to vessels and to refrigerated tank storage, drying and cooling activities are accompanied by safety procedures such as nitrogen purging in case of furnace failure; control of utilities; Automatic control of fire system through jockey pumps; visual control of torch operation and quality control of inputs for burning in furnace.

Liquefied gas purification is achieved by removing water in drying through dehumidification columns. This purification is complemented by removal of solids through filtration and removal of gas through torch to avoid instability in compression and in thermal cycle for refrigeration. Moisture removal in towers is done through desorption, heating by furnaces, cooling through exchangers, using gas in thermal expansion and compression cycles. For safety reasons, it is important maintain control in refrigerated tank by maintaining vaporization through compression and return of condensed vapors. The critical factors are in process variables, operation, and results maybe in maintenance aspects. Some features from the team and logistic operations are important too.

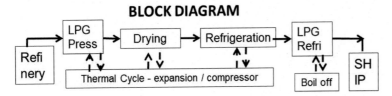

FIGURE 7.26 Simplified processes of the liquefied gas company plant.

Abnormalities and human performance

Human performance factors influence the task, including considering that technology is part of industry with job stability. The supplier has a plant in a supply chain of energy products, when common shareholders, due to the possibility of situations such as leakage from the industry, as what happened with New Mexico City (Pietersen, 1988; Shariff et al., 2016). Part of the operations is carried out at high pressure and part at low temperature, which can influence the occurrence of safety events. High pressure increases gas leakage, and low temperature, in the presence of water, can induce the formation of hydrates, making transfer difficult.

Logistics operations, remote or local, depend on task planning and involve an inventory, pressurized fluid transport, and time for each operation. Failures in motorized valves and other events keep the maintenance backlog high due to contaminants from the supplier, as solids and moisture, which impact on the pumps and vessels.

Efficiency in the transport, processing and storage of gases depends on productivity and the nonoccurrence of unsafe events. Productivity requires continuous communication with the supplier about operations. Safety is the result of good quality of raw material received continuity of operations and level of attention of operators to keep normality.

The best production control is performed with process automation and procedures for alignment operation of compressors and pumps. These operations require maintenance and safety care. Production team has the knowledge and insight to visualize onset of failure in continuous processes and intermittent procedures. Automation and visualization of supervisory screen operations indicate control modes and their failure modes.

The operator's commitment to the task may be low due to job stability. This commitment can cause safety deviance and/or delays in the production. Low rotation of the team can cause human errors due to the existence of workplace addictions, especially excessive self-confidence. Risks of rapid pressurization of liquefied gases with consequent leaks that are eventually not alarmed to the operator's perception can cause discomfort or even public complaints.

The logistics of this inflammable liquefied gas is important to the investigation to avoid quality (moisture, solids, and light gas) and production problems (for example, delay due to pumping). Production program problems are availability of critical equipment as columns, compressors, or vessels. Operation problems can be failure of furnace or boil-off abnormality in the refrigerated tank. Moreover, other logistic problems are about shipment connections that can cause delay or accident, due to pressure conditions.

Experience in the plant

From experience in a period of 4 months in the company's work environment, carried out during shift hours, the following characteristics and events were noted:

1. Opening of operation database, without restrictions, demonstrating interest in the theme of HRA by operators and supervisors.
2. Supervision and coordination available to decision sharing with new operators. Where a large part of them was prepared to discuss this subject.
3. Interest of a large part of the team about understanding what the research work is about, but at time of data collection and processing there were no formal interviews. The informal meetings were done at:
   a. Entrances of the control panel.
   b. Lunchtime.
   c. In area visits with operation follow-up.
4. Existence of many announcements on the walls relating to local and general information.
5. Presentation of policies posted on bulletin boards located in corridors and near the place of withdrawal of the Work Permit (WP).
6. "Excess WPs" in the first hours and complaints about not performing scheduled maintenance services due to *backlog*.
7. The focal points of social relations that are not directly involved with the job are found in the company's cafeteria.
8. The relationship with maintenance is only contractual, with characteristics from price-based competition to contracted services.
9. The coordination seeks to isolate the risks and is aware of the possibility of human error that may cause the event of serious accidents.

In this work, initial "informality" accelerated access to information and to understanding situations that depend on the shift group experience. Management acted less as "overly centralized leader" and acted more as "participatory management." The aspects about job stability and behavior were considered in this investigation. If "excessively participatory management" would occur, situations of low productivity or lack of responsibility for critical procedures could arise.

Union relations in the suppliers also influence these inadequate situations because there is no possibility of dismissals for improper behavior. Thus to motivate change it is necessary to conquer certain focal points of resistance, and this can be done through attractiveness of work in human reliability. During the period of experience, there were resistances located in some long-standing operators and in the staff, such as, difficulties in accessing the areas.

In the current economic situation, human errors need to be addressed to avoid process losses and reduced profits. It is vetoed to operate with lost time, low levels of productivity and even abnormalities that may cause accidents and operational discontinuity. It is known that many primary operational errors occur at facilities of chemical processes (e.g., cavitation). It is also notorious that safety indicators based on number of accidents do not always meet expectations of the safety, environment and health areas.

Recognition of social rituals and organizational culture (PA13, 15)

The company has a *reduced staff* to support the operation of chemical processes and the operation of equipment. Few engineers and specialists work in the company and its main team is formed by operators who need knowledge in logistics, compression processes, refrigeration, and safety in the logistics of liquefied gas.

The previous stability of oil prices at high levels favored cash reserves and the payment of profit sharing in the company. With the current economic situation of decrease in international price of oil and the reduction of its derivatives prices, cash restrictions and consequent flattening of results began to occur, thus reducing the profit sharing to employees. Thus there is a need to change the technical–operational culture to prevent production failures or interruptions that cause process or productivity losses. Hence the importance of working to increase reliability of equipment, processes, operational and human reliability.

7.3.2 Survey of technical, human and social data

Data collection is done at different points on the process in the industry (T1) and from analysis of records of health, safety and environment (HSE) Sectors (PA14). The query basis for processing data is the shift register. We also consult operational procedures, process variable incident and accident records.

A discussion on techniques for data collection relates numbers to facts within the activity stream presented in Figs. 7.3 and 7.4. Data collected in the shift register (T1) are functionally classified:

1. Statisticians and chronological.
2. Division of labor in tasks.
3. Task descriptors. Because of the processing of this data, we define the following information for the analysis:
    a. Requirements for task control.
    b. Analysis of failure warnings and signs of abnormalities.
    c. Classification of operational factors.
    d. Classification of operational failure.

The survey of the social and human type of the company's employees (T7 to T10) also involves data collection in the HR department, including inventory of knowledge, individual and group behaviors of the shift team. Data collection is also done from surveys, interviews and dynamics applied on groupings of operators in the shift, in this case in an internal Workshop. This data is processed and resulting information will allow identification of structural, nondeterministic characteristics that may lead the individual or group to commit failures. To identify types of team behavior, it is necessary to analyze routine in different classes regarding items: competency, tendency to slip and tendency not to follow established rules.

7.3.2.1 Routine, task, process and safety

Data collected in the shift register (T1)

The information collected and treated in this research involves the following aspects: (1) statistical and temporal identifier; (2) division of labor identifier; (3) task description identifier; (4) task control identifier; (5) analysis of signs abnormalities; (6) classification of operational factors; and (7) classification of operational failure. Details are discussed below:

1. Statistical and time identifier. Entry data: month (M), day (D), shift class (Tt) and time (Ho). Reasons:
    a. Identify the trends of abnormality and seasonality of event in the timeline, in the class and in the shift time.
    b. Analyze variables of physical environment, that is, work clock: morning/night, winter/summer, Monday/Saturday.
    c. Analyze failures in shift changes or caused by inadequate supervision profile.

2. Identifier of the division of labor. Input data: number of operators (Op), number of operators in double shifts (OpD), percentage of double shifts ((OpD/Op)*100) and number of abnormal event signals (Sean). Reasons:
   a. Analyze whether the number of abnormal events can be caused by physical or cognitive stress resulting from percentage of double shifts.
3. Task description identifier. Input data: process section (Spc), tasks or operations (Tf), place of origin of the task (OT), task destination location (DT), processed volume (Vl), time to perform task (t), process quality (Qpc, e.g., temperature), product quality (Qpd; e.g., % contaminant), effluent quality (Qef; e.g., ppm of contamination). Reasons:
   a. Check the conditions for performing tasks and operations.
   b. Check the level of control from times and qualities in the process, product and effluent.
4. Task control identifier. Calculate: number of hours per operation (Nhor); number of times that task is performed (nTf); total time spent per operation (Tti = Nhor*nTf); total time spent for all operations (Tt = Σ Tti); hours available per shift (Td = Op * 8); average hours spent per operator for equipment and people safety (Hseg); total hours with occupancy (Htot = Tt + Hseg * Op); percentage (%) (Ocp = Htot/Td * 100); and percentage of technical idleness (Oci = 100 − Ocp (%)). Reasons:
   a. Check the productivity and occupancy rate when performing routine tasks.
   b. Reverse analysis of occupation operation safely from facilities and processes.
5. Analysis of signs of abnormalities. Description of abnormal events (Desc), number of signs of abnormal events (Sean), maintenance specialty (Ms = Mec, Ins, Cald, El, Pr). Thus Mec = mechanical; INS = Instrumentation, Cald = boiler services, El = Electric, Pr = Process. Mechanical events: disarmament (des), sealing (Seal) and lubrication (lub) and others. Instrumentation events: control valve (CV), solenoid valve (SV), temperature (T), pressure (P), level (L) and amperage (A) and others. Boiler events: obstruction (Ob), holes (Fu), overflow (Tb), leakage (Vz), problems with motorized valves (Vmotor) and others. Electrical events: short circuit (Cc), crumpled electro duct (Elt), lighting (Il) and others. This survey motivated the preparation of map of abnormal events (MEA) to build the chains of abnormalities and prepare the interpretation on the dynamic failure in the task.
6. Classification operational factors. Root cause (Cr), secondary cause (cs), primary to quaternary consequence (cq: p/s/t/q), failure signal (Si). Factors that potentially increases or decreases effect problem (Fp), Failure Cycle Factor (FRc), with motivation to enable preparation of the chain of abnormalities, following signs of abnormalities extracted from the shift register and the process and operation data.
7. Classification of operational failure. Technical error (etc): information (info), material (mat), movement (mov) and tool type (Fer). Task Error (ETa): planning (Pl), perception, observation (PO), and interpretation (In), execution (Exe) state measurement (ME). Human Error (EHu): memory, attention, knowledge, rules, and commitment. Management error (Ge): decision (Dc), guideline (Di), style (Es) and communication (Co). Environment influence (IAmb): public (Com), organizational culture (CO), economic position (PE) and social insertion (IS). With motivation to facilitate mooring of managerial and technical mechanisms for correction failures and facilitate taking of action to neutralize harmful work environments to accomplishment of task.

Data collected in the operational procedure for task analysis (PA1)

Data were collected to enable the investigation of the tasks in the gas processing, transfer and storage units, and all tasks of the following types were investigated: (1) Safety and utilities; (2) Liquefied Gas transfer; (3) Storage control (tank pressure control); and (4) Processing (drying, cooling and regeneration).

Data collected to enable the investigation of the tasks in the gas processing, transfer and storage units, and all tasks of the following types were investigated: (1) Safety and utilities; (2) Liquefied Gas transfer; (3) Storage control (tank pressure control); and (4) Processing (drying, cooling and regeneration).

Study of task control considered the following aspects: 1. Classification of function in the process (emergency, normal operation, stop and interlock); 2. Steps of the procedure; 3. Safety-related items such as use of PSVs; 4. Closed control meshes (flow, pressure, temperature, and level); and 5. Aspects related to alignments in the area by task.

Data collected on process variables for use in AEP (T3)

Control and verification items important to this investigation using statistic descriptive tools are drying (large volume of processing and occupation of operators in this activity); liquefied gas pumps at entrance and exit of plant (transfer problems, obstructions, malfunction of control, and failure at motorized valves); and flash control.

The most used statistical parameters for analysis of abnormal processes and events are mean, median, standard deviation, and maximum value. Data considered on the statistics of 5 days on average. For analysis of trends in process data related to logistics operations, which involves transfers and intermittent operations, some derived variables can indicate efficiency in tank pressure control, transfer and cooling. Thus the sum of standard deviations of discharge pressure, oil pressure and compressor current allows to stop machine compensation with operating machines being possible to trend analysis.

After consensus with the company staff and analysis of the process variables, it was concluded that those indicate release of failure energy in some process areas where we cite some parameters and variable important to be followed:

1. In drying and cooling operations, 1. Pressure at the coolant outlet and control valve; 2. Pressure in the thermal cycle and control valve; 3. Sum of standard deviations from compressor discharge pressure; and 4. Plant output temperature after cooling and valve.
2. In the tank pressure control in liquefied gases: 5. Mean and standard deviation sum at tank pressure control compressor amperage; 6. Amperage Sum of tank pressure control compressors; 7. Tank pressure control pressure and valve; 8. Pressure of tank pressure control and Sum of Pressure compressor lubrication; 9. Discharge pressure and the pressure of the tank pressure control compressor lubricating oil, sum.

Data collected in safety records—HSE (PA14)

Occupational Safety and Health management practices in the workplace follow voluntary standards, but mandatory for requirements of the international product market, thus the importance of ISO-14000 and 18000 is great for export strategies. Even companies that do not export are obliged to meet regulatory standards, therefore with the legal requirement of creating an Internal Commission for Accident Prevention to prevent accidents and implement programs for intervention, reducing environmental risk through Environmental Risk Prevention Program.

The data processed were title, days for implementation of actions, date of occurrence, area of occurrence, keyword, type of deviance, degree, level of reality, initial or not, descriptive, analysis and action. Keywords are shortcuts in colloquial language that quickly identify the event. The type of occurrence is classified as: accident, accident with injury, deviance, critical deviance, systemic deviance, systemic operational failure, incident, high potential incident, critical incident, noncompliance, abnormal occurrence and complaint. The major incidents are those that cause the most severe consequence.

After statistical evaluation of the company's HSE data, the difference was verified between the aspects involved in the recorded deviances (infrastructure and administration) in relation to the signs of abnormalities identified in the research that indicate chains of abnormalities.

7.3.2.2 Human and social data collection (T7 to T10)

The discussion about the confidentiality of human and social data caused the researcher to work with group and nonindividual data, avoiding identifying names, the shift group, thus being able to preserve the operators' identity. Another important aspect is that the collected data do not provide actions based on certainty, thus leading to the definition of trends. The level of certainty of human behavior depends on the validation of data from different sources.

It is noted that observations were made about the experience in the work environment (T12), showing the trends of culture and types of worker behavior based on aspects of regional, global and organizational culture. Observations were also made about economic positioning (PA16) and the new paradigms faced by the industrial public company and its affiliates, in this case, a gas-processing company.

The analysis of the relationships that define the behavior of the operator at work depends on the composition of the behavior based on data and information collected from the following sources:

1. Primary data from HR, interviews, questionnaires, and dynamics (T10).
2. Information resulting from data processing and relationships (behavior, body reading, history, resulting from culture, competence—T7).
3. Positioning regarding: the analysis of reward for the company and the employee, the acceptance of rules and authority, the inventory of competencies (PA5), the level of commitment (PA12).
4. Analysis of the type of behaviors in the work routine (T9 and PA11): inclusion, affection, creativity, attention, memory, discipline, organization and justice.
5. Aspects related to the accomplishment of the task: leadership, control, decision, cooperation and group roles (PA3, PA10 and PA11).

Presented below is a greater detail in the data collected in HR and in the workshop focused on this purpose. Data regarding the identification of the technical culture and referring to the analysis of the stress team are treated in items 7.3.4.3 and 7.3.5.4.

Social and individual operator data (T7)

The self-assessment of the worker is divided into two blocks of information: the social data, which indicate the level of insertion in the company or in the group; and the human data, which deal with self-assessment of operators that indicates the reflection of their individual experience about their body in areas of health, cognition and behavior.

The social data of the survey were divided into cultural information, influence of parents, family aspects, and aspects about public itself, food and social customs, bonds of friendship, leisure and hobby, actions in society and in the family.

A brief discussion on analysis of these data indicates conclusions regarding human and social types. Religion is inserted in the cultural movement and may indicate a relationship of critical or blind fidelity; it is not a deterministic factor, but along with other information, it may indicate cultural trends. The same can be said in relation to the sport or football team, knowing that supporting and playing sports can indicate "oxygenation in thoughts," following two trends: preserving leisure for quality of life and body health that leads to the balanced mind.

Family relationships in Brazilian culture are preserved by maintaining the proximity of parents, although the globalized trend is the nonaffective bond after a certain time, which may mean placing family members in the nursing home. Thus the presence of living parents who are financially dependent on company employees can mean an emotional bond or, at the other extreme, emotional imbalance. The position in the family indicates level of bond adopted. Emotional balance is complete or incomplete, it depends on constitution of the family and its harmonization; thus data such as marital status, number of times married (egocentrism and building affective bonds), number of children and age are collected.

The performance in the public is self-assessed by the worker who responds about politics, ethics, social action and form of participation. Depending on the answer and the other traits raised, there may be no sincerity in these answers, due to the preferred behavior, but not always reasonably practiced "socially." Aspects related to the dominant culture explain the customs of eating, living with parents, parties, drinking and smoking. The answers may indicate dependence or even the adoption of a globalized or mass culture.

Finally, keeping friends after establishing yourself in the work environment and equalizing with family and public means being active; how far is it possible to reach this state? From how many hours of occupation is leisure considered a vice? What is the intensity of participation in the family and society that can be considered reasonable for the current culture?

The aspects of coexistence in society, at work and in the family generate behaviors resulting from impositions of these means for social inclusion/financial survival or generate behaviors formulated from strategies for the defense of this social inclusion/financial survival. As a result of imposition or preformulated strategy, both behaviors can be discussed based on the behavior of the operator and family in relation to health, self-image, cognitive self-assessment, and behavior.

In the self-assessment poll, information on family health; operator's childhood; the type of behavior with primary information on diseases, weight, height; aspects of personal hygiene and esthetics; aspects related to human cognition, which somehow influences the construction of a mind map for the execution of the task at work.

A brief discussion is made to guide the use of data. Items related to family health may indicate the propensity not to bear stress due to cardiovascular diseases or diseases related to excessive authority or control.

The relationship between weight and height indicates nutritional stability or even regional customs that can prevail over the globalized custom. Heart disease related to blood pressure may indicate greater sensitivity to stress, traits not interesting in the current world of work, but which are very present.

Respiratory diseases may indicate dependencies when related to family or infant aspects. Diseases of the intestinal tract may be characteristic of reaction to authorities, being indicative of the level of fidelity. Some symptoms involving skin diseases, according to psychosomatic medicine, may involve unresolved guilt processes and affective bonds in social environments (Haynal et al., 2001).

As for aspects related to esthetics, they may indicate confrontation in relation to society, where the noncleaning of the own body may be associated with discredit in aspects about "process losses" in the work environment. Cognition is a primary factor in the work environment, and we analyzed positions on: attention, memory, sense of perception, thought, language and intelligence are questioned.

Questionnaire data on individual behavior in the group (T8)

When answering the questionnaire about individual behaviors, it is expected to locate the individual and the group about the questions of sociability. Questions are asked with yes and no answers. Each respondent has several expected responses based on the results of this "supposed" pattern. The mean, standard deviation, variability factor and difference in relation to the expected pattern are calculated.

1. Mental attitudes that contribute to sociability: Kindness (tendency to union), compassion (affect competitiveness), affectivity (sensitivity toward third parties), optimism (moderate to work with detachment), self-mastery (attending to emergency situations); respect (sympathy or empathy for the group).
2. Unfavorable mental attitudes for sociability: arrogance (belittling others), selfishness (exaggerated opinion of oneself and disparages self-love of others), excessive criticism of defects (fend off co-workers), self-centrism (focusing on self rather than colleagues), angry temperament (low self-control and discussion with colleagues).

Questionnaire on group behavior: cooperation and leadership (T9)

According to Schutz (1966), the FIROB method analyzes the behavioral orientation of fundamental interpersonal relationships. It states that there are three fundamental areas for understanding and predicting interpersonal behavior. Although many other factors influence a person's attitudes, by knowing their position in three dimensions, significant inferences can be made about their behavior. These three areas are: inclusion, control, and affection.

Inclusion is causally related to sociability, thus indicating the tendency to be grouped or isolated. Inclusion, degree in which the person associates with their mother is evaluated. According to Jung (2002), concepts of "introversion" and "extroversion" are related to inclusion in society or social groups. On the other hand, the maintenance of social situations in control is related to rationality and may lead to greater comfort for certain more centralizing leaders during emergency decisions. Control measures the degree in which the person takes responsibility, makes decisions, or dominates other people. In relation to affection, emotional balance can impact cognitive processing and respective organizational analyses (reward, commitment, biases in decision-making, justice, and application of competencies).

Human resource or human factor data (T10)

The data from the Department of Personnel or Human Resources should consist of a sample from 20%–40%, randomly, on the team of operators and leaders.

The employee's origin, whether from the logistics in the company or another company; date of birth and age; the level of instruction with several hours indicating level of knowledge; the number of dependents and the age of the employee per dependent. In addition, the period between current time and admission compared to the time calculated by exact date. The change of positions, how many levels have been acquired so far, age by years of work, years of work per dependent, what salary progressions (levels per year of work and by period), time of leave in days and leave per year.

Some experienced employees from oil and gas have greater preparation, with skills regarding emergency and routine situations. The worker's older age may represent more experience; thus this analysis requires care and validation from the individual identification of the technical–operational culture, while the "younger ones," have less knowledge and maybe take initiatives without thinking.

The level of education may not mean greater knowledge, given the quality of knowledge acquired today, but, taking into account the possibility of increasing complexity in thinking, it can facilitate the interpretation of complex problems. Tests and dynamics with stress can indicate which trend, whether combined, with available knowledge, or following the trends of globalization.

Number of dependents may be linked to the need to build bonds or eventualities in the life of couples or even in practicality.

Period of admission may indicate cultural changes in the company regarding safety aspects, regarding proximity or not in interpersonal relationships and regarding management and union issues that can occur. Time at home can indicate people who are disbelieving (old in silence) or energetic (new and willing to change).

Number of levels acquired in relation to time in the Company may represent a greater commitment and identity of the company in relation to these employees. It can represent "patronage," leading to a centralizing managerial profile with a low sense of justice.

Some important numbers are: (1) ages per years of work, which may indicate permanence as dependent for longer; (2) years of work per dependent, indicating tendency to build emotional bonds or not adopting differences in dominant culture (global vs regional); (3) absence due to accident per year of work, indicating trends in safety culture or social culture.

7.3.3 Abnormal event mapping, signs, and failure mode (MEA and FMEA)

From shift register for a period of 2 months, occurrences that indicate abnormalities were collected. Abnormal events mapping prepared through consultation and investigation of occurrences recording in tree form and encoded according to connections of gas processing. Relationships between these occurrences construct chains of abnormalities that are discussed in abnormal events report (REA). Validation of REA through testing is an important part of the stabilization process. Some identified in the preparation of qualitative mapping that help to define with statistical indicators of abnormal events (MEA) are:

- Unavailability of Raw Material in the supplier.
- Unavailability of a ship to receive from refrigerated tanks.
- Decreased compression capacity due to a percentage of light gas.
- Failures in some automation instruments (pressure, levels and flows).
- Failure in some rotating equipment, in motorized valves, and filtration systems.
- Large number of logistics and safety procedures performed by operator and per shift, could inducing to human error.

The abnormalities recorded in the form of a causal chain need to be validated through discussion with the staff and with the shift. Some abnormalities in Table 7.9 are classified by type. Part of these abnormalities is connected to each other, and chain 22 and 23 represent abnormalities that affect the entire gas-processing plant.

TABLE 7.9 Sample of chains of abnormalities in oil and gas processing.

C #	Event	Description
C1	Fire system	The fire system unit must be available when it is required, has configuration in independent rings for the vessels and for refrigerated tanks.
C2	Dirt coalescer and filter—water	The signs of dirty filter and presence of water should be analyzed to avoid the consequences of the presence of hydrates and high humidity affecting drying efficiency. Routine drainages.
C5	Cooling efficiency	Liquefied gas should not stand still in pipes or vessels too long due to expansion possibilities (high pressure) and leaks.
C6	Events in the compressor	The events that may involve the refrigeration compressor: (a) refrigerant quality (light gas), (b) demand for thermal cycle, (c) difficulties with the load valve, (d) overload in compression system.
C7	Regeneration failure 1	Humidity meters indicate the moment (time) of saturation (with water) of the column—insulation and regeneration. If the column is not regenerated properly, water and hydrates are sent, thus causing obstructions due to freezing.
C8	Regeneration failure 2	The liquefied gases pump for the caved furnace due to instrumentation failures for safety. The noncirculation of this used to regenerate the berth causes backpressure in the furnace with risk of explosion, high pressure.
C9	Furnace failure, liquefied gases flow	At the present time we will be concerned with showing signs of abnormality that may be involved with the circuit (a) referring to the part of the burn that provides heat to the circuit (b).
C11	Pilot failure in furnace safety	The safety system does not act properly causing the pilot to lose control to keep burning in the furnace.
C12	Fault burner furnace safety	The very high temperature in regeneration can cause damage to the fixed berth, having its stability compromised and may fragment by sending waste from the berth to regeneration.
C15	Compressor stop	The cold assigned in refrigeration using the thermal cycle of expansion and compression of gas depends on the load of the drying process. The same thermal cycle that cools liquefied gases cools in drying and regeneration.
C19	Pump failed	Reasons for low yield or failure in the vessel pump: pressure in suction, amperage, low level in pump (well) and in the relief line or breath, pressure in discharge. In addition, the existence of relief valve that sends to the torch.
C22	Liquefied gas contamination and effect on ship load	At this time, the possible effects of liquefied gases contamination for processing gas originated in the are present. These effects cause disturbances throughout the plant with different impacts. There are signs and consequences that mark the path and that need to be analyzed to take preventive and corrective actions.
C24	Abnormalities in HV valves	The motorized valve decreases human effort in alignments for the transfer of large volumes of fluids. Therefore, to facilitate operations, the valve is operated remotely. In gas-processing facilities, many valves have constant failures indicating the need for displacement and greater effort to perform alignments.

(Abnormality Chain—AC 1/3) production support systems (fire, compressed air).
(AC4) communication with the supervisory, electrical control centers, LCP (Logical Control Panel), CVTV (Internal TV Network).
(AC 2/20/21/22/23) quality of the raw material for the presence of solids, excess water, excess light gas, or corrosive (R_2S_2).
(AC 9/10/11/12/13/14) failures in the regeneration furnace (in load, safety system for pilot/burn and control).
(AC 7/8) failures in the regeneration bed (saturation).
(AC 15/20) failures in refrigeration due to light or other contaminants, causing equipment (thermal cycle compressors and refrigerated tank top).
(AC 16/19) product and process pump failures (product and condensate pumps).
(AC 17/18) failures in the control of the refrigerated tank.

> *(Conclusion PCA1/16—abnormalities) The presence of solids in the feeding circuit affects valves, pumps and vessel. High pressure caused by low operability of control valves and motorized valves. The improper compressed air problem potentiates failures, but the main cause may be due to design errors. Contamination by moisture and sulfurous compounds accelerates problems.*

7.3.4 Process analysis, logistics, operations, maintenance and safety (T3—AEP, T4—EVA)

The technical aspects discussed here are related to tools already developed and applied in the industrial plants of aliphatic amines in fine chemistry, aromatic amines in polyurethane market, and polymers as in the case of polycarbonate (Ávila, 2004), related to statistical analysis based on classes and temporal aspects. The dynamic analyses of the process named as AEP or statistical process monitoring-control, discussed in Section 7.1.2. The discrete data analysis, on a chronological or class basis, named as EVA tool, here statistics of abnormal event and referring to logistics, operations, maintenance, and safety aspects, treated in items 7.1.2 to 7.1.4.

7.3.4.1 Process analysis

The company's liquefied gas site is capable of stocking pressurized gas from supplier in vessels; refrigerated and liquefied Gas resulting from process this installation; and loading/unloading, from ships. The transfer of pressurized gas from the supplier to the liquefied gas site of the terminal takes place through ducts that feed the vessels.

Main operations and processes have important equipment to purify this liquefied gas as compressors and absorption columns. These equipment including compressors with different functions: boil-off in refrigerated tanks; tank pressure control compressor; and refrigeration compressors.

The dry process and tank pressure control are monitored to avoid production delays and safety events. Gas-processing unit based on logistic operations, work to meet demand of production. Staff need to avoid high charge to equipment through implementation of stabilization processes Program and human reliability. The pressure may be the result of load for drying or flow controls, due to restriction on refrigeration or receiving refrigerated tank.

During these months of analyses the plant had controlled pressure stability including the pressure in the discharge of tank pressure control vessels, located at the top of refrigerated tanks, to control vaporization and keep liquefied gases refrigerated.

Any fluctuations in the control of drying pressure may or may not affect the control of the 3rd stage pressure in the refrigeration compressor. At this moment, as shown in Fig. 7.27, it is intended to evaluate the sum of standard deviation of parameters "indication" and "valve opening."

With liquefied gases receiving and scheduled output, rotating equipment could work with continuity and greater reliability. This variation in discharge pressure can be caused by the presence of light gas in dry gas. Thus it is necessary to monitor quality and quantity received.

The temperature after refrigeration and its respective opening of control valve has direct relationship with the drying load and the efficiency of the thermal cycle. Both work well if the coolant is dry, ethane-free and liquefied gases is water-free. Fig. 7.28 shows large oscillations in the valve opening, with an average of 50%—a saw effect with temperature oscillations, which reaches levels: 20°C, 10°C and −5°C.

All tank pressure control compressors work with average current in saw or zigzag effect, that is, they are oscillating. The scheduling of receipt and consequent processing of gas is carried out as production unit sends liquefied gases; this

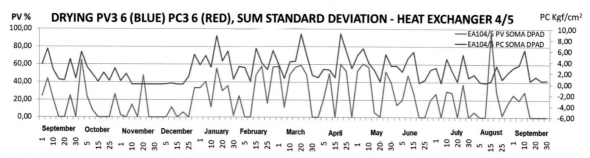

FIGURE 7.27 Sum of standard deviations of discharge pressure/thermal cycle and control valve. *From Ávila, S.F., 2010. Etiology of Operational Abnormalities in the Industry: Modeling for Learning (Thesis—Doctor's degree in Chemical and Biochemical Process Technology). Federal University of Rio de Janeiro, School of Chemistry, Rio de Janeiro, 296 p.*

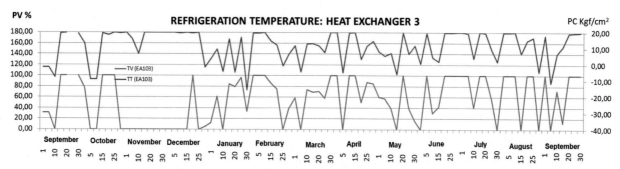

FIGURE 7.28 Temperature at the refrigerated outlet. *From Ávila, S.F., 2010. Etiology of Operational Abnormalities in the Industry: Modeling for Learning (Thesis—Doctor's degree in Chemical and Biochemical Process Technology). Federal University of Rio de Janeiro, School of Chemistry, Rio de Janeiro, 296 p.*

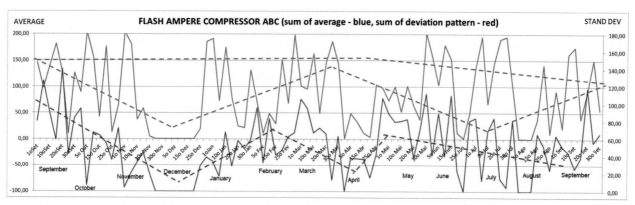

FIGURE 7.29 Sum of amperage averages and tank pressure control compressor standard deviations. *From Ávila, S.F., 2010. Etiology of Operational Abnormalities in the Industry: Modeling for Learning (Thesis—Doctor's degree in Chemical and Biochemical Process Technology). Federal University of Rio de Janeiro, School of Chemistry, Rio de Janeiro, 296 p.*

means that steam loads for refrigerated tank can be oscillating depending on demand. The steam loads are due to the vaporization of power in the inlet exchanger; thermal energy of the refrigerated pump; heat received on transfer line to piers; heat received by the ship tank; tank boil-off in operation; displaced steam from the liquefied oil gas. It is important to estimate each load to decrease the oscillation as far as possible.

With reference to Fig. 7.29, although the sum of the standard current deviation of the tank pressure control compressor is oscillating in a "saw" effect, it does not follow the same trends as the sum of the amperage mean. At first, the reduction of the sum of the amperages was equivalent to the reduction of the standard deviation, but then the trends were diverse, showing that, possibly the steam load of different origins may be causing variation in compressor performance.

The maximum amperage of each compressor is added and represents a parameter, there was a stabilization of the curve, despite operations with alternating compressors, that is, operates and stop. In Fig. 7.30, at the beginning of the period this sum reached 400 amps with great efforts from there for 100 amps, indicating low load and returning to 200 amps with normal efforts. Thus ascending peaks to 300 amps or descendants for 100 amps indicate events that correspond to varying amounts of steam in the circuit leading to the need for preventive and corrective action.

The opening of the pressure control valve follows the reverse direction of the average pressure in the *tank pressure control* compressor. So, when the valve tends to open, the pressure tends to drop. The interesting thing is that every thermal cycle system operates with the reverse direction between average and valve opening, probably also occurs in the regeneration conditions. liquefied gases quality is related to moisture and then causing uncontrolled situations.

The lubricating oil pressure fluctuates between 2 and 15 kgf/cm^2, more smoothly, and indicates overload over the system as shown in Fig. 7.31. The discharge pressure sum oscillate, which may indicate increased load, or change in the quality of the finished product.

> *(CONCLUSION PCA2/16—process) The pressure indication PC3, PC6 (standard deviation) can indicate process and safety trends, the PV regulates though in manual. The maximum discharge pressure monitoring of pressurized Site pumps is discussed, but the sum of the standard deviation is more significant. The sum of the standard deviations of the amperage of the ABC Tank pressure control Compressor contributes to the knowledge, the sum of the amperage maximums in this Compressor and the sum of the lubrication oil pressure of the tank pressure control contribute to the knowledge about pattern and failure in this system.*

7.3.4.2 Logistics analysis

Ship loading projection on the pier defines production planning and control to meet liquefied gases product sales area. Existence of product in tanks/vessels and availability of operational area for purification (equipment and people) enables the loading of ships. On the supplier's side, demand for product is seasonal and ships that make their transport are in a limited number indicating that at certain times, gas processing must operate, thus giving priority to the

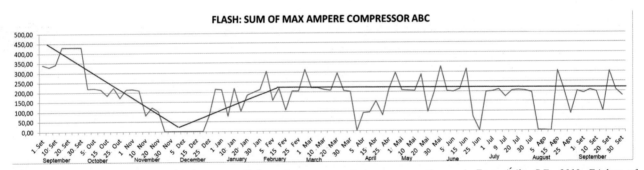

FIGURE 7.30 Sum of maximums electric current of *the tank pressure control compressors (ampere)*. From Ávila, S.F., 2010. Etiology of Operational Abnormalities in the Industry: Modeling for Learning (Thesis—Doctor's degree in Chemical and Biochemical Process Technology). Federal University of Rio de Janeiro, School of Chemistry, Rio de Janeiro, 296 p.

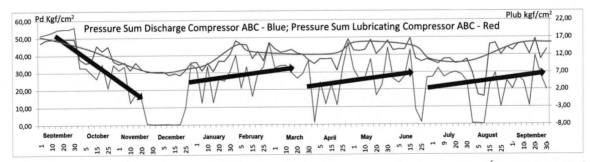

FIGURE 7.31 Sum discharge pressure and oil lubrication pressure of the *tank pressure control* compressor. From Ávila, S.F., 2010. Etiology of Operational Abnormalities in the Industry: Modeling for Learning (Thesis—Doctor's degree in Chemical and Biochemical Process Technology). Federal University of Rio de Janeiro, School of Chemistry, Rio de Janeiro, 296 p.

availability of a ship that is already docked on the pier and may even stop for a short period other transfer operations to meet this priority. In this way, it becomes more difficult to increase the productivity of loading and gas processing, based on waiting time for the arrival of the next ship.

The monitoring in logistics is carried out as listed below, in which the comments made are about increase in productivity in loading and transfer logistics: (1) Number of loads and load hours for shipment; (2) Final cargo per shift schedule per ship; (3) Number of days between shiploads; and (4) Shift schedule.

> *(PCA3/16 Conclusion—Logistics) As logistics operations are tied to delivery priorities of processed material to meet ship and liquefied gases availability in supply, block-based activities do not bring much numerical information but discrete decisions about the action to be performed.*

7.3.4.3 Operations analysis

The following productivity control of the production team affects the cost and the availability of Product: average number, daily, double shifts, average number of operators per day for a shift (operators), average daily percentage of team double shifts, and number of abnormalities per day. In addition, we consider daily processed volume, number of operations, and team performance.

This performance includes following indicators: hours per operator, hours per operation, and the team's technical idleness level. This technical term (idleness) in the team refers to the inverse of occupancy rate for activities related to the operation function in the liquefied gases unit.

Some variables influence these parameters, for example, liquefied gases availability for processing, vessel, or ship availability. It is based on the premise, in this research, that technical idleness is not a characteristic that depends on man. It depends on liquefied gas availability and ship availability.

Operations: Average number of operators per day

As shown in Fig. 7.32, in percentage terms, the following classes were detected: no double shifts (18.6%), 1 double (27.1%), 2 doubles (25.4%), 3 doubles (22.1%), 4 doubles (3.4%) and 5 doubles (3.4%). The team (on internal or company demand) performed around 2 doubles, but this high number is considered for human body clock. Ideally, there would be one double per shift to avoid fatigue or stress effects on operator. The average per daily shift of working operators is 4, ranging from at least 3 to up to 4.4 operators per shift. Although double shifts mentioned above have been programmed in advance, repetition can lead to physical fatigue and worsening cognitive condition in development of task.

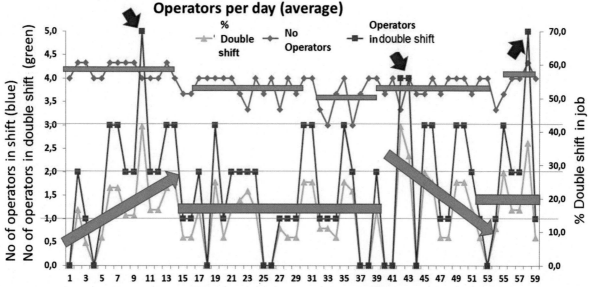

FIGURE 7.32 Number of operators per day, double shifts, and percentage of double shifts. *From Ávila, S.F., 2010. Etiology of Operational Abnormalities in the Industry: Modeling for Learning (Thesis—Doctor's degree in Chemical and Biochemical Process Technology). Federal University of Rio de Janeiro, School of Chemistry, Rio de Janeiro, 296 p.*

Operations: Sum of abnormality signs per day

Based on technology and manual procedures, certain shift records are considered signs of abnormality that need to be validated through testing. Causality is tested and validated by shift team. While abnormalities were extracted from of the register occurrences of the shift during two-month period at the beginning of the year. On the other hand, patterns differentiate normal state from abnormal, suggested by the operation guide, in which, any sign of abnormality must be presented. It is known that certain signs, although not considered abnormalities, may indicate their existence, thus being valued in identifying the cause in latent failure. The sum of abnormalities (or signs) was analyzed to indicate events per week or per day, in addition to trends in the period.

The recording of signs of abnormalities can be "false" if absolute value is considered, due to aspects of omissions in the writing of the report. The lack of more detailed information about abnormalities can be considered as a cultural trait and is influenced by managerial profile and its acceptance in the shift team. The intention is to analyze the trends indicated in Fig. 7.33 of this indicator and may enable the analysis of traces of the technical–operational culture. It is interesting that the problem that occurred in late January in this gas plant has fewer signs of abnormality than at the beginning of the analysis period, thus indicating that during the stop period, there is a decrease in the signs of abnormality when compared to full operation.

Operations: Processed volume and number of operations

Fig. 7.34 analyzes three variables in operations at the plant: processed volume, number of operations (multiplied by 5 to improve visualization), and number of hours of operation. This information corresponds to a sum and is not always

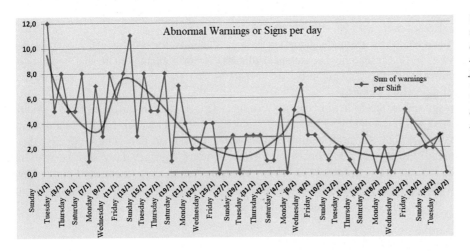

FIGURE 7.33 Number of abnormality signs or warnings per day. *From Ávila, S.F., 2010. Etiology of Operational Abnormalities in the Industry: Modeling for Learning (Thesis—Doctor's degree in Chemical and Biochemical Process Technology). Federal University of Rio de Janeiro, School of Chemistry, Rio de Janeiro, 296 p.*

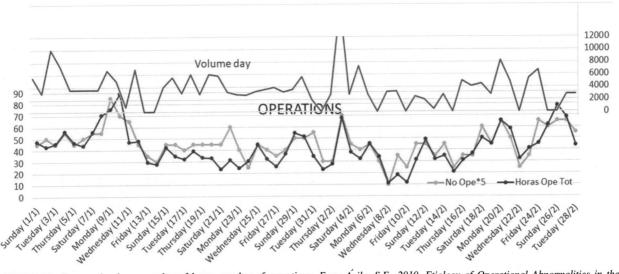

FIGURE 7.34 Processed volume, number of hours, number of operations. *From Ávila, S.F., 2010. Etiology of Operational Abnormalities in the Industry: Modeling for Learning (Thesis—Doctor's degree in Chemical and Biochemical Process Technology). Federal University of Rio de Janeiro, School of Chemistry, Rio de Janeiro, 296 p.*

reported in the shift report, both about volume and hours of operation, as well as the number of operations carried out on the day.

Operations: Team performance: number of hours per operator and per operation

Team productivity is also analyzed through the number of hours developed for each operation and the number of hours per operator as shown in Fig. 7.35. In addition, the number of hours each operator spends on each operation. The database of this analysis is the shift period (eight hours) and the average is calculated for the daily activities. Noted similarities indicate correlations between the variables. The characteristic of job organization in operation indicates that 4 operators can conclude from 5 to 6 operations, thus maintaining the ratio between 1.25 and 1.5 operator per operation.

The type of operation can favor operator's performance, which can be simple operations or parallel operations in continuous mode, without interruptions. The average values considered are 3 hours per operator performing operations, 4.5 hours per operation performed and 1.2 hours per operator per average daily operation.

Operations: Technical idleness by schedule and per class per week (Fig. 7.36)

The premise adopted is that the data collected in the shift register are correct. It is known in literature (Hollnagel, 1993; Rasmussen, 1997) that omission is a cause of human error and occurs in the execution of the task. In addition, that, based on the practice of work in the operation in general, the shift register has omissions, problem indicated by Reason (1990) and discussed in chemical process industries, therefore, the interpretations used for decision-making should be analyzed.

Class E had the lowest percentage of technical idleness (16%), thus indicating the highest capacity to be busy. Class B had the highest average percentage of technical idleness (21%), thus with difficulties in being busy on the shift. The greatest variation in technical idleness is in classes A and E, with 15% for standard deviation. The week of highest technical idleness was 6, and the lowest week was week 8. The groups that remained cohesive about this percentage were respectively: week 1—EAB; 2—ACDE; 3—BCDE; 4—ACD; 5—ABCDE; 6—BCDE; 7—ABD; 8—BCD. In this cohesion, the class that is most repeated is D, while the ones that are least present with downward trends, are classes A and E.

> (PCA4/16 Conclusion—operations) The percentage of double shifts may be an important factor to indicate propensity to error, low process volume and increased signs of abnormalities. This sum of signs of abnormalities brings important meanings for the analysis and serves as an indicator of organizational efficiency, as well as the processed volume of liquefied gases although the data, in the occurrence register, had low information quality. The number of hours per operation and per operator indicates complexity of the task and commitment to perform and may be an indicator of motivation and competence applied in the accomplishment of the task. The organizations of reinforcement should be considered in the shift for plant operability, such as the presence of redundant operator in the issue of Work Permits.

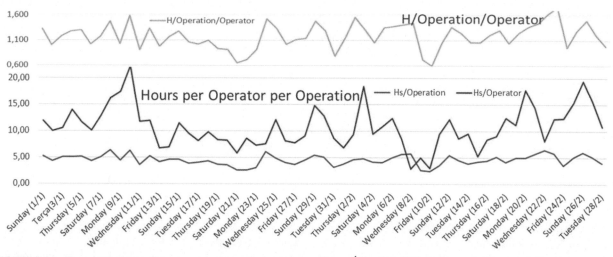

FIGURE 7.35 Team performance, hours per operator and per operation. *From Ávila, S.F., 2010. Etiology of Operational Abnormalities in the Industry: Modeling for Learning (Thesis—Doctor's degree in Chemical and Biochemical Process Technology). Federal University of Rio de Janeiro, School of Chemistry, Rio de Janeiro, 296 p.*

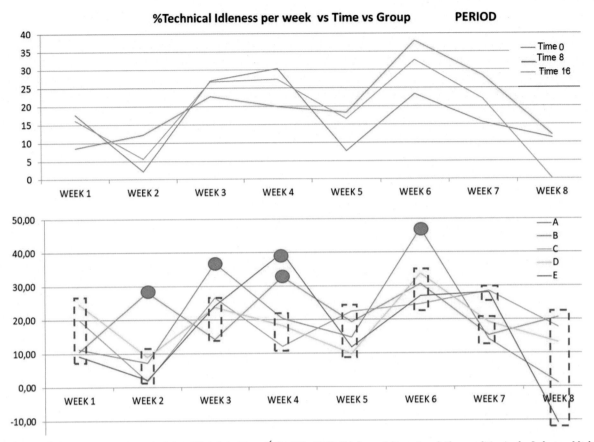

FIGURE 7.36 Technical idleness by schedule and by class. *From Ávila, S.F., 2010. Etiology of Operational Abnormalities in the Industry: Modeling for Learning (Thesis—Doctor's degree in Chemical and Biochemical Process Technology). Federal University of Rio de Janeiro, School of Chemistry, Rio de Janeiro, 296 p.*

7.3.4.4 Maintenance, safety and environmental analysis

In Fig. 7.37, in a pie graph of abnormalities by maintenance and occupational specialty, it is noted that most of the abnormalities recorded in the shift register of occurrences are related to safety and instrumentation. Many of the citations on electricity are general issues, such as lighting and faulty electro ducts. Regarding environmental impact, small leaks with liquefied gases is most cited. This logistics industry involves compressed and refrigerated gases, all flammable, therefore, the number of operations with alignments, compression and cooling cycle is large. In general terms, when a classification is verified, it is noted that valves, pumps, and compressors are the main citations responsible in the nonconformities discussed in the case book.

It is important to comment that events occur with the neighboring public and that are a risk due to the public not being aware of safety procedures such as the operation team, so weight seven in the table and Fig. 7.38. Another weight correction was the multiplication by two, high weight, of the events involving compressors due to the high-pressure risks in the plant.

The bar chart in Fig. 7.38 shows events related mainly to valves with 28%, including stuck and inverted signals and compressors in the order of 23%. The main occurrences in the pumps are leaks and problems in the automatic activation of the motors in general. Events involving the public, such as vehicle crash outside the factory walls, fire in the trash transit of strange people on the site. In addition, events relating to process issues are ahead of events relating tanks, lines, filters and coalescers.

After statistical analyses, some items are chosen to be investigated, grouped as (1) Motorized valves, control valves and solenoids; (2) Filters and events with corrosion; and (3) liquefied gases pump and vessels, in addition to pressurization of lines and systems. The other groups are (4) Fire system pumps; (5) Refrigerated liquefied gases tanking; (6) Drying and regeneration, including the operation of the furnaces and their solenoid valves; (7) Tank pressure control system, compression and tank pressure control pump. Finally, (8) Cooling system, compression.

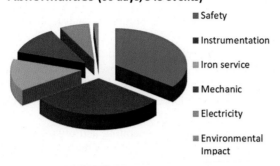

FIGURE 7.37 Abnormality by maintenance specialty and health, safety and environment. *From Ávila, S.F., 2010. Etiology of Operational Abnormalities in the Industry: Modeling for Learning (Thesis—Doctor's degree in Chemical and Biochemical Process Technology). Federal University of Rio de Janeiro, School of Chemistry, Rio de Janeiro, 296 p.*

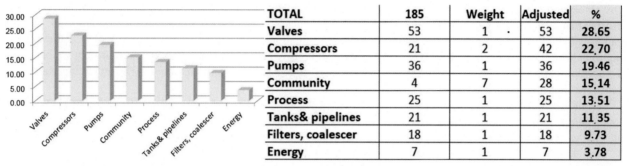

TOTAL	185	Weight	Adjusted	%
Valves	53	1	53	28.65
Compressors	21	2	42	22.70
Pumps	36	1	36	19.46
Community	4	7	28	15.14
Process	25	1	25	13.51
Tanks & pipelines	21	1	21	11.35
Filters, coalescer	18	1	18	9.73
Energy	7	1	7	3.78

FIGURE 7.38 Abnormality by type of event in equipment or in general. *From Ávila, S.F., 2010. Etiology of Operational Abnormalities in the Industry: Modeling for Learning (Thesis—Doctor's degree in Chemical and Biochemical Process Technology). Federal University of Rio de Janeiro, School of Chemistry, Rio de Janeiro, 296 p.*

> (PCA5/16 Completion—Maintenance) It is important, as already commented in the abnormalities, to include events that gather three blocks and that appear here in maintenance as well: events involving valves (22% of citations); events involving filters, coalescers (11%); and pressurized and vessel pumps, which may be related to the possibility of high pressure.

The main factor to be controlled in the refrigeration unit is pressure, to initially avoid leaks that can cause fires and explosion. All other factors are connected to the issue of pressure, as in the case of control of the production process through valves, lines and tanks. There may be a reduction in the load for refrigeration from the liquefied gas itself also has a content; thus compressors do not condense the entire liquefied gases, and have material loss for torch. Presence of gases as derivatives of mercaptans with liquefied gases, facilitates the occurrence of corrosion. The presence of hydrates, leftovers from molecular sieve, presence of materials from corrosion hinder operation of motorized valves and solenoid valves, mobilizing operations in alternative valves in the field. The result from all these events can generate pressure, which impairs cooling systems, tank pressure control, affecting performance of compressors.

The control of the furnace depends on the functioning solenoid valves, which may be impaired by mixing, thus generating soot or oscillations of liquefied gases quality, regarding the presence of water, and difficulties in condensation of this water, sending back to the molecular sieve column, causing its saturation even after regeneration.

Events with ducts may indicate maintenance error or behavior, with risks for occurrence of site in flammable gas area. The law of least effort avoids additional activities avoided, thus worsening the pattern of the operation. Meanwhile, some abnormal events may result from interaction with the neighboring public, such as the accident caused on the company's vehicle, or even a fire put into garbage near the gas pipe. These events require "latent failure reading," and care as to the public's failure not to accept this processing.

Finally, it is noted that, despite the importance of the pier and the number of operations performed and risks of pressure nonrelation, in the case of liquefied gases loading, there are not many events recorded. It is important to examine whether the monitoring for these operations is satisfactory, with their *checklists*. This by the potential risk of fuel gas.

In the analysis of equipment related to safety issues, as indicated in Fig. 7.39, events involving vessels, tanks and equipment, such as the furnace. Most of this equipment is protected by PSvs It was observed that the operators (apparently) avoided transiting near the tanks, especially near the vessels. In other industrial sulfuric acid processing plants, operators felt inhibited for action in the field due to the risks of burn accident. Thus it is important to analyze this indicator of behavior that may be reflecting in lack of water drainages in the vessels.

The record of incidents, deviances and accidents is analyzed at this time, incorporating certain aspects already discussed in the technology behavior (or technical type). Themes of these nonconformities are diverse, involving aspects related to technical systems, such as the operation of equipment and instruments, which can lead to short circuits, leaks or accidents. Some behavioral aspects of body of work are analyzed in relation to safety and public itself, which, being a neighbor is active in living with company.

The technical aspects can be originated from process problems (quality) of the way of operating and organizing the shift, in addition to maintenance errors in the delivery of these technical systems. An important factor that was repeated in the events recorded and investigated in the HSE.

Due to the degree of impact, it is noted that 32.5% of the cases have a higher degree, indicating high criticality of gas-processing systems that can be increased if only events related to the process are considered. The technical team affirms that 80% of safety cases analyzed are real (not hypotheses). Less than 20% was raised without proof. The indications of abnormalities of the process, show that the real situation would be reversed (with 80% of hypotheses), leading to believe that records do not analyze latent failures, but only real failures after they happen, or very visible factors. To corroborate the previous statements, events have a high percentage of repetition, above 40%, indicating that the root cause was not resolved.

As shown in Fig. 7.40, general events involve the entire process area and make up 18.8% of the total events with local, sanitary, sampling, identification, and allocation of waste in the plant as a whole. The communication item is at 16.3%, due to sudden signal loss due to logical, contact devices or even physical issues, such as severed optic

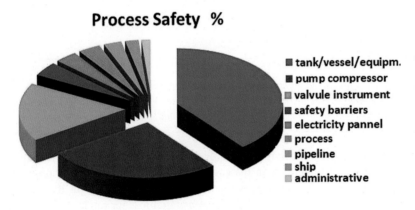

FIGURE 7.39 Equipment and functions in safety process. *From Ávila, S.F., 2010. Etiology of Operational Abnormalities in the Industry: Modeling for Learning (Thesis—Doctor's degree in Chemical and Biochemical Process Technology). Federal University of Rio de Janeiro, School of Chemistry, Rio de Janeiro, 296 p.*

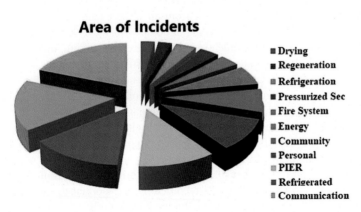

FIGURE 7.40 Investigation of recorded incidents and accidents. *From Ávila, S.F., 2010. Etiology of Operational Abnormalities in the Industry: Modeling for Learning (Thesis—Doctor's degree in Chemical and Biochemical Process Technology). Federal University of Rio de Janeiro, School of Chemistry, Rio de Janeiro, 296 p.*

fiber. In the refrigerated area 13.8% of the cases cited for losses by leakage or sending to torch. Events involving the pier have 11.3% of the cases reported. Then are the subjects related to personnel with 10%, indicating reinforcement in training and greater commitment to the safety of the routine. The public, or events involving interaction with the local public, are around 7.5%. Electrical events, such as short circuits and events with electric motors are in 6.3% of cases.

The facilities were generally mentioned in incidents recorded by HSE, indicating that it needs repair. The same can be said in relation to the reoccurrence of incident problems in communication with supervisory and interconnection systems.

> *(PCA6/16 Conclusion—Process Safety) This item reaffirmed the importance of analyzing high pressure in the system. Events with the furnace fire and interlock system can be indicators that lead to process safety failure in prediction, so failures in testing should be included. In HSE records, attention is drawn to electrical events and communication-related events for process control.*

7.3.5 Analysis of the task and results

The purpose of this analysis is to validate the techniques developed for Task Analysis. In this research applied to the industry, in which it is intended to identify the origin of the failure in execution and planning of task. Techniques presented in item 7.2.5.1 and Table 7.6 are applied for the case of liquefied gas processing, both in routine and in emergency situations. Some criteria are considered to evaluate where to assess critical tasks: high temperature; high flow rate; safety failure possibility; presence of hydrates in cooling; high risks of operational problems; critical inventory, vessel or tank; and history of events in safety in area of furnaces. As a result, the liquefied gas regeneration area was chosen for analysis and, since the beginning of regeneration depends on the drying operation, it is also analyzed.

7.3.5.1 Types of task analysis

In Table 7.10, we are listing the techniques of task analysis, goals, and objectives in addition to the benefits of each technique per step. Next, we detail the steps of analysis of Standard Routine Procedures (7.3.5.2), Failure Task Analysis (7.3.5.3) and Emergency Behavior Analysis (7.3.5.4). The tools listed in Table 7.10 and applied during task assessment could be in standardization time (A) and failure analyzes time (B).

7.3.5.2 Routine standard procedures

The regeneration task is intermittent and is only triggered when at least one column is saturated after a certain drying time, so the drying batch of the molecular sieve columns depends on the amount of moisture that enters the process with the liquefied gases of refining.

Decision heuristics (informal rules) are constructed through verification, through interviews and field surveys, on how decision-making happens and important parameters for monitoring task. In current case, Task 3, column regeneration, is summarized as follows: **Decision 1**—in maintenance planning, perform tests in advance to put regeneration to operate; **Decision 2**—with realization that tower is saturated, it is needed to start regeneration; **Decision 3**—to circulate, heat and cool the liquefied gas in the process circuit. This decision depends on temperature and the presence of water in the boot of separator vessel; **Decision 4**—after confirming that all water has been removed from the separator boot, lowering the temperature until the furnace is switched off and the drying is switched off.

Task project analysis

(A1) Task architecture

Task Architecture (Fig. 7.41) helps organize the Manual of Best Practices following recommendations from Embrey (2000), Wreathall et al. (2003). For the sake of safety, the steps of nitrogen purge, control of utilities and safety are planned. Next, raw material is transferred from the refining unit to the company. The liquefied gas already stored in the vessel initiates purification, removal of water, solids and light gas. The desorption and cooling can promote condensation and drainage of water. To maintain process and safety controls, it is essential to: control the torch and control furnace with fuel gas flow and pressure.

When receiving the liquefied product in the refrigerated tank, there are pressure and temperature controls to be maintained with respective automatic controls. The vaporization control is carried out through the boil-off compressors in the refrigerated tank. Other heat input reasons require a higher compression capability through the tank pressure control. The thermal cycle maintained in refrigeration is responsible for the low temperatures of liquefied gases. Finally, we discussed loading and unloading on pier of pressurized gas.

TABLE 7.10 Tools for task evaluation and planning.

Assessment type	Target	Benefits
Consensus analyzes of task—CARMAN—**David Embrey**	Develop best practice guideline of critical tasks based on consensus in relation to practices in routine	Avoid work with informality and simplify the procedures toward common practices giving safety to routine
Standards and procedures in the operation—PADOP—**Salvador Ávila Filho**	Assess efficacy and effectiveness of the standards and critical procedures in the operation	Discuss application of resources using environment assessment, cognitive processing, and way to review the tasks
PADOP: IAT—Environment assessment	**PADOP: PCET—Cognitive processing and task execution**	**PADOP: ATEE—Task assessment based on efficacy and effectiveness**
Planning the task by safety process: **Frank Lees**	Task assessment by planning cycle based on knowledge. Task review based on skills.	Discuss task analyzes investigating interface man machine issues and operation competences
(A) Planning the task by human factors	Project the tasks based on expected standard of product, equipment and human factors	Keep cost and quality standards in product, production, environment, keep team motivated
(A1) Task architecture (good practice guideline): HTA, **Embrey, Ávila, Lees**	Organize order for tasks, procedures, and steps in the production administration, similar HTA	Verify level of complexity in topography or layout of tasks
(A2) Task Design based on requirements, steps and targets: **Hollnagel**	Organize the task by targets, requirements, steps and permissions, including cognitive limitations (alarms), automation, operation and safety	Verify if the steps can be achieved and keep task safety in relation of accident probability
General assessment	**Specific steps assessment**	**Functional assessment: safety and process**
(A3) ST risk dominance and parallelism in technical and organizational factors	Organize the task steps in order of safety priority, sequence, who execute in the same time two or more activities	Help to organize choice tasks and steps in decision-making process, help to motivate cooperative work
ST risk assessment	**Risk dominance assessment**	**Parallelism of sequence action**
(A4) Cognitive and physical enforcement assessment based on complexity of task	Verify the need of managerial tools and systems to control process Physical and cognitive quantity of work and respective complexity	Optimize resources and do quick changes to avoid human error in the task
Complexity assessment	**Physical enforcement assessment**	**Cognitive enforcement assessment**
(A5) Classification of task steps by type and place	The distance between places that the task is been done indicate new necessities in workstation and the type indicate demand for new competences	Optimize resources and do quick changes to avoid human error in the task
Place	**Functions involved**	**Type of step**
(A6) Chronology and function assessment of task: Embrey	Analyzes of chronology and function using simplified symbols and signs (debottlenecking)	Verify need data for decision about actions considering what to do in how much time…
(B) Dynamic fault assessment in the task	Know the abnormal chains in the operation during routine of task execution	Define barriers to avoid abnormalities and failure in task
(B1) Failure phenomenology in the task, body and definition of control volume	Delimit physically where occur the task fault and study the route of fault body through the equipment and process	After identification of abnormalities in operation routine during task execution is possible to study and measure the fault energy
(B2) Logical analyzes of task by type of factor and by sociotechnical dimension	Define the logical functions of task to verify the respective impact and what are the current and best barriers, the connection between operational factors	Help to identify root cause and the best place to install the barriers of safety indicating where and how review the task safety

(Continued)

TABLE 7.10 (Continued)

(B3) Fault chronology assessment	Identify times that happen the fault operational factors	Help to measure the energy of fault avoiding accident
(B4) Materialization point	Analyze the point of migration from material body to cognitive or informational body in task	Help communication tools and work cultural environment in the organization
(B5) Connectivity assessment	Verify points where the fault is in connection with other faults giving complexity	Increase the visibility of failure process becoming easier the measure and take preventive action
(B6) Critical analyzes for locate barriers in the task	Using the architecture diagram, chronology diagram, logic, connectivity we can assess the efficacy of barriers (technical, organizational and human tools).	Help to review barriers of safety installed and to be installed, help to design devices to avoid failures and human errors

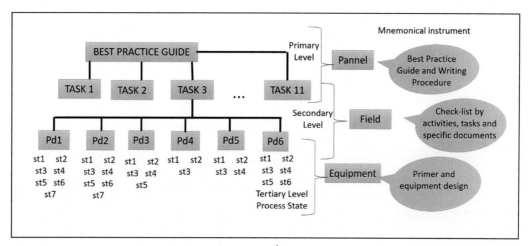

FIGURE 7.41 Architecture of tasks with drying and regeneration. *From Ávila, S.F., 2010. Etiology of Operational Abnormalities in the Industry: Modeling for Learning (Thesis—Doctor's degree in Chemical and Biochemical Process Technology). Federal University of Rio de Janeiro, School of Chemistry, Rio de Janeiro, 296 p.*

The detailed task of liquefied gases drying and regeneration needs six specific procedures, in which there is a logical order of realization. The plant is available for operation and needs to remove water through drying. If the vessel is not inerted we need to perform Procedure 1. When the dehumidification column is saturated, we need to regenerate this equipment using Procedure 2. The start-up of the furnace and respective process control depend on the implementation of Procedure 3. The maintenance of operating regeneration depends on Procedure 4. The decision to stop the liquefied gases heating through the furnace, the water drainage stop, the cooling of the regeneration systems, and the availability of the tower for operation is discussed in Procedure 5. Finally, shutdown of the plant follows the instructions of procedure 6. Starting procedure 1 requires sequential, nonparallel steps. Procedures 2, 3 and 5 have sequential and parallel steps, indicating complexity in the distribution of work between operators. Procedures 4 and 6 are parallel, requiring high control and communication to avoid taking improper action.

(A3) Analysis of the greatest risks of failure coming from organizational factor and parallel actions

In the specific case of a processing company, the need for dominance analysis is considered to control safety. Another aspect to be analyzed is how the steps of the procedure are sequential or parallel, involving communication needs between different jobs. For each stage of the task related to drying and regeneration, dominance analysis was performed in terms of safety, sequence of activities, parallelism and visual idea regarding cognitive and physical load of the task. An example of applying this technique is presented in Section 2.1.4 of this book.

(A4) Analysis of physical and cognitive effort based on the complexity of the task

The discussion about complexity is exemplified through the results of Table 7.6. Although the reduction of physical and cognitive efforts can be achieved through automation, the more automated the process is, the more complex its control becomes and the greater the attention to data collection in the supervisory.

Risk dominance in a specific step of the task indicates the need for greater attention to the signs and events that induce the accident. It is important to establish barriers to avoid communication noises in these critical steps. One of the barriers is to decrease the physical-cognitive effort to provide good quality of attention and memory.

Auxiliary systems of equipment and processes with high complexity require greater care to avoid the occurrence of events not visible in the operation and that can occur in parallel with the main activities. Thus it is important to draw appropriate safeguard so as not to reduce concentration in relation to the main activities and the safety area.

Task 3, procedure 2, regeneration, have features of complexity discussed in Table 7.11. The low level of automation (13%), the low level of risk dominance demand low attention (5%); the high use of main system (92%), excellent commitment with written procedure (100%), low level of physical and cognitive effort, indicate low complexity procedure.

Then, low demand for greater attention effort without deviation of attention by parallel activities compared to actions for the main systems. The activities are accomplished according to schedule, in the same way in which they are written. The steps in the procedure are all from search for the location and act on devices and equipment, that is, SA (Search and Action).

TABLE 7.11 Complexity analysis—procedure 2—preparation for regeneration.

Stage	Physical effort	Cognitive effort	Task complexity			
			Auto / M	Dom / =	Pri / Aux	Of / alt
Align the available tower, SA (Search and Action)	1	0	0	0	100	100
Isolate the tower to be regenerated, SA	1	0	10	0	100	100
Establish drainage and equalize vessel, SA	1	1	0	10	100	100
Direct and align for vessel or for drying (drain), SA	1	1	0	0	100	100
Align dry liquified gases for furnace coil (lights up after establishing flow), SA	1	1	0	0	100	100
Establish furnace liquified gases recirculation, exchangers, vessel, drying, SA	2	1	0	0	90	100
Close the bypass valve of the exchangers	1	0	0	0	100	100
Lighting the pilot, N2 alignments, SA	1	1	100	30	50	100
AVERAGE			13	5	92	100

(A5) Analysis of task steps by type and location (Table 7.7)

Search and action (SA) activities are mechanical and require care to avoid skipping steps. In the search and action, there is the focus of going straight to a certain place to perform a certain activity. Surveillance activities require increased attention and keen perception to identify abnormal situations. On the other hand, planning and action (PA) activities require some attention regarding the activation of long-term knowledge to assist in the decision.

In Procedure 1 assessment, start-up and maintenance of drying column, some concerns discussed are: location, the start is carried out in three different locations, requiring displacement and speed, a characteristic not always available in the shift team. Most of the activities take place near the drying towers. The level of automation then could be a point to improve safety process. Displacement takes place as follows from, 1 (pressurized) to 2, 3, 4 (in drying) and from there to 5 (in the trombone) and ending in 6, 7 (in drying).

Regarding the type of stage, the activities are distributed in this procedure, in which more than 50% in the search and action that simplifies cognitive effort and allows increasing physical effort. On the other hand, 30% of the activities require planning and decision-making, indicating the need for greater cooperation and greater assertiveness in actions. Only 15% of the activities are in the surveillance line with greater sense of perception, verifying the need for auxiliary tools to perceive abnormalities. Steps 1, 2, 3 are search and action, action 4 requires planning and decision, then in step 5 search and action activities, and again returning to planning in step 6. In step 7, the procedure is concluded with surveillance activity.

> *(PCA7/16 Conclusion—Task Analysis) The tools in this stage of the project will help in the choice of preventive actions to improve the position of the artificial variable resulting from the groupings of variables about the task: cognitive effort (respect the team's offer), dominance in safety, parallelism, and complexity of the task, in addition to the need for changes in the profile to finalize field activities within a certain flexibility (good practice and competencies in failure processes).*

7.3.5.3 Failure task analysis (PA2)

The default task, standard

Task Control related to liquefied gas from the oil process requires efficiency analysis in terms of product quality, operating performance, safety risks and pressure control in refrigerated systems. The quality of the liquefied gases, raw material, influences the yield of operations and depends on the supplier's receiving schedule.

The quality of the product, result of gas processing, has density as a control item. Upon approval, it is sent to a tank or ship. The density of the refrigerated gas depends on temperature and pressure conditions and affects its transfer flow and consequent loading of the ship.

The yield for removal of impurities depends on equipment and processes such as, filtration to remove solid, drying of liquefied gases and cooling operations during regeneration that authorizes separation of water and its drainage. In specific cases, the presence of corrosion caused by raw material out of specification generates solids and requires system insulation and material return to the supplier. Drying of the gas and its regeneration, with furnace operation, requires safety and cooling care at low temperatures. Thermal cycles are used to maintain control of refrigerated gas sent to tank and ship.

(B1) Phenomenology of task failure and definition of control volume

In liquefied gas sites, the physical failure bodies travel through processes and equipment and take different forms. The presence of moisture and air in the vessel storage of pressurized liquefied gas and for burning is considered failure body too as, the solids resulting from soot (at burning) and flow of liquefied gases (process) in the heating of berth (origin of residues) through the furnace.

On the other hand, the presence of ice and hydrates in cooling and separation of water from heated liquefied gases; and control of tank pressure, which, at specific times, can limit drying flows, are also bodies travel with hazard through process.

The study of phenomenology of failure requires identification of bodies with hazard, which may be technical or social. In this study, gas purification includes technical or physical bodies (hazard carriers) such as: solids, water or moisture, hydrates, flow and pressure are discussed. Control volume is the area delimited by the paths traversed in the circulation of this body that carry hazard energy. In this control volume, there are the equipment and related operations.

Process solids are in the sections of drying and cooling. Solid residue can be deposited in the furnace (volume of control 2, V2) impairing the photocell that detects the flame in interlocking system. This dirt can be generated from soot by incomplete burning or residual refractory material due to improper maintenance. There is still the possibility of solids coming from the vessel to the intermediate vessel. The formation of hydrates in the regeneration and fines of drying towers (V3) affect indication of level of the vessels, thus causing difficulties in controlling flow and pressure.

The presence of excess moisture (V4) can saturate drying towers by decreasing the time on operating cycle and requiring in short time further regeneration. If there is too much moisture, there will be water in the drain of the decanter boot, in the regeneration section, consequently, difficulty in controlling water with return to the drying towers. Moisture can be caused by production, no drainage of the vessels by human error, or low coalescer efficiency. In this case, this moisture ends up not being completely removed and increases the formation of hydrates (V3) at low temperature, impairing the control of pressure and flow in refrigeration.

Ideally, the liquefied gas flowing from the furnace serpentine entering the cooling is the same to the gas flowing out of the vessels. If this flow is out of control, the vessel overflows or the pump cavitates. The localized overheating of in furnace serpentine due to momentary "stop" of circulation can cause ignition (V2, V3).

Liquefied gas circulates through serpentine with normal flow and adequate pressure until fragility of the serpentine and internal circulation stop occurs. Excess of this product due to the interlocking of the burner, stops the flame, increasing the presence of this flammable gas with air in the furnace leading to leakage into atmosphere. The fire can be originated externally, and for some reason of failure in the interlock, the pilot kept turned on, causing explosion. The fragile serpentine opens and causes fire in equipment, piping and atmosphere.

The Control volumes are identified to explain phenomenon of failure in regeneration and drying sections as shown in Fig. 7.42. Each crash of hazard body in its journey is included in one or more control volumes. The circuit of the furnace starts in the vessel (V1) passes through the coalescer then goes to the drying column that operate (V4). It passes through the furnace to heat (V2), through the saturated column (V4) to remove moisture and from there to cooling, separation and drainage of the water returning and closing the cycle to the coalescer at the entrance or to the vessel.

In another process, the firing circuit in the furnace, where it exits the vessel (V1) to feed the accumulator vessel (V5), and then proceeds to the burning in the furnace (V2) and then to the atmosphere. There may eventually be high pressure and send the liquefied gases from the accumulator vessel to the torch. The possibility of serpentine explosion (V4) occurs due to liquefied gases contaminated with air from the (V1) after maintenance, which causes the explosion of the furnace (V4), and hence, by fire channel, "in domino effect" to the other vessels of the process.

(B2) Logical analysis of the task by factor type and by technical socio-dimension.

In the logical analysis of the failure in the task (discussed in Chapter 6: Human Reliability: SPAR-H Cases, and beginning of this chapter) we seek to define the causal link indicating probable causes, intermediate operational factors, consequences and interconnection between chains of different abnormalities through common factors. This analysis seeks to define the regions of installation of barriers to prevent the failure factors from continuing to work, producing consequences, and generating major incidents or accidents.

In the present case, the analysis refers to the chains related to the task of drying and regeneration. The abnormalities analyzed are part of the investigation already carried out, in which the logical diagram and other tools are used to mitigate or eliminate the failure. Note that some of the flaws have their origin in organizational issues.

The construction of a logic diagram to install barriers must follow some rules. This was performed to drying and regeneration's reliability assessment. Each color represents a differentiated source of failure where C indicates cause, Q

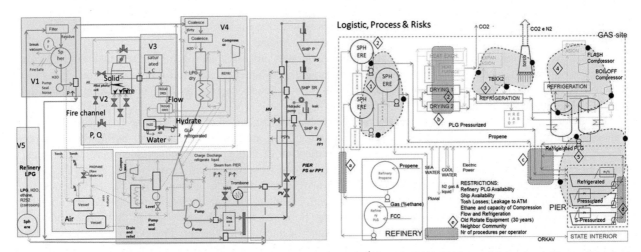

FIGURE 7.42 Control volume failure and hazard body for study. *Adapted from Ávila, S.F., 2010. Etiology of Operational Abnormalities in the Industry: Modeling for Learning (Thesis—Doctor's degree in Chemical and Biochemical Process Technology). Federal University of Rio de Janeiro, School of Chemistry, Rio de Janeiro, 296 p.*

consequence, F indicates operational factor. The number on the left is the causal nexus number and number on right is order number of the operating factor. Barriers to avoid bad events in operational failure are in the center with single color. Continuing the signs, n symbolizes networking, receiving, or sending failure energy from more than one operational factor, B means safety or failure barrier, R is redundancy, p is primary, s is secondary, and t is tertiary, number range (example 1–7) means that it receives energy from factors linked to various causal nexus.

In Fig. 7.43, factors are coded, facilitating the visualization of relationships and their interpretation. The fatigue of Man resulting from the performance of excessive manual services, with high risks of working with liquefied gases, ends up returning and refeeding abnormalities. The consequences are loss of productivity (or processing time) in case of consecutive stops due to furnace interruptions and in higher impact event, the explosion of the furnace with liquefied gases -rich internal atmosphere and the firing of the furnace. If failure processes are isolated through preventive actions in: 1Fn4, 7F3, 78nBp, 9/10mB, 23C1, six aforementioned events are simply deactivated at the same time.

Fig. 7.44 shows causative factors by dimension and location in relation to the phases of the operational failure. It is noted that in the inertial phase (with low visualization) the following factors are found: 1Fn4 (does not allow pilot lighting), 7F3 (extremely high process temperature) and 23C1 (improper maintenance). It is noted that in the development phase the factor 78nBp (inadequate flow control) is located and in the risk phase the operational factor 9/10mB (passing water for drying or vessel). As for the size of these operational factors, it is found that: 1Fn4, 7F3, 78nBp and 9/10mB are technical factors, while 23C1 is a factor related to the task.

The root causes are described as follows: maintenance contract, commitment of ask, improper operation and maintenance, nonoperation of instruments in fa102, inadequate Cv (Valve Flow Coefficient) of control valves since project (requiring studies on reasons), load oscillation in refrigeration, fatigue of operators and maintenance technicians (these are also final consequence that return as the first cause).

(B3) Failure chronology analysis

The failure scenario analyzed by the chronology technique has been described earlier and intends to check: times, operational factors that increase or decrease failure energy, and the failure format over time. This exercise is carried out by simulating a situation prepared based on events measured with the company's staff, without joint occurrence of the factors, and with the times estimated by a specialist.

The events that occurred in the drying and regeneration sections were chosen to construct a simulation of an explosion situation, from the signs of abnormalities collected in the shift register, such as the absence of flames, corrective

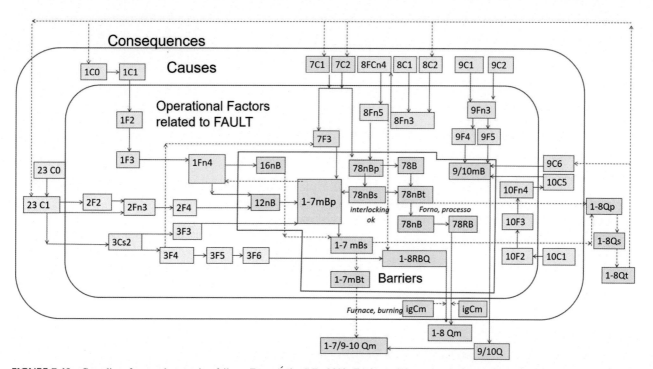

FIGURE 7.43 Causality of network operating failure. *From Ávila, S.F., 2010. Etiology of Operational Abnormalities in the Industry: Modeling for Learning (Thesis—Doctor's degree in Chemical and Biochemical Process Technology). Federal University of Rio de Janeiro, School of Chemistry, Rio de Janeiro, 296 p.*

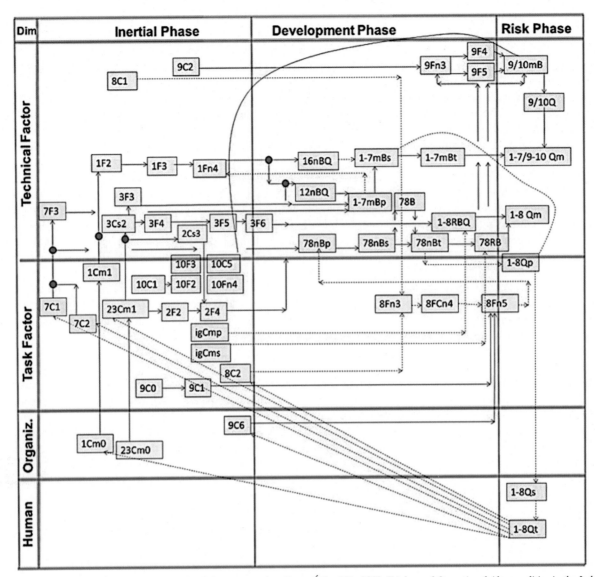

FIGURE 7.44 Logical diagram of the failure involving regeneration. *From Ávila, S.F., 2010. Etiology of Operational Abnormalities in the Industry: Modeling for Learning (Thesis—Doctor's degree in Chemical and Biochemical Process Technology). Federal University of Rio de Janeiro, School of Chemistry, Rio de Janeiro, 296 p.*

measures to decrease the failure energy, then reaching the explosion of the furnace and equipment by "domino effect." These events were based on the OD and on hypotheses about the continuation of abnormalities.

Accident or crisis behavior is a consequence of chain reaction of events, in a short period of time, the failure energy (or released danger energy) increases geometrically. If you consider the entire chronology of the birth of the failure, it goes from the classification of signals to process uncontrol to then error in the task, and hence medium and serious events involving safety aspects, and such as sinister, or disaster caused by a BLEVE (*Boiling Liquid Expanding Vapor Explosion*).

To facilitate the construction of graph, time and hazard energy, parameters were used from experience and research on events in history. We had to use numbers resulting from Napierian logarithm for both failure and time (ln) as shown in Table 7.12 and Fig. 7.45.

Assumptions for validation:

- No corrective actions were taken in the phase of signs of abnormalities.
- Actions were taken only after insistently confirmed the absence of flame, but the actions were about consequences and not on the root cause.
- The 6th to the 15th operational factors occurred in the period of up to 24 hours.

TABLE 7.12 Simulation: Illustrative analysis of fault chronology.

Factor	Description	Time	Accumul. Time	Ln of time	Days	Hazard energy	Accumulated energy	Ln energy
1	Absence of furnace flame	1	1	0		10	10	2.303
2	Absence of furnace flame	80	81	4.394	3.3	10	20	2.996
3	Absence of furnace flame	48	129	4.860	5.4	10	30	3.401
4	Open fault solenoid	20	149	5.004	6.2	30	60	4.094
5	Pilot entry with gas	12	161	5.081	6.7	90	150	5.011
6	Test system safety and fixes	10	171	5.142	7.1	−120	30	3.401
7	Operational safety and fault	12	173	5.153	7.2	150	180	5.193
8	Overheating in burning	7	180	5.193	7.5	1800	1980	7.591
9	Furnace stop for 4 h	4	184	5.215	7.7	−400	1580	7.365
10	LV and pump failures, flow	2	186	5.226	7.8	2200	3780	8.238
11	Furnace explosion	3	189	5.242	7.9	5000	8780	9.080
12	Domino effect equipment	2	191	5.252	8.0	7500	16,280	9.698
13	Cooling of equipment	2	193	5.262	8.0	-1500	14,780	9.601
15	Bleve	2	195	5.273	8.1	15,000	29,780	10.302

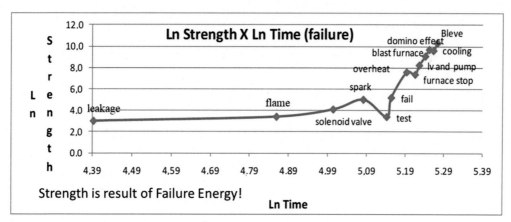

FIGURE 7.45 Time versus failure energy (bi, ln). *From Ávila, S.F., 2010. Etiology of Operational Abnormalities in the Industry: Modeling for Learning (Thesis—Doctor's degree in Chemical and Biochemical Process Technology). Federal University of Rio de Janeiro, School of Chemistry, Rio de Janeiro, 296 p.*

- Corrective actions were factor 6 (interlock tests), factor 9 (furnace stop), and factor 13 (equipment cooling).
- There have been two major failure energy changes (from 150 to 1800), with PSV and task failure, and the second moment (7500 to 15,000), coming out of the blast range of the furnace to the Bleve of the accumulator vessel that feeds the furnace.

(B4) Analysis of social/technical energy migration in the task (materialization)

In the global market, where the maintenance contract with the supplier is the result of bidding for the lowest price, if some service providers want to reduce costs, they can remove items that affect safety or quality from the formulation of their offer, with possibilities for inadequate quality services. Thus considering that process control systems are serviced by design and maintenance service with contract, some indicator instruments, or even controllers, may have reduced performance.

Difficulties in flow control, pressure control and level control, probably because of valve capacity (CV), or maintenance, cause the control to be manual. The dirty photocell causes furnace to stop. These control items affect flow on the process side. This situation contributes to incomplete burning, with soot, reducing availability of furnace.

As a result of this simulation and application of tools, the care in Maintenance Management to prevent the occurrence of the fire vessel is: (1) maintenance contract based on competence and leadership in process safety and not based on minimum cost; (2) living with gas leakage depends on the assumed safety standard; and (3) the calibration plan of PSVs should be well studied and with maximum tests, to avoid surprises.

(B5) Failure connectivity analysis

Connectivity Analysis was discussed in Chapter 3, Worker Performing Task, and Chapter 5, Learned Lessons: Human Factor Assessment in Task. The possible points of connection between failures in different abnormality chains are due to the common use of liquefied gas for process and usefulness of failure energy transfer. In the drying-regeneration furnace, the gas is being used for combustion and for moisture removal. In the refrigeration system participates in the process and eventually is part of the thermal cycle. High pressure can cause leaks in various operations and parts of the process. High humidity integrates the relationship with supplier (production unit) with all process steps. The motorized valve and compressed air system are also factoring that can connect equipment and processes to different failures. In the production was analyzed the following connections that can complex, interconnecting different abnormalities: (1) The common liquefied gases between burning and regeneration; (2) Common liquefied gases between drying and regeneration; (3) Relation of the Furnace and its heating coil; (4) Instrumentation and Interlocking; (5) Measurement of failure behavior; (6) Maximum point or success of the failure.

(B6) Critical analysis of safety barriers in the task.

The absence of an in-line moisture meter or even the absence of moisture measurement in liquefied gases indicates possible failures in drying control, leading to the loss of time due to production stops, hydrates in valves and rotating systems. We need to investigate the common causes and the root cause of these events.

The existence of *fire-safety* prevents the formation of fire flow channel. The need for positive pressure requires system installation with a vacuum-breaking line that previously utilized vessel liquefied gases. The pump is of fundamental importance to ensure the flow of process through the furnace, requiring attention to the safety systems that can stop the flow. Torch pressure control and quality for burning are part of the care in torch operation. The furnace interlock system has important functions that cannot fail mainly about the photocell that detects the flame. In the control of burning, in addition to the issue of air intake, comes the issue of the operation of the valves, including solenoids and temperature control. The calibration plan needs to be revised, with real high-pressure tests. Attention is drawn to the following PSVs: accumulator vessel for furnace; vessel and lines; vessel in the cooling process; other vessels; and changer.

> *(PCA8/16 Completion—Task Failure Analysis) The analysis of failure in the task has the main usefulness of preparing training material for the formation of procedural skills on the failure. In the choice of the area of furnaces, attention was drawn to the risks involved, although there is no information that meets the need of this instrument since the probability of such an event is low and is not a latent socio-technical problem, apparently. It is a more technological and localized issue.*

7.3.5.4 Risk and behavior analysis in emergency or extreme stress (PA3/PA7)

A fault tree (Fig. 7.46) is designed in this work of accident simulation after the application of an accident situation with liquefied gases, includes fire, explosion, and gas leak with fire pool fire. This simulated event ends up turning into a chain reaction with explosion of various equipment until it reaches the vessels.

This fault tree includes LOPA concepts, except for layers that are not independent. We call this tool LODA (Stress Progressive Layers Operational Decision Analysis during Operational Management of a Crisis). To elaborate this scenario, the techniques of analysis of the OD and staff were applied, and the information of the MEA was used. After quantitative Fault Tree Analysis, we estimate the probability for each event based on LOPA numbers from Dow

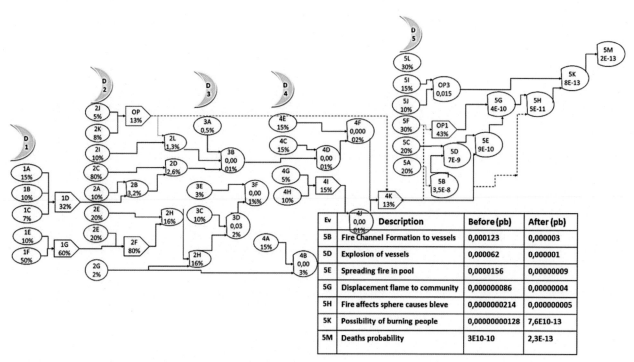

FIGURE 7.46 Fault tree per operational routine-based layer (FTAh) and results of operational, managerial, and process safety devices. *From Ávila, S.F., 2010. Etiology of Operational Abnormalities in the Industry: Modeling for Learning (Thesis—Doctor's degree in Chemical and Biochemical Process Technology). Federal University of Rio de Janeiro, School of Chemistry, Rio de Janeiro, 296 p.*

Chemical and for the greater consequence of the accident, 40 fatalities in the neighboring community, as what happened in New Mexico (Pietersen, 1988; Shariff et al., 2016) and Rio de Janeiro. The exercise continues with the installation of defense mechanisms suggested by the operation teams and new probability calculations.

In the risk analysis, the responses of the operation team that participated in the workshop were considered, in which data on human factors were collected. This group attended the presentation of the Scenario with progressive stress and proposed actions performing a HAZOP. Key words were used, like the HAZOP applied to understand the risks (formation of fire channel, explosion of vessels, spreading of fire in puddle, possibility of fire affecting the vessel, burning people and fatality occur in the accident).

The steps indicated in the procedure were performed and a fault tree constructed by decision levels of the operation: D1, on the signs of abnormality; D2 on the uncontrolled of processes; D3 about errors in the task and equipment failures; D4 on safety aspects of the company's people, equipment, and assets; and D5 on the events involved with the accident with impact on the public.

The main recommendations of the operation teams that were in the workshop, and who intended to avoid the accident with liquefied gases are presented in Table 7.13 at the same decision levels as the failure tree in Fig. 7.46. The scenario of the accident with the physical location of the equipment involved and the drag direction of the fire, the location of the control room and the public are presented in Fig. 7.47. The operation team that was present at the Workshop was divided into five groups to participate in stress dynamics (LODA). In this event we designed the explosion situation of the liquefied gas drying furnace, collected the proposed actions through HAZOP and observed behavior of the leadership and groups that were summarized and presented in Table 13. The characteristics of human behavior, which were observed and analyzed in this emergency scenario were: leadership and type, physical movement, commitment to the task, communication, new ideas and stress level, description on Table 7.14.

(CONCLUSION PCA9/16—emergency behavior) The same previous conclusion serves for the current situation, in which the main technical objective of this analysis is learning. In the dynamic, there is a still "warm" environment in terms of stress, drawing attention to aspects in the area of commitment and competence formation that are in the corrections to be made to make the artificial variable to be within the best organizational efficiency.

TABLE 7.13 Recommendations at the 5 decision levels of the simulated sinister operation—liquefied gases.

Level	Theme	Keyword	What?	To avoid what?
Decision 1	Abnormalities or deviances in the process	Check	Filter	Obstruction
		Check, Adjust	Controllers/PIC/FI	Possibility of unmanageability
		Check	Alignments, flanges, rackets and valves	Flow, high pressure
			Interlock in the furnace, flame	Possibility of failure, N2
Decision 2	Unmanageability of the process	Check	Inlet and outlet flow, solenoids	Flow
		Cut	liquefied gases and drain tower	Remove inventories
		Isolate	Vessel and vessels	Remove inventory
Decision 3	Equipment and task	Check	N2 alignment	Inerting furnace
		Lock	Feed from vessel to furnace	Remove inventory
		Open drains	Align liquefied gases for *flare* controlled	Remove inventory
Decision 4	Safety flaws	Start	Alarm and evacuation of area, fire brigade	Fire control and Avoid accidents
		Communicate	Coordination and management	Managerial control
		Cooling	Equipment nearby	Avoiding explosions (Fig. 7.23)
Decision 5	Failures in the accident control	Communicate	Neighboring public	Avoid accidents
		Start	Containment curtains	Cool equipment
		Confirm	Contingency plan	Handle accident

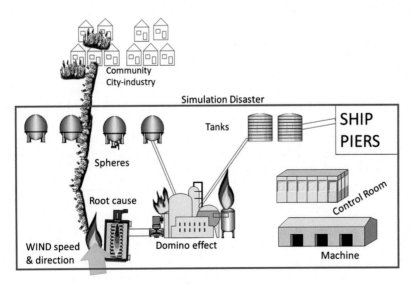

FIGURE 7.47 Scenario of simulation. *Adapted from Ávila, S.F., 2010. Etiology of Operational Abnormalities in the Industry: Modeling for Learning (Thesis—Doctor's degree in Chemical and Biochemical Process Technology). Federal University of Rio de Janeiro, School of Chemistry, Rio de Janeiro, 296 p.*

TABLE 7.14 Results of dynamics under stress of the simulated scenario of explosion accident in oil and gas industry.

General	The participants of the dynamic were receptive to the Workshop and the integration was high. They answered all the polls, identified the technical culture and participated in the dynamics effectively. There was a good climate to discuss events that happened in the past, that is, safety was transmitted for this to happen. Participation was small in quantity, but with the effective participation of all aspects of human reliability, its role for society and operation were clarified. The participants talked about the subject of the Workshop, which showed the interest of the participants and the clarification about it. There was nonexplicit resistance of a component and the presentation of leadership "inhibition" in one of the cases.
Quality of dynamics—workshop	
Leadership	With a small number of people and perhaps because there are not many formal leaders either, the leadership was not rotative, it was fixed on each team. There was an established leader in group 3 and in the other teams the leadership was less evident.
Physical movement	There were no excessive physical movements, indicating that the stress level of the team was moderate, except for person 3.
Commitment to the task	The commitment to the task was quite reasonable (in 91%). In decision 4 and 5 the involvement was lower, and by team 3, it was not attributed importance to the most critical situation, as commented, we just have to leave the area.
Communication	Communication was reasonable between the operators, except for 20% of the team that did not participate, and a team that was resistant.
New ideas	33% of the team did not present new ideas to treat the stress situation. Team 2 in the decisions of signs in the routine, process uncontrol and performance in the task / equipment practically did not present new ideas. There were new ideas from all teams for decision 4 and 5.
Stress level	33% with moderate stress and 67% with high stress. Team 3 did not become apparently involved like the others; it was at moderate stress level.

7.3.6 Analysis of human and social data in the work environment

The data collected in this survey is based on a 20%—40% sampling of the shift team. We were careful to avoid exposing individual situations. This research on human and technical typology (Ávila, 2010) intends, through the processing of information, to indicate the paths for the change of behavior and adjustment of the Operational—Technical Culture (TC).

7.3.6.1 About social data

As shown in Fig. 7.48, there are indications for the adoption of patterns and conservative view in 70% of cases and sedentary lifestyle in 70% of cases. The responding operators are first-born in 56% of the cases, increasing the responsibility on family members, confirming a relationship of dependence of the parents by 67%. The situations raised indicate that the family is a fundamental part of the emotional balance of the operator, in a ratio of 65%—75%.

The division of affection in living with partners brings increased emotional stability. In this sampling of the operation team, following the current economic culture, 78% of the interviewees have up to one child. Most of the group are older than 40 years, which indicates some maturity; on the other hand, by age and routine, work can also be considered a necessity and not a challenge. Political action in the proportion of 62.5% is interesting, and it should be confirmed whether it is only labor policy or also involves public issues.

Regarding the culture, the operators identified themselves in a balanced way, with 50% adoption of the culture alternated between local and global. The emphasis in regional culture has been reduced, and adoption of the global culture is widespread. In this case, although there is the presence of the family, as indicated above, there are possibilities of behavior fluctuation. The operator is sociable, although in the cooperation test shows the restrictions on inviting elements into personal space. It is customary to spend the weekend with the stable family at 45%, although 22% consider themselves in unstable relations. There are indications of behavior fluctuation of around 30%, with friendships in more than one tribe.

> (Conclusion PCA10/16—emotional balance, sociability) The comments linked to the conclusions of 10 to 14 feed correction programs, the tool that uses the PCA. In the social data presented directly, there was disagreement with the questionnaires indirectly. Despite the differences, there is a need to increase the organizational and social integrations promoted by the company.

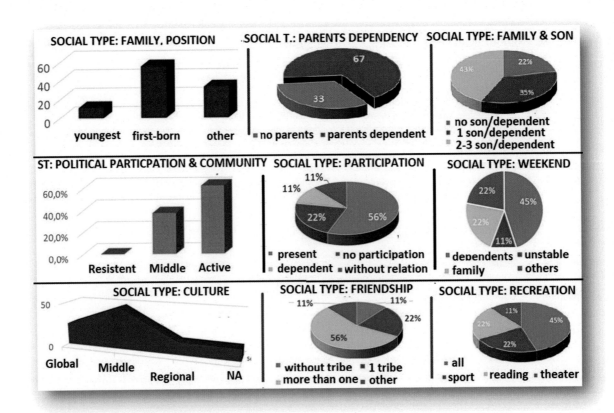

FIGURE 7.48 Data to identify the social typology of the operation team. *From Ávila, S.F., 2010. Etiology of Operational Abnormalities in the Industry: Modeling for Learning (Thesis—Doctor's degree in Chemical and Biochemical Process Technology). Federal University of Rio de Janeiro, School of Chemistry, Rio de Janeiro, 296 p.*

7.3.6.2 About individual data

There are references in psychosomatic medicine that reveal that at least 85% of the causes of certain diseases can be emotional in nature; thus the poll was designed to analyze the possibility of imbalance. As shown in Fig. 7.49, the altered pressure is indicating a state of melancholy and coincides with the proportion of the cardiovascular stage. This behavior may be related to the lack of affective bonds, indicating no major commitments in 11% to 22%. Tiredness can be a symptom of discouragement, with possibilities of depressive condition (traits); it is necessary to motivate these 20% of the staff to seek new challenges within the work environment. Part of the team presents a picture for constipation, which may indicate resistance to leadership (20%), or fear of prospects—diarrhea (10%).

The cognitive system is the "home" where the main human errors are installed in the accomplishment of the task. With 11% of the answers indicating changes in memory that, based on other results and those who did not answered, can reach 15%. Regarding attention, same can be affirmed, where 20% are classified as an intermediary between inattentive and attentive, this indicates the need to activate the attention of team members, perhaps motivating with something visual, with campaigns that move people in search of focus, attention and commitment. In spatial perception, 30% of the team confirmed difficulties, indicating need for better identification of areas and search for simulated exercises to avoid automatism of activities. Under times of stress, language is impaired to 10% (reasonable), and cooperation between colleagues decreases the negative impact of this characteristic. More than 50% of the team is excessively rational, leading, by extrapolation, to obsessive pictures, which were not confirmed by pressure. Thus social integration is a necessity for the operation team and that is carried out through leisure. It is proposed that the company perform rituals of social integration (external activities) to incorporate team feeling, thus avoiding compulsion to work, avoiding over rationality and developing affective bonds with work.

A joint analysis is indicated in Fig. 7.50, hexagon of emotional balance, in which aspects related to culture, digestive behavior and blood pressure are interpreted together with weight and reflex in self-esteem, in addition to memory and

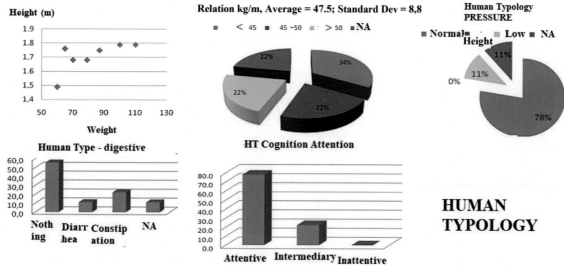

FIGURE 7.49 Data to identify the individual human typology of the operation team. *From Ávila, S.F., 2010. Etiology of Operational Abnormalities in the Industry: Modeling for Learning (Thesis—Doctor's degree in Chemical and Biochemical Process Technology). Federal University of Rio de Janeiro, School of Chemistry, Rio de Janeiro, 296 p.*

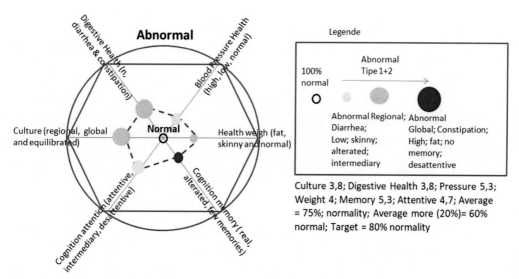

FIGURE 7.50 Hexagon of emotional balance. *From Ávila, S.F., 2010. Etiology of Operational Abnormalities in the Industry: Modeling for Learning (Thesis—Doctor's degree in Chemical and Biochemical Process Technology). Federal University of Rio de Janeiro, School of Chemistry, Rio de Janeiro, 296 p.*

attention (reasons for human errors in the execution of the task). In the abnormality there are two situations: confrontation and acceptance and is present by the color of the flag in the emotional balance ruler. The center indicates the state of normality, while the extremities represent the state of maximum abnormality. The highest normality values are in pressure and memory with 88%, and the lower values with 63% of normality are digestive symptoms and culture.

The average normality reached values of 75% that are close to the target value of 80%. However, part of the data was lost due to operators who, despite participating in the Workshop, did not answer the questionnaire. This occurred in the order of magnitude of 15%. Thus to be conservative, a correction of 70% over 15% is made and applies to a reduction in the achieved rates of normality, finally reaching 70% of normality.

For this work and definition of management programs, goals of 80% of normality are sought, thus indicating the need to build inclusion programs for the correction of 10% in emotional balance. The behavior indicated for the above discussion is in the hexagon of emotional balance, represented in Fig. 7.50.

> (Conclusion PCA11/16—emotional balance, health) Following the interpretation presented above, there are indications of a need for adjustment in the profile in terms of emotional balance that leads to attention, good memory use, construction of satisfactory mind map, good level of communications and good spatial location. Thus in the PCA, staff relocation and training in specific skills and abilities can be performance correction instruments.

7.3.6.3 About the individual's behavior data for cooperation

Depending on the age, regional culture or organizational phase that the company goes through, it may happen that acceptable patterns of behavior are changed or, which is certainly the case, the measured responses are changed, indicating behavior change. As these are behavioral questions, it's important to be done in group sampling, at different times and with different questions. Due to cultural changes in society, level of selfishness and egocentrism are factors that hinder the achievement of ideal standards for the formation of cohesive groups, being challenges for the modern manager, being recommended to review ancient and deterministic forms, through intuitive aspects of administration.

Although not conclusive, there are more disaggregating forces than aggregators, and risks for the treatment of small emergencies as shown in Fig. 7.51. The increase in measurement in relation to the target is interesting for the hexagon and for the emergency bar, opposite is not, due to pentagon with disaggregating tendencies. Extreme values are not common, thus leading to loss of individuality, for example. Even with low values in relation to the mean values, emergence, criticality and selfishness are human behaviors that prevail as disaggregation. There are large variations in emergency situations in routine behavior, indicating lack of preparation, apparently, in the work environment. Statistical data of this survey are presented in Fig. 7.52, indicating difference in relation to the target.

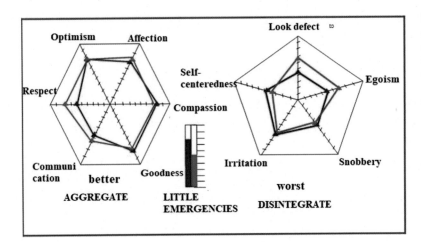

FIGURE 7.51 Group aggregation indication. *From Ávila, S.F., 2010. Etiology of Operational Abnormalities in the Industry: Modeling for Learning (Thesis—Doctor's degree in Chemical and Biochemical Process Technology). Federal University of Rio de Janeiro, School of Chemistry, Rio de Janeiro, 296 p.*

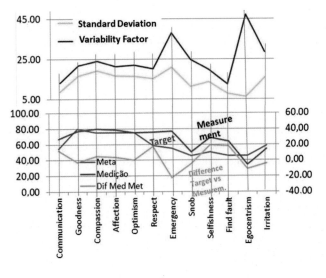

FIGURE 7.52 Characteristics and behaviors of the individual by group. *Adapted from Ávila, S.F., 2010. Etiology of Operational Abnormalities in the Industry: Modeling for Learning (Thesis—Doctor's degree in Chemical and Biochemical Process Technology). Federal University of Rio de Janeiro, School of Chemistry, Rio de Janeiro, 296 p.*

> *(Conclusion PCA12/16—group)* In the measurement performed, it is noted that the mean value is very close to the desired pattern, but isolated cases lead to the following observations. Social ties and the comparison between individual and organizational values indicate that work that can be done to increase man's bond with the organization. This conclusion added to the previous ones and confirmed when the movements (signals from) disaggregators are greater than the team aggregators are.

7.3.6.4 About group behavior and leadership

In general terms, the team tends to be socially flexible (66.7% of cases), and presents a slight tendency toward rebellious character, which coincides with critical behavior and the issue of selfishness. In a proportion of 55% people have little initiative, are more passive and do not want other people to relate to it; in the other 45% are those who have normal initiative with traces of rebel, as stated before. Still in terms of sociability, the participants are very selective of their intimacy, this in a proportion of 44%. Around 33% of the group prefers not to relate to other people, the rest is considered normal. The group is emotionally passive, in which 66% prefer to receive affection rather than to offer. They are not sure how to deal with small emergencies, which was the main reason for difference in relation to the general goal of the items analyzed. It was found that 56% of the team is normal regarding responsibility for the task, 44% avoid responsibility.

Much of the sample, about 90%, avoids frequent relationships and is selective, making it difficult to work with mutual trust, which is a worrisome fact for cooperativism. Passivity is confirmed by the following data collected: 44% of operators prefer to be invited while 22% prefer to invite; and 44% of people prefer to take orders. Table 7.15 shows sociability, sense of responsibility, importance to affection and social interaction had a standard deviation between 14% and 15.5% of the mean value, indicating that this group varies in the same way. The importance to people already has a higher variation equivalent to 18% of the average. There are large variations in modulus in the differences of control, inclusion, and affection, with, 78%, 39% and 32%, respectively. The maximum and minimum values were preferably in profile 4 and 5, which differed in relation to the rest of team.

> *(Conclusion PCA13/16—leadership)* There is conflicting information in this item, but which shows the working potential for adequacy of leadership profiles. The need for inclusion is confirmed, that of team aggregation as well, but the environment is conducive to the necessary changes. Leaders need adjustment to bring to company greater organizational efficiency: volume processed safely and with minimal technical idleness. In the PCA enters as a correction factor to adapt organizational efficiency.

7.3.6.5 About the social data collected in the company's human resources

Wage progression is divided into two classes: above R$ 1 per year (symbolic unit) and below R$ 1 per year. The highest progression mainly includes employees migrated from the affiliate (supplier). Wage progression is an indicator that there has been a greater commitment; accident is an indicator of problems that may be related to lack of commitment or technology.

Data related to work, such as salary progression and leave, are presented in Table 7.16 with sampling until 40% (15 operators of a group of 40). In the Absence, four classes were detected with leave and an average of 53 days of leave for each employee. Class 1, below 5 days, indicates that 1/3 of the team performs well in safety: class 2, with 5 to 10 absences, total 24 days of absence, equivalent to low percentage values in relation to the total, 3%. In class 3, where 4 operators were away for 10 to 80 days, with 145 days of leave equivalent to 18% of the total: and finally, in class 4, where 3 of the 5 oldest employees of the company and totaling 616 days of leave, almost 80% of the total. The leave per year of work is a relative number, indicating that there may be greater leave with more time of service. It was divided into 3 classes of leave per year of work: class 1, from 0 to 1 leave per year (40%), usually the newest; class 2, between 1 and 3 absences per year (33%); and class 3, above 4 absences per year, including employees from the affiliate (Supplier), who constitute at least 60%.

For the calculation of the commitment, it is considered that the cultural issues that indicate stability regarding family planning and the beginning of work earlier are favorable to the increase of commitment and will be used as a reference. In this case, we propose a way to calculate the commitment where the fundamental parts are the wage progression and the number of days away per year. Cultural factors are time-per-dependent, in which the current pattern of comparison is 5, whereas in the old pattern, the default time per dependent is around 10. A greater the number of dependents indicates less planning in the family, that is, less tendency to stability. Another multiplicative factor is the working time by

TABLE 7.15 Analysis of leadership characteristics and groups.

Responder	N activity	N need	Importance to people	Responsibility task	Affection importance	Social interaction	Invitation/is invited	Obey/give orders	Receive and give affection
1	11.0	10.8	10.2	7.4	4.1	21.8	−1.6	0.6	1.2
2	12.9	12.0	13.0	7.7	4.2	24.9	−1.9	1.7	1.1
3	9.3	10.6	10.3	5.7	3.9	19.9	−2.8	0.3	1.2
4	14.2	12.7	13.1	8.2	5.6	26.9	−2.7	1.8	2.4
5	9.2	8.1	7.0	6.3	4.0	17.3	−0.8	0.3	1.6
6	12.8	11.3	12.4	7.0	4.7	24.1	−1.8	2.3	0.9
7	9.6	9.3	9.3	5.1	4.4	18.9	−1.8	0.7	1.3
8	11.2	8.9	10.0	5.8	4.3	20.1	−0.9	1.3	1.9
9	11.8	12.3	11.3	7.2	5.6	24.1	−2.4	0.1	1.8
Avera	11.3	10.7	10.8	6.7	4.5	22.0	−1.8	1.0	1.5
StdDv	1.76	1.60	1.97	1.05	0.63	3.18	0.71	0.78	0.48
Fatvar	0.155	0.150	0.183	0.156	0.138	0.144	−0.387	0.775	0.320
Max	14.2	12.7	13.1	8.2	5.6	26.9	−0.8	2.3	2.4
Min	9.2	8.1	7.0	5.1	3.9	17.3	−2.8	0.1	0.9

TABLE 7.16 Operator performance and characteristics—human resource data.

Age/Year	Time/dependent	Years of Work	Wage Progression	Absences	Absences per Year
7.4	1.5	5	1.20	46	9.2
5.1	11.8	7	0.71	10	1.4
7.9	3.0	7	0.86	2	0.3
6.4	3.4	8	0.75	2	0.3
6.1	6.8	8	0.75	1	0.1
5.4	3.4	8	0.75	1	0.1
5.0	3.4	8	0.75	7	0.9
5.4	2.3	8	0.75	2	0.3
5.6	1.7	8	0.75	9	1.1
4.7	3.4	8	0.75	8	1.0
2.2	7.4	23	0.78	60	2.6
2.1	11.5	24	1.33	96	4.0
2.0	12.0	25	0.64	376	15.0
2.2	11.6	24	1.33	29	1.2
2.1	11.0	23	1.04	144	6.3

age, indicating that currently, in a more unstable situation, work begins at a later age. After applying the formula, 40% of the team, for these criteria have higher commitment and 33% have less commitment. Commitment = function (salary progression, leave per year, working time by age and working time per dependent).

(Conclusion PCA14/16—commitment) The data and information analyzed (leave of absence, progression, dependents) give indications of team formation with different characteristics and commitments as well. The main conclusion is the search for homogenization that is considered action for organizational efficiency adjustment after artificial measurement in the PCA.

7.3.7 Competence analysis and results

Organizational psychologists such as Muchinsky (2004) have found difficulties to ensure the correlation between the psychological profile of the worker and performance in the work environment. The difficulties continue when the development of competency is programmed through training and, after execution, it is not possible to achieve the program.

The ritual of entering the worker into the organizational environment includes a process of "acculturation" and it is necessary to adopt concepts about cognitive gaps that have not existed until then. Thus the adaptation of the competency is made, shaping the ideal professional for the company. This auxiliary adequacy of competencies indicates that in the selection we need a confirmation regarding the performance effectiveness of the contracted profile.

The competency analysis was performed from the survey of demands of the technology in operation compared with the offer of the contracted team, bringing as a result success or failure. Competency analysis indicates Technical–operational Culture installed in production. Human types are classified regarding the application of competency in the accomplishment of task. The confirmation of behaviors is with evaluation of knowledge and interpretations related to failures that occurred in the company's routine.

7.3.7.1 Demand based on technology

From the experience, study of technology, analysis of procedures and the mapping of processes that indicates human errors and routine failures, themes and situations were chosen to identify demand for competencies in liquefied gas processing: fluid properties, alignments, contaminants, logistics, safety and compressor operation, operation and maintenance.

Liquefied gas properties in various proportions of gases alter cooling balances, leading to differentiated conditions in heat exchangers and compressors. In addition to alignment problems, these conditions can result in high-low pressures in such a way that leads to the opening of PSV to the torch. On the other hand, low temperature conditions lead to high flows in the transfer to the ship, indicating the need to maintain the safety of reducing this flow rate through operation. These properties can also alter the viscosity in the liquid state and slightly displace the curve of the pump.

In the fluid properties we found aspects about alignments, contaminants, and logistics; safety system including instrumentation; actions with the public; and compressors, pumps and intermittent operations.

Competency map.

The systematic training program intend to hold conferences regarding level of knowledge, skills and mind map formation for problem solving. Competency is classified into five levels, shown in Fig. 7.53 and its details and interconnections are: (1) understanding and control of the task; (2) concepts to accomplish the task; (3) analysis of the failure in the task; (4) construction of skills in the routine; and (5) concepts for process control and safety.

Level (a) Understanding and control of the task.

To understand and control task it is necessary to have a good level of abstraction, being able to plan task, imagining how it would be performed by the team, and enabling interpretation for the taking of action. At this stage, facilities are developed to elaborate the mind map. Still at this moment, logical knowledge of causality is needed to facilitate treatment of problems and to perform simple calculations to assist decision-making. The relationships of preference, dominance, and pressure as they result in the fluid flow.

Level (b) Concepts for the task.

Some basic concepts are required to perform the operator's work in this gas logistics unit. Thus how combustion works, what are the principles behind the change of phases and the mixture of gases, what the properties of the fluids,

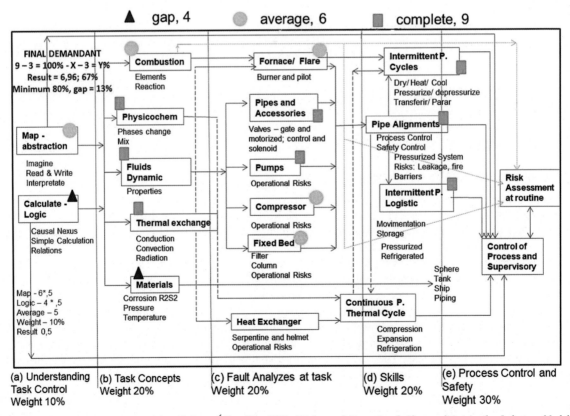

FIGURE 7.53 Demanded competence. *Adapted from Ávila, S.F., 2010. Etiology of Operational Abnormalities in the Industry: Modeling for Learning (Thesis—Doctor's degree in Chemical and Biochemical Process Technology). Federal University of Rio de Janeiro, School of Chemistry, Rio de Janeiro, 296 p.*

principles for thermal exchange and the resistance of the materials present in the equipment to process the gases in relation to corrosion, temperature and pressure.

Level (c). Task failure analysis.

This knowledge about task failure is closer to skills and related to the operator's routine in performing the tasks. Some equipment investigated are the focus of attention, furnace and flare depend on pilot operation and burner efficiency, at the same time, the pipelines work using valves to control flow, pressure and temperature. The rotative equipment (pump and compressors) are more complex but some static equipment can cause big problems as the dehumidification columns. The heat exchangers, filters and coalescer depend on liquefied gases quality, where border conditions of operation must be avoided.

Level (d). Skills.

Field and panel skills, which are important to avoid accidents in critical operations in this oil and gas site, depend on devices and controls to organize intermittent operations with times and cycles for each type of material being processed. Thus drying, heating, cooling, pressurizing, depressurizing, transferring and stopping require attention. Several alignments between equipment are for process control and safety. Part of these involving pressurized systems, possibility of leakage and fire events. The clarification of barriers avoids human error and failures, out of the system. Inventory, production control and availability of equipment keep intermittent and continuous processes under control.

Level (e). Process and safety control.

At this moment, control is achieved in the gas-processing unit and process control is achieved with supervisory and systematic communication. Safety control being envisioned through routine risk analysis methods. Refrigeration operations are intensively developed when compared to liquefied gases transfer operations from vendor to processing company.

It can be considered that the number of hours spent in refrigeration is more than double the number of hours spent transferring from the supplier to the company, mainly motivated by difference inflows.

Considering the requirement of specific knowledge for each activity, it's important in refrigeration to know how to operate under pressure and low temperatures, compression, pumping, fixed berth operations and furnaces/torch. In the case of transfers from the supplier to processor: communication, monitoring of quality, transfer pressure, possibility of solids, automatic/manual and semi-automatic alignments (motorized valves) become important, as well as area inspection to detect leaks. In the loading of ships, there are specific issues of equalization of pressures, in addition to the connections to the ships.

7.3.7.2 Competence offering—dynamics and identification of technical culture

The demand for knowledge to operate the technology is defined by process mapping after at least 1 year of continuous operation. MEA, a tool that analyzes the writing of the occurrence register, is aided by inclusion of process and safety data and targeted interviews with coordination and operational staff. Latent faults and other visible flaws were mapped in the chain of abnormalities, completed from the experience of the interviewees and from there, the skills map necessary to operate the flammable gas-processing plant was designed. So, the Competency Map is based on the Abnormal Event Map.

The technical skill map is the result of the competency analysis tool or, symbolically, personality ruler represented in Fig. 7.14. The preparation of the competency analysis requires measurement of inventories of knowledge installed and identification of the technical–operational culture from the point of view of an operators' group (40%) representing the entire team of the operation. The information to make up this competency map depends on questions about the gaps identified through historical human errors and that return through examination applied in workshop.

7.3.7.3 Examination for technical–operational culture identification

The technical culture identification exam, as well as of installed skills, based on abnormal event mapping and human errors that may be in knowledge, decision, communication and cooperation.

Technical–operational culture examination

The process of preparing this exam was not simple. After identifying abnormalities and interconnecting through their logic and connection, it was possible to classify and analyze the type of human error performed in the processing of flammable gas. Additionally, the task step or activity in which the failure was initiated was identified. Based on the tasks, 70 questions were prepared for 4 months of analysis of the shift register and process and production variables.

Knowing the limitations of production by making operators available for training, we filtered the most important questions to analyze the Technical—Operational Culture. In limiting training time to only 4 hours, culture exam was reduced to 15 questions in Fig. 7.54.

Discussion

The team has the knowledge of historical events, but that is not suitable for failure process training, there is no discussion about documented cause and effect. Additional effort is required to prepare training materials and, from there, prepare multipliers in failure processes. We recommend the tool of the Five Whys in the operational routine. It is a good strategy to avoid the complacency of getting used to standard response. FMEA helps to define the cause and consequence of problems in more detail. This interpretation is valid in the discussion about the event of contaminants in liquefied gases, for example.

The operation and vision of the production process can facilitate the interpretation of certain situations of contamination in the liquefied gases. In the current case, contamination with sulfur compounds, water or sodium needs to be discussed in a shared way with the supplier. After these supplier considerations, it is important to insert this into the abnormal event map if new factors also explain the events.

Motorized valves are considered the third equipment impacted by solids, after the vessel and the coalescer. The technical—operational culture indicates the possibilities of solids to generate problems in the motorized valves. There is great variability in the ICT responses, thus indicating that knowledge is internalized differently for each operator.

There was great variability of the processes in the period investigated. It is necessary to standardize information about the sequence of steps in the tasks. It is important to retrieve the events that occurred to find when, and at what stage, the task runs away from the expected routine. What may actually have happened and was not recorded clearly and logically. Lack of information makes it difficult to set priorities and may affect decision-making.

The responses related to the accident caused by BLEVE in the culture exam were coherent, but there were no other options to avoid BLEVE, which may indicate difficulties in constructing the mind map for complex events related to extremely high risk. Most chose the furnace event as the riskiest, with 33%, and then equally divided: leakage into the atmosphere, high pressure in the compressors and pressure in the refrigerated tank. All approached coherently about the

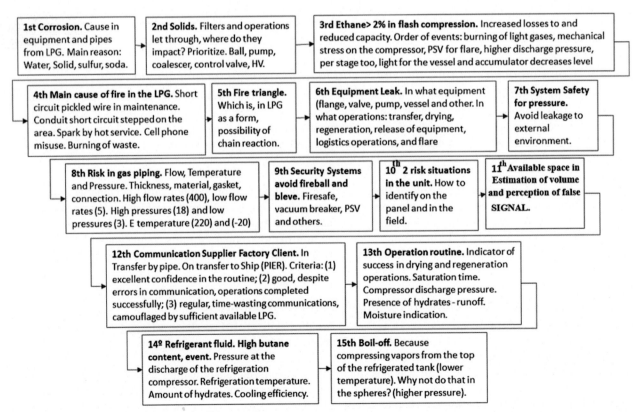

FIGURE 7.54 Examination of technical and operational culture.

signals in the panel and field, in which they issued their opinion on the problem. The identification of risk situation was agreed among the respondents.

> *(Conclusion PCA15/16—competency) Depending on the grouping that needs to be corrected, it can be sought to adapt to the installed competency based on the identifications already performed. The identification of these situations activates the need for training on failure processes; bringing as an indirect consequence a greater integration of the team (this training uses the leaders themselves as multipliers of knowledge).*

7.3.8 Standard for behavior analysis

It is important to define a pattern of comparison for analyzing behavior of operators and leaders in the work routine. This important part of the research allows management to compose its team by harmonizing the disaggregating energies. This discussion is divided into performance in the workplace (7.3.8.1), behavior analysis (7.3.8.2), human error in the task (7.3.8.3), and type of human and social behavior (7.3.8.4).

7.3.8.1 Performance in the workplace

The operator's behavior during the execution of the task in the jobs is the result of relationships that involve aspects related to individual behavior and group relationship. Some characteristics are important for the execution of the task by operation team, such as leadership, task control, decision-making, action in progress, cooperative relationships of team members and the roles or functions of operators in routine and emergency actions, on a fixed or rotating basis.

Leadership discusses the advantages of informality versus formalization, leadership direction, turnover, and power relations. Depending on leadership guidance, in the direction of blame or just culture, and influence on groups (delegates or not), formal organization chart can be merged with the informal, presenting what is to be called organizational gelatin as indicated in Fig. 6.1. It is important to detect which of these movements creates environments not appropriate for human reliability and, from there, take actions in the formal or informal field to neutralize these trends.

Other topics discussed in the workplace are the control of the task and the moment of decision-making in the operation and in the shift, where it is detailed: type of situation and speed in the decision, what is the person's profile responsible for the decision, harmonization of ideas in the decision, criteria for decision and analysis of consequences. It is also important to evaluate the action chosen after the decision, time, steps, devices, tools, type of risk, communication and documents. Also, analyze the action in behavioral terms: natural or planned action; studied, impulsive, consulted action, team, individual, managerial or supervisory. Dependencies or requirements should be preevaluated and made available for immediate use.

Still in the job it is important to analyze the level of cooperation of the team members. The work environment must be involved in a climate of cooperation for the team to work unified, in which there are no leaders directing their activities to different points. Team cooperation is measured when standards are unique and uniform in the team. This pattern in written form should be easily reproduced to the oral form because it is easy to assemble a mind map of the activity to be performed.

In the group, the existence of conflicts is healthy, during discussions and interpretations about the activities to be carried out, in the investigation stage for the decision, with the behavioral conflict and the nonacceptance of the best decision being unhealthy. In the social process of cooperation, people are aware of the needs of group action and do not assume egocentric or selfish postures, some experts call it an inflated ego. Usually, people who are concerned about the overall outcome and who perceive group differences are in favor of cooperation in the workplace. The level of team cooperation depends on important aspects known to shift leader and production coordinator: social inclusion of the operator in the group and in society; organizational integration in the shift and in the company; the affective bond with work and with the leader; and to the sense of justice of leaders and the company.

7.3.8.2 Analysis of behavior in the workplace

Workers perform the task as they move between the area and the control panel. In the area, depending on the operation being performed, they may be close to some equipment or even close to PLC, to perform control activities. At all these times there is an expected default behavior and limits to behavior oscillations, to avoid human error or operational failure.

At this moment, behavioral issues of the individual will be discussed, as an operator, in relation to leadership, process control and operations, group or individual decision-making, team cooperation, and specific roles in day-to-day.

Knowledge of the standard behavior allows evaluating the oscillations of the behavior and possible motivations for this. As they speak of individuals in the work environment, a rational model is always sought, although in this workplace the affective aspects and individual values circulate in the decisions. Unfortunately, to hinder quick solutions, part of these studies is subjective, involving emotional balance and stresses that activate inadequate and often unconscious behaviors, leading to human error and accident.

7.3.8.3 Human error in the task

The internal processes of the individual in the work environment promote conscious or unconscious changes that affect behavior, then generating human error. These internal processes are related to: (1) value analysis or reward analysis; (2) analysis of competence and commitments in its implementation; (3) neutral behavior or influenced by cultural issues; (4) intrinsic preferences in the social relationship and in the type of environment. Other internal processes are: (5) family breakdown with real losses (disease or separation) or with symbolic losses (depression for no apparent reason) and (6) lack of knowledge of himself/herself, leading to illogical behaviors, but that do not make it impossible to perform the work.

In some cases, these internal movements do not fulfill the expectation after performing the job (7) where competency or emotional experience applied is low in relation to the requirements of the situation. In the case of cognitive work processes where no (8) recognition occurs after comparing the lowest level of competency, or (9) when fear of developing certain activities in the work environment dominate the occasion. Other problems happens when there is no (10) recognition by the team and mainly by the leader when the worker feels left out of the group (no group insertion). Other process possibilities that affect operator behavior can generate failure situations.

Some individual behaviors may be inappropriate for performing the group task at work. These behaviors may be causing the appearance of human errors and with possibilities of turning into team error, managerial error or technical failure related to the equipment and or task.

Some inappropriate behavior and human error in the task performance are the sense of low social inclusion; poor organizational integration in plant stops; lack of organizational integration and affective stability; nonfulfillment of roles in the family and in other groups. In other cases, inhibition of creativity; lack of attention, memory failure and nonacceptance of the rules; or opposite, with excess discipline; with the lack of organization (identification of requirements and areas) and the lack of a sense of justice in decision-making.

By recognizing the existence of these aspects, it is possible to work with the identification of human failure and raise the hypothesis of which environments promote human error. With these hypotheses, dynamics and tests are then performed for confirmation and only posteriorly working on environmental and competency changes. Accordingly, it becomes possible to link a program to increase human reliability.

7.3.8.4 Analyses for the type of human and social behavior

Human factors in the work environment have their own dynamics that transit from the structure of the individual to the shift group and from the shift to the organization and back again, similarly to the communicative processes indicated in Fig. 7.6 in the organizational pyramid model. The information collected for the analysis of human behavior at work indicates, after correct analysis: the reward balance; differences in standards and level of acceptance; the consequence analysis; the availability of competency for the development of the task and level of commitments. This information circulates through individual, group, and organizational structures, and can be influenced by the profile of the leader and by organizational type and the company's safety culture.

Certain repetitive forms of behavior indicate an addictive tendency to compulsion, resulting from myths and rituals imposed by regional and/or organizational culture. With the awareness of these movements, operational routines can be protected to avoid the oscillation of behavior toward the addictive vector originated in cultures and directed to the individuals and the group.

Competency building results in a different construct with theoretical knowledge as a base for future complexities; general practical knowledge that is not necessarily related to the end activity; skills resulting from the experience in the end activity, which some authors name as working memory; and (4) knowledge and experience in interpersonal relationships, reflecting a greater emotional balance.

In the competency analysis, it is also important to understand the differences between supply and demand, indicating the need for adjustment of competencies. The employee's commitment to the company allows the transfer of knowledge inventory to its application in a partial-total way.

Reward analysis, the key to relations

The swap relations in the society, economy and work try to be just giving correct value for each service or product. In the case of work at operation, it is not different, this relation has an expectation of reward for the good work performed. It reflects not only the job partners but also, in the life of a couple, public relations and others. The operator performs a reward analysis to understand if the given mission fulfills the expectations of the company, for example. The company compares the balance between the results achieved after application of operator competency with financial return. The emotional balance resulting from organizational inclusion, where nonfinancial values come into balance between the organization and individual is important. Moreover, analysis of the company regarding the balance between consequence of the decisions taken and initial expectation of work in the operation.

Management profile

The analysis of the managerial profile allows characterizing the type of leadership and influence on the decisions made in the routine of the operation. The profile may indicate trends to informality, internal motivations toward the delegation or even careerist professional carrying vices (negative) and vast skills in the end activity of the team (positive).

Economic-social utility relations at the workplace

In Fig. 7.55, we indicate the data analyzed and care through Motivational Programs to favor the various economic relations integrated to the sociotechnical process:

1. Reward analysis: home time, salary progression, and employee level reflect this reward analysis, but mainly productivity.
2. Commitment analysis: commitment is related to the comfort of being in the role and fidelity toward leaders and organizations.
3. Analysis of allocation in the workplace: ergonomics, cognitive processing facilities, availability of the procedure and instruments for task control are important aspects in this case.
4. Analysis of competency application: the relationship of base knowledge (course), specific knowledge (operator training), skills (time of experience) and interpersonal relationships (mood and time of experience).

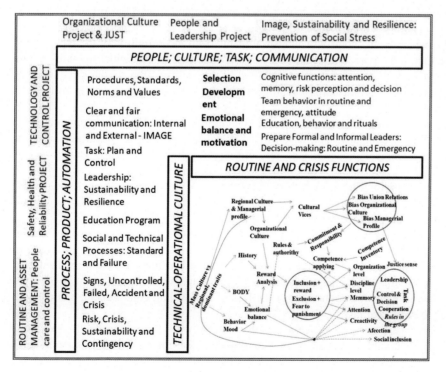

FIGURE 7.55 Relationships for reliability in task execution. *Adapted from Ávila, S.F., 2010. Etiology of Operational Abnormalities in the Industry: Modeling for Learning (Thesis—Doctor's degree in Chemical and Biochemical Process Technology). Federal University of Rio de Janeiro, School of Chemistry, Rio de Janeiro, 296 p.*

5. Emotional balance analysis: the relationship of the worker in inclusion in society and in the family results in a good emotional balance. Emotional balance keeps psychological functions tied and functioning properly, including perception, memory, and cognitive processing. Emotional balance reflects on the body through symptoms or diseases.
6. Balance analysis of decisions: the influence of the biases of the economy, society, regional culture, organizational culture, and direct managerial profile affect the balance in decision-making. In direct terms, union relations, job stability, being "owner of the area," aspects of individual behavior that reflect culture affect the appearance of human errors.
7. Balance analysis in the sense of justice: balancing the attentions of the various demands of the team and the organization, without allowing the power relationship to overlap with decisions, demonstrates an established and balanced sense of justice.
8. Dominant culture analysis: global versus local.

Behavior and cognition

Behavior is influenced by dominant culture and can affect the work environment despite organizational procedures that try to inhibit negative influences. Cognitive memory functions, attention, risk perception, acceptance and understanding of rules and procedures can be altered individually or in groups.

1. Analysis of the dominant human type. Based on the coordinator's response on human traits and based on the personal responses of the operator, we seek to classify in a maximum of 4 profiles, showing a tendency toward certain behaviors and how to inhibit unsafe behaviors.
2. Analysis of organizational and managerial solutions: verify the issues that most impact the company and the leader about the team's behavior, and that can cause human error.

What is sought as corrective, preventive and mitigating action to reduce human errors resulting from inadequate relationships between human, social and organizational factors?

Discussing barriers to undue and unsafe behavior in Fig. 7.52

The identification of relationships that can promote operational control or human error indicates the need to draw barriers to avoid inadequate and unsafe behavior. This action is strategic and involves all departments of the company and the procedures during the design and operation of industrial facilities. The design of the projects is a primary phase not to accumulate future problems. On the one hand, objective projects involving the processing of materials and data for transformation into a product or service: technological (process and product); safety, health, environment and reliability; and routine and asset management. On the other hand, most subjective projects include: fair organizational culture; people and leadership; image, sustainability and resilience.

These projects define the level of automation and control of industrial plants from the definition of the types of process, product, people, and culture resulting in the operation of the company. To this end, it is essential to build communication and task control programs. Operational, occupational, process and health safety is the result of requirements and actions entered in organizational and behavior standards operationalized through operating limits in procedures, standards and values.

The tools for planning, executing, and adjusting internal and EC is in the culture project. From this communication, we seek a guarantee of positive image in the company. The task will be designed based on technical and cognitive requirements respecting social processes, technical processes and resulting application of contracted knowledge and skill in the operational routine.

Leadership plays an important role in calibrating the team with keen perception of difficult-to-visualize sociotechnical failure processes. The efficiency of the operational team is demonstrated through cost control and quality in the routine and the ability to perceive and solve chains of abnormality that can lead to the accident.

One of the most important and most subjective parts is the establishment of organizational resilience and sustainability of processes in routine and in periods of degradation of structures. The importance of knowing how to select and develop people will ensure future of the business. Thus the main product of analysis of human factors is to establish cognitive and work behavior criteria for routine and emergency moments, although the emergency or crisis is not expected. Leadership development should trigger protections against the loss of motivation, commitment, and cooperation in the functioning of teams during execution of task.

In conclusion, we will only be able to ensure sense of justice and motivation for work from a technical—operational culture where leaders (formal or formal) are humble in behavior, in which there is determination in appropriate actions and tools of intervention guaranteed that provisional solutions of unresolved cases are investigated to develop a stable format of the process. Labor functions and behaviors are stabilized from the trust established between the organization, external public, stakeholders, operations team and leaders.

7.3.9 Abnormal event cluster analysis (T14)

Based on the assumption that there is "hidden" information between the events raised in the shift register, by equipment and by process, the mapping seeks to identify trends and general rules between these events following a chronological rule. In 7.2.9.1 the analysis is done for groupings of events in process and in 7.2.9.2 the analysis is done for groupings of events by equipment.

7.3.9.1 Cluster analysis and relationships in process abnormalities

Fig. 7.56 analyzes the groupings of data that can bring information when properly processed by statistical and visual tools. Each information group is unique by number, frequency, and indicates probable events in occurrence. This information is analyzed indicating a pattern of repetition and different relationships, thus believing in the existence of chronology in the events. The period of Cyclic events is 10 days where the principal phenomena involved is the corrosion and white water in PLG, we don't know exactly... Is it cause or consequence? The events are grouped in physical and social aspects.

Data groupings are gathered through rectangles that circle the series of events indicated in Fig. 7.53. Grouping 1 involves vessels (EF), pumps (B03), filters (FDCORR) in liquefied gas feed, and corrosion events (CORRL). In the grouping of information 2, there is repetition of pattern, but with a certain dispersion involving air (KAR) and refrigeration (KR) compressors, which would be indirectly related by the operation of the control valves. In group 3, several events such as high pressure in the systems (Palta) are related to: pressure control (FLASH), tank pressure control pump (condensates—pump), refrigerated tank, and pump (TQ72, pump), indicating that pressure oscillations in the plant can cause oscillations in the pressure control circuit and refrigerated tank. In cluster 4, where general (GERAL) and electrical problem related (ELDUTO) events are repeated.

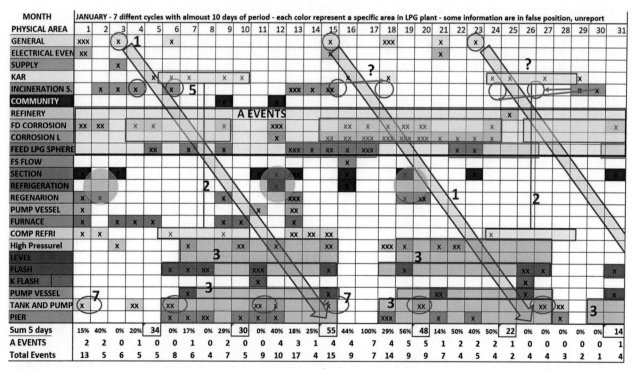

FIGURE 7.56 Cluster analysis in process abnormalities. *Adapted from Ávila, S.F., 2010. Etiology of Operational Abnormalities in the Industry: Modeling for Learning (Thesis—Doctor's degree in Chemical and Biochemical Process Technology). Federal University of Rio de Janeiro, School of Chemistry, Rio de Janeiro, 296 p.*

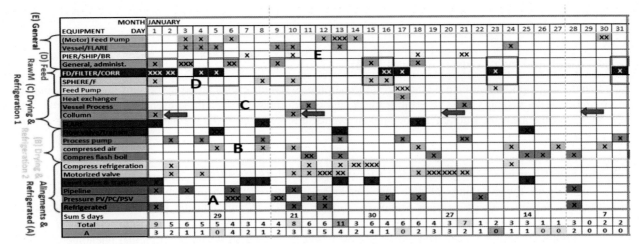

FIGURE 7.57 Cluster analysis and relationships in equipment abnormalities. *From Ávila, S.F., 2010. Etiology of Operational Abnormalities in the Industry: Modeling for Learning (Thesis—Doctor's degree in Chemical and Biochemical Process Technology). Federal University of Rio de Janeiro, School of Chemistry, Rio de Janeiro, 296 p.*

From the blue arrow in Fig. 7.56, it is noted that every 12 days, certain events are repeated with a tendency to decrease between January and February 2006. Data groupings in 5, 6, and 8 indicate the appearance of events in clusters but not repeated at a certain frequency. In 5 represents events with fire system (SISINC). In 6 events, events involving refrigeration (REFRI) are indicated, and in 8 events on the pie. Due to the higher density of data and due to indicators related to plant performance, the conclusion is for the choice of the cluster that involves filters, or filtering, including events with sphere and pumps.

7.3.9.2 Cluster analysis and relationships in equipment abnormalities

The events related to equipment were launched in the grouping worksheet, and in some cases, several related equipment is part of the same group, with sufficient data density for interpretation about the clusters. The behaviors identified in the general class, relating to fire system, furnaces, pier, general and administrative. In the feed section, there are pumping systems, filters, and vessel, drying and cooling section, events with exchangers, tower and vessels. In refrigeration events with control valves and pumps are present. In tank pressure control and refrigerated units, there are events with motorized valves, level, compressor, pipelines and refrigerated tank. Although there was the search for equipment following, on average, process direction, the repeated pattern shifted the tank pressure control compressor item close to refrigeration and the refrigerated compressor item close to the refrigerated.

Finally, as shown in Fig. 7.57, the clusters are about equipment in logic of process. In A, coincident with alignments and refrigerated equipment, presenting higher data density and consistent with events related to high pressure, motorized valves, and events in refrigerated area. In B, coincident with drying and cooling, grouping 2, where the events of compressed air and pumps are distributed in the period. In C, coincident with drying and cooling, grouping 1, where events with process vessels and drying towers can be analyzed. In D, coincident with the feed, where events involving filter, pump and vessel are grouped. Apparently, cluster A is the most significant, both by the amount of data and repetition, and within these grouping events aligned through HV valves and involved with pressure control. Thus there are some options for analyzing discrete data behavior in the case of equipment: (1) all events listed or (2) events with motorized (HV valves) and involving pressure, or (3) events of cluster A.

> (PCA16/16 Conclusion—clusters) Choice of data in the following order: feed (filters + pumps + vessels), alignments (HV valves with pressure).

7.4 Qualitative results: chemical industry cases

With the implementation of the methodology in an industrial gas-processing environment, following the logical and chronological order indicated in Figs. 7.2, 7.3 and 7.4, it was possible to corroborate concepts, techniques and procedures discussed in this research. In this application, there was concern in testing the methods and statistically validating the hypotheses. In the chemical industries where the technical typology identification step was applied, it was possible to measure the results after field tests and the change of patterns. These results are cited as a reference as to what can be achieved with the methodology. The results of the research are divided into two types: (7.4.1) managerial aspects for decision-making that includes the preparation of programs and training for the adequacy of competencies and (7.4.2) development of a tool for analysis of failure energy through a procedure for mathematical calculation.

7.4.1 Management aspects for decision

The identification of the technical, human and social typology allows decision-making toward the greater organizational efficiency, which is represented by the objective function of the project, gross profit and image. Process losses are decreased through management programs in the areas of process stabilization and human reliability.

Initially, it is important to define the possible parameters for measuring gross profit margin (7.4.1.1), thus indicating how to measure the results of the implemented programs. Next, it is important to compare the previous work carried out in the chemical industry with the comparison between cases (7.4.1.2). Thus the presentation of the results in a comparative way between chemical plants and this case of LIQUEFIED GASES makes it possible to understand and present management programs and opportunities to increase human reliability and stabilization of industrial processes.

7.4.1.1 Objective function: gross profit margin

The gross profit margin consists of production cost components and components related to the company's image. Various items related to process losses are directly and indirectly influenced by both aspects the. In the case of components linked to the company's image, it can be affirmed that after accidents and serious incidents, there is an increase in the costs of renewing insurance, borrowing costs in banks, and even reducing the stock market value, leading in the end to the reduction of the company's revenue.

Table 7.17 describes the objective function components that can be measured for result analysis in programs related to process losses as shown in Fig. 7.2 of this study. The gross profit margin increases when costs decrease in relation to past situations, thus higher revenue and lower cost indicate the increase in gross profit. The difference in revenue (ΔFAT) is influenced by the loss of production due to shutdown (E); loss of revenue due to products with reduced specification (H); by loss in the delivery time to customers (I); by low production load (M); and by to reduced credit because of negative image (R/S).

The cost difference based on the acquisition of materials ($\Delta COST_{rm,\ is,\ uts,\ mat}$) is mainly due to reprocessing (G); reaction failure (K); low efficiency in separation operations (L); increase in consumption of utilities (N) and spare parts (J). In terms of costs related to production and team ($\Delta COST_{opm}$) it is mainly due to yield in the work of the operation (A); performance in maintenance work (B), rework in operation and maintenance (C/D); fixed cost increases due to unscheduled stop (F); and raising the reprocessing cost. Also, in the analysis of production cost there are aspects related to the company's image, thus ($\Delta COST_{Ef,\ IMA,\ IM)}$: effluent cost (O); fines on effluent, labor and the customer (P/Q); and insurance renewal costs (T). Finally, it is necessary to analyze the increase in cost from low plant yield due to inadequate projects (U).

7.4.1.2 Comparison between cases already applied

In the period between 1992 and 1998, cases 1, 2 and 3 were developed in the chemical industry with a focus on identifying the behavior of technology for Process Stabilization. The implementation of changes in standards and procedures, in addition to the investment in instruments and equipment, enabled the achievement of increase in gross profit. In 2008/2009, work was carried out with the application of the identification of technical, human and social aspects in an energy company to define management programs in the areas of process stabilization and human reliability.

TABLE 7.17 Objective Function in loss investigation.

NoObF	Area	Objective function	Equation
A	Operation	Loss of yield in operating work	$(\Delta\eta_{hh}*CF_o)$
B	Maintenance	Loss of yield in maintenance work	$(\Delta\eta_{hh}*CF_m)$
I—GM, gross margin	FAT—COST$_{rm,is,util}$—COST$_{om}$—C$_{mat}$—C$_{ef,IMA,IM}$—C$_{div}$		
C	Operation	Rework in operation	$(N_{hh}*P_{rto})$
D	Maintenance	Maintenance rework	$(N_{hh}*P_{rtm})$
E	Production	Loss of production per stop	$(P/h)*h_{stops}$
II—ΔREV, revenue variation	$\Sigma (E+H+I+M+R+S)_{today}-\Sigma (E+H+I+M+R+S)_{yesterday}$		
F	Production	Increment—fixed cost unscheduled stop	$(H_{stop}*CF_{om})$
G	Production	Cost increase per reprocess	$(Q_{rep}*CV_{util}*R_q)$
III—ΔCOST$_{rm, is, util}$	$\Sigma (G+K+L+N)_{today}-\Sigma (G+K+L+N)_{yesterday}$		
H	Sales	Price loss, $P_{Offspec}$ revenue	$(\Delta P_{offspec}*Q_{offspec})$
I	Sales	Customer loss, revenue, term, Quality$_{Offspec}$ and price	$(\Delta = C_{fat}-C_{losses})$
J	Procurement	Increase in cost of supply due to failure	$(\Delta = C_{cps}-C_{fails})$
IV—ΔCOST$_{opm}$	$\Sigma (A+B+C+D+F+G)_{today}-\Sigma (A+B+C+D+F+G)_{yesterday}$		
K	Process	Loss of raw material and inputs by failure to react	$(P_{MPI}*Q_{loss})$
L	Operation	Loss of inputs by low efficiency in operations	$(P_{is}*Q_{loss})$
V—ΔCOST$_{mat}$, materials	$\Sigma (G+J)_{today}-\Sigma (G+J)_{yesterday}$		
M	Production	Loss of revenue by low load	$(\Delta C*IC_{mpi}*P_{mpi})$
N	Production	Increased utility consumption due to failure	$(\Delta C*IC_{util}*P_{util})$
O	Effluent and safety	Increase in effluent cost due to failure	(ΔC_{eflu})
P	Effluent and safety	Fines for out-of-standard effluent	(M_{eflu})
Q	Effluent and safety	Fines for labor—man, management, organization	(M_{MMO})
R	Effluent and safety	Loss of revenue, credit or value—negative image—accident or environmental event	$(\Delta F_{imag}-\Delta C\ end_{imag})\ IAMB$
S	Product	Loss of revenue, credit or value—negative image—product application	$(\Delta F_{imag}-\Delta C\ fin_{imag})_{prod}$
T	Reinsurance	Increased risk due to IMA and SEG events and labor causes—failure	$(\Delta R_{faults} * f_{power} * R_{Normal})$
VI—ΔCOST$_{ef, IMA, IM}$	$\Sigma (O+P+Q+R+S+T)_{today}-\Sigma (O+P+Q+R+S+T)_{yesterday}$		
U	Project	Increase in cost per inadequate project and failure	(ΔC_{proj})
VII—ΔPROJECT$_{COST}$	$\Delta Cdiv = \Sigma (U)_{today}-\Sigma (U)_{yesterday}$		

As shown in Table 7.18, the cases of the chemical industry are from private companies around production of materials for plastics (polycarbonates and polyurethanes) and for pesticides (alkyl amines), different from the case of the energy industry, that is a liquefied gases processing and logistics company. The main objective of the study was different in the four cases, in which case 1 was cost reduction, in case 2 increase of continuity, in case 3 decrease of solvent loss and in the case of gases, decrease in the risks of accidents with the public. All cases are related to process losses. The identification/analysis of human behavior and competency analysis was developed with greater detail in the case of gas processing.

Technical, human, and social results are presented in Tables 7.19, 7.20 and 7.21. In comparative terms, in the case of liquefied gas, it is made in conjunction with cases of chemical industry, thus indicating possibilities of actual results.

Technical results of process stabilization (Table 7.19)

The definition of the process variables for monitoring, providing support for change of procedures and standards was the main result of this work in the liquefied gas company, while suitability in 5 sigmas, thus increasing the capability of the processes in case 2 and 3, was the practical result achieved. In the technical program to stabilize processes, actions involving the presence of solids and the operation of motorized valves were identified with great possibility of increasing the availability of processes (goal is reached in cases 2 and 3). At the same time, in these cases of the chemical industry, the consumption of spare parts was reduced, a target not applied in the gas company. In terms of people, there was a reduction of overtime in cases of the chemical industry and identification of factors such as % of double shifts and technical idleness, which are important indicators for increasing human reliability. In the area of accidents and incidents, appropriate techniques were developed for Liquefied Gases, while in case 2 there was an explicit gain in accident reduction with revision of tasks. In the direct analysis of process losses, the great prospect of increasing load, reducing unscheduled shutdown, and reducing consumption due to reduced pressure and sending liquefied gases to *flare* is noted. The validated site of the root cause in case 2 is established in the recycle line; in case 3 in inadequate technical and managerial standards, meanwhile; in the case of gases, the site is in the low quality of raw material, inadequate maintenance contracts and the need for competency adjustments. These root cause sites in the case of liquefied gases will be validated with the implementation of technical and managerial programs.

Technical results of task analysis (Table 7.20)

The methods for analysis of the standard task, in failure and in emergency situations, were developed and applied in the liquefied gases Company. In comparative terms, without following established procedures for analyzing the task, 19 procedures were revised in case 2, 14 procedures in case 3 and 6 procedures in the Liquefied Gas Case. The operational failure analysis with impact on the operation was applied to cases 2 (12 chains of abnormalities), 3 (23 chains of abnormalities) and was applied in the operation and systematized to safety area in the Liquefied Gas Case (24 chains of abnormalities in the operation and 8 chains of abnormalities related to safety). The techniques of risk analysis and stress dynamics that consider human factor in the decision of actions to increase human reliability were developed and applied in the case of gases.

TABLE 7.18 Comparison between cases with partial application.

Characteristics	Case 1	Case 2	Case 3	Gases
Product and use	Alkyl amines; pesticide	Methyl diphenyl Amine. Polyurethane	Polycarbonate; engineering plastic	Liquefied gases
Type	Private	Private	Private	Private
Activity	Industry	Industry	Industry	Industry and service
Objective	Reduce cost	Increase continuity	Decrease solvent loss	Reduce risk accident process
TT Tool (technical typology)	MEA AER EVA	MEA REA EVA AEP defaults	MEA REA EVA AEP defaults	MEA REA EVA AEP fault standard and emergency task
HST Tool (human and social typology)	Not applicable	Team pattern, aggregation	Team pattern, addictions	Behavior and pattern in routine and stress
Competence	Not applicable	Development, self-management	Training	Competency analysis
MAT Tool (mathematical or statistical)	Statistics	Statistics	Statistics	Statistics, PCA, heuristics, failure

TABLE 7.19 Results of process stabilization programs.

Results	Case 2	Case 3	Gases
Controlled process variables	70%	80%	12 variables stable
Equipment unavailability (minor)	60%	70%	Solid and motorized valve
Reduce spare parts consumption	20%	40%	Not applicable
Overtime reduction (target 0.5%)	1.5%	1%	Double shifts and technical idleness
Accident (absence/year) and incident	1 accident per year, 4 incident per year	Not applicable	FTA fire study and BLEVE
Effluent and emissions reduction	50% effluent	50% effluent	Reduce pressure and <flare
Increased load, less discontinuity	+15% load, 50% discontinuity	+10% load, −30% discontinuity	Not applicable
Decrease consumption index, reduce losses	Raw Material—20%	Input—15%	Lower Pressure, Less Loss
Root cause	Recycle, raw material flow	Technical and managerial standards	Raw Material Quality, Maintenance Contract, Competence

TABLE 7.20 Results in the application of task analysis.

Task	Case 2	Case 3	Gases
Task review	16 task, 2 laboratory and 1 process	6 operation, 5 process, 3 managerial	Drying/regeneration
Task project	Not applicable	No applicable	All systematized tasks, focus on drying and regeneration
Failure analysis in the chains of abnormalities	12	23	24 chains of abnormalities, 8 with accident failure analysis
Risk analysis	Not applicable	Not applicable	Adapted LOPA technique (layers) with human factor and FTA

TABLE 7.21 Results in the identification of elements for analysis of human reliability and socioeconomic environments.

Aspects	Case 2	Case 3	Gases
Motivation (voluntary participation)	Social events (75%)	Social events (65%)	Good participation in the workshop for human reliability (over 40%)
Behavior pattern	Applied standards and manual preparation	Not applicable	Standard in application, manuals and various subjects in scheduling.
Analysis and diagnostics	Self-management	Training and functions	Emotional balance; aggregation of teams; role of leaders; commitment; performance under stress

Social human results (Table 7.21)

The methodology developed and applied in human reliability is of rapid acceptance and results in broad spectrum due to working with the behavior of the workforce intrinsically, when applied in the bottom-up mode. The first diagnosis is the identification of the failure anatomy (chain of abnormalities) to facilitate decision and identification of technical culture to analyze competency. Behavior is a consequence of social relations, the insertion of the manager's discourse in the workforce and the risks of incidents and accidents in the work environment. Thus the use of routine data and the change of patterns and procedures in consensus with shift team, present positive results in terms of motivation indicated by greater participation in external social activities and awareness-raising work in human reliability. The identification of patterns of social and technical behavior is essential for conducting training for multipliers in shift classes. These manuals, the training of leaders, the development of work from the operational—technical culture allowed to achieve the level of self-management in the company in case 2. In the case of gases, diagnoses of emotional balance in group, competencies analysis, aggregation of teams, level of commitment and processes in leadership were developed. To finalize, the procedure of behavior under stress (LODA) allows to develop critical sense in the operation team and enables the participation of groups in risk analysis.

7.4.1.3 Management programs to increase human reliability

With the implementation of identification methods of human, social and technical behaviors in the face of failure in the task, it becomes possible to propose management programs to adjust coexistence of the team. The important thing is to gather information, develop indicators, and prepare procedures that are intended to harmonize teams and promote their development in search of self-management. To this end, it is important to develop leaders and knowledge multipliers. The management programs suggested in this chapter have as a result the increase in human reliability.

Increased human reliability

Provide management programs to increase human reliability calculated/estimated directly or, indirectly, together with tasks in the field. The measurement of the increase in human reliability is given either by reducing the rate of human failures in technical activities or by increasing the time between human and operational errors. Increased reliability can increase the volume processed, improve the individual's occupation rate at work and decrease the number of signs of abnormality in the shift. The increase in human reliability is based on direct action through programs on the identified elements.

Analysis of critical tasks in routine and emergency

Based on the architecture of task and global dominance in terms of safety, it is necessary to analyze critical tasks. The techniques used for standard task analysis, task failure analysis and failure (emergency) analysis, are selected based on the criterion: application time, complexity, and results. Identify the roadmap to analyze the task and what the function of each tool, allowing for the choice of applicable ones and developing procedure for application. Indicate Action Plan to deploy the recommendations indicated in the task failure analysis. Prioritize chains of abnormality and techniques for analysis of critical tasks in routine and emergency. Prepare a roadmap to formulate the Manual of Good Practices in a simplified way and with the corresponding field booklets.

Behavior patterns

Behavior patterns should be transformed into guidelines in the work routine. At least one component of the shift class, which may be the formal or informal leader, is trained as a multiplier on this subject. In this training instructions are given on understanding about human relationship in the workplace, the expected behaviors, and behaviors not appropriate in the routine, how inappropriate behavior can generate human error, which human types are most prone to certain errors, how to measure critical situations in the company's social body and what actions will be taken.

Classification of human error

The classification procedure needs to be created/reviewed together with the production team and people through *the workshop* that will also address behavior patterns and database formation. This procedure will guide on how to build the database to deal with human reliability issues and process stabilization. Remembering that in the procedure, the process signals, the OD and the manager's discourse are given, helping in the investigation of human error or technical failure. Part of this data already exists in production, health-safety-environment, people, being then necessary to add other maintenance data, for example, systematized to assist in the classification.

Adequacy of competences

The development of skills is a priority of the Company, given the result of the analysis of competencies. In the diagnosis of the competencies required for the technology, there are topics to be worked on to avoid failure. During a three-year period, training of all teams on failure mode and safety and control programs as analyzed and application of exams and dynamics for purpose of selecting shift leaders based on best competencies, including interpersonal relationships.

Integration: rituals of inclusion of operation and maintenance

Inclusion rituals serve to increase the cohesion of groups but must respect local cultural aspects and sometimes clash with certain organizational norms. Informality is an important factor to facilitate the inclusion of dispersed people in relation to larger institution, that is, the organization. The proposed activities require authorization from coordination and go beyond the formal environment and cognitive processes, entering areas of the informal environments and the affective/interpersonal relationships.

Management paradigms: productivity and safety

In routine management, team motivation should be sought around overcoming difficulties and paradigms established by company's culture or global culture. Among paradigms is the environmental sustainability, the need to open the company in relation to external bench markings of productivity, reducing technical idleness, incorporating investigation of accident risks from operation routine and development of skills to manage conflicts between people.

Communication instruments: occurrences on the shift.

Some of the communication tools can help in the aggregation of groups and deserve attention in the type of language, in the codes to enable communication and in the tools that help the transmission of the message. In written communication of shift occurrences, omission is a problem, operator may not write or perform what was requested. The other problem is the commission, in which the operator does not perform and/or does not register in the requested way, being able to skip steps or insert new steps.

Team adjustments and leadership.

Leaders who manage to maintain the growth of the team despite dynamic and threatening environments are in tune with the self-motivated team and practice self-management, thus an increase of productivity can effectively be achieved and then promoted organizational resilience. Attitudes during the operational routine and the difficulties of planning the team indicate leadership deficiencies that hinder the definition of priorities and the realization of actions. The aspects that may be involved are several: emotional balance, aggregation, organization, skills, training, conflicts, manuals and documents. It is necessary to harmonize the team and more than that, develop formal leaders to meet this important need, self-management.

7.4.1.4 Stabilization program for processes, opportunities and suggested techniques

With the identification of process variables, failure energy measurement items and events that are involved with chains of abnormalities, it is possible to establish a statistical process monitoring program and start application of tests in normal operation for revising patterns and tasks. The opportunities are diverse in the technical area, which includes the adequacy of control systems with emphasis on air system instruments, supervisory screens, and better risk management through new parameters to be monitored and the application of analysis and sociotechnical system audits.

Process stabilization program

Prepare a program for statistical monitoring based on statistical analysis of processes and events carried out based on the period of year 2006. In this program, tools are proposed for correlations, temporal analysis and automatic construction of graphs. Initially, the chosen process variables are linked to liquefied gas transfer between sections, compressors, refrigeration and tank pressure control. The variables that have predictable behavior for correction due to direct cause-consequence relationship are, apparently, pressure, flow, temperature; and the statistical parameters indicated for the analysis are mean, standard deviation, maximum values and sum of parameters. The operation variables discussed are linked to productivity, such as technical idleness and sum of signs of abnormality. Maintenance variables are important for defining control volume, failure region, including problems with rotating equipment and valves. Production variables are processed volume and daily electricity consumption. In process safety, events involving high pressure and equipment/processes that may leak liquefied gases. From the Department of Health Safety and Environment, data are analyzed leading to conclusions about the behaviors of processes and people. The objectives in the implementation of the process stabilization program are to increase the capacity of the process with increased load, the availability of equipment, the reduction of unscheduled downtime and the reduction of material losses through *the flare*.

Adequacy in the supervisory and in the field

It is known that there are technical adjustments to be made in the control modes of gas processing. They are proposed for the preparation of simplified screens in the function's quality, safety and logistics. On the other hand, many of the control meshes must have their operation improved, same happens in terms of remote-control valves, so a program to audit the operation of control valves/motorized/solenoids is important in industrial installations.

Better risk management, safety.
With the analysis of the failure in the task, the analysis of emergencies/contingencies and the statistical analysis of events/processes/equipment allow the proposition of technical and competency adjustments to deal with failures. Data are collected in the Health-Safety sector and in the operation, thus assisting in the construction of a program for process safety, in which audits in the calibration of PSV; inspection of the operation of critical equipment; and for the operation of safety systems.

7.5 Quantitative results: oil and gas case

7.5.1 Failure energy analysis

The research in human and operational reliability with interventions in human behavior and operational–technical culture aims to validate diagnoses with mathematical tools and field tests. Among the activities for the design of the mathematical method, are prepare a conceptual algorithm to apply the PCA; apply scatter plots to analyze failure energy; and in future studies, correct performance factors using pertinent models (FUZZY), correcting the factors to achieve better human performance.

The objective of this research is also to calculate the failure energy through predictive methods of behavior of the operation routine inserted in an organizational environment. This calculation results in a cloudy region that after being compared with an optimal behavior pattern (or maximum organizational efficiency) presents a difference in behaviors and indicates a tendency for both error and correction. In the case of a trend toward error, we seek to change the trajectory in the dispersion chart for situations of greater comfort through managerial adjustments about environments, tools, people, groups and tasks.

The PCA method is simple and can be used automatically through different software. In the current case, Minitab was used. PCA is a method that analyzes the behavior of covariance of large amounts of process variables with the intention of reducing this amount of data by generating artificial numbers. The artificial numbers generated represent so many other numbers that measure operational failure, that is, they are indicators of organizational efficiency. Thus in the scatter plot, it is important to record the sequence of points and the measurement of organizational efficiency for each point. This mapping of efficiencies allows for the analysis of behavior of artificial variables indicating level curves of organizational efficiencies, topologically established, or even data clouds representing three dimensions. The regions' maximum organizational efficiency indicates target points to be reached by the behavior of the process. With the measurement during the prediction of behavior, it is possible to indicate needs of adjustments in process, people, and tasks.

According to Lindsay (2002), PCA is a way to identify patterns in data and express data to evidence similarities. The input data are numerous and present a different nature. They can be difficult to detect, in which case, we studied a broader group of data through the PCA. One advantage of PCA is that once the patterns for input and data are found, these can be compressed, thus reducing the amount with minimal loss of information.

According to Lindsay, the steps to reduce the amount of data are: (1) Calculating covariance matrix between data that resembles and intends to start compression; (2) Calculation of vectors and (Eigen) values of the covariance matrix; (3) Choosing components and forming a characteristic vector; (4) Deriving a new data series.

Other important information that characterizes the PCA tool is (Lindsay, 2002; Shlens, 2009) covariance, which is a degree of linearity between two variables; high positive covariance value indicates possibility of correlated data; linearity formats the problem as a base change; several research areas seek to extend these views to nonlinear regime. These considerations also encompass the belief that data have high SNR (as the signals indicate, it can be changed by a noise).

7.5.1.1 Rules of technical behavior, environment, and individuals

After analyzing the partial conclusions in 7.3.2.3, a series of behavior rules can be listed that can be used in the operational routine and in the manager's routine. These rules are suggested and after a period of coexistence with the production classes and the application of techniques for identifying the sociotechnical typology, adjustments or corrections are made for the application of the new rules in the management tool, seeking improvements in the performance of shift groups.

Proposal for rules of conduct
With the identification of signs of latent failure, with its reordering and its statistical validation, the aim is to build situations away from riskier scenarios mapped by Department of Health, Safety and Environment in addition to operation of the company. Signals are detected by the MEA, based on technology and task.

Type of technical, social and human behavior

The intention of identifying a social type is to verify how the environments (organizational and socioeconomic) that influence, or that are part of organizations, cause instabilities in sociotechnical behavior. After identifying the technical behavior based on characteristics of technology and the accomplishment of tasks, the OD is read to map the abnormalities, discovering the codes between multidimensional factors, through the causal and organic nexus. We consider as organic nexus, a term discussed earlier, the time when the individual's imaginary is involved with intuitive and emotional aspects, hindering the logic of the simple causal nexus and, therefore, the chronological and physical rule in the investigation loses value.

Thus in the periods of difficulty in plant operation, we must identify the organizational and social changes in this macro period and what are the critical parameters to be analyzed statistically and qualitatively. This work aims to identify types of behavior of technology and possible relationships with organizational and social rules.

Each operational factor or complete event can be identified as technical, but it always involves human and social factors. The factors that make up the chain of abnormalities are statistically analyzed, being suited for the maintenance of specialties and potential activities for environmental incidents or accidents with absence.

The characterization of the type of technical behavior is completed with the statistical interpretation of process variables and intermittent liquefied gas transfer and storage operations. Remembering that this logistics activity involves processes of refrigeration, compression, drying and incineration of light gases, contaminants of liquefied gases.

Analysis of operational failure in complex systems

Gas transfer operations are simple and depend on the availability of lines, vessels, tanks and people to perform operations. The safety risks are due to the handling of flammable liquid and gas in a place of human presence and active politically in the neighboring public. The complexity involved in the logistics systems for the transfer of flammable gases lies in the various possibilities of alignments and the unguaranteed quality of liquefied gases then causing instabilities in processes and operations. Another difficulty is the lack of knowledge of the chains of abnormalities with their various signs, making it impossible to analyze critical scenarios for fire and explosion and that can affect the public.

The presence of contaminants (solids, moisture, and organic compounds with sulfur) in the gas circuit and improper maintenance can cause many interventions in the motorized valves, leading to critical operating conditions. It is worth emphasizing the importance of the political aspects about the operational failure, among these, the wage campaign of the union, the possible social benefits of company's presence and relationship with employees in the work routine. Operational failure analysis is done using the tools discussed here, failure logic, chronology, task sequence analysis, connectivity, and the materialization of the operating failure energy carrying body.

Hybrid process and PCA: failure energy

The choice of parameters related to the main process constraints allows using up to three characteristic groups that describe the failure energy. Initially we designed the following parameters cited in Table 7.22, in Group 1, gas pressure (Pg) and presence of contaminants that increases the control volume of the investigation (% contaminants) also, the number of nonconformities with compressors (Nc). In Group 2, events with filters and coalescer (Nf), number of nonconformities with motorized valve, with control valve and with solenoid valve (Nm), average occupancy with safety actions in the installations (Hs). Moreover, in Group 3, gas processing load in the plant (Ca) and average time with critical operations (t).

This measurement is made considering different profiles in the shift class, considering level of applied competence (knowledge, skill, and commitment); dispersion level, leading to loss of attention and slippage (low, medium); possibility of not following rules.

Each group develops a result in the PCA chart that, according to environmental influence (union relations and organizational pressures), seeks to correct organizational efficiency goals in relation to level of stress in the work environment. If this correction is not satisfactory in relation to usefulness of company in society, it seeks to correct human performance factors.

Environmental influence

Identifying whether the environment is stressful depends on the following aspects (Table 7.2): time of year regarding the wage campaign (Tcs), percentage (%) in double shifts in the period (%d) and availability of products for processing (% Load). The more stressful the environment, the more the resulting PCA moves away from the high-performance goal in the routine. Idle environments also have the effect of moving away from the (high-performance) goal in the routine.

Heuristics in cognitive processing (Table 7.2)

The speed, sharpness and cooperation of the team in decision-making depends on the human or group type present. Thus two types of behavior are analyzed, in terms of the speed with which the failure can emit signals, intensifying the possibility of reaching its maximum value. The team (1) that has good visibility, allowing the construction of the causal link, and where inadequate characteristics are covered by qualities in the shift. The team (2) that does not have good visibility, thus promoting failure situation, and people have porous memory characteristic, leading to situations in which the group cannot reduce the possibility of failure of individuals.

If team 1 is working, the result is of satisfactory effectiveness in the construction of mind maps and schemes, leading to the planning of the satisfactory task. If team 2 is acting, the resulting one is the unsatisfactory effectiveness in the planning of the task, then contributing to the increase of the failure energy and the increase in the possibility of the larger event to happen, that is, the failure exceeds the function reference value.

Decision-making and critical task

Planning the task means verifying the realization environment, rescuing information and tools necessary for its execution, verifying requirements, assigning functions, deciding the most appropriate way among the alternatives of performance and enable its execution. This planning needs to be transformed as a production program published for group validation.

Another level of task analysis is to verify what the worker's behaviors are at the time of decision-making. In the case of gas processing, the groups of operators are divided according to the causal link resulting for decision-making. Group A has a broad view of the alternatives, knowing their consequences and promoting agility and assertiveness around the decision—effectiveness of this group depends on successful identification of the causal link previously evaluated. Group B: has a restricted view of the alternatives, because it has low long-term memory rescuing (knowledge base) and reduced access to short-term memory (work), leading to faltering and incorrect decisions, or mixed decisions, due to compensation in the teams, where assertive and correct decisions also occur. In this Group B, there is high possibility of construction or increase of operational failure.

After identifying the basic types of behavior in the work environment, it is possible to suggest individual and group actions that indicate the human response to the execution of the task. In the operator and process behavior, parameters resulting from the actions in the task are chosen, which make up the PCA. This calculation, when it indicates a tendency to uncontrol, in that the points are in the opposite direction to the goal, leads to the transfer of records to the memory of the organization as indicated in the Algorithm described in Fig. 7.16. These memories return to the operational routine, influencing the behavior of process. Past memory influences the resulting control (PCA) in production systems.

Fix for applying behavior rules.

Variables of environmental influence were suggested in the Heuristics mentioned in 7.2.3 in PA8.1 (analysis procedure). There was a lack of data to infer about the classes in the Heuristics that initially dealt with the following subjects: time of year regarding the wage campaign; double shifts (% d), adopted in the research; and availability of products for processing, variable reflected indirectly through processed volume that participated in the objective function.

In cognitive processing, as indicated in Table 7.2, behaviors chosen to measure the speed, sharpness, and cooperation of the team in decision-making were maintained. Team 1 has good visibility, allowing construction of causal link, and which inadequate characteristics are covered by qualities in the shift. Team 2 does not have good visibility, promoting situations of failure, and people have a porous memory characteristic, leading to situations in which the group cannot reduce the possibility of failure of individuals.

The result of this cognitive processing was maintained. While Team 1 is working, there is satisfactory efficacy in the construction of maps and mental schemes, leading to the planning of the satisfactory task. If Team 2 is acting, the result then is the unsatisfactory efficacy in the planning of the task, contributing to the increase of the failure energy and the increase in the possibility of the larger event, that is, the failure exceeds the reference value of the objective function. It was not possible to confirm aspects about decision-making, being evaluated later for the construction of the failure prediction tool.

7.5.1.2 Criteria and selection for measuring organizational efficiency

Based on the stated objective of the company focus of the study, on impressions regarding its role in society, in interviews with staff of the logistics function with liquefied gases, and on the availability of data positioning indicators were chosen regarding organizational efficiency and that are related to gross profit and image of the company.

Among the objectives declared by the company, there is the respect for the demands of society in general and specifically to local communities. Thus the company is committed to maintaining communications and adding value with its presence in the local public. Aspects related to safety implicitly correspond to the satisfaction of the public by the presence of a safe industrial unit and that eventually has management tools for predicting risk situations in their work routine. Thus signs that indicate abnormalities should propose preventive actions that avoid risk situations.

The company under study, which is inserted in globalized situations, internally promotes the discussion regarding the paradigms of environmental sustainability, thus seeking new technical and social standards. Productivity becomes primordial, regardless of the origin of capital. Thus it should adopt indexes that indicate team productivity, as well as the adequacy of tools and work techniques.

After analyzing the arguments presented, organizational efficiency is measured from:

1. The volume processed, which indicates compliance with production schedule.
2. The technical idleness or occupancy rate of the operator, which indicates the commitment and the proper application of this commitment in the task.
3. In safety processes and operations, using important signals already identified in the techniques of qualitative and quantitative analyses of abnormalities, as number of deviances or warnings of failure, or signs that indicate presence of fault.

In addition to the listed items (volume, technical idleness, and process safety), it would be economically important to include cost of production through electricity consumption and liquefied gases quality in relation to light gas; however, at this time, unavailability of information does not allow inclusion of these items.

7.5.1.3 Analysis and choice of principal component analysis groupings: partial conclusions

The partial conclusions of each analysis tool and procedures used led to observations on the application of the PCA methodology, thus it was necessary to group according to discussion below.

Abnormalities (conclusion PCA1/16).

The presence of solids in liquefied gas feed circuit affects valves, pumps, and vessel. High pressure can be caused by low operability of control valves and motorized valves. The problem of improper compressed air may be related to the presence of excess or moisture fluctuations and contamination by sulfurous compounds. It is necessary to detail and test abnormalities that potentialize failures.

Process (conclusion PCA2/16).

The indication of pressure before (average) and after, in refrigeration (sum of standard deviation), can indicate trends in the quality and safety of the process. The maximum discharge pressure monitoring of pressurized liquefied gases pumps is discussed, but the sum of the standard deviation is more significant. For knowledge about the failure, it is important to analyze the sum of the standard deviations of the amperage of the compressors; the sum of the amperage maximums in this compressor; and the sum of the lubricating oil pressure of the tank pressure control.

Logistics (conclusion PCA3/16).

Logistics operations are tied to delivery priorities for processed material, to meet liquefied gases availability in supply and ship on the pier. The activities are carried out discreetly and do not bring much numerical information. This activity has discrete decisions about the action to be performed.

Operations (conclusion PCA4/16).

The percentage of double shift working hours may be an important factor in indicating propensity to error, the low volume processed, and the increase in signs of abnormalities. The sum of signs of abnormality brings important meanings for the analysis, and serves as an indicator of organizational efficiency, as well as is related to the processed volume of liquefied gases (care must be taken due to the quality of this information in the occurrence book). The number of hours per operation and per operator indicates the complexity of the task and commitment to its performance and can be an indicator of motivation and competence applied in the accomplishment of the task. It should be considered both reinforcement in the shift for plant operability, as well as the presence of redundant operator in the emission of Work Permits.

Maintenance (conclusion PCA5/16).

It is important, as already commented in the abnormalities, to include events that gather three blocks and that appear here in maintenance as well: events involving valves (22% of citations); events involving filters, coalescer (11%); and pressurized and vessel pumps, which may be related to the possibility of high pressure.

Process safety (conclusion PCA6/16).

This item reaffirmed the importance of analyzing high pressure in the system. Events with fire and interlock system can be indicators that lead to process safety failure in prediction. In the Safety and Environment records, attention is drawn to events with short circuit in electrical systems and events related to communication for process control.

Task analysis (conclusion PCA7/16).

The techniques identified in this stage of the project will help in the choice of preventive actions to improve position of the artificial variable resulting from groupings of variables, about the task. Cognitive effort (respect the limits in the team's offer), dominance in safety to project barriers in the steps of task, parallelism of steps thinking about the efficiency of communications, and complexity of the task. In addition to the need for changes in the operator profile to finalize field activities within a certain flexibility due to the availability of control devices in the operational routine (good practices and competencies in failure processes).

Task failure analysis (conclusion PCA8/16).

The failure analysis of the task has the main utility of preparing the training material for formation of procedural skills on the failure. In the choice of the area of furnaces, attention was drawn to the risks involved, although there is no information that meets the need of this instrument, since probability of such an event is low and is not a latent sociotechnical problem. Apparently, it's a more technological and localized issue. Following new process safety standards, although it's a rare event, it may happen.

Emergency behavior (conclusion PCA9/16).

The same previous conclusion is useful for current situation, in which the main technical objective of this analysis is learning. In the dynamics, there is a still "warm" environment in terms of stress, drawing attention to aspects in commitment and competence formation that are in the corrections to be made to frame the artificial variable within the best organizational efficiency.

Emotional balance, sociability (conclusion PCA10/16).

The comments linked to the conclusions from 10 to 14 do not refer to data to make up the rules of human and social behavior, but rather they are intended for the preparation of programs that will correct aspects of the task, thus reviewing the results measured and released in the PCA. There was no consensus in the answers of the social data presented directly in relation to the questionnaires that deal with the subject indirectly. Despite the differences, there is a need to increase the organizational and social integrations promoted by the company. Current Programs needs to be revised.

Emotional balance, health (conclusion PCA11/16).

There are indications of need to adjust the profile regarding emotional balance. The balanced worker has as a result the correct application of cognitive faculties, are attentive, have good and available memory, and has satisfactory mind map construction, good level of communication and good spatial location. Thus in the PCA, the relocation of personnel and training in specific skills and abilities can be performance correction instruments. The level of stress that can change behavior in the routine may be related to the percentage of double shifts at work.

Group (conclusion PCA12/16).

In the measurement performed, it is noted that the mean value is very close to the desired pattern, but isolated cases lead to detecting noise in the formation of social bonds and after the comparison between individual and organizational values, it is recommended to perform work to increase the bond of the individual with the organization. This conclusion is added to previous ones and confirmed when movements indicated by disaggregating signals are greater than the team aggregators.

Leadership (conclusion PCA13/16).

There is conflicting information in this item, but it shows working potential for matching leadership profiles. The need for inclusion is confirmed, as well as that of team aggregation, but the environment is conducive to the necessary changes. Leaders need to be adjusted to bring greater organizational efficiency to the company: increased volume processed safely and with minimal technical idleness. Leadership adjustments are used to correct artificial variables after changes in leadership, thus adjusting organizational efficiency.

Commitment (conclusion PCA14/16).

The data and information analyzed (distance, progression, dependents etc.) suggest the formation of the team with differentiated commitments and characteristics. The main conclusion is the need for "positive" homogenization considered as corrective action for the adjustment of organizational efficiency by adjusting artificial measurement in the PCA.

Competence (conclusion PCA15/16).

Depending on the grouping that needs to be corrected, one can seek to adapt the competency installed based on the identifications already performed. The identification of these situations activates the need for training on failure processes, bringing as an indirect consequence the greater integration of the team (this training uses the leaders themselves as multipliers of knowledge).

Clusters or clustering study (conclusion PCA16/16).

From the analysis of the figures that visually indicate seasonal movements from the signs collected from the shift register, the data chosen for the analysis organization. This feeding in the plant including the following equipment: filters, pumps and vessels. It also includes alignment operations that include motorized valves for high-pressure systems (HV valve with pressure).

7.5.1.4 Choice of input data and organizational efficiency for application

Discussion of input data in the PCA

The choice of discrete variables for the calculation of PCA was based on the criteria of: (1) performance in the graphs of number of abnormalities per maintenance item, equipment, and processes; (2) interpretations of the chains of abnormalities and suggested control items; and (3) grouping of relational events or repetition in discrete diagram by process and equipment. The discrete and continuous data chosen for data entry in PCA are described in Table 7.22.

The Components classified to perform the PCA; each one is composed of two variables. Component 1, C1, valve faults, HV and events with pressure. Component 2, C2, indication of mean pressure at the entrance, standard deviation at the plant outlet, and sum of the standard deviation in the discharge of the pressurized pump. Component 2, C2, sum of the standard deviations of the amperage and sum of the maximum sums of the oil pressure, tank pressure control compressor. Component 3, C3, % of double shifts, sum of signs of abnormalities, number of hours per operation and per operator. Component 4, C4, events with filters, pumps and vessel. Some experimental tests were performed for the final choice of the analyzed components.

Discussion of output data in the PCA: organizational efficiency measurement

Based on the availability of information, we had chosen for calculation of organizational efficiency the parameters presented in Table 7.23. Volume of gas processing, indicates billing, equipment reliability and process availability. Technical idleness of the team that is related to the application of safety barriers in the routine and to the execution of tasks according to planning. In addition, number of signs of abnormalities that is related to safety standards, commitment to the organization, installed skills and level of cooperation of the team.

Depending on the organization performance and the variables chosen, organizational efficiency can be in the form of "clouds—3D" or "ranges of level curves—2D." Organizational efficiency can be identified in a graph of two (x, y) or three (x, y, z) dimensions, can be identified by item type to be measured in isolation or can be a percentage weighted sum of the items (chosen here), as indicated in Table 7.24.

The established weight was defined by the safety criterion as Priority and then Production. In the processed volume, the established weight follows its data quality, where it has 40% of the total in relation to the target volume. In technical idleness, the established value is 30% of the total. Based on the signs of abnormality, the other 30% are considered.

TABLE 7.22 Data entry for principal component analysis (PCA).

Task	Case 2	Case 3	Gases
Task review	16 task, 2 laboratory and 1 process	6 operation, 5 process, 3 managerial	Drying/Regeneration
Task project	Not applicable	No applicable	All systematized tasks, focus on drying and regeneration
Failure analysis in the chains of abnormalities	12	23	24 chains of abnormalities, 8 with accident failure analysis
Risk Analysis	Not applicable	Not applicable	Adapted LOPA technique (layers) with Human Factor and FTA

TABLE 7.23 Analysis of measurements for organizational efficiency in principal component analysis (PCA).

Data Type		Measured variable	Relationship with	Availability for PCA use	Keyword	PCA Test
Disc	Cont					
		Processed volume	Production, quantity produced	Available	Billing	E1
		Technical idleness	Productivity	Available	Cost	E2
		Number of signs of abnormality	Process safety	Available	Image	E3
		Electricity consumption	Production cost	Unavailable	Cost	

TABLE 7.24 Criteria and weights for calculating organizational efficiency.

Measured item	Result	Track	Correction	Weight
Processed volume	Class 1	0 to 3000	60%	40%
	Class 2	3000 to 5000	90%	
	Class 3	Over 5000	120%	
Technical idleness	Class 1	Over 20%	60%	20%
	Class 2	10% to 20%	90%	
	Class 3	Up to 10%	120%	
Signs of abnormality (on the day)	Class 1	Over 5	60%	40%
	Class 2	3 to 5	90%	
	Class 3	0 to 2	120%	

Y = (volume/technical idleness/abnormalities); Y = f (X1, X2, X3).

Discussion around programs for environmental adjustments and social systems

The activities in Table 7.25 discussed in this investigation can be applied to adjust human factors for increasing operational and human reliability.

7.5.1.5 Data processing, results

Data processing activities involve the use of software *(Excel* and *Minitab)* and the handling of data generated from technical, human and social typology. *Excel* is used in the organization and processing of data before being processed in *Minitab*, right after using *Minitab* for multivariate analysis or PCA, and finally, with the release of the artificial variables generated, the scatter plot is reached to form the organizational efficiency curves and facilitate the interpretation of the measurement for prediction of operational failure.

With the analysis of the failure energy, adjustment of the technical parameters of the task and human and social behaviors are started using the rules already assembled in the form *of fuzzy mechanics*, also resulting from the partial conclusions presented in 7.2.3. The script for data processing follows instructions 1 through 5 and is discussed in Fig. 7.58 through Fig. 7.61.

TABLE 7.25 Technical partner adjustments for production scheduling.

Type of information	Action program	Relationship with	Availability for PCA use	Keyword	PCA Test
Socio-technical adjustment	Database preparation and debugging for input (variables) and output (yields)	Routine management	In formulation	Staff	
Socio-technical adjustment	Specific training on failure processes	Increased critical sense and reflection on productivity	In formulation	Staff	AST1
Socio-technical adjustment	Training and work for conceptual base review	Interpersonal relationship and negotiation for leaders	In formulation	People	
Socio-technical adjustment	Training in specific skills and abilities	Routine management	In formulation	Staff	AST2
Socio-technical adjustment	Organizational integration campaigns	Increased emotional balance and commitment	In formulation	People	AST3
Socio-technical adjustment	Social integration campaigns	Increased emotional balance and psychological contract	In formulation	People	
Socio-technical adjustment	Leadership development for team aggregation	Routine management	In formulation	People	AST4
Socio-technical adjustment	Revise of standards in technical and social systems	Behavior and transparency of objectives	In formulation	CEO	AST5
Socio-technical adjustment	Revise of organizational policies	Commitment analysis and psychological contract	In formulation	CEO	

1. Build an *Excel* table with the primary data to be used in the PCA as indicated in Fig. 7.58, differentiating entries and measurements of organizational efficiency with the respective criteria for classification.
2. Build an *Excel* table with selected and preprocessed data for PCA analysis as shown in Fig. 7.59.
3. Post the data to Minitab and rotate the data as shown in Fig. 7.60.
4. Finally, structure the responses generated by Minitab in the PCA part *(Eigen values)* as shown in Fig. 7.61—PCA Results, and record the *Eigens* numbers resulting from the PCA, preparing the correction to launch in the chart.
5. Prepare an *Excel* chart and post to level curves or data clouds as shown in Fig. 7.61 for discussion of the results.

			Components: C1A, C1B, C2A											
	Day		Presssure C1A					C1B Pressure			Quality C2A Current ampere		FLASH	Compressor flash
			No of Abnormalities in valves and pressure						Refrigerat input	Input Refri Output Refri Output	A B C			Current Ampere
No	Dat	Week day	control Number	pressure Number	motorized Number	solenoid Number	Sum	Pressure Average	Stand Devi	Stand Devi Sum SD	Stand Dev Stand Dev Stand Dev			Sum SD
1	1	Sunday	2	3	1	0	6	9.55	1.95	1,54 3.49	17.74 0 16.84			34,58

Components: C2A, C3, C4											
C2B Oil Lub Comp				Compromisso e Competência C3 Productivity		Quality and ICT Tec Oper Culture C4 Number of Abnormalities					
A max	B max	C max	Sum max	% double time	No hours per operation	Filter+coal+cor Number	Pumps pres Number	Sphere Number	Sum		GLOBAL ORGANIZATIONAL EFFICIENCY
5.53	0	4.53	10.06	0.0	1.325	3	0	1	4		
WEIGHT 0.4				WEIGHT 0.2		WEIGHT 0.4			1		
E1 Volume processado m3/h	1 - 0 -3000 2 - 3000 - 5000 3 -> 5000 Class			0,6 E2 0,9 IDLENESS 1,2 TECHNICAL %	1 - until10% 2 - 10 - 20% 3 -> 20% Class	1,2 E3 0.9 Warnings or Sins 0.6 abnormalities Number	1 - 0 - 2 02/03/2005 3 -> 6 Class		1.2 0.9 0.6 Em % Value		Global Efficiency
	5500		3	1,2	12.815	2	0.9	12	3	0.6	90,00%

Row labels (leftmost column of lower block): WEIGHT, OBJECTIVE FUNCTION

FIGURE 7.58 Primary data for the principal component analysis. *From Ávila, S.F., 2010. Etiology of Operational Abnormalities in the Industry: Modeling for Learning (Thesis—Doctor's degree in Chemical and Biochemical Process Technology). Federal University of Rio de Janeiro, School of Chemistry, Rio de Janeiro, 296 p.*

As part of the data has subjective basis and the collection was made in incident reports, there may be errors that reflect on the quality of the data. Thus the exercise performed to reach dispersion plot involving artificial variables of the PCA needs future corrections.

Better detailing the steps

Step 1. Release of primary data

Fig. 7.59 presents data related to pressure (C1), quality (C2) and commitment to the application of competencies (C3) in the shift routine. Other data were released in this table, but were not used in the PCA exercise, as is the case of the C4 grouping related to events in the vessel, in the feed pump and in equipment for the removal of water and solids from liquefied gases.

Pressure-related data can be an indirect consequence of valve failures, work with high pressure being named C1A, and numerically the sum of these four data is made. They are included in valves, events with motorized, control and solenoid valves. Pressure-related data can be linked to the continuous indication of pressure or variation of the opening of the control valve, being C1B. In the present case, the mean pressure at the refrigeration inlet and the sum of the standard deviations at the input and output are considered.

When discussing quality, the performance of liquefied gases in compression is mainly worked regarding the behavior resulting from the levels of gases. The most representative data regarding the signs of failures caused by quality, with fluctuation and measurement of trends, are in the current (C2A) and particularly in the sum of the standard deviations of the three tank pressure control compressors. Another important indicator is the maximum pressure of the lubricating oil and in particular the sum of these pressures (C2B).

Although released in the table in Fig. 7.58, the C3 data referring to productivity used in the data entry for the PCA ended up not being used in the generation of the dispersion plot in Fig. 7.60. The C4 data referring to quality regarding solids and humidity were not used even to generate artificial variables.

The indexes that represent organizational efficiency are presented in Fig. 7.58. The volume in m3/day is rated at three levels and weight of 40%; the technical idleness also divided into three classes with above 20% in the worst situation and weight also of 20%, while the number of signs of abnormality with weight of 40% has, in the worst situation, above 6 abnormal signs or events.

Step 2. Data processed for PCA.

Fig. 7.59 shows the data separated and processed for direct application in the PCA and its overall organizational efficiency as a percentage. Thus in the first grouping (C1) representing the pressure, there are 3 data to be processed: sum of signals with pressure and valves; average pressure at the refrigerated inlet; and sum of the standard deviations of coolant inlet and outlet. In the second grouping (C2) are the continuous data of: sum of the standard deviations of the tank pressure control current compressor and the sum of the maximum lubrication pressure also of these compressors. The third grouping, although organized, was not used for PCA that are data related to quality and technical culture, thus the percentage of shift work and hours of work per operation and per operator.

Day week	Pressure C1A Valves Pressure Sum	C1B Refrigeration Pressure Input Refri Average	Quality C2A AmpereFLASH Input+Output A/B/C Sum Stand Dev	C2B Oil Press Flash A/B/C Sum Stand Dev	Commitment and Competence C3 Productivity % Sum MAX	Double time	Quality and Tec Oper Culture C3 hs operation operato Sum	ORGANIZATIONAL EFFICIENCY SPECIFIC M3/H	%Ildness	No abnorm	GLOBAL
1 Sunday	6	9.55	3.49	34.58	10,06	0	1,325	5500	12.815	12	90.0%
2 Monday	2	10.77	3.53	10.72	9.53	16.66667	1.001538	3000	20.62667	5	84.0%
3 Tuesday	1	9.76	2.42	19.68	9.44	6.666667	1.179487	10000	19,06333	8	90.0%
4 Wednesday	2	9.4	4.17	56.35	9.69	0	1,277273	7300	3,956667	5	108.0%
5 Thursday	1	9,63	5.14	17,49	7.52	8.333333	1.3	3600	13,75	5	90.0%
6 Friday	6	8.63	4,89	54,19	10	23,33333	1.015385	3570	20,41667	8	72.0%
7 Saturday	1	13.99	1.46	60.81	9.72	23.33333	1.170629	3600	8.396667	1	108.0%

FIGURE 7.59 Data for principal component analysis application. *From Ávila, S.F., 2010. Etiology of Operational Abnormalities in the Industry: Modeling for Learning (Thesis—Doctor's degree in Chemical and Biochemical Process Technology). Federal University of Rio de Janeiro, School of Chemistry, Rio de Janeiro, 296 p.*

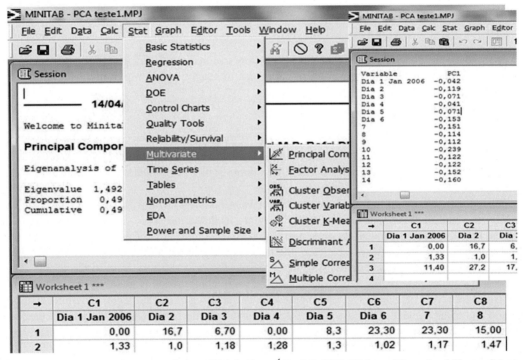

FIGURE 7.60 Running the principal component analysis on Minitab. *From Ávila, S.F., 2010. Etiology of Operational Abnormalities in the Industry: Modeling for Learning (Thesis—Doctor's degree in Chemical and Biochemical Process Technology). Federal University of Rio de Janeiro, School of Chemistry, Rio de Janeiro, 296 p.*

Step 3. Running PCA on Minitab

At this point, as indicated in Fig. 7.60, the data is entered transversally inverted, date on the title row and the variables will be put in the columns. By running a statistical mode, multivariate analysis, PCA, marking covariance option and requesting generation of only representative value generated, a low is reached. It is noted that most of the values generated are negative and small.

Step 4. Adjust the data for posting to scatter diagram.

As shown in Fig. 7.61, the data are organized, in which C1, an artificial variable that represents pressure, will position itself on the X axis; and which C2, an artificial variable that represents quality, will be on the Y axis. Artificial values are low and negative, and correction must be performed by multiplying by (−) 100. Overall efficiency is posted, and the sequence is also for identification at each point on the scatter plot.

	Day	1	2	3	4	5
C1 Pressure	Valv+P	6	2	1	2	1
	Aver input	9.55	10.77	9.76	9.4	9.63
mult * -100	SDEV INP+OUT	3.49	3.53	2.42	4.17	5.14
-100	Eigen	-0.085	-0.18	-0.18	-0.148	-0,166
X	corrector	8.5	18	18	14,8	16.6
Quality						
C2 Comp	amp sum STAND D	34.58	10.72	19.68	56.35	17.49
	lub sum mx	10.06	9,53	9.44	9.69	7.52
	Eigen	-0.053	-0.003	-0.022	-0.101	-0.022
Y	corrector	5.3	0,3	2.2	10.1	2.2
Commitm	% double	0	16.66667	6.666667	0	8.333333
Tech Ope C	hours p operat/opt	1,325	1.001538	1.179487	1.277273	1.3
C3	No abnormalid	11.385	27.19821	17.28615	10.96727	17.15333
	Eigen	-0.042	-0.119	-0.071	-0.041	-0.071
	corrector	4,2	11,9	7,1	4,1	7,1
Global Efficiency		90.0%	84.0%	90.0%	108,0%	90.0%

FIGURE 7.61 Principal component analysis results. *From Ávila, S.F., 2010. Etiology of Operational Abnormalities in the Industry: Modeling for Learning (Thesis—Doctor's degree in Chemical and Biochemical Process Technology). Federal University of Rio de Janeiro, School of Chemistry, Rio de Janeiro, 296 p.*

7.5.1.6 Mapping organizational efficiency and simulating situation: failure energy

As shown in Fig. 7.62, with the preparation of the dispersion plot, organizational efficiencies are spread following the 60 days of liquefied gases production. In more detail, organizational efficiencies are classified into three main ranges: at red lines, low yield ranges (70% to 84%); at the green circle, the average income range (100%); and at the yellow circle in the center, possibility of achieving 120% organizational efficiency.

It is noted that several points are outside the ranges defined by curves or circles leading to the belief of the need to improve the quality of data or data processing. There are many ways to improve the quality of results: increase the quality of the data collected; choose other data already related to test in the analysis that was done in two dimensions; perform the analysis with three variables; and analyze the results considering other compositions of organizational efficiency.

As shown in Fig. 7.62, in the center of mass of the yellow circle, is marked as the point of maximum efficiency and toward the highest density of the points, straight to the red curve that represents the worst organizational efficiency region. This height connects the point of maximum efficiency (maximum volume, no abnormalities and low technical idleness) to the minimum that means uncomfortable situation or almost major accident.

The maximum height, in the current exercise, is shown in Fig. 7.62 on the black double arrow as 12.5. When measuring the process, already with the mapping of the organizational efficiencies launched, for the exercise effect it is noted that point 25 with 96% organizational efficiency has traces of failure and could be closer to the range of 120% organizational efficiency.

After measuring the height of this point, we trace a perpendicular straight line in the center of mass of the maximum h point and taking the purple line to point 25. Noted that the failure is 40% of vectorial distance to achieve the almost accident (as indicated in the cylinder of the fault), we can consider 60% percent of sociotechnical reliability, leading the organization to invest more intensely in the investigation to find a solution for the problems.

For testing procedure and knowing that this is a method for analysis of sociotechnical systems, it was considered that there was success in the demonstration. The mapping of sociotechnical behavior after the adoption of simulation data for production planning allows locating future performance in regions of organizational efficiency. Therefore it is possible to change the production schedule.

7.5.1.7 Final observations for the application of the principal component analysis

In the application of the scatter plot after principal component analysis, it is necessary to correct the quality of the data to decrease the variability between the points. There is a lot of data that is not part of organizational efficiency curves indicating the need for adjustments.

The direction of the data for maximum height of the failure cylinder is defined by the center of mass of the highest efficiency and the direction of the highest data density. The measurement is made through a parallel of this line that

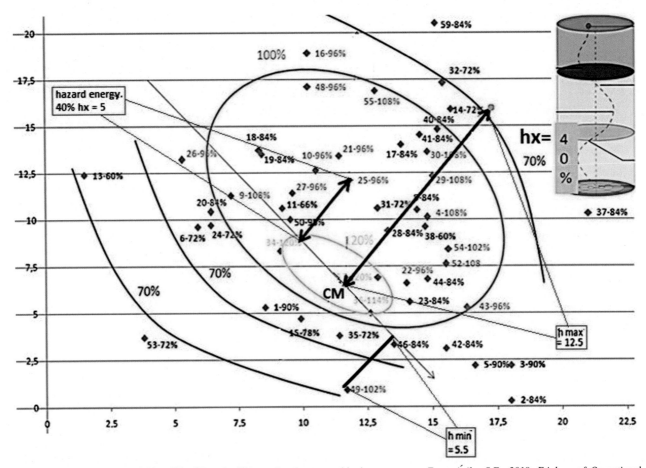

FIGURE 7.62 Artificial variables (C1, C2) and efficiency level curves with time sequence. *From Ávila, S.F., 2010. Etiology of Operational Abnormalities in the Industry: Modeling for Learning (Thesis—Doctor's degree in Chemical and Biochemical Process Technology). Federal University of Rio de Janeiro, School of Chemistry, Rio de Janeiro, 296 p.*

represents the maximum height until it finds the measured point. Thus it becomes possible to estimate the failure energy in the learning model.

7.6 Future work: task cross-assessment based on particle swarm model

Literature review is adapted in this book to reflect natural reality that is multidisciplinary, much discussed but not highly practiced. It is not possible to verify progress of sciences in the composition of methods that do not require cross-sectional limits or fixed contents. The convincing of entrepreneurs and managers about the importance of applying this bottom-up method opens the possibilities of validating programs in the chemical, energy and maybe i in the areas of transportation and public health.

As commented by ABIQUIM in meetings about this methodology, most of the concepts, techniques and procedures developed and applied here are useful for defining the project criterion, also about investment in instrumented safety systems that is influenced by human factors. Thus it is in the interests of project companies and industries in general to develop these important parameters for decision-making.

The procedures for harmonization of the production team are dispersed in social and technical applications. It is intended in future research to gather this list of information and procedures to facilitate decision-making at the strategic, managerial, operational and emergency levels.

Ávila and Dias (2017) discussed the theoretical model of bee swarm at ESREL 2017. An interpretation of the factors that affect worker productivity and performance in the task is done using a map that compares a hive and a routine operational activity in the industry.

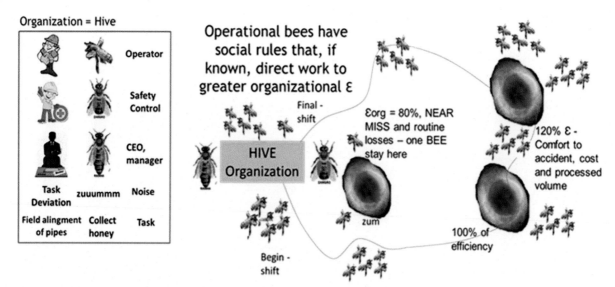

FIGURE 7.63 Task at the hive workplace and honey collection (Ávila and Dionizio, 2017). *From Ávila, S.F., Dionizio, J., 2017. Systemic fault analysis to calculate the approximation of the top event (near miss evaluation system): NEMESYS. Safety and Reliability. Theory and Applications, ESREL, Portoroz, Slovenia, June 18–20, 2017.*

The work of bees in nature is like the work of industry operators. Both have schedules to perform activity cycles and require control for a task. The collection, transportation and storage of honey requires a preestablished route with environmental risks or task scheduling that can reduce the efficiency of the organization, as occurs in industrial activities. This is group work directed by senior management (queen bee) and supervised by areas of EHS as indicated in Fig. 7.63.

It is important to consider that a company established only through policies (top down) is not always successful. When an organization is inserted into the culture of guilt, the lack of feedback to correct activities can cause loss of efficiency. In this reasoning, a safety team assumes the role of supervision in industrial activities, following the rules of the culture of guilt, consequently, with omissions and difficulties in communicating occurrences. Contrary to what usually happens, both senior management and safety must adopt a learning posture to build a just culture.

This is how we elaborated from the case of liquefied gases an interpretation of what would be the swarm of particles (operators) in carrying out their task of cooling and transfer of flammable gas. Fig. 7.64 describes the movements with approximations and distances from regions with better organizational efficiency. Direction of particle displacement, in the collective, depends on the sequence of points, density of the data and the approximation to accident regions, indicates the existence of noise in the task or change of environmental pressures. There are points that do not fit the displacement indicating that there are different collectivity rules or different processes, thus requiring debugging data and techniques. This is a future work that in conjunction with human factor analysis will join the current work to improve critical operations.

7.6.1 The path of workers

Each point in the scatter plot in Fig. 7.64 has an organizational efficiency that represents the results of that week or that day. After the first day, comes the second day, and finally reaching the 60th working day, which is our case of application. The points on chart indicate the following movements:

(Region A, 80%) Beginning of the study period positioned close to the red zone of 60%–80% of organizational efficiency, but closer to the green zone, therefore, 80%.
(Region ABC Hypothesis): Due to organizational guidelines and better product quality and resources, the group performed its tasks better.
(Region B, 90%): The group of operators went toward maximum efficiency (yellow zone) but due to production factors, passed on the external border to green region, we will call 90%, and with greater efforts to enter and exceed 100% organizational efficiency.

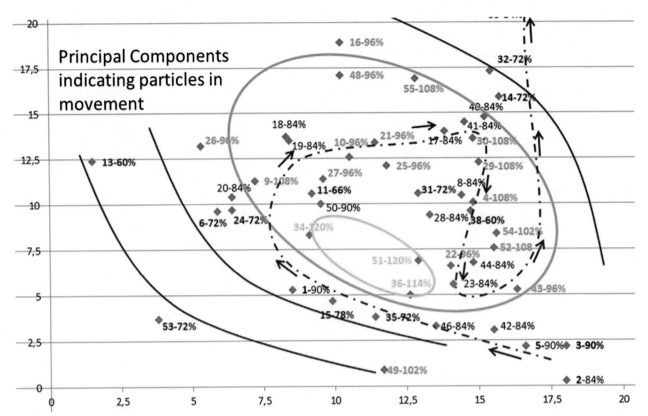

FIGURE 7.64 Swarm of particles or operator bees swarm—LIQUEFIED GASES site. *From Ávila, S.F., 2010. Etiology of Operational Abnormalities in the Industry: Modeling for Learning (Thesis—Doctor's degree in Chemical and Biochemical Process Technology). Federal University of Rio de Janeiro, School of Chemistry, Rio de Janeiro, 296 p.*

(Region C, 105%): The group enter inside 100% region fleeing beside target region of 120%.
(Region CDE hypothesis): Excess bending, high stress, workload, ship arrival and manager complaints drove group away from optimal point.
(Region D, 103%): The group passed this region.
(Region E, 100%): The group passed this region.
(FGH hypothesis): the reinforcement of the teams to avoid the loss of control improved the result to 100%, but after 10 days with the return of normality and the return of stress and corrosion problems coming along with the supplier's liquefied gases drove result to 50% and plant stop.
(Region F, 105%): The group passed this region.
(Region G, 99%): The group passed this region.
(Region H, 50%): The worst are of organizational efficiency that the group of workers passed on; we need urgent correction actions and need to understand the root cause to take important decisions.

7.6.2 The bridge to the future

The great challenge of this work is to perform an exercise of sociotechnical reliability based on the signs of the operational routine and the variables that represent the technical and social processes. Somehow, these variables are interconnected through phenomena that are not yet under discussion based on the concepts of engineering or psychology. The interconnection of discourse, emotional state, and regional rules with technical control actions is intended to achieve the goal of product production, within a technical–operational culture that meets the requirements of cost, revenue, and image.

In the midst of this discussion, we realized the importance of how to mine social and technical data for use in predicting future failures. We created new concepts from subjective relationship for performing a successful exercise in the liquefied petroleum plant.

The concepts, tools, and procedures created are discussed and exercised in Chapters 1–6. In this chapter, we studied the case of the liquefied gases plant by analyzing human reliability, and proposed the management programs to be applied.

This area is new and intends to indicate methods for achieving self-management and organizational resilience. We included in this work a glossary that clarifies part of the concepts. We also included an abbreviation list to unify the terms used in the book. In continuity, we will be adding in the annex a tutorial on how to use the methodology indicated in the book, polls, methods, interviews, and others.

The modeling of the subjective along with the objective is complex, but it presents a scenario closer to reality in the industrial plant. In this way, it is possible to avoid major accidents and surprises in installations of high complexity and risk.

References

ABIQUIM, Responsible Care (Action) Management System. 2021. São Paulo. Available from: <https://abiquim.org.br/programas/sistemaGestaoAr> (accessed November, 30, 2021).

Alvarenga, M.A.B., Fruticose, P.F., Fonseca, R.A., 2009. A review of the models for evaluating organizational factors in human reliability analysis. In: Proceedings of the International Nuclear Atlantic Conference, INAC, Rio de Janeiro, Brazil, October 2009.

Ávila, S.F., 2004. Methodology to Minimize Effluents at Source From the Investigation of Operational Abnormalities: Case of the Chemical Industry (Dissertation—Professional Master's degree in Management and Environmental Technologies in the Productive Process). Federal University of Bahia, Teclim, 115 p.

Ávila, S.F., 2010. Etiology of Operational Abnormalities in the Industry: Modeling for Learning (Thesis—Doctor's degree in Chemical and Biochemical Process Technology). Federal University of Rio de Janeiro, School of Chemistry, Rio de Janeiro, 296 p.

Ávila, S.F., 2011. Dependent layer of operation decision analyzes (LODA) to calculate human factor, a simulated case with LPG event. In: 7th GCPS Global Conference on Process Safety, Chicago, IL.

Ávila, S.F., 2012. The exergy of operational failure, a philosophy discussion with practice base. In: Proceedings of 19th Brazilian Chemical Engineering Congress—COBEQ, Búzios, Rio de Janeiro.

Ávila, S.F., Dias, C., 2017. Reliability research to design barriers of sociotechnical failure. In: 27th European Safety and Reliability Conference on Safety and Reliability. Theory and Applications, ESREL, Portoroz, Slovenia, June 18–20, 2017.

Ávila, S.F., Dionizio, J., 2017. Systemic fault analysis to calculate the approximation of the top event (near miss evaluation system): NEMESYS. Safety and Reliability. Theory and Applications, ESREL, Portoroz, Slovenia, June 18–20, 2017.

Ávila, S.F., Pessoa, F.P., Andrade J.C., 2006. Dynamic analysis of human reliability. In: Proceedings of Brazilian Congress of Chemical Engineering, 16, UNICAMP, Campinas, p. 115.

Ávila, S.F., Pessoa, F.P, Andrade J.C., Figueiroa, C., 2008a. Worker character/personality classification and human error possibilities in procedures execution. In: Proceedings of CISAP3–3rd International Conference on Safety and Environment in Process Industry, Rome, pp. 279–286.

Ávila, S.F., Pessoa, F.P, Andrade J.C., Figueiroa, C., 2008b. Human reliability risk management system (SGRCH). In: Proceedings of Brazilian Congress of Chemical Engineering, ABEQ, Recife.

Ávila, S.F., Cerqueira, I.C., Ferreira, J.F.M.G., Drigo, E., 2018. Task & communication in just culture, oil offshore exercise. In: Proceedings of 2018 Spring Meeting and 14th Global Congress on Process Safety, EUA, Orlando, FL, April 22–27, 2018.

Ávila, S.F., Pessoa, F.P., Andrade J.C., 2006. Dynamic analysis of human reliability. In: Proceedings of Brazilian Congress of Chemical Engineering, 16, UNICAMP, Campinas, p. 115.

Cerqueira I., Ávila, S.F., Nascimento, A.J., 2017. Project for the implementation of the research and development center for disaster skills at UFBA. In: UFBA Congress, Salvador, Brazil, October 16–18, 2017.

Dalgalarrondo, P., 2000. Psychopathology and Semiology of Mental Transforms. Artmed, Porto Alegre, 271 p.

Deming, W.E., 1990. Quality: The Administration Revolution Marques Saraiva.

Dodsworth, M., Connelly, K., Ellett, C., Sharratt, P., 2007. Organizational climate metrics as safety, health, and environment performance indicators and to relative risk ranking within industry. Process Safety and Environmental Protection 85 (1), 59–69.

Drigo, E.S., Ávila, S.F., 2017. Communicative skills and the training of the collaborative operator. In: Proceedings of 8th International Conference on Applied Human Factors and Ergonomics (AHFE 2017), Los Angeles, CA, USA, July, 17–21, 2017.

Embrey, D., 2000. Preventing Human Error: Developing a Consensus Led Safety Culture Based on Best Practice. Human Reliability Associates Ltd, London, p. 14.

Falconi, V.C., 2014. TQC—Total Quality Control in Japanese Style. Falconi.

Fadiman, J., Frager, R., 2002. Theory of personality. HARBRA, São Paulo, 2002, 393 p.

Freud, Sigmund. 1989. The ego and the id (1923). TACD Journal, v. 17, n. 1, p. 5-22.

Freud, S., 1923. The ego and the id. TACD Journal 17 (1), 5–22.

Haynal, A., Pasini, W., Archinard, M., 2001. Psychosomatic Medicine: Psychosocial Concerns. Medsi, Rio de Janeiro, 342 p.
Hollnagel, E., 1993. Human Reliability Analysis Context and Control: Computers and People Series. Academic Press Inc, San Diego, CA.
Jung, C.G., 2000. Archetypes and the Collective Unconscious. Vozes, Petrópolis, Rio de Janeiro.
Jung, C.G., 2002. Psychic Energy, eighth ed. Vozes, Petrópolis, Rio de Janeiro, p. 95.
Lees, F.P., 1996. Loss Prevention in the Process Industries: Hazard Identification, Assessment, and Control, second ed. Butterworth-Heinemann, Great Britain, GB, vol. 1–3, 3.680 p.
Louart, P., Maslow, A., Herzberg, F. 2002. Et les theories du content motivational. Les cahiers de la recherche, pp. 1–18 (in French).
Leveson, N.G., 2011. Applying systems thinking to analyze and learn from events. Safety Science 49 (1), 55–64.
Lindsay, S., 2002. Tutorial on principal components analysis. <http://www.cs.otago.ac.nz/cosc453/student_tutorials/principal_components.pdf> (accessed 05.06.10).
Lorenzo, D.K., 2001. API770—A Manager's Guide to Reducing Human Errors, Improving Human in the Process Industries. API Publishing Services, Washington, DC.
Muchinsky, P.M., 2004. Organizational Psychology, seventh ed. Pioneira Thomson Learning, São Paulo, p. 508.
Perrow, C., 1984. Normal Accidents: Living With High-Risk Technologies. Basic Books, New York, NY, p. 453.
Pietersen, C.M., 1988. Analysis of the LPG-disaster in Mexico City. Journal of Hazardous Materials 20, 85–107.
Rasmussen, J., 1997. Risk Management in a Dynamic Society: A Modeling Problem. Elsevier Safety Science, London, vol. 27, n. 2/3, pp. 183–213.
Reason, James., 1990. Human error. Cambridge university press.
Shariff, A.M., Wahab, N.A., Rusli, R., 2016. Assessing the hazards from a BLEVE and minimizing its impacts using the inherent safety concept. Journal of Loss Prevention in the Process Industries 41, 303–314.
Shönbeck, M., 2007. Human and Organizational Factors in the Operational Phase of Safety Instrumented Systems: A New Approach (Master's dissertation). Eindhoven University of Technology (TU/e), Trondheim, 58 p.
Schutz, W., 1966. The Interpersonal Underworld. Science & Behavior Books, Palo Alto, p. 168.
Shlens, J.A., 2009. Tutorial on Principal Component Analysis. Center for Neural Science, Version 3.01, New York, 12 p. Available from: <http://www.snl.salk.edu/~shlens/pca.pdf> (accessed 25.05. 10).
Sternberg Jr, R., 2008. Cognitive Psychology. Artmed, Porto Alegre, p. 582.
Vygotsky, L.S., 1987. Thought and Language. Martins Fontes, São Paulo.
Yantovski, E., 2004. Zero Emissions Membrane Piston Engine System for a bus. In: Proceedings of VAFSEP.
Wreathall, J., Roth, E. M., Bley, D., & Multer, J., 2003. Human reliability analysis in support of risk assessment for positive train control (No. DOT-VNTSC-FRA-03–03), United States. Federal Railroad Administration. HUMAN reliability Analysis in Support of Risk Assessment for Positive Train Control: Human Factors in Railroad Operations. National Technical Information Service. Springfield, VA, 140 p. Available from: <http://www.fra.dot.gov> (accessed 09.05.07).

Chapter 8

Conclusion and products

8.1 Conclusion

The human reliability analysis (HRA) began, historically, through demands from the nuclear energy area, repeating the study of human error based on the premise of nonvariability of the cognitive machine during the task, and THERP is considered as a 1st Generation tool. In the 2nd generation, environmental variations were included, with the classic view of human error, such as equipment failure with a constant failure rate and specific moments of change. In the third generation, the premise adopted is that the environment can change the way of performing tasks, differing from the standards, thus bringing latent rules of work, which alter individual or group work. The 3rd generation (HRA) works with informal rules that admit a dynamic risk, and the organization tries to keep human factors under control, which is not possible through discreet actions. Ávila (2010) mapped regions of organizational efficiency and defined rules to correct human factors and the respective organizational efficiency, targeting regions without accidents, considering that it is part of the concepts of FUZZY mechanics. The subjectivities that involve human at work and in society demand adaptive methodologies that accompany cultural changes in society, in the region, in the place, and in sectors of the risk industry.

The conclusions after application of this quantitative–qualitative analysis methods for the calculation of human reliability, including process stabilization work and adjustments of human factors, are presented and possible applications of future work are discussed.

The issue of the root cause of abnormalities in the industry is based on the need for sociotechnical studies to increase the competitiveness of the oil and chemical sectors in Brazil and worldwide. For the academy, this topic opens horizons by working with shop floor data, thus researching productivity indicators in the operational technical culture of the industry and allowing advances on social data, previously not processed. For the industrial sector, it is possible to add value to the operational database by transforming it into a management tool for decision making.

The paradigms are many and deserve attention to enable the analysis of the sociotechnical system. The chemical industry, that has its mature technology in terms of process and equipment, with advances in automation, instrumentation and control, still has not succeeded in transforming new projects into consistent results, lacking analyses of themes related to operational excellence in productivity and safety. The top-down management programs applied by the industry such as "responsible care," that treat the performance to avoid environmental accidents and ISO-14000, do not bring about expected and effective results, because they still do not direct the production team to self-management.

Through this research, a new type of investigation was presented on the type of technical, human, and social behavior. The key that allows the door to open this investigation is in the type of research, in the bottom/up direction, in which formulations are prepared together with the production team as well as the validations of the respective interpretations and diagnoses. Subsequently authorizations from the board are obtained, to implement these programs, with the objective of reducing losses in the production process, reducing accidents and improving the image.

A group of concepts was designed to enable the transformation of ideas into techniques and from there into applicable analysis procedures. The superficial treatment so far attributed to this subject, boosted the researcher in search of methods that translated the operator discourse into information enabling "better decision making," which until then, in the obscurity of latent failures, were positioned in the apparent "best place."

The main concept used is based on the extraction of abnormalities and signals that indicate process deviances, presented by the shift register together with statistical performance of the process and production. The mapping of these signals and the interpretation of causality generate the possibilities of identifying root causes of latent failures.

The difficulties encountered and announced by articles and public reports that analyze aspects related to the safety and productivity of the task refer mainly to the impossibility of predicting human failure in technical-technological systems, citing as the main reason, the variation of human behavior, and the low perception of risk. By being able to work on the signs of abnormality, and by reaching clarifications about complex process relationships from the operator's discourse, traces of behavior are forcibly included that are clarified by variations in environmental aspects, such as economic positioning, new legislation with the demands of society, traces of dominant social culture in behavior, and organizational culture thinking.

The production manager, who had long sought tools to clarify the root cause of latent failures, starts to act safely in decisions when there is the reading of the operational technical culture defined by the typology or type of operational activities in the industry. Failure can be predicted both analytically and in a timely manner, through mathematical tools that follow dynamic premises of environments, causing structural oscillations in the individual, equipment and processes.

The workers' activities are affected by increasingly uncertain environments in the globalized world and permeated with fundamental issues for their sustainability. The origin of the data in this area leaves managers and engineers, who have classical training, in a situation of discomfort. These professionals are accustomed to the vision of concepts and the right answers to each question.

In terms of application, the effectiveness of the operator's discourse analysis to build the abnormal event map (MEA), in Chapter 5 (Section 5.3.1), is performed and, as a result, we found the relation of causal nexus to the root cause in the failure birth phase in the factory operation. The benefits were measured in process losses, mainly in production cost and process stability. The mapping was successfully tested, mainly in terms of learning about the birth and development of operational failure, in a public logistics company that processes flammable gases.

Safety activities, measurements of operation indexes and logistics activities had direct visualization benefits about the failure in the operation and the possibilities of human error in the execution of the task. By locating points where failure energy based on its "body" can be measured, it is possible to develop alarms as to its presence, based on the causal link identified through the MEA.

The location of the regions to be protected in terms of safety (defense mechanisms) depends on the analysis of factors that make up the operational failure, a subject also discussed herein. The methodology that searches for the operational failure origin and human error helps three important sectors for the management of the sustainability of the company: production management, that deals with the operational routine; people management, that analyzes aspects of selection and development of competencies; and strategic management, that deals with the projects and the vision of the company's future. These managers, when equipped with programs based on the mapping of abnormalities, and, on the discourse of the operator, have awareness on how to adapt the environments (organizational policies), human structures (worker and manager profile), social structures (group formation and cohesion maintenance around organizational efficiency), and especially, learning processes about failure. Operational failures change shape and "do not give warnings," leaving management efforts in a real chaos.

While routine management has the benefit of knowing how to apply task analysis in the design step and during operation, it can also visualize the necessary adjustments on human-machine-process systems. These adjustments were previously spread in a wide variety of latent failures in coexistence with the operation since the start of the plant, that is, the production team, not knowing how to identify the root cause, experience the inefficiencies in the routine of the operation.

Production management also awakes to the need to migrate from the form of training, from classic basic concepts to information about the process of failure in equipment, processes, and groups of people. The safety area gains, through the study of the shift register, the inclusion of human and social factors (operational technical culture) in risk analysis techniques. This is "because" knowledge allows one to be more effective in the corrections of safety devices or in the design of new devices to be installed in the industrial area.

It is important to demonstrate that the behavior of the operator during an emergency is very different in relation to the routine, and the development of techniques to act with the stress of the emergency allows, in a more assertive way, to act in the correction of profiles in the workplace. By quoting Embrey (2000) with the information gathered in this learning model, it becomes possible to prepare a good practice manual with modules based on the birth of the failure in this sociotechnical system.

In the area of strategic management, contract management must expand its performance to the overall result by inserting the support activities as participants in the good results and in the processes of failure in the operation (to acquire greater visibility). It is also necessary to review communication instruments internally (equalizing languages of many cultures) and externally (reviewing from signs coexistence with the public, organizational behavior).

The board realizes the need to plan procedures that confirm the desired competencies for production in the industry with measurement results in the routine of organizational efficiency. People management, which combines matters with strategic management, seeks to achieve states of social inclusion, organizational integration, and adequacy of competencies, consensus regarding patterns of workers' behavior, and criteria for the identification and development of leaders in the operational routine.

As being a bottom-up methodology and with dynamic format, the application of these concepts and procedures, in partnership with the logistics and industrialization activities of liquified gases, allowed the opening of strategies to prevent operational failure, demonstrating the real possibility of joint action between the university and companies.

By creating procedures to identify behaviors of operators and technologies on the shop floor, it defines productivity indicators inscribed with social aspects in this technical and operational culture. The strategy of disseminating concepts in development, as well as the respective techniques and procedures, enables the assimilation of the demands for knowledge from the operational routine environment.

It is worth mentioning that with the application of these new concepts and methods in chemical and energy industries, the possibility of identifying the root causes of latent failures is increased. Although still in progress, it accesses information from the company's technical culture, allowing the application of tools that help decision making in the areas of production (failure analysis and actions), in the strategic area with criteria for new projects, sustainability based on the reduction of losses, and in people management, with the harmonization of teams and the analysis of the "real" competencies installed.

The creation of an algorithm for predicting failure, the elaboration of rules for adjusting sociotechnical behaviors, and the exercise of mathematical tests using the PCA tool and dispersion diagrams demonstrate the potential for failure prevention, all of which has been validated by situations in the chemical industry and by research carried out in the liquified gases processing plant. The sociotechnical systems until then had restrictions on the use of mathematical tools that help decision making in the routine of production. With the present work, research is used to predict situations of failure and thus program the production more effectively and effectively.

The great differential in the method presented in this research is in the way of "mining" the data, in which the observations of techniques and procedures allow the best choice of data. Another great differential is in discussing the failure or release of hazard energy for indirect calculation of human error probability. Mathematical models become dynamic with corrections based on formal rules resulting from environments and technical, social and human typology. The relative measurement of failure energy indicates which one is best set up for the comfort of the production campaign. This measurement follows the organizational resilience goals that are part of cultural analysis, trying to adjust the organization's culture to the best organizational efficiency option.

This nondeterministic method, which begins with the identification of signals to measure energy of emerging or growing failure, can also measure chains of parallel abnormalities. Etiology of failure uses a robust statistical technique, which analyzes the main components and dispersion diagrams for visualization purposes, once the data used to identify variables to be analyzed is in in relation to long periods, with due robustness, thus bringing a greater certainty as to level of hits.

The transformation of a large amount of data into smaller quantities with the PCA tool allowed the graphical view of approaching optimal points of organizational efficiency in the scatter diagram. The adjustment of human factors (PSF) is the missing element to be inserted in future researches to enable the correction of environments, tasks and groups.

The definition of the formal or heuristic rules of behavior (Table 7.3.2.1) in a generic procedure, inserted in the algorithm in Fig. 7.15, allowed the presentation of a list of team behavior rules, social and economic organizational environments, and people's coexistence, the initial period of experience of the researcher in the liquified gas industry. These rules were corrected after the analysis of the partial conclusions leading to observations that helped in the choice of components to be reduced.

Despite the rules of behavior analysis, the final part of the algorithm that refers to the correction of production performance has not been experienced. Then, from rules and weights of alternatives, there would be corrections seeking new positions of organizational efficiency. Also, during the investigation of tools for social behavior analysis, there was the particle swarm technique, discussed in Future Work (Section 7.4.2), intended to adapt, in the future, the management programs for each company leading to the suggestion of generic program for human reliability and process stabilization. The intention is also to validate the mathematical tools and expand knowledge in the area of competency analysis, task analysis, commitment, emotional balance in the workplace and analysis of team-building factors.

8.2 Future book: human factor routine and emergency analysis

The cases of the chemical and oil and gas industries that focused actions on the analysis of technical culture (Chapter 7) to identify the regions of organizational efficiency, show that corrections must be made in the performance factors to avoid a future accident. These factors initially presented by API770 (Chapter 3) and later discussed in practical cases by SPARH (Chapter 6) were part of this discussion in the current book on human reliability.

We started with the discussion on cognitive quality from the concepts presented in Chapter 2 (Human cognition and reliability), Chapter 3 (task analysis and human and social typologies) and Chapter 8, when we discussed product 2 and 3, respectively, task analysis and cognitive quality. The development and validation of tools to identify patterns and to measure the quality of behavior related to C4t (communication, cooperation, competence, and commitment) will be part of the next book to be published. The development and validation of tools to address critical management issues (leadership for process security and crisis control) and technology (complexity and risk) will also be part of this next book.

We present herein the theoretical concepts and models on the design elements and human factors involved with failure and accident. As a starting point, following the Reason (1990) and Rasmussen (1997) models and algorithms to define a network of human-organizational-social and managerial factors and elements, trying to avoid the chain of events that may lead to an accident or disaster. Thus the following subjects will be addressed:

1. Principles and concepts for the hazard energy model.
2. Technical culture—operational and informal team rules.
3. Social-operational context, human factors and technologies.
4. Cognitive quality in the task, lessons learned and the unusual.
5. Performance model and algorithms for the behavior of the leader and the emergency team.
6. Analysis of cognitive gaps in past accidents.
7. Human reliability organizational project.
8. Measurement, management, and analysis of elements
   human factor, pattern, relationship, and interaction for arresting the increase of failure energy.
9. Identification of human elements in the dimensions of management—technology—behavior.
10. Characterization of human factors through technology and culture
    Triggers and breakdown of loss events in the dimensions of culture, social phenomena, company, management, individual groups, operational control, failure-accident-disaster control.
11. Cross task evaluation based on the particle Swarm model.
12. Crisis management from human elements and factors.

We intend to execute this book based on the validation of researches carried out in Brazil and expanding the discussion with research leaders from the United States and Europe. Therefore it will be a book with international authorships and with cases from different countries.

8.3 Products in general

The products described are applicable to any production or service system that has the possibility of loss in terms of affecting corporate sustainability. The company's value reflects its performance toward society, therefore accidents, environmental impacts, high production costs, low compliance are losses to be mapped in order to define the direction and priority level of action for the application of financial resources and technological solutions, and there may be a need for changes in organizational culture, safety, and reliability.

In this book, we adopt the construction of thinking based on the concept and then confirm through practice. Therefore the chapters are built based on the appropriate concepts for the implementation of solutions in practice. In Chapter 8, we will deal with the products, using concepts and practical examples mentioned in the book, therefore the products will follow the line of preliminary diagnosis, validations, detailed diagnoses, interventions, and adjustments to reach the global goal. Thus we present the following products: (1) mapping of process losses; (2) task analysis; (3) analysis of cognitive quality; (4) sparh application and calibration; (5) sociotechnical reliability; (6) diagnostics and interventions based on the identification of technical culture. These product issues indicate parameters for the decision on human performance factors, for analysis of human reliability. Respecting previous attempts at calculating human reliability, this methodology intends (P1—CH4) to map losses indicating regions and priorities, based on human errors, equipment and process failures, all during execution, planning and diagnosis of critical tasks (P2—CH3).

The premise adopted is that the system under evaluation already has standards and procedures, requiring a new analysis of the task, carried out by specialists based on experience and knowledge. Then, if there is a finding of frequent and impacting human errors, (P3—CH2) study of the cognitive quality for the task is proposed, considering whether there are possible gaps hindering the processing of information and the respective action. The reader of the book, understanding cognitive quality, has knowledge about human factors, which makes it easier to calculate human reliability with the SPARH procedure (P4—CH6), originally applied in the Nuclear Industry and which may need to be calibrated for chemical industry installations, for example.

Based on the mapping, the analysis of the task and cognitive quality, there are high differences between the performance of the equipment, process, operation, and human action. Thus a new indicator, already validated in the manufacturing and continuous processes industry, is suggested: the analysis of the reliability in the sociotechnical system—CSST (P5—CH6), according to the level of complexity.

So far, the methods adopt the specialist's discourse with the manager's discourse guidelines, but these fail to direct efforts to the main and current region where latent and emerging failures occur. It is then necessary to apply methods that analyze the technical operational culture (P6—CH7) of the factory floor. From the Operator's Discourse (OD) that can be taken from the occurrence register, from workshops for the analysis of accident scenarios, and from courses directed at human factors. The OD and that of the Manager (MD), in their respective records, indicate chained abnormalities. This scenario, together with the process and production variables, presents a high number of sociotechnical data. Data mining and multivariate statistics indicate standard qualitative and quantitative behaviors that represent cyclical or even emerging events. The construction of curves that indicate historical organizational performance enables the interpretation of workers' movements that result in organizational performance.

The back-casting analysis, with the team's past efficiency curves over 2 months, for example, and the list of informal rules in the routine, shape the likely paths for the future production campaign.

The mapped informal rules allow directing actions to points of connection between GHOST factors (management—humans—organizational—social—technological) in a complex sociotechnical system. Descriptive, multivariate statistical techniques, and fuzzy mechanics facilitate measuring the level of approximation of the next production campaign in relation to the near accident.

The applications for the mapping of process losses were carried out in the chemical and metallurgical industries, SPARH (NUREG 6883, 2005) was applied in the chemical and refining industries, among others, and the CSST (reliability of sociotechnical system) was applied in the refining industry. The identification of the technical operational culture was studied in the LPG processing industry and the actions carried out, among others, promoted behavior change in the fertilizer industry (change of rituals, educational programs, and training). The results after the applications were reduction of the accident probability with LPG, applying LODA tool (Ávila, 2011a) to emergency team; increasing reliability in amine and sulfuric acid plants; in reducing methylene chloride losses in the polycarbonate industry; and in the reduction of ammoniacal load for the fertilizer plant effluent (Fig. 8.1).

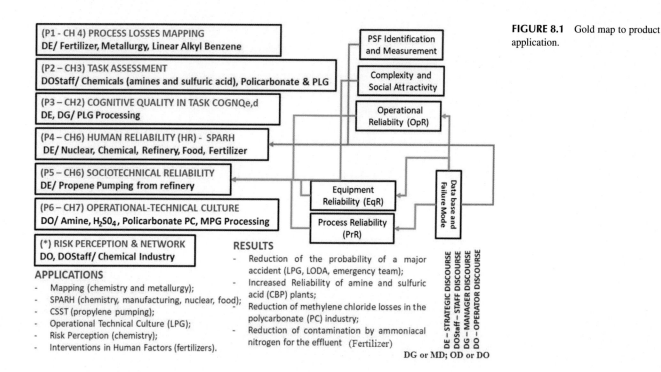

FIGURE 8.1 Gold map to product application.

8.4 Product 1 (Chapter 4)—process loss mapping

8.4.1 Introduction and methods

The generation of losses in industrial processes reduces productivity, increases company costs, and decreases team motivation. In Chapter 4, the main types of losses were discussed in terms of their nature and type, the most well-known of which are: loss due to overproduction; loss due to waiting time; transport loss; loss from over-processing; inventory loss; loss by movement and loss by defects. However, currently there are many other losses not only related to what was known in the literature by Ohno (2000) and Shingo (1996), but also losses related to the company's image (social and environmental issues) and accidents / incidents (Ávila, 2004).

8.4.2 Methods

For the diagnosis and treatment of industrial process losses, some techniques are presented in Chapter 4 and indicated in Fig. 8.2, such as abnormal events map (MEA); statistical analysis of abnormal events; audit; process loss mapping; and mass balance.

The search for solutions to avoid process losses starts with the observation of the continuous routine, with mathematical methods, and then continues with the analysis of the technological, managerial, and behavior dimensions, based on good practices. In this way, a system was developed to demonstrate functions already presented in the example of metallurgy and the chemical industry on the loss diagnosis methodology, Chapter 4 (Section 4.4). Organizational system for Sustainability Management—RESOTECH (Fig. 8.3) presented at the onshore Oil Production Congress in China (Ávila et al., 2020), has three main functions: design, operation and management. In RESOTECH we indicate the need for Management and Control of water and energy resources in production, the control of CO_2 emissions from improved combustion efficiency, the energy efficiency of thermal equipment and systems, desalination to increase water availability, effluent reuse for the process and utilities, loss control from its mapping. In the operation, it involves the environmental discussion, the cost of production resulting from energy savings and the nonwaste of time. The reason for building this system is to demonstrate that the implementation of sustainability management control brings benefits to the Image, the reduction of costs (such as losses), and the participation of the global objective of reducing CO_2 emissions and saving on water consumption and energy.

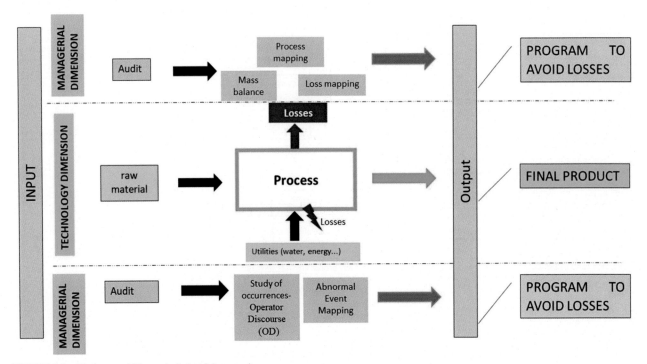

FIGURE 8.2 Tools to avoid losses in industrial processing.

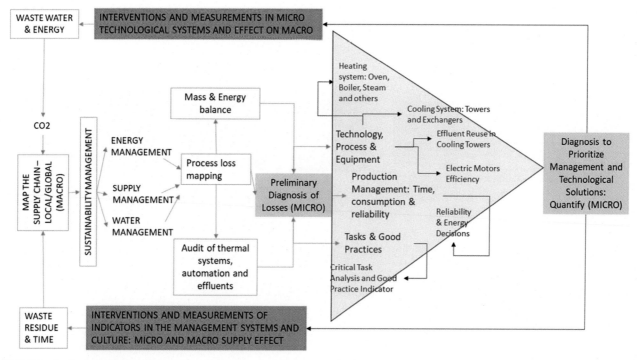

FIGURE 8.3 Tools to identify and correct process losses.

The RESOTECH Method (Ávila et al., 2020) proposes a management system that guarantees the company's economic and environmental sustainability. In this intention, it develops three levels of management, hydro-energy and supply management, but it is also causally related to the project, routine, and asset management. To simplify, we will divide this work into: (1) Identification of the management system to avoid losses; (2) Design of mapping of process, water-energy and effluent losses through mass balance and field audits; (3) Preparation of preliminary diagnosis with immediate managerial actions; (4) Research on Technologies, Processes and Equipment to reduce losses of energy, water and waste generation; (5) Reliability Management based on the recognition of impacts on Energy and Water with the Plant downtime; (6) Analysis of Critical Tasks to avoid losses of Energy and Control of Practices based on Indicators. With these investigations, it is possible to elaborate (7) Diagnosis of Losses with the prioritization of Technological, Managerial and Behavioral Systems. RESOTECH methodology concludes, on the one hand, with interventions and measurements in the MICRO production system and which influences the MACRO system, in relation to the supply chain. In addition, it proposes revisions in the Project and in the Management, it is important to establish a strong Culture to avoid losses of processes through the Good Practices of the team.

8.4.3 Discussion

To objectify the product of process loss mapping with an investigation of easy application, low cost and use of few resources, the dynamics of the process loss mapping methodology is presented in Fig. 8.4. This methodology has already been presented and demonstrated in the case of metallurgy (Section 4.5.2). The mapping allows the investigation from the type of loss to the area of concentration, then allowing to calculate and identify the priority of loss, that is, the critical loss of the process.

To know the failure scenario, or the identification of failures or losses in the process, it is necessary to initially investigate how the task routine works. The operational routine must be regulated in procedures and standards accessible to everyone on the team. In this stage, too, products, technology used and the functioning of the processes to perform the task are discussed. In this way, critical environments are initially evaluated through routine monitoring.

Routine monitoring adds technical knowledge to theoretical knowledge, allowing to verify the real need and effectiveness of the processes.

In routine monitoring, interviews were initially carried out with the management team (MD) on the processes and technology; however, it is interesting to also include the OD in the methodology.

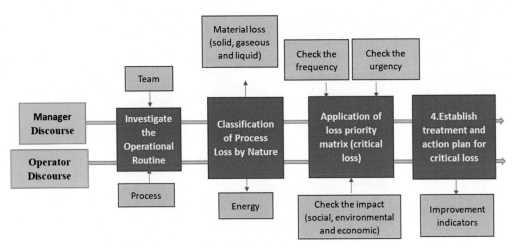

FIGURE 8.4 Block diagram operator and manager discourse solutions (OD and MD).

8.4.4 Calculation based on metallurgy case (Section 4.5.2)

To carry out the mapping of process losses in the metallurgy case (Section 4.5.2), a base form (Fig. 9.4) was prepared for completion during the investigation where losses are divided into two types of nature: loss of material nature and loss of nature of energy. Material losses can be in the solid, liquid, or gaseous state, such as: particulate material, effluents, scrap, slag, gaseous emissions, and among others. Energy loss is analyzed through consumption, whether it is unnecessary consumption, such as rework or lack of process optimization.

The prioritization by mapping process losses is due to competence to identify the loss prioritization (Pp), through the FIU (frequency impact urgency) matrix, Chapter 4 (Section 4.4.5). In this discussion, we demonstrate the calculation method below:

$$P_p = F \times I \times U$$

To achieve the priority, aspects were indicated: frequency, impact, and urgency. In the mapping, the loss can be related to the main operational problem or to the failures that occur in the task and that result in its greatest generation.

The process of filling in the mapping was done initially with the management, but it can also be carried out, in parallel, with the operation team (OD). In this way, you can allow the applicator/researcher to calculate information in the priority matrix to be crossed.

A scoring scale was created for this matrix (FIU), presented in Chapter 4 (Section 4.4.5), which allows to fit each aspect to the current state of the loss.

In the case, the main loss pointed out was Slag, with a score of 27 points in the FIU matrix (Table 4.9—Loss mapping in metallurgy). This loss was routine in the process, but its volume and frequency were increasing. Thus after the mapping, an investigation for the causes of generation was carried out, with some engineering and quality tools, such as: mass balance, Ishikawa diagram, OD, Likert Scale and Fuzzy Logic. With the known causes, an action plan was developed to treat and improve the failures that occurred in the task and resulted in an increase in the generation of critical loss.

8.5 Product 2—task assessment—PADOP

8.5.1 Introduction

The plants operate under normal condition when the lead planners translate technological assumptions of control in the form of a procedure and when they encode the operator's mental maps for the action, in the format of written communication with auxiliary memories, in the appropriate format of oral communication and respecting different hierarchical-functional levels. The improvement of operational and human reliability depends on the operational context and skills installed in this scenario. Products 2 and 3 of this work are intrinsically related and directly influence the performance of industrial plants. When considering that this book has prioritized the improvement of the performance of the plants in operation, it is considered that there is a work pattern in the routine of operations. From the mapping of losses, it is noted that this pattern may be fallible or may be inserted in a dynamic sociotechnical scenario.

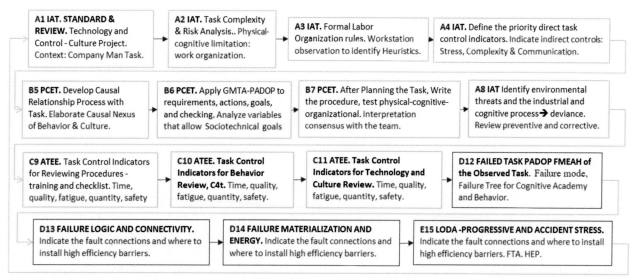

FIGURE 8.5 PADOP procedures, patterns and failure assessment.

Prior to the organizational requirements to design the procedure steps is the cognitive and physical environments for carrying out the activities from the point of view of the company, the task, and man (IAT). Thus there are initial indicators to be calculated and processed to classify the level of risk, the level of complexity, business priorities, the knowledge applied to the actions under discussion, and the comfort of doing something already experienced by operators. This condition of work availability for the activity depends on the organizational action in terms of selection, development of the teams, and dynamic capacity to respond by acute demand to actions that require emotional balance and intuitive perception. We will divide the dimensions for the analysis of the task in design of standard, analysis of failure in the activity with revision and interventions in the technological project and in the organizational culture.

Product 2 for Chapter 3 of the book focuses on the analysis of the standard task and the need to revise the task due to failure. The indicators to aid in the elaboration and adjustment of procedures are built from questions regarding the task environment, the ability to develop standards, the cognitive recognition of actions, and the control of the task.

The indicators are based on a survey of routine and technology. The questions listed in Chapter 3 suggest rules for triggering review of environmental requirements, preparation of auxiliary memory, review of intermediate and final control parameters, understanding of the effectiveness and priority of critical tasks.

In the guidance on how to apply the PADOP Product, Product 2, we indicate the following diagram of the simplified block, in Fig. 8.5, with activities in the planning of procedures for the cases of the chemical process industry: refinery logistics services; starting procedures at the sulfuric acid plant; task failure in aromatic amine plants; and LPG processing.

8.5.2 PADOP standard and review

8.5.2.1 Environment assessment—IAT

8.5.2.1.1 A1 context

In Fig. 3.18, item 3.2.2. we asked about context focused on human, company and the task, 27 questions. After this we classify as satisfactory or not based on standard.

8.5.2.1.2 Context PLG facility

A flammable product, the culture of a public company with job stability, procedures with more than 45 stages, processing with high variability due to the logistical capacity and due to the quality of the raw material, history of fire events, spheres with drainage operations and presence of hydrates presents a scenario of low reliability in the context of Man, the company and the task.

A2 risk and complexity The risk analysis of the task indicates the possibility of loss of control of the technology with an impact on the business economy, Man (occupational), neighbors (process safety), property (material damage) and the environment. But, if we think about the context of anthropometric, cognitive, and work limitations, it is important

TABLE 8.1 Evaluation of the measurement aspects of the complexity of the task / process.

Aspects	Measurement
(1) Compliance with the task	0–1
(2) Type of technology used in process control	0–1
(3) Type of equipment with installed control instruments	0–1
(4) Number of steps in the process	0–1
(5) Degree of attention required to perform the task by the operator	0–1

to analyze the risk of low quality of communication, difficulties in accessing equipment-processes, limits of human perception for signs of the process, and complexity involved in the process—cognitive processing, all for carrying out actions to control the task. An important aspect that relates to all the human factors in the operation, and all the human elements in the project, is stress during the task, which alters the quality of the cognitive apparatus and affects the employee's motivation attitude.

The analysis of risk and task complexity goes beyond the classic process and equipment and increases the resilience of organizational systems. The PRA-SH (Ávila and Barroso, 2012a) in Chemical Plant, Social HAZOP (Ávila et al., 2013a) in a case of separator drainage at a Refinery, LODA (Ávila, 2011a) in PLG furnace and Cognitive-Social Matrix in the Interfaces (Ávila and Pessoa, 2015c) of air product plant tools join the complexity calculation discussed widely in this book and presented in item 8.8.2., Table 8.1. In this case we classify the risk from 0 to 6 and the complexity from 3 to 1000.

Risk PLG facility In Chapter 3 (Fig 3.36), Chapter 7 (operational context), and in the thesis of Ávila (2010), we evaluate ways derived from verifying the risk of events with flammable gas in the oil industry. We note that the risk level classification may be related to the number of failure warnings and the number of double jobs in the shift. In the case of oil and gas, this risk was classified as a high result of over 20% double work load and over 6 failure warnings in the 8-hour shift. The risk is high, with a value of 5 within a scale upto 6. Knowing the number of activities performed by the operator in the shift, we can define the probable safeguard, in this specific case, the best solution is organizational change in the production leadership and process security.

Complexity refinery facility The complexity was calculated by level of attention, memory, process control, and others. In this way, the case of the forklift accident in a refinery, yields a complexity of 79 between 3 to 1000. This parameter is used to calculate the social technical reliability. It indicates a low level of complexity, but, if we investigate aspects the rules to perform contract of logistic, the conclusion is that it is not possible to guarantee the safety rules.

8.5.2.1.3 A3 labor organization rules

Rules and heuristics sulfuric acid plant The design change with increased capacity of the main plant, toluene diisocyanate (TDI), brought consequences for the nitration reaction and the subsequent separation of sulfuric acid in relation to the dinitro toluene (DNT) product. Sulfuric acid after being separated from DNT came with an excess of contaminants (NOX and DNT) to the glass plant, causing problems of back pressure and equipment breakdown. Due to a power relation where the main plant (TDI) commanded the strategic actions and due to the investigation of the incomplete problem, looking only for the immediate causes, without analyzing the problem in depth and in the sociotechnical dimension, the low safety standard instituted the rules in the operational routine.

Two phrases compose the heuristics of the behavior in this sulfuric acid reconcentration plant: (1) might is right, and the sensible obey; (2) I do not go to the area for fear of dying; (3) there is no point in talking since the power is in the hands of the supplier.

Meanwhile, Security believes that when providing personal protective equipment, it does not need to revise standards. In fact, the team unsuccessfully signals that the acid on the PPEs and can cause an accident to the operator.

This conflicting situation with continuous breaks and stops creates the conflict between policies and practices and internal accusations. Shift work enters a cycle of nonquality bringing discredit and dissatisfaction. Thus other consequences, such as delay in starting the plant, excessive double shifts, high maintenance backlog, all hinder the work organization. A lot of risky activity is carried out over the weekend to preserve the administrative team from possible accidents.

8.5.2.1.4 A4 task control indicators

Control indicators in PLG column dryer When analyzing the context in the analysis of the environment specificities of the task, Man and of the company are included. When mapping the production processes with the product and process properties, each task has a main function with the respective requirements and checks, as well a secondary function, also with the respective requirements and checks. These secondary functions serve the context of the activity.

In the case of LPG drying, the main function is to remove moisture through the drying column equipment with a product flow and appropriate temperatures. In addition, perform regeneration operations in constant cycles.

The main indicators of the task considered are in meeting the objective of removing moisture within a standardized time cycle, in a certain temperature and flow range. We must not forget to include the auxiliary indicators: no accidents, no controlled cost, no environmental impact, no maintenance problems, and no health problems for the worker. We currently include activities with minimal CO_2 emissions, reduced energy, and water expenditure, avoiding events that go beyond the factory boundaries and having feedback from the operation group regarding quality of life and level of job satisfaction.

8.5.2.2 Cognitive processing and execution—PCET

In Chapter 3 we showed questions about the second phase of PADOP, PCET or cognitive processing and execution phase. In this step there are five activities to be performed: task planning (Table 3.3); written procedure preparation (Table 3.4); workstation design (Table 3.5); chronology and criticality analysis; and task execution. These questions and explanations guide the investigator to look for the cause of low operational continuity of the critical equipment. The cause mode is built if we understand the phenomena behind the process in the industry.

8.5.2.2.1 B5 PCET. develop causal relationship process with task

In practice, PCET activities are possible only after understanding the relation between the industrial process, product transformation, and each step of the task, then we must classify each task, subtask, actions as a hierarchy architecture based on Risk and Complexity. In this situation, we need to understand the normality level of product, equipment, environment, human and process, step by step, to avoid future problems or failures.

Causal nexus—equipment, process and operation in caprolactam industry The definition of hierarchization in the tasks and failures, to define level of priority depends on the causal nexus between the normal phenomena (project) or in cause mode of a failure. In a high complexity chemical plant, where the process control depends on laboratory and maintenance services, the communication between the areas needs to be clear to discuss actual problems.

The production of caprolactam to supply raw material to produce nylon fiber 66 depends on cyclohexane facility in first step of production. The plant has a high level of automation to control equipment and processes. The functioning of hydraulic system to control turbine speed and the prioritization of services and lab results depends on clear communication about the possible problem.

Breakage event on glass pipeline in a sulfuric acid plant On the other hand, to understand impossible events considering the physical law of cause and consequence, we need to construct several hypotheses and validated one by one. A visual assessment, drawing the problem as a scenario, helps us to understand the movements of a Teflon piece in the case of vacuum instantaneous breakage in the sulfuric acid plant.

To design patterns or to investigate causal mode of a fault we need to develop causal nexus as a learning for the future task and process of production or for the accidents that happened.

8.5.2.2.2 B6 PCET. Apply GMTA-PADOP to requirements, actions, goals and checking

After understanding the several phenomena in chemical process to the production of products with good quality, we can understand probable causes of accidents in this technology. To keep production without surprises we need to reproduce the standard activities, in written documents, for the performance of the steps of a task by the operations team. The information that is part of this document can show, to the operators' team, the requirements before the actions, the goals and targets giving correct direction for each activity, the actions with progressive grade of complexity done by a person or a group, with interface or not with other departments. Finally, the group needs to check the level of success of the task (mind map) or of the procedure (written).

Task planning is described through the following activities indicated in the questionnaire in Table 3.3 in Chapter 3: explaining the task through the causal nexus, restrictions, company objectives, and the team's cognitive limitations;

structure the task by differentiating environmental restrictions, task history, and legislation requirements; specific steps with requirements and confirmations and with the final check of the task.

Task planning also includes analyses of: the type of activity to define cognitive-motor difficulties (search and action; surveillance; diagnosis; and planning); recording of routine observations; monitoring of task performance indicators (how to measure, where to register, to whom to communicate); inside the task context discuss again complexity and risk; define kind of auxiliary memory; define automatic tool (statistical and math) to help the task perform; and task hierarchy assessment.

Operational instruction to clean acid tank in aromatic amine site The design of each stage of the tank cleaning service and the respective checks must be made for the analysis of risk, complexity and for the scheduling of activities in the service agenda. Then the planning of the activities is the result of a discussion between the planners, with documents and an analysis of the task environment, in the context of Man and the company. In this way, as the objective is to clean the tank without accidents and at a controlled cost, we have an idea of the requirements for this accomplishment.

The prescribed activity is recorded in the document, step by step, as well as the confirmation of success, recorded in the checklist. Both are created from a mind map resulting from consulting drawings, documents, experiences and investigations. The written operating instruction, after being validated by an expert, is sent to the task participants. At that moment, they return to the mental map mode for physical preparation in carrying out this instruction.

This instruction is not normally done, therefore there may be additional precautions when there is no previous history of performance. The health consequences and the possibility of identifying damage to the tank can change the instruction steps.

8.5.2.2.3 B7 PCET. After planning task, write procedure, test physical-cognitive-organizational

The cost in task planning is based on hours of specialist work, based on knowledge or practice. Sometimes the industry prefers to delegate the development of new procedures to new operators or engineers. This can be a serious mistake if the lack of practice or concept can cause a failure or accident.

After planning the task, we try to write the procedure and checklist, which is a challenge due to possible linguistic flaws or difficulties in transferring a mind map to writing with the guarantee of composing the same or similar mind map. After reading the procedure or from the oral speech, this mind map will become an action.

The writing of the procedure described in Table 3.4 is composed of the following activities: general actions to elaborate the procedure; writing of the procedure and linguistic understanding with sampling of the operator's interpretation; validation of written activity through field confirmations about what was prescribed; and design of safeguards or barriers to prevent failure, including fault assessment of barriers.

At this moment, it is important to draw attention to other human interfaces with the control room, with field notices, in oral communication to establish hierarchies, and in the speed of motor action and communication. Parallel activities, with several operators related to the same task, or even with several activities being done at the same time by the same operator, must be reported in chronological order and in the opposite direction to the progression of risk.

Experts in linguistics in cross-culture environments have argued that the best configuration of the procedure in high-risk activities and industry, for the new generation of operators and engineers, adopts less written words and more symbols and figures. This perception was presented at the Global Congress on Process Safety.

It is recommended to take the mind map test, manager for planner as the context of the company, planner for writing as a scenario generated by an expert, written procedure for an operator as a mind map in comparison with the expert map, and an operator mind map compared to the result activities. This test indicates the quality of the task planning but also indicates the level of approximation of the just culture in Security and the Organization.

Based on SKR Models from Reason (1990) and based on discussions from Rasmussen (1997) we remember to recover the Table 3.5, from Chapter 3, that presented Cognitive Workstation criteria for the Industrial Project. These criteria are being updated in task assessment aspects here in topics B7 and A8. What are the questions to be remembered: a safe-alert-resilient behavior; a good commitment between the worker and the plant; the adequate application of knowledge and skills in critical task; and the understanding that the formal documents are important to operational discipline, important characteristic to keep safety in complex processes.

Procedure revision in startup of a sulfuric acid stripping column The sulfuric acid plant is at high risk due to the product (burn damage to man), the process (high temperature), the material (mechanical or thermal impact can cause leakage), and people (poor team performance and low plant performance) reliability leads to low motivation.

The requirements for operation are related to flow and temperature at the start of the plant. The stages involve these two actions in a gradual and ascending manner until reaching the plant's load standard, productive capacity. Communication is done in a simple way, from the field to the panel, from the plant to the supplier, and from the plant to the customer.

An investigation work on the abnormalities in the shift based on the operator's discourse demanded a review of the procedures of the sulfuric acid reconcentration plant. The big change is regarding the minimum quality of the raw material for starting purification. Additional requirements for plant materials and pipe flange tightening have also been adopted. The training for these tasks was revised, presenting safety, quality, and maintenance items.

There was a managerial shift from doing cognitive tests on each new procedure being carried out in sampling by field and panel operators. Some alarms were considered priority and every plant shutdown is discussed for burn accidents and or low reliability in production.

8.5.2.2.4 A8 IAT identify environmental threats and the industrial and cognitive process→ deviance

The natural, physical, social and cognitive environments are in continuous change, often not visible to the human eye. These changes can influence the performance of the task. It is probably not visible due to the characteristics of complexity and risk aversion. In fact, the lack of Strategic Intelligence to build future scenarios means that fragments of information, signals, or spurious data of natural and social phenomena are not given value. What we see, where attention is calibrated, is related to direct losses such as plant shutdown and serious accidents with loss of time and image.

In B7, when it comes to cognitive and physical tests for interpretation, what is done is a conference of perception of the stages of carrying out the task, in fact the prescribed activity is being confirmed. But, before performing the task, it is necessary to confirm the current environment, the current scenario, thus whether the prescribed activity can be performed on the equipment, processes, people and environments in the current scenario. In this case, we have two cognitive processes, the quality of the mental map in the prewritten activity and the sense of reality for the construction of a scenario or environment for carrying out the task. Environmental threats arise through invisible signs and the team must be prepared to perceive and value this set of hidden information.

Self-management, barrier in organization and technology polycarbonate plant The improvements made to the polycarbonate plant followed the methodology of identifying the technical-operational culture according to the subject of Chapter 7 and product 6. The main objective was to reduce losses of methylene chloride, a chlorinated polymerization solvent. From the defragmentation of the operator's discourse, in conjunction with the analysis of process, production and maintenance variables, it was possible to map chains of abnormality that we name as systemic failure. Tests regarding the cause of these abnormalities indicated the necessary changes in equipment, processes, tasks and behavioral adjustments.

These operational research activities were carried out in normal operations of the plant and we discovered the following side events that were treated: (1) control of the reaction from the dosage of solid—raw material, from the variation of CO pressure and also obstruction in the caustic soda pump; (2) analysis of the cause of failures in the SABROE cold system to reduce the temperature of methylene chloride integrated with the mass balance and losses of this solvent, in this case sealing an alternative pump, re-entering the process; (3) start-up of the plant without proper learning due to the centralization of the director who was an operator in the past; (4) stop organization to correct equipment and systems; (5) best operating standard for the final solvent adjustment column for the effluent; (6) study of the failure mode of equipment (kneader) and process (vacuum) in the purification; and (7) review of procedures, adjustment of equipment, process changes and training of staff at the level of concepts, skills and management.

The global indicators for the start-up and normal operation tasks of the plant were: consumption of methylene chloride per ton of polycarbonate, on spec, produced; losses of methylene chloride to the effluent and the environment. We sometimes noticed that the organizational practices did not coincide with the intention of the improvement project, where the board addressed the immediate task of avoiding preventive stops "ordering" to produce even without environmental quality.

Thus to keep the standard achieved in the execution and cognitive processing of the team intact, and knowing the decision errors of the centralized management, the most correct measure can be ensured by promoting a state of self-management directing the sense of shared responsibility for the formal and informal leaders.

The conclusion reached is that Task Analysis, its control and improvements only occur when it is combined with changes in organizational culture, team behavior, leadership style and the way of maintaining improvements. Therefore the need to reach the state of self-management is confirmed, only in this way do we achieve organizational resilience in the face of the dynamism of current events.

8.5.2.3 Task efficiency control—ATEE

The objective function in carrying out the task is composed of different control items from the various areas of impact of this task. Some areas are part of the main functions, without which the company stops billing. If the product is not specified and production is expensive, it is difficult to operate. Therefore quality, quantity, reliability (time) and cost are major parts of the objective function. The secondary parts, but no less important than the main ones, are part of the consequences of production activities and which affect the image and, indirectly, also affect the value of the business in the capital market and the intellectual value in the labor market.

Depending on environmental, labor, health, and safety legislation that respond to societal pressures reacting to local and global impacting events, we can list other component parts of the objective function (FO) with the accomplishment of the task: environmental, occupational, and specific characteristics in the task control. Remembering that these secondary parts can also be at a level of degradation with the threat of plant shutdown.

The weights of these parts of the FO vary according to the period of the project's cycle, which may be in the stages of assembly, start of operation, maturity, material degradation, organizational degradation, accidents, and loss of control with the possibility of chain reaction. These FO weights can also vary according to external demands (economy and natural threats) and internal demands (social and leadership behavior) in the organization.

8.5.2.3.1 Objective function weight calibration—the case of PLG plant

In calibrating the objective function, we must understand the risks and complexity of the task. Which procedure is most critical? How to keep the plant in continuous operation, by maintaining controlled flow and pressure? Or whether we need to transfer the product, considering the possible quality of the process and the attention of the operation. We compared two procedures: (1) normal drying operation and temperature reduction in relation to, (2) transfer of raw material from the refinery sphere to the gas unit sphere or transferring the finished product from the refrigerated tank to the ship.

As two tasks are at risk of leaking processes, the transfer procedures are related to intermittent activities whereas to maintain normal operation, adjustments are made to a standard condition. Although the transfer is a less complex control, it is riskier. The biggest cost of production is knowing how to control a normal operation.

8.5.2.3.2 C9 ATEE. Task control indicators for reviewing procedures—manuals, training and checklist

The task indicators will guide as to what kind of improvement should be made. These improvements may seek to return current process state to the expected standard after correcting the causes of equipment or process failure. One can, for example, seek to reduce water consumption, trying to reach a new standard in times of water scarcity. There is still the possibility of seeking safeguards to avoid loss of quality in the process or product, thus increasing stability time and improving the company's results in terms of reliability, availability and maintainability. Observation in the field and in the control room, allows for calculating the indicators and monitoring the phenomenon trying to make a causal correlation to know where to perform the intervention.

8.5.2.3.3 Process control and procedure review, aromatic amines

As the indicators are monitored for task control, we notice a change in the mean, standard deviation, or measures derived from these indicators. In this case, we are talking about the following procedures of the aromatic amine and isocyanate plant: cleaning dispersers in the feed of raw materials; flow control for the amination reaction; and the control of the process variables in the purification, where the return of aniline occurs or, incorrectly, the return of MDA to the aniline tank. The experience accumulated and the investigation based on operator discourse show the necessity of change in the process, the equipment and the procedure. We need to answer these questions:

What are the principal impacts? What parts of equipment, materials, or process need changes? What kind of investigation is required to detect and to follow the indicators after the change? What is the change in procedure? What kind of new manual or guide is required? What training do we need to address the new needs? What checklists need to be provided?

8.5.2.3.4 C10 ATEE. Task control indicators for behavior review, C4t

Some properties necessary for sociability at work and for carrying out the task were classified as human elements, or safeguards installed in the project to prevent the accident. Conducting resilient and sustainable production campaigns requires monitoring of these properties, quality of communication, cooperation and commitment. In addition, we seek to measure the competencies demanded by technology and those offered by the team to measure cognitive gaps. The level of stress is accompanied by derived indicators and the leadership must fit the typical scenario in living with the task within normality and outside, in crisis situations.

Measuring behavior for change—LPG industry Ávila thesys (2010) We understand that it is possible to manage subjective aspects of worker behavior, treating the human elements in C4t step (Cerqueira et al., 2018), part of SARS Tool (Ávila et al., 2018). Some indicators were tested in the Oil & Gas and Industry. We understand that emotional state is based on cultural aspects, cognitive quality and possibility of psychosomatic diseases, giving in the Gas Industry a final value of 7 out of 10 and standard 8. In case of commitment, we consider worker information such as time working at the company, whether married, how many children, and other personal information. We tested and the result in 3 groups was of 30% in high commitment level. The level of cooperation is based on aggregation (6.8 from standard 8) and disaggregation (5 from standard mean of 5.4) as parallel movements. The competency analysis is based on the human errors during long period of shift, 2–6 months, and task results. In this period, we can elaborate cognitive and behavior gaps and create an exam that will be applied in operational group and indicate training and behavior necessities to avoid accidents. Finally, Dos Santos Drigo and Avila (2017) elaborated Organizational Communication Model to measure the quality (omissions and commissions) of ascendant, horizontal and descendent communication.

8.5.2.3.5 C11 ATEE. Task control indicators for technology and culture review

The integration between Organizational Culture and Technology in operation indicates the possibility of achieving and maintaining organizational resilience in the face of external threatening environments that can impact the stability of the task. This discussion took place in 2004 when Ávila presented the Operator Discourse Analysis to Paul Sharratt at UMIST, with the intention of investigating deviations and abnormalities not visible in the operational routine, the subject of a possible thesis at this University, and which ended up being held at UFRJ, in Rio de Janeiro, Brazil.

Until that moment, 2004, Organizational Culture and Task did not communicate, statements made by the research leader in process security at UMIST. And Ávila (2010), with the discovery of the operator's discourse analysis technique, seeks to break this silence and presents options to prevent future failure considering an environment of high connectivity: Social Influences, Managerial Preferences, Technologies, and the resultant Operational Practices.

Prospecting just culture—a case of the metallurgy industry A family-owned company in the metallurgy sector opened its capital to the stock market, with the intention of increasing the size of the business and turnover. No analysis of organizational change was carried out and technological transformations started, and transformations to meet legal requirements in relation to the environment, work and occupational health and safety.

Due to the arc casting technology, which requires a lot of energy, the company was integrated with the biomass-cellulose area and currently with the production of energy from wind farms. Thus with regional distances, the company grew, it was previously in mining and metallurgy and introduced cellulose and a wind farm.

The growth brought Labor Market Managers and other partner facilities, specialists, living with professionals trained on the spot through schools maintained by the original family that owns the factory. This situation of organizational change with new management systems clashed with the customs adopted while the company was family-owned with local regional cultures.

Although the economic results were satisfactory, there were still many accidents with fatalities and generation of excess slag with mountains of material to be reprocessed. The lack of integration of the new security and organizational culture with the task ends up creating conflicts that make it difficult to reach the standards expected by the open market.

The lack of comfort in the workplace with high temperatures, distrust or organizational discredit, or even the nonidentity of a linguistic code to make the task possible removes the cognitive perception for the manual action of separating the scum from the league bringing negative economic and occupational consequences.

In this case, it's necessary to study the past to advance to the future, review the criteria at the workstation to reduce losses, standardize procedures, in stages using, visual communication to facilitate the commitment. Finally, trying to migrate fragmented information to chart a path toward a fair culture, transparent communication.

It is confirmed that cultural transformation is necessary to achieve effectiveness in the Task. Without which there will be no guarantees of organizational efficiency.

8.5.3 Task failure assessment

The Task Failure Analysis in Complex Systems depends on the researcher's knowledge requirements: (1) knowing how to differentiate normality from abnormality and its intermediate states; (2) identify and differentiate deviations, failures, incidents and accidents to adjust the task; (3) verify communication noises that affect the decisions and cause the failure to perform and diagnose procedure; (4) conduct failure mode analysis including human and organizational factors; (5) have active listening, observe the post and analyze the history to build mental maps about the phenomenon; (6) knowing how to separate production systems into processes, then into equipment, then components, to analyze the failure mode; and (7) when investigating open your mind to discuss new possibilities for failure, such as human error, bad habits and incorrect decisions.

Ávila (2010, 2012b) when investigating complex systems presents ways to analyze the risk (Ávila, 2011b) but also to investigate the connections of different dimensions of the failure, here named as systemic failure (Ávila and Dias, 2017). The tools developed involve investigation of machine failures, analysis of complex systems and their different functions and connections, failure energy and chronology during the accident, communication noise being transformed into equipment failure, and many others.

8.5.3.1 D12 PADOP, FMEAH, FTAH of the observed failed task

The monitoring of the quality of the processes, product and effluent tries to prevent the state of abnormality from occurring; tries to ensure that the operation does not get used to deviations and that failures do not start; thus the importance of studying the task and monitoring the reliability of critical systems. Observation of the indicators and variables that represent the process state is followed by the monitoring of field signals, contamination, noise, odors, vibrations and visual signals that can show the beginning of a failure.

The analytical view of the task, accompanied by the recording of abnormalities in a disciplined manner and the observations and knowledge of critical equipment, is the base material for the initial investigations of the failure, which will be represented through graphs and tables described in the FMEA and FTA. When there is a strong influence of human factors, we also use FTAH and FMEAH graphs and tables.

8.5.3.1.1 Operational continuity in the fertilizer and caprolactam industry

Research on operational continuity-reliability in the chemical industry depends on the application of appropriate techniques for root cause investigation. Failures in the productive task can be the result of human errors and material degradation. Chemical plants for urea production and caprolactam production had serious reliability problems with the possibility of environmental impact. In the case of the fertilizer industry, incorrect decisions on the shift caused operational abnormalities and caused contamination of the effluent with ammoniacal nitrogen. In the caprolactam industry, plant shutdown events, resulting from compressor shutdown, in addition to impacting economic loss could contaminate the environment with benzene or cyclohexane.

Although the weekly Operating Instructions, the procedures for stopping and starting the plant, the respective checklists are followed in a standardized manner, the operational reliability is low, indicating a high cost of production, what to do? Review cognitive gaps that impact the functioning of technology with the effect of time and centralized management.

The Nitrogenated Fertilizer Company, despite the size and the production time, unfortunately only managed to produce 85% of the program, applied low standard solutions in the project, used old methane compressor in the CO compressor function, experienced unscheduled stops due to the lack of ammonia (compressor) and still has a negative environmental image due to continuous contamination of the effluent. The team does not believe in the company and the company changes its manager every two years but does not maintain a continuous production line. Operators are not committed and have job stability. The procedures are extensive and due to the high cost of production, the solutions are of a low standard. The technological update in Process Safety (new control panel and instrumented safety system) did not prevent explosion of ovens due to improper maintenance on the plant and incorrect strategic decisions. Some of the critical procedures were informal, without certification. The staff team was aware of the difficulties but was not involved with the operation, communication did not flow, the methods did not include analysis of human and social factors.

The caprolactam plant, on the other hand, was not a public company, it had a certain level of work organization, it had standards for critical processes and equipment, but it had a very high level of complexity and the product was in low competitiveness in the global market. Finally, the technology was at high risk for flammability and toxicity with a history of international accidents in Europe. We noticed that the team was motivated, but the staff did not know how to analyze the failure when it included loss of perception for systems acquired as a black box, such as the turbine speed that provided work for the hydrogen compressor in the transformation of benzene to cyclohexane.

8.5.3.2 D13 failure logic and connectivity

Since the project, the start-up of the plant and the operation, decision errors occurred, which hid or accumulated its effect, and instead of the Board acting on the root cause, it decided to purchase end-of-line equipment such as the cold chloride box methylene to kill gases in the exhaust gas system. The identification of the thread in these cases past ten years is done through observation at the workplace and the hypotheses elaborated from the operator's discourse (Ávila, 2010).

Once again, we confirm that the complex format of the systemic failure is built with inadequate corrective actions to make production possible immediately due to sustainability not being a problem at the time of the emergence of the failure. Thus the routine is to invest little money and resources in cases of deviations and failures in the most recent period of the plant.

An intricate relationship is built between the failure modes of processes-equipment including the social, human and organizational dimension are, unfortunately, protected by the BLAME culture where might is right, and the sensible obey.

If someone asked me how to solve this, I would say through excellent quality communication, upward, downward and across. In addition, through the mastery of Technology with well-planned tasks considering all dimensions of control and a team based on trust, commitment and leaders who seek through fierce determination to avoid accidents and provide time and tools for the investigation of the root cause, even if the cause emerged ten years before and may be related to incorrect decisions in the management of operations.

Unfortunately, this ideality is difficult to achieve due to the Operational Culture being "linked" in the EROS domain, where leaders consider that "these bad things" do not happen in my plant, therefore latent, or complex flaws that have high connectivity are not investigated.

8.5.3.2.1 Reduction of methylene chloride loss in the polycarbonate industry

The polycarbonate plant was small but had a cohesive operation team making it easy to solve some of the immediate problems. The losses with methylene chloride were giving the plant leaders "headaches." These losses were measured by general indicators and therefore difficult to indicate where the main causes are. Certainly, the losses were scattered in incorrect decisions and hidden in time. With an "apparent" good level of organization and an "apparent" goodwill of the team, it was difficult to locate the *bad boys* in the production theater. Therefore as you know, the analysis of the operator's discourse was applied to understand the format of the systemic failure and finally open the "Pandora's Box."

Around 20 chains of abnormalities were found from the investigation of the shift register and we focused on 6 to study the loss of methylene chloride. We then applied the fault logic and connectivity diagrams to complex systems. Here we present some considerations valuing the stage of study of the failure in PADOP to avoid that it from becoming an accident or crisis. Some of the main failure events with high connectivity and multiple dimensions are:

1. Reaction control affects stability in purification and effluent quality.
2. The plant's start-up can accumulate problems due to off-spec product and deficiencies in methylene chloride vaporization.
3. Living with old problems in alternative Alfa Laval pumps makes it difficult to control the use of the solvent, which in this case is an external sealing fluid.
4. Overconfidence in the operational leaders, in parallel with the excess of centralization at the plant's start, has the effect of group omission.
5. Incorrect strategic decisions to prioritize production in cases of unscheduled shutdown, reactor bore or kneader obstruction.
6. Trace steam to vaporize methylene chloride in the effluent would bring strategic problems and the transfer of blame to the staff.

These actions did not impact the immediate sustainability of the plant but brought the disease that degrades and kills the patient from one moment to the next through a major process accident with an impact on the effluent (7 tons of solvent for the external secondary handler), for the neighboring community (borehole in a poorly corrected reactor causing a big event with phosgene) or in occupational terms (man gets into equipment and dies melted with polycarbonate).

8.5.3.3 D14 failure materialization and energy

The tools indicated by PADOP following the guidance of operational and dynamic risk management and, therefore with the inclusion of human-cognitive-social factors, and applied in actual cases of the chemical process industry are:

1. Risk in Project and Task Review: PHA-SH (Ávila and Barroso, 2012a); Social HAZOP (Ávila et al., 2013a); and Socio-Human-Interface Matrix (Ávila and Pessoa, 2015c).
2. Failure and Accident Analysis: (Ávila et al., 2012c); FTAH (Ávila, 2011b); Bowtie (Ávila et al., 2016); LODA (Ávila, 2011a); Chronology and Danger Energy (Ávila, 2010, 2012b).
3. Decision under progressive stress: LODA (Ávila, 2011a).
4. Analysis of failure and human errors in the task: executive function and task failure (Ávila, 2015a); archetype analysis (Ávila and Menezes, 2015d); classification of human derived error (Ávila, 2010).
5. Analysis of the operator's discourse (Ávila, 2010).
6. Communication quality: organizational communication pyramid (dos Santos Drigo and Ávila, 2017).
7. Competency analysis: personality rule; and examination of technical and operational culture (Ávila, 2010).
8. Complexity analysis: logical diagram of the task failure and connectivity diagram (Ávila, 2010, 2012b).
9. Failure and mind map analysis: materialization and the body with danger (Ávila, 2010, 2012b).
10. Criticality: analysis of type, parallelism and risk in the actions that initiate the failure (Ávila, 2010, Ávila and Dias, 2017).

Here, in this topic, we will discuss experiences in the analysis of the mind map with the materialization of the failure in the event of an accident at the LPG plant.

8.5.3.3.1 Cause mode and solid in PLG plant

The materialization tool and chronology versus danger energy released with the fault have already been presented in Chapters 3 and 7 in the case of LPG. The decisions in the operation task brought physical consequences (pressure variation or appearance of solids) and release of energies in each uncontrolled phase until the disaster occurred.

Now when the presence of sulfur in the refinery's LPG purification is not resolved, for reasons hidden within the operation history of this project, there is an oscillation in its quality with the presence of variable humidity, making cooling difficult due to the generation of hydrates.

This hydrate cannot be retained in the drying columns and affects the quality of the cooling process. The materialization diagram indicates where the source of the noise is in the mental map (powers between supplier and customer as to who is responsible for the failure) and when that noise turns into physical uncontrolled process, hydrate. In this way, it makes it easier to indicate where to install safeguards to interrupt the failure process at the source or halfway.

8.5.3.3.2 Accident with fire ball, in the LPG industry

The events identified since the deviation stage, through the uncontrolled process (flow, temperature), passing through the phase where the coil in the furnace weakens with the failure of materials, the fire comes into direct contact with the process's LPG causing the explosion release of large amounts of LPG into the atmosphere. Unfortunately, the failures can come in chains, and apparently, the inertization after the maintenance of the sphere was poorly done, facilitating the transfer of fire to vessels in the process with series explosions. This event was built from the survey of the operator and staff discourse and completed with the investigation of historical accidents, causes and consequences. Dynamic analysis of events over a long period of time is plotted on a scatter plot against the energy release estimate, based on the discourse, and then we plotted the data on a biln plot.

This tool allows you to check from what point the event is totally uncontrolled and where the trigger for the disaster is given.

8.5.4 Task emergency assessment

The first experiences in industrial production require exploration of the best alternative for the stages of a task that had not yet been written, thus being the beginning of learning from practice. The objective is known and defined, at the end of the day, it could be verified that it was the same as the task accomplished. Once the best way to operate is confirmed, given the objective function and the control items suggested during task planning, the standardization process begins, which ends when the standard is validated through tests and with the consensus of the operation team. Thus the writing of the procedure, the annexes, and even the checklist is prepared. The auxiliary memory and cognitive processing tools are developed in the form of documents and software, then, developed and applied for training the team.

This standard must be revised at stipulated periods, but, as the risk is dynamic, to respond to feedback from the operation team, there may be a revision of the procedure due to spontaneous demand. The critical Observer-Facilitator of the task will confirm the need for an immediate review or within the normal validation process.

Due to influences from the western culture, risk aversion and "link with EROS," ignoring Thanatos, it may happen that the operating team has problems of lack of operational discipline, which may be the dimension of the degraded technical-operational culture. This cultural disease takes the form of normalizing the deviation in the operation (CCPS, 2019).

The accumulation of investigations not carried out on the root cause of problems can lead the industrial plant to a dangerous state of release from danger on a large scale, or, the event of the accident.

The researchers from the Research Group on operational and dynamic risks, understanding this phenomenon, know that stress can instigate cognitive processing to the point of greatly increasing the occurrence of human errors, with a probability of 10% in normal routine and reaching 70% with maximum stress (Ávila, 2011a).

This group has focused on important research such as the effect of stress on the decision and the body, it has also studied the various phases and events in the occurrence of the disaster. The researchers have tried to quantify the hazard energy to compose the network of human factors. Thus we discussed the progressive stress in the phases and committees organized to deal with the crisis: strategic intelligence; integrated management and decisions; local operations command; and organizational resilience.

8.5.4.1 E15 LODA—progressive and accident stress

In the article that discusses stress in oil platform embedded activities (Ávila et al., 2021a) we note that stress control is a capillary barrier as to the possibility of failure along with technological and cognitive control of complexity and control of communication and feedback.

We also note that the operating team that performs routine tasks must have characteristics that favor cognitive quality (Ávila et al., 2021b) to avoid failure in the critical task. So, having proper leadership in a high-risk environment helps a lot.

Having the necessary competency to solve problems, knowing how to work cooperatively in the team facilitates the redundancy of activities, and having the commitment so that the task reaches the business objectives.

Know how to listen and communicate in any hierarchy and at any time. Having the perception of events during the task and knowing how to work all the stress vectors: cognitive, organizational, physical, social, and unknown.

8.6 Product 3—cognitive quality

8.6.1 Introduction

The Operational Standard, as well as the failure, follows its own causality that can be altered by elements of the organizational (management), physical (material), or natural (climate) environment. The difference between the standard and the failure is that, while the standard procedure seeks to repeat activities to add value to the product or service, the failure disaggregates the company's value, decreasing the probability of being sustainable. The importance of choosing the best variables and indicators for quality, cost, and to preserve the positive image, differentiates companies in terms of business continuity.

From the point of view of sociotechnical systems and following the Swiss cheese model (Reason, 1997), the release of danger energy in the organizational environment can cause the accident from failures in the barriers or design safeguards. This danger energy migrates through its carrier body classified as technical or social. This interpretation is clearly seen when analyzing the failure scenario in complex systems, during the operator's communication between the field and the panel, in activities of the chemical industry supervisory. The communication noise uses the operator as the bearer of the mind map and this danger energy is transferred to the equipment through, for example, pressure loss. The discussion of cognitive quality is found at the interface between the operator's intrinsic ability in terms of competency and cognitive quality, versus, the process control technology supported by computer monitors and the respective closed loop controllers.

Product 3 of Chapter 2 will be based on this cognitive ability to respond to the demand for technology control. In this way, the evaluation of primary psychological functions, of learning functions in the dimension of concepts and practices, and of functions related to cognitive processing for decision, will be made to adjust the leadership and the team in management processes and technological interfaces.

This cognitive quality analysis is performed in a static-individual or dynamic-collective format. In the collective exam, a simulated exercise of a real scenario is done together with the cognitive analysis through techniques that characterize the Executive Function, that create the progressive dynamic of stress and that include decision processes with intrinsic variables. In this case, factors that occur during the event of failures and even accidents are considered, and are related to emerging movements, at first unknown but that change the human and social typology, thus requiring new adjustments in the organizational and safety culture. Behavior changes and unusual technological scenarios must be included in the analysis of cognitive ability that is the object of this product.

Competency assessment for unusual situations includes the ability to recognize the new scenario and to be able to develop solutions with the interpolation and extrapolation exercises and the principles of pseudo-concept by Vygotsky (1987). The Analysis of Scenarios of high stress and cognitive load is made based on the principles of operating groups and the LODA tool discussed in Chapter 2.

The assessment of cognitive quality establishes standards of comparison that depend on the job title, technological requirements (process, product, and equipment) and organizational requirements, leadership and stress level. Emotional balance makes cognitive stability possible.

8.6.2 Cognitive quality elements, functions and subfunctions

The elements that are part of the cognitive processing were discussed in the models and practical cases of Chapter 2 through the referenced authors (Ávila, 2015; Hollnagel, 1993, Muchinsky, 2004; Reason, 1990; Sternberg, 2008). Here in this evaluation work we divide into three types of elements based on psychological functions: primary, learning, and decision-action. These elements are subdivided and interact with technological-organizational and external threats. Each element is built during the professional's previous experience and learning and must match the current job position.

The performance characteristics of each function element indicates characterization of subfunctions. We will present here how deviations caused by threats occur from the application of subfunctions, preferably in real cases already presented.

The cognitive primary functions (Sternberg, 2008) are perception, attention and memory as already discussed in Chapter 2 (Section 2.2.1). Failed perception may indicate that the constructed image of the scenario is not complete. The image may be contaminated with behavioral aspects from the beginning, causing difficulties early in the decision. Often, the perceived scenario does not represent reality, causing delay reactions, or not being appropriate to the situation. The expected pattern is to be able to perceive reality without disturbance. The tests are related to the feeling of reality and urgency level.

Attention is a primary function that needs to be calibrated according to the indicated priority and tested after perception. Attention is a function required in the work environment and can be of two types, both important, for each demand or situation. Attention can then be focused or dispersed. The focused attention is important for the diagnostic activity and the dispersed attention for the general control of the process and the initial and final checking of the shift work.

The work of interpreting the event depends on memory operations with coding, repetition, deposit, and redemption. In addition, memory is related to the decision through operations of comparison, interpolation, extrapolation and noncognitive analyzes required during the emergency.

Short-term memory or working memory is related to learning through the installation of skills. Long-term memory can be classified as a knowledge base. Thus defects in memorization directly affect the competency of the professional and inner and outer threats alter emotional balance and rescue low quality memories, also depositing false perceptions. Therefore presenting the model that characterizes memory as conjugated and human error as a derivative.

The type of memory demand depends on the speed of the situation. Therefore the recovery of parallel memories depends on the practice of simulations, as there is not enough time for linear and consecutive operations.

The big problem is that preconditioning hinders the development of unusual scenarios due to impulsive action. This situation indicates the need to review training for emergency teams.

Reason (1997) indicates in his model the need for this practical, theoretical knowledge and for organizational standards to make the functioning of industrial processes feasible. Although a great indicator of learning is the practical and theoretical examination of skills, if there is an organizational threat that induces the worker to low commitment, the hours of learning are deflated, decreasing applied SKR as discussed in Chapter 2 (Section 2.2.6) in the Fig. 2.22.

The perception in Chapter 2 (Section 2.1) of the unusual does not depend only on SKR, it is important to have within the group people who project new causal sociotechnical relationships, and to create an appropriate environment for these interpretations. There is no point in training the team for nuclear fission emergencies if the problem is in the redundant utilities in TMI—Three Mile Island, Llory (1999).

The human mistakes made in the decision and action come from the primary cognitive and learning functions. In principle, failures in equipment and processes are therefore related to skill and knowledge. Those related to disruption of security principles are related to the lack of understanding of the importance of the agreed rules. And system failures caused by overconfidence may be directly involved with patterns that do not represent the real picture. Finally, installed competency indicates a lack, excess or inadequacy in the specific competency.

In decision and action functions, intrinsically related to the previous ones, it is important to have a level of commitment and an understanding of cooperation in group work to avoid human errors, bad habits and incorrect decisions. The subfunctions that must be preserved and trained are construction of a mental map for causality analysis, elaboration and control of communication interfaces, projection, and interpretation of future scenarios; choice of the best alternative based on the impact of future scenarios, disruption of the inertia with action at the right time keeping the activity under control.

Cognitive functions in Chapter 2 (Section 2.2.7) in the Fig. 2.24 are influenced by the organizational environment through external threats that are part of the project and are also influenced by the internal environment, intrinsic to man and the role of the group of operators. These influences alter decisions and cause human errors. Organizational Culture and Technology are stabilized based on barriers here named as human elements that combat threats to the project.

Visible threats such as conflict of priorities, mental burden resulting from the high risk and complexity of the task and the product, and mental burden resulting from external situations (market, climate, scarcity of resources), cause situations of low organizational efficiency. In this way, threats promote the diversion of primary cognitive functions, learning functions and functions related to decision and action.

When technology and organizational climate are known and dominated, deviations do not happen with enough intensity to cause human errors, incorrect decisions and bad team habits. This model is named as static cognitive quality (COGNQe) shown in Fig. 8.6, where events are known, and patterns can be controlled more easily.

In the situation where individual and group behaviors are unknown social emergent, it is difficult to diagnose and define barriers to avoid the accident. This model is named as dynamic cognitive quality shown in Fig. 8.8.

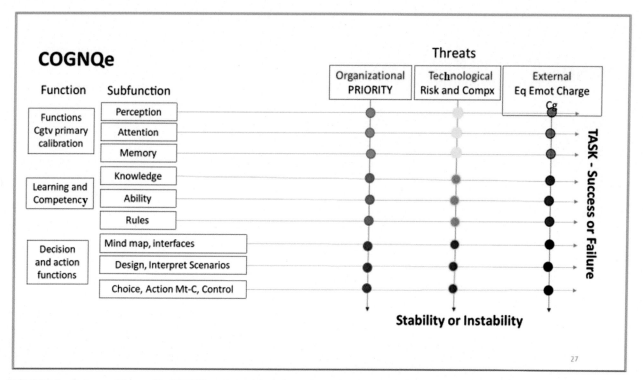

FIGURE 8.6 Static cognition quality COGNQe author original.

8.6.3 Static cognitive quality—COGNQe

This model intends to direct algorithms with procedures for data collection and calculations of cognitive qualities, standard and after routine threats. The definition of standard quality depends on the technological-organizational project to carry out the task within an organizational efficiency that meets the sustainability objective. The stability of barriers to address threats makes the initial design robust. The complexity of the functions follows the downward line (primary → learning → decision) and the color that signals the influence of the threat on the functions gradually increases in tone. Tests, questionnaires and workshops are carried out to identify the level of cognitive quality (Fig. 8.6).

In the book, we index the subjects for quantification as follows: Primary cognitive functions: perception of attention, memory in Chapter 2 (Sections 2.2.1; 2.2.2) in the Fig. 2.17. Learning functions and skills training: knowledge in Chapter 2 (Section 2.2.3) and Fig. 2.18, skills and rules. Decision and action functions: mind map and interfaces in Chapter 2 (Section 2.2.5) in the Fig. 2.19.

The classification of cognitive quality depends on the calculation (Fig. 8.7) as to the level of modification of pattern (functions) in relation to threats by type of technology and team. The questions asked about leaders and team will demonstrate that the average operator lives in the virtual world, thus changing reality, transforming it into more subjective than objective. This aspect affects the priority when a focused attention cannot be reached to make important decisions. Finally, it has the defect of not being able to encode this memory, reducing the quality of the current work. The application of questionnaires and the monitoring of the workplace during nonadministrative hours indicates that the final cognitive quality is 63%, below the minimum expected standard, which is 75%.

8.6.4 Discussion about dynamic cognitive quality—COGNQd

The study of static cognitive quality considers organizational, technological and external threats that affect primary, learning and decision-making functions. This investigation indicates human performance in execution, planning and diagnosis in the work environment. The threats to cognitive stability in COGNQe are more visible than when considering sociocultural emergent and the resulting social phenomena. Therefore it is difficult to detect failures through static tests on cognitive processing for cases that affect behavior resulting from the following events: gender conflict, generation conflict, multiculturalism, addictions, bad habits, and aspects of social culture, in regional terms or globally.

For cases related to behavior change in the face of dynamic culture, we suggest the application of COGNQd that works with the projection of an accident with the inserted cultural characteristics and following the LODA technique

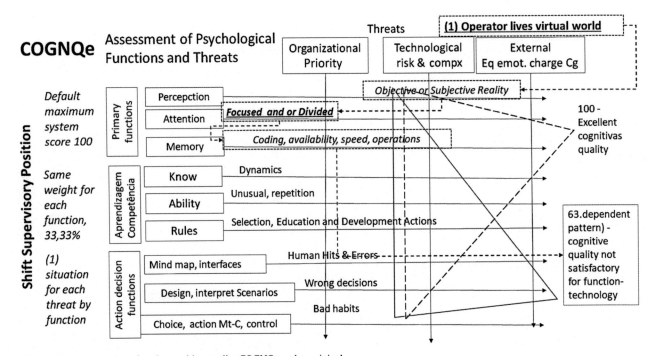

FIGURE 8.7 Calculation of static cognition quality COGNQe author original.

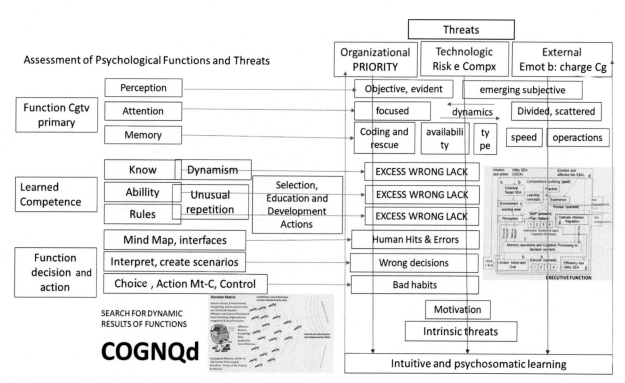

FIGURE 8.8 Dynamic cognition quality COGNQd author original.

(Ávila, 2011a) adapted from Pichon Riviere, Operative Group. The dynamics, through the projection of the accident in progressive states of stress, cause the individual protections of personality-behavior to fall, making explicit new characteristics in an emergency scenario. The elements, functions and subfunctions are the same as those of the COGNQe model (Fig. 8.5) adding in the methodology the description of an accident following the chronology and through an emotional discourse regarding the high impact that includes fatalities in the community. Fig. 8.8 presents the COGNQd model that measures primary and secondary cognitive functions in the same way as in COGNQe, but in the middle of the warm simulated exercise (emotional stamp). At this time, dynamic decision analysis is also carried out, where the priority changes according to the related social phenomenon and the impact on the team. For this purpose, the priority orbital model (Ávila, 2010) is used to review the dynamic influence of the external world and intrinsic behavioral changes.

Another tool included in the COGNQd, and already discussed in this book, is the Executive Function Analysis to dynamically assess, following the story of the accident, the human errors resulting from cognitive failures and cultural issues. This methodology will work on the relationship between the worker, his level of motivation and the work environment involved in the aforementioned social phenomena. Therefore there are latent threats here named as intrinsic that explain strange elements in the investigation of the accident, where the impact has direct contact without barriers, with a high level of danger energy and it seems that the cognitive apparatus is turned off.

This is reminiscent of the human learning process that indicates that in the previous stage verbalization equals the behavior of a child who has the same traits of schizophrenia. Some accidents are happening primarily.

8.7 Product 4—human reliability SPARH

To apply the SPAR-H method, knowledge of the task, processes, losses, and staff is required by the researcher, as already seen in the description of P1, P2 and P3. The SPAR-H method was initially created to serve the nuclear industry; however, as seen in Chapter 6 it can be adapted for other types of industries, requiring the calibration of performance factors (PSF's). Knowing about this calibration demand, a general flowchart was elaborated on the methodology proposed in this book, being followed in any industry segment (Fig. 8.9). It is worth mentioning that the Calibration step (4) must be carried out by specialists in the area, since both the calibration of the NHEP and some PSF can directly impact the results of HEP and Chu (human reliability calculation). Calibration is a practice that this book proposes to the applicator of the traditional method. The data collection step (2) needs to be applied together with the entire team

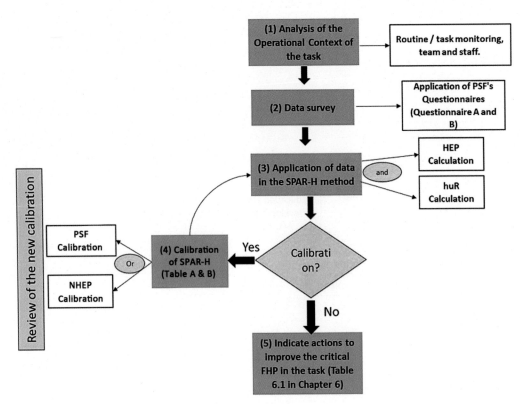

FIGURE 8.9 New methodology to SPAR-H assessment author original.

involved in the task. After the indications of actions (5) it is necessary to reassess and apply the SPAR-H method again in a certain programmed period to verify its effectiveness.

8.7.1 Operational context

The calibration of the SPAR-H method was presented in Chapter 6 in the release service for batch transport in the manufacture of sporting goods (Section 6.3.7). The case presents a problematic scenario, with a high rate of human and technical errors.

Initially, the original SPAR-H methodology was applied with NHEP and PSF values original to the method. From the results, HEP = 114%, values were found totally out of context, requiring a new calibration in the parameters used. For this, a real and ideal survey of the probability of error of this task was necessary, being quite different from the activities of the nuclear or chemical industry. We conclude, then, that this analysis of human reliability by the original SPAR-H calibration was inadequate for the reality of the task, as well as if we admit a 114% probability in the task would be to state that the occurrence of the failure is certain, and its frequency is high.

From this, a new calibration was proposed, using the method Fig. 8.10. This new calibration was performed by the person responsible for the PCP and based on the important points raised in the company (Section 6.3.7), the following changes were defined:

1. The PSF procedure was recalibrated, reducing the score from nonexistent to 20.
2. The PSF training/experience was also recalibrated for the diagnostic task, due to the alignment of action task values and diagnostic tasks (reduction of error severity), leaving 5 for diagnosis and maintaining 3 for action.
3. And the NHEP values were reset to: NHEP for diagnosis would be 0.001 and NHEP for action task would be = 0.01.

8.7.2 Performance factors assessment

Human performance factors at work can have their criteria weights modified due to different or new technologies, and due to cultural changes, that alter behaviors for safety and productivity. The indicative values consider the application

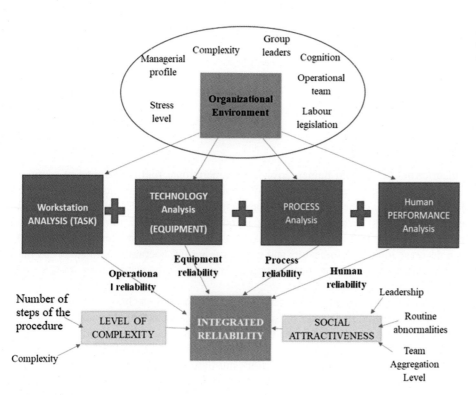

FIGURE 8.10 Practical method of sociotechnical reliability. *Author's Original.*

of SPARH for the Nuclear Industry that has in its Management and Project redundancies that are not repeated in relation to the chemical process industry. On the other hand, new technologies can alter characteristics that modify the risk and complexity involved in critical tasks, processes, and equipment.

In the case of changes caused by the Global or Regional Social Culture, which affect the operational routine, causing local social phenomena or even events related to the company's values in the globalized world, this can generate conflicts in the prioritization of decisions, in Organizational Belief and at work cooperative to carry out procedures.

The change in behavior during the work is caused by the change in the leaders' sense of justice, the great difference between policies and practices, leading to alter the perception of the organization's image toward society and the group. Culture (social and organizational) can affect the belief about the company and its leadership by modifying project characteristics. Design safeguards or human elements that resist the burden of cultural change are: stress control; quality in communication and feedback tools; elaboration and control of critical procedures; level of team cooperation; application of skills in routine and emergency; establish commitment to sustainability and organizational resilience; know how to plan the task in complex systems; leadership development for the emergency; and analysis of complexities and their dynamics in the operational routine.

A more assertive calibration needs real criteria to modify the probability of human error in the action task, as well as for the diagnostic task.

8.7.3 Calculation and calibration

After defining the new values for NHEP and PSF (Table 6.9), they were applied again in the calculation of the SPAR-H method, using the formulas:

$$\text{PSF}_{composto} = \sum \text{PSFr} \tag{8.1}$$

$$\text{HEP} = \frac{\text{NHEP} \cdot \text{PSF}_{composto}}{\text{NHEP} \cdot (\text{PSF}_{composto} - 1) + 1} \tag{8.2}$$

$$\text{HEP}_{Total} = \text{HEP}_{Diag} + \text{HEP}_{Ação} \tag{8.3}$$

$$C_{hu} = 1 - \text{HEP}_{Total} \tag{8.4}$$

With the calibration of the items, it was possible to adapt the HEP and Chu to be closer to the company's reality. Total HEP resulted in 71% and Chu in 29%. Human reliability is still extremely low, but it is justified by the frequency of mistakes made daily. The company urgently needs new procedures and task readjustment.

New PSF's can be updated over time, so that it can be adapted to the reality of the scenario. Calibration is flexible, provided it is done and validated by specialists in the analyzed area. The questionnaires, Figs. 8.9 and 8.10, can be applied both in the original method and in the recalibration of PSF's and NHEP.

8.8 Product 5—social-technical reliability

The main cause of accidents is human error. Along with human error in the operational routine, failures of routine or contingency management, inefficiency of leadership and communication tools, failures in maintenance of equipment and design failures, such as inadequate technology for use or technology adjusted to start the plant with low quality changes. These problems increase the likelihood of greater loss of process and accidents. The control of the production process tries to avoid human action directly in the task, increasing the risk with equipment and processes. The low integration of activities or functions can decrease reliability indicators and transform normal operations into abnormal ones. These events may be related to the inappropriate use of communication tools that affect the status of the process causing stoppages and accidents.

Product 5 was prepared using the methodology proposed by Ávila (2013b), named as sociotechnical reliability, or also called integrated reliability. This content was covered in Chapter 6 (Section 6.5). The sociotechnical reliability methodology appears with the demand to perform an analysis and quantify the integration between the technical and social reliability. The methodology is ideal for verification, knowledge and control of complex sociotechnical relations, which are not easily known or are intertwined. The investigation of sociotechnical relations must arise from the analysis of the profile of the leadership and the operational team with the type of technology and task proposed. It is important to emphasize that the skills of the leader are fundamental to define the nature, direction and impact of this management. On the other hand, it is considered that the leader—in his thinking, feeling and responses—is also shaped by organizational culture. Thus it is understood that the ability to understand and maintain culture is a prerequisite for having good leadership.

The Method (Fig. 8.10) that constitutes product 6 involves the following steps: identification of the operational context; mapping of functions, processes and activities that have a significant influence on reliability; system complexity analysis; individual reliability calculations, which are: human, equipment, operational and process reliability; and finally, the results of the previous steps are combined to estimate the reliability of the integrated system. After executing the methodology, it is important to carry out the validation based on the real economy with field data, map the reliability and carry out an intervention plan with improvement actions. Culture and manager profile, complexity and parametric equations, individuals reliability calculations, and sociotechnical reliability calculation with an example of hydrochloric acid are discussed.

8.8.1 Culture and manager profile

The approximation of policies in relation to practices brings the intended Organizational Culture closer to the installed Technical-Operational Culture. The informal rules of operation of the team take the set of industry operators toward the organizational objective. This result comes from the Managers' understanding of the importance of controlling human performance factors to maintain reliability indicators at high values. However, these Managers admit the multidisciplinary importance in solving problems and of active listening for the construction of future scenarios. In the discussion on Sociotechnical Reliability, we tried to classify the managerial profile and the organizational culture by the social attractiveness indicator demonstrated in the reliability calculation.

8.8.2 Complexity of parametric equations

According to Ávila (2013b), Perrow (1984) and dos Santos Drigo and Ávila (2017), complexity can be associated with aspects that cause noise in the mental map, hindering decision making in the task. To analyze the complexity, some points of the process must be analyzed, such as: the level of automation, equipment control functions, the question of cognition to perform the task and the verification of procedures. Table 8.1 was prepared for information on the aspects and the unit of measurement used to quantify the complexity in the Integrated Reliability methodology (Chapter 6.6).

To calculate the complexity, the score of items (1–5) of Table 8.1 in Eq. (8.5) is used (Chapter 6.6)

$$\text{Complexity level} = (1) \times (2) \times (3) \times (4) \times (5) \times 100,000 \qquad (8.5)$$

8.8.3 Individual reliabilities

The integrated reliability methodology aims to analyze the reliability of both technical and human aspects in a task/process, involving the following reliabilities: (1) equipment; (2) operational; (3) process and (4) human (Chapter 6.6).

8.8.3.1 Equipment reliability (Req)

The reliability of the equipment should not be analyzed in isolation, but in conjunction with other sectors of a production system for greater efficiency and better results. Accordingly, the reliability of each sector or function is interconnected and depends on the level of automation, type of process, management style and organizational environment.

The equipment reliability calculation (Req) is related to the capacity of the studied system to remain operational during its use. Following the methodology proposed by Lafraia (2001), an exponential distribution was used, which considers that the equipment has a constant failure rate and whose only dependent variable is the Failure Rate (λ). Reliability is a Function of Time; it is not a predefined number as indicated by the formula:

$$R(t) = e^{-\lambda t} \tag{8.6}$$

To know the reliability of two or more equipment that work together in the process, the reliability of the system must be evaluated. In the serial system, reliability decreases, because if any component of the system fails, the entire production process will stop. In the case of parallel systems, the reliability increases, since there are other components that will serve as a back-up in the event of a failure. That is, the production process does not stop. To know the reliability of two or more equipment, we have the reliability for a series system in Eq. (8.7) and for a parallel system in Eq. (8.8):

$$R(t) = R1 \times R2 \times R3 \times \ldots Rn \tag{8.7}$$

$$R(t) = 1 - (1 - R1) \times (1 - R2) \times (1 - R3) \times \ldots (1 - Rn) \tag{8.8}$$

8.8.3.2 Operational reliability (Rop)

The calculation proposed by Ávila (2013b) for operational reliability was performed based on the complexity of the task [Eq. (8.5)].

$$R_{op}(\%) = \left(100 - \left(\frac{\text{LOG(complexity)}}{4}\right)\right) \times 100 \tag{8.9}$$

8.8.3.3 Process reliability (Rpr)

To assess process reliability, it is necessary to know, analyze and identify the process, the product and the task indicating which main process variables cause the plant to stop. The issue of process conformity and product quality must be analyzed. Process reliability is a result of process capability. As we are not going to discuss here in the book the reliability of processes, we indicate the premise that because they are current chemical processes and technologies, we managed to reach 5 Sigma in the variables that cause plant shutdown, that is, 99%.

8.8.3.4 Human reliability (Rhu)

To present the Integrated Reliability product, the SPAR-H method for human reliability was used. The formulas are already known and demonstrated in Chapter 6 (Section 6.2.2). The product 4—human reliability SPAR-H, also demonstrates the SPAR-H methodology for applying the original method and for cases that need recalibration.

It is important to remember that, using the formulas of the probability of human failure (HEP) through Eqs. (8.10) and (8.11), occurs when there are less than 3 negative PSF in the operator's performance, with more negative PSF, ideally Eqs. (8.12) and (8.13). Eq. (8.14) will be used for both HEP situations.

$$\text{HEP}_{\text{Diag.}} = \text{NHEP}_{\text{Diag.}} \times \prod \text{PSF} \tag{8.10}$$

$$\text{HEP}_{\text{Action}} = \text{NHEP}_{\text{Action}} \times \prod \text{PSF} \tag{8.11}$$

$$\text{PSF}_{\text{composed}} = \sum \text{PSFr} \tag{8.12}$$

$$\text{HEP} = \frac{\text{NHEP.PSF}_{composed}}{\text{NHEP.}(\text{PSF}_{composed}-1) + 1} \tag{8.13}$$

$$\text{HEP}_{Total} = \text{HEP}_{Diag} + \text{HEP}_{Action} \tag{8.14}$$

After the identification of the Probability of human failure, the result is applied in Eq. (8.15) to know the human reliability of the task.

$$R_{hu.} = 1 - \text{HEP} \tag{8.15}$$

8.8.4 Social-technical reliability calculation

Integrated reliability depends on the level of complexity and social attractiveness. The level of complexity of the process / task is determined from the evaluation of experts in the field. To determine the level, one must analyze stages of the processes, procedures, technology used and manufactured product. The equations for the respective levels were presented in Chapter 6 (Section 6.5.1), but will be recalled below:

$$\text{High complexity} \to \text{IR} = \left(R_{pr}\right)^2 \times R_{eq} \times \left(R_{hu} \times R_{op}\right)^{1/2} \tag{16}$$

$$\text{Average complexity} \to \text{IR} = R_{pr} \times R_{eq} \times \left(1.5 \times R_{hu} \times R_{op}\right) \tag{17}$$

$$\text{Low complexity} \to \text{IR} = \left(R_{eq}\right)^{1/3} \times R_{pr} \times \left(1.7 \times R_{hu} \times R_{op}\right)^2 \tag{18}$$

For the calculation of social attractiveness, one can mention the relationship between monitoring situations involving culture (organizational, security, regional, and global) with human errors, such as: conflicting priorities between security and production; inadequate communication or feedback; conflict between policy and practice; violation of the stereotype population; standardization of deviation; cultural flexibility; style in team coordination; organizational change and external environmental factors resulting from social phenomena. These human-organizational factors define the level of social attractiveness that must be estimated in this method to correct the calculated sociotechnical reliability.

$$\text{IR}_{final} = \text{Attractiveness} \times \text{IR} \tag{19}$$

For Ávila (2015b) this case, social attractiveness can be classified into:

1. Very good → 1.2
2. Medium → 1
3. Low → 0.90

8.8.4.1 Hydrochloric acid case

The sociotechnical reliability method was applied to the case of a chemical industry, in the task of loading hydrochloric acid (HCl), already presented in the context of the case in Chapter 6 (Section 6.5). Considered as equipment analysis, two pumps that participate in the pumping of HCl and that are in parallel.

About the process involving HCl, the temperature and concentration conditions guarantee control of the process and are maintained through it, in addition to the redundancies and safeguards of the system, being framed in the 5 SIGMAS, resulting in the reliability of the process (Rpr) in 99%. Human reliability already calculated in item 6.3.3 resulted in 74.6%. Equipment Reliability was performed on pumps, with the reliability of each pump 71% and the system reliability considering two parallel pumps 92%. Operational Reliability resulted in 80.5%. From these results, and considering the average social attractiveness (value = 1) and the low complexity [Eq. (8.18)], there is [IR] final = 92.6%.

8.9 Product 6—operational-technical culture and prediction

8.9.1 Introduction

Product 6 indicated in Fig. 8.11, analysis of the technical-operational culture and prediction of future production campaigns is the result of Chapter 7 of this book. The identification of the human, social and technical, or technological

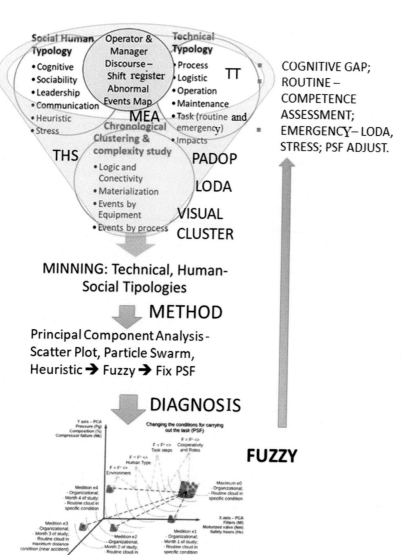

FIGURE 8.11 3rd generation operational culture identification and near miss position NEMESIS.

typology, together with the mapping of Abnormalities in the shift work (collected from the occurrences recorded by the operator or the control system) will feed a descriptive analysis of data to identify critical region, where the main causes of the accident are probably located.

The descriptive statistics of the variables and parameters of the process, logistics and operation, the maintenance, safety and environmental impact indicators, the types of maintenance specialty involved in the accident indicate cyclical phenomena and saturation points that cause plant shutdown. Or, these variables may indicate spurious events resulting from new technological or social phenomena.

A confirmation or validation of the datamining result based on this descriptive statistic is made from the chaining of events by date and following the logic of the process and equipment failure. We will do a visual analysis of abnormality cycles for this confirmation.

The selected variables of process, maintenance, operation, task (PADOP and LODA) join with the human typology through the analysis of main components, generating a dispersion graph that shows the tendency of stability or instability of the sociotechnical system. The mapping of the operation history and the informal rules in the routine indicate which technical-human and social factors are to be corrected. After addressing the cognitive or work organization gap, after adjusting the factors that indicate abnormalities in the scatter plot, based on organizational efficiency, we will have a program with interventions to be carried out on the team, management, and technology.

8.9.2 Methodology and products

The methodology indicated in Chapter 7 was built based on new concepts detailed in previous chapters and some discussed here. The sociotechnical system influenced by nonvisible rules affect the company's results. These rules do not have an explicit physical relationship between the aspects related to technology, people, the task, and the culture that involve it, but there are informal rules not visible between these factors of production. We must learn how these rules work on technology, equipment, and processes.

We will review the methodology indicating its objectives, inputs and outputs, briefly discussing the tools in application in the sociotechnical system, in the current case for the gas processing industry. The algorithms to identify the technical, human and social typologies were essential and possible from the introduction of these new concepts.

We conducted this discussion in Fig. 8.12, where we proposed the application of programs and tools to correct the culture, revise the task, change the worker's behavior, improve practices in the work routine and, as a result, increase the equipment efficiency. We showed all the tools, inputs, outputs and some important concerns.

The activities for elaboration of the Method are listed and continued with proposition of Barriers to contain the flow of Hazard Energy in Production, in our specific case, Liquefied Petroleum Gas processing. In Fig. 8.12, we retrieve tools used so far in this Chapter 7, show the resulting analysis tools discussed in Section 7.3, and anticipate the discussion about tools that unfold to this methodology to achieve the results.

The 24 subproducts and steps of the methodology or research activity to identify the Operational Culture with important PSFs and the correct barriers are:

1. The technology and task: know the technology, the task for competency analysis, and the expected behavior.
2. Experience: acquire experience to identify the heuristics of behavior in the routine—workplace.
3. New concepts: elaboration of concepts in sociotechnical systems that reproduce history: tests in industry or service.
4. Data collection tools: testing and proposition of tools for data collection and statistical processing.
5. Procedure analysis: prepare procedures and test for competency, behavior, commitment, cooperation, task.
6. Abnormalities investigation: investigation about abnormalities in the shift.
7. Research to mine for data: data mining representing the technical and operational culture: the variables chosen from the MEA (mapping of abnormal events), EVA, AEP (statistical process analysis) are analyzed and indicated for PCA.

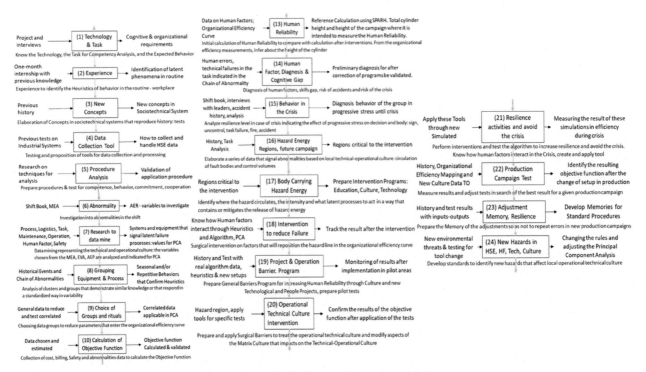

FIGURE 8.12 Tools for a resilient and sustainable company, finding future accident author original.

8. Grouping techniques: analysis of clusters and groups that demonstrate similar knowledge or that respond in a standardized way in variability.
9. Choice of groups and rituals: choosing data groups to reduce parameters that enter the organizational efficiency curve.
10. Calculation of objective function: collection of cost, billing, safety and abnormalities data to calculate the objective function.
11. Identify process losses and data for PCA: build organizational efficiency curves indicating how variable groups move in relation to the objective function (organizational efficiency).
12. Efficiency and objective function analysis: analysis of responses in the objective function and representing organizational efficiency.
13. Human reliability: initial calculation of human reliability to compare with calculation after interventions. From the organizational efficiency measurements, infer about the height of the cylinder.
14. Human factor, diagnosis and cognitive gap: diagnosis of human factors, skill gap, risk of accidents and risk of the crisis.
15. Behavior in the crisis: analyze resilience level in case of crisis indicating the effect of progressive stress on decision and body: sign, uncontrol, task failure, fire, accident.
16. Hazard energy regions, future campaign: elaborate a series of data that signal abnormalities based on local technical-operational culture: circulation of fault bodies and control volumes.
17. Body carrying hazard energy: identify where the hazard circulates, the intensity and what latent processes to act in a way that contains or mitigates the release of hazard energy.
18. Intervention to reduce failure: surgical intervention on factors that will reposition the hazard line in the organizational efficiency curve.
19. Project and operation barriers program: prepare general barriers program for increasing human reliability through culture and new technological and people projects, prepare pilot tests.
20. Operational technical culture intervention: prepare and apply surgical barriers to treat the operational technical culture and modify aspects of the matrix culture that impacts on the technical-operational culture.
21. Resilience activities and avoiding crisis: perform interventions and test the algorithm to increase resilience and avoid crisis. Know how human factors interact in the crisis, create and apply tool.
22. Production campaign test: measure results and adjust tests in search of the best result for a given production campaign.
23. Adjustment memory, resilience: prepare the memory of the adjustments so as not to repeat errors in the new production campaigns.
24. New hazards in HSE and production: develop standards to identify new hazards that affect local operational technical culture.

8.9.3 Management programs for human and sociotechnical reliability

The management program to increase human and sociotechnical reliability was divided into 4 sequential-parallel modules or subprograms with implementation planned for 3 years. They are:

1. Program 1: Indicators of tendency to human error and technical failure
2. Program 2: Routine management for environment favorable human reliability
3. Program 3: Managerial to validate the failure analysis model
4. Program 4: Opportunities for logistics and safety operations management

8.9.3.1 Program 1: human error and technical failure trend indicators

The objective of this program is to gather information from the technical, human and social typology indicating procedures to change environments, structures, people and management processes in order to enable the increase of human and operational reliability in the industrial plant.

Through these procedures, it is intended to direct the routine management to the maximum points of organizational efficiency and safety in the company. By grouping scattered knowledge about latent failure processes and making them available for the development of competencies, and by motivating agglutinating tendencies in the shift team, it becomes possible to increase Human Reliability.

The steps of the program are: (1) Implement Program to Increase Human Reliability; (2) Analyze critical tasks in routine and emergency situations; (3) Patterns of behavior with elements of the Human Reliability; and (4) Classification of Human Error.

The important work of mapping and identifying technical, human and social typologies, complemented with competency analysis provides information with enough clarity for the formulation of programs and procedures.

8.9.3.1.1 Implement program to increase human reliability

The implementation of this stage of the first program is to provide management programs to increase the human reliability calculated/estimated directly, or indirectly, together with tasks in the field. The measurement of the increase in human reliability is given either by reducing the rate of human failures in technical activities or by increasing the time between human and operational errors. This number can also be composed of the volume processed, the rate of occupation of the worker (or the reverse, technical idleness) and the number of signs of abnormality in the shift.

The rate of human failure in field action tasks, surveillance, problem diagnosis and decision making should be reduced initially by 15%. The time between human and operational errors should be over 72 hours, fluid alignment errors requiring rework. The processed product volume will be increased by at least 20% with human reliability work. The rate of occupation of man in the task of at least 90% and the number of signs of abnormality per day, recorded in the occurrence register or detected in the shift, computed at a maximum of 3 per day. These signals can be seal leaks, dirty filter, equipment disempower, failed safety test and others. This program is premised on the existence of an organized database that involves aspects of operation, safety, and people.

With the confirmation of typologies and the relationship with the human performance factors presented by API770 (Lorenzo, 2001) and Hollnagel (1993), it becomes possible to implement a program for Human Reliability. With the direction defined, for the program to enter the routine management, it is important to systematize database in the routine and control items to measure the results of the program.

The human reliability Program includes these stages of implementation; initially validate the human and social typology with specific workshop including operational leaders. Then, relate performance factors and social-organizational environments to be changed based on API770 considerations. The stages also include deployment of database in routine for HRA, develop, and apply of procedures to measure targets through operational indicators. With the analysis of critical tasks, we can implement adjustments in the shift team and leaders, and we can develop skills and leadership.

8.9.3.1.2 Analyze critical tasks in routine and emergency situations

This step aims to identify the roadmap to analyze the task based on the function of each tool, then allowing the choice of the applicable tools. In addition to indicating an Action Plan to implement the recommendations for task failure analysis, prioritize the chains of abnormality and techniques for the analysis of critical tasks in routine and emergency and prepare a roadmap to formulate the best practice manual in a simplified way and with the corresponding field booklets.

The analysis of critical tasks is essential to prevent operational and human failure. The levels of complexity of the tasks are different from each other and it is therefore important to analyze which techniques will be used for each case, without which it would be expensive to analyze all the tasks or even review them. Certainly, the most complex tasks will be analyzed in detail.

Knowledge of the task by the operation and availability of important aspects of failure for the learning of the shift enables the implementation of self-management.

The steps for analysis of the critical task are standardize roadmap for analyzing the task that depends on the complexity and risks involved; implement recommendations in the failure analysis of asking; prioritize chains of abnormality in task for analysis; and present a roadmap to prepare the Good Practice Manual.

8.9.3.1.3 Defining behavior patterns with elements of human reliability

In this program, it is intended to define behavior patterns for leaders and operators, in addition to defining which elements guide the behavior toward increasing Human Reliability. This discussion of behavior patterns should be available in two formats: primer and document used for training.

Knowledge of expected operating patterns allows adequacy of behaviors and induces employee satisfaction. Receiving feedback of the standard expected by organization and management regarding competency, motivation and attitude, brings an improvement in the quality of the service provided.

Communication with the group also clearly differentiates the role of the leader in relation to the role of the operator. In this way, leadership is valued to achieve behavior change. The use of behavior booklets facilitates the implementation in the routine because it is written with figures and words used in the daily routine.

This subject little discussed should be democratized with the team and reviewed periodically (minimum 2 years) to avoid surprises of cultural changes imposing behavior alteration without which, stereotypes in the team would result. The purpose is to avoid conflicts in the shift and decrease the effect of behavior fluctuation.

At certain times the team will demand special attention from leader for behavioral change campaign, this being the subject of the Future Operation Manager.

The steps for elaboration, dissemination and implementation of behavior patterns are: performance in workplace; workplace behaviors; human error in the task; analysis for Human and Social Typology; primary measurements and relationship analyses; and deployment program.

8.9.3.1.4 Human error classification

Abnormality identification chains and task through critical analysis clears environments, for verification of the latent failure and for identification of root causes. With the deployment of a database for human reliability, it is possible to classify human error and direct the investigation of cause. This procedure aims to identify type of human error by class and indicate environments to be corrected.

The complexity of this work requires an adequate methodology that separates possibilities of environment, the type of task, the group relationship, and the operator's cognitive relationship, causing errors about the productive systems. There is still the possibility of equipment and its interfaces, as well as design aspects related to classification. The management of a database of human failures and errors depend on its classification and chronological, physical and communicational relationships.

The goal of the activity is to include this classification in shift control through automatic data entry systems, not increasing assignments for the operation.

With the classification of human error and social and human typology, in addition to the procedure for standard behavior, it becomes possible to take preventive actions.

The classification of human error initially prioritizes the investigation of failure and human error. In this way, it can classify human error including the composition of respective derived human error. We recommend constructing a database management to classify this derived human error. We can classify each abnormality by risk and application of financial and human resources. Then, the manager needs to identify the critical areas, functions, workers, leaders, and equipment. The organizational level of the failure includes strategic, tactical, operation, and emergency concerns. Finally, with the investigation about the informal rules of human behavior, we can write the activities necessary for the correction of human performance factors that implement these activities and update human reliability program with the respective indicators.

8.9.3.2 Program 2—routine management for environment favorable to human reliability

Routine management program aims to develop a favorable environment for increasing human reliability and consequently operational reliability, dealing with routine human and social factors. In this program, intended to suggest procedures, diagnoses, campaigns, and standards that will support gain of human reliability. The survey of social and human typology and respective statistic treatment mixed with technical analysis of process, help to indicate deficiencies in several areas.

1. Technical skills need to focus their actions based on the history of the number of failures.
2. The integration of groups must be dynamic in so as not to crystallize positions without resilience to threats.
3. The integration is very important for the operation, but maintenance must also be included in this commitment.
4. Digital communication tools should be reviewed to avoid copy-paste.
5. A critical reading of unresolved pending issues should be made to avoid critical vision loss.
6. Review communication tools, including procedures and booklets.
7. In the routine management, a real interest of the team in search of productivity and safety should be developed, and it is necessary to cause the de-structuring and search for an appropriate balance for the "new times."
8. The operation shift classes are "marked" and need to be reviewed seeking new centering and adjustments in the operator's behavior.
9. Leaders need to be trained to understand the need for this new centering, both informal and formal.
10. Some suggestions to adjust the management profile-seeking adequacy in relation to paradigm breakage.

The implementation of the reliability program is complemented by the adjustments proposed herein and that need to be reviewed by "neutral auditor" of behavior and analysis of results in the social, human, technical and safety areas. The suggested programs always have pilot stage, validation and standardization, considering that they are suggestions based on nondeterministic methods. It is not understood that revisions or nondeployments as new standards of certain systems are considered errors, but rather learning in search of operational and safety excellence.

In this context, we present the tools to be implemented in program 2, routine management: competency adequacy program; integration campaigns with rituals of inclusion operation and maintenance; review of communication tools for occurrences in shift; routine management to break productivity and safety paradigms; team adjustments leadership training; and suggestions for adjustments to manager profile. These program 2 target will be developed in the future, after the discussion of human organizational and social factors. This subject is addressed in a future book.

8.9.3.3 Program 3—management to validate the failure analysis model

Diagnosis on human reliability indicates the need for new technical tools to support the prevention of failure. This proposed a new management tool to implement statistical monitoring of processes and events aiming at identifying changes in variables and pattern reviews, rapid identification of changes in technical and operational culture based on routine warnings and the definition of which human factors should be adjusted to avoid the accident, and standardization of failure prediction tools.

Although there have been interviews, questionnaires, application of exams and meetings with the operation team, supervisors, staff and the coordinator, it is necessary to revise and validate for application, with a higher level of certainty, the recommendations made in this research work explained in the book.

Technical-social and human diagnoses are sufficient to initiate efforts to increase human reliability, and thus increase guarantees on specific programs, especially technical ones. The plant as well as human degrades, and it is necessary to revalidate technical-human-social typologies after a time interval of 3 years. These program 3 subjects will be developed after discussion on tools to analyze the degradation of identified sociotechnical systems. This subject will be approached in a future book.

8.9.3.4 Program 4—opportunities for logistics and safety operations management

Process control supervisors can have built-in design flaws that lead to human error. Improving process reliability requires changes to the control screen and maintenance management. The revised maintenance program should include new rules in calibration of safety valves, new tools for the analysis of the consequence of failures involving product quality, for example, moisture and corrosion. These new criteria in area of human reliability demand changes in reliability program of equipment and instruments.

The oil and gas industries might receive equipment and instruments that do not meet the final demand of the operation: a valve that does not work by CV, poorly sized moisture instrument, filter that does not work by failure in pneumatic system design and many others. It is important to discuss the relationship between projects and adequacy of use in the operation.

These program 4 subjects will be developed with the introduction of human reliability criteria in the project area. We call attention to changes in supervisory and maintenance management.

In the supervisory area, we can mention new logistics screen and quality to avoid human error; new safety screen to avoid the accident by monitoring routine deviances; and more criteria for the work in the design, in the specification of operating conditions to avoid failures with control meshes and analyzers.

Review the procedures for PSV audit is important in maintenance management, for the quality control of liquified gases and petroleum products, negotiations with the supplier to adjust the quality of the raw material; improvement of electrical systems, valves, compressors and filters; and improved reliability of critical instruments and equipment in the fire system and in the safety control of the furnaces.

References

Ávila, S.F., 2010. Etiology of operational abnormalities in the industry: modeling for learning. 296. (Ph.D. thesis in Chemical and Biochemical Process Technology)—Federal University of Rio de Janeiro, School of Chemistry, Rio de Janeiro.

Ávila, S.F., 2011a. Dependent layer of operation decision analyzes (LODA) to calculate human factor, a simulated case with PLG event. In: Proceedings of the Seventh GCPS Global Conference on Process Safety, Chicago.

Ávila, S., 2011b. Assessment of human elements to avoid accidents and failures in task perform, cognitive and intuitive schemes. In: Proceedings of the Seventh Global Congress on Process Safety, Chicago.

Ávila F.S., 2012b. Failure analysis in complex processes. In: Proceedings of the Nineteenth Brazilian Chemical Engineering Congress—COBEQ, Búzios.

Ávila F.S., 2013b. Review of risk analysis and accident on the routine operations in the oil industry. In: Proceedings of the Fifth CCPS Latin America Conference on Process Safety. Cartagena das India's.

Ávila, S.F., 2004. Methodology to minimize effluents at source from the investigation of operational abnormalities: case of the chemical industry. Dissertation (Professional Master's degree in Management and Environmental Technologies in the Productive Process). In: Teclim, Federal University of Bahia, pp. 115–p.

Ávila, S.F., Silva C.C. 2015a. Analysis of cognitive deficit in routine task, as a strategy to reduce accidents and increase industrial production. European Safety and Reliability Conference—ESREL. Zurich.

Ávila, S.F., 2015b. Reliability analysis for socio-technical system, case propene pumping. Engineering Failure Analysis 56, 177–184.

Ávila, S.F., & Barroso, M.P., 2012a. Human and social preliminary risk analysis (HS-PRA), a case at LPG site. In: Proceedings of the Eighth Global Congress on Process Safety—GCPS, Houston.

Ávila, S.F., Mendes P. C. F., Carvalho V., Amaral J. 2012c. Human factors analysis in turbocharger equipment failure FMEAH in chemical plant. Proceedings of 27th Brazilian Maintenance Congress. Rio de Janeiro.

Ávila, S.F.; Dias, C., 2017. Reliability research to design barriers of sociotechnical failure. In: Proceedings of the Annual European Safety and Reliability Conference, Portorož.

Ávila, F.S., & Menezes, M.L.A. 2015d. Influence of local archetypes on the operability & usability of instruments in control rooms. In: Proceedings of European Safety and Reliability Conference—ESREL, Zurich.

Ávila, S.F., Pessoa, F.L.P., 2015c. Proposition of review in EEMUA 201 & ISO Standard 11064 based on cultural aspects in labor team, LNG case. Procedia Manufacturing 3, 6101–6108.

Ávila, S.F., Pessoa, F.L.P., Andrade, J.C.S., 2013a. Social HAZOP at an oil refinery. Process Safety Progress 32, 17–21.

Ávila, S., Ferreira, J., Sousa, C., & Kalid, R., 2016. Dynamics operational risk management in organizational design, the challenge for sustainability. In: Proceedings of the Twelfth Global Congress on Process Safety, Houston.

Ávila, S., Cerqueira, I., & Drigo, E. 2018, July. Cognitive, intuitive and educational intervention strategies for behavior change in high-risk activities- SARS. In: Proceedings of the International Conference on Applied Human Factors and Ergonomics, pp. 367–377, Springer, Cham.

Ávila S.F., Ávila R.C.S., Ávila J.S., 2020. Dynamic and operational risk at oil exploration and related services, human reliability and energy efficiency— ReSoTech management system. In: Proceedings of the IFEDC International Field Exploration and Development Conference 2020, Chengdu.

Ávila et al. 2021a. Stress–complexity–communication on oil platform & mental map gaps, situations of routine and emergencies of JBH, the operator. In: Proceedings of the Virtual Global Congress for Process Safety.

Ávila et al. 2021b. Continuous stress and consequence on C4t, complexity, risk, and necessity of leadership level 5. In: Proceedings of the Applied Human Factor and Ergonomics Conference, New York.

CCPS, 2019. Center for chemical process safety. In: Recognizing and Responding to Normalization of Deviance. ISBN 978–1–119–50670-6.

Cerqueira, I., Drigo, E., Ávila, S., & Gagliano, M., 2018, July. C4t: safe behavior performance tool. In: Proceedings of the International Conference on Applied Human Factors and Ergonomics, pp. 343–353, Springer, Cham.

dos Santos Drigo, E., Ávila, F.S., 2017. Organizational communication: discussion of pyramid model application in shift records. Advances in Human Factors, Business Management, Training and Education. Springer, Cham, pp. 739–750.

Embrey, D., 2000. Preventing human error: developing a consensus led safety culture based on best practice. Human Reliability Associates Ltd., London, p. 14p.

Hollnagel, E., 1993. The phenotype of erroneous actions. International Journal of Man-Machine Studies 39 (1), p.1–32.

Lafraia, J.R., 2001. Manual of reliability, maintainability: and availability.

Llory, M., 1999. Industrial Accidents, the Cost of Silence. Editorial Mediation, Rio de Janeiro.

Lorenzo, D.K., 2001. API770—A Manager's Guide to Reducing Human Errors, Improving Human in the Process Industries. API Publishing Services, Washington.

Muchinsky, P.M., 2004. Organizational Psychology, seventh ed. Pioneer Thomson Learning, São Paulo.

NUREG 6883, 2005. United States Nuclear Regulatory Commission Regulation—NUREG—the SPAR-H Human Reliability Analysis Method. United States Department of Energy, Washington.

Ohno, H., 2000. Analysis and modeling of human driving behaviors using adaptive cruise control. In: Proceedings of the Industrial Electronics Society IECON Twenty-Sixth Annual Conference of the IEEE 2000, vol. 4, pp. 2803–2808.

Perrow, C., 1984. Normal Accidents: Living With High-Risk Technologies. New York, NY: Basic.

Rasmussen, J., 1997. Risk management in a dynamic society: a modeling problem. London: Elsevier Safety Science. England 27 (2/3), 183–213.

Reason, J., 1990. Human Error. Cambridge University Press.

Reason, J.T., 1997. Managing the risks of organizational accidents. Ashgate Publishing Limited, Aldershot, ISBN, (13), 978. 252 p.

Shingo, S., 1996. The Toyota Production System from the point of view of Production Engineering. Bookman, Porto Alegre.

Sternberg Junior, R., 2008. Cognitive Psychology. Artmed, Porto Alegre.

Vygotsky, L.S., 1987. Thought and Language. Martins Fontes, São Paulo.

Annex

Table A.1 Chapter, products, and annex, OTC operational–technical culture.

Annex		Context	Chapter/product	Subject
A. Process lossing map		Model for loss mapping	Chapter 4/Product 01	Loss of process
B. Calibration of spar-h method	B1-Operational context SPAR-H	Questionnaire for the analysis of the operational context, for operational team & staff/managers.	Chapter 6/Product 4	SPAR-H
	B2-PSF calibration SPAR-H	Parameters for calibrating the PSF's, for action and diagnostic task	Chapter 6/Product 4	SPAR-H
C. Task assessment	C1—Standard task data collection—PADOP	Task data collection form	Chapter 3/Product 2	Task analysis: PADOP
	C2—Collection of failure data for task assessment	3-part form for collecting data on task failure and abnormal events	Chapter 3/Product 2	Task analysis: PADOP; MEA
	C3—Data collection for emergency assessment	Form for task emergency assessment	Chapter 3/Product 2	Task analysis: PADOP
D. Technical survey	D1—Technical data collection for OTC	Form for collecting technical data of the process	Chapter 7, Product 6	OTC
	D2—API 770 Data & OTC	Form for analysis of API 770 components and operational–technical culture	Chapter 3 and 7/Product 6	OTC; API 770
	D3—Continuous process technology requirement	Form for assessing technology requirements in continuous processes	Chapter 7, Product 6	OTC
	D4—Routine data collection for OTC	Information board for routine data collection for operational–technical culture	Chapter 7, Product 6	OTC
	D5—Process and production variables collection	Form for evaluating process and production variables	Chapter 7, Product 6	OTC
E. Social typology	E1—Data collection to political aspects & practices	Form to evaluate aspects and practices of the company	Chapter 3	Social typology
	E2—Data collection for social relations	Data collection form—social typology	Chapter 3	Social typology
F. Human typology	F1—Data collection to emotional Imbalance	Data collection form—human typology	Chapter 3	Human typology
	F2—Data collection for cognition quality & OTC	Approach to assess cognitive quality and OTC	Chapter 3; Chapter 7, Product 2 e 6	Human typology; OTC
	F3—Data collection for cooperation level of operational team	Indication of themes to assess the level of cooperation for OTC	Chapter 3; Chapter 7; Product 6	Human typology. OTC
	F4—Data collection for leadership assessment	Approach to assess the type of leadership for OTC	Chapter 3; Chapter 7; Product 6	Human typology; OTC
	F5—Data collection for commitment assessment	Form to assess the level of commitment for OTC	Chapter 3; Chapter 7; Product 6	Human typology; OTC
G—Exercise	G1—*Exercises*—Task assessment in sulfuric acid facility	Exercise about task analysis	Chapter 3	Task analysis
	G2-Exercise API770	Exercises about API 770	Chapter 3	API 770

A.1 Process loss map (Product 01)

To assist in the mapping of losses, Product 1 [8.4], a base model was built, Annex A, filling-in by type of waste, location, and an evaluation by the FIC Matrix [4.4]. This model is developed and indicated for the researcher to know the losses that exist in the work area, recognizing its social, environmental, and economic impacts.

In this mapping model, the following definitions of residues are considered: solid residues are residues in the solid state such as particulate material, scrap, and process residues. Liquid wastes are the wastes generated during/in the process in the liquid state such as industrial effluents, oils (derived from leaks), and others. Gaseous wastes are generated during the production process such as fugitive emissions.

(B) Loss of Energy Consumption	(A) Industrial Losses			Loss Classification	BASE MODEL FOR LOSS MAPPING
	GAS	LIQUID	SOLID	Type of Waste	
				Losses	
				Communication - Action Progm	
				Location Production Area	
				Frequency	
				IMPACT = I = S * E * Ec — Social S	
				Environ on E	
				Economy Ec	
				Criticality - C	
				Total = F * I * C	
				Priority	

A.2 Calibration of the SPAR-H method

A.2.1 Survey about operational context—operators/staff

To assist in the investigation of product application 6 [8.7], referring to the SPAR-H method, a questionnaire was prepared to be applied to the operational public and the staff. Quiz A, indicated to be applied to operators, investigates the operational routine, work conditions, communication between the teams, the operators' cognition, and understanding of the task performed. Quiz B, indicated to be applied to the staff, investigates the work environment through the perception of management, with analysis of the investigation modes for events (accidents) or failures, validation of procedures and reviews, knowledge analysis of the team profile for the task, and among others.

Quiz (A)
Analysis of the Operational Context of the task—Operators Interview
Position: _____ Experience in the role: _____
Company time: _____

GUIDING QUESTIONS FOR THE INTERVIEW
1. Are incidents or abnormalities discussed at meetings with the participation of operators? Soon after, are the initial causes of the occurrence indicated with an action plan to prevent recurrence?
2. Is the work environment (lighting, temperature, noise, general cleaning, etc.) kept within comfortable limits?
3. Are there any layout-related issues that make it difficult for operators to perform their tasks correctly?
4. Are the forms of communication in the change of shift effective to describe the necessary information about the state of the process (process/product conditions, process deviations, out-of-service equipment, work permits, and others.)?
5. Are the shift reports correctly filled in, so that a history is built on the signs of abnormalities in the process and are clearly identified/located when necessary?
6. Do operators actively participate in the identification of situations susceptible to errors in existing procedures?
7. Do operators participate in the review of operational procedures, such as indicating improvements in the most critical tasks or improving the presentation of writing (step-by-step format; diagrams, photographs, drawings)?
8. Do the specific activities of the operation strictly follow the guidelines contained in the procedures?
9. Do you believe that the operating procedures are clear and easily accessible?
10. Are operational procedures frequently consulted? How many times on average do you consult the procedure in the week?
11. Are critical tasks carried out following a step-by-step checklist?
12. How often do training or retraining critical tasks take place?

Quiz (B)
Analysis of the Operational Context of the task—Operators Interview
Position: _____ Experience in the role: _____
Company time: _____

GUIDING QUESTIONS FOR THE INTERVIEW
1. Are all equipment failure and intervention events duly recorded and analyzed?
2. Are incidents or abnormalities discussed at meetings? Right after, is an action plan suggested to prevent recurrence?
3. Is a root cause analysis carried out whenever possible taking into account technical, human, and organizational factors?
4. Are the shift reports correctly filled in, so that a history is built on the signs of abnormalities in the process and are clearly identified/located when necessary?
5. Are the mental and physical aspects analyzed for both routine and emergency activities?
6. Is the work environment (temperature, noise, lighting, general cleaning, etc.) kept within comfortable limits?
7. Does the process design and writing process meet the requirements of ergonomics, communication, layout, human machine interface?
8. Do operators participate in the identification of situations susceptible to errors in existing projects/procedures?
9. Do operators get involved in the evaluation of new projects/procedures for critical tasks?
10. Are the operational procedures available, clear and easily accessible?
11. What is the frequency of review/update of the procedures?
12. Does the use of specific procedures for the diagnosis of abnormalities assist in the identification of faults?
13. Could the inclusion of human factors alerts in operational procedures, such as the need for greater physical fitness or greater concentration, reduce failures in carrying out activities?
14. Are critical tasks carried out following a step-by-step (checklist)?
15. Did the selection of workers consider the appropriate criteria based on physical skills, aptitudes, and experience, among others?

A.2.2 PSF calibration

To assist in the new calibration of the SPAR-H method, Product 6 [8.7], a base model for directing the adjustment of the PSF's calibration for action and diagnosis task was elaborated, according to experts' guidelines. It is worth mentioning that this consists of an orientation of values, being subject to change according to the industrial scenario.

PSF action task

PSF	Level	Calibration criteria	Multiplier for action	Calibration adjustment
Available time	( ) Inadequate time. ( ) Time available = time required. ( ) Nominal time. ( ) Time ≥ 5X the required time. ( ) Time ≥ 50X the required time. ( ) Insufficient information.	The time PSF relates to other PSFs as the complexity of the task, the stress in the execution. The way of managing also causes differences in the criterion of time available. Therefore, one must evaluate these questions to reevaluate the multiplier. Processes with automated tasks <Calibration tends to reduce, or equal, the original multiplier> Processes with non-automated or semi-automated tasks (metallurgy, manufacturing) <Calibration tends to increase the original multiplier>	P (failure) = 1 10 1 0.1 0.01 1	
Stress/stressing agent	( ) Extreme. ( ) High. ( ) Nominal. ( ) Insufficient Information	Behavioral subjectivities must be analyzed before assigning discrete values to PSF Stress. The individual characteristics and the respective consequences of high stress for the body and the mind may be different due to: stage of the enterprise cycle, emotional stability of critical functions at work, availability of memory aid and cognitive processing, and quality of the simulated for situations of high risk and possibility of accident. Some cultural characteristics related to guilt and communication can also alter the criteria of multiplier values for the	5 2 1 1	

		impact of stress on the likelihood of human error. Processes with automated tasks (car assembly industry and others) <Calibration tends to reduce, or equal, the original multiplier> Processes with automated tasks, but with flammable products (chemical, petrochemical, and others) <Calibration tends to increase, or equal, the original multiplier> Processes with non-automated or semi-automated tasks (metallurgy, steel, manufacturing) <Calibration tends to increase the original multiplier>	
Complexity	( ) Highly complex. () Moderately complex. () Nominal () Insufficient information.	In the case of complexity, the ability to analyze, decide and act with the complexity of the situation must be compared, the existence of auxiliary systems, the level of risk for accidents, the number of requirements in the task and the respective number of steps. It is also important to analyze the mode of operation and the technology of the product, process and safety. Processes with automated tasks and few steps (chemical, petrochemical) <Calibration tends to reduce, or equal, the original multiplier> Processes with non-automated or semi-automated tasks (manufacturing, surgery) <Calibration tends to increase the original multiplier>	5 2 1 1

Training/Experience	( ) Low. ( ) Nominal ( ) High. ( ) Insufficient information.	The calibration to be done in the training case is related to the level of commitment and the need for specific skills and knowledge training. Processes with automated tasks (chemical, petrochemical) <Calibration tends to reduce, or equal, the original multiplier> Processes with non-automated or semi-automated tasks (food, manufacturing) <Calibration tends to increase the original multiplier>	3 1 0.5 1
Standard and Procedure	( ) Nonexistent ( ) Incomplete ( ) Available, but poorly designed ( ) Nominal ( ) Insufficient information	The calibration of multiplier values for standards and procedure in the application of SPARH depends on the culture and technology that the company has. Processes with automated tasks (chemical, petrochemical) <Calibration tends to increase, or equal, the multiplier> Processes with non-automated or semi-automated tasks (food, manufacturing) <Calibration tends to reduce the multiplier>	50 20 5 1 1
Instrumentation and ergonomics	( ) Nonexistent/poorly designed ( ) Bad ( ) Nominal ( ) Good ( ) Insufficient Information	Depending on the process control technology, the Gross Work, which involves a high physical and logistical effort, the need arises to calibrate this PSF. Processes with automated tasks (chemical, petrochemical) <Calibration tends to reduce, or equal, the original multiplier> Processes with non-automated or semi-automated tasks (food, manufacturing) <Calibration tends to increase, or equal, the original multiplier>	50 10 1 0.5 1

Fitness for service	( ) Not fit ( ) Degraded fitness ( ) Normal ( ) Insufficient information	Operational discipline, commitment, human and social typology are causally related to physical and cognitive aptitude for work. Processes with automated tasks (chemical, petrochemical) <Calibration tends to reduce, or equal, the multiplier> Processes with non-automated or semi-automated tasks (metallurgy, manufacturing) <Calibration tends to increase, or equal, the original multiplier>	P (failure) = 1 5 1 1	
Work relationships	( ) Bad ( ) Normal ( ) Good ( ) Insufficient Information	The coefficient assigned to relationships at work should be treated as an average in assessing the behavior of the teams, but, in some critical functions, the behavior of these operators must be considered. Processes with automated tasks (chemical, petrochemical) <Calibration tends to reduce, or equal, the multiplier> Processes with non-automated or semi-automated tasks (metallurgy, manufacturing, coconut industry) <Calibration tends to increase the original multiplier>	5 1 0.5 1	

PSF diagnoses task

PSF	Level	Calibration criteria	Multiplier for action	Calibration adjustment
Available time	( ) Inadequate time. ( ) Time close to nominal (~2/3). ( ) Nominal time. ( ) Extra time (between 1 and 2X the nominal time and > 30 min).	The time PSF relates to other PSFs as the complexity of the task, the stress in the execution. The way of managing also causes differences in the criterion of time available. Therefore,	P(failure) = 1 10 1 0.1 0.01	

	( ) Expansive time (> 2X required time and > 30 min). ( ) Insufficient Information	one must evaluate these questions to reevaluate the multiplier. Processes with automated tasks <Calibration tends to reduce, or equal, the original multiplier> Processes with non-automated or semi-automated tasks (metallurgy, manufacturing and others) <Calibration tends to increase the original multiplier>	1
Stress/Stressing Agent	( ) Extreme. ( ) High. ( ) Nominal. ( ) Insufficient information.	Behavioral subjectivities must be analyzed before assigning discrete values to PSF Stress. The individual characteristics and the respective consequences of high stress for the body and the mind may be different due to stage of the enterprise cycle, emotional stability of critical functions at work, availability of memory aid and cognitive processing, and quality of the simulated for situations of high risk and possibility of accident. Some cultural characteristics related to guilt and communication can also alter the criteria of multiplier values for the impact of stress on the likelihood of human error. Processes with automated tasks (car assembly industry and others) <Calibration tends to reduce, or equal, the original multiplier> Processes with automated tasks, but with flammable products (chemical, petrochemical, and others) <Calibration tends to increase, or equal, the original multiplier> Processes with non-automated or semi-	5 2 1 1

Complexity	( ) Highly complex. ( ) Moderately complex. ( ) Nominal ( ) Obvious diagnosis	automated tasks (metallurgy, steel, manufacturing) <Calibration tends to increase the original multiplier> In the case of complexity, the ability to analyze, decide and act with the complexity of the situation must be compared, the existence of auxiliary systems, the level of risk for accidents, the number of requirements in the task and the respective number of steps. It is also important to analyze the mode of operation and the technology of the product, process and safety. Processes with automated tasks and few steps (chemical, petrochemical) <Calibration tends to reduce, or equal, the original multiplier> Processes with non-automated or semi-automated tasks (manufacturing, surgery) <Calibration tends to increase the original multiplier>	5 2 1 0.1
Training/Experience	( ) Low ( ) Nominal ( ) High ( ) Insufficient information.	The calibration to be done in the training case is related to the level of commitment and the need for specific skills and knowledge training. Processes with automated tasks (chemical, petrochemical) <Calibration tends to reduce, or equal, the original multiplier> Processes with non-automated or semi-automated tasks (food, manufacturing) <Calibration tends to increase the original multiplier>	10 1 0.5 1
Standard and Procedure	( ) Nonexistent ( ) Incomplete ( ) Available, but poorly designed ( ) Nominal	The calibration of multiplier values for standard and procedure in the application of SPARH depends on the	50 20 5 1

	( ) Diagnostic/symptom-oriented ( ) Insufficient information.	culture and technology that the company has. Processes with automated tasks (chemical, petrochemical) <Calibration tends to increase, or equal, the multiplier> Processes with non-automated or semi-automated tasks (food, manufacturing) <Calibration tends to reduce the multiplier>	0.5 1
Instrumentation and Ergonomics	( ) Nonexistent/poorly designed ( ) Bad ( ) Nominal ( ) Good ( ) Insufficient information.	Depending on the process control technology, the Gross Work, which involves a high physical and logistical effort, the need arises to calibrate this PSF. Processes with automated tasks (chemical, petrochemical) <Calibration tends to reduce, or equal, the original multiplier> Processes with non-automated or semi-automated tasks (food, manufacturing) <Calibration tends to increase, or equal, the original multiplier>	50 10 1 0.5 1
Fitness for service	( ) Not fit ( ) Degraded fitness ( ) Normal ( ) Insufficient Information.	Operational discipline, commitment, human and social typology are related to physical and cognitive aptitude for work. Processes with automated tasks (chemical, petrochemical) <Calibration tends to reduce, or equal, the multiplier> Processes with non-automated or semi-automated tasks (metallurgy, manufacturing) <Calibration tends to increase, or equal, the original multiplier>	P (failure) = 1 5 1 1
Work relationships	( ) Bad ( ) Normal ( ) Good ( ) Insufficient information.	The coefficient assigned to relationships at work should be treated as an average in assessing the behavior of the teams, but, in some critical	2 1 0.8 1

	functions, the behavior of these operators must be considered. Processes with automated tasks (chemical, petrochemical) <Calibration tends to reduce, or equal, the multiplier> Processes with non-automated or semi-automated tasks (metallurgy, manufacturing, coconut industry) <Calibration tends to increase the original multiplier>		

A.3 Task assessment

A.3.1 Standard task data collection—PADOP

When investigating the task, it is important to know the component items. For an initial analysis, a survey of the parts that make up the task was suggested, such as identification of the task, identification of the procedure; plan actions in the procedure; in carrying out the task, the safety items involved in the task; and the question of cognition and complexity of the task. In Chapter 3/Product 2, content was addressed to carry out this analysis, which resulted in the design of a form for the initial collection of task data.

Identify the task for planning Identify architecture Sequence of procedures and hierarchy	Task title			
	Numerical goal			
	Normal process state	Process deviations	Intermediate state and gradations	Abnormal process state
	Requirements			
Identify the procedure after planning the task to generate a written document	Title of the procedure Numerical goal State of process/equipment/people			
Planning actions to perform procedure	Requirement Phases Sequence Time Type (surveillance, action, planning-deciding) Location (distance between actions) General consultation (field/panel) Specific query (field/panel) Authorization			
Perform the task, the procedure	Task automation Automatic operations Manual operations Monitoring			
Task security	Security barriers Management barriers Sequential human action (step) Parallel human action (steps) Multiple human actions (at the same time) High impacts (procedures and steps)			

	Task complexity Physical effort Cognitive effort Automation level: auto/manual Main/auxiliary Done/written Divided attention/focused attention Communication type and efficacy
Auxiliary memory	Written procedure Record Check list Design Booklet
Auxiliary processing	Calculations

A.3.2 Collection of failure data for task assessment—MEA (abnormal events)

Task failures [3.2.4] are indications of process malfunction. This problem must be analyzed carefully using tools such as PADOP and MEA. These tools were discussed in Chapter 3 [3.2.2] and Chapter 7 [7.3.3] as options for operational routine investigation and abnormal events. To assist in the investigation of flaws in the work environment, the ANNEX C2form was suggested. This form allows a survey of data from the occurrence of the event to the failure's connectivity to the environment and the task's execution team.

PART I—Abnormal events MEA assessment					
Identification of abnormal occurrence and causal relationships for investigation	Fault description and location Frequency Process Equipment People				
	Position and function	Failure warning, failure signal, deviation				
		Causes				
		Root or main		Secondary	Tertiary
		intermediaries Consequence				
		Principal	Secondary	Tertiary	Quaternary
	Previous factor (defragment and reorder) Posterior factor (defragment and reorder)				
	Process loss involved. Impact on the management system and on the business	Time Environmental Occupational Energy Financial Image			
	Severity				
Related variables	Process Product Wastewater Production				
Anatomy of failure Abnormal events map (MEA)	Identification	Title Place or region Period (critical, from 2 until 5 months) Warning, signal Root cause Secondary cause Event			
	Direct causal relationship					

Anatomy of failure Abnormal events map (MEA)	Influence relationship	Consequence *p/s/t/q* Top event Disaster Organizational culture Social culture Work environment False factor Anticipated factor
	Time correction factors	
Validation		Potentiating factor (positive or negative) Failure control volume Danger-bearing body
		Occurrence identification Related variables Failure anatomy Procedure test

A.3.3 Collection of failure data for task assessment—PADOP

Understanding about failure in complex process, task and social relations requires information for applying different diagrams: logic, chronology & hazard energy, materialization and connectivity, each one is represented by surveys Part II, and III. In Part II, we are collecting data about failure logic, chronology & hazard energy. It is important to remember that all these diagrams are based on DO [2.1.5] and MEA Assessment. The diagrams in Part II and III were discussed in Chapter 3 [3.2.4] and applied in Chapter 7, when discussing the cases in the oil & gas and in chemical plants [7.3.5.3]; and presented in Chapter 8 in Product 2, [8.5.3.2] and [8.5.3.3].

PART II—Logic, chronology-hazard energy, and failure materialization		
Failure logic	Identification	Title Place or region Period
	Sociotechnical classification	Task Managerial Organizational Human Technical
	Failure zone classification	Inertial Development Risk Top event
	Ramification	Feedback Feed-forward
	Retention time	
	Actions	Corrective Preventive or safeguard
Failure chronology and hazard energy	Identification	Title Place Period
	Failure zone classification	Inertial Development Risk Top event
	Temporal failure sequencing (anatomy)	
	Time by factor	Estimated real (hour) Relative (Ln time)
	Hazard energy	Real estimated (without dimension unit of energy) Relative (Ln energy failure or hazard energy released)

Failure materialization	Identification	Title
		Place or region
		Period
	Sociotechnical classification	Task
		Managerial
		Organizational
		Human
		Technical
	Failure zone classification	Inertial
		Development
		Risk
		Top event
	Bad actors in failure	Equipment
		Human
		Product
		Process
	Failure energy migration	Information (imaginary) for matter/energy
		Matter/energy for information (imaginary)

A.3.4 Collection of failure data for task assessment

The understanding about failure in complex process, task and social relations requires information for applying different diagrams: logic, chronology & hazard energy, materialization, and connectivity, each one is represented by Surveys Part II, and III. In Part II we are collecting data about Failure logic, chronology & hazard energy. It is important to remember that all these diagrams are based on DO [2.1.5] and MEA Assessment. The Diagrams in Part II and III were discussed in Chapter 3, [3.2.4] and applied in Chapter 7, when discussing the cases in oil & gas and in chemical plants [7.3.5.3]; and presented in Chapter 8 in Product 2, [8.5.3.2] and [8.5.3.3].

	PART III—Connectivity	
Failure connectivity Failure connectivity	Identification	Title
		Place or region
		Period
	Failure type	Simple causal—linear
		Organic causal—complex
	Factor type	Causal
		Effect
		Signal or warning
		False
		Anticipated
		Potentiating factor
		Top event
	Cause type	Root
		Secondary
	Consequence type	Primary
		Secondary
		Tertiary
		Quaternary
	Failure feedback points Failure energy measurement point Top event	
	Main system causality Auxiliary system causality (diagram)	Equipment (diagram)
		Equipment
		Utilities
		Human
	Diagram identification	Failure measurement
		Equipment
		Product flow
		Feedback point
	Construction of the connectivity diagrams in parallel faults (ex: fatigue, cavitation and lubrication) Connectivity point in parallel faults or parallel blades	

A.3.5 Data collection for emergency assessment

In Annex C3 we suggest a survey to collect data to design a risk assessment in progressive levels of Stress. The intention is to: (1) Construct a Disaster from the level of deviance; (2) Perform a Dynamic Simulation based on DO; 3. Observe the reaction in this Simulation and Perform Behavior Assessment; 4. Prepare and FTA Analyses; 5. Calculate Human Factor Probability in this Simulated Event; and 6. Locate Technical and Human Barriers. This discussion was discussed in Chapter 3 [3.2.5], was applied in Chapter 7, [7.1.2.3] and in Chapter 8 Task Assessment Product 2 [8.5.4].

In the assessment of emergency situations in the task, it is necessary to collect data based on the events occurred, on the work environment and on the Operator's Discourse, as already discussed in Chapter 3 [3.2.5]. The behavior of the team in an emergency differs from normal work moments. Thus, it is important to know the team, the task, the work environment, and its connections with the emergency event. For emergency simulations, Annex C.3 was indicated, a form, to be filled out together with the operational team. From this it is possible to know and anticipate paths and solutions not foreseen in the company's procedure.

1. Anatomy of the accident	Act with the staff	
	Beta Anatomy Construction based on MEA: signs for intermediate state (DO)—SI	
	Survey of Accident or Serious Accident—AG	
	Hypothesis of factors not mentioned to link SI—AG	
	Validation of Hypotheses	
	Construction of a new Anatomy 1 for dynamics	
Description of the progression of events at levels	Signs of abnormality, deviance	
	Uncontrolled process	
	Unstable equipment and failed task	
	Unsafe situations and events	
	Major accident or event	
2. Dynamic simulation based on DO	Prepare dynamics with events and scenarios	
	Split groups	
	Low stress scenario—signs	Display scenario
	Medium stress scenario—uncontrolled process	Delimit time
	High stress scenario—unstable equipment and task at fault	Design operational actions by group
	Extremely high stress scenario—unsafe but controlled situation (ex: fire)	Display scenario
	Extreme stress scenario—accident with several deaths	Delimit time
	Gather teams' programs (from 3 to 5)	Design operational actions by group
	Analyze suggestions for technical and human barriers.	Display scenario
	Prepare dynamics with events and scenarios	Delimit time
		Design operational actions by group
	Split groups	Display scenario
	Low stress scenario—signs	Delimit time
	Medium stress scenario—uncontrolled process	Design operational actions by group
	High stress scenario—unstable equipment and task at fault	Display scenario
	Extremely high stress scenario—unsafe but controlled situation (ex: fire)	Delimit time
	Extreme stress scenario—accident with several deaths	Design operational actions by group
	Gather teams' programs (from 3 to 5)	
	Analyze suggestions for technical and human barriers	
	Prepare Anatomy 2 with improvements	
3. Behavior analysis	Train behavior watchers to record people's reactions to stress	
	Check key points related to stress	
		Leadership
		Organization
		Persistence
		Linear Thinking

			Creativity
			Knowledge
			Ability
			Social relationships
			Clarity
4. Failure tree analysis	Act with the staff		
	Interpret individual and team behavior		
	Act with the staff		
	Prepare anatomy 1 (no improvements)		
	Define logical operators		
	Define positive influence operator		Much
			Little
	Define positive influence operator		Much
			Little
5. Human factor valuation	Measure the probabilities of failure in Technical Devices		
	Estimate probability of failure in Human and Organizational Factors		
	Calculate the probability of claim 1		
	Correct glaring aspects and calculate the probability of the accident 2		
	Act with the staff		
	Estimate probability target value for the claim		
	Compare the likelihood of claim 2 with target value		
	Correct logical operators (if necessary)		
	Correct probability values for human errors		
	Analyze the probability assignment for technical failures		
	Recalculate		
	To correct		
	Probability of claim n = target value		
6. Technical and human barriers	Act with the staff		
	Estimate goal to be achieved with improvements		
	Estimate reduction in the likelihood of events if improvement occurs		
	Correct logical operators (if necessary)		
	Recalculate		

A.4 Technical survey

The technical survey suggested in this Annex D is to try to understand the restriction to operate a continuous plant. These data are collected from book of technology, process description, specification of critical equipment and information about occurrences from the routine.

A.4.1 Data collection of technical: OTC O&G—general

Product type	Yes	No	What? Which?
Toxicity			
Flammability			
Reactivity			
Physical state			
Process type			
Temperature			
Pressure			
Automation level and equipment			
Control valves			
Rotating equipment			
Static equipment			
Production			
Safety			
Frequency/severity of incidents			
Frequency/severity of accidents			

Voluntary participation in the brigade People Number of employees Number of formal leaders Average age at jobs Proportion of contractors Tasks and human error Compliance with written procedure steps in relation to practice Accidents initiated or developed from human errors Number of nonconformities Product quality Effluent quality Environmental impact Occupational impact			

A.4.2 API 770 data & operational–technical culture

Analysis of the technical-operational culture, discussed in Chapter 7, and in product 6 in Chapter 8. points out that identification of the Human, Social and Technological Typology and mapping of critical regions, because of the collection of data from operators, should contribute to the improvement campaigns for the production area, and thus, it is possible to know the main causes of accidents. This form contains variables that stand out for the quantitative understanding of what is possible to do to avoid possible accidents.

Policy enforcement	Type 1 Culture	Power Papers Assignment People	
	Type 2 Culture	Fault Learn	
	Reaction level	Proactive Active Reactive	
	Knowledge	Techniques Policies Practices	
	Clarity & credit Deviation record	Amount Veracity	
	Hazard/Risk identified		
Human factor engineering Knowledge and training Cognitive processing Cognitive functions Skills Knowledge level Human error type & mechanism Behavior and workstation Behavior under stress Performance factors Emotional balance and impact on behavior Workstation commitment Standards review			

Worker	Circadian cycle Skill Experience Fitness Training Concept Failure process Principles Social relationships Health Stress level Emotional balance
Task analysis	Standards review Methods to identify the root cause, signs and latent failure Communication tools Guidelines Instruction Procedures Occurrence Task failure analysis Control volume Post and pre-accident Routine tasks Critical tasks Emergencies Division of labor Work load Rotation level Scheduled folds Extra hour Re-service Knowledge critical situations
Procedure	Availability Revision Unknown events Complexity Application level Low risk flexibility Work permission Assertive Level of detail Parallel actions Draft procedure Phases Sequence Commitments Requirements Goals Impacts
Human machine interface design	Improper posture Inadequate movement Console usability and access Traffic on the panel and risk Alarm signals per person Normal routine Normal shutdown Emergency shutdown Security devices Human decision level Security barriers Fear of risks

A.4.3 Continuous process technology requirements

In the case of Chemical Processes, we prepared a Survey to ask about requirements to operate and which, if not attended, can cause failures.

Equipment materials and accessories	Types of equipment and products (corrosion and erosion)	
	Types of seals, materials, and products (liquid leakage)	
	Types of gaskets, materials, and products (liquid leakage)	
	Type of seals and gaskets (fugitive gas emissions)	
	Mechanical impacts with cracking or cracking of materials	
	Thermal shock with cracking or cracking of materials	
Technology	Critical chemical reactions (wastes and controls)	
	Critical physical separation processes (waste and controls)	
	Critical physical transport processes (accidents and waste)	
	Critical manufacturing processes (accidents and losses)	
Equipment inventory	Critical inventory for production maintenance	
	Critical inventory for effluent control	
Rotating equipment and interconnections	Movements in rotating equipment with fatigue.	
	Rotary lubrication characteristics.	
	Connection between pipes with pressure or where it passes	
	Connection between pipes passes toxic product or even	
	Connection between pipes at high temperature.	
Process characteristics	Pressure fluctuation	
	Residence time and corrosion	
	Production cycles.	
Environmental aspects	Excessive rain and equipment	
	Excessive and effluent rain	
	Insolation of equipment	
Fluid flow characteristics	Speed and dirt	
	Viscosity and heating	
	Flow and vibration regime	
Contamination possibilities	Finished product	
	Effluent	
Low efficiency separation	Emulsion formation	
	Nearby densities	
Precipitation of solids	Process disorders	
	Equipment disturbances.	
Process disorders	In heating, steam, quality and quantity	
	On cooling, water, dirt, temperature	
Emissions to the atmosphere	Inadequate drainage of liquids	
	Inadequate venting of gases	
Process controllers.	In manual	
	100% open	
	0% open	
	Cool by-pass	

A.4.4 Routine data collection

Based on the application in an oil & gas facility, we suggest collecting and statistical treatment of these data. Chapter 7 [7.1.2.2] [7.3.2.1].

Event class	Data, processing, and information	Parameters and analysis criteria
Statistical and time identifier	Month (M), Day (D), Class of shift (Tt), Time (Ho), specific events, stops and others.	Seasonality Physical environment (day/night) Physical environment (winter/summer) Physical environment (Monday/Sun) Change turn Leadership profile
Identify division of labor	Number of operators (Op), number of operators in double working time (OpD), % double working time ((OpD/Op) * 100), and no signs of abnormal events (Sean).	Relationships between fold% Division of work Physical and cognitive effort Number of abnormal events
Task description identifier	Process section (Spc), Tasks or operations (Tf), Task location (OT), Task destination location (DT), Processed volume (Vl), Time to perform the task (t), Quality of work process (Qpc, e.g.: temperature), Product quality (Qpd, eg: % contaminant), Effluent quality (Qef, eg: ppm of contamination)	Check the condition of the tasks Check operating conditions Check the level of control times Check control—which process Check the control—which in the product Check the control—which is not effluent.
Task control identifier	Number of hours per operation (Nhor), number of times the task is performed (nTf), total time spent per operation (Tti = Nhor * nTf), total time spent for all operations (Tt = ΣTti), hours available per shift (Td = Op * 8), average hours spent per operator for the safety of equipment and people (Hseg), total hours with occupation (Htot = Tt + Hseg * Op),% occupation (Ocp = Htot/Td*100), % idleness (Oci = 100-Ocp (%)).	Productivity per operation Task frequency Term compliance analysis Technical idleness Occupancy rate Security time
Analysis of abnormal signs	Descriptive of abnormal events (Desc), number of signs of abnormal events (Sean), specialty of maintenance (EM = Mec, Ins, Cald, El, Pr). Where, Mec = mechanics; Ins = Instrumentation, Cald = boiler, El = Electrical, Pr = Process. Mechanical events that can be: disarming (dis), sealing (Seal) and lubrication (lub) and others. Instrumentation events that can be: control valve (CV), solenoid valve (SV), temperature (T), pressure (P), level (L) and current (A) and others. Boiler making events: obstruction (Ob), holes (Fu), overflow (Tb), leakage (Vz), problems with motorized valves (V motor) and others. Electrical events: short circuit (Cc), dented conduit (Elt), lighting (Il).	Causality Maintenance classes: mechanics; instrumentation; boiler making; electric; process control. Mechanics: disarm; sealing; lubrication Instrumentation: control valve; solenoid valve; T, P, L, A. Boiler making: Obstruction, hole, transshipment, leak and motorized valve Electrical: Short circuit

A.4.5 Process and production variables collection

In this case, it is only possible to treat information from operational routine that indicates deviation, abnormalities, in the investigation from MEA in chemical process industries, cited on Chapter 7.

Type of data	Details			Variables
Quality	Product quality (contaminant, physical property) Quality matter-p. (contaminant, physical property) Process quality (T, P) Relationships between qualities			
Process variables	Critical flows Valve opening Critical pressures Valve opening Critical temperatures Valve opening Critical pressures Valve opening Amperage Pressures Pressure delta Sum of standard deviation of pressure Sum of maximum pressures Moisture Others			
Production reports	Operation incidents Maintenance incidents SMS incidents Personnel incidents			
Automation security analysis	Operation stage (stop, start, normal, emergency) Number of automated operations Number of manual operations Number of control loops Number of priority alarms and risks Automatic safety (SIS, interlock)			
Management report		Standard	Measuring	Δ
	Customer relationship Supplier relationship Production performance SMS performance Personal performance			

A.5 Social typology

A.5.1 Data collection for political aspects and practices

For mapping and identification of social typology, it is necessary to collect information of the technical, human, and social typologes, in this Annex, only the social one, procedures are identified to change environments, structures, people and management processes that contribute to increase human reliability and operation in the industrial plant. This form can be filled out individually so that there is enough information for a well-founded descriptive statistical analysis. Chapter 3, product 6, of the program 8.9.3 Chapter 8.

Cultural aspects	
Key elements of security policy	
Key elements in organizational policy	
Level of acceptance of procedures	
Level of compliance with procedures	
Expectation of compliance with procedures	
Perception of the image	
Routine practices	
Operational routine management aspects	
Type of company	
Business technologies	
Economic stability ratio	
Level of delegation	
Insufflation level of leaders' ego	
Insufflation level of the followers' ego	
Technical culture stability	
Existence of vanity bonfires	
Relationship of territory and power	
Types of addictions	
Decision model	
Impulsive action maps	
Interpretation of addictions	
Interpretation of bias in the routine	

A.5.2 Data collection for social relations

This survey was used in the case of an oil and gas facility. Chapter 7.

Annex E.2 aims to carry out a survey on the social relationships in the organization. Social typology, which has already been discussed in Chapter 3 [3.1.2], aims at a broader approach on social relations that can positively and negatively impact the performance of the operator/team. The knowledge of desirable profiles in workgroups can avoid human error by automatism, which can compromise the stability of the organization, derived from individuals who have intrinsic behaviors under the influence of a social environment.

Human-social typology 1			
1 General			
1.1 Religion	Catholic	Evangelical	Other
1.2 Sport	Yes	No	Other
1.3 Football team	A	B	Other
2 Parents			
2.1 Living with parents?	Yes	No	
2.2 Position in the family (brothers)	Youngest	First-born	Other
2.3 Financial dependency	Yes	No	
3 Family			
3.1 Marital status	Married	Not married	
3.2 Remarried	2nd time	3rd time	Above
3.3 Number of children	1	2 to 3	Above 4
3.4 Ages	25–35	36–55	Over 56
4 Community			
4.1 Opinion on politics	Resistant	Half term	Active and constructive
4.2 Opinion on ethics	Influenceable	Follow the law	Ethical
4.3 Opinion on social action	Does not participate	It depends on the advantages	Unstable, power, and physical
4.4 Type of participation	Does not participate	Only speech	Speech and action
5 Food and personal customs			
5.1 Normal customs	Emphatic regionalization	Between global and local	Globalized customs
5.2 Weekend Customs	Stable, family	Depends on the week	Participate
5.3 Party customs	Does not participate	Depends on the week	Participate
5.4 Drink	Does not drink	Sometimes	Drink
5.5 Smoking	Does not smoke	Sometimes	Smoke

3.6 Friendship 6.1 Class	1 to 2 friends	2 to 5 friends	Above 5 friends
6.2. Group	No specific tribe/group	1 Tribe/specific group	More than 1 specific tribe/group
7 Leisure/hobby 7.1 Hours per week	0	2–4	Above 4 h
7.2 Type	Sport	Theater, cinema, other	Reading
8 Actions in society and family 8.1 Intensity	Low (once/m)	Average (every week)	High (almost every day)

A.6 Human typology

A.6.1 Data collection to emotional imbalance

The purpose of this form is to collect personal and behavioral data for the analysis of emotional balance, a factor that stands out in any accident scenario. This Annex corresponds to Chapter 3/7, Product 3 [8.6.4].

Human-social typology 2			
1 Family			
1.1 Health	0 Serious illness	1–3 Serious illnesses	Over 3 diseases
1.2 Customs	Excessive care	Basic care	Only when sick
1.3 Cleaning habits	Excessive	Normal	No worries
1.4 Symptoms	Excessive	Normal	No worries
1.5 Children's diseases	Asthma	The flu	Others
1.6 Wives' diseases	Gynecological	Sinusitis/Migraine/Pain	Skin
1.7 Relatives' diseases	Thrombosis	Diabetes	Heart
2 Childhood of the operator			
2.1 Diseases	Asthma	Diarrhea/constipation	The flu
2.2 Surgeries performed	By impact	Internal	No surgeries
2.3 Medications	Do not use	Light	Intense
3 Human typology			
3.1 Weight			
3.2 Height			
3.3 Weight-to-height ratio			
3.4 Cardio-vascular stage (heart, blood vessels . . .)	Serious	Sometimes	No occurrences
3.5 Blood Pressure	High	Low	Normal
3.6 Breathing: asthma and others	Asthma	Tiredness	No registry
3.7 Digestive: constipation, diarrhea and others	Cold	Diarrhea	No occurrences
3.8 Nutritional	Obese	Skinny	Normal
3.9 Skin	Yea	No	
3.10 Serious illnesses (cancer and others)	Yea	No	
3.11 Sleep (light, heavy, medium)	Light	Heavy	Normal
3.12 Endocrine (pituitary, thyroid, adrenal, reproductive)	Yes	No	
3.13 Gynecological	Yes	No	
4 Esthetics and features			
4.1 Personal care	Confront	Oscillates	If accepted
4.2 Hygiene	Careless	Depends on the time	Takes care
4.3 Costumes	Over-care	No care	Normal

A.6.2 Data collection for cognition quality & CTO

For analysis of cognitive quality, the mapping was carried out through a survey to observe the results obtained. To understand the application, in chapter 8, product 3 [8.6.2].

Mapping cognition variables			
WARNING	Real memory	Intermediate	Inattentive
MEMORY	Attentive	Changes memory	No memories
Sense of perception	Sharp perception	Intermediary	Don't notice
Thought	Very rational	Normal logic flow	Difficulties in thinking
Language	Good	Average	Difficulty speaking
Intelligence	Good	Average	Reasonable

A.6.3 Data collection for cooperation level of operational team

Annex 6.3 indicates the topics discussed to calculate the level of cooperation of the Aggregation and Disaggregation Assessment for an operational–technical culture, Chapter 3 [3.1.1] and Chapter 7 [7.1.2], Product 6 [8.9]. Themes discussed to calculate Level of cooperation from Aggregation, Disaggregation Assessment.

Questions 1 to 26—Sociability and Communication. Affirmative responses indicate tact.

Questions 27 to 42—Mental attitudes that contribute to sociability naturally—Kindness. The greater the number of affirmative responses, the greater the dose of goodness in their formation.

Questions 43 to 54—Mental attitudes that contribute to sociability naturally—Compassion.

Questions 55 to 66—Affectivity. Affectionate people are sensitive to the manifestations of others.

Questions 67 to 82—Optimism. People with optimism find pleasant ways to say things.

Questions 83 to 95—Ability to respond to emergencies. Self-control, when things are bad, is essential for touch.

Questions 96 to 107—Respect. People with less are likely to appear to be important. People with higher marks help others to feel important, this is the essence of tact.

Questions 108 to 117—Unfavorable mental attitudes. We will now deal with several series of questions that revolve around qualities that are unfavorable to the touch. The inclinations for snobbery come first. Snobbery.

Questions 118 to 127—Selfishness. The fewer affirmative responses, the better. The operator is harmed if he forms an exaggerated opinion about himself or if he tries to maintain his conviction by disparaging the self-love of others.

Questions 128 to 138—Search for defects. Defect-seeking is common among people who try to preserve their self-love by disparaging others.

Questions 139 to 150—Egocentrism. The average score is five affirmative responses. More than that means tendencies to focus on yourself rather than on other colleagues.

Questions 151 to 162—Angry temper. Do you get irritated, want to make feline remarks, get your muscles tense or feel like hitting someone when:...

A.6.4 Data collection for leadership assessment

Themes discussed to classification of type of leadership.

Inclusion, Control and Affection issues that will diagnose leadership profiles:

Importance of people; Take responsibilities; Importance of affection; Social interaction

They also analyze specific profiles in the inclusion questions such as:

The dusky twilight; The inhibited individual; The social flexible; Hidden inhibitions.

They also analyze specific profiles in the control questions such as:

Openly dependent; Dependency vs. independence conflict; The rebel; Participative management; Mission Impossible; Impossible mission with narcissistic tendency

They also analyze specific profiles in questions about affection such as:

Optimistic; Pessimistic; Cautious lover; Image of intimacy

A.6.5 Data collection for commitment assessment

Themes are discussed to calculate level of commitment. The survey of data about items that demonstrate the individual's level of commitment should be seen as a priority aspect for the composition of the profile of the operational team, Chapter 3 [3.1.1] and Chapter 7 [7.1.2], Product 6 [8.9]. Annex F.5 discusses some of the topics used to calculate the level of Commitment.

Primary data	Information
Home company	
Year of birth	
Age	
Degree of education	
Salary level	
Admission	
	Salary progression
Dependents	
	Age/dependent
Change of posts	
Levels	
	Company age/time
	company/dependent time
Years of work	
Clearances	
	Leave/years of work
Voluntary participation	
Exam results	

A.7 Exercise

A.7.1 Exercises task assessment in sulfuric acid facility

First divide people in your team into groups with basic knowledge in the areas of chemistry, processes, and operations. Analyze the starting task (procedure 1) of the sulfuric acid stripping column (flowchart 2) considering the techniques presented for planning the standard task GMTA and PADOP, Chapter 3 [3.2]. Create failure situations to describe the items listed in checklist E.

Documents

A. Sulfuric acid plant start-up procedure
B. Routine information—shift occurrences
C. Description of the sulfuric acid process
D. Items in the standard PADOP worksheet
E. Task analysis checklist

A. Procedure: Start of the sulfuric acid reconcentration plant

1 Purpose
Guide the departure of the operational units responsible for the reconcentration of spent acid from the DNT plant. Allow the plant to operate continuously under the appropriate operating conditions.
2 Responsibilities
2.1 It is the responsibility of the process engineering of the DNT plant to ensure that the spent acid arrives in the proper conditions.
2.2 It is the responsibility of the Process Engineering of Operational Management II to update and disseminate this procedure in addition to monitoring compliance.
3 Definitions
3.1 **AMG**: waste acid
3.2 **ASP**: waste sulfuric acid
3.3 **QUENCH COOLER**: cooling from thermal shock

3.4 Sulfuric acid reconcentration unit

4 General conditions

4.1 This standard establishes the requirements to make the acid reconcentration unit operational continuously.

4.2 Hot AMG from the DNT plant can be pumped to the acid plant as long as it is within specifications.

4.3 The hot acid to be reconcentrated must be characterized in compliance with the provisions of Operational Instruction.

5 Specific operation and process conditions

5.1 The preheating of the AMG in the exchanger requires that the valves at the inlet be properly aligned.

5.2 Check that the temperature at the inlet of the inlet exchanger is less than 100₀C.

5.3 Check that the valve at the inlet of the second exchanger is aligned.

5.4 Check valve alignment at the changer inlet.

5.5 Check if the steam flow to the stripper column is adequate and monitor the temperature up to 160°C.

5.6 Analysis of the top vapors of the stripping column and monitoring of the temperature when leaving the exchanger for reduction in the quench cooler.

5.7 Collect sample in the stripping column. A for analysis of the acid concentration according to the recommendations of IO 14.3.1.

5.8 Observe level indicators on the changer at the top of the stripping column.

5.9 Adjust valve at the changer outlet.

5.10 Adjust all controllers.

5.11 Observe if the currents are within the specification.

5.12 Check the sprays for operation.

6 Operational care

6.1 Temperature adjustments must be gradual.

6.2 As the column is filled, the temperature and pressure must be adjusted.

6.3 There must be no water in the gutting column.

6.4 The presence of solids at the start should be avoided (avoid suction of solids that hinder the flow and can be accumulated)

6.5 Check the plant for leaks.

7. Check list

(a) Circulating AGR water

(b) Place the meshes in Manual FC and TC

(c) Place the pump to operate according to the pump starting procedure

(d) Inventory in the preheater up to 50% level

(e) Gradually increase the opening of the CT, by 5 in 5%

(f) After reaching 120°C, align to the tower's inventory, with a flow rate of 1500 kg/h.

(g) Inventory the tower up to 20% and align steam8. Exposure control and Individual Protection

8. Material handling care

8.1. Engineering control

Good general ventilation must be ensured to maintain product vapor concentrations below levels of exposure. It may be necessary to use vapor absorbers or mist eliminators.

8.2. Individual protection equipment

Keep the following PPE available and use it under the conditions of exposure: Chemical splash goggles, face shield (with chemical splash goggles), long-sleeved PVC gloves, acid-resistant apron and boots, shirt long sleeves made of polyester or acrylic fibers, acid-resistant hooded clothing and respiratory protection.

8.3. Occupational exposure limits

ACGIH—TLV/TWA: 0.2 mg/m^3 .... (Base 2004).

9. Neutralizing agents

Dilute with water in large quantities and then neutralize with calcium oxides (lime) and sodium carbonate (soda ash).

10. Material handling cautions

Heat shock caused by water in acid or acid in water, glass.

Thermal shock caused by rain on hot glass.

Stress caused by over-tightening of glazed flange.

Acid burn

Mechanical shock of foreign materials (nuts, stones) on glass
Obstructions with organic acid inlet
Obstructions with iron sulfate at the acid outlet
Presence of NOx in mono spent acid
Inadequate mixing of effluents

B. Routine information—shift occurrences

This plant has a history of constant shutdowns due to cracking in glass equipment and various leaks of sulfuric acid. The processes are simple, but the contaminants DNT (solid) and NOx (gaseous) fluctuate a lot and can cause several obstructions.

The lack of glass pieces due to the low agility in the import causes the stop of one of the sides (A or B) hindering and making the production of the client/supplier that is producer of DNT to produce TDI more difficult and expensive.

Operators are afraid to go to the area due to the high number of events with equipment breakdowns and leaks with sulfuric acid causing low quality in the process.

The vacuum has difficulties in framing, the stripper has a history of filling with preferential paths and the DNT at the entrance temporarily obstructs exchangers, which can cause internal impacts on the equipment.

There is a large amount of iron sulphate in the finished product tank (pump seal and carbon steel holes) and large quantities of DNT in the raw material tank (pump seal, overflow level and obstructions in pipes and exchangers).

The vacuum difficulties are constant and due to the sudden presence of DNT, the break of the big pipeline that reunites vaporized waters of the process can happen.

The access to the plant goes through a gravel yard and the protection of the first-floor equipment, pure glass, is made of plastic, there are flaws in the shutters and torrential rains.

The effluent, when mixed with the organic effluent from the isocyanate plant, reacts generating waste called "gum". DNT is considered waste due to the amount and difficulty of collection, although it is a raw material for TDI.

The managerial profile is centralized, and the maintainers have their origins in a bank culture. There is a lot of environmental problem and the operators complain a lot. Engineers are not in the habit of collecting historical data for investigating events. The plant builder constantly visits technological support.

C. Description of the sulfuric acid process

The reconcentrated acid, 96% sulfuric acid, is obtained from the evaporation of the water contained in the mono waste acid in two stages of vacuum evaporation. The spent hot AMG acid generated at the DNT plant is pumped directly to the URAS, using a line coated internally with TEFLON. Thus, eliminating the possibility of obstructions. The current is divided between units A and B by means of the respective flow controls.

Initially the AMG is preheated in heat exchangers where it will exchange heat with the purified sulfuric acid ASP stream, from the bottom of the stripping column, or with the 96% reconcentrated acid stream raising the temperature to about 100°C. The option chosen will depend on the amount of Mono Spent Acid to be processed and the capacity of the concentration stages. Then the AMG passes through the final heater which operates by the thermo-siphon principle and uses steam in the hull, where its temperature will reach 160°C, before feeding the NOx stripping. This column is provided with superheated steam injection in the bottom to assist gutting.

The vapors at the top of the column and exchanger made up of Organic Matter (DNT), nitrous oxides (NOx) and nitric acid HNO_3 will be annihilated in the direct contact condenser where they are cooled to 70°C through the injection of water. The vapors not yet condensed will go to the final condenser, which will operate in a common SWITCH regime for units A and B, that is, while one operates normally the other is regenerated with steam for later replacement. The non-condensable are finally sent to the NOx absorption column while the condensates from the quench cooler and exchanger and other contaminated liquid effluents from the units go to the waste water collection vessel, where they are kept at 70°C and transferred through the pumps to the water plant. DNT, thus reincorporating the organic matter into the process, as well as reusing this water in the washing of the DNT. The transfer line is equipped with external heating to avoid possible obstructions.

The bottom product of the stripping column, free of DNT, HNO_3, NOx, now called sulfuric acid purified at 160°C, is then fed to the first evaporation stage, in a vacuum flash tank. The acid circulation is made with a pump through a heater to heat the acid and the mixture of the feed in the flash tank. Preconcentrated acid, 87% sulfuric acid, overflows from the lung tank to feed the second evaporation stage. In the final concentration, the second evaporation stage, the same process as the first stage occurs, but under different conditions of temperature and pressure (vacuum). The preconcentrated acid is fed into the flash tank of the second evaporation stage through a wash column. The reconcentrated acid, 96% sulfuric acid, overflows from the lung tank to the vessel, then it is cooled and sent to storage.

D. Items in the standard PADOP worksheet

Objective
Meta State
General requirement
General risk
Cognitive and motor requirement
Human risk
Architecture of procedures in the task with hierarchy
Procedure 1
Specific requirement 1
Specific requirement 2
Carrying out the procedure
Steps 1 through 35 (for example)
Step 1
Step 2
... Step 35
Meta state
Barriers
Application
Care
Alignments
Automatic control
Measurement and control instrument
Manual controls
Scratches
Authorization

E. Task analysis checklist

1. History of events
2. Psychological conditions (attention and intensity of knowledge)
3. Time available (event horizon, history, synchronize task and dynamic process)
4. Availability of plans/tasks/heuristics/routines: people must know what to do and use in the procedures
5. Number of simultaneous goals (number of tasks or number of lines of action), indicates the workload
6. Mode of execution (verbalization and interaction between team members, automatic or step by step?)
7. Process status (alarms and indicators)
8. Adequacy of the human machine interface and operational support (lack of support, incomplete/insufficient procedures)
9. Suitability of the organization (roles, responsibilities, assistance, communication, security, emergency instructions)
10. Control mode, human/machine interface

A.7.2 Exercise API770

To verify the application difficulties and what are the contents present in the API770 questionnaires, Chapter 3 [3.3], it is intended to apply the questionnaire on 4 groups, each group will answer different topics like this, group 1 (policy), group 2 (work), group 3 (human machine interface), group 4 (procedure).

The total group should have between 10 and 20 participants, being divided as follows:

At least 2 participants per group and a maximum of 5

Each group will have a copywriter, an exhibitor, a debater and analysts to critique the responses of competing groups ...

The writer writes the work and passes it off to deliver as an evaluation.
The exhibitor will present the work to the other groups.
The debater will be entitled to a question opposing an exhibitor's positions.
Analysts will critically analyze the exhibition and the debate.
The following tasks will be performed:

1. Answer the questionnaire: yes or no, and why?
2. Take files about limitations or strengths in your company regarding the questions
    Team structuring
    Knowledge
    Managerial involvement
    Communication
    Commitment
3. Answer supplementary questions
- What are the management and educational processes?
- What are the mandatory structures?
- What are the responsible human factors?
- What physical and psychological stressors do they influence over?
- What does stress influence?
- What are the performance factors?
- What are the hypotheses that most cause human error?
4. Prepare abstracts: speaker, panelist on questions

Questions for group 1
API770 Questionnaire—policies

1. Is the commitment of top management clear? Do workers understand policies and are they convinced of the sincerity of top management? Is there evidence?

2. Do supervisors and workers believe that security has a high (or at least equal) status compared to other business objectives?

4. Is management of workers' health and safety essential in management activities?

5. Are occupational health and safety issues discussed at management meetings at all levels?

6. Has top management established policies for human factor engineering? How did the authorities remedy human factors from engineering deficiencies?

7. In the areas of research, design, assembly, contracts, operations, maintenance, management and others, are procedures clearly defined to assess engineering aspects of human factors of: new and modified processes?

8. Are human factor engineering resources available in the organization and are they available to help resolve design and procedural issues?

9. Are the allocated times and resources suitable for engineering human factors in projects?

10. Do workers help to identify similar situations with errors in procedures and projects?

11. Are workers encouraged to discuss potential human errors and near loss with their supervisors? How are process designs modified to prevent deficiencies in future projects?

12. Are supervisors trained and encouraged to identify similar situations of errors, unsafe behaviors and personnel problems that could adversely affect the worker's performance? What actions are taken if a problem is identified?

13. Is the data collected on human errors available to managers? Has the data been used as a basis for making any decisions?

Complementary questions

- What are the management and educational processes?
- What are the necessary structures?
- What are the human factors involved?
- What physical and psychological stressors do they influence over?
- What does stress influence?
- What are the performance factors?
- What are the situations that most cause human error?

Questions for group 2
API770 Questionnaire—task and human machine interface
Work and task issues

14. Have critical jobs and tasks been identified? Have the mental and physical aspects of these works been analyzed for both routine and emergency activities?

15. Have the jobs and tasks been designed to keep the worker interested and involved? Are responsibilities rotating to avoid overload and increase the worker's experience? How have activities been emphasized in relation to security?

16. Are tasks that require intense, repetitive or rare monitored abnormalities assigned to machines whenever possible?

17. Are the individual responsibilities of workers clearly defined? How are these responsibilities distributed across the team?

Issues of the human machine interface (HMI)

18. Has the HMI suggested revisions in human factor engineering?

19. Is the information adequate about normal and abnormal process conditions signaled in the control room?

20. Are workers provided with information for diagnosing non-compliance when an alarm sounds? Are the signs adequately visible from all relevant work positions?

21. Does the layout of control panels reflect functional aspects of the process or equipment? Are related signals and controls grouped together?

22. Are all controls accessible and easily distinguishable? Are the controls easy to use in the right format and difficult to use in the wrong format?

23. Is there a formal mechanism to correct deficiencies in human engineering identified by the worker? Have workers made changes to signs, controls or equipment to make it better and more appropriate for their needs at work?

Complementary questions

- What are the management and educational processes?
- What are the necessary structures?
- What are the human factors involved?
- What physical and psychological stressors do they have influence over?
- What does stress influence?
- What are the performance factors?
- What are the situations that most cause human error?

Questions for group 3
API770 Questionnaire—man-machine interface.

25. Are instruments, signals, and controls readily repaired after malfunction? How much time? What priority criteria?

26. Are the work environment (temperature, noise, lighting, cleanliness, etc.) kept within comfortable limits?

27. Are the right tools available and used when needed? Are all required special devices available to perform tasks safely?

28. Is there adequate access for the routine operation and maintenance of all equipment? And of all the critical equipment?

30. Are all important equipment (vessels, pipes, valves, instruments, controls) clearly identified (unambiguous)?

31. Are specific needs for communication and team cooperation analyzed in the workplace project? How do different shifts communicate the process status for each one? What is the procedure for communication between departments?

32. Are there critical procedures for communications, during emergencies between workers and emergency personnel response, plant management, corporation management, and public authorities?

33. Are workers encouraged to ask supervisors for assistance?

35. Is the shift rotation schedule prepared to minimize the break in workers' circadian rhythms?

Complementary questions

- What are the management and educational processes?
- What are the necessary structures?
- What are the human factors involved?
- What physical and psychological stressors do they influence over?

- What does stress influence?
- What are the performance factors?
- What are the situations that most cause human error?

Questions for group 4
API770 Questionnaire—procedures and worker

36. Is there an updated and complete series of procedures available for the use of workers? How are specific procedures kept up to date?

37. Are the procedures written for the correct level of knowledge and understanding of the workers?

38. Do worker practices coincide with written procedures? How are differences detected and resolved?

39. Are work permit systems used correctly? How are outsourced workers included in these systems? Are PT's requirements differentiated in relation to hazards and risks?

40. Are emergency procedures clearly described? Are they practiced regularly? How many immediate actions are required?

41. Are checklists used for critical procedures? Is there only one action specified for each numbered step?

42. Does a human factor specialist help to develop the worker's competence?

43. Is there a training policy applicable to all workers, including those hired?

44. Are training records maintained? How are training needs raised?

45. Are screening and health journal evaluations carried out for workers who meet and maintain defined medical standards? Is the worker's health assessed before he is allowed to return to work after an illness?

46. Are there any programs to identify and help substance abuse workers or mental health professionals?

Complementary questions

- What are the management and educational processes?
- What are the necessary structures?
- What are the human factors involved?
- What physical and psychological stressors do they have influence over?
- What does stress influence?
- What are the performance factors?
- What are the situations that most cause human error?

List of Abbreviations

ABIQUIM	Brazilian Chemical Industry Association
ABRACOPEL	Brazilian Association for Awareness of the Hazards of Electricity
BLEVE	Boiling Liquid Expanding Vapor Explosion
CV	Valve Capacity
TMBF	Mean Time Between Failures
CCPS	Center for Chemical Process Safety
CCg	Cognitive Communication
CH	Human Reliability
CIDEM	International Congress on Disasters and Emergencies
CLT	Consolidation of Labor Laws
COE	External Organizational Communication
COI	Internal Organizational Communication
COCOM	Contextual Control Model
CPA	Principal Component Analysis
CREAM	Cognitive Reliability and Error Analysis Method
CSNI	Committee on the Safety of Nuclear Installations
CRM	Reliability Centered Maintenance
DG or MD	Manager's Speech or Manager's Discourse
DIPEA	Process Dissection and Abnormal Events
DO or OD	Operator's Speech or Operator Discourse
EDUC	Educational Programs.
EVA	Statistic of Abnormal Events
EVOP	Test in Normal Operation and Operational Events;
FIROB	Fundamental Interpersonal Relations Orientation
FMEA	Failure Mode and Effect Analysis
FTA	Failure Tree Analyses
GPLAT	Platform Manager
HPF	Human Factors of Performance
HAZOP	Hazard and Operability Study
HEP	Probabilities of Human Error
HSE	Health Security and Environment
HMI	Human Machine Interface
HTA	Hierarchical Task Analysis
IBAMA	Brazilian Institute for Environment and Renewable Natural Resources Protection
ISO	International Organization for Standardization
ICT	Operational-Technical Culture
JBH	Johnny Big Head
LODA	Dependent Layers of Operation Decision Assessment
LOPA	Layer of Protection Analysis
LPG	Liquified Petroleum Gas

List of Abbreviations

MDA	Methyl Diphenylamine
MEA	Abnormal Event Mapping
MTBF	Mean Time Between Failures
MSA	Maintenance, Safety, and Environment Analysis
MULTI	Multiplier Agents.
OSHA	Occupational Safety and Health Administration
PADOP	Standards and Procedures in Operation
PLCs	Programmable Logic Controllers
PSF	Performance Shaping Factors
PSV	Pressure Safety Valve
PCL	Programmable Logic Controllers
RHA	Human Reliability Analysis
RH	Human Resource
RCM	Reliability Centered Maintenance
ROTLIM	Clean Routines
SAP	Statistical Analysis of Processes
SMS	Safety, Health, and Environment Sector
SPM	Statistical Process Monitoring
ST-FTA	Sociotechnical Failure Tree Analysis.
OSH	Occupational Safety and Health
OTC	Operational-Technical Culture
TDC	Thinking, Deciding, and Acting
TREINOP	Training for Operation.
TQC	Total Quality Control
TPM	Total Productive Maintenance

Glossary

Abnormal events map is a mapping technique that involves the extraction of abnormalities or signs of abnormalities recorded in the shift register for a map that gathers the history of the plant for at least two months. This map also includes product and effluent quality information and eventually the process. The map is used to facilitate the construction of causality through the chain of abnormal events through codes and symbols.

Abnormality refers to signals emitted during operational events, which, at first, do not affect the established operating standards, but it may indicate a tendency to change the state of the process or people, from normal to abnormal. In other words, if the signals, or if the abnormalities still fall within the operational standards, these warnings are not necessarily considered to be of high risk, but some of them can bring important information to avoid the trigger of the failure from the latent or inertial phase.

Bottom-up technique is the technique used in the mapping of abnormalities that privileges the experience of the routine and the analysis of the occurrence records (bottom) to define the appropriate route in search of human reliability and stabilization of the processes (up) and that after the Board approval returns in the form of a management program. These types of techniques facilitate the installation of self management in production.

Causal link is the formation of a chain of simple linear abnormalities with the direct appearance of cause and consequence.

Chain of abnormalities are a series of signs of abnormalities or operational factors, which, after validation are classified as cause, consequence, false, potentializing, anticipated, corrective, preventive, and other factors, which continue to be codenamed as signs of failure. Several chains of abnormalities can happen in parallel with common operational factors.

Cluster analysis of data is a verification of the trends of increment, reduction or cyclic fluctuation of processes, and equipment from the abnormal signs extracted from the routine through the mapping of abnormal events (MEA).

Competency analysis is performed based on the identification of the technical–operational culture that results from the mapping of operational abnormalities. The analysis of competencies is made by comparing the demand for operations to achieve organizational efficiency with the supply of competencies based on the examination carried out by the shift team.

Conjugated memory is composed of events that happened, gaps in unremembered experiences, and the replacement of memories by others that are more favorable to the physical and psychological survival of the worker. In the same way, the organization's memory is built with transformations involving cancellation and substitution.

Environments in the failure study indicate traits that affect the history of operational and human failures. Environments such as organizational culture (organizational chart and company structure), social culture (features in decision-making and emotional stability), economic situation (growth and risks), and occupational legislation (resulting from society's demands for industry) are considered.

Etiology is the study of the origin that in the case of research on human and operational failure allows identifying the root cause by reading the operator's discourse (OD) and the manager's discourse (MD).

Failure hook is the term used for the characteristics in human, social, and technical structures composing organizational environments and jobs. The mitigation or inhibition of these characteristics is one of the objectives of management programs for human reliability and process stabilization.

Failure informalization is a process that involves the migration of part or all of the failure energy originally inserted in equipment (solids or pressure fluctuation) and destined for the social environment through individual fatigue or conflicts in the group.

Failure life cycle or failure life is the set of all operational factors from the moment of zero energy of the failure to the maximum point where the maximum loss of process occurs. The characteristics of technical and human structures are not the focus of failure life cycle analysis (learning model).

Failure materialization is the process that involves the migration of part or all of the failure energy originally inserted in the social environment (leaders in conflict) or in the organizational environment (policies do not match the practices) and is intended for equipment and processes through the lack of control in the execution of the task.

Goal state is the state of the process, equipment, and product achieved after the task is performed. In this state, motivational aspects and workload are analyzed.

Human reliability is the increase in the psychological quality of the individual and the group leading to a reduction in the frequency and impact resulting from operational failures. It includes task analysis activities and team improvements in leadership and cooperativeness.

Latent failure is part of the life cycle of the failure, and this is at the base of the learning cylinder and occurs when the failure is close to the axial axis indicating low visibility. After passing through the inertial zone (in the cylinder) where the fault is considered latent, the fault's growth phase begins, which at this moment becomes active and apparent and the abnormal event and its various factors can then be characterized.

Normal state is when the process works as indicated by the project or as indicated by the project in normal operation. Technical culture may erroneously claim that routine signals such as a dirty filter or bomb trip are included in the normal state. In this work of stabilizing processes and human reliability, it is necessary to recalibrate the balance of the scale that classifies normality in the industrial process.

Organic nexus is the formation of a chain of complex abnormalities with skins of energy feedback from the failure or sharing the failure in new dimensions (for example, from organizational to technical structures or from individual to group).

Operational factors are items that make up the anatomy of the failure or accident and can be, after their functional understanding, cause, consequence, or other functions. At the beginning of the study, all operational factors are considered as signs of failure and as they are understood, their functions are defined.

Personality ruler is a technique used to facilitate the analysis of competencies where it measures the competency offered by the team and compares the competencies required by technology. In practice, it means comparing the needs for technical / interpersonal knowledge and skills to perform the task with the provision of available knowledge and skills by the team. This work is carried out after the identification of the technical culture.

Potentiation (or failure potentiation) is a process of increasing (decreasing) the failure energy, increasing (decreasing) also your risk.

Potentiating event is one that increases the failure energy in a geometric way, implying reaching the risk zone in the cylindrical learning model. It may happen that the event geometrically reduces the failure energy, being a decrease of potent event.

Potentiating factor (depotentializing) is a human, technical, or environmental event that geometrically increases (or decreases) the failure energy.

Process stabilization is the state in which minimal losses and team motivation occur when compared to historical and worldwide company standards.

Starting event often cited by Reason and Hollnagel is also considered the root cause of latent failures or a common cause in active failures.

Root cause is the cause that originates or initiates the abnormality where, after passing the low visibility zone, it starts to indicate the existence of the abnormal event. The root cause can be very hidden, being identified after tests and adjustments accompanied by standards and procedures. A root cause can initiate several chains of abnormalities.

Structures in the study of failure are individuals (human), groups (social) and equipment / processes (physical, technical) responsible for enabling industrial economic activity and services.

Task analysis includes verification of the standards and procedures applied, verification of the behavior of the failure in the task including aspects raised from the mapping of abnormalities and verification of the team result in situations with high stress including failure analysis with layered operational decisions and calculation of probability considering human factors.

Technical idleness or non occupancy rate for activities related to the function, being a reference number due to the multiple variables in this case, for example, availability of LPG for processing, availability of sphere, or of ship - starting from the premise, in this research that non occupancy rate is not an intrinsic trait of man.

Technical–operational culture is the felt behavior, analyzed, and measured resulting from the identification techniques for the human, social and technical type established in the workplace and in the technology in operation.

Top-down technique is the technique used by management systems centralized in people representing top management (top) such as ISO-9000/14000, TQC, TPM, 5S where the use of pre-formatted techniques to be applied in the areas of factory, especially in the production area (down). As there is no prior consultation with the local technical or organizational culture, there may be difficulties in implementation.

Type of human behavior or human typology is the way in which individuals act in isolation, allowing inferences about emotional balance and level of commitment based on the identity between human and organizational values. To identify this behavior, specialized techniques are used in dynamics, surveys, and historical data statistics. In the identification of the typology, it is possible to recognize signs of "hooks" of operational failure and human error.

Type of social behavior or social typology is the investigation of how the groups of operators / workers act in carrying out the task and the possibilities of being influenced by environmental issues (social, economic, and organizational). To identify this behavior, specialized techniques are used in dynamics, surveys, experience, and statistics of historical data.

Type of technical behavior or technical typology is the way in which the equipment, products, and processes behave during the operation of the operation in factory production. The technical structures may have defects, "hooks" or inappropriate characteristics and that can eventually cause operational failure or human error, after an initiating event, or root cause, such as dissatisfaction caused by inadequate commitment resulting from a maintenance contract purely for the price.

Index

Note: Page numbers followed by "*f*" and "*t*" refer to figures and tables, respectively.

A

A8 IAT
  identifying environmental threats and industrial and cognitive process, 485–486
ABIQUIM, 361
Abnormal event cluster analysis, 449–450
  cluster analysis and relationships in equipment abnormalities, 450
  cluster analysis and relationships in process abnormalities, 449–450
Abnormal event qualitative-quantitative map based on operator's discourse, 186–189
  description of failure & history resulting in process loss, 189
  operational factors, 189
  period for investigation, 187
  process abnormalities, 187–188
Abnormal events map, 372, 478
  abnormal event mapping, signs, and failure mode, 412–414
  and chain of abnormalities, 196–197
Abnormalities inventory, 392–393
Abnormality logic in complex processes, 393–395
Abnormality mapping and operator discourse, 43–45
  complexity, 43–44
  failure logic in complex systems, 44
  influence of technology on human error, 44
  normal state of process, 44–45
  operational factors for failure, 44
  process disorders, 45
Absences at work, 183
Accident cases in contractors, 227–228
  accident with death during service in TDI tank, 227
  accident with death when handling forklifts, 227–228
Accident Sequence Assessment (ASEP), 31
Accidents, incidents and morale changes, 183–184
Acid plant, 216
Acrylates, proposal for chemical industry of, 284–286
Alcohol, 165
Aliphatic amines, 395–396
  EVA, 395–396
  operational context, 395
  technology, 395

Alkylamines, 279
American anthropometrical ergonomics, 7
American Nuclear Industry Regulatory Commission (NRC), 310
Antecedent, behavior, consequent tool (ABC tool), 299
Anthropometry, 7
Anticipated factor, 46
API770, 85, 91, 143–148, 222
  analysis, 144–148
    assignments in tasks, 147
    human machine interface, 147
    policy analysis and guidelines, 146–147
    worker, 148
    written procedures, 147–148
  concepts and assessment from, 143–144
    managers, 143–144
  data & operational–technical culture, 525–526
  exercise, 536–539
Aptitude for work, 322–323
Archetypes, 88–90, 364
Aromatic amines, 396–401
  industry, 279
  operational context, 396–399
  operational diagnosis on standard, 399–401
  statistical monitoring of processes, 399
  technology, 396
Assessment of task, efficacy, and effectiveness (ATEE), 121–122
Audit for production and actions, 202–204
Automatism, 55–56
Available time, 316–317

B

B5 PCET, 483
B6 PCET, 483–484
B7 PCET, 484–485
Back-casting analysis, 477
Behavior analysis. *See also* Failure analysis
  rules of, 475
  standard for, 445–449
    analyses for type of human and social behavior, 446–449
    analysis of behavior in workplace, 445–446
    human error in task, 446
    performance in workplace, 445
Behavior in emergency task, 377

Behavior patterns, defining, 504–505
Benzene, Toluene and Xylene (BTX), 286
Biodiesel, 164–165
Biofuels industry, process losses in, 164–166
Blast-explosive industry, 281
Bottom-medium-up movement, 367
Bottom-up methodology, 475
Brazilian culture, 273–275
Business environment, 11–12

C

Calibration, 315–324
  of SPAR-H method, 510–519
    PSF calibration, 512–519
    survey about operational context, 510–511
Caprolactam industry, 279
CARMAN Tool, 320–321
Center for Chemical Process Safety (CCPS), 15
Chemical and polymer case, 195–207
  abnormal event map and chain of abnormalities, 196–197
  audit for production and actions, 202–204
  results, 204–207
  statistical analysis of
    abnormal event, 198–201
    process variable and effluent, 198
Chemical industry, 331–333
  application of SPAR-H method, 331–333
  case application, 391–404, 451–457
    abnormalities inventory, 392–393
    abnormality logic in complex processes, 393–395
    aliphatic amines, 395–396
    aromatic amines, 396–401
    polycarbonates, 401–404
  considerations, 333
  and electricity distribution, 278–286
    blast-explosive industry, 281
    cases, 278–279
    human reliability research proposal, 281–284
    proposal for chemical industry of acrylates, 284–286
    real cases of chemical industry demand and electricity distribution demand, 280–281
  industrial context, 331
  management aspects for decision, 451–457

545

Chemical industry (Continued)
  cases, 451–455
    management programs to increase human reliability, 455–456
    objective function, 451
    stabilization program for processes, opportunities and suggested techniques, 456–457
  process analysis, 331
  process losses in, 167
  in United States, 8
Chemical plant, 216
Chicken industry, 325–328
  application of SPAR-H method, 328
  considerations, 328
  equipment, 327
  industrial context, 325–326
  process analysis, 326–327
Chicken manufacturing, 296–299
  real cases of demand, 298
Chronology, 118
Client, 18
Cluster analysis and relationships
  in equipment abnormalities, 450
  in process abnormalities, 449–450
CO compressors, 261–269
Coconut industry, 342–344
  application of SPAR-H method, 343
  considerations, 343–344
  operational context, 342
  task, 343
COCOOM model, 106, 108
Cognitive and behavioral academy, 237–248
  comments, 248
  diseases and vaccines in study of behavioral aspects, 239–241
  fertilizer case, public industry, 241–243
  polycarbonate case, 244–247
  program, 241
  PROGRAM friends of emergency pool's activities description, 244
Cognitive efforts, 123
Cognitive ergonomics, 6–7
Cognitive models about human error in Nordic countries, 7
Cognitive processing, 52–82
  cases, failure, and skill knowledgement, rules, 52–60
    conflict of generations and supervision of shifts, 57
    culture biases and population stereotypes, 59
    defragmenting operator's discourse to investigate accident, 55
    divided attention in oil industry, 53
    emergency, Three Mile Island, and Fukushima, 59–60
    failure analysis with cognitive nature cause, 58–59
    gender conflict and ethnicity prejudice in shift operator accident, 56
    human error, automatism and excessive self-confidence, 55–56
    memory in workplace, 53–54
    mind map, perception and mnemogram, 55
    primary memory capacity and memory with association, 54
    skills, knowledge, and rules, 60
    slips, organizational and individual values, 55
  cognitive functions and decision processes, 60–64
    routine of field operator in chemical industry, 63–64
  cognitive model, 71–74
    cognition and memory operations, 72–73
    decision model for observer and controller, 73–74
    rules, gaps, and questions, 71–72
    simple model for cognition, 74
  decision, and action model, 364
  decision-making process, 66–71
  and execution, 483–486
    A8 IAT identify environmental threats and industrial and cognitive process, 485–486
    B5 PCET, 483
    B6 PCET, 483–484
    B7 PCET, 484–485
  human behavior dynamics in company, 74–82
  learning and skill, 64–65
  motivation and decision, 65–66
Cognitive quality, 491–495
  dynamic cognitive quality, 494–495
  elements, functions and subfunctions, 492–493
  static cognitive quality, 494
Cognitive Reliability & Error Analysis Method (CREAM), 34
Collection workshop, 375
Commission, 42
Commitment, 230–231
Committee on the Safety of Nuclear Installations (CSNI), 5
Communication, 232
  chronological description of occurrences with HCl in chemical installations, 252–254
  considerations, 254
  normality status, 252
  in routine, 251–254
Companies, 1
Comparative analysis, SPAR-H, 347–349
  recommendations, 349
Competence
  analysis and results, 441–445
    competence offering, 443
    demand based on technology, 441–443
    examination for technical–operational culture identification, 443–445
  assessment, 380
Competency
  analysis, 2–3
  map, 442
Conceptual failure energy model, 368
Conjugated memory construction, 364
Connectivity analysis, principles and concepts of, 134
Consumption index, 180, 182
Continuity, 181
Continuous process technology requirements, 527
Contract management, 474
Cooling system, 134–135
Corrective action, 46
Costs from inadequate projects, 184
Criticality analysis, 118
Culture, 228–230
  biases, 59
  corporate policy, management profile, and conflicting priorities, 229
  cultural vices, 228–229
  in metallurgy and oil industries, 273–277
    Brazilian culture, 273–275
    considerations, 277
    fair, organizational and safety culture in metallurgy, 275–276
    fair, organizational and safety culture on OFFSHORE oil platform, 276–277
  new technical-operational culture, 229–230
  of safety and guilt, 229
  of safety demands, 277–304
Cyclic individual, 87
Cylindrical models for operational failure, 127

D

D12 PADOP, 488–489
D13 failure logic and connectivity, 489–490
D14 failure materialization and energy, 490
Data collection, processing, and information, procedure for, 384
Decision, 65–66
  analysis under stress, 136–138
    organization of techniques for task review, 137–138
  decision-making process, 66–71
  decision–action cycle, 101–102
  model for observer and controller, 73–74
Depressed individual, 87
Deviance normalization, 11
Dinitro toluene (DNT), 482
Dissecting the Process to Investigate Abnormal Events (DIPEA), 128–129
Divided attention in oil industry, 53
Dynamic(s)
  cognitive quality, 494–495
  model in production system, 367
  of operational factors in failure process, 368
  risk management, 41–42

E

E15 LODA, 491
Education, 14–15, 231
Effluent
  quantity and quality, 180, 182, 184
  and waste treatment, 184
Effluent treatment section (STE), 195
Electricity distribution in Brazil, 279

Emergency, 59–60
Environmental, Health and Safety (EHS), guilt and information for, 222–223
Environments in human and operational reliability, 104–105
Equipment reliability, 354, 499
ER3 assessment, 215–216
Ergonomics, 321–322
  studies, 2
Ethnicity prejudice in shift operator accident, 56
Executive function (EF), 90
  analysis, 102–104
Exergy of operational failure and relationship with productive system, 370
External communication (EC), 367
External organizational communication (EOC), 367
Extra hour/overtime, 182
EVents in abnormal state (EVA), 395–396

F

Failure analysis
  of CO compressor in ammonia unit, fertilizers, 266–269
    case study, 268–269
    human reliability, 267–268
    Ishikawa diagram and components reliability, 267
    operational context and task analysis, 266–267
    preliminary analysis, 266
  with cognitive nature cause, 58–59
  management to validate failure analysis model, 506
  in production systems, 46–48
    moment of failure materialization, 47–48
Failure cylinder, 389
Failure energy analysis, 457–468
  analysis and choice of principal component analysis groupings, 460–462
  choice of input data and organizational efficiency for application, 462–463
  criteria and selection for measuring organizational efficiency, 459–460
  data processing, 463–466
  mapping organizational efficiency and simulating situation, 467
  observations, 467–468
  rules of technical behavior, environment, and individuals, 457–459
Failure energy feedback, 46
Failure Mode and Effects Analysis (FMEA), 31
Failure task analysis, 427–432
False factor, 46
Fault logic diagram, 129
Fault tree, 432
Fault tree analysis (FTA), 31
Federal Inspection Service (FIS), 326
Fertilizer case, public industry, 241–243
Fertilizer industry, 126, 337–342
  application of SPAR-H method, 339–341
  considerations, 341–342
  equipment, 339
  general context, 337–338
  process analysis, 338
  and refining units, 294–296
    real cases of demand, 295–296
  task, 338
Field operator in chemical industry, 63–64
Fines, 178–179, 184
FMEAH, 488–489
FO. See Objective function (FO)
Fordism in United States, 6
Forklift in oil industry, 271–273
Free-thinking (FT), 77
French ergonomics, 6
FTAH, 488–489
Fukushima, 59–60
Functional Resonance Analysis Method, 35
Fundamental Interpersonal Relations Orientation-Behavior technique (FIROB technique), 4
Fuzzy logic, 2
  to change performance factors, 35–36
Fuzzy mechanics in human reliability, 39–40

G

Gas industry for energy, process losses in, 163–164
Gender conflict, 56
Generation conflict, 57
Generic algorithm, 49
Globalization, 97–99
GMTA method, 111–112
GR6 assessment, 216–218
GRODIN, 298, 301
Gross profit margin, 451
Group behavior, 439
Group work, 232

H

Health, safety and environment (HSE), 11, 17, 408
Health risks, 231
Heuristic rules for human reliability, 385
Heuristics, application of, 388
HR department data, 376
Human activity, 9
Human behavior dynamics in company, 74–82
  Johnny Big Head's life cycle in organization, 80–82
  movements, 77–78
  organizational processes, 79–80
    moment of causal and organic nexus, 80
    moment of elaboration of causal nexus, 79
    moment of fantasies and unconscious processes, 79
    moment of information processing, memory and action, 80
    moment of organic nexus, 79–80
    moment of reality and cognitive processes, 79
    thought and fantasy, 79
  psychological/psychic functions in informational process, 76–77
Human error, 1–2, 5, 8–19, 25, 30, 55–56, 104–105
  analysis in US aviation, 7
  classification, 90–93, 93f, 380
  cognitive models, 7
  in context of critical activities, 309–310
  human error-rate prediction technique, 32–33
  probability of, 2
  study of, 3
  in task, 446
  and technical failure trend indicators, 503–505
    analyzing critical tasks in routine and emergency situations, 504
    classification, 505
    defining behavior patterns with elements of human reliability, 504–505
    implementing program to increase human reliability, 504
    management to validate failure analysis model, 506
    opportunities for logistics and safety operations management, 506
    routine management for environment favorable to human reliability, 505–506
  types, 3
Human error probability (HEP), 268, 310
  calculation, 314
Human factors (HF), 85
  algorithm for correction of, 386
  analysis and diagnosis of, 3–4
  routine and emergency analysis, 475–476
Human reliability (HR), 25–52, 277–304, 354, 499–500
  chemical industry case application, 391–404
  classic concepts, 28–31
  and cognitive processing, 52–82
  conceptual and mathematical models in, 1–2
  in cultural link, 38–40
    database, 39
    fuzzy mechanics, 39–40
    human design and reliability, 38
    operational failure exergy, 38–39
    structures, processes, and environments, 38
  and efficiency, 99–100
  environments in, 104–105
  historical vision about schools related to origin, 5–8
  methodology description, 359–391
    concept, tool, and procedure for technical, social, environment, and human typologies, 361–391
    guiding algorithms to apply technological tools and social environment, 359–361
  modeling, 31–36
    exercise, 33–35
    first generation, 32–33
    second generation, 34
    third generation, 35–36
  in nuclear power plants, 7
  oil and gas case application, 404–450
  operator discourse, 41–48

548 Index

Human reliability (HR) (*Continued*)
   qualitative results, 451–457
   quantitative results, 457–468
   research proposal, 281–284
   SPARH, 495–498
      calculation and calibration, 497–498
      operational context, 496
      performance factors assessment, 496–497
   standardized plant analysis. risk-human reliability analysis method and case in chemical facility, 36–38
   study, 25–28
      management analysis, skill, perception, knowledge, and rule, 27t
      root cause and elements for building accidents, 28f
   task cross-assessment based on particle swarm model, 468–471
   third-generation application in calculation of organizational efficiency, Oil & Gas, 48–52
      generic algorithm, 49
      method, 48–49
      situation simulation, 49–52
Human reliability analysis (HRA), 29, 95–100, 143, 307, 310, 359, 473
   influence on workplace, 99
   social culture, globalization, and human types of society for work, 97–99
Human types of society for work, 97–99
Human typology, 86–88, 531–533
   data collection for
      cognition quality &CTO, 532
      commitment assessment, 533
      cooperation level of operational team, 532
      to emotional imbalance, 531
      leadership assessment, 532
   types of organizational behavior, 86–88
Human-machine interaction, 321–322
Human–machine interface (HMI), 4, 35, 93, 265
Hydrochloric acid (HCl), 331, 353
Hydrochloric acid case, 500
Hydrogen compressors (H_2 compressors), 261–269
   in benzene hydrogenation unit, 261–263
   failure analysis in HDT unit, refinery, 263–266

I
IAT (Identification of Environment Task), 113–115
Idaho National Laboratory (INL), 310
Image, 179
   on workplace, 183
Individual and social data about worker, 374
Individual data, 436–438
Individual's behavior data for cooperation, 438–439
Industrial plant load and continuity, 180
Industry alarms and shutdown, 261–269
Insurance, 179

Integrated reliability, 349–356. *See also* Human reliability (HR)
   control and metrics to achieve integrated or sociotechnical reliability, 351–352
   results, 354–356
   simulation, 353–354
      complexity, 354
      equipment reliability, 354
      human reliability, 354
      illustrative questions from experts, 353–354
      operational reliability, 354
      process reliability, 354
Intentional violation, 224–225
Intermediate State of Abnormality (ISA), 252
Intermediate State of Normality (ISN), 252
Internal organizational communication (IOC), 367
Isocyanate industry, 279

J
Job and society, worker role in, 8–19
Johnny Big Head (JBH), 26
   life cycle in organization, 80–82
   operator, 26

K
Knowledge, 60, 230–231
   base and skill in planning and adjusting task, 111
   and investigation of causal link, 230
   of production process, 172
   of technology and experience in process, 406–408

L
Labor relations, 323
Latent failure, concepts and investigation of, 93–95
Leadership, 15, 439
   human typologies, 87
Learning, 64–65
LESHA tool, 137
Lessons learned, 213
   culture in metallurgy and oil industries, 273–277
   human reliability, sociotechnical reliability, culture of safety demands, 277–304
   research *vs.* society's demand, 304
   routine, environments, human types, and class of errors, 213–228
      accident cases in contractors, 227–228
      emergency case, 215–216
      inappropriate design and operation, 223
      operational safety case, 221–223
      organizational change, 223–224
      practical skills case, 216
      problem analysis, 219–220
      process control, 218–219
      routine management, technical-operational culture, 224–226

      routine management case, 216–218
      routine management case GR, 214
      technical and operational culture, 215
   routine learning, 228–235
   and validation of guidelines, 235–277
      application of tools for archetype analysis and executive function in industry, 248–251
      cognitive and behavioral academy, 237–248
      communication in routine, 251–254
      industry alarms and shutdown, 261–269
      investigation of technical failure and human error in sulfuric acid plant, 255–261
      task complexity, low efficiency, and accident investigation, 269–273
Liquefied gas purification, 406
Load, 181
LODA Tool, 122, 137, 432–433
Logistics, 17–18
   opportunities for, 506
Loss mapping through manager's speech, 207–210
Losses in process industries, 157–168
Low Risk Perception (LRP), 280

M
Maintainability and availability of critical equipment, 180
Maintenance re-work, 181
Manager's discourse (MD), 186, 365, 367
Mapping losses, 189–190
Maritime transportation in United Kingdom, 8
Mass balance, 190–191
   for materials, 210–211
Mass production in United Kingdom, 6
Master's degree proposal for Bahia firefighters department, 2018–19, 302–304
Material loss risk, 19–22
Mathematical models, 475
Mean time between failures (MTBF), 2, 29
Means and goals in task planning, 111–112
Medical error in US health services, 8
Memory
   operations, 72–73
   in workplace, 53–54
Metallurgical industry, 296–299
   real cases of demand, 298
Metallurgy
   based on manager discourse and technical issues, 207–211
      loss mapping through manager's speech, 207–210
      mass balance for materials, 210–211
   fair, organizational and safety culture in, 275–276
   mining and cellulose industry proposal, 2018–19, 298–299
   process losses in metallurgy industry, 168
Methyl diphenylamine (MDA), 215–216
Mind map, 55

Mnemogram, 55
Motivation, 65–66, 183
Movements, 77–78
MPBA. *See* Public Ministry of the State of Bahia (MPBA)

N

Nondeterministic method, 475
Normal State (NS), 252
NRC. *See* American Nuclear Industry Regulatory Commission (NRC)
NS. *See* Normal State (NS)

O

Objective function (FO), 451, 486
Obsessive individuals, 87
Occupational health risk, 221
OFFSHORE oil platform, fair, organizational and safety culture on, 276–277
Oil and gas case application, 404–450, 457–468
   abnormal event cluster analysis, 449–450
   analysis of human and social data in work environment, 435–441
      group behavior and leadership, 439
      individual data, 436–438
      individual's behavior data for cooperation, 438–439
      social data, 435
      social data in company's human resources, 439–441
   analysis of task and results, 423–434
      failure task analysis, 427–432
      risk and behavior analysis in emergency or extreme stress, 432–434
      routine standard procedures, 423–427
      types of task analysis, 423
   competence analysis and results, 441–445
   context identification, 405–408
      application of methodology, 405–406
      knowledge of technology and experience in process, 406–408
   failure energy analysis, 457–468
   logistics analysis, 416–417
   maintenance, safety and environmental analysis, 420–423
   operations analysis, 417–419
   process analysis, 414–416
   standard for behavior analysis, 445–449
   survey of technical, human and social data, 408–412
      abnormal event mapping, signs, and failure mode, 412–414
      human and social data collection, 410–412
      routine, task, process and safety, 408–410
Oil and gas processing, 126, 129–134
Oil and gas site, 377
Oil industry, 105–106
   process losses in, 159–163
   in United States, 8
Omission, 42

Onshore offshore oil and gas industry, 287–294
   proposal for offshore, Brazil, 2010 public industry, 289
   proposal for offshore, Brazil, 2016–17 public O&G industry, 291–293
   proposal for offshore, USA, 2018–19, 293–294
   proposal for onshore, Brazil, 2008 public industry, 288–289
   proposal for onshore, Brazil, 2012 private industry, 289–290
   proposal for onshore, Brazil, 2017–18 public industry, research, 293
   real cases of demand for research and services, 287–288
Operational failure
   birth and development of, 369
   connectivity, 134
   exergy, 38–39
Operational reliability, 354, 499
   conceptual and mathematical models in, 1–2
   environments in, 104–105
Operational-technical culture and prediction, 500–506
   management programs for human and sociotechnical reliability, 503–506
   methodology and products, 502–503
Operator action tree, or decision and action mind map (OAET), 31
Operator discourse (OD), 36, 41–48, 186, 279, 361, 477
   abnormality mapping and, 43–45
   assumptions for investigation, 42–43
      fluctuation of behavior, 43
      multicultural, 43
      omission and commission, 42
      operational mode differences for task analysis, 43
      patterns and culture, 43
      process interconnects, 43
      skills analysis, 43
   dynamic risk management and, 41–42
   failure analysis in production systems, 46–48
   identification of types of operational factors in failure, 45–46
      cause, signal and consequence, 45–46
      corrective action and preventive action, 46
      failure energy feedback, 46
      potentiating factor, false factor, and anticipated factor, 46
   importance, 365
Organizational culture, 13, 99
Organizational efficiency
   choice of input data and organizational efficiency for application, 462–463
   criteria and selection for measuring, 459–460
Organizational environment, 99–100
Organizational integration, 232
Overtime, 181

P

Packing list services in manufacture of sports products
   considerations, 347
   general context, 344
   process analysis, 344
   scenario with original calibration, 345–347
   SPAR-H application, 345
   task, 345
PADOP, 113–122
   application of PADOP for sulfuric acid plant case, 138–143
   ATEE, 121–122
   cognitive processing and execution, 483–486
   environment assessment, 481–483
      A1 context, 481
      A3 labor organization rules, 482
      A4 task control indicators, 483
      context PLG facility, 481–482
   IAT, 113–115
   implementation process, 122
   method, 376
   PCET
      cognitive processing, 115–119
      task execution, 119–121
   standard and review, 481–488
   task efficiency control, 486–488
Paranoid individual, 88
Particle swarm model, task cross-assessment based on, 468–471
   bridge to future, 470–471
   path of workers, 469–470
PCET, 115–119
People management, 14
Perception, 55
Performance shaping factor (PSFs), 310, 315–324, 335
   calibration, 512–519
Personal protective equipment (PPE), 221
   failure of, 221
Petrochemical industry, 286–287
   real cases of petrochemical industry, demand for services, 286–287
Petrochemical industry, process losses in, 166–167
Phobic individuals, 87–88
Physical efforts, 123
Place and task type, 124
Polycarbonates, 401–404
   case, 244–247
   industry, 279
   operational context, 402
   statistical monitoring of processes, 402–404
   technology, 401–402
Polymer industry, process losses in, 167–168
Population stereotypes, 59
Porous individual, 88
Potentiating factor, 46
Pressure safety valve (PSV), 406
Preventive action, 46
Primary memory capacity and memory with association, 54

Principal component analysis (PCA), 49, 361
  analysis and choice of principal component analysis groupings, 460–462
  application of, 388
Priority Orbital Decision, 68
Proactive behavior, 87
Procedures, 231
Process and efficiency of operating procedures, 182
Process and production variables collection, 529
Process data, safety, logistics, operation, and maintenance, 374
Process industries, 325
Process loss(es), 6–7, 19, 26, 29f, 152
  in biofuels industry, 164–166
  in chemical industry, 167
  competencies to assess, 155–157
    premises and competencies, 155–157
  context, 151–155
  diagnosis, 169–194
    collecting data, 172–176
    commercial, 176
    institutional relations area, 176
    integrated management, occupational, and environmental area, 175–176
    knowledge of production process, 172
    measuring results, 176–184
    personnel area, 176
    production area, 174–175
    project and process engineering area, 175
  diagnostics with quantitative and qualitative analysis, 194–211
  in gas industry for energy, 163–164
  losses in process industries, 157–168
  map(ping), 478–480, 510
    calculation based on metallurgy case, 480
    methods, 478
  metallurgy based on manager discourse and technical issues, 207–211
  in metallurgy industry, 168
  in oil industry, 159–163
  in petrochemical industry, 166–167
  in polymer industry, 167–168
  risk of loss due to technology, 168
  tools and methods, 185–194
Process reliability, 354, 499
Process stabilization, 360–361
Process variables, 179
Product quality, 180, 182
Production, 181–182
Production audits, 193–194
Production management, 474
  profile, 13–14
Production manager, 474
Production Planning and Control team (PPC team), 345
Products, 476–477
  cognitive quality, 491–495
  human reliability SPARH, 495–498
  operational-technical culture and prediction, 500–506
  process loss mapping, 478–480
  social-technical reliability, 498–500
  task assessment, 480–491

PROGRAM friends of emergency pool's activities description, 244
Projects, 12–13
Propylene, 334
Psychoanalysis, 98
Psychodynamics at work in France and England, 6
Psychological/psychic functions in informational process, 76–77
Psychology, 6
Public Ministry of the State of Bahia (MPBA), 300
Public security agencies, 300–304
  real cases of demand, 300–302
Purification section (SP), 195, 393
Pyramid of organizational communication, 366

Q

Quantitative analysis of failure approximation, 385
Quantitative–qualitative analysis methods, 473
Quantity and reprocessing costs, 182
Questionnaire on individual and group behavior, 375

R

Rail transportation in United States, 8
Reaction section (SR), 195
Reactive behavior, 86–87
Real cases, 295–296, 298, 300–302
Refining industry, 333–337
  application of SPAR-H method, 335–337
  considerations, 337
  equipment, 335
  industrial context, 333
  process analysis, 333–334
  task, 334
Reliability Centered Maintenance (RCM), 5
RESOTECH method, 478–479
Responsible care, 473
Revenue, 178
Risk
  analysis in critical systems, cost or security decisions, 226
  and behavior analysis in emergency or extreme stress, 432–434
  dominance, 123
  of loss due to technology, 168
  management
    in complex processes and environments, 2
    material loss risk, 19–22
  on material losses and operations, 19–22
Routine data collection, 528
Routine learning, 228–235
  culture, 228–230
  knowledge, commitment, and standards, 230–231
  operation, project, maintenance, and contractors, 232–234
    contractors and commitments, 233–234
    operation and design standard, 232–233
    task and shutdown planning, 233

route of human, group and organizational error, 234–235
social relations at work, 232
Routine management
  case GR, 214
  for environment favorable to human reliability, 505–506
Routine task standard, 377
Rules, 60

S

Safety, 474
  culture, 275–277
  operations management, 506
  of work, 6
Sales amount, 178
Satisfaction and loss of customers, 178
Schools, 5–8
  historical vision about schools related to origin of human reliability, 5–8
  American anthropometrical ergonomics, 7
  chemical industry and oil industry in United States, 8
  cognitive ergonomics, 6–7
  cognitive models about human error in Nordic countries, 7
  human error analysis in US aviation, 7
  human reliability in nuclear power plants, 7
  maritime transportation in United Kingdom, 8
  mass production in United Kingdom and Fordism in United States, 6
  medical error in US health services, 8
  psychodynamics at work in France and England, 6
  psychology and safety of work, 6
  rail transportation in United States, 8
  task analysis in England, Manchester, and Sheffield, 7
  Toyota, total quality control and total productivity maintenance in Japan, 6
Search and action (SA), 427
Self-confidence, excessive, 55–56
Separation section (SS), 195
Simplified plant analysis risk-human reliability analysis (SPAR-H), 307
  calibration, 310–324
  and case in chemical facility, 36–38
  case studies, 324–347
  comparative analysis, 347–349
  human error probability calculation, 314
  human errors in context of critical activities, 309–310
  integrated reliability, 349–356
  operational context, 311–314
    continuous process industry, 312–313
    culture and operational safety, 313–314
    manufacturing or intermittent process industry, 312
    performance shaping factors and calibration, 315–324
Skills, 60, 64–65, 230–231

Slips, 55
Social culture, 97–99
Social data, 435
   in company's human resources, 439–441
Social demands, 1
Social environment, organization in, 1
Social inclusion, 232
Social relations at work, 232
   group work and communication, 232
   organizational integration and social inclusion, 232
Social typology, 88–90, 529–531
   data collection for political aspects and practices, 529–530
   data collection for social relations, 530–531
Social-technical reliability, 498–500
   calculation, 500
      hydrochloric acid case, 500
   complexity of parametric equations, 498
   culture and manager profile, 498
   individual reliabilities, 499–500
Socio-functional relationship at workplace and behavior diagnosis, 379
Socioeconomic environment, 95–100
Socioeconomic-affective utility (SEAU), 101
   cycle of, 101
Sociotechnical reliability, 277–304
Sociotechnical system, 368
Spare parts costs, 180–181
Stabilization program for processes, opportunities and suggested techniques, 456–457
Standard/procedures, 320–321
Standards, 230–231
State of abnormality (SA), 252
Static cognitive quality, 493–494
Statistical analysis
   of abnormal events, 192–193, 198–201, 374
   of process and effluent variables, 191–192
   of process variable and effluent, 198
Statistical process assessment (SPA), 245
Statistical process control/AEP, 373
STE. See Effluent treatment section (STE)
Step parallelism, 123
Stress, 4
   stress/stress-causing agents, 317–318
Submissive individual, 87
Sugar, 165
Sulfuric acid, 105–106
   industry, 279
   investigation of technical failure and human error in sulfuric acid plant, 255–261
      activities, 257–261
      culture of operational safety, 255–256
      investigation of abnormalities, 257
      process description and main events, 255
      work and people, 256–257
Summarizing, 18
Supervision of shifts, 57
Survival, 77
System reliability, 31

T

Task analysis in England, Manchester, and Sheffield, 7
Task architecture, 123, 423
Task assessment, 106–143, 480–491, 519–524
   application of PADOP for sulfuric acid plant case, 138–143
      critical analysis, 141–143
      operational context of case of sulfuric acid facility, 139–140
      task and barriers analysis, 140–141
   collection of failure data for, 520–522
   data collection for emergency assessment, 523–524
   decision analysis under stress, 136–138
   knowledge base and skill in planning and adjusting task, 111
   means and goals in task planning, 111–112
   methods in task analysis, 109–111
   PADOP, 113–122
      standard and review, 481–488
   requirements for task analysis, 111
   standard task data collection, 519–520
   in sulfuric acid facility exercise, 533–536
   task control complexity, 109
   task failure assessment, 488–490
   tools for planning standard task, 122–135
Task complexity, 124, 318–319
   low efficiency, and accident investigation, 269–273
      case study of forklift in oil industry, 271–273
      critical task complexity, forklift accident, 269–271
      qualitative approach for risk and task complexity, 271
Task cross-assessment based on particle swarm model, 468–471
Task emergency assessment, 490–491
   E15 LODA, 491
Task failure
   assessment, 377, 488–490
      D12 PADOP, FMEAH, FTAH, 488–489
      D13 failure logic and connectivity, 489–490
      D14 failure materialization and energy, 490
      task emergency assessment, 490–491
   phenomenology of, 128
Task in failure, dynamic investigation of, 126–129
Task planning, 2–3
Task project, 123
   analysis, 423
Technical and socio-human data collection, 372
Technical survey, 524–529
   API 770 data & operational–technical culture, 525–526
   continuous process technology requirements, 527
   data collection of technical, 524–525
   process and production variables collection, 529
   routine data collection, 528
Technical–operational culture (TC), 375–376
   examination for technical–operational culture identification, 443–445
Technique Empirica Stima Errori Operatori Tool (TESEO Tool), 32, 34
Technique for Human Error-Rate Prediction method (THERP method), 30, 32
   exercise, 34t
   probability data, 32t
Technology and projects, 12–13
Thinking cycle, 101
Three Mile Island, 59–60
Timeline, 5–8
   analysis, 126
Toluene diisocyanate (TDI), 482
Top-down management programs, 473
Total productivity maintenance (TPM), 6
Total quality control (TQC), 6
Toyota, 6
Training, 14–15
Training and experience, 319–320
Transport and storage, 165
Tray packing machine, 327
Turnover, 182–183

U

Unit operations efficiency and chemical reaction, 179
Uranium industry, 328–331
   application of SPAR-H method, 330
   considerations for uranium ore extraction industry, 331
   industrial context, 329
   process analysis, 329–330

V

Violation, 225–226

W

Worker performing task
   classification of human error, 90–93, 93f
   concepts and investigation of latent failure, 93–95
   diseases, bad habits and cognitive academy, 105–106
   environments in human and operational reliability, 104–105
   executive function analysis, 102–104
   human typology, 86–88
   organizational environment, 99–100
   social typology and archetypes, 88–90
   socioeconomic environment, 95–100
   socioeconomic-affective cycle and cycle of thinking and decision making, 101–102
   task assessment, 106–143
Worker role in job and society, 8–19
   departments, functions, and human reliability, 11–19

Worker role in job and society (*Continued*)
  business environment, 11–12
  client, 18
  education and training, 14–15
  example of possibility of same error, 18–19
  health, security and environment profile, 17
  laboratory profile, 17
  leadership, 15
  logistics, 17–18
  maintenance profile, 16–17
  operation profile, 15–16
  organizational culture, 13
  people management, 14
  production management profile, 13–14
  selection by function, 14
  summarizing, 18
  technology and projects, 12–13
  human activity, 9
  role at work and society, 9–11
Workstation, 118
  questionnaire for, 120*t*

Y

Yield, 179